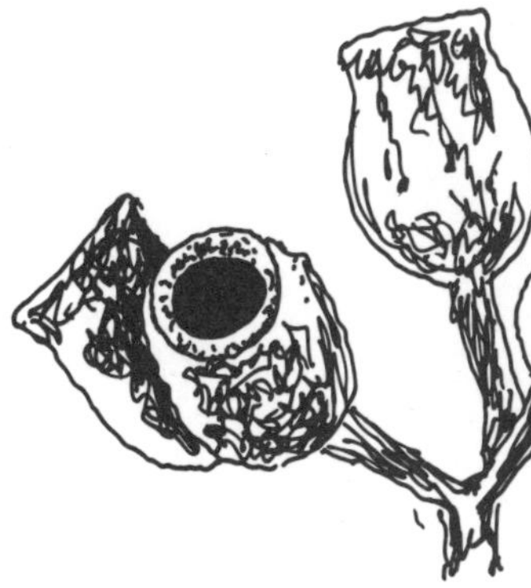

Pink Bloodwood
Eucalyptus intermedia

A medium sized to tall tree, with a straight trunk. The heart wood of this tree is deep pink to dark red-brown. The species occurs as one of the dominant trees of open forests, often bordering rainforests.

Maned Duck
Chenonetta jubata

Maned ducks are true ducks despite superficial similarity to geese. They feed frequently on grass, hence have benefited from European settlement. They nest in holes in trees, possibly mating for life.

pecies
rea
t is
with
es
ple

AF680783

Plaesiorhinus sp.

This small weevil lays it's eggs in the fruit of some legumes. when the larvae emerge they pupate in the soil. Little is known of the adult life of this species.

Conserving Biodiversity:
Threats and Solutions

PLATE 1 *(above)*. The Barmah forest is a Red Gum *Eucalyptus camaldulensis* forest which receives water from the Murray River. It is an important wetland area as well as being habitat for birds such as the rare Regent Parrot *Polytelis anthopeplus*. Logging of the red gums and changed regimes of flooding due to irrigation have the potential to alter the structure of the forest with wide ranging effects on resident plants and animals. How can the resources of the river and forest be used while maintaining resident biota? *(Photograph by D. Eastburn.)*

PLATE 2 *(left)*. A water sample from the Darling River at Bourke, New South Wales in 1991, illustrating a high concentration of blue-green algae. Recent outbreaks of these algal blooms in the inland waterways of eastern Australia have focused attention on the complex issues of water usage by humans. Both people and biodiversity are dependent on a ready supply of clean water in inland Australia. *(Photograph by Kathie Atkinson.)*

Conserving Biodiversity: Threats and Solutions

Edited by

R. A. Bradstock, T. D. Auld, D. A. Keith, R. T. Kingsford,
D. Lunney and D. P. Sivertsen

Published by

Surrey Beatty & Sons Pty Limited
In association with
NSW National Parks and Wildlife Service

This book is copyright. Apart from any fair dealing for purposes of private study, research, criticism or review, as permitted under the Copyright Act, no part may be reproduced by any process without written permission. Enquiries should be made to the publisher.

Surrey Beatty & Sons Pty Limited is a member of Copyright Agency Limited (CAL) in Australia and is a respondent to overseas collecting societies.

The National Library of Australia
Cataloguing-in-Publication entry:

Conserving biodiversity: threats and solutions.

Bibliography.

Includes Index.

ISBN 0 949324 59 0.

1. Biological diversity conservation — Australia. 2. Biological diversity conservation — New South Wales. 3. Biological diversity — Australia. 4. Biological diversity — New South Wales. 5. Ecosystem management — Australia. 6. Ecosystem management — New South Wales. I. Bradstock, Ross Andrew. II. New South Wales. National Parks and Wildlife Service.

333.95160994

Published July, 1995

NSW
NATIONAL
PARKS AND
WILDLIFE
SERVICE

AUSTRALIA

Department of the
ENVIRONMENT
SPORT and
TERRITORIES

PRINTED IN AUSTRALIA BY
SURREY BEATTY & SONS PTY LIMITED
43 Rickard Road, Chipping Norton, NSW 2170

Preface

THIS book provides a record of a similarly titled conference held at the University of Sydney in June 1993, organized by the NSW National Parks and Wildlife Service. The conference was an integral part of the process of deriving a Biodiversity Conservation Strategy for New South Wales, the state with longest history of occupation by Europeans on the Australian continent.

The impetus behind both the conference and the Strategy was to evaluate threats to biodiversity both within a state and national context and to review and foreshadow the range of measures needed to counter these threats. We think that a review of the effects of these threatening processes is a pre-requisite for action. While it is not hard to list the processes that threaten biodiversity, it is significantly more difficult to agree about which elements of biodiversity are affected by each threat and to what degree. A systematic attempt has been made here to cover the current and future threats to biodiversity. Threatening processes are reflected as themes in the book.

Any scientific review of threats, no matter how comprehensive, is ultimately of little use unless these can be practically addressed. As a prime objective, this book highlights the range of solutions needed to conserve biodiversity in the light of current knowledge. But solutions must also be implemented, and we have fostered discussion and debate about who, within our society, is responsible for conserving biodiversity. What are the roles of the law, landholders, industry and lobby groups? How does a government agency charged with the task of conservation see its current and future role? While the views presented here are not exhaustive, we consider them indicative of the important elements in the debate.

Ultimately we must face the reality that conservation of biodiversity is not going to be achieved solely by reservation of land in national parks or other kinds of conservation reserves. Management of threats to biodiversity must take place across the spectrum of land tenure, but it is essential that these threats are dealt with in a systematic and co-ordinated way. Given this need, we contend that conservation in the future is only possible when a partnership of responsibility is accepted among all individuals, communities, institutions and governments of our society.

While this book and the conference have been fostered by the government agency charged with conserving biodiversity in the state of New South Wales, the information and ideas are of far wider significance. In many instances an explicit, national perspective is presented. Where the focus appears to be more parochial, we argue that, by example, the material has broad relevance beyond the context of one state or even of one group of biota.

Finally we draw attention to some omissions. A book of this kind cannot cover everything, nor can it cover all shades of opinion. In this respect the content reflects the biases and attitudes of the conference organizers, editors and associates. We do, however, specifically apologise for the omission of two invited review papers, covering history of land clearance and climate change and a further paper dealing with water management policies. While these papers were presented at the conference they were not submitted for publication. Adequate substitutes, dealing with climate change and water policies, could not be arranged in time for publication.

Conference organization and subsequent publication of papers has consumed much time and energy. But this effort is miniscule compared with what is needed to conserve the biodiversity of Australia. We hope that our effort encourages others to see the issues more clearly and to play a part in conserving the species and ecosystems that are part of the fabric of this continent and the life supports for the human species.

ACKNOWLEDGEMENTS

Thanks are due to many people who assisted with the smooth organization and running of the conference and the production of this book. Most noteworthy of these are: the Commonwealth Department of Environment, Sport and Territories (DEST) who provided generous financial sponsorship; Mark Ooi who handled most of the unglamorous work of production with characteristic coolness, humour and vigour; Lynda Wild whose organizational skills, efficiency and unflappability made the conference, and ultimately this book, happen; the many National Parks and Wildlife Service staff who most capably assisted with the running of the conference; Grant Bywater, author of the Biodiversity Strategy, for valuable and friendly discussion of ideas and material that are integral to this book; Ivor Beatty and Jim Johnson of Surrey Beatty and Sons who provided valuable advice and assistance in a friendly and professional manner; the numerous referees who have contributed greatly to the high quality of material presented, and, finally, to all conference participants, including speakers at affiliated sessions not covered here, and the registrants who have patiently waited for this publication.

Ross Bradstock
Tony Auld
David Keith
Richard Kingsford
Dan Lunney
Dominic Sivertsen

March 1995

Contents

Preface .. v
List of Colour Plates .. viii

CONSERVING BIODIVERSITY .. 1
Why conserve biodiversity? By A. J. Beattie .. 3
Practical methods for conserving biodiversity. By A. A. Burbidge and K. J. Wallace .. 11

HABITAT LOSS .. 27
Habitat loss — its nature and effects (including case studies from New South Wales). By D. P. Sivertsen .. 29
Farmers as conservators of a sustainable production base. By R. L. Davidson .. 43
Urbanization and transport. By P. Adam .. 55
Mining in relation to habitat loss and reconstruction. By D. Woodside and D. O'Neill .. 76
The responses of some selected arthropods and the frog *Crinia signifera*, to habitat fragmentation. By C. R. Margules, K. F. Davies, J. A. Meyers and G. A. Milkovits .. 94
Habitat reconstruction: the revegetation imperative. By D. A. Saunders and R. J. Hobbs .. 104

DEGRADATION AND POLUTION OF WATER RESOURCES .. 113
Water resources and biodiversity: past, present and future problems and solutions. By P. Cullen and P. S. Lake .. 115
The impact of salinization and sedimentation on aquatic biota. By L. Metzeling, T. Doeg and W. O'Connor .. 126
Water pollution, biodiversity and the effectiveness of water quality criteria. By R. T. Buckney 137
Ecological effects of river management in New South Wales. By R. T. Kingsford .. 144
Managing water resources: seeking unity in the interests of diversity. By P. Crabb .. 162

WEEDS AND FERAL ANIMALS .. 175
Conserving biodiversity: impact and management of exotic organisms. By M. D. Fox .. 177
The threat of environmental weeds to biodiversity in Australia: a search for solutions. By R. J. Adair 184
Solutions to feral animal problems: ecological and economic principles. By M. Bomford, A. Newsome and P. O'Brien .. 202
Solutions to weed problems in urban bushland. By R. Buchanan .. 210
Weeds and feral animals: management of future threats. By F. D. Panetta and J. K. Scott .. 220

COMMERCIAL USE OF NATIVE BIOTA .. 229
Resource harvesting and biodiversity: the contribution of ecologists to policy decisions. By I. R. Noble .. 231
Sustainable forestry — the urgency to make the myth a reality. By T. W. Norton and J. B. Kirkpatrick .. 240
How does grazing of native pastures affect their biodiversity? By M. H. Friedel and C. D. James 249
Harvesting wildlife: kangaroos and waterfowl. By S. C. Cairns and R. T. Kingsford .. 260
Australia's native wildlife trade: scale, trends and impacts. By D. J. Callister and L. M. Williams 273
The practical application of population viability analysis for conservation planning. By H. P. Possingham .. 292
Biodiversity and fisheries management: the implications of extracting maximum yields from interactive ecosystems. By R. E. Kearney .. 300

CHANGES TO FIRE REGIMES .. 307
Extinction of biota by fires. By A. M. Gill and R. A. Bradstock .. 309
Fire and conservation: imperatives and constraints on managing for diversity. By R. A. Bradstock, D. A. Keith and T. D. Auld .. 323

CAN GOVERNMENTS SOLVE THE PROBLEMS? .. 335
Policy instruments for conserving biodiversity on private land. By D. Farrier .. 337
Some socio-economic and ecological implications of alternative biological diversity conservation strategy options. By M. D. Young .. 360
How well does the EIA process protect biodiversity? By R. Buckley .. 365
The role of non-government organizations in the development of international and national policies for the conservation of biological diversity. By R. C. Nias .. 375
The role of activists. By R. Breckwoldt .. 380
Conserving biodiversity — a viewpoint on the role of the landholder. By M. Tooth .. 387
Biodiversity conservation in New South Wales: the role of the National Parks and Wildlife Service. By D. Papps and P. Wilson .. 391
Conserving biodiversity — threats and solutions: Overview. By R. Purdie .. 410

Index .. 417

List of Colour Plates

Plate 1. Barmah Forest. Photo by D. Eastburn .. .. Opposite Title Page
Plate 2. Blue-green algae. Photo by Kathie Atkinson .. .. Opposite Title Page
Plate 3. Orb Spider. Photo by Kathie Atkinson .. .. Page 2
Plate 4. Ants on carcass. Photo by Kathie Atkinson .. .. Page 2
Plate 5. Land clearance. Photo courtesy NSW Department of Conservation and land Management .. .. Page 28
Plate 6. Usage over time. Photo courtesy NSW Department of Conservation and Land Management .. .. Page 28
Plate 7. Macquarie Marshes. Photo by D. Eastburn .. .. Page 114
Plate 8. Burrendong Dam. Photo by Richard Kingsford .. .. Page 114
Plate 9. Cotton farming. Photo by Richard Kingsford .. .. Page 114
Plate 10. Aquatic invertebrates. Photo courtesy Museum of Victoria .. .. Page 174
Plate 11. Stream-bank damage. Photo by Leon Metzling .. .. Page 174
Plate 12. Fox baiting. Photo by David Priddel .. .. Page 176
Plate 13. Cat with bird. Photo courtesy NSW NP&WS .. .. Page 176
Plate 14. Damage to acacia by rabbits. Photo by Tony Auld .. .. Page 176
Plate 15. Bitou bush. Photo by Kathie Atkinson .. .. Page 176
Plate 16. Logging coupe. Photo by Dan Lunney .. .. Page 230
Plate 17. Galah smuggling. Photo courtesy Australian Customs Service .. .. Page 230
Plate 18. Duck hunting. Photo by Wayne Lawler .. .. Page 306
Plate 19. Black duck. Photo by Wayne Lawler .. .. Page 306
Plate 20. High intensity fire. Photo by Tony Auld .. .. Page 308
Plate 21. Low intensity fire. Photo by Ross Bradstock .. .. Page 308
Plate 22. Fire regime map. Photo by Ross Bradstock .. .. Page 334
Plate 23. Houses in bush. Photo by Ross Bradstock .. .. Page 334
Plate 24. Rural bush remnants. Photo by Dan Lunney .. .. Page 336
Plate 25. Woodchip protest. Photo by Roland Breckwoldt .. .. Page 336

Conserving Biodiversity

PLATES 3 *(left)* and 4 *(below)*. Invertebrates serve to remind us of the qualities of beauty, utility (e.g., *left*, the web of the orb weaving spider may form the basis of new fibres) and services (e.g. *below*, ants as agents of organic decomposition) that humans derive from biodiversity. Yet invertebrates are often overlooked in many assessments of biodiversity. *(Photographs by Kathie Atkinson.)*

Why conserve biodiversity?

A. J. BEATTIE[1]

There are three overlapping reasons for conserving biodiversity: beauty, utility and profit. Beauty can be the most difficult to appreciate but some simple analogies are remarkably illuminating and should not be neglected. Utilitarian reasons chiefly concern the maintenance of ecosystem function and, although obvious and cohesive with all-embracing economic and political realities, have not been widely recognized until recently. Reasons of profit have been controversial but are increasingly accepted and are likely to be potent and effective in the inevitable context of a world marketplace.

These reasons (i) Demonstrate the importance of conserving all levels of biodiversity, genes, species and ecosystems, each of which has unique properties that contribute profoundly to human welfare and economics in very different ways. (ii) Emphasize that the task of exploring biodiversity and discovering its multiple resource values, while massive should not be avoided. Their skilful use could increase funding horizons by orders of magnitude to place biodiversity among the select handful of fields known as big science. (iii) Incorporate the concept of intergenerational equity. Thus, the quality of life for an expanding human population with massive and increasing demands for biodiversity-based services and resources will be determined, in large part, by contemporary conservation policy and effort.

INTRODUCTION

THERE are many reasons for conserving biodiversity but I have organized mine around just three words: beauty, function and profit. The concepts they symbolize undoubtedly overlap and many would probably choose other words, but these are the ones upon which I shall attempt to build my arguments.

Beauty

By beauty I mean, *A quality that is present in a thing or person giving intense aesthetic pleasure or deep satisfaction to the senses or the mind* (Stein 1984). It is commonplace to hear people ask why this or that species should be conserved. I sympathize with this kind of question, especially when the species concerned is a beetle less than 1 mm long, or a scorpion that has just stung someone. To answer it I examine my own experience that has been repeated again and again, especially when I was young. At first a new organism would seem repulsive or even dangerous. However, as I got to know it, read about it, kept it as a pet and watched it in the garden, my interest and respect grew until I would become positively protective. Of course, the key word here is *know* and knowledge is a word that, at least in this context, seems intimately associated with beauty.

What I am suggesting here is that there is usually something beautiful about an organism, but you may have to get to know it before that beauty is evident. Knowledge is likely to increase the accessibility or appreciation of beauty. Thus, the beauty of a habitat or a community may be found in its appearance, sounds, smells and colours, variety of forms and shapes, its organization, or its innate ability to make us relax. Similarly, the beauty of a species may be found in its sound, colour, form, smell, behaviour, adaptations, ability to survive, and its interactions.

In this light, even that scorpion can be genuinely admired. This is where David Attenborough and a few like him succeed so well. They take an obscure, ugly or even repulsive organism, examine it closely and show us the beauty of its adaptations and life style. If the beauty of the organism is not immediately self-evident, they may provide us with sufficient knowledge to find it.

A parallel with human relationships seems to be useful: Most of us have a kind of generic caring about a remote group of people that are in trouble that leads us to a feeling of unease when we hear of their plight. However, when the anthropologist or war correspondent documents their lives and hardships we begin to sympathize with their circumstances. We may show this by donating to the Red Cross. However, when they get off the plane as refugees and see their faces and hear their stories we become more active and call up to offer clothing, furniture or even a job. Finally, when we meet them at the refugee centre we may cry with them and take their children into our homes. Knowledge brings with it an increasing level of involvement. It helps us to appreciate the beauty of those people who once seemed so strange on the TV screen but now receive our active concern and respect.

While some organisms strike us as immediately beautiful, the majority probably do not. However, in either case, knowledge of an organism develops our interest, respect, concern and our sense of its beauty. As this sense grows and

[1]Research Unit for Biodiversity and Bioresources, School of Biological Sciences, Macquarie University, New South Wales, Australia 2109.

Pages 3–10 *in* CONSERVING BIODIVERSITY: THREATS AND SOLUTIONS, ed by R. A. Bradstock, T. D. Auld, D. A. Keith, R. T. Kingsford, D. Lunney and D. P. Sivertsen. Surrey Beatty & Sons. 1995.

matures, our motivation to conserve is increased. These ideas are summarized in Figure 1. The two curves represent the motivation to conserve. The curve A is for those species that are immediately appealing or beautiful. It could be called the koala curve. In this case, the steepest portion is to the right indicating that koalas, in general, have our interest and respect but it is not until knowledge levels are high that we appreciate they are endangered and the motivation to conserve them rises significantly. The curve B is for the vast majority of species that are not immediately appealing or beautiful. It might be called the weevil curve. Here, motivation for conservation is only generated by knowledge and it is uphill all the way. However, the beauty of some aspect of the organism does come into play at some point, perhaps increasing the steepness of the motivation curve.

Some of my colleagues have suggested that I should make knowledge the key word rather than beauty. Others think that the word education would be better. Perhaps they are right but the idea that the motivation to conserve originates from an appreciation of beauty and not necessarily because of knowledge has strong appeal. This is because many things strike us as beautiful spontaneously, without knowledge, and because knowledge may be incomplete.

I am grateful to my colleague Mark Westoby for expanding on the theme of knowledge and conservation: His analogy is books and libraries. Like books, living species contain a vast amount of information and therefore also merit preservation, maintenance and study. Also like books, their value to humanity is not only an immediate or instrumental one. Many books languish unread in libraries for years but become cherished when they provide new insights for readers perhaps centuries after the death of the author. Few question the value of maintaining libraries and all the books they contain on a permanent basis and when a library is destroyed by warfare or fire, there is a general appreciation of the disaster it represents. Books and species may have immediate or future utility, or they may simply enhance human life and culture by their availability for study, interpretation and revelation.

Function

By function I mean ecosystem function. The connection with biodiversity is the millions of species whose metabolism and interactions cumulatively produce the end products, the emergent properties, that we call ecosystem functions. They include primary and secondary productivity, gas exchange, decomposition,

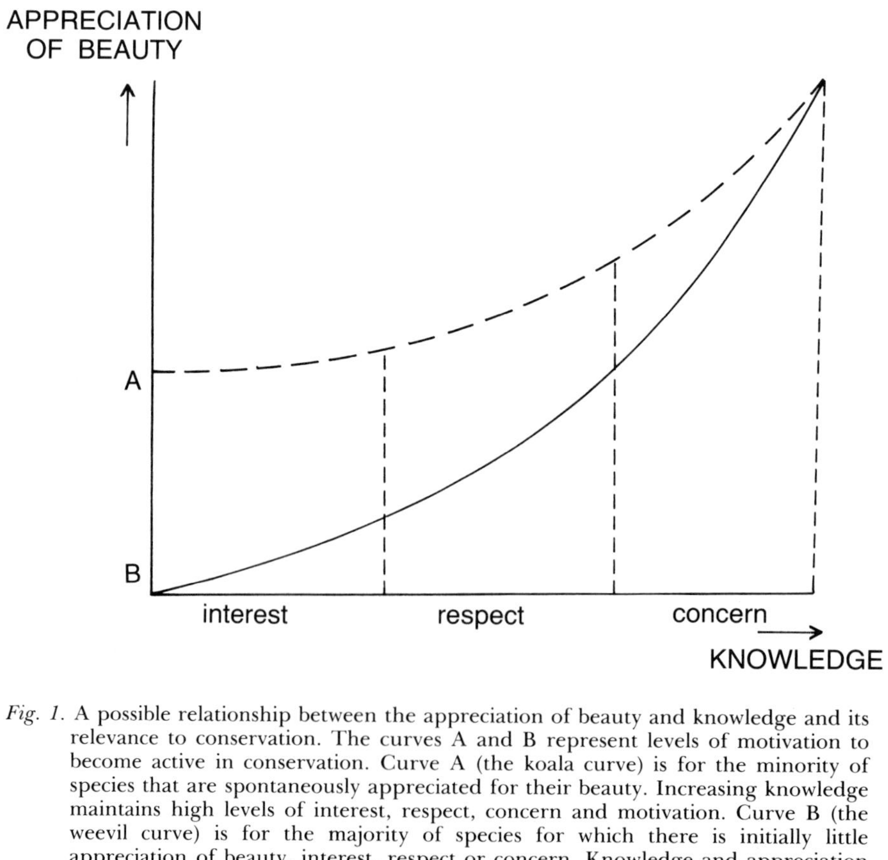

Fig. 1. A possible relationship between the appreciation of beauty and knowledge and its relevance to conservation. The curves A and B represent levels of motivation to become active in conservation. Curve A (the koala curve) is for the minority of species that are spontaneously appreciated for their beauty. Increasing knowledge maintains high levels of interest, respect, concern and motivation. Curve B (the weevil curve) is for the majority of species for which there is initially little appreciation of beauty, interest, respect or concern. Knowledge and appreciation of beauty may work synergistically to raise levels of motivation.

nutrient cycling, energy flow, herbivory, pollination and seed dispersal. When we interpret ecosystem functions as processes that support human survival and civilization, we call them ecosystem services. Thus, functions that may have highly technical names in the ecological literature become more familiar, for example, the maintenance of air and water quality, the maintenance of soil fertility, and natural pest control. I am not going to elaborate on the values of ecosystem services as writers such as the Ehrlichs have written about them for many years (Ehrlich and Ehrlich 1991). Their articulate and continuous output has repeatedly demonstrated that human beings, whether we exist as nomads, rural, urban or city dwellers, are interacting parts of the ecosystems we live in. While this may appear self-evident, the message has been remarkably difficult to get across.

We will take a brief look at a two examples: waste disposal and the degradation of pollutants and pest control. Many city dwellers have simply never thought about what happens to their bathwater, toilet water or garbage once it goes down the gurgler. It may or may not be treated but, sooner or later, it is dumped into an ecosystem. Rural dwellers may simply never have thought about what happens to the fertilizers and herbicides sprayed on crops and paddocks. These chemicals, just like city wastes, also enter ecosystems.

Let us be a little more explicit here — those city and rural wastes enter the habitats of hundreds of thousands of species, that is, the biodiversity that constitutes the living part of the ecosystem and, whether we live in the city or the country, all of us hope that the metabolism and interactions of those species function to degrade, detoxify, or sequester these wastes and perhaps even recycle the useful components. We all hope that the functions these organisms generate will do the job for us. Of course, when they do not, these wastes may quickly begin swilling around our doorsteps.

In another arena, for decades we have believed that pesticides control pests and indeed they do in some circumstances. However, the Volterra effect in which pesticide applications actually *increase* pest levels was predicted on theoretical grounds six decades ago (Volterra 1931). This has happened in practice so many times that people scarcely bother to think about it any more and respond, unthinking, by increasing pesticide applications still further. Yet there is a growing recognition in many countries, increasingly in the third world, that most pest control is, or can be, a function of biodiversity, that is, the diversity of predators, parasites and pathogens. Major trials have shown that by allowing these components of biodiversity access to crops and forests, pest control may be achieved at economically harmonious levels, with little or no requirement for pesticides (Joyce 1988; Sattaur 1988; DeBach and Rosen 1991).

These and other functions or services are widely regarded as free or, to put it another way, it is generally thought that their value cannot be assigned in monetary terms. Economists ask how you can place a value on a suite of fifty bacterial species that detoxify a factory waste, or guilds of parasitoid wasps, spiders and entomogenous fungi that control a crop pest?

Such questions are disingenuous because we perform the experiments to find out values such as these on a regular basis. These experiments are mostly inadvertent but sometimes quite well designed as there are before and after data for comparison. We perform these experiments every time we overload an ecosystem and poison some vital components of its biodiversity, for example, when sewage washes up on shore, a river is taken over by a toxic algal bloom, or pests increase in density following pesticide application. We then find out precisely what it costs to avoid overloading the inshore ecosystem, clean up the river, and the loss of the crop to pests.

These costs are a measure of the value of ecosystem functions. It is not hard to figure them out. The world spent $US15.9 billion on pesticides in 1985 (Postel 1987). Some part of this sum could be defrayed by biodiversity. For example, recent experience in several provinces of Indonesia suggest ways of calculating what might be saved by curtailing pesticide applications in favour of control by natural assemblages of predators and parasites. In some areas losses of the rice crop during 1986–88 were estimated at $US1.5 billion. Following the banning of several pesticides and massive reductions in pesticide applications, these losses were recouped in 3–5 years, depending on the location (McNeely 1988). The value of Indonesian insect predators and parasites which subsequently achieved economically viable levels of pest control is thus, very approximately, $US1.5 billion. I say "approximately" because, in general, savings are actually much greater: savings on fuel for pesticide applicators, savings on hospital bills from farm workers affected by pesticides, and savings on other crop yields reduced when pollinators are killed (Pimentel and Perkins 1980). Myers (1983) calculated that biological control projects saved Californian agriculture $US987 million between 1928 and 1979. A programme of biological control of the Rhodesgrass mealybug cost $US200,000 with estimated savings to the economy of Texas of US$177 million a year (DeBach and Rosen 1991).

Closer to home, what are the ecosystems and biodiversity off the Sydney coastline worth? When our beaches were plagued with human wastes it was decided to pump these wastes to areas many kilometres off shore. The Sydney Water Board informs me that this cost $307 million. This is a starting figure for the calculation of the net worth of the marine biodiversity around Sydney. However, sewage disposal is only one function; the supply of stock for commercial and sport fishing is another and, more recently, tourism has benefited greatly from the revived pastime of whale-watching.

The concept of ecosystem services is a profound one and extends further than many realize. Thus, photosynthesis and evolutionary processes are also ecosystem services. To some this may appear an extreme point of view but it is important to reflect on the basic fact that the sun is the sole source of energy for the planet and that photosynthesis is the only mechanism that enables us to take advantage of it. It is a monopoly and no viable alternative has ever been contemplated. From this vantage point, photosynthesis is the most fundamental ecosystem service of all. Evolutionary processes are driven by the energy derived from photosynthesis and produce the genomes that provide us with species for food, clothing, medicines and a countless other products. Evolutionary processes are the mechanism for testing the effectiveness of genes in literally millions of combinations. These tests yield the genomes that produce ecosystem function. Anyone who believes that genetic engineering and biotechnology can replace these processes is profoundly misinformed. While these wonderful new techniques enable us to tinker with genes and genomes in clever new ways, they are not within a million years — literally — of achieving the complexity, flexibility and resilience of an evolutionary product. I shall follow this up in my final section when I discuss the commercial use of evolutionary processes as a reason for conserving biodiversity.

It is important at this point to enter the debate about the conservation of function versus the conservation of species. Here I agree wholeheartedly with those who believe that conservation programmes concentrating on the maintenance of individual species, whether they are rare, endangered, charismatic, keystone or indicator species, are likely to encounter serious problems (Franklin 1993; Orians 1993; Walker 1989). However, this kind of reasoning can go too far. It has been suggested that the most wise course of action may be the conservation of ecosystem function. Further, it is assumed that some species are more essential to ecosystem function than others and that these should be, therefore, the primary subjects for conservation resources. Further still, not all species are equally important. Indeed, some, even many, may be redundant (Walker 1992; Lawton and Brown 1993).

I wonder if this line of argument is a little too static for modern conservation and ecological thinking. I agree that the conservation of habitats, communities and ecosystems, as opposed to individual species, is the best strategy. However, the question arises as to what, precisely, must be conserved to maintain function. The implication is that some sub-set of species will do this and others will not. However, the claim that some species are more important than others, or that some species are redundant, seems very premature, for several reasons.

First, most often we do not know which species are the most important to a given function and this will be very difficult to find out even for the better known guilds. The task appears nigh impossible for guilds of bacteria and fungi involved in, for example, essential nutrient cycles. If it is not impossible, it is going to take a very long time to find out.

Second, once some key species have been identified, how are the essential and the redundant species disentangled in a conservation strategy?

Third, we know that ecosystem functions change in both time and space. This is a deeply vexatious question in applied ecology, perhaps epitomized by the experience in East African game parks (Botkin 1990; Walker 1989). It seems that change is just about all ecologists can predict with certainty for ecosystems and most of the time we do not even know if the change is cyclical, periodic, chaotic or some progressive, open-ended process. Often it seems to be the latter and, as a result, we may say that there is no such thing as "balance" in nature, or that species composition is not in "equilibrium". Thus, it becomes difficult or perhaps impossible to predict which species will be more or less essential to the ecosystem, and whether or not they are or will be redundant, when ecosystem functions are likely to change widely in space and inexorably through time.

Fourth, as I wish to expand on in a moment, I believe we have a too conservative view of the utility of species. This may be because we have not yet evolved a rigorous discipline for the discovery of commercially useful species. My final discussion calls into question those who dismiss the future discovery of useful species as only a very small element of the conservation imperative.

In summary, I am concerned about the concentration of strategy on ecosystem function firstly, because of the constant change in ecosystems, and secondly because of the potential utility of species. In short, right now, we have little knowledge of which species will be most important to evolving ecosystems or which species will be valuable resources for future generations. The conservation of ecosystem function and species is inseparable for these reasons.

Utility

My final reason for the conservation of biodiversity is utility. This refers to the biological resources derived from biodiversity. There are three major categories of biological resources: ecosystem services, conventional bioresources and novel bioresources. As I have already spent some time on ecosystem services I will move straight on to conventional and novel bioresources.

Conventional bioresources are those that humankind utilizes for maintenance and trade. This utilization may have continued for many thousands of years. The species include: many species of marine and freshwater fish, molluscs, honeybees, cattle, sheep, goats, pigs, horses, chickens, turkeys, game birds, wheat, rice, corn, oats, barley, millet, apples, citrus fruits, bananas, many plant species for medicines, timber, clothing and building materials. Reflection on a list such as this reveals that, on the face of it, very few corners of biodiversity have been left unexplored. When the plant, animal and microbial resources of indigenous peoples are included with those of large and expansive civilizations, the inventory includes many thousands of species (Myers 1983).

While this list is long it should be borne in mind how few species have been described by science, perhaps 1.7 million out of 10, 30 or 50 million (Erwin 1991). This means that the potential for the discovery of new biological resources is great. Most national governments and many industries are, of course, aware of this and the major questions at the Rio de Janeiro Summit on Biodiversity were resource questions such as: Who owns them? Who has access to them? How will they be valued and paid for?

At the same time we are not being very clever about finding them. Major new programmes and initiatives for finding new pharmaceuticals from tropical plants, coral reef organisms or soil fungi are being announced around the world. Most of the time they involve massive and expensive screening programmes. Thousands of species are processed in the hope of finding something profitable. The emphasis is invariably on high technology such as new instrumentation for chemical analysis, new ways for assaying products, and sophisticated software to create designer drugs.

It seems to me that evolutionary biology is more incisive than screening and has worked very well so far to reveal new biological resources. Its application to a far wider array of fields than pharmaceuticals seems the most reasonable way to proceed.

Some of my earlier remarks were about evolutionary processes shaping the organisms with which we share the planet. Natural selection has moulded genomes, shifting the frequency and pattern of genes into units of variety that we call species. Therefore, each species is a kind of triumphal survivor with a set of genes that codes for a suite of adaptations. As we have shared much of the same selection pressures as other organisms, we can ask: Are these adaptations our solutions? This provides a new perspective on the search for biological resources. We can look for problems analogous to the human ones in biodiversity and examine how they have been solved by the species involved. We may be able to use any aspect of the natural adaptations that solved the problem: the genes, natural products, cells, tissues, behaviour or interactions.

My own research for new sources of antibiotics is an example. Antibiotic resistance in human pathogens is an increasingly frightening prospect and the search for an entirely new generation of antibiotics is in progress. In my lab we applied the evolutionary paradigm and asked: *Under what conditions would you expect antibiotics to evolve?* We know they have evolved in soils as fungi compete for resources by poisoning each other with the chemicals we call antibiotics. However, the question opens up many other possibilities. For example, antibiotics may be expected to have evolved: (i) where animals aggregate in large numbers, and (ii) where contagious diseases are therefore likely to be disastrous. This opens up many further possibilities, for example, breeding colonies, feeding flocks and overwintering aggregations. However, these are so diverse that more limited conditions are necessary to provide focus. Thus, the reasoning can be continued: (iii) where individuals live in close proximity, (iv) where this is obligate, as in societies, (v) where the societies are highly organized, and (vi) where the juvenile stages are kept together. This points us to the social insects, the bees, wasps, and ants. With the help of the Australian Research Council and a multi-national pharmaceutical company we have found antibiotics in ants which are proving to be both potent and unusual (Veal *et al.* 1992).

We are now taking the evolutionary paradigm even further by selecting those ant species whose life-history traits are arguably the most promising such as South American ants that cultivate a fungus, that is, a monoculture, actively suppressing the hundreds of bacterial and fungal species that would otherwise invade their fungal gardens.

The problem held in common is contagious disease. Organisms that in many ways are like us have evolved adaptations to solve the problem. In the case of ants, they have been at it about 60 million years longer than we have. Perhaps their adaptations can be adapted for our solutions.

At the moment then, ants are genuine biological resources. We do not yet know which species. There are between 4 000 and 8 000 species in Australia alone. Further, ants have had other proven commercial applications: as a source of new anti-termite chemicals and for insect pest control (Perfecto 1991; Augereau 1988). These examples have so far come from overseas, but which species might possess similar resource possibilities in Australia?

Having made a pitch for the evolutionary paradigm and the potential commercial importance of obscure components of biodiversity such as ants, it is now legitimate for you to wonder how generalizable this is. I will present just two examples out of the many available (Beattie 1994). They are good examples because they are (1) fascinating, (2) they illustrate the power of evolutionary biology, and (3) because they remind us sharply that you never know what species or groups of organisms will turn out to be useful.

The first is the case of a group of materials scientists in the USA searching for a natural paradigm for the ultrastructural design of an unlikely material, flexible concrete. Knowledge of natural history and the reasoning of evolutionary biology led them to the shells of deep sea molluscs. These shells must withstand many atmospheres of pressure and have evolved both toughness and flexibility. Shells were brought into the lab, sectioned and examined under the electron microscope. Their ultrastructure has subsequently formed the basis for new man-made materials now used in the manufacture of car parts and, before long, in construction projects (Birchall 1989; Webb *et al.* 1991).

The second example is the demand for high-tensile, ultra-strong, light-weight fibres. One application is stopping high speed projectiles such as bullets. The basic question is: Where would the evolution of fibres with the following properties be expected? Fibres adapted to (i) absorbing large amounts of kinetic energy, (ii) without damage to the projectile, while (iii) minimizing damage to the matrix of fibres and (iv) with only a gentle rebound?

Many kinds of natural fibres spring to mind, for example, hemp, flax, sixal, tendons and hair, but none of these has the required adaptations. However, a select group of spiders, the orb-web builders, have identical requirements. Here the adaptations seem very precise. Fast flying projectiles must be stopped on the spot (they cannot crash through and then fall to the ground). The fibres must absorb the energy of the projectile (the prey) so that it is not damaged. This is because these spiders can ingest only fluids and a damaged, leaking insect would lose much of its nutritional value. Finally it is an advantage for the fibres (web) to be reusable. Spider silk is being tested for many applications, for example, the filling of bullet-proof vests. Several research institutions are now working on obtaining commercially feasible quantities of silk from selected spider species for a variety of industrial uses (Helton 1990; Beard 1992).

Ten years ago who would have anticipated that species of ants, molluscs and spiders would be of commercial importance? Who would have anticipated that ant, mollusc and spider diversity was a resource? A world survey of similar proven or potential projects leaves little of biodiversity untouched (Beattie 1994).

The point is that evolution by natural selection has produced a vast variety of species that contain a massive store of problem-solving information of immediate or future utility. In addition to the ants, molluscs and spiders discussed here, commercial products, processes or services developed in the last few years involve many species of bacteria and fungi, algae and flowering plants, termites, wasps, lace bugs, nematodes and many other kinds of worms, protozoans, and fish.

Evolutionary biology and its associated disciplines: natural history, ecology, systematics and genetics recognize and understand biodiversity and provide a strong centralized theory of its origins, distribution, and function. They are practical tools for the discovery of resources. They provide an infrastructure whereby we can move away from screening and, through careful logic, identify those environments, families, species, tissues or natural products most likely to make commercial investigation profitable.

This brings me to my final point. We have discussed the importance of the conservation of biodiversity through the conservation of habitats, communities and ecosystems. I have suggested that there is currently little rational basis to claim that some species are more important than others, and that, on functional grounds, all species should be conserved. Here, I add that species should be conserved as potential resources. It is important to emphasize once again that it is the species that counts. The search for biological control agents is a good way of illustrating this. It has often been shown that biological control is successful only when *precisely* the right species for the job has been found. This is no surprise. Each species, by definition, has a unique genome, a unique assemblage of genes. We know as a result of bitter experience, that most often only one genome meets the highly specific requirements of the job (Room 1990). Only one set of genes codes for that precise behaviour, life-history or bioactive compound that is required to effectively combat a particular pest.

To return to the title of the talk: Why conserve biodiversity? I suggest one answer is because we are only just beginning to explore its commercial potential. We are only just inventing ways of doing this and the next several generations are likely to have developed them beyond our imagination. The basic resources are species or rather their genetic information, and all of its morphological, physiological, biochemical and behavioural derivatives. For this reason I agree with E. O. Wilson (Wilson 1989) and others that we should make every effort to inventory species. Erwin (1991) said that determining the number of species on Earth is like reaching for the stars. I wonder if he intended his analogy to be so apropos. Internationally, non-defence scientific efforts are massively concentrated on the stars, the extra-terrestrial. Why are all these resources and funds directed away from the concrete problems right here, like finding out how many species there are on Earth and what useful adaptations they harbour?

The answer may be because the aerospace industry and its captive sciences have persuaded entire governments and much of the public that cosmic exploration is important. I have to confess that I doubt the veracity of their claims of finding the origin of the universe while simultaneously improving the lot of the third world with spin-off technologies. I am not alone in questioning the value of the so-called spin-off technologies (Slakey 1993). What might the international conservation agencies do with the kinds of funds assigned to space craft, space stations and space telescopes? We might do well to examine the strategy. I get impatient with biologists who claim we will never know the number of species on Earth or never be able to inventory them. We know this is a gargantuan task but then so do the space agencies know that *theirs* is a gargantuan task. It has not stopped *them* from going all out for funding and channelling billions of dollars into space projects.

It is worth aiming at total inventory because we need to know what the components of ecosystem functions are and because we are most likely sitting on the edge of a vast array of new resources. Thus, there are hundreds of frontiers left on Earth and all manner of exploration to be carried out. Perhaps what we need is the equivalent of an aerospace industry. Those industries that develop around ecosystem management and restoration , and around new natural resources, might just be the boost that is needed.

We conserve biodiversity because it is beautiful, because it is the functional foundation of civilization, and because it undoubtedly harbours an astonishing array of resources. We all think of biodiversity as being integrated over at last three levels — genes, species and ecosystems. Accordingly, it is across all three levels that it should be conserved.

REFERENCES

Augereau, J. F., 1988. Natural insecticide derived from ants. *The Guardian*, July 17, p. 14.

Beard, J., 1992. Warding off bullets by a spider's thread. *New Scientist* **136:** 18.

Beattie, A. J., 1994. Conservation, evolutionary biology and the discovery of future biological resources. Pp. 305–12 in *Conservation biology in Australia and Oceania* ed by C. Moritz and J. Kikkawa. Surrey Beatty & Sons: Chipping Norton.

Birchall, J. D., 1989. The importance of the study of minerals to materials technology. Pp. 491–509 in *Biomineralization: chemical and biochemical perspective* ed by S. Mann, J. Webb and R. J. P. Williams. Verlagsgesellschaft: Hamburg.

Botkin, D. B., 1990. *Discordant harmonies. A new ecology for the twenty-first century.* Oxford University Press: New York.

Debach, P. and Rosen, D., 1991. *Biological control by natural enemies.* Cambridge University Press: New York.

Ehrlich, P. R. and Ehrlich, A. H., 1991. *Healing the Planet.* Addison Wesley: Menlo Park, California.

Erwin, T. L., 1991. How many species are there? Revisited. *Cons. Biol.* **5:** 330–33.

Franklin, J. F., 1993. Preserving biodiversity; species, ecosystems or landscapes. *Ecol. App.* **3:** 202–05.

Helton, D., 1990. Web of Steel. In *Wildlife Tales.* M. Birkhead (Producer), B. Green (Narrator) (audiovisual). ABC/Kane Productions: New York, NY.

Joyce, C., 1988. Nature helps Indonesia to ant its pesticide bill. *New Scientist* **118:** 35.

Lawton, J. H. and Brown, V. K., 1993. Redundancy in ecosystems. Pp. 255–70 in *Biodiversity and ecosystem function* ed by E. D. Schulze and H. A. Mooney. Springer-Verlag: New York.

Mcneely, J. A., 1988. *Economics and biological diversity.* I.U.C.N., Gland: Switzerland.

Myers, N., 1983. *A wealth of wild species.* Westview Press: Boulder Colorado.

Orians, G. H., 1993. Endangered at what level? *Ecol. Appl.* **3:** 206–08.

Perfecto, I., 1991. Ants as natural control agents of pests in irrigated maize in Nicaragua. *Jour. Eco. Entomol.* **84:** 65–70.

Postel, S., 1987. *Defusing the toxic threat: controlling pesticides and industrial waste.* Worldwatch Paper No. 79. Worldwatch Institute: Washington, DC.

Pimentel, D. and Perkins, J. H., 1980. *Pest control: cultural and environmental aspects.* AAAS Selected Symposium 43. Westview Press: Boulder, Colorado.

Room, P. M., 1990. Ecology of a simple plant-herbivore system: biological control of *Salvinia. Trends in Ecology and Evolution* **5:** 74–79.

Sattaur, O., 1988. A new crop of pest controls. *New Scientist* **119:** 48–54

Slakey, F., 1993. Rationales for research. *New Scientist* **139:** 42–43.

Stein, J. (Ed), 1984. *Random House Dictionary of the English Language*. Random House: New York.

Veal, D. A., Trimble, J. E. and Beattie, A. J., 1992. Antimicrobial properties of secretions of the metapleural glands of *Myrmecia gulosa. Jour. App. Bac.* **72:** 188–94.

Volterra, V., 1931. *Lecons sur la théorie mathématique de la lutte pour la vie.* Gauthier-Villars: Paris.

Walker, B., 1989. Diversity and stability in ecosystem conservation. Pp. 121–32 in *Conservation for the twenty-first century* ed by D. Western and M. C. Pearl. Oxford University Press: New York.

Walker, B., 1992. Biodiversity and ecological redundancy. *Cons. Biol.* **6:** 18–23.

Webb, J., Evans, L. A., St. Pierre, G. and Macey, D. J., 1991. Biominerals: source and inspiration for novel advanced markets. *Search* **22:** 137–39.

Wilson, E. O., 1989. Conservation: the next hundred years. In *Conservation for the twenty-first century* ed by D. Western and M. C. Pearl. Oxford University Press: Oxford.

Practical methods for conserving biodiversity

A. A. BURBIDGE[1] and K. J. WALLACE[2]

Biodiversity can be affected by almost every human activity and, accordingly, if biodiversity conservation is to be achieved, it must be integrated with the everyday activities of individuals, businesses, industries and governments, as envisaged by the ecologically sustainable development process. We discuss practical methods for conserving biodiversity that can be implemented now under three broad groupings: social and cultural issues, conservation biology and resources. Actions are suggested for developing a community commitment to conservation, developing a better protected area system, improving off reserve conservation, recovering threatened species and threatened ecological communities, controlling alien and problem indigenous species, and the provision of resources. Priorities for research are also proposed. The adoption and implementation of ESD principles, including biodiversity conservation, are enormous challenges facing our society. Unless we accept these challenges and are successful in solving the associated problems, we may well leave an impoverished world to future generations. Implementing biodiversity conservation is possible. Already, we know enough about the problems and solutions to be able to make rapid progress. The challenge is to convert ideas, information and strategies into effective action.

INTRODUCTION

RECENTLY, much has been said and written about the benefits of and need for the conservation of biological diversity (herein termed biodiversity conservation). Internationally, the recently concluded Convention on Biological Diversity, which Australia has signed (along with over 150 other nations) and ratified, has highlighted the contribution of biological diversity to human well-being and economies, and promoted the conservation of ecosystems, species and genetic diversity.

Within Australia, biodiversity conservation is one of three core objectives of the National Strategy for Ecologically Sustainable Development (Commonwealth of Australia 1992). This strategy was endorsed by the Council of Australian Governments in December 1992. Under the Inter-Governmental Agreement on the Environment (IGAE), national strategies for biodiversity conservation and for the conservation of threatened species and ecological communities have been prepared.

While much has been said, written and published about biodiversity conservation, the ability of most government conservation agencies to actually do things is limited by the recession and its aftermath, which have resulted in fewer resources. Conservation biologists and managers, who must implement the strategies, could be excused for concluding that the rhetoric about biodiversity outstrips their capacity to actually conserve and manage it, and far outstrips the political commitment to allocate the necessary resources. Australia has many excellent conservation biologists, but not all are employed.

It is easy to set down some of the impediments to effective biodiversity conservation. For example, many people do not understand what biodiversity is or why it is important, let alone why society should spend money to conserve it. Conservation biologists know little about the functioning of ecosystems, have only a general idea of how many species of organisms there are on earth and know little about genetic diversity or its conservation. We in Australia could easily set up expensive research programmes that might, after several decades, produce more information about some of these subjects. But, if we wait for more research data before acting, it may be too late for major components of biodiversity — we must act now and, if necessary, learn from our mistakes. Individuals and land managers (of both public and private lands) do want to know what they can do now. They want advice on practical, achievable actions that can be implemented without too much additional cost. In this paper we attempt to provide this guidance.

Biodiversity is affected by almost every activity carried out by humans and, accordingly, if biodiversity conservation is to be achieved, it must be integrated with the everyday activities of individuals, businesses, industries and governments. ESD principles provide a theoretical basis for such integration, but need to be applied so as to provide actions at the local level.

We will discuss biodiversity conservation under three broad groupings — and it may surprise many that a paper written by two biologists starts by concentrating on social and cultural issues and on information and education. This is intended to emphasize that

[1]Department of Conservation and Land Management, Western Australia, Threatened Species and Communities Unit, P.O. Box 51, Wanneroo, Western Australia, Australia 6065.
[2]Department of Conservation and Land Management, Western Australia, Wheatbelt Region, P.O. Box 100, Narrogin, Western Australia, Australia 6312.

Pages 11–26 *in* CONSERVING BIODIVERSITY: THREATS AND SOLUTIONS, ed by R. A. Bradstock, T. D. Auld, D. A. Keith, R. T. Kingsford, D. Lunney and D. P. Sivertsen. Surrey Beatty & Sons. 1995.

biodiversity conservation depends most on the attitudes of people and that the attitudes of all sectors of society are important. It is one thing to have support for conservation in the big cities — this may influence governments to pass legislation, produce strategies and provide money — it is another thing to have the support of economists, business-people and country people. Farmers neighbouring a national park have much more influence on it and on its management than most people who visit occasionally from a distant city.

While we wish to emphasize the importance of social and cultural issues, we will not ignore conservation biology, which is dealt with in the second section. In the third section we discuss resources — the issue that will, in the long term, determine success. Throughout, we will concentrate on practical measures that can be implemented now by governments, business and industry, and individuals. We will also, as requested by the conference organizers, provide examples from our experience in Western Australia.

THE NON-BIOLOGICAL PREREQUISITE — SOCIAL AND CULTURAL ISSUES

In western societies such as Australia, the goal of conservation of biological diversity will be achieved only if most individuals accept, for whatever reason, that this goal is in their personal best interests. The combination of private ownership of businesses and agricultural lands, individual long-term leases of pastoral lands, western concepts of "individual rights", and the cultural importance of material wealth, make this essential.

Given that problems such as land degradation, over-use of resources and pollution continue to threaten biodiversity conservation, it is apparent that there is as yet insufficient commitment at the individual level. Developing a commitment to conservation largely involves socio-cultural issues, therefore it is essential to discuss these when considering practical aspects of conserving biodiversity. In this section the development of individual commitment is discussed, followed by a discussion of several important strategies.

Individual commitment to conservation

There are three closely related ways of developing personal commitment to biodiversity conservation. These are:

1. the cosmological imperative;
2. the ethical imperative; and
3. the selfish imperative.

Each of these are inter-dependent — any significant change in one is likely to affect the others. They are considered below together with relevant action steps.

The Cosmological Imperative

A number of writers have stated or inferred that, to protect and maintain our environment, the cosmology of society must be consistent with conservation goals (Knudtson and Suzuki 1992; Nash 1990; Skolimowski 1988). (Cosmology is defined as the human view of the universe; the branch of philosophy that concerns itself with the origin and general structure of the universe, its parts, elements, and laws.)

For example, according to mythologist Joseph Campbell, Judaeo-Christian beliefs are based on a mythology which has separated humans from nature and is thus inherently anti-nature (Campbell with Moyers 1988). This has encouraged a utilitarian view of nature typified by the philosophy that plants and animals were put on this earth solely for human use and benefit. Indeed, "taming the wilderness" has long been viewed as a Christian task (Nash 1982). Although this view is not shared by all Christians, where this interpretation of Christian cosmology has prevailed over that promoted by St Francis of Assisi and others, it has frequently proved inimical with conservation goals.

However one views these issues, there can be no doubt that cosmological beliefs profoundly influence the way in which humans relate to their environment. As any professional manager of conservation lands knows only too well, logic and persuasion are weak forces of attitudinal change unless they are consistent with underlying value systems. Thus while cosmological issues may seem removed from everyday action, they are in fact frequently at the heart of land management conflicts.

Therefore it is very important to recognize the effects of cosmologies and their associated value systems on conservation of biodiversity and to operate accordingly.

Although not necessarily related to cosmological beliefs, it is relevant to mention here the dangers inherent in different disciplines and organizations developing quite different cultures and ways of knowing. Norgaard (1992) has pointed out that, within disciplines or organizations, belief systems are simplified and incapable of encompassing the complexity of reality. In contrast "Interdisciplinary scholars move between disparate cultures and become acutely conscious of the simplicity of each discipline's belief system and the conflicts that arise from attempting to overlap them" (p. 210). The clash of organizational and disciplinary cultures results in significant inefficiencies in resource use, and is a significant impediment to conserving biodiversity.

In our current circumstances practical action steps include:

- Providing conservation professionals with an understanding of the relationships between cosmologies, value systems and management. This can be achieved through professional training and by promoting debate within institutions and government agencies (for example, see Saunders and Burbidge 1988).
- Encouraging discussion and adoption of cosmologies and value systems that are consistent with conservation of biodiversity. In pursuing this goal, it must be recognized that there is no *one* appropriate cosmology. The key issue is whether or not a particular cosmology is consistent with conservation of biodiversity.
- All conservation professionals tithing to informing, educating and exposing the public, particularly younger people, to value systems consistent with conservation goals. Tithing should be directed mainly towards groups that will benefit from a better understanding of the issues — preaching only to the converted is a waste of time. Preaching to the converted is, however, worthwhile; if the converted (most of whom are non-professionals) can be stimulated to tithe some of their time to teaching the unconverted, the task of the professional is made easier.
- Facilitating staff transfers between agencies and organizations (including industrial and international organizations) to improve the ability of professionals to work across cultures and value systems. Staff of large organizations tend to become introverted and unable to see other points of view.
- Developing clear visions and goals for organizations which reflect as appropriate community values and aspirations, and ensuring that these are, as far as practicable, consistent with those of other organizations. It is essential that conservation organizations develop a coherent operational philosophy and educate their employees about it.

Although the expenditure of resources in this area, where results are generally long term, may seem wasteful to bureaucrats concerned with balancing budgets, not to do so will inevitably prove very costly.

The Ethical Imperative

The success of human social systems is dependent on people, at least at some critical, minimum level, giving respect to others and co-operating for the mutual benefit of all. Such respect and co-operation are framed in a set of ethics or laws, for example prohibitions on murder, which are based on cosmologies and mythological systems, or on systems of logic, or on some combination of both.

Leopold (1949) was among the first to argue a detailed case for ethics to be extended to land and the wildlife living on it and Nash (1990) has reviewed the case for the "rights of nature".

It is now widely recognized that if we do not look after our environment, then we will eventually suffer a dramatic decline in living standards and an associated decline in health. Pollution of the air we breathe or of drinking water supplies, degradation of agricultural land and loss of biotic and landscape diversity are some examples of detrimental environmental changes. We will also irretrievably lose the potential to benignly exploit natural genetic resources, e.g., for pharmacology. If we are to counter these trends, we must look after the environment. Conservation of biodiversity has, therefore, sufficient personal importance to warrant expression in a set of appropriate social ethics that formalize respect for nature.

Logically, then, one can show the need, couched in terms of individual human survival, for an ethic which incorporates conservation of biodiversity. Such an ethic has been defined by the Western Australian Department of Conservation and Land management (CALM) as "the duty of humans to recognize that, for their own benefit, they must have a symbiotic relationship with their environment, and must care for that environment" (CALM 1990).

However, one can also arrive at a similar ethic based on spiritual beliefs — for example, by asserting that all life is sacred and must therefore be protected (although this statement must be qualified to the extent that natural extinctions do occur and that most religions would expect other species to be sacrificed before humans). Nash (1990) describes many of the issues involved.

Thus the social laws expressed as ethics are a combination of Cosmological and Selfish Imperatives that, some may argue, are themselves two expressions of the complex body of human needs.

If we really want people to conserve biodiversity, then ultimately this must be seen as an ethical consideration of equivalent value to accepted cultural prohibitions on crimes against people, even, in extreme cases such as extinction, to murder. This is the only means, in our society, by which conservation of biodiversity can be implemented in day-to-day operations. Even then there will be environmental "murders", just as there continues to be real murders despite the strength of ethical and legal barriers.

Accepting that in the short term conservation of biodiversity cannot await a cosmological transformation, humans must be persuaded to an ethic which is in sympathy with their own needs and conservation of biodiversity. Acceptance of such an ethic will therefore rely on actions listed under The Selfish Imperative, however, in the medium to longer term those listed under The Cosmological Imperative are also essential.

The Selfish Imperative

If we are to really convince most people that conservation of biodiversity is worthwhile in the short term, then we must unequivocally demonstrate that conservation of biodiversity is of direct and immediate relevance to the quality of life of each individual human. That a recent workshop on biodiversity in Mediterranean climatic regions of Australia (Hobbs 1992) was unable to conclude that maintenance of biodiversity is essential for ecosystem functioning suggests that collectively we need to do some very quick and incisive thinking and research. Those of us who are "biodiversity conservation professionals" will not be convincing in this most important of issues if we are publicly equivocal.

The values of biodiversity to human survival and well-being are becoming better documented and are dealt with in another paper in this conference. These values must be clearly stated by those concerned with biodiversity conservation.

It is not enough to describe the values of biodiversity conservation in terms of economics, ecosystem resilience, human health and comfort, aesthetics and so on. Nor is it enough to ensure that a significant percentage of the community understand and express support for these values in either emotional or intellectual terms. What is required is that the community understands these values and considers them sufficiently highly to act for them at some personal cost in the short term. This point is crucial to evoke the Selfish Imperative, and is therefore explored further.

Difficulties in this area are underlined by Dunlap's (1987) work in which United States of America voting patterns seemed to be at variance with expressed public concern for the environment. There is also evidence that individuals expressing concern for the environment are less likely to support management actions to protect the environment which impinge directly on their lives (Carr and Tait 1991; Noe and Hammitt 1992). Even if the expressed attitudes are favourable, they may not translate into commitment and appropriate action. A related phenomenon is a variation on "nimbyism" (the "not in my back yard" syndrome) when people express very high concern for things happening close to home, but do not care when it happens elsewhere.

More recent research by Dunlap (1991) shows that, at least in the United States, public concern about environmental issues has increased. However, the translation of this concern into effective action has not occurred to any significant degree.

For adequate resources to be provided, it is essential that individual support for environmental issues translates into action. We should constantly remind ourselves of Frazier's (1990 p. 394) comments that "conservation of natural resources — particularly in the international arena — has much less to do with biology than we ecologists and wildlife managers would like; key decisions are in the hands of politicians, economists, developers and that ever-growing tribe of international experts. A week's activities in foreign aid, or international development, or international debt, or political pressures or social conflict, have much more impact on the use of natural resources than do decades of careful ecological studies and wildlife management projects, whether we like it or not".

If truly committed to conservation of biodiversity, then conservation managers and researchers must be prepared to contribute in the socio-political domain as well as the biological. Complaining that lobby groups have got their "facts" wrong is a waste of time if you are not willing to contribute to the debate.

Actions that can be taken now include:

- Ensuring that the case for conservation of biodiversity and its linkage to the well-being of each individual is clearly established in terms which are relevant to, and accepted by, the community. This acceptance must be sufficient that actions are in accord with expressed values.
- Explaining biodiversity values. Nature conservation professionals themselves must be clear on the values of conservation of biodiversity, and be prepared to explain them to any and all segments of the community including politicians and industry. In our 15 years of interviewing potential professional land managers, we have noted that questions concerning the explanation of nature conservation to farmers have been consistently poorly answered by job applicants, and there appears to have been only marginal improvement over the period.

- Tithing. As discussed above, the concept of professionals devoting a significant proportion of their time in helping the public and politicians understand nature conservation (see for example Saunders *et al.* 1987) is very important. Conservation professionals form perhaps the largest, informed group capable of delivering the conservation message to a wide audience. It is essential that professionals contribute to this task and do so both as individuals and through their professional societies. The latter is especially important as it provides a voice to those who are otherwise restricted, because of their employment, from being involved.

- Researching cultural change. We have little concept of the current beliefs within the community, how these relate to conservation of biodiversity, or how community views might be altered without resort to indoctrination and other unacceptable techniques (see Prior 1992 for some ideas). These are major gaps in our current knowledge. Machlis (1992) concludes that analysis of social and political issues, combined with research of the human causes of biodiversity loss and the formulation of solutions, are all possible and necessary. Sociology and the other social sciences can make an important contribution to conservation biology.

- Providing leadership and living examples. Governments and committed individuals can provide demonstration areas where ESD actually works and can be evaluated in economic as well as environmental terms.

- Many action steps listed under other sections (e.g., Protected Areas) are also relevant here. Some, such as those involving shared management, should encourage empowerment of local people so that they implement conservation and provide political pressure for resources.

Strategies for change

A wide range of approaches are available to bring about the social changes discussed above. Important means include legislation and enforcement, education and access to information. The following discussion is not intended to be comprehensive, but rather, to highlight practices that are particularly important in conservation of biodiversity.

Legislation and Enforcement

Although it is theoretically possible to achieve conservation of biodiversity by enforcement, somewhat as occurred within the hunting parks of England, this is not an acceptable approach in our society. However, while not a feasible means of achieving conservation of biodiversity by itself, it has an important role, particularly when combined with other means such as education.

Empowerment of local communities — helping people to help themselves — is a crucial component of achieving many Landcare goals (Hartley *et al.* 1992) and is equally important in conservation of biodiversity. However, while empowerment, education and extension will remain important and popular conservation strategies, the proper use of legislation and enforcement is also an important tool for assisting social change and compliance.

In a review of soil conservation legislation in Australia and overseas, Bradsen (1992a) concluded (among other things) that while a sympathetic land ethic is important to ensure effective legislation is enacted, the latter is essential to prevent land degradation — a land ethic alone is not enough. Elsewhere, Bradsen (1992b) has emphasized the importance of using an effective mix of species and habitat legislation to ensure conservation of biodiversity.

As might be expected, the answer is not empowerment versus enforcement, it is the effective mix of appropriate techniques for specific situations. The successful mix of restrictions through police power, incentives based on self interest, and personal ethics, to achieve a conservation goal is well described in Blanchard and Monroe's (1990) paper concerning the reversal of seabird population declines in Quebec.

It should be stressed that regulation does not have to be a negative process; for example Porter (1991) has argued that environmental regulations in industry can enhance rather than hinder competitive advantage. Two ways in which this occurs are through direct savings from pollution control and the profitable export of environmental technologies. Such viewpoints must be adequately explained and demonstrated to industry, politicians and the community.

Action steps include:

- Reviewing the efficacy of various legislative and enforcement alternatives in achieving conservation of biodiversity. This would involve a number of projects, but could readily begin with appropriate literature reviews and dissemination of results.

- Through research and training, improving use of enforcement as a management tool.

- Improving integration of enforcement and education so that they act synergistically to bring about desirable conservation outcomes.

- Demonstrating to industry that regulation can enhance their economic position.

Education and information

Given the importance of land management to human survival, and the extent and pace of social and cultural change since World War II, it is remarkable that social aspects of land management have attracted so little research. As stated previously, a better understanding of how social change occurs, and of public attitudes and values referring to conservation, are important tools for achieving conservation of biodiversity. The work cited previously by Blanchard and Monroe (1990) provides an informative example of education and an understanding of social context being used in combination with other management techniques to achieve a positive change for wildlife conservation.

While qualitative techniques — such as talking to client groups, analysing the contents of letters received, and so on — may be used to assess the knowledge and needs of target groups during information programmes for conservation of biodiversity, we are not aware of market research being used as a basis for preparing and monitoring the effectiveness of extension information in Australia. Given the significant expenditure of resources in this area, one would expect that funds would be allocated to client research so that information programmes achieved maximum effect.

Duda *et al.* (1989) describe a wildlife education programme in Florida based on social and biological research and marketing techniques. While the evaluation phase had not been undertaken at the time the article was written, this was built in to the planning process, and the programme was being implemented. A similar approach could be used with great value in Australia.

Unfortunately most information products are assessed by peer comment rather than client comment, and the desire to use resources to produce tangible, physical products, rather than assess client needs, is a powerful motivating force on funding bodies and conservation managers. These attitudes must change if we are to properly use our limited information resources. If we cannot properly use such resources, then they should be re-allocated to implement some of the action steps listed here.

Similar comments apply to other extension activities. The importance of understanding the sociology of extension methods and ensuring effective communication is well documented (for example McDowell *et al.* 1989; Hartley *et al.* 1992).

Communication barriers exist within professional ranks as well as "out there". Magill (1991) has described a number of barriers within natural resource professionals which interfere with effective communication. These include attitudes of "knowing best", "being above politics", using a special, excluding language and not dealing well with issues of values and ethics. Wallace (1992) also describes a range of issues, including historical matters and organizational cultures, which provide barriers to effective communication by land conservation advisory services.

The barriers organizational cultures provide to effective integration of ideas and information are well described by Norgaard (1992). For example, he cites an instance where differences in world view and organizational culture contribute to the failure of the US Department of Agriculture and Environmental Protection Agency to develop an effective joint programme for the reduction of pesticide use. Similar inefficiencies occur within Australia, and we can ill afford the waste of resources which results.

Action steps include:

- Conducting and using social research to improve communication of information and promote positive attitudinal change.
- Properly identifying target groups for information programmes, and establishing what they want, know and need before implementing an information strategy (proper client research). We all know we should do this, but there is little evidence that it happens.
- Ensuring that information and education programmes are implemented only where demonstrably needed, and not just because they seem like a good idea or to conform with professional fashion.
- Training natural resource professionals to be effective communicators. Recher (1992) and Ehrlich (1993) discuss many of the important issues and their solution.
- Identifying organizational constraints to effective extension and removing or decreasing them.

With regard to educational programmes in general, they are too often prone to justifying the activities of organizations and agencies, worthy as these may be, rather than developing a community which is informed on land management issues. This is very understandable. If you are getting bad press or need community support (and we all do), the normal response is to spend time and money on information programmes that emphasize an organization's importance. Although such programmes have their place, a cynical public often perceives them as propaganda, and the community may remain uninformed of the complexity of biodiversity and land management decisions. Too often we assume a level of community knowledge and understanding which greatly exceeds the reality.

Effective land management and land use decisions — and these are crucial to conservation of biodiversity — require a good understanding of the full range of land use options together with their values, inter-relationships, environmental impacts and socio-cultural context. These issues must be better explained. Although a few managers still suffer from the misapprehension that an informed public is less comfortable to deal with, most now accept that criticisms and interactions will be more valuable if everyone understands the basis and detail of land use decisions. As well, the public will better accept the reversals of land use decisions that inevitably prove necessary with further research and more experience (e.g., prescribed burning in some US forests instead of fire exclusion as promoted by "Smoky Bear").

In the specific case of schools, we often have limited understanding of the pressures on teachers and inflated expectations of what teachers actually know concerning the environment. There are many expensive teachers' activity kits and programmes which are never used. While there have been successful school environmental programmes, the development of an understanding of land use and management in schools could not be considered adequate given that these topics are fundamental to life. Perhaps we have not fully recognized that the fourth "R", responsibility towards the environment, should be a significant part of the content used to teach the other three "Rs".

Practical actions include:

- Improving liaison between organizations involved in land use and thus encouraging collaboration to produce educational material and information which better explains land use and management choices and decisions.
- Developing educational programmes which take a broad view of land management issues and biodiversity conservation.
- Developing programmes with teachers which meet their needs in environmental education, and those of the students, rather than those of funding organizations.

There are manuals for farmers and manuals for revegetation, but manuals for managing biodiversity are very, very scarce. One of us was recently involved in producing a manual for managing remnant bushland in southern Western Australia (Hussey and Wallace 1993), and found that while there is an enormous amount of relevant information and research, it is exceedingly poorly integrated into forms which provide land managers with either a conceptual basis for land management decisions or a review of the facts with "best bet" methods. Reviews such as those provided in Saunders *et al.* (1993) and manuals such as Hussey and Wallace (1993) also help researchers to identify high priorities and avoid replicating work which has already been done.

While there are important reviews which do help — for example, those by Main (1987) and the papers in Hobbs and Saunders (1993) — they are not common, nor are they necessarily in a form which is readily accessible to land managers.

Reviewing and integrating information into forms which provide insights across disciplines, and which are accessible and helpful to land managers are important challenges. Except in the areas of bibliographic indexing and abstracting, the expectations stimulated by computer-based technologies have not been met, and there is, it seems, no substitute for someone actually sitting down and doing the hard slog. Unfortunately, there is little reward for this in the research/academic community, and most managers find it exceedingly difficult to find the time to read the literature. Managers can and do, however, implement prescriptions that have been produced by others and the development of prescriptions produced by multi-disciplinary teams should be promoted. Terminology must be kept simple and, if necessary, explained. Nothing puts the lay-person off so fast as inexplicable jargon and acronyms.

Finally, the amount of relevant unpublished and anecdotal information that is undocumented and is slowly being lost is frightening. Reviews that tap into unpublished data, including oral histories, will often advance knowledge more than experimental studies that, to some extent, "re-invent the wheel" because previous work is unpublished.

Actions include:

- Rewarding researchers who integrate information into a form that is useful and accessible to managers and other researchers. With proper qualification, there is scope in such reviews for proposing new ideas and concepts to provide managers with models they can test.
- Resourcing oral histories which document biodiversity changes and natural histories before those with memories of (in some areas) the initial impact of European settlement are dead.
- Improving networking between managers and researchers. While there are many managers and researchers who successfully network to the advantage of both research

and management, it is not enough. In particular, the importance of researchers "kicking the dust" with managers is insufficiently recognized.

- Encouraging the production of management plans, guidelines and prescriptions which not only provide methods, but also the reasoning which underlies these. The latter is essential if managers are to understand techniques, adjust them for local situations, and provide effective feedback to planners and researchers.
- Providing effective and easy to use means for managers to document information and management practices applicable to biodiversity conservation. Many interesting observations and "experiments" go unrecorded, and monitoring techniques should be more effectively designed and implemented.
- Developing, making available and promoting systems of information retrieval which allow managers and researchers to quickly and easily access published information and other documents.

Social and cultural issues are vital in changing human attitudes to biodiversity conservation. The next step is to ensure that conservation biologists provide the knowledge and cost-effective techniques to prevent extinctions.

CONSERVATION BIOLOGY

The aim of conservation biology is to prevent the loss of biodiversity. Strategies to achieve this aim are (after Western 1989):

1. To identify biological conservation problems.
2. To develop scientific conservation principles.
3. To establish management procedures.
4. To bridge science and management by making scientists responsive to the problems of practising conservation in a real world and managers responsive to scientific issues.

To us, conservation biology includes the political, social, economic and cultural issues, as well as the science and management, of biodiversity conservation.

Conservation biology is a new, broad science, combining several fields of scientific endeavour some of which, on their own, may not necessarily be directed towards its aim. Because it is so broad, it is necessary to define priorities. Our priorities follow.

Protected areas

It is a basic tenet of conservation biology that the vast majority of species and ecological communities can not survive in ecosystems that have been destroyed or significantly modified. Thus, for many decades, one of the most important techniques applied for biodiversity conservation has been the creation of systems of protected areas, where harmful actions are banned or controlled. In Australia, the existing system of nature conservation reserves is far from representative and a priority for the nation is to make it more so (House of Representatives Standing Committee on Environment, Recreation and the Arts 1993). Many areas of Australia have not had even a cursory biogeographic survey and the conservation status of many taxa, even those in relatively well known groups such as vertebrate animals and vascular plants, is completely unknown. Australia is probably the only "western" country where new mammal species are being described every year and the conservation status of many mammal and even bird species is not known. Nevertheless, we do have enough information to design a much better system than currently exists and recommendations for additional areas and better systems abound — the problems are not so much scientific as political.

In Australia we have developed the existing protected area network in competition with agriculture and pastoralism (on land) and fishing (in marine areas), and legislation and practice clearly reflect the view that the above land uses are largely incompatible (however, this does not mean they can not be integrated; see Hobbs and Saunders 1993). To extend the reserve system in much of Australia we must now develop an approach to land protection that dovetails as far as possible with the aspirations of Aborigines and the needs of industry, especially the mining industry, and for remnant vegetation in particular, agriculture. At present the aspirations of these land users are preventing the declaration of many biologically important areas.

Multiple-use is a "dirty word" to some conservationists, reflecting a view that uses other than conservation are incompatible or will always have priority in multiple-use areas. However, this need not be the case and in some parts of the country progress has been made with various modifications of the concept. For example, Western Australian State forests have been managed on a multiple-use basis for many years (CALM 1992). There is no doubt that, outside high priority protected areas such as national parks and nature reserves, the multiple-use concept must be used more widely if biodiversity is to be retained. Those espousing and using multiple-use must ensure that uses are not prioritized in such a way that significantly disadvantages biodiversity values.

Protected areas must be managed if they are to retain their biodiversity. Many protected areas are too small for them to retain some of their species — indeed, many are small remnants of formerly widespread communities occurring on what is now agricultural land. Much research and management effort is now being applied in Australia to the management of remnants (for example, Saunders *et al.* 1990). Society must recognize that protected areas will not retain the values for which they were reserved in the first place unless scientifically based land management follows.

Actions that can be taken now include:

- Arranging biogeographic surveys of regions, including areas of land use conflict, that involve all interested parties, such as local government, industry, farmers and conservation groups. State conservation agencies should co-ordinate such work to ensure that information is collected in a manner which allows different areas to be compared.
- Continuing to conserve representative samples of ecosystems by declaring protected areas from public lands and, where needed, by purchasing private land.
- Working out ways of protecting areas that have multiple resource values, so that their biodiversity can be protected and managed. We should promote experiments with tenure and management control. Government agencies are not the only ones with expertise in this area and promoting the involvement of locals will help counteract the "locked up" view of protected areas.
- Developing appropriate land tenures and management schemes for areas of value to Aboriginal society so that Aborigines and conservation agencies can work together to achieve common goals.
- Providing better descriptions of the community values of protected areas, e.g., their contribution to flood mitigation, groundwater quality, and so on.
- Developing improved information and decision-making systems for managers (see below).
- Working towards increased political and community support for protected areas.
- Involving local communities more in the management of vegetation remnants in urban, agricultural and pastoral landscapes, particularly where management is not only protecting biodiversity, but also contributing to community and individual quality of life. For example, rehabilitating remnants so that they better conserve local biodiversity and pump increased volumes of groundwater can help to protect agriculture. Other examples include managing remnant vegetation as a seed source for community revegetation projects and using remnants as "outdoor classrooms" and as parts of heritage trails.
- Improving community understanding of values at risk, e.g., even the scientific community underrates the biodiversity of the Western Australian wheatbelt and this probably applies to other agricultural and pastoral regions.
- Directing scientific research effort towards managing those factors that are currently threatening the biodiversity of protected areas, in particular disturbance (especially fire) regimes, exotic species, plant diseases and changing landscape hydrology.
- Striking a balance between expenditure on recreation and conservation; in most protected areas the balance is tilted far too much towards managing people.
- Training and educating managers in managing for conservation of biodiversity.
- Involving volunteers and community groups in research into conservation biology.

Improved off-reserve conservation

The protected area system will never, on its own, be capable of conserving all Australia's biodiversity. For example, of the 1 386 populations of the 238 taxa of declared threatened flora in Western Australia in 1989, only 30 per cent of the populations occurred within conservation reserves, while 16 per cent were found only on freehold land (Hopper *et al.* 1990). A considerable proportion of biodiversity can be conserved only on land and waters allocated primarily for uses such as agriculture, pastoralism, timber harvesting, mining, fishing, military training, human habitation or on land owned or reserved for Aboriginal purposes.

Many community groups are grappling with the need for nature conservation outside protected areas, while maintaining economic activity. Off-reserve conservation is being tackled in a variety of ways, e.g., through Land Conservation District Committees, catchment management groups and Farmers' organizations. Basic to the success of these ventures is the involvement of local people and the demonstration that good conservation practice benefits local productivity.

Many of the actions relevant to public lands are applicable here. Additional actions that can be taken now include:

- Improving collaboration between government agencies.

- Reviewing, at least Australia-wide, the history of land clearing controls and other mechanisms for protecting areas of biodiversity on freehold and leasehold lands. Are there general principles, or is the divergence of approaches in Australia a necessary reflection of different circumstances?
- Reviewing the efficacy of the Land Care approach to biodiversity conservation. A lot of faith has been placed in this approach, but there is some evidence that it will not be enough by itself (Bradsen 1992a) and it is taking a lot of resources. For example, revegetation is often limited unless commercially driven (Shea *et al.* 1992) and catchment management groups may attract only those landholders most at risk from hydrological changes.
- Reviewing the use and application of planning and zoning as a mechanism for promoting nature conservation with extractive uses of lands and waters.
- Establishing to what degree biodiversity is essential to sustainable land use in particular circumstances.
- Providing practical revegetation prescriptions that help land holders integrate conservation and production.
- Providing financial assistance or incentives to promote the retention and management of biodiversity.
- Promoting the productive use of "problem" indigenous animals.
- Developing commercial industries based on local species as a means of preventing introductions and encouraging revegetation sympathetic to maintaining biodiversity.

The recovery of threatened species and ecological communities

The conservation of threatened species of vertebrate animals and vascular plants has received increasing attention in recent years. The concept of conserving ecological communities, integrated assemblages of native species that inhabit a particular area in nature, is much newer, but equally or more important. This is often termed the landscape management approach.

There is currently rapid progress in this area. A national strategy has been prepared for signature by the Prime Minister and Premiers, the Commonwealth has entered the field and is providing significant extra resources to the States and Territories, most State and Territory conservation agencies are contributing an increasing proportion of their resources to the development and implementation of Recovery Plans, and non-government organizations, with WWF Australia at the forefront, are lobbying for action and providing assistance.

The "recovery process" provides guidance for the conservation of threatened species and communities:

1. review the conservation status of taxa and communities;
2. prepare priority lists of threatened taxa and communities;
3. conduct the necessary research;
4. produce costed Recovery Plans; and, for each Recovery Plan:
5. obtain funding;
6. implement;
7. monitor and review.

This process can be followed at national, state or local levels. A vital part of the process is the setting up of Recovery Teams to prepare and implement each Recovery Plan. A Recovery Team should be small and effective and include those who have a stake in or are affected by the recovery of the species.

Increasingly, there are examples of the effectiveness of the recovery process for threatened species. Success stories include work on the Western Swamp Tortoise *Pseudemydura umbrina* (Burbidge and Kuchling 1994), Noisy Scrub-bird *Atrichornis clamosus* (Danks *et al.*, in press) and Eastern Barred Bandicoot *Perameles gunnii* (mainland race) (Seebeck 1990). Many Recovery Plans are being prepared or getting underway.

Recovery Plans and Teams for threatened ecological communities are a newer concept, but are starting to gain momentum. The Recovery Plan for Toolibin Lake in Western Australia is an example (Bowman Bishaw Gorham *et al.* 1992). We cannot attain the goal of conserving biological diversity if we focus only on species — there are simply too many species and such an approach will soon exhaust the resources available. While a species approach is appropriate for high profile animal and plant species and will conserve many other species and communities (the "umbrella effect"), most species can only be conserved by ensuring that ecological communities are conserved within functioning ecosystems.

Actions that can be taken now include:

- Preparing and reviewing priorities for work on threatened species and ecological communities on a national, state and local basis.

- Setting up Recovery Teams for the most threatened species and ecological communities.
- Preparing and implementing Recovery Plans on a priority basis.
- Providing factual information to land holders and the public about threatened species and communities.

Alien species and problem indigenous species

Alien species

Australia has enormous problems with alien species. Much of the loss of biodiversity outside cleared areas, including much of the loss within protected areas, is associated with introduced species — predators such as the feral cat and fox, herbivores such as rabbits and pigs, weeds such as bitou bush and plant diseases such as *Phytophthora* spp. Much more needs to be done to control aliens.

Progress with the understanding and control of the European Red Fox *Vulpes vulpes* is a good example of what can be done. While many mammalogists warned that exotic predators could be a cause of decline and extinction in Australia (e.g., Calaby 1971; Christensen 1980) such warnings were largely ignored by policy makers and managers because of a lack of proof, statements in ecological texts that predators could not eliminate prey species on a continental scale (e.g., Frankel and Soulé 1981) and the lack of suitable fox control techniques. It took the pioneering work of Kinnear *et al.* (1988) to prove that the fox could be controlled and that control resulted in the recovery of remnant Black-footed Rock-wallaby *Petrogale lateralis* populations. In Western Australia this quickly led to similar results with other species, e.g., the Numbat *Myrmecobius fasciatus* (Friend 1990).

Following Kinnear's work CALM has carried out fox control, using meat and egg baits with "1080" (an environmentally-friendly toxin in Western Australia, King and Kinnear 1991), around remnant populations of a variety of threatened species. As well as Black-footed Rock-wallabies and Numbats, dramatic recovery has now been demonstrated in Common Brushtail Possums *Trichosurus vulpecula,* Woylies *Bettongia penicillata,* Rothschild's Rock-wallaby *Petrogale rothschildi,* Tammar Wallabies *Macropus eugenii* and Western Brush Wallabies *Macropus irma*. Kinnear's seminal work has led to a reappraisal of the role of exotic predators in Australia and to the new Co-operative Research Centre on the Control of Vertebrate Pest Populations commencing research into the biological control of foxes as well as rabbits. But foxes are not the only problem introduced predator. As similar projects develop in arid areas it is becoming abundantly clear that, in some areas, feral cats are not controlled adequately by standard fox control technology and that research into cat control techniques is urgently needed.

Research into other "biodiversity pests and diseases", such as *Phytophthora* (Shearer *et al.* 1991), pigs and the many bushland weeds that are invading native vegetation (Humphries *et al.* 1991) must continue.

Problem native species

Some indigenous species also pose problems for land managers. Although in most cases these affect the agricultural, horticultural and pastoral industries, they also create difficulties for managers of protected lands.

For example, kangaroos and scarab beetles may significantly damage revegetation and regeneration, and the expansion in range and numbers of species such as the galah is undoubtedly impinging on the breeding success of less common, local species and often damaging native vegetation remaining on farms. Problems which occur are not only economic, they are also social. The negative feelings towards wildlife which are stimulated in some segments of the rural community are detrimental to achieving biodiversity conservation, and the views expressed by some conservation and animal rights groups may exacerbate feelings.

Although the immediate management issues — such as damage licences and ensuring species are not excessively culled — are generally very well handled, it would contribute to conservation if the sustainable economic use of problem indigenous species was better developed.

Actions that can be taken now include:

- Where cost effective techniques exist, controlling "biodiversity pests and diseases" in protected areas and on other land valuable for biodiversity conservation. Such pests include foxes, rabbits, goats, feral horses, donkeys and pigs.
- Where cost effective control techniques do not exist (e.g., for *Phytophthora* spp.), ensuring that management actions are designed to minimize their spread.
- Preventing the import and release of new aliens that may damage the environment and lead to a loss of biodiversity — basic to this is the better use of local species of plants and animals for commercial production, land conservation works and landscaping.

- Promoting research into the biological control of major pests, weeds and diseases.
- Ensuring that pests, weeds and diseases are controlled at the local level.
- Developing sustainable use of problem indigenous species.

Priorities for research

Conservation of biodiversity is a complex subject and conservation biology as a science is in its infancy. Australia must ensure that conservation biology is a national research priority — it is much more important to the future of humans than some of the things we spend enormous amounts of money researching at present, e.g., radio astronomy.

Research priorities need to be strongly influenced by the users. In conservation biology the users include land managers (including pastoralists and farmers), industry, policy makers and the voluntary conservation movement. Scientists must seek input from users when setting goals and seeking funds.

As a starting point for debate, we propose that the following areas are research priorities for Australia (after Burbidge 1994) (these are not in order of priority; all are needed):

- Improving alpha taxonomy. We believe that the first task is to classify vascular plants and vertebrates, and selected groups of invertebrates and non-vascular plants, so that their conservation can be addressed with reasonable confidence of taxonomic boundaries. Invertebrate and non-vascular groups selected should be those that:
  - (i) are beautiful, unusual or spectacular (or even dangerous) (because the public will empathize with or be interested in them); or
  - (ii) show local endemism and low dispersability so that fine-grained patterns in the biota can be addressed in protected area system design (e.g., Solem and McKenzie 1991, McKenzie and Belbin 1991); or
  - (iii) are likely to be most affected by current environmental change; or
  - (iv) are likely to contribute to land conservation and commercial production.
- Co-ordinating and developing biogeographic surveys and surveys of threatened taxa.
- Establishing a representative protected area system.
- Developing management systems applicable at a landscape or catchment level that integrate nature conservation and sustainable primary production.
- Developing and testing methods for monitoring change in ecosystems, both managed and unmanaged; these methods should be implementable by semi-skilled land management staff.
- Protecting and managing islands which, largely untouched by the major disturbances that have affected the mainland, now contain ecosystems and species extinct or threatened on the mainland (Burbidge 1989).

  For example, 13 Australian islands harbour populations of 13 endangered mammal taxa on the 1991 Australian and New Zealand Environment and Conservation Council (ANZECC) List of Australian Endangered Vertebrates (three populations are introduced) (ANPWS 1991), while 19 islands harbour populations of five vulnerable mammal taxa (two populations are introduced). Eleven species of Australian mammals occur only on islands and all except two of these occurred on the mainland in historic times.
- Preventing extinctions among vascular plants and vertebrates and developing cost-effective techniques that can be widely applied to the recovery of endangered species.
- Working on control or eradication of the major threatening exotics: foxes, feral cats, rabbits, *Phytophthora* spp. and those bushland weeds that are threatening native species and whole ecosystems (e.g., Ward's Weed *Carrichtera annua* in the Nullarbor Plain, McKenzie *et al.* 1991).
- Developing fire regimes that protect human and conservation values within major Australian biomes.
- Researching social and cultural issues to provide a firm basis for a range of operational and policy activities. More specific information is given above.
- Developing better tools for decision-making and integration, collation and access of information. This will involve inter-disciplinary research — we must be prepared to learn from other fields including industrial research.
- Researching methods for re-establishing favourable hydrological regimes in agricultural areas.
- Expanding our understanding of key life cycles (autecology) so that we can develop management models and principles which enhance biodiversity conservation. Such research would also provide a better basis for calculating minimum viable population sizes across a range of organisms, rather than the current more theoretical approach.

- Developing useful organizing concepts, which improve the structuring of knowledge and theoretical development. Ensuring that such concepts are real; for example, the keystone species concept may have done little to promote conservation (Mills *et al.* 1993).

PROVISION OF RESOURCES

If Australia is to overcome the extinction crisis and conserve its biodiversity, it will cost money. It will not cost a lot in comparison with the expense of defending ourselves against the rest of the human race (compare the cost of one jet fighter with the cost of all nature conservation work in Australia), but we should not kid ourselves that we can do it by fiddling around at the edges. The cost should be seen as an investment in the future — timely conservation action now will allow this generation to pass biological diversity on to future generations and will save far greater expenditure in the future — extinctions are irreversible and recovering threatened species and communities is more expensive than preventing them from becoming threatened in the first place.

Community issues

Governments will only provide more resources if the people who vote for them demand it, and business and industry will integrate biodiversity conservation into their activities only if they are not disadvantaged and/or market forces demand it. Thus, resource allocation is very much dependent on community attitudes affecting voting patterns and markets via buying power, discussed in the first section above.

Furthermore, successful implementation of actions listed in the first section will decrease management costs, most obviously in relation to enforcement and vandalism. Such gains would release funds for other important work.

Political issues

Those wanting to promote biodiversity conservation cannot expect political parties to pick up the issue in a big way unless they participate in the political process and convince politicians that biodiversity conservation is an issue that attracts votes. Again, resource allocation is very much dependent on community attitudes, discussed in the first section above.

Private individuals and organizations

While governments must take the lead, there are many other ways that resources can be provided or augmented. Australians are infamous for expecting "the government" to do everything, while at the same time whingeing about the level of taxation. Other countries have taken different paths and have demonstrated that private people, either individually or collectively, can have an enormous impact. Organizations such as the Nature Conservancy in the United States of America, the National Trust and Royal Society for the Protection of Birds in the United Kingdom and the World Wide Fund for Nature (which operates in many countries, including Australia) have shown the way.

Committed conservationists need to develop similar levels of private funding in Australia. Surely the idea of paying less tax to a nebulous Government Treasury and making up the difference by donating money to a cause you believe in should be more attractive than it currently appears to be.

Much, also, can be done by large and small companies and corporations to sponsor research into conservation biology and support management action.

Volunteers are already making a significant contribution to practical biodiversity conservation and can make more if organizations continue to change their culture to promote volunteer input.

As noted above, the work by Dunlap (1991) suggests that there is strong public support for solving environmental issues, but this has not been converted into public action. This is an important challenge for all conservation organizations.

Decision-making systems

While decision guidelines are relatively common for specific physical operations, such as fire-fighting or herbicide use, or management of particular threatened species which have been studied intensively, and management plans exist for some protected areas, there is little guidance for managers on how to make decisions and implement actions for managing biodiversity on a district, regional, state or national basis. Nor is there adequate direction for research, an issue covered above.

To implement cost-effective management and research, it is imperative that better decisions are made. Main (1992) is one of the few Australian researchers to tackle this issue, and other issues are outlined in workshop notes by Wallace (1988). Unless decision-making systems are improved, the costs to management will be high.

The early promise that computer-based systems would lead to a new age of sound decision-making has not been realized. While a limited number of geographic information systems and expert systems are in use — for example at Kosciusko, Blue Mountains, and Kakadu — their use is not widespread, and their cost relative to some other methods of conserving biodiversity is often high.

Julien *et al.* (1992) note the limited success of expert systems in environmental evaluation despite their promise, and suggest that knowledge acquisition is the major problem, with the principal difficulty being representing in the system how experts state their knowledge. Similar issues pertain to the use of expert systems in biodiversity management. For example Moffat (1990), in a review of potential values and problems associated with information technology notes that the promise of geographical information systems, remote sensing systems and expert systems has not been realized. In some cases the financial and personnel costs of establishing systems prohibit their successful use. Ventura and Giampetroni (1993) document this and other barriers to implementing geographical information systems for soil conservation workers in Wisconsin.

When one examines the range of technologies that have provided so much promise — expert systems, geographic information systems, remote sensing and so on — and then examines the actual benefit these have provided to conservation of biodiversity, it is difficult not to be disappointed and suspect that, sometimes, resources could have been more effectively allocated. Too often people seem to find a new technological toy, play with it, then ask the questions "How can I get money to keep playing with this? How can this be used?" Rather than "What are the things we really need to do? What technology can we use and how can it be developed?" This is not a Luddite argument, merely a plea for conservation professionals to get their priorities clear and not to lose sight of the bigger picture.

We suggest that decision-making systems are important, and they need to be explored. However, perhaps we also need to better include socio-cultural factors into the decision-making process, and concern ourselves more with identifying and describing the important biological principles we may use as decision-making tools.

Action steps include:

- Better describing the socio-cultural factors and biological principles which should drive the decision-making process.
- Bringing together researchers, managers and those who have developed decision-making systems in other disciplines to develop systems which are effective for conservation of biodiversity.
- Ensuring that expenditure on technological innovations, including development of computer systems, is driven by biodiversity conservation goals and priorities, rather than other, more personal goals.

Land management organizations

In addition to government nature conservation agencies, there are already large land management-oriented organizations all over Australia: Agriculture Departments, Soil Conservation Services, Lands Departments, and so on. They possess more resources than do nature conservation agencies and have a profound influence on how land is managed and how biodiversity is treated.

A change in philosophy by these organizations, within the ESD framework, could have a significant effect on biodiversity conservation. This can happen, either by decisions at the political level or the high management level in those organizations, or by networking at the operational level. All three should be occurring and have the potential to save resources which can then be used effectively to implement other actions proposed in this paper.

CONCLUSION

Much effort has been put into the development of the principles of Ecologically Sustainable Development, both in Australia and elsewhere. Now, Australians have the more difficult task of ensuring that the Strategies are more than mere words on paper, destined for a dusty shelf.

Quite simply, the implementation of ESD principles, including biodiversity conservation, is an enormous challenge now facing our society. Unless we accept the challenge and are successful in solving the associated problems, we will leave an impoverished world to future generations.

Implementing biodiversity conservation is possible. Already, we know enough about the problems and solutions to be able to make rapid progress. The challenge is to convert ideas, information and strategies into effective action.

ACKNOWLEDGEMENTS

We thank Richard Hobbs, David Mitchell, Denis Saunders, Tony Start and Roger Underwood for constructive comments on a draft of this paper and Harry Recher and an anonymous referee for their reviews.

REFERENCES

ANPWS, 1991. List of Australian endangered vertebrate fauna (April 1991). Australian National Parks and Wildlife Service: Canberra.

Blanchard, K. A. and Monroe, M. C., 1990. Effective educational strategies for reversing population declines in seabirds. Pp. 108–17 in *Transactions of the 55th North American Wildlife and Natural Resources Conference* ed by R. E. McCabe. Wildlife Management Institute: Washington DC.

Bowman Bishaw Gorham, Jim Davies and Associates and Rural Planning, 1992. Recovery Plan for Lake Toolibin and surrounding nature reserves. Report RI2147, prepared for the Department of Conservation and Land Management, unpublished.

Bradsen, J., 1992a. A review of soil conservation legislation in Australia. Pp. 21–35 in *Proceedings of the 5th Australian Soil Conservation Conference, Volume 1, Central Conference* ed by G. J. Hamilton, K. M. Howes and R. Attwater. Department of Agriculture: Perth.

Bradsen, J., 1992b. Biodiversity legislation: species, vegetation, habitat. *Environ. Plan. and Law Jour.* **9**(3): 175–80.

Burbidge, A. A. (ed.), 1989. *Australian and New Zealand islands: nature conservation values and management.* Occasional Paper 2/89. Department of Conservation and Land Management: Perth.

Burbidge, A. A., 1994. Conservation biology in Australia: where should it be heading, will it be applied? Pp. 27–37 in *Conservation biology in Australia and Oceania* ed by C. Moritz and J. Kikkawa. Surrey Beatty & Sons: Chipping Norton.

Burbidge, A. A. and Kuchling, G., 1994. Recovery Plan for the Western Swamp Tortoise *Pseudemydura umbrina.* Western Australian Wildlife Management Program No. 11. Department of Conservation and Land Management: Perth.

Calaby, J. H., 1971. The current status of Australian macropodidae. *Aust. Zool.* **16:** 17–29.

CALM, 1990. CALM's role in management of native vegetation in rural areas. Policy Statement No. 27. Department of Conservation and Land Management: Perth.

CALM, 1992. Mangement strategies for the south-west forests of Western Australia, a review. Draft for public comment, February 1992. Department of Conservation and Land Management: Perth.

Campbell, J. with Moyers, B., 1988. *The power of myth.* Doubleday: New York.

Carr, S. and Tait, J., 1991. Differences in the attitudes of farmers and conservationists and their implications. *Jour. Environ. Man.* **32:** 281–94.

Christensen, P., 1980. A sad day for native fauna. *Forest Focus* No. 23. Forests Department: Perth.

Commonwealth of Australia, 1992. National strategy for ecologically sustainable development. Australian Government Publishing Service: Canberra.

Danks, A., Burbidge, A. A., Burbidge, A. H. and Smith, G. T. (in press). Noisy Scrub-Bird Recovery Plan. Western Australian Wildlife Management Program No. 12. Department of Conservation and Land Management: Perth.

Duda, M. D., Cerulean, S. I. and Gillan, J. A., 1989. Comprehensive wildlife education planning in Florida: the value of human dimensions research. Pp. 455–67 in *Transactions of the 54th North American Wildlife and Natural Resources Conference* ed by R. E. McCabe. Wildlife Management Institute: Washington DC.

Dunlap, R. E., 1987. Polls, pollution, and politics revisited: public opinion on the environment in the Reagan era. *Environment* **29**(6): 6–37.

Dunlap, R. E., 1991. Public opinion in the 1980s: clear consensus, ambiguous commitment. *Environment* **33**(8): 10–37.

Ehrlich, P., 1993. Communication: how can ecologists get their message out? Pp. 295–301 in *Nature conservation 3: reconstruction of fragmented ecosystems — global and regional perspectives* ed by D. A. Saunders, R. J. Hobbs and P. R. Ehrlich. Surrey Beatty & Sons in association with Western Australian Laboratory of the CSIRO, Division of Wildlife and Ecology; Centre for Conservation Biology, Stanford University; Tammin Land Conservation District Committee: Chipping Norton.

Frankel, O. H. and Soulé, M. E., 1981. *Conservation and evolution.* Cambridge University Press: Cambridge.

Frazier, J. G., 1990. International resource conservation: thoughts on the challenge. Pp. 384–95 in *Transactions of the 55th North American Wildlife and Natural Resources Conference* ed by R. E. McCabe. Wildlife Management Institute: Washington DC.

Friend, A. J., 1990. The numbat *Myrmecobius fasciatus* (Myrmecobiidae): history of decline and potential for recovery. *Proc. Ecol. Soc. Aust.* **16:** 369–77.

Hartley, R. E. R., Riches, J. R. H., and Davis, J. K., 1992. A systems approach for landcare. Pp. 217–22 in *7th ISCO conference Sydney — people protecting their land. Proceedings Volume 1* ed by P. G. Haskins and B. M. Murphy. International Soil Conservation Organization under the auspices of the Department of Conservation and Land Management: Sydney.

Hobbs, R., 1992. Is biodiversity important for ecosystem functioning? Implications for research and management. Pp. 211—29 in *Biodiversity of Mediterranean ecosystems in Australia* ed by R. J. Hobbs. Surrey Beatty & Sons: Chipping Norton.

Hobbs, R. J and Saunders, D. A. (eds) 1993. *Reintegrating fragmented landscapes. Towards sustainable production and nature conservation.* Springer-Verlag: New York.

Hopper, S. D., van Leeuwen, S., Brown, A. P. and Patrick, S. J., 1990. *Western Australia's endangered flora and other plants under consideration for declaration.* Department of Conservation and Land Management: Perth.

House of Representatives Standing Committee on Environment, Recreation and the Arts, 1993. *Biodiversity: the role of protected areas.* Australian Government Publishing Service: Canberra.

Humphries, S. E., Groves, R. H. and Mitchell, D. S., 1991. Plant invasions of Australian ecosystems: a status review and management directions. Kowari 2, Part 1. Australian National Parks and Wildlife Service: Canberra.

Hussey, B. M. J. and Wallace, K. J., 1993. *Managing your bushland.* Department of Conservation and Land Management: Perth.

Julien, B., Fenves, S. J. and Small, M. J., 1992. Knowledge acquisition methods for environmental evaluation. *AI Applications* **6**(1): 1–20.

King, D. and Kinnear, J., 1991. 1080: the toxic paradox. *Landscope* **6**(4): 14–19.

Kinnear, J. E., Onus, M. L. and Bromilow, R. N., 1988. Fox control and rock wallaby population dynamics. *Aust. Wild. Res.* **15:** 435–50.

Knudtson, P. and Suzuki, D., 1992. *Wisdom of the Elders.* Allen and Unwin: North Sydney.

Leopold, A., 1949. *A sand country almanac and sketches here and there.* Oxford University Press: New York.

Machlis, G. E., 1992. The contribution of sociology to biodiversity research and management. *Biol. Cons.* **62:** 161–70.

Magill, A. W., 1991. Barriers to effective public interaction. *Jour. For.* **89**(10): 16–18.

Main, A. R., 1987. Management of remnants of native vegetation — a review of the problems and the development of an approach with reference to the wheatbelt of Western Australia. Pp. 1–13 in *Nature conservation: the role of remnants of native vegetation* ed by D. A. Saunders, G. W. Arnold, A. A. Burbidge and A. J. M. Hopkins. Surrey Beatty & Sons in association with CSIRO and CALM: Chipping Norton.

Main, A. R., 1992. Management to retain biodiversity in the face of uncertainty. Pp. 193–209 in *Biodiversity of Mediterranean ecosystems in Australia* ed by R. J. Hobbs. Surrey Beatty & Sons: Chipping Norton.

McDowell, C., Sparks, R., Grindley, J. and Moll, E., 1989. Persuading the landowner to conserve natural ecosystems through effective communication. *Jour. Envir. Man.* **28:** 211–25.

McKenzie, N. L. and Belbin, L., 1991. Kimberley rainforest communities; reserve recommendations and management considerations. Pp. 453–68 in *Kimberley rainforests of Australia* ed by N. L. McKenzie, R. B. Johnston and P. G. Kendrick. Surrey Beatty & Sons: Chipping Norton.

McKenzie, N. L., Robinson, A. C. and Belbin, L., 1991. Biogeographic survey of the Nullarbor District, Australia. Pp. 109–26 in *Nature conservation: cost effective biological surveys and data analysis* ed by C. R. Margules and M. P. Austin. CSIRO Australia: Melbourne.

Mills, L. S., Soulé, M. E. and Doak, D. F., 1993. The keystone-species concept in ecology and conservation. *BioScience* **43**(4): 219–24.

Moffat, I., 1990. The potentialities and problems associated with applying information technology to environmental management. *Jour. Envir. Man.* **30:** 209–20.

Nash, R., 1982. *Wilderness and the American mind.* 3rd Edn. Yale University: New Haven.

Nash, R., 1990. *The rights of nature.* Primavera Press: Leichhardt.

Noe, F. P. and Hammitt, W. E., 1992. Environmental attitudes and the personal relevance of management actions in a park setting. *Jour. Envir. Man.* **35:** 205–16.

Norgaard, R. B., 1992. Co-ordinating disciplinary and organizational ways of knowing. *Agri. Ecosys. Envir.* **42:** 205–16.

Porter, M. E., 1991. Essay — America's green strategy. *Scientific American* **264**(4): 96.

Prior, M., 1992. Responding to threat: psychology and environmental issues. *Habitat* **20**(2): 12–14.

Recher, H. F., 1992. Simple journalists *or* simple scientists?: Are environmental issues too complex for the media? *Aust. Zool.* **28**(1–4): 19–23.

Saunders, D. A. and Burbidge, A. A., 1988. Ecological theory and biological management of ecosystems: workshop summary and discussion. Pp. 77–85 in *Ecological theory and biological management of ecosystems* ed by D. A. Saunders and A. A. Burbidge. Occasional Paper No. 1/88. Department of Conservation and Land Management: Perth.

Saunders, D. A., Arnold, G. W., Burbidge, A. A., and Hopkins, A. J. M., 1987. The role of remnants of native vegetation in nature conservation : future directions. Pp. 387–92 in *Nature conservation: the role of remnants of native vegetation* ed by D. A Saunders, G. W. Arnold, A. A. Burbidge and A. J. M. Hopkins. Surrey Beatty & Sons in association with CSIRO and CALM: Chipping Norton.

Saunders, D. A., Hobbs, R. J. and Ehrlich, P., 1993. Reconstruction of fragmented ecosystems: problems and possibilities. Pp. 305–13 in *Nature conservation 3: reconstruction of fragmented ecosystems — global and regional perspectives* ed by D. A. Saunders, R. J. Hobbs and P. R. Ehrlich. Surrey Beatty and Sons in association with Western Australian Laboratory of the CSIRO, Division of Wildlife and Ecology; Centre for Conservation Biology, Stanford University; Tammin Land Conservation District Committee: Chipping Norton.

Saunders, D. A., Hobbs, R. J. and Margules, C. R., 1990. Biological consequences of ecosystem fragmentation: a review. *Cons. Biol.* **5:** 18–32.

Seebeck, J. S., 1990. Recovery management of the eastern barred bandicoot in Victoria: statewide strategy. Pp. 165–77 in *Management and conservation of small populations* ed by T. W. Clark and J. H. Seebeck. Chicago Zoological Society: Brookfield, Illinois.

Shea, S., Bartle, J. and Inions, G., 1992. Tree crops for farms. *Landscope* **8**(2): 47–53.

Shearer, B., Wills, R. and Stukely, M., 1991. Wildlflower killers. *Landscope* **7**(1): 28–34.

Solem, A. and McKenzie, N. L., 1991. The composition of land snail assemblages in the Kimberley rainforests. Pp. 247–63 in *Kimberley rainforests of Australia* ed by N. L. McKenzie, R. B. Johnston and P. G. Kendrick. Surrey Beatty & Sons: Chipping Norton.

Skolimowski, H., 1988. Eco-philosophy and deep ecology. *The Ecologist* **18**(4/5): 124–27.

Ventura, S. J. and Giampetroni, D. A., 1993. Wisconsin conservationists respond to field office overload. *Jour. Soil Water Con.* **48**(2): 83–89.

Wallace, K. J., 1988. Priorities of research and management and the setting of objectives. Pp. 73–76 in *Ecological theory and biological management of ecosystems* ed by D. A. Saunders and A. A. Burbidge. Department of Conservation and Land Management: Perth.

Wallace, K. J., 1992. Services — a barrier to achieving land conservation objectives? Proceedings Volume 2. Pp. 459–66 in *7th ISCO conference Sydney — people protecting their land* ed by P. G. Haskins and B. M. Murphy. International Soil Conservation Organization under the auspices of the Department of Conservation and Land Management: Sydney.

Western, D., 1989. Conservation biology. Pp. 31–36 in *Conservation for the twenty-first century* ed by D. Western and M. C. Pearl. Oxford University Press: New York.

Habitat Loss

PLATES 5 and 6. An example of progressive habitat loss in the New South Wales wheatbelt over the period 1981–91. *(Photographs courtesy of NSW Department of Conservation and Land Management.)*

Habitat loss — its nature and effects (including case studies from New South Wales)

D. P. SIVERTSEN[1]

The problem of habitat loss sits at the very heart of biodiversity conservation.

Whilst we may legitimately be concerned about and discuss such things as the effects of pollution, hunting, exotic species and even climate change, they will cease to be important when habitat for our native species has disappeared. There is a rapidly increasing body of literature which deals with habitat loss and species decline all over the world. However, the factual information for these trends in Australia is generally poor and, apart from our record number of mammalian extinctions, is not evident in international digests, e.g., Hammond (1994).

In this paper I have cited a number of case studies to illustrate the extent of habitat loss in the local context. In each of these studies the percentage of habitat that has already been cleared is very high. It is also possible to establish that clearing is continuing in many instances. The main reason for past and present habitat loss is clearing for agriculture. In each case study clearing of native vegetation equates to loss of biodiversity at the ecosystem, population, species and genetic levels. Loss of habitat for one species often means an increase in habitat for another. The changing balance in Australia's native biota is examined briefly.

The inescapable conclusion to be drawn from these case studies is that time is short and that, whilst it may be possible to halt clearing through legislation, a fundamental change in attitude to our use of the land is required in the long term.

Hammond, A. L., 1994. *World resources 1994–95.* Oxford University Press: Oxford.

INTRODUCTION

IT is reasonable to suggest that something like half of all terrestrial species are likely to become extinct over the next 50 years, if current trends continue (May, in press). These trends include loss of habitat.

In this paper my aim is to describe and illustrate habitat loss using examples from New South Wales whilst looking at the consequences for biodiversity conservation in broader national and global contexts.

The concept of habitat loss and its effect on biodiversity is of acute concern globally. From the equator to the poles, loss of habitat is an issue. Rainforests are being felled at the rate of about 200 000 km^2/year (Western 1989) for agricultural expansion and woodchipping. The area of the world's deserts is increasing by about 100 000 km^2/year (Western 1989) again as a result of expanding agricultural and pastoral activities. In temperate lands also, native vegetation is being progressively cleared for agricultural and pastoral expansion (e.g., Saunders *et al.* 1985; Sivertsen 1994). Even in the Antarctic, an island which supports vast penguin colonies has been destroyed to make an airstrip.

Habitat loss is not a new phenomenon. A very vivid image of historical habitat loss is conjured by the television personality David Attenborough sitting on a dock in the ancient port city of Carthage pointing to the modern Mediterranean coastline, now just visible on the horizon, many kilometres to the north. In Carthaginian hinterlands the native forests were cleared for agricultural expansion and timber production, this area was one of the "bread baskets" of the Roman Empire. Ancient Carthage is no longer a coastal city with a richly timbered and biologically diverse hinterland, but an ancient ruin being buried deeper and deeper under shifting sands some kilometres from the coast. Another example of historical habitat loss is conjured by Heroditus (420 B.C.). This author documents the Persian campaign to conquer the Greek states around 489 B.C.; the Persian fleet alone in these wars consisted of about of 3 000 wooden ships. Construction, and subsequent destruction, of wooden vessels was a huge drain on timber resources. In the battle of Lepanto in 1571, for example, one galley required 50 beech trees for its oars, 300 pines and firs for its planks and spars and over 300 mature oaks for its hull. This battle in which only about 500 ships took part is estimated to have "cost" over 250 000 trees (Attenborough 1987). Millions of trees were felled to build the Persian and Greek fleets and the floating bridges across the Bosphorus for the Persian army's crossing in the Persian campaign. Nearly all of this timber construction was destroyed in the ensuing battles (Heroditus 420 B.C.). The eastern Mediterranean hinterlands now contain bare stony deserts where once there were magnificent forests; at least in part a direct result of this campaign.

In both cases the effects of habitat loss are still being felt some 2 000 or more years later!

[1]NSW National Parks and Wildlife Service, P.O. Box 1967, Hurstville, New South Wales, Australia 2220.

Pages 29–42 *in* CONSERVING BIODIVERSITY: THREATS AND SOLUTIONS, ed by R. A. Bradstock, T. D. Auld, D. A. Keith, R. T. Kingsford, D. Lunney and D. P. Sivertsen. Surrey Beatty & Sons. 1995.

DEFINING THE PROBLEM

Habitat loss is often cited as the main cause of species decline and extinction; "most extinct, endangered or vulnerable species are the unintended casualties of **habitat destruction** caused by expanding human populations and activities" (Endangered Species Advisory Committee 1992).

So what is "Habitat Loss"?

This term will obviously mean different things to different people and, as ever, one definition will not satisfy everyone. Habitat may be broadly defined as: the places where biota live, the processes which sustain them, and species interactions occurring there. Habitat loss may therefore be defined, again broadly, as anything that alters the habitat so that critical resources (e.g., food, shelter, mates) are no longer available to some or all of the biota. Under this definition habitat loss could have a very broad application. In this paper I intend to focus the discussion on habitat loss associated with clearing of native vegetation and to some of the implications of this process for biodiversity conservation. In a number of the case studies I will be examining, clearing and grazing are inextricably linked. In part therefore, the discussion will also relate to habitat loss and biodiversity decline due to grazing.

It is often difficult to define precisely the point at which habitat is "lost". For the purposes of this discussion, forest, woodland and shrubland communities which have been reduced to a crown cover (*sensu* Walker and Hopkins 1984) of 5 per cent or less are considered to have been cleared. Herbaceous strata and communities which have been ploughed, extensively or intensively grazed and which have suffered declines in native species (*sensu* Adamson and Fox 1982) are also considered to have been cleared. These parameters are a guide only; in examining historical records, for example, such definitions are unknown.

The discussion will focus mainly on New South Wales where a number of case studies will be used to illustrate habitat loss and the attendant decline in biodivesity. The New South Wales experience will then be viewed in a national and global context.

HABITAT LOSS CASE STUDIES

The Sydney Region

The eastern seaboard is the most densely populated part of New South Wales; here urban areas and infrastructure flourish and continue to grow. The effects of urbanization are to be explored in detail in a later chapter (Adams, this volume), suffice it to say here that associated habitat loss from this cause is extensive, although usually not complete, but is very difficult to rehabilitate. The narrow Illawarra coastal plain, the Sydney Basin and the lower Hunter Valley must rate as among the most extensively and irrevocably altered landscapes in Australia. The Cumberland Plain in the Sydney Basin, for example, currently supports about 6 400 ha of Cumberland Plain Woodlands, about 6 per cent of the area originally occupied by this community (Benson and Howell 1990) (Table 1). In just 200 years we have reduced this widespread native vegetation type by some 94 per cent! Other vegetation types of the Sydney Basin have fared even worse; the Blue Gum High Forests have been reduced to about 1 per cent of their former area, and the Turpentine-Ironbark woodlands to less than 0.5 per cent (Benson and Howell 1990).

Evidence also suggests that a number of vertebrate species have been driven to extinction in the Sydney region since European settlement, these are shown in Table 2. Recher *et al.* (1993) document declines in a wide variety of species. However, the status of many species mentioned in this work remain unclear. Recher *et al.* (1993) also point out that many terrestrial and aquatic vertebrate species, other than the mammals shown in Table 2, have declined in the Nepean and Hawkesbury catchments in the past 200 years. Urbanization has not halted. The native bushland of suburbs like Kellyville, Vineyard, Castle Hill, Frenchs Forest, Killarney Heights, Jannali, Como and Sutherland have been given over to low density housing over the past forty years or so. Precise figures on loss of habitat due to urban expansion are not readily available. However, the aerial photographic record attests to it, (see Benson and Howell (1990) for examples).

Table 1. Habitat loss due to clearing of native vegetation in selected case study areas.

Area or map name	Percentage cleared	Hectares remaining	Hectares cleared
Cumberland Plain woodland	94	6 400	100 300
Bega Valley forests and woodlands	92	7 150	82 200
Central West woodlands	76	1 560 000	4 940 000
Wheatbelt — St George (1:250 000)	68	105 380	224 000
Goondiwindi (1:250 000)	81	60 220	258 000
Moree (1:250 000)	68	461 000	980 000
Condobolin (1:100 000)	86	36 600	225 000
Cargelligo (1:250 000)	81	233 200	982 000

Table 2. Status of selected native mammal species in each of the case study areas. Codings in the table refer to the particular case study area only. Nomenclature is from Strahan (1983). The following subjective codes are used: X — extinct; O — not recorded in historical times; R — rare; U — uncommon; C — common; V — vulnerable on a state-wide basis; T — threatened on a state-wide basis; ? — status not reported in the literature; NA — not applicable, the species is not known to occur in the area. Different authors tend to represent status/threat in different ways; some of the papers cited do not explicitly state a level of threat or a current status. The sources of information for each area are: General: Strahan (1983), Lunney *et al.* (submitted); Sydney: Recher *et al.* (1993); South Coast: Lunney and Barker (1987) and Lunney and Leary (1988); Central West: Goldney and Bowie (1987); Wheatbelt: Dickman (1994); and Western Region: Dickman *et al.* (1993).

		Case study areas					
Common name	Scientific name	Sydney	South Coast	Central West	Wheatbelt	Western Division	Status in New South Wales
DASYURIDS							
Spotted-tailed Quoll	*Dasyurus maculatus*	R	R	U	U	U	V
Eastern Quoll	*Dasyurus viverrinus*	X	X	X	?	NA	T
Western Quoll	*Dasyurus geoffroii*	X	X	X	X	X	X
Mulgara	*Dasyurus cristicauda*	NA	NA	NA	NA	O	X
Brush-tail Phascogale	*Phascogale tapoatafa*	X	X	X	?	U	T
Red-tailed Phascogale	*Phascogale calura*	NA	NA	NA	NA	O	T
Yellow-footed Antechinus	*Antechinus flavipes*	?	?	C	?	C	
Brown Antechinus	*Antechinus stuartii*	U	C	C	NA	NA	
Dusky Antechinus	*Antechinus swainsonii*	R	U	C	NA	NA	
Common Dunnart	*Sminthopsis murina*	R	?	C	?	C	
White-footed Dunnart	*Sminthopsis leucopus*	?	U	NA	NA	NA	V
Fat-tailed Dunnart	*Sminthopsis crassicaudata*	NA	NA	C	?	U	
Stripe-faced Dunnart	*Sminthopsis macroura*	NA	NA	NA	?	R	V
Southern Ningaui	*Ningaui yvonneae*	NA	NA	?	?	O	V
Kultarr	*Antechinomys laniger*	NA	NA	X	?	R	T
Narrow-nosed Planigale	*Planigale tenuirostris*	NA	NA	?	?	U	
Paucident Planigale	*Planigale gilesi*	NA	NA	?	?	R	
NUMBATS							
Numbat	*Myrmecobius fasciatus*	NA	NA	NA	O	O	X
BANDICOOTS							
Southern Brown Bandicoot	*Isoodon obesulus*	R	X?	R	NA	NA	T
Northern Brown Bandicoot	*Isoodon macrourus*	?	NA	R	?	NA	
Golden Bandicoot	*Isoodon auratus*	NA	NA	NA	NA	O	X
Long-nosed Bandicoot	*Perameles nasuta*	R	U	R	NA	NA	
Western Barred Bandicoot	*Perameles bougainville*	NA	NA	X	X	X	X
Pig Footed Bandicoot	*Chaeropus ecaudatus*	NA	NA	NA	NA	O	X
BILBIES							
Bilby	*Macrotis lagotis*	NA	NA	X	X	X	X
KOALAS							
Koala	*Phascolarctos cinereus*	R	R	C	?	C	V
WOMBATS							
Common Wombat	*Vombatus ursinus*	R	C	C	?	NA	
Northern Hairy-nosed Wombat	*Lasiorhinus krefftii*	NA	NA	NA	X	X	X
POSSUMS AND GLIDERS							
Ringtail Possum	*Pseudocherius peregrinus*	C	U	C	?	C	
Greater Glider	*Petauroides volans*	U	U	U	?	NA	
Yellow-bellied Glider	*Petaurus australis*	U	U	R	NA	NA	V
Sugar Glider	*Petaurus breviceps*	U	C	C	?	C	
Squirrel Glider	*Petaurus norfolcensis*	NA	NA	R	?	C	V
Brushtail Possum	*Trichosurus vulpecula*	C	U	C	C	C	
Mountain Brushtail Possum	*Trichosurus caninus*	NA	NA	U	NA	NA	
Eastern Pigmy Possum	*Cercartetus nanus*	U	R	U	NA	NA	
Feathertail Glider	*Acrobates pygmaeus*	U	R	C	?	C	
POTOROOS AND BETTONGS							
Long-nosed Potoroo	*Potorous tridactylus*	X?	R	NA	NA	NA	V
Brush-tailed Bettong	*Bettongia penicillata*	NA	NA	X	X	X	X
Tasmanian Bettong	*Bettongia gaimardi*	X	X	X	X	NA	X
Burrowing Bettong	*Bettongia lesueur*	NA	NA	X	X	X	X
Tropical Bettong	*Bettongia tropica*	NA	NA	NA	X	X	X
Rufous Bettong	*Aepyprymnus rufescens*	X	NA	X	?	NA	V

Table 2 — continued

Common name	Scientific name	Sydney	South Coast	Central West	Wheatbelt	Western Division	Status in New South Wales
		Case study areas					
HARE-WALLABIES							
Eastern Hare-wallaby	*Lagorchestes conspicillatus*	NA	NA	X	X	X	X
NAILTAIL WALLABIES							
Bridled Nailtail Wallaby	*Onychogalea fraenata*	NA	NA	X	X	X	X
Crescent Nailtail Wallaby	*Onychogalea lunata*	NA	NA	NA	NA	X	X
ROCK WALLABIES							
Brush-tailed Rock Wallaby	*Petrogale penicillata*	X	NA	R	?	R	V
Yellow-footed Rock Wallaby	*Petrogale xanthopus*	NA	NA	NA	NA	R	T
PADEMELONS							
Red-necked Pademelon	*Thylogale thetis*	X?	C	?	NA	NA	
MACROPODS							
Parma Wallaby	*Macropus parma*	X	X	NA	NA	NA	V
Black-striped Wallaby	*Macropus dorsalis*	NA	NA	NA	R	?	T
Red-necked Wallaby	*Macropus rufogriseus*	U	C	C	?	NA	
Eastern Grey Kangaroo	*Macropus giganteus*	U	U	C	C	C	
Western Grey Kangaroo	*Macropus fuliginosus*	NA	NA	C	C	C	
Walleroo	*Macropus robustus*	?	C	C	?	C	
Red Kangaroo	*Macropus rufus*	NA	NA	C	?	C	
Swamp Wallaby	*Wallabia bicolor*	?	C	C	?	R	
FRUIT BATS							
Grey-headed Flying Fox	*Pteropus poliocephalus*	C	R	C	?	NA	
Little Red Flying Fox	*Pteropus scapulatus*	C	R	C	?	U	
RODENTS							
Water-rat	*Hydromys chrysogaster*	R	R	C	?	U	
White-footed Rabbit-rat	*Conilurus albipes*	X	X	X	X	X	X
Plains Rat	*Pseudomys australis*	NA	NA	NA	X	X	X
Gould's Mouse	*Pseudomys gouldii*	NA	NA	X	X	X	X
Hastings River Mouse	*Pseudomys oralis*	X?	?	X	NA	NA	T
Eastern Chestnut Mouse	*Pseudomys gracilicaudulus*	R	NA	NA	NA	NA	V
Desert Mouse	*Pseudomys desertor*	NA	NA	NA	NA	X	X
Sandy Inland Mouse	*Pseudomys hermannsburgensis*	NA	NA	NA	NA	U	V
New Holland Mouse	*Pseudomys novaehollandiae*	R	?	R	NA	NA	
Silky Mouse	*Pseudomys apodemoides*	NA	NA	NA	NA	R	T
Smoky Mouse	*Pseudomys fumeus*	X?	X?	NA	NA	NA	T
Bolams Mouse	*Pseudomys bolami*	NA	NA	NA	?	R	T
Forrests Mouse	*Leggadina forresti*	NA	NA	NA	NA	U	V
Greater Stick-nest Rat	*Leporillus conditor*	NA	NA	NA	NA	X	X
Lesser Stick-nest Rat	*Leporillus apicalis*	NA	NA	NA	NA	X	X
Mitchell's Hopping-mouse	*Notomys mitchelli*	NA	NA	NA	NA	R	X
Dusky Hopping-mouse	*Notomys fuscus*	NA	NA	NA	NA	X	X
Fawn Hopping-mouse	*Notomys cervinus*	NA	NA	NA	NA	R	X
Long-tailed Hopping-mouse	*Notomys longicaudatus*	NA	NA	NA	NA	X	X
Bush Rat	*Rattus fuscipes*	R	C	C	NA	NA	
Swamp Rat	*Rattus lutreolus*	R	R	U	NA	NA	
Long-haired Rat	*Rattus villosissimus*	NA	NA	NA	?	U	V
Pale Field Rat	*Rattus tunneyi*	NA	NA	NA	?	X	
DINGO							
Dingo	*Canis familiaris dingo*	R	R	?	?	U	

The Bega (South Coast) Region

A similar trend has been observed in the Bega district on the far south coast of New South Wales (Lunney and Leary 1988). According to these authors the district was first settled and grazed by cattle in 1829 and by 1866 it was reported that: "The native grasses have nearly disappeared; their place has been taken by clover (*Trifolium* sp.), prairie grass (*Bromus* sp.) and rye-grass (*Lolium* sp.)" (Braine 1866). By the 1870s clearing had begun in earnest and by the first decade of this century nearly all of the Bega Valley and surrounding arable lands were cleared, with only small patches of forest and scattered trees remaining (Keith, in press) (Table 1). This loss of native vegetation cover was accompanied by rapid increases in weed species such as South American Thistle *Carduus* sp., Bathurst Burr, *Xanthium spinosum,* Blackberry *Rubus fruticosus* and Spotted Cabbage Thistle *Sylybum marianum.* Lunney and Leary (1988) go on to document the progressive decline of many native mammals in the Bega district from about 1900 onwards. The groups in which they have documented declines include: possums, gliders, quolls, bandicoots, koalas, small macropods ("pademelons" and "rat-kangaroos"), wallabies, kangaroos and flying foxes (see Table 2).

The Central West

To the north-west the Bathurst to Forbes area (the Central West) comprises some of the longest established agricultural lands in Australia. Bathurst was proclaimed a town in 1815 and "development" of the Central West proceeded rapidly from that time. Habitat loss here is both extensive and ongoing. Work currently underway by Dr David Goldney indicates that some 76 per cent of the native vegetation has been cleared from the Central West (Goldney, unpubl. data) (Table 1). What remains is divided amongst some 3 500 individual remnants, 90 per cent of which are less than 5 km^2 in area (Goldney and Bowle 1986). In the Central West about one-third of the mammal fauna have been driven to extinction (at least regionally) since European settlement, a further third have lost more than 80 per cent of their former range and are close to extinction (Goldney and Bowie 1986). Table 2 shows the native mammal species known to have disappeared from the region as well as the status of many others.

The New South Wales Wheatbelt

A similar situation exists further west in the wheatbelt of New South Wales. In some parts of this region as much as 95 per cent of the native vegetation has been cleared in the past 150 years according to the Murray Darling Basin Ministerial Council (1987). In the north of the state, in the St George, Goondiwindi and Moree areas (Fig. 1), for example, an average of 55 per cent of the remaining native vegetation was cleared during the 1970s and 1980s (Sivertsen 1994). Similarly to the south, on the area covered by the Condobolin 1:100 000 map sheet (Fig. 2), 61 per cent of the remaining native vegetation was cleared in the 15 year period between 1974 and 1989 (Sivertsen 1993). Table 1 shows the total extent of clearing for these areas at the time of mapping. These are not small areas of habitat loss. Table 3 shows the rates of clearing in selected areas of the New South Wales wheatbelt. There is no reason to regard these rates of clearing as atypical. Walker *et al.* (1993) estimate that between 12 and 15 billion trees have been cleared from the Murray-Darling Basin since European settlement!

As with the urban areas of the eastern seaboard, habitat loss in our agricultural heartlands has not stopped. Although the data are not yet available, direct observation indicates that clearing continues apace. We should remember, however, that clearing is not something that goes on at the same rate year after year (the danger of dealing with averages!). Clearing rates are driven by social, economic and technological factors, the influences of which tend to be patchy in space and time. Gasteen (1989), for example, points out that it was not until the advent of large and powerful machinery in the 1960s that clearing of brigalow began in earnest. The "Brigalow Belt" once ran from central Queensland into central northern New South Wales, to about Narrabri; brigalow is now considered a rare community in this state. The

Table 3. Clearing rates for selected areas of the New South Wales wheatbelt.

Mapping area (dates of air photographs used)	Annual average clearing rate (ha)	Annual average clearing per 1 000 ha of mapped area (ha)	Total area cleared in the period (ha)
St George (1977 and 1985)	13 440	36	107 512
Goondiwindi (1973 and 1984)	17 720	57	124 060
Moree (1977 and 1985)	49 440	34	460 920
Condobolin (1974 and 1989)	3 800	10.5	57 400

Fig. 1a

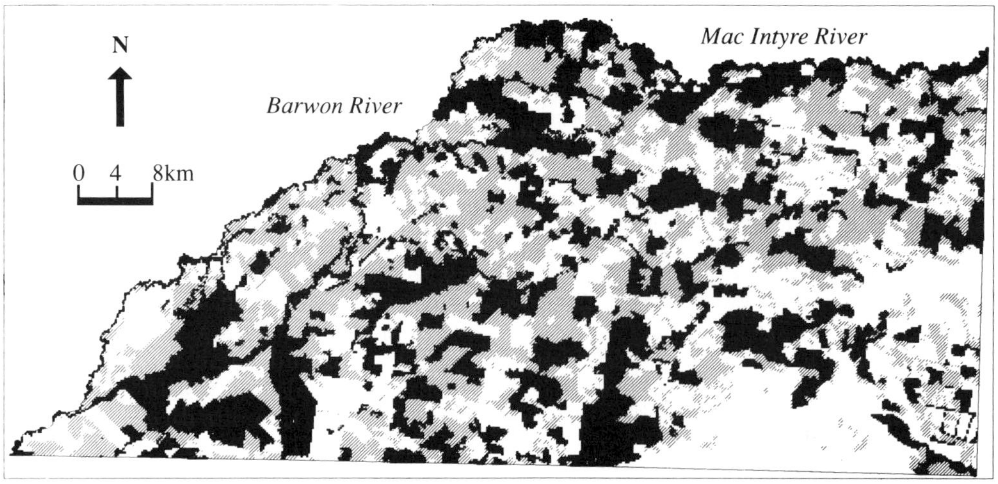

Fig. 1b

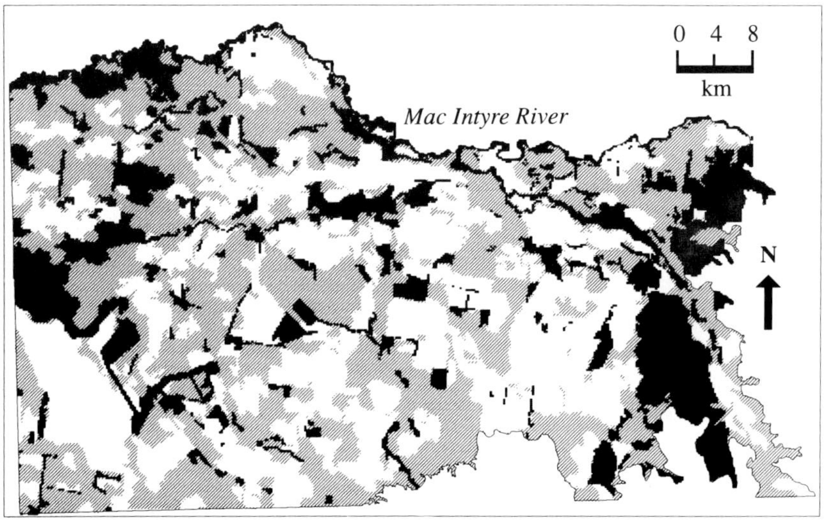

Fig. 1c

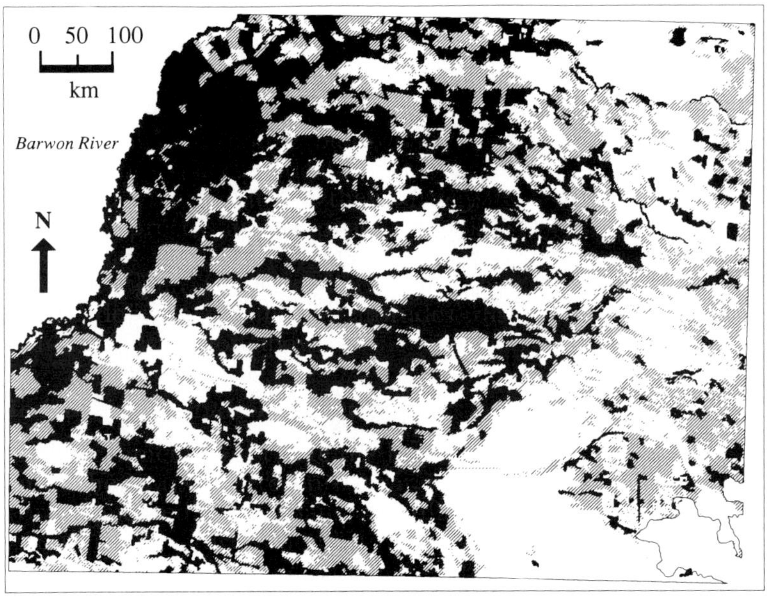

possibility of very high returns on cotton is the current incentive for extensive clearing of native vegetation in some parts of the New South Wales wheatbelt. The area sown to cotton in New South Wales rose by 43 per cent between 1988/89 and 1990/91 (Farrell 1993). Then again, small scale, piecemeal clearing continues throughout our agricultural lands as individuals clear a paddock here and there to boost production a little or to "improve" their holding.

Saunders (1994) and Saunders *et al.* (1985) have documented an analogous situation in the wheatbelt of Western Australia. Here, too, clearing for agriculture has been extensive and has left behind a multitude of native vegetation fragments which are not ecologically sustainable (Saunders 1994).

The status of mammalian fauna in the New South Wales wheatbelt is very poorly documented. Where possible the status of mammals known to occur in the wheatbelt have been given in Table 2.

The Western Division

In far western New South Wales, in the arid to semi-arid Western Division of the state, habitat loss through clearing is a relatively recent phenomenon. Here, the land is predominantly leasehold, and licenses to clear and/or crop must be obtained from the Department of Conservation and Land Management. More than 640 000 ha of land in the Western Division of New South Wales was licensed for clearing (although not necessarily cleared) between 1984 and 1990, and there was a tenfold increase in the area devoted to dryland cropping in the same region between 1965 and 1980 (Pressey 1990). Most of this clearing and cropping activity occurred along the eastern boundary of the Western Division, close to service centres, major rivers and to the existing wheatbelt. Unlike the wheatbelt, the status of mammalian fauna in the Western Division has been investigated and well documented (Dickman *et al.* 1993) (Table 2).

Lessons from the case studies

What we see in these case studies is a trend of habitat loss due to clearing from east to west across New South Wales. What we are also seeing is a profound change to the ecology of the state, as evidenced by the documented declines in native mammal fauna. The available habitat for much of our native biota is being reduced at an alarming rate and individual populations of species, whether plant or animal, are eliminated in the process. This alone implies a diminution of biodiversity at the population and genetic scales.

Not only is the habitat area reduced but the quality of that habitat is also being altered dramatically. Clearing leaves behind fragments of habitat of varying sizes and degrees of isolation (Plates 1 and 2). Fragmentation substantially changes the processes which helped to define that habitat. The radiation balance, wind penetration and turbulence and water balance are all altered in isolated habitat remnants. The degree of alteration together with remnant size, shape and degree of isolation will greatly influence species survival in those native habitat remnants (Saunders *et al.* 1991).

Plate 1. Many remnants are small and isolated by surrounding cultivation. (Photo J. Benson.)

Plate 2. Wheatbelt remnants may be interconnected by corridors, by scattered trees and by pastures containing native species. All remnants are under threat from clearing, see foreground. (Photo S. King.)

Facing page — Fig. 1. Clearing of native vegetation in the New South Wales wheatbelt: (a) the St George mapping area in 1977 (▨ and ■) and that remaining in 1988 (■); (b) the Goondiwindi mapping area in 1973 (▨ and ■) and that remaining in 1984 (■); (c) the Moree mapping area in 1977 (▨ and ■) and that remaining in 1985 (■). Sivertsen (1994).

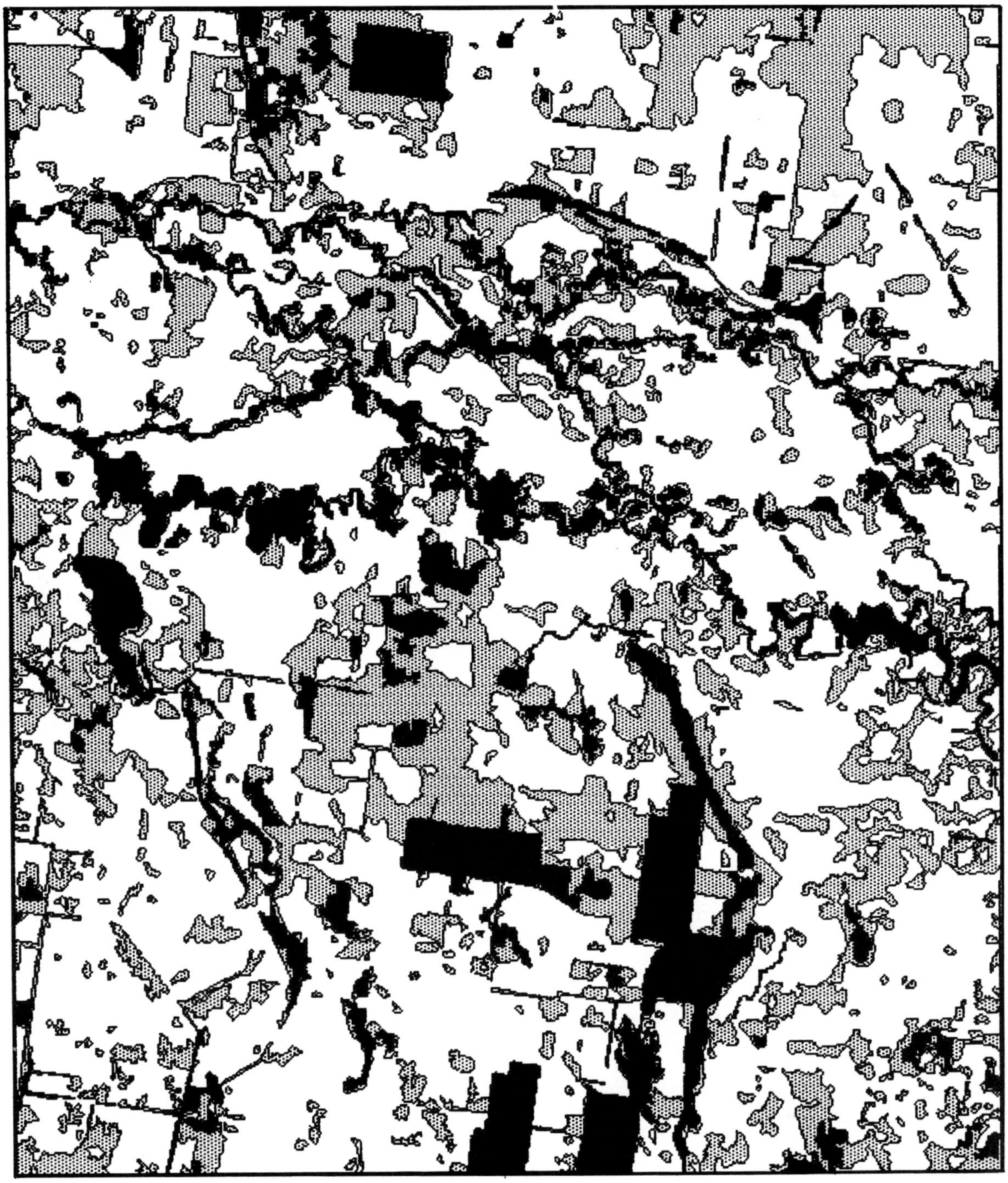

Fig. 2. Clearing of native vegetation in the New South Wales wheatbelt, the Condobolin 1:100 000 map sheet in 1974 (▒ and ■) and that remaining in 1989 (■), Sivertsen (1993).

It is important to stress that clearing alone is not the sum total of much of the habitat loss described above. Grazing has removed understorey habitat; it has been the primary cause of decline in many native grass and palatable shrub species (Adamson and Fox 1982; Denny 1987) and is the cause of non-recruitment amongst many woody species (Auld 1990; Pickard 1991). Grazing by sheep at the turn of the century, is responsible for the type of habitat loss to be seen along Marra Creek in western New South Wales today (Plate 3). Reduced fauna populations are also more suseptable to predation; yet another process which contributes to declines in native biodiversity.

Clearing and grazing together provide the disturbance necessary for invasion by many exotic species. Remnant vegetation in the wheatbelt of New South Wales, most of which is heavily grazed, may comprise as much as 40 per cent exotic plant species (Sivertsen submitted). The

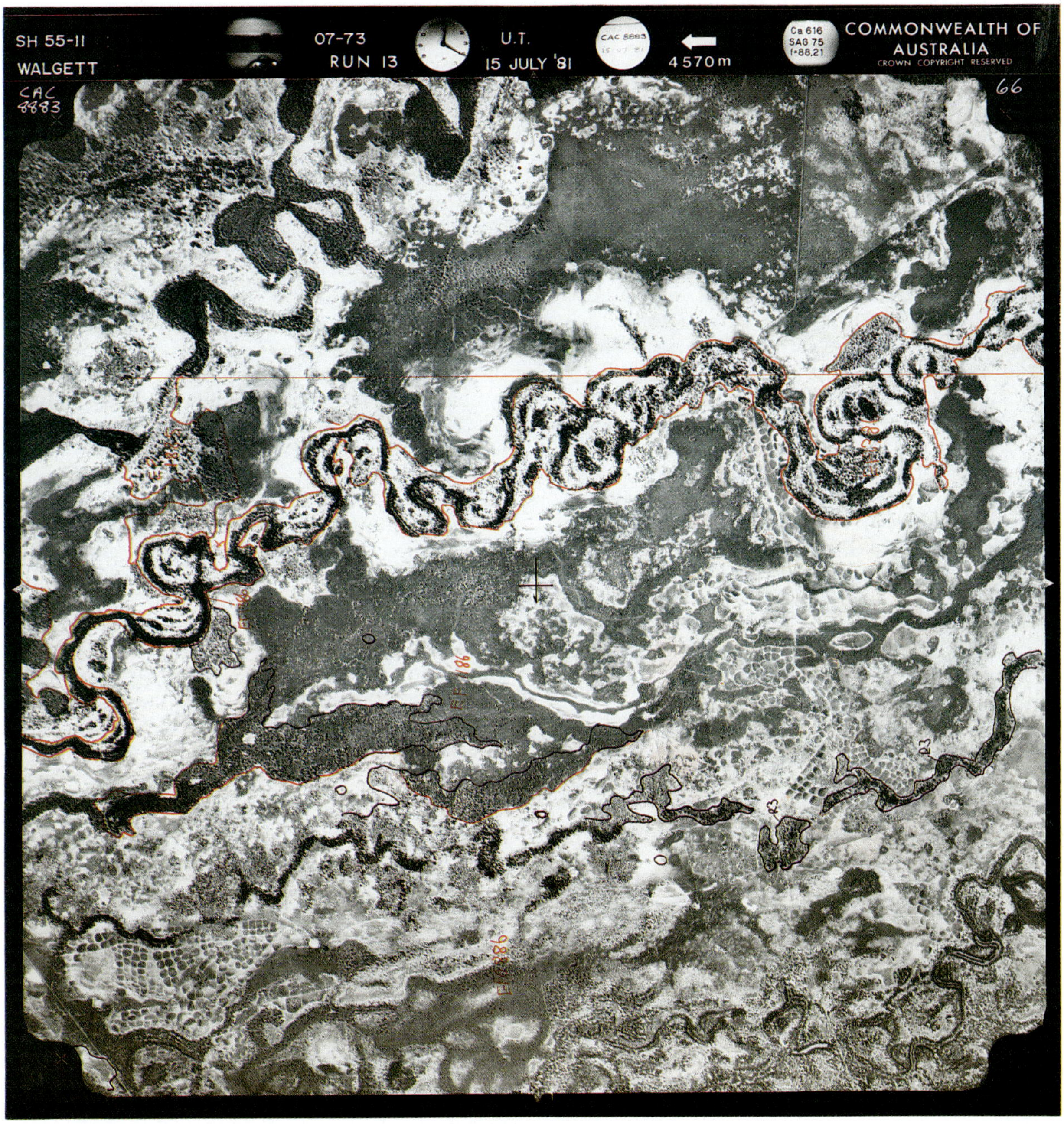

Plate 3. Habitat loss in Chenopod Sheublands. The effects of sheep grazing along Marra Creek, in western New South Wales, early this century can still be seen today as white patches on the aerial photographs.

effects of grazing on native fauna have not been well documented to date. However, Friedel and James (this volume), indicate that there are significantly less reptiles, birds and selected invertebrates in heavily grazed arid and semi-arid rangelands when compared with adjacent ungrazed areas. In the rural landscape at least, clearing and grazing must be considered together in determining habitat loss.

IMPLICATIONS FOR BIODIVERSITY

The main implications of habitat loss for biodiversity will be addressed in a number of chapters in this volume. Habitat loss, in the case studies cited, can be quantified in the clearing of hundreds of thousands of hectares of native vegetation state-wide (Table 1); clearing which continues apace today. In this paper I have called upon data for the Sydney Basin, the Bega district, the Central West, the Wheatbelt and the Western Division as examples. These areas are in no way atypical and together represent most of the area of New South Wales; more atypical are those areas of the Great Dividing Range where large tracts of native vegetation have been retained, more or less intact, by virtue of rugged terrain and reserved Crown Lands.

Given this vast amount of clearing, what are the implications for biodiversity? Loss of biodiversity can take a number of forms. What follows is a very brief, and necessarily, cursory exploration of loss of diversity due to habitat loss.

Ecosytem diversity

The whole thrust of this paper is to describe loss as the ecosystem level. By clearing such vast amounts of our native vegetation, and altering even more through grazing, we are successfully depleting our native ecosystem diversity considerably. The habitat loss discussed, impinges on many ecosystem types in the state. Other examples, which could have been examined equally well, include the south-east forests (where logging for the woodchip industry is an issue), the Northern Coastal Plains (where the Big Scrub has been reduced to tiny fragments), the Brigalow lands and, the New England Tablelands (where the agricultural and pastoral industries rely on extensive clearing of native vegetation). There are very few recognized ecosystems in Australia that have not suffered depletion, simplification and loss of diversity.

Population diversity

We can also recognize biodiversity at the population level. On the Cumberland Plain we have removed 94 per cent of the original woodland vegetation, in the wheatbelt we have removed somewhere between 80 per cent and 95 per cent, in the Bega Valley we have removed 92 per cent of the native vegetation on arable lands. In achieving this level of clearing we have succeeded in destroying whole populations of individual species even though most of those species have not yet become extinct. At the population level, therefore, there has also been a net loss of biodiversity. It is safe to assume, despite the paucity of data, that these trends of biodiversity loss are as true for invertebrates and amphibians, reptiles and birds as they are for the plants which comprise the vegetation communities being cleared and the mammals documented in Table 2.

Genetic diversity

The term "genetic diversity" describes a dynamic equilibrium. The total genetic code is diminished when any individual is lost from the system and is added to when new individuals come into the system. A dynamic balance exists when births and deaths (losses and gains) are more or less equal over time. Loss of habitat and, by definition, destruction of large numbers of individuals (plants and animals) results in an upset of this balance and a net loss of genetic diversity results. Take for example the Turpentine-Ironbark forests of the Sydney Basin. This community has been reduced to 0.5 per cent of its original area; there has been a net loss of individuals of many species and therefore, a net loss of biodiversity at the genetic level, even though none of the component species has yet become extinct (Benson and Howell 1990).

Species diversity

Although loss of biodiversity does not begin and end with extinction, this remains the most important measure of such loss. Populations may recover and species my survive well enough with diminished genetic resources, but when a species is extinct, it is gone forever. As early as 1948 Fairfield Osborn maintained that; "The tide of Earth's population is rising, the reservoir of Earth's living resources is falling. . . ." (Osborn 1948). Expansion of the human population and sphere of influence and the attendant loss of habitat has grave implications for species extinctions. Modern extinction rates, calculated on a per species basis, may be as high as those which existed in the "great extinction events" of the Paleozoic and Mesozoic (Wilson 1989). This means a rate of extinction not seen on Earth for 65 million years, and all due to the activities of one species — *Homo sapiens*. Wilson (1989) also points out one very important difference between the ancient and modern extinction events. In the past these events have tended to affect mainly faunal species, and usually particular taxonomic groups (echinoderms, cephalopods and reptiles), whilst the modern event is universal in its application — plants, invertebrates and vertebrates are all affected. "The ultimate result of this is impossible to predict, but it is not something, I think, with which humanity will want to gamble." (Wilson 1989).

Australia has the dubious honour of having the highest rate of mammalian extinctions in modern times; 50 per cent of all the mammal species that have gone extinct world-wide in the past 200 years were from Australia (Endangered Species Advisory Committee 1992). Twenty Australian mammal species have been driven to extinction in just 200 years. In the same period some 97 species of plants have been pushed to extinction Australia-wide (Briggs and Leigh 1988). In New South Wales 532 species of plants are regarded as rare or threatened (Briggs and Leigh 1988). In the Eden Forests Region of New South Wales, which includes the Bega Valley, about 30 per cent (432 taxa) of the flora is considered to be worthy of special conservation planning (Keith and Ashby 1992) although only 41 of these are listed as nationally rare or threatened (Briggs and Leigh 1988). Many of these 432 taxa have been listed because they have suffered declines in range and population size due to habitat loss. In the Western Division of New South Wales Bowen and Pressey (1993) list 239 plant taxa with restricted distributions and which are therefore worthy of special attention in conservation planning. In the wheatbelt 40 per cent of the native plant taxa recorded appeared only in one or two of 1 200 quadrat sites recorded in a recent survey (Sivertsen 1993).

Whether we are looking at global extinction trends or at documented extinctions nationally or at the trend towards extinction regionally the message is clear: human activity, specifically destruction of habitat, has dire consequences for biodiversity at all levels.

Loss of habitat has seen the extinction of many species already and, if ecologists such as May and Wilson are to be believed, we stand on the brink of extinction events equal to the largest in Earth's multimillion year history.

Changing the balance

Habitat loss is not just about depletion of biodiversity, however. Imbalances, which can be engendered by the habitat loss, are also significant. A number of native species have been considerably advantaged by the changes wrought since European settlement. Red Kangaroos and Western Grey Kangaroos in western New South Wales have been advantaged by more open country with many permanent watering points and a reduction in the numbers of dingos "inside" the dingo fence (Friedel and James, this volume). In contrast habitat destruction in this same region has contributed to the extinction of 28 species of mammal in the same period (Dickman 1994). Loss of woodland habitat in the Western Australian wheatbelt has contributed to the decline of Carnaby's Black Cockatoo *Calyptorhynchus funereus latirostris* whilst the agricultural industries and landscapes that replaced that habitat have given rise to increasing numbers and breeding success of Galahs *Cacatua roseicapilla* (Saunders and Ingram 1987). Similarly in the agricultural landscape of the eastern Australian wheatbelt species such as Galahs, Little Corellas and Sulphur-crested Cockatoos have increased in number whilst clearing of native forests and woodlands for agriculture has brought about declines in species such as the Superb Parrot *Polytelis swainsonii* (Webster 1988) and the Regent Honeyeater *Xanthomyza phrygia* (Franklin *et al.* 1988).

Habitat loss is bringing about profound changes to the ecological balances of our nation. The long-term effects of such shifts in the balance are as hard to gauge as Wilson's (1989) mass extinctions; and perhaps as unwelcome.

WHERE TO FROM HERE?

Graetz *et al.* (1992) and Hobbs and Hopkins (1990), among others, establish that trends in habitat loss, similar to those described above, are occurring Australia wide. Osborn (1948), Weston and Pearl (1989), Wilson (1989) and many others, establish beyond any doubt that similar trends are occurring world wide. It is interesting, and depressing, to note in passing that an increasing proportion of the ecological literature is devoted to "fragmented landscapes"; what Noss (1990) refers to as the "voluminous literature on habitat fragmentation". Dunning *et al.* (1992) in examining "ecological processes that affect populations in complex landscapes" assume a fragmented landscape. Habitat loss is a reality and an ongoing problem.

In New South Wales, and indeed in Australia generally, most habitat loss is the result of clearing and grazing, which is associated with our agricultural and pastoral industries. Calling a halt to clearing will not be simple. In Australia many of our rural, social and economic paradigms include sustained growth and expansion. On the ground these cannot be maintained without continued clearing and "putting into production" of more and more land. Altering such paradigms implies huge social and economic change as well as adoption of a new, environmentally sensitive ethos. Added to this, biodiversity decline will not stop unaided; this implies an urgent need to expand our knowledge of all aspects of biodiversity. Western (1989) identifies seven broad areas which are likely to dominate the conservation movement in the twenty-first century.

1. A complete biotic survey.
2. *Ex situ* conservation.
3. Combining conservation and economic development.
4. Pressures from assistance and lending agencies.
5. Restoration ecology.
6. Engagement by the social sciences.
7 Aesthetic and moral reasoning.

Some of these already form an important part of our thinking; biotic survey, *ex situ* conservation and restoration ecology are three that spring to mind. Another, that Western has not identified, is autecological research which will remain an important part of our thinking for the remainder of this century at least.

Authors such as Hobbs and Hopkins (1990), Young *et al.* (1984), Burbidge and Wallace (this volume) and Saunders and Hobbs (this volume) provide well reasoned and well argued methods for minimizing loss of our native plant and animal diversity. The proposals put forward by such authors begin to challenge some of our social and economic paradigms. Movements such as Landcare, Greening Australia and Total Catchment Management are also beginning to challenge accepted thinking and accepted attitudes to land use. The question immediately arises, "Will this process of change be fast enough to address the problem?" Unfortunately, the short answer is probably "No!" However, this needs to be qualified to some extent. A recent conference in Geraldton, Western Australia (The role of Networks in Nature Conservation, Saunders, in prep.) contained many individual examples where the answer is "Yes!" (e.g., Ahmad, in prep.; Chaudhry, in prep.; Williams, in prep.; Newmark *et al.*, in prep.; Gamassa, in prep. and Doley, in prep.). In these cases, highly dedicated individuals working with local human communities have managed to change people's thinking and actions enough to turn the tide of environmental degradation at a local scale. However, we are a long way from universal acceptance of such thinking.

This then raises the question of legislative control which, in turn, **must** raise the question of management. We may be able to stop clearing with legislation but, if the land is not managed sympathetically and continues to be overgrazed and degraded then its long-term future is just as bleak as if there was no clearing control. Woody species in vegetation remnants eventually become senescent and die. If grazing pressure is such that there is no recruitment in these remnants, they will gradually fade as individuals die and eventually disappear just as surely as if they had been cleared (Pickard 1991).

The plight of some nature reserves in the Western Australian wheatbelt serve to illustrate this point in another way. There are cases where these reserves, isolated remnants, are underlain by very saline and rapidly rising watertables. It is estimated that it will be only a matter of a decade or so before all the native vegetation in such reserves is killed by that saline groundwater (Blyth *et al.*, in prep.). Prevention of this implies a significant shift in local landuse practises and in the social and economic expectations of landholders.

The paths lie clearly before us. On the one hand we can continue as we are now, on the other we can change our thinking and our ethos. On the first path we face mass extinctions with who knows what ultimate consequences. The alternative path requires a huge effort to bring about social change and entails, not just undertaking the science which allows us to understand ecology, decline and restoration, but embracing the themes put forward by Western (1989) in our capacity as scientists and conservationists.

Overcoming habitat loss, however, is not the job of scientists or governments or landholders working in isolation. This very large and difficult undertaking requires the active support and cooperation of all sections of the human community if success is to be a viable option.

ACKNOWLEDGEMENTS

I wish to thank my two referees Richard Major and Andrew Taplin for their critical review and helpful suggestions for improving the text; David Keith who offered many helpful comments on an earlier draft; Lisa Metcalfe who devoted many hours to preparation of the Plates and Figures; and Save the Bush and the NPWS who have joint funded my work in the wheatbelt of New South Wales.

REFERENCES

Adams, P., this volume. Urbanization and Transport. In *Conserving biodiversity: threats and Solutions* ed by R. A. Bradstock, T. D. Auld, D. A. Keith, R. T. Kingsford, D. Lunney and D. P. Sivertsen. Surrey Beatty & Sons: Chipping Norton.

Adamson, D. A. and Fox, M. D., 1982. Change in Australasian vegetation since European settlement. Pp. 109–60 in *A history of Australasian vegetation* ed by J. M. B. Smith. McGraw Hill: Sydney.

Ahmad, A., in prep. Community based conservation in Pakistan. In *Nature conservation: the role of networks* ed by D. A. Saunders. Surrey Beatty & Sons: Chipping Norton.

Attenborough, D., 1987. *The first Eden: the Mediterranean world and man.* Little, Brown and Co.: Boston.

Auld, T. D., 1990. Regeneration in populations of the arid zone plants *Acacia carnei* and *A. oswaldii*. *Proc. Ecol. Soc. Aust.* **16:** 267–72.

Benson, D. and Howell, J., 1990. *Taken for granted: the bushland of Sydney and its suburbs*. Kangaroo Press: Sydney.

Bowen, P. F. and Pressey, R. L., 1993. Localities and Habitats of Plants with Restricted Distributions in the Western Division of New South Wales. *Occasional Paper* No. 17, NPWS: Sydney.

Blyth, J. D., Brown, A. P. and Burbidge, A. A., in prep. Community networks and recovery plans. In *Nature conservation: the role of networks ed* by D. A. Saunders. Surrey Beatty & Sons: Chipping Norton.

Braine, W. H., 1866. Notes on Bega and people in Bega: from the diary of W. H. Braine, Unpublished manuscript, Mitchell Library, Sydney. Quoted in Lunney and Leary, 1988.

Briggs, J. D. and Leigh, J. H., 1988. Rare or threatened Australian plants. Special Publication 14. ANPWS: Canberra.

Burbidge, A. A. and Wàllace, K. J., this volume. Practical methods for conserving biodiversity. In *Conserving biodiversity: threats and solutions* ed by R. A. Bradstock, T. D. Auld, D. A. Keith, R. T. Kingsford, D. Lunney and D. P. Sivertsen. Surrey Beatty & Sons: Chipping Norton.

Chaudhry, I., in prep. Rural communities and conservation of bear species. In *Nature conservation: the role of networks* ed by D. A. Saunders. Surrey Beatty & Sons: Chipping Norton.

Denny, M., 1987. Historical and ecological study of the effects of European settlement in inland New South Wales. Report to the New South Wales Nature Conservation Council and National Estate Committee of the Heritage Council of New South Wales. Nature Conservation Council: Sydney.

Dickman, C. R., 1994. Native mammals of western New South Wales: past neglect, future rehabilitation? Pp. 81–91 in *Future of the fauna of western New South Wales* ed by D. Lunney, S. Hand, P. Reed and D. Butcher. Transactions of the Australian Zoological Society: Sydney.

Dickman, C. R., Pressey, R. L., Lim, L. and Parnaby, H. E., 1993. Mammals of particular conservation concern in the Western Division of New South Wales. *Biol. Conserv.* **65:** 219–48.

Doley, A., in prep. Linkages needed by individual farmers: a personal perspective from the northern wheatbelt of Western Australia. In *Nature conservation: the role of networks* ed by D. A. Saunders. Surrey Beatty & Sons: Chipping Norton.

Dunning, J. B., Danielson, B. J. and Pulliam, H. R., 1992. Ecological processes that affect populations in complex landscapes. *Oikos* **65**(1)**:** 169–75.

Endangered Species Advisory Committee, 1992. An Australian national strategy for the conservation of Australian species and communities threatened with extinction. Aust. National Parks and Wildlife Service: Canberra.

Farrell, D., 1993. New South Wales Year Book No 73 1993. Australian Bureau of Statistics: New South Wales.

Franklin, D. C., Menkhorst, P. W. and Robinson, J. L., 1988. Ecology of the Regent Honeyeater *Xanthomyza phrygia*. *Emu* **89:** 140–45.

Friedel, M. H. and James, C. D., this volume. How does grazing of native pastures affect their biodiversity? In *Conserving biodiversity: threats and solutions* ed by R. A. Bradstock, T. D. Auld, D. A. Keith, R. T. Kingsford, D. Lunney and D. P. Sivertsen. Surrey Beatty & Sons: Chipping Norton.

Gamassa, D.-G. M., in prep. Indigenous peoples I conservation: the pastoral Maasai and wildlife conservation in Tanzania. In *Nature conservation: the role of networks* ed by D. A. Saunders. Surrey Beatty & Sons: Chipping Norton.

Gasteen, J., 1989. The changing face of the Brigalow lands. *Wildl. Aust.* Vol. **26**(1)**:** 10–13.

Goldney, D. C. and Bowie, I. J. S. (eds), 1986. The National Trust (NSW) scenic and scientific survey of the Central West (Draft Report), National Trust of New South Wales.

Heroditus, 430 B.C. *The histories*. Translated by Aubrey de Silincourt. Penguin Books: London.

Hobbs, R. J. and Hopkins, A. J. M., 1990. From frontier to fragments: European impact on Australia's vegetation. *Proc. Ecol. Soc. Aust.* **16:** 93–114.

Keith, D. A., in press. The role of the science in community-driven conservation: a case study in the Bega Valley, New South Wales. In *Nature conservation: the role of networks* ed by D. A. Saunders. Surrey Beatty & Sons: Chipping Norton.

Keith, D. and Ashby, E., 1992. Vascular plants of conservation significance in the south-east forests of New South Wales — June 1992. *Occasional Paper* No. 11, NPWS: Sydney.

Lunney, D. and Barker, J., 1987. Mammals of the coastal forests near Bega, New South Wales. II. Annotated checklist. *Aust. Zool.* **23**(3)**:** 41–49.

Lunney, D., Curtin, A., Ayres, D., Cogger, H. G. and Dickman, C. R., submitted. Identifying the endangered fauna of New South Wales: an ecological approach to a systematic evaluation of the status of all species. *Pac. Cons. Biol.*

Lunney, D. and Leary, T., 1988. The impact on native mammals of land-use changes and exotic species in the Bega district, New South Wales, since settlement. *Aust. J. Ecol.* **13:** 67–92.

May, R. M., in press. Past efforts and future prospects towards understanding how many species there are. In *Biological diversity and global change*. A symposium at the 24th General Assembly of IUBS, Amsterdam.

Murray-Darling Basin Ministerial Council, 1987. Murray-Darling Basin Environmental Resources Study. SPCC: Sydney.

Newmark, D. M., Sariko, H. I. and Gamassa, D.-G. M., in prep. Conservation attitudes of local people living adjacent to five protected areas in Tanzania. In *Nature conservation: the role of networks* ed by D. A. Saunders. Surrey Beatty & Sons: Chipping Norton.

Noss, R. F., 1990. Indicators for monitoring biodiversity: a hierarchical approach. *Cons. Biol.* **4**(4)**:** 355–64.

Osborn, F., 1948. *Our plundered planet*. Little, Brown and Co.: Boston.

Pickard, J., 1991. Sheep and rabbits — the biological chainsaws. *Search* **22**(2): 48–50.

Pressey, R. L., 1990. Clearing and conservation in the Western Division. *Nat. Parks Jour.* **34**(6)**:** 16–24.

Recher, H. F., Hutchings, P. A. and Rosen, S., 1993. The biota of the Hawkesbury-Nepean catchment: reconstruction and restoration. *Aust. Zool.* **29:** 3–42.

Saunders, D. A. (senior editor), in prep. *Nature conservation; the role of networks.* Surrey Beatty & Sons: Chipping Norton.

Saunders, D. A., Hobbs, R. J. and Margules, C. R., 1991. Biological consequences of ecosystem fragmentation. *Cons. Biol.* **5:** 18–32.

Saunders, D. A. and Hobbs, R. J., this volume. Habitat reconstruction: the revegetation imperative. In *Conserving biodiversity: threats and solutions* ed by R. A. Bradstock, T. D. Auld, D. A. Keith, R. T. Kingsford, D. Lunney and D. P. Sivertsen. Surrey Beatty & Sons: Chipping Norton.

Saunders, D. A., 1994. The effects of habitat reduction and fragmentation on the mammals and birds of the Western Australian central wheatbelt: lessons for western New South Wales. Pp. 99–106 in *The future of the fauna of western New South Wales* ed by D. Lunney, S. Hand, P. Reed and D. Butcher. Royal Zoological Society of NSW: Mosman.

Saunders, D. A. and Ingram, J. A., 1987. Factors affecting survival of breeding populations of Carnaby's Cockatoo *Calyptorhynchus funereus latirostris* in remnants of native vegetation. Pp. 249–58 in *Nature conservation: the role of remnants of native vegetation* ed by D. A. Saunders, G. Arnold, A. A. Burbidge and A. Hopkins. Surrey Beatty & Sons: Chipping Norton.

Saunders, D. A., Rowley, I. and Smith, G. T., 1985. The effects of clearing for agriculture on the distribution of cockatoos in the south-west of Western Australia. Pp. 309–21 in *Birds of Eucalypt forests and woodlands: ecology, conservation and management* ed by A. Keast, H. F. Recher, H. Ford and D. Saunders. Surrey Beatty & Sons: Chipping Norton.

Sivertsen, D. P., 1993. Conservation of remnant vegetation in the box and ironbark lands of New South Wales. *Vic. Nat.* **110**(1)**:** 24–29.

Sivertsen, D. P., 1994. The native vegetation crisis in the wheat belt of New South Wales. *Search* **25:** 5–8.

Sivertsen, D. P., submitted. Natural vegetation remnants of the southern wheat-belt (Forbes and Cargelligo 1:250,000 map sheets). *Cunninghamia.*

Strahan, R. (ed), 1983. *The Australian Museum complete book of Australian mammals.* Angus and Robertson: Sydney.

Walker, J. and Hopkins, M. S., 1984. Vegetation. In *Australian soil and land survey field handbook* ed by R. C. McDonald, R. F. Isbell, J. G. Speight, J. Walker and M. S. Hopkins. Inkata Press: Melbourne.

Walker, J., Bullen, F. and Williams, B. G., 1993. Ecohydrological changes in the Murray-Darling Basin. I. The number of trees cleared over two centuries. *J. App. Ecol.* **30**(2)**:** 265–73.

Webster, R., 1988. The Superb Parrot; a survey of the breeding distribution and habitat requirements. Report Series 12. Australian National Parks and Wildlife Service: Canberra.

Western, D., 1989. Overview. Pp. xi–xv in *Conservation for the twenty-first century.* Oxford University Press: New York.

Western, D. and Pearl, M. R., 1989. *Conservation for the twenty-first century.* Oxford University Press: New York.

Williams, S., in prep. Malleefowl as a flagship for on-farm conservation in the northern Murray mallee of South Australia. In *Nature conservation: the role of networks* ed by D. A. Saunders. Surrey Beatty & Sons: Chipping Norton.

Wilson, E. O., 1989. Conservation: The next hundred years. Pp. 1–7 in *Conservation for the twenty-first century* ed by D. Western and M. Pearl. Oxford University Press: New York.

Young, M. D., Wilson, A. D and Harrington, G. N., 1984. Management aims, objectives and responsibilities. Pp. 15–22 in *Management of Australia's rangelands* ed by G. N. Harrington, A. D. Wilson and M. D. Young. CSIRO: Melbourne.

Farmers as conservators of a sustainable production base

R. L. DAVIDSON[1]

Farmers are vaguely aware that they farm within ecosystems, but essential information about agricultural ecosystems has not been made available to farmers. The reasons for this poor communication are briefly examined. The unique value of native bushland on farms is discussed, in the context of recent developments in Landcare and Total Catchment Management Committees.

It is essential that sustainable agriculture be discussed in agricultural terms. Production processes such as nutrient cycling, soil improvement, pest control (in crops, pastures and trees), prevention and amelioration of dryland salinity, protection from wind, cold and heat stress and many other ecosystems processes require large populations of a wide diversity of plant and animal species.

Farmers have always been in the forefront in controlling feral pests and weeds in the vast areas that they manage in Australia. The National Farmers Federation and all the state associations of farmers and graziers have espoused the conservation ethic. Farmers will take the lead in promoting biodiversity when they are assured of its value.

INTRODUCTION

THE 125 615 farmers in Australia manage 500 million ha, or two-thirds of the land surface of the whole country (ABS 1992; Plumb 1982). This is an awesome responsibility borne by each individual farmer and corporate manager in their daily decisions and in their long-term planning. However, the opinions of farmers depend on the information available to them, and regrettably this technical advice to farmers is fragmented into specialist scientific disciplines, agriculture and conservation departments, economics and technology services (Vanclay 1992). Farmers have to pick their way through a minefield of conflicting advice.

Land use should be seen in historical perspective. Clearing land of forest, woodland, mallee and shrubland for cultivation of crops and for pastures was essential for production of food and fibre. Land clearing was a condition of title in the early days, and from 1936 to 1983 clearing was encouraged by tax concessions. As early as the 1950s and 60s intensification of agriculture was driven by the politics of development (Lines 1991).

Throughout this period an ethos of farming was developed, which relegated native plants and animals of the original ecosystem to a position of irrelevance or pest status. To their credit many farmers and especially graziers retained bushland on steep slopes, shallow soil and riverbanks, but almost invariably they failed to exclude livestock. There was a strong ethic of tidiness which resulted in cleaning up and burning the fallen timber and dead trees. In effect farmers deliberately destroyed the habitat of thousands of birds, mammals, reptiles and insects. In the cropping areas and to a lesser extent in the grazing lands the number of plant species has declined dramatically over the past century. The plant and animal species which have almost disappeared from farmed land are mostly common species which are essential for the existence of stable bushland ecosystems. Most people on the land do not know how these plants and animals contribute to sustainable agriculture.

Farmers and graziers have developed this ethos with pride in their individual achievements. They have been intelligent and hard-working in developing a unique style of farming in a strange continent. No one could doubt the outstanding success of Australian broadacre productivity. Farmers paid levies on wool, meat and grain to fund agricultural research and marketing. Their sons and daughters went to agricultural colleges and universities, but learned nothing about natural ecosystems.

Agricultural systems were defined in terms of cattle, sheep and pasture plant species. Arable farming was simply crops, weeds and insect pests. In the 1970s the conquest of nature was exemplified by technology rather than science. Rather than seeking the causes of problems, short term research sought the solution in high tech fixes. I was doing research in CSIRO at this time, and was disturbed by the lack of support for long-term field experimentation in plant and animal production. Of course there were exceptions, and in the 1980s a few farmers and some ecologists examined the role of bushland in farming, stimulated by a growing uneasiness about rural tree decline (rural dieback).

[1]14 Burgess Street, Armidale, New South Wales, Australia 2350.

Pages 43–54 *in* CONSERVING BIODIVERSITY: THREATS AND SOLUTIONS, ed by R. A. Bradstock, T. D. Auld, D. A. Keith, R. T. Kingsford, D. Lunney and D. P. Sivertsen. Surrey Beatty & Sons. 1995.

Plate 1a. Remnant woodland on a sheep property where regeneration is inhibited by heavy grazing. There is no understorey of shrubs and herbs to provide habitat for birds, wasps and other useful animals.

Plate 1b. The inevitable consequence of heavy grazing is a loss of native ecosystems and a bleak, eroded farm landscape.

Plate 1c. Light intermittent grazing will maintain viable woodland with understorey shrubs. Species diversity will not be achieved unless corridors link farms to nearby undisturbed woodland with a rich flora and fauna.

RURAL TREE DECLINE

Although clearing of native vegetation mainly by heavy machinery and herbicides continues to this day in New South Wales and elsewhere, a more insidious loss of natural ecosystems on farms results from rural dieback. There is widespread misunderstanding about the causes of this tree decline (Landsberg and Wylie 1988; Beckmann and Davidson 1990; Davidson and Davidson 1992). All the causes of tree mortality such as aging, disease, drought, flooding, insect damage, lightning, hares and rabbits should be seen in relation to the rate of regeneration from seed. The increase in grazing intensity and trampling by cattle and sheep since about 1960 (Fig. 1) has eliminated this natural replacement of trees and shrubs because livestock were not fenced out of farm bushland remnants retained on most farms. Plate 1a shows a stand of trees on a sheep property, where there is no understorey, no herb layer and no regeneration of eucalypts from seedlings. The decline of many of the farm remnants of natural ecosystems is inevitable (Plate 1b), unless livestock are excluded for most of the year, and the remnants are managed to promote regeneration (Plate 1c).

Two examples of this insidious decline in native vegetation in New South Wales are given. Figure 2 shows the decline in native vegetation in the wheat/sheep region of western New South Wales at Condobolin between 1974 and 1989 (Sivertsen 1993). In Figure 3 the bushland remaining on the northern tablelands in 1962 after clearing, but before the increase in livestock numbers with improved pastures, is shown. Apart from the gorge country, which is still covered with forest and woodland, the farmland is either open grassland or has scattered trees (2–25 trees per ha). Blocks of woodland (too small to show on this scale) were retained on larger farms, but many have deteriorated over the past 30 years because of neglect. Two basic explanations for rural tree decline are firstly a public misconception that healthy native bushland is incompatible with modern agriculture, and secondly that agricultural research has failed to convince farmers that native bushland has a unique value in profitable sustainable agriculture.

THE UNIQUE VALUE OF BUSHLAND ON FARMS

The term bushland is used to encompass remnant vegetation and the native mammals, birds, reptiles, amphibians, pond-life and soil organisms which may be found in the bushland on farms. Being a word in common use, it does not imply a pristine ecosystem unaffected by fire and agriculture. Indeed bushland on farms is a goal for the regeneration of diverse plant/animal communities on selected portions of every farm.

There are several reasons why trees are of value to farmers, but the unique value of an ecosystem of trees, shrubs, grasses and forbs as well

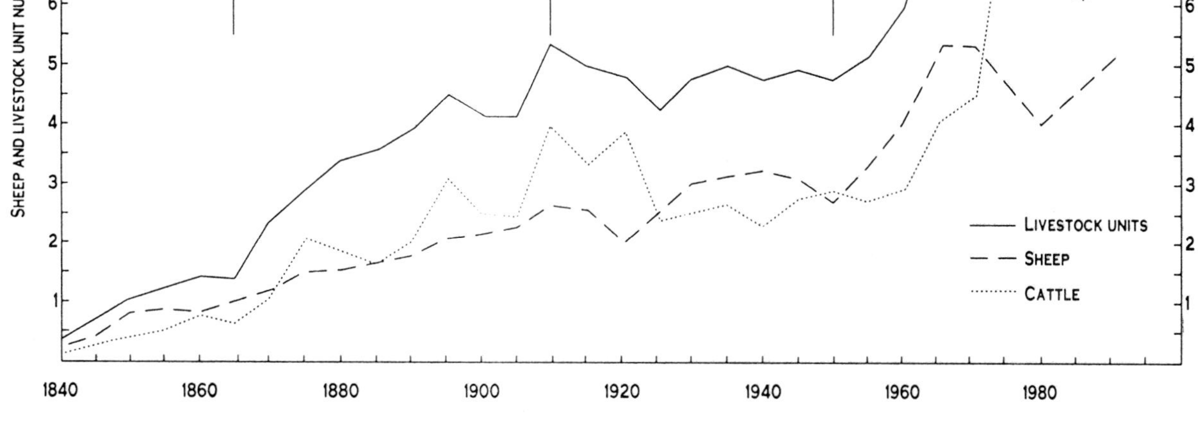

Fig. 1. Numbers of sheep, cattle and livestock units (cattle equivalent to seven sheep) on the northern tablelands, in relation to historical development. From McDonald (1968) amended, by courtesy of the Geographical Society of New South Wales Inc. and the Australian Bureau of Statistics. *Reproduced with permission from* Bushland on farms: do you have a choice? *by Rob and Steve Davidson. AGPS Press 1992.*

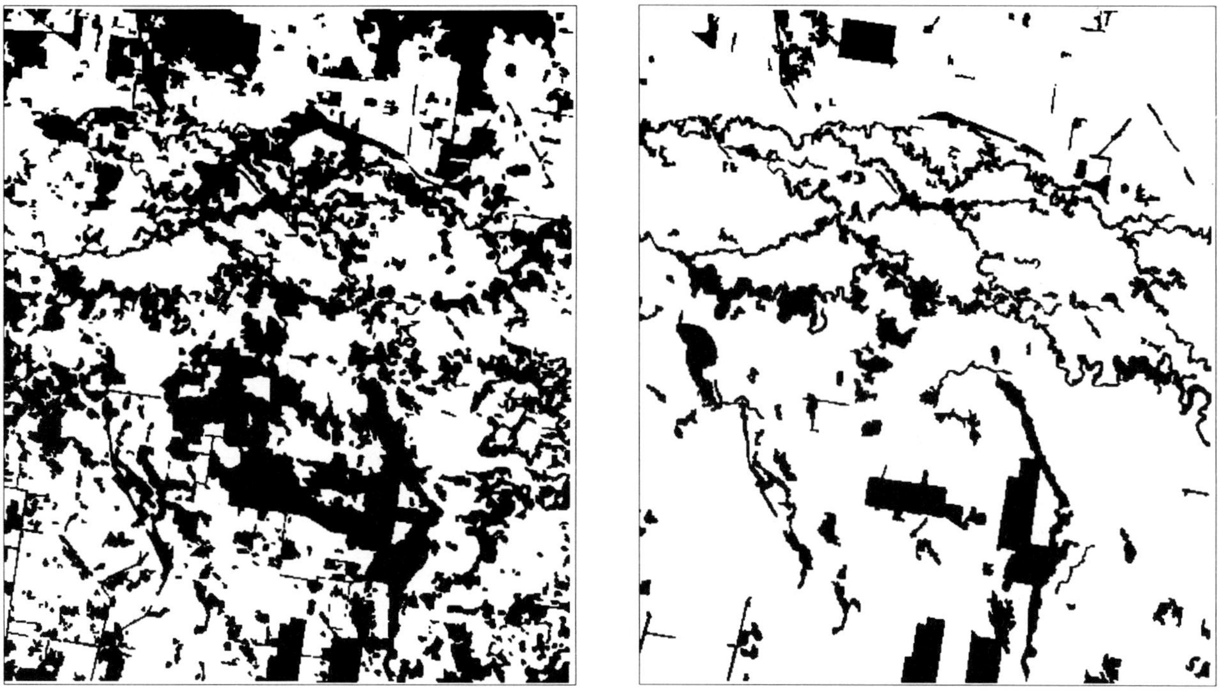

Condobolin 1:100 000 Sheet
(a) Woody vegetation in 1974

Condobolin 1:100 000 Sheet
(b) Woody vegetation in 1989

Fig. 2. The native woody vegetation around Condobolin has declined 61 per cent between 1974 (a) and 1988 (b). From Sivertsen (1993). *By courtesy of the* Victorian Naturalist *and NSW National Parks and Wildlife Service.*

as the associated animals is outstanding. Bushland remnants on farms, provided they are not too small or too isolated, have an awesome array of predators and parasites which help to control insect pests in pastures and crops (Davidson 1984).

In the book *Bushland on farms: do you have a choice?* (Davidson and Davidson 1992) pasture scarabs, crickets, grasshoppers and the insects damaging trees are described as examples of pests which are kept to economically acceptable levels of infestation by the combination of climate, disease, predators and parasites. In Table 1 the natural mortalities of vertebrate and invertebrate animals that are necessary to maintain stable populations are shown to be between 90 per cent in rabbits to 99.9 per cent in the black field cricket. When farmers realize that in most crop and pasture pests there are natural mortalities of around 90 per cent they may wish to enquire further into nature's way of pest control. Death strikes down insects at every turn. Nature's weapons are climate, disease, predators and parasites. The eggs are desiccated in spells of dry weather or flooded by rain or killed by disease. Predators like spiders, lizards and beetles feed on the eggs, and wasps and flies lay their eggs in them. The larvae and nymphs (mostly soft-bodied) are sought after by hordes of hungry predators, like birds, robber flies, dragon flies, echidnas, lizards and spiders. It is with deep regret that I say that most people have little idea of the useful animals that occur in bushland.

These natural controls are what farmers stand to lose unless they familiarize themselves with the habitat requirements of these useful aids to pest management. Birds will require nesting sites, whether hollows in trees, or prickly shrubs like blackthorn and wattle, or even tall grasses. Wasps and flies will need nectar at the appropriate time of year (see Table 2). Birds, which do not hibernate, will require nectar throughout the year, and insects when feeding young. Some birds like magpies and choughs will need insects throughout the year, and logs on the ground will provide wet sites for these insects in droughts. Small birds (which have disappeared from most farms) will need the insects like scale and lerps and caterpillars on eucalypts to satisfy their needs (see Table 3).

Of course, pesticide salesman and the literature they provide will tell farmers frightening stories of how fast the pests breed, and how much they eat. Worst of all they will fail to tell them that pest outbreaks are infrequent and localized, but instead will advise that spraying be done by the calendar, even if pest numbers are below the economic threshold. Such preventive spraying by the calendar is insisted upon in orchard crops and some vegetable crops, especially those grown under contract to canneries. This practice is justified by the excuse that monitoring pest numbers is time consuming. However, a decision support computer programme (SIRATAC) developed by CSIRO and New South Wales Agriculture for

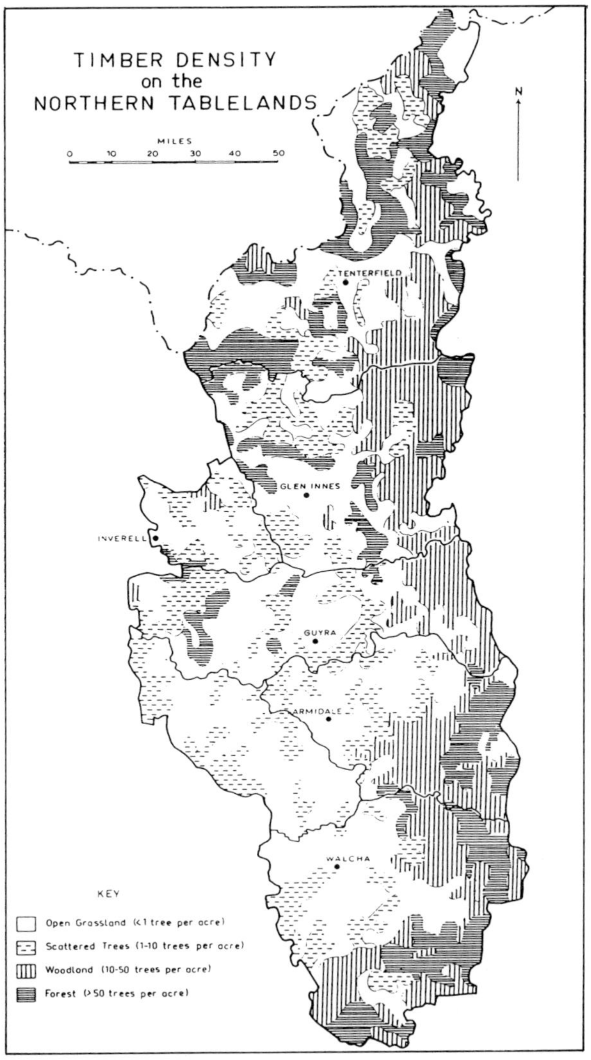

Fig. 3. Map of timber density on the northern tablelands before dieback became series. McDonald (1968) mapped these broad classes of timber density from aerial photographs taken in 1962 by the New South Wales Department of Lands. *By courtesy of The Geographical Society of New South Wales Inc.*

the cotton industry has proved that cost savings are possible in pesticide use (Ralph 1990; Considine 1992). The centralized SIRATAC is being replaced with a decision support system called EntomoLOGIC available for personal computers. Farmers will make their own estimates of pest numbers in their cotton fields (McKewan *et al.* 1992). The trend to wasteful use of farm chemicals must be reversed if agriculture is to become economically viable (see Fig. 4), and ecologically sustainable.

LOOKING AHEAD TO INTEGRATED PEST MANAGEMENT

Farmers understandably do not ask for advice about pest control until they face a pest outbreak. They would not be impressed if an extension officer advised them to plant a tree to encourage the magpies to nest there, or to plant shrubs to encourage parasitic wasps. In 1982 I presented a paper to my scientific colleagues entitled "Correcting past mistakes — loss of habitat for predators and parasites of pasture pests" (Davidson 1982b). Goodyer (1985) wrote an Agfact on scarab grubs in northern tablelands pastures which recommended to farmers the retention of bushland from which cattle and sheep were excluded for the development of the necessary diversity of plants and animals to support natural predator populations. No chemical is registered for use against scarabs on tablelands pastures in New South Wales.

The savings to farmers may be judged by the cost of chemical control of scarabs in New Zealand pastures — between $48 and $200 per hectare for each application excluding labour and machinery costs. These organophosphates are active for only a year, and treated pasture can not be grazed for several weeks. New Zealand's improved pastures, with radiata pine and cypress windbreaks for shelter, are disturbed, unnatural ecosystems which are not conducive to natural enemy populations (Pottinger 1977).

Once broadacre spraying of pastures is practised the chances of natural control are diminished. Firstly, predators and parasites will be killed by the pesticide and their numbers will be temporarily reduced. Secondly the farmer will inevitably neglect to provide habitat essential for the increase of predator and parasite populations.

No decision about the use of pesticide should be made on one season's calculation of crop or pasture losses. In New South Wales long-term measures are available of pasture scarab populations (Roberts *et al.* 1982), wingless grasshoppers (Baker 1993) and Australian plague locusts (APL Commission).

Scarabs are present in temperate pastures at all times, but high densities are localized and sporadic; loss of animal production occurs about once in six years (Fig. 5). Considering that only a part of any property is affected in any one year, the damage spread over six years is negligible. When outbreaks occur birds like crows, magpies, magpie-larks and ibises can be attracted by a light harrowing or trampling by livestock.

Wingless grasshoppers are always present in small numbers, and only increase to high densities as a result of high survival of young hoppers during very dry springs, especially on warm, north-facing ridges without tree cover, where heavy grazing has occurred. The intermittent outbreaks of this pest could be much reduced by planting perennial grasses (without clover or weeds) and trees in the source areas (Farrow and Baker 1993).

Table 1. Number of progeny produced in a lifetime of each mated pair in species considered pests.

Species	No. of young (per female)	Lifespan (months)	Percentage mortality (a) For stable population	(b) To double population
Eastern Grey Kangaroo	12	100	83.0	67.0
Rabbit	20	24	90.0	80.0
Dingo	40	120	95.0	90.0
Pasture Scarab	30	12–24	93.0	87.0
Bush Fly	100	2	98.0	96.0
Cabbage Butterfly	350	2	99.4	98.9
Oncopera Moth	2 000	12	99.9	99.8
Field Cricket	2 500	12	99.9	99.8

Reproduced with permission from Bushland on farms: do you have a choice? *by Rob and Steve Davidson. AGPS Press. 1992.*

Table 2. Flowering periods of native trees, shrubs and herbs that are a source of nectar and pollen.

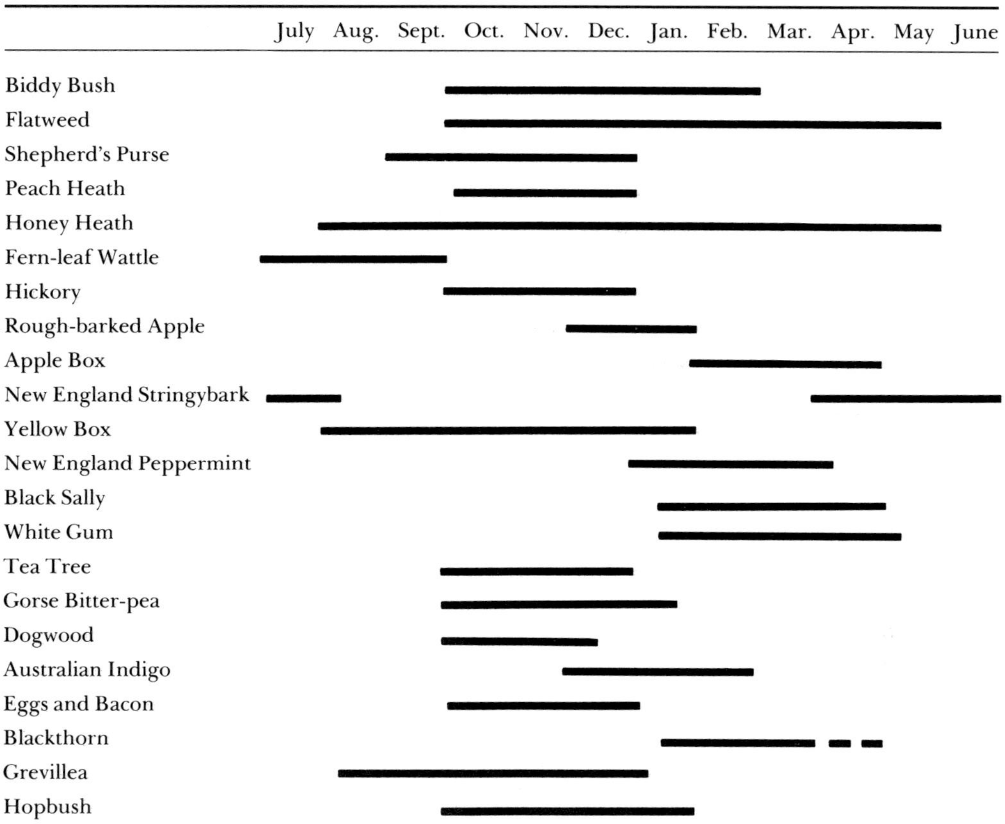

Source: Sue Bowen, pers. comm., in Davidson (1982b). *By courtesy of CSIRO Australia.*

Reproduced with permission from Bushland on farms: do you have a choice? *by Rob and Steve Davidson. AGPS Press. 1992.*

Outbreaks of wingless grasshoppers seem to have increased with increase in improved pastures, clear felling and heavy grazing. When the eggs hatch, strategic spraying only of these limited areas is recommended. Spraying should not be done after the hoppers disperse, that is not after early December in the northern tablelands and mid-December in the central and southern tablelands. Natural predators and parasites of the grasshoppers bring these outbreaks to an end, and broadacre spraying is not advocated (Baker 1981). There are many other grasshoppers occurring together, including two wingless species and several winged species. All these potential pests in pastures including scarabs, crickets and many caterpillars could become more serious if widespread insecticidal control were practised against any one species — with disastrous economic consequences.

THRESHOLDS OF DAMAGE

Farmers and graziers will be persuaded to practise natural methods of pest control only if convincing evidence is available that light infestations are innocuous and that even

Table 3. Seasonal prevalence of insects on trees as a source of food for predators and parasites.

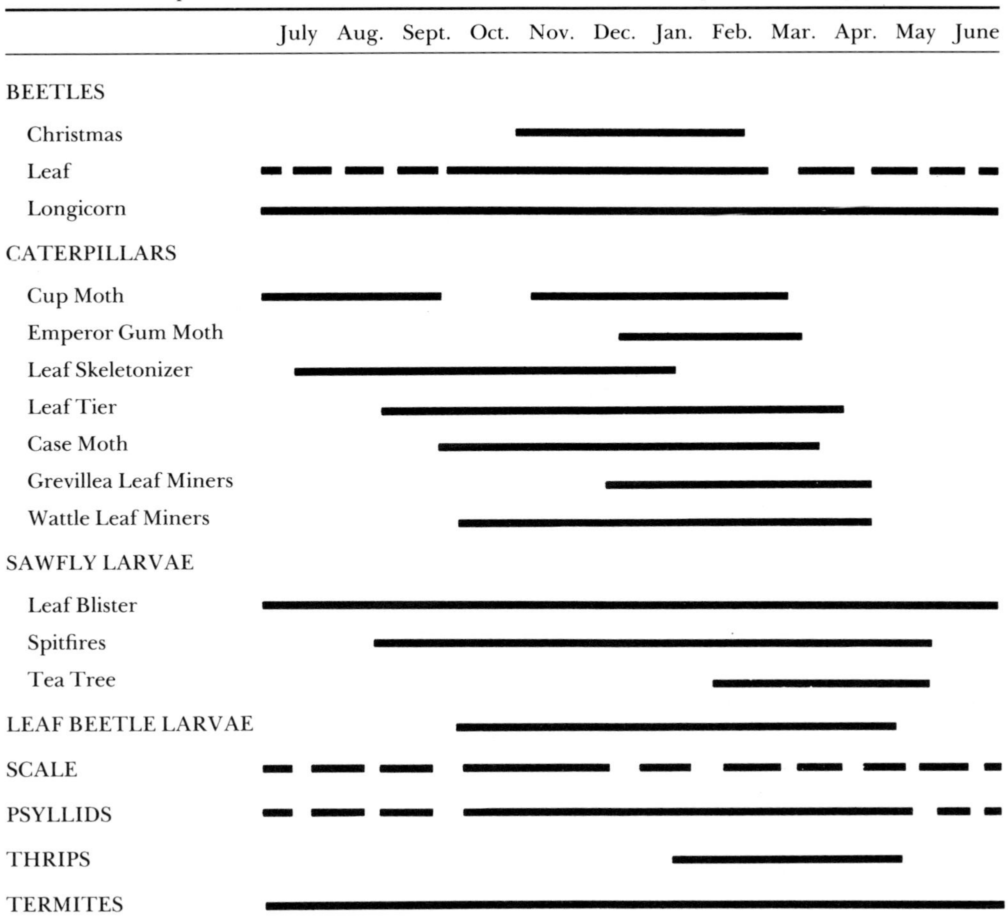

Source: Hadlington and Johnston (1988). *Reproduced by courtesy of New South Wales University Press. Reproduced with permission from* Bushland on farms: do you have a choice? *by Rob and Steve Davidson. AGPS Press. 1992.*

temporary heavy infestations cause less economic loss than the cost of pesticides. Predators and parasites need time to increase their numbers to match host populations.

Figure 5 shows the number of scarabs of all species over 18 years on the northern tablelands estimated from light-trap records (Roberts *et al.* 1982) at the CSIRO Pastoral Research Laboratory. Light infestations cause no damage, and may be beneficial in root-bound pastures. Increasing numbers of grubs cause reduction in roots of pastures, and still greater numbers cause a reduction in foliage growth. However, animal intake is not affected until the amount of standing green herbage falls below 1.2 tonnes per hectare. In pastures stocked at 12 sheep/ha on the tablelands this usually occurs only in winter and early spring. From the diagram, sheep production is affected in three years in 18 years, and spread over 18 years the loss is tolerable. Dense aggregations of scarabs are localized. If only one fifth of a property is affected, three outbreaks in 18 years overall means that any one paddock will suffer economic damage just one year in 30 years, and the loss of production lasts for only a few months. For scarabs in pastures it is clear that the cost of insecticidal control at $48 to $200 per hectare can never be justified in Australia.

Wingless grasshopper outbreaks on the northern tablelands over the past 50 years (Baker 1993) have occurred in only eight years (Fig. 6), and two of these were not serious. In the last outbreak in 1982, the grasshoppers were killed by nematodes and bee-fly larvae. On the northern tablelands there is an enlightened attitude to pest control, because of the evidence available to graziers. Many landholders have for some years been regenerating bushland which is the habitat for natural controls of pasture pests. They have learned to roll with the punches, and regard occasional pest outbreaks as no different from occasional droughts (Fig. 7).

REGENERATING NATURAL ECOSYSTEMS ON FARMS

Assuming that a group of farmers has been persuaded that they should regenerate the

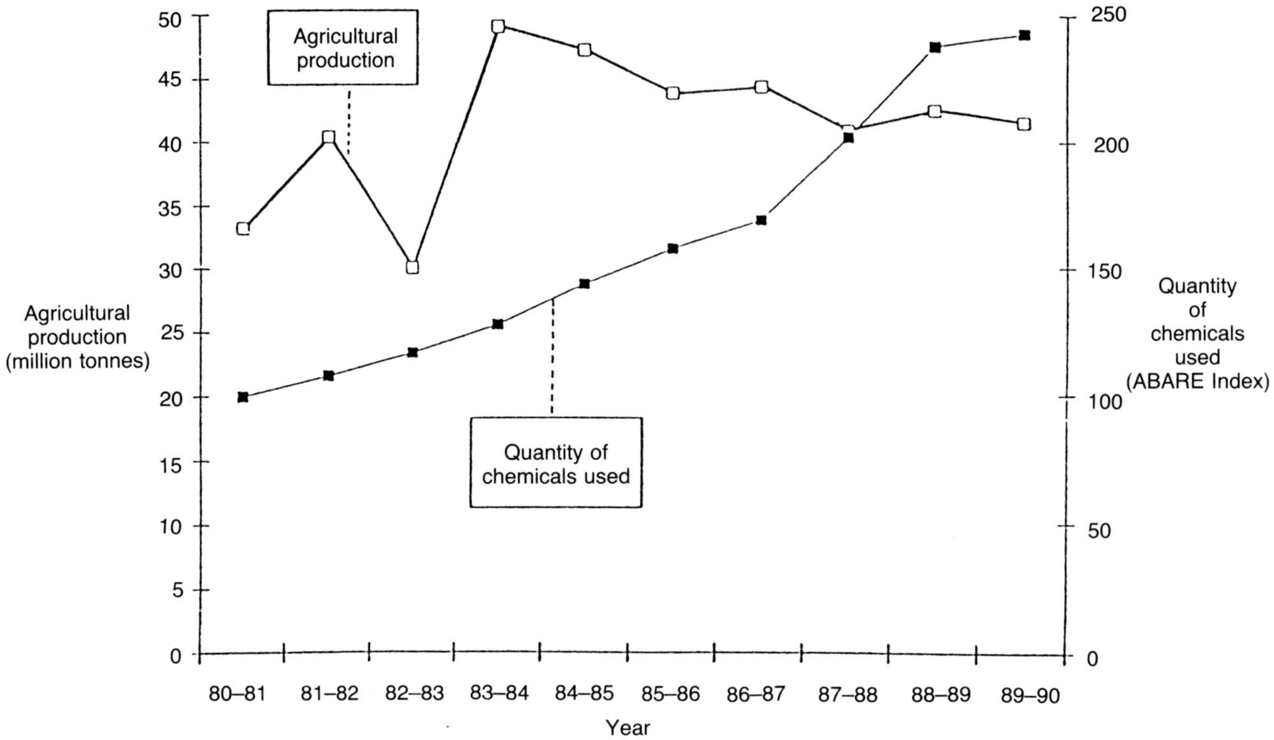

Fig. 4. Use of chemicals in agricultural production, excluding fertilizer. From Reeve and Brouwer (1992). *Courtesy of New South Wales Agriculture Tocal.*

bushland remnants on their properties, what are the difficulties they face?

Discussing this in agricultural terms, they first might consider where they need shelter from wind and sun to protect sheep off shears and new-born lambs. Bird (1984) gives examples of the value of shelter for livestock. Wool production at a high stocking rate was increased 31 per cent by shelter compared with unprotected sheep. Studies in southeastern Australia have shown that lamb deaths can be halved by providing adequate shelter from wind in wet and cold weather. In March 1983 in Victoria 11 per cent of sheep in unsheltered paddocks died after shearing, while all survived in sheltered paddocks. In northern Queensland shade trees resulted in 10–16 per cent more lambs, and cattle using shade were in better condition, with calving percentage increased because of fewer deaths.

Secondly the group of farmers might look for the upper slopes of hills to consider where to regenerate woodland that will prevent rising water tables in low lying parts of their properties that might cause salinity problems. At the same time the woodland would prevent erosion of steep slopes, and the north-facing slopes would not be used as egg-laying sites for wingless grasshoppers.

Thirdly the group would certainly need to map the existing remnants of woodland on their adjacent properties, and if these are in poor condition they might have to look further afield and persuade another farmer to join in the planning. They should be aware that birds and other predators and parasites of pests in crops and pastures can not be planted like trees or shrubs. These animal components of ecosystems can only be attracted by planting shrubs that flower regularly each year, like tea-trees, wattles, grevilleas, bottlebrushes, banksias and blackthorns. Corridors will be necessary between the proposed areas to be regenerated, and these may be useful as windbreaks for livestock. Riverbanks and drainage lines make good corridors, especially if they are fenced off from livestock.

As the estimated cost of fencing mounts, the group might consider forming a Landcare group, and enlist the support of a landcare coordinator to submit a proposal for financial assistance.

All the planting of trees and shrubs need not be done at one time. Indeed there is virtue in planting trees further apart than usual, and drilling or broadcasting seed of local trees and shrubs between them. Seeding along fence-lines with native grasses like wallaby grass, kangaroo grass and snow grass is a good idea to suppress weeds, and provide grazing for stock in times of drought. Tall grass is a valuable habitat for spiders and many useful insects.

There may be difficulty in deciding what species to plant or seed. Greening Australia will provide lists of tree and shrub species appropriate for most areas. Regional plans have been prepared for some regions, e.g., "Land

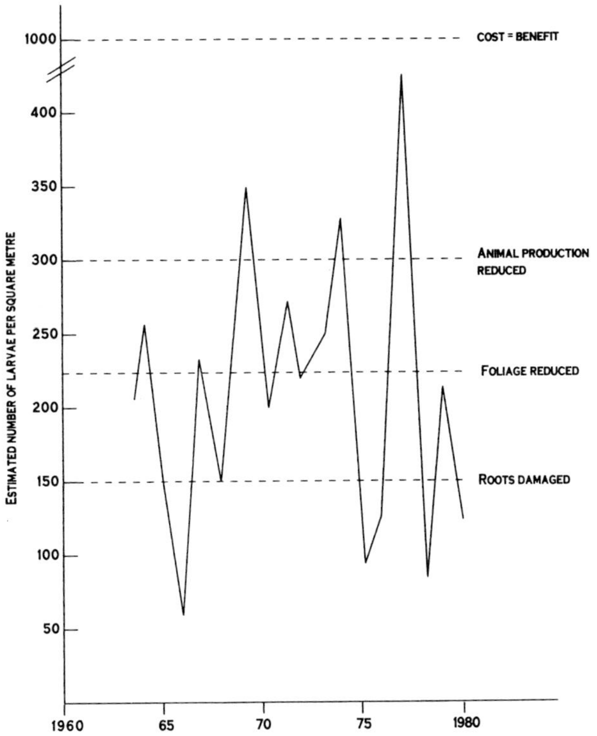

Fig. 5. Estimated numbers of pasture scarab larvae in spring at "Chiswick" 1963–80 and the approximate thresholds for root damage, foliage reduced, animal production reduced and animal production losses equal to cost of insecticide treatment. Actual thresholds will depend on species and pasture growth. From Davidson (1982a). *By courtesy of CSIRO Australia. Reproduced with permission from* Bushland on farms: do you have a choice? *by Rob and Steve Davidson. AGPS Press. 1992.*

conservation for the New England tableland — a regional plan" (Morgan and Terrey 1990) to be published by The World Wide Fund for Nature Australia. This lists the tree species found in each natural region (soil type, topography, etc.), but does not list the shrub and herb species that are essential for functional ecosystems. Local botanists can help, and National Parks and Wildlife staff will help. The Society for Growing Australian Plants and voluntary conservation groups will assist. Local provenances of plant species should be used, and the collection of seed from herbs and small shrubs is a task for farmers, volunteer conservationists and local nurserymen.

Farmers can be reassured that after 5–10 years trees and shrubs will flower and set seed, providing bushland with plants of all ages. If there are clumps of old trees or even isolated trees on the property, they could witness natural regeneration from seed and lignotubers if livestock are excluded. Where the soil under trees has been compacted by trampling, or the soil has become water-repellent because of manure, soil preparation may be necessary.

The development of biodiversity on farms is only a part of the planning for sustainable agriculture. For the reasons given in the introduction it is the most difficult to achieve. In the excellent home study unit on sustainable agriculture produced by New South Wales Agriculture and the Murray Darling Basin Natural Resources Management Programme (Reeve and Brouwer 1992) the other changes needed in farming practices are clearly and convincingly explained. Landcare and Total Catchment Committees are providing integrated advice to local farming communities.

BIODIVERSITY ON FARMS

In agricultural terms there are two dimensions of biodiversity. The first dimension is the number of taxa (species, subspecies, varieties and forms) of plants and animals on the farm. Many will say that this is unlikely to excite a commercial farmer. Yet our experience suggests otherwise. The Clark family at "Bimbadeen" near Taroom in Central Queensland has preserved all the original vegetation types and some parts of the property are set aside purely for wildlife habitat (Lawrie 1989; Voller 1992). Ooline *Cadellia pentastylis* was once reasonably common in the area, but over-clearing in Queensland and New South Wales has made it rare. Adam Clark has protected

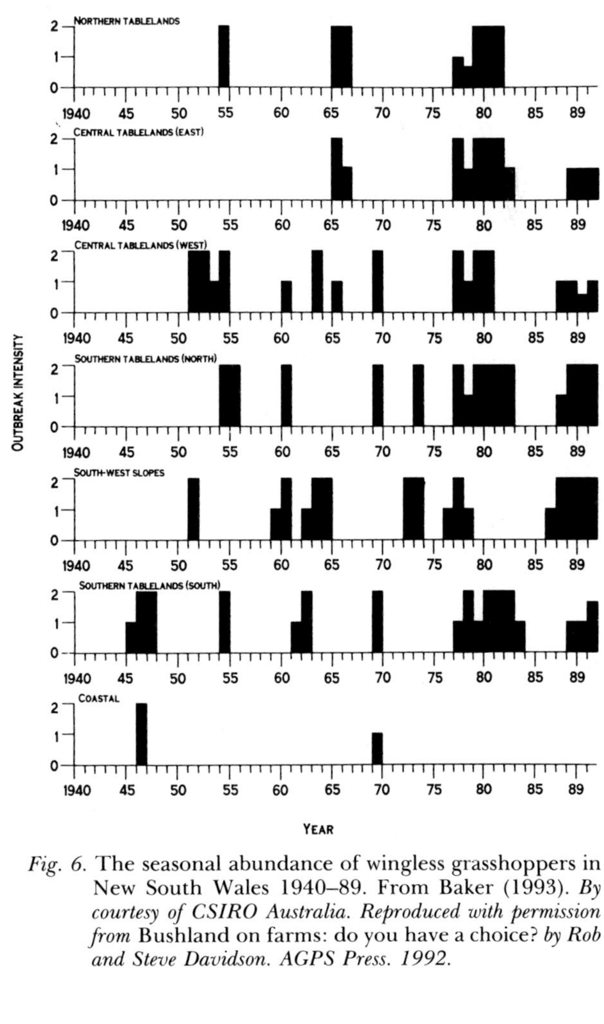

Fig. 6. The seasonal abundance of wingless grasshoppers in New South Wales 1940–89. From Baker (1993). *By courtesy of CSIRO Australia. Reproduced with permission from* Bushland on farms: do you have a choice? *by Rob and Steve Davidson. AGPS Press. 1992.*

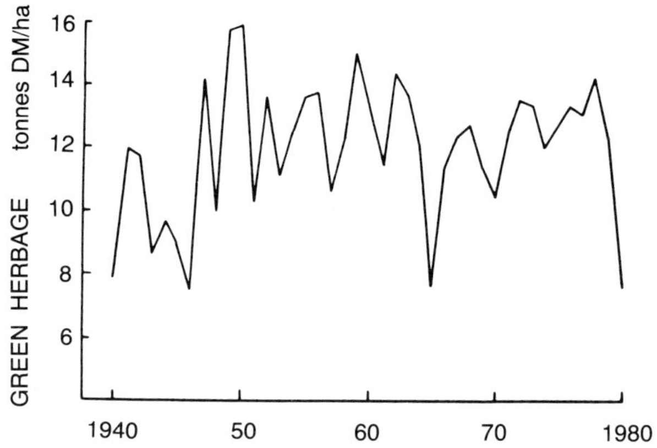

Fig. 7. Simulated annual herbage production in improved pastures, Armidale, from computer simulation (Blair *et al.* 1976), using weather data for the period 1940–80, and assuming fertilized sown pastures.

a patch of softwood scrub where Ooline grows. Joy and Don Williams are wheat and sheep farmers on the sandplains of Badgingarra in Western Australia. They have set aside 373 ha of their 1984 ha property as a nature reserve, preserving seven very rare plant species including four endangered eucalypts (Reading 1992).

Leigh and Briggs (1992) provide estimates of the effects of agriculture and grazing on extinction and threats to endangered species of plants in Australia. Of the 83 extinct species, 34 were caused by grazing and 44 by other agricultural activities. Of the 232 plant species that are endangered, 51 are threatened by grazing and 112 by agriculture. In future 55 more are still threatened with extinction by grazing and 53 by agriculture. Again putting this in an agricultural context, I am sure that any farmer to whom an appeal is made for preservation of endangered species of plants and animals, will respond positively. Better communication is sorely needed, which is surprising considering the technology available through television, radio and personal computer links to data bases.

The second dimension of biodiversity on farms is the number of individual plants and animals of each species present. When considering ecosystem processes such as pest control, nutrient cycling, improvement of soil structure, organic matter in soils, water tables, shelter and so on, the abundance of the plants and animals that are useful to farmers is paramount. The collection of data on fluctuating populations is time consuming, and trends over time are hard to measure. Conservationists and field biologists could greatly assist farmers by contributing to the meagre information on the populations of spiders, scorpions, insects, lizards, birds, nematodes, fungi and bacterial and virus diseases of pests. If these animals could be related to the habitat on farms, we could talk with confidence about sustainable agriculture.

On most farms the threads from which the ecosystem pattern was woven have unravelled as one species after another has dwindled. There is very little comparative information on plant and animal populations in grassland, cropland, woodland and partially cleared woodland. Ford and Bell (1982) and Ford (1989) have counted birds in such contrasted habitats around Armidale, and the results are striking. In Eastwood State Forest there were 21 birds per ha in August/September; during the same period there were ten birds per ha in open woodland while in a dieback area at Newholme there were two birds per ha and treeless grassland had less than one bird per ha. The number of bird species recorded in each of these habitats was respectively 46, 35, 8 and 6.

Some people have said that the natural ecosystems are lost forever from farms, and one has accused those who try to establish such systems as a faction that plays God with landscapes. Be this as it may, there are many examples of farmers who have replanted or rehabilitated bushland. Cicely and John Fenton at Branxholme in the Western District of Victoria have transformed 683 ha of their property from a virtually treeless, heavily grazed farm to a low maintenance productive property (Powell 1992). Wetlands which had been drained and grazed have been refilled (6%), plantations (10%) and agroforestry (5%) established, and laneways and buildings surrounded with amenity plantings (2%). Bird species have increased from 35 to 147 species, of which 44 are breeding there. The pasture retained on 77 per cent of the land is profitable because insecticides and sprays have not been used for 25 years, and fertilizer application has been very light. The lighter than usual stocking intensity has provided trouble-free management. Land valuation has increased and the farming enterprise is sustainable. John and Cicely Fenton are conservators of a sustainable production base.

CONCLUSIONS

I am aware that Steve and I together with the seven contributors from all States to *Bushland on farms: do you have a choice?* are asking Australian farmers to do what no other country has succeeded in doing, namely regenerating natural ecosystems on land cleared for agriculture — retaining and rejuvenating on farms a web of woodland, mallee and heathland comprising about one-tenth the area of freehold and leasehold land. My experience over 15 years talking with farmers has convinced me that the tide has already turned. Sufficient landholders have already achieved natural ecosystems on commercial farms in every State and Territory to prove that it is possible everywhere, no matter how bleak the countryside was before. These farmers have taken the lead in the conservation movement. If each one of them persuaded only one other farmer to follow their lead each year, in six years *all* of the 125 000 farm families Australia-wide would have begun to revegetate 50 million ha of farmland to make agriculture sustainable.

Michael Tooth in the last Session: "Can Government Solve the Problems?" will tell you about the role of landholders in conserving biodiversity. We are agreed that farmers are addressing the wider problems of agriculture, and need more information. But a teacher with a big stick is just a bully. I would never advocate legislation to compel farmers to cease clearing or to restrict their use of pesticides or compel them to lighten stocking rates or even repair eroded land. Peer pressure alone will shame the farmers who drag their feet, in the same way that peer pressure shamed them into clearing their land and burning the logs and litter in the first place.

This is not the first time farmers have been advised to farm within ecosystems. In Australia Illingworth (1921) recommended that sugar cane growers leave remnants of forest between the cane farms to provide habitat for wasp parasites to control cane grubs. Instead the government introduced the cane toad.

In the USA Rachael Carson (1962) wrote "Silent Spring" to warn of the consequences of pesticide overuse. What has happened in the USA since then (van den Bosch 1980) is a salutary lesson for Australian farmers as they strive for biodiversity and sustainable agriculture.

REFERENCES

Australian Bureau of Statistics (ABS), 1992. Characteristics of Australian farms. Pub. 7102.0 Australian Bureau Statistics.

Baker, G., 1981. Wingless grasshopper. New South Wales Agriculture Agfact AE.I.

Baker, G., 1993. Current outbreak and pest status of wingless grasshopper, *Phaulacridium vittatum*, in southeastern Australia. Pp. 121–27 in *Pests and pastures"* ed by E. Delfosse. CSIRO: Australia.

Beckmann, R. and Davidson, S., 1990. Reversing rural tree decline. *Rural Res.* **146:** 19–26.

Bird, P, 1984. Effects of trees on agricultural productivity. Pp. 13–35 in *Focus on farm trees II: reversing the decline of trees* ed by J. Hofler. Canberra Publishing and Printing: Fyshwick.

Blair, G., Till, A. and Smith, R., 1976. The phosphorous cycle — what are the sensitive areas? Pp. 9–19 in *Reviews in rural science III* ed by G. Blair. University of New England: Armidale.

Carson, R., 1962. *Silent Spring*. Houghton Mifflin Co.: Boston.

Considine, M., 1992. Coming pest forecasting for cotton. *Rural Res.* **156:** 24–27.

Davidson, R., 1982a. An ecological view of pasture pests on the Northern Tablelands of New South Wales. Pp. 67–79 in *Proceedings 3rd Australasian Conference Grassland Invertebrate Ecology* ed by K. Lee. South Australian Government Printer: Adelaide.

Davidson, R., 1982b. Correcting past mistakes — loss of habitat for predators and parasites of pasture pests. Pp. 199–206 in *Proceedings 3rd Australasian Conference Grassland Invertebrate Ecology* ed by K. Lee. South Australian Government Printer: Adelaide.

Davidson, R., 1984. The value of habitat. Pp. 39–51 in *Focus on farm trees II: reversing the decline of trees* ed by J. Hofler. Canberra Publishing and Printing: Fyshwick.

Davidson, R. and Davidson, S., 1992. *Bushland on farms: do you have a choice?* AGPS Press: Canberra.

Farrow, R. and Baker, G., 1993. The reduction of wingless grasshopper infestations by sound farm management practices. Pp. 202–05 in *Pests of pastures* ed by E. Delfosse. CSIRO: Australia.

Ford, H., 1989. *Ecology of birds: an Australian perspective.* Surrey Beatty & Sons: Chipping Norton.

Ford, H. and Bell, H., 1982. Density of birds in eucalypt woodland affected to varying degrees by dieback. *Emu* **82:** 202–08.

Goodyer, G., 1985. Scarab grubs in northern tableland pastures. New South Wales Agriculture Agfact P2. AE.1.

Hadlington, P. and Johnston, J., 1988. *Australian trees: a guide to their care and cure.* New South Wales University Press: Sydney.

Illingworth, J., 1921. Queensland Bureau Sugar Experiment Stations. Pp. 5–47 in *Division Entomology Bull.* **13.** Government Printer: Brisbane.

Landsberg, J. and Wylie, F., 1988. Dieback of rural trees in Australia. *Geojournal* **17:** 231–37.

Lawrie, B., 1989. Land development and wildlife conservation: seeking common ground. In *The management of eucalypt country in North Queensland.* Dalrymple Landcare Committee: Charters Towers.

Leigh, J. and Briggs, J., (eds) 1992. *Threatened Australian plants: overview and case studies.* Australia National Parks and Wildlife Service: Commonwealth of Australia, Canberra.

Lines, W., 1991. *Taming the great southland: a history of the conquest of nature in Australia.* Allen and Unwin: Sydney.

McDonald, G., 1968. Recent pasture development on the Northern Tablelands of New South Wales. *Australian Geographer* **10:** 382–91.

McKewen, L., McFarlane, S. and Maddern, W., 1992. EntomoLOGIC and HydroLOGIC — what's happening? In *Proceedings 6th Australia Cotton Conference.*

Morgan, G. and Terrey, J., 1990. Land conservation for the New England tableland: a regional plan. World Wide Fund for Nature, Australia. Unpublished report.

Plumb, T., 1982. *Atlas of Australian Resources 3rd Series, Vol III Agriculture.* Division of National Mapping: Canberra.

Pottinger, P., 1977. Pests of grassland and small seeds crops. In *Pastures and pasture plants* ed by R. Langer. Reed: Wellington, NZ.

Powell, P., 1992. John and Cicely Fenton: "Lanark". Pp. 172–80 in *Bushland on farms: do you have a choice?* AGPS Press: Canberra.

Ralph, W., 1990. SIRATAC: death and rebirth. *Rural Res.* **147:** 8–12.

Reading, J., 1992. Conservation "on the land". Pp. 184–90 in *Bushland on farms: do you have a choice?*. AGPS Press: Canberra.

Reeve, I. and Brouwer, D., 1992. Your farm's future: how to preserve its sustainability. New South Wales Agriculture and MDB Natural Resources Management Strategy. C.B. Alexander Agricultural College: Tocal.

Roberts, R., Ridsdill Smith, J., Porter, M. and Sawtell, N., 1982. Fluctuations in abundance of pasture scarabs over an 18-year period of light-trapping. Pp.75–79 in *Proceedings 3rd Australasian Conference Grassland Invertebrate Ecology* ed by K. Lee. South Australian Government Printer.

Sivertsen, D., 1993. Conservation of remnant vegetation in the box and ironbark lands of New South Wales. Vic. Nat. **110:** 24–29.

Vanclay, F., 1992. The social context of farmers' adoption of environmentally sound farming practices. Pp. 94–121 in *Agriculture, environment and society* ed by G. Lawrence, F. Vanclay and B. Furze. Macmillan Australia: Crows Nest.

van den Bosch, R., 1980. *The pesticide conspiracy*. Prism Press: Dorchester.

Voller, P., 1992. An old motto at work in land management: Adam Clark's story. Pp. 191–94 in *Bushland on farms: do you have a choice?*. AGPS Press: Canberra.

Urbanization and Transport

P. ADAM[1]

Urban development inevitably involves habitat destruction. Although the area of cities in Australia is a small proportion of the total land surface the distribution of major cities in coastal regions has resulted in a disproportionate loss of some coastal and estuarine habitat types. Urbanization also results in serious degradation of surviving habitats, while some impacts extend far from the city. The nature of degrading factors is described and the conservation management implications of the continuance of degrading processes are discussed.

The impact of urbanization on the various components of biodiversity is discussed. Although the adverse impacts of urbanization on the environment can be identified, it is difficult to quantify the effects on biodiversity given the lack of data on both the pre and post urbanization environment.

Transportation is also responsible for both habitat destruction and degradation. Although, on balance, transportation has had a negative effect on biodiversity, the verge of roads and railway embankments may function as habitat corridors and also provide opportunities for the conservation of species and communities which may otherwise have been lost.

Australian cities will continue to grow. In addition, there will be movement away from the current metropolitan centres to rapidly developing urban areas on the coast. The impacts of urbanization in the future will be more extensive then at present but there are opportunities to learn from the past and to implement planning and development controls which might minimize environmental damage and maximize conservation opportunities.

URBANIZATION

URBANIZATION has a dramatic impact on the environment. Nevertheless within Australia, urbanization has directly affected only a very small proportion of the land surface (the exact extent depending on the definition of urban, conventional statistics treat any settlement of more than a 1 000 people as urban, but even smaller communities have impacts on the environment similar in scope if not in scale to larger communities).

The definition of "urban" in official statistics is complex and areas classified as urban may nevertheless incorporate natural areas and agricultural land. After a census, "a boundary was finalized for each population cluster of 1 000 or more people and for known holiday resorts of less population if they contained 250 or more dwellings (*not* including caravans, etc. in caravan parks) of which at least 100 were occupied. These clusters are known as urban centres and the populations and dwellings enumerated in them are classified as urban for statistical purposes" (Australian Bureau of Statistics 1986a). Capital city statistical divisions are "predominantly urban in character and the boundaries are delineated to contain the anticipated urban development of the capital cities (and associated urban centres) for a period of at least 20 years." . . . "A capital city statistical division boundary should delineate an area which is now, or is expected to be socially and economically oriented towards the metropolitan area in question" (Australian Bureau of Statistics 1986b).

Australia is a highly urbanized country. In 1986 an estimated 85 per cent of the population lived in urban areas (as against a global average for developed countries of 73%), with 63 per cent living in the 12 major areas with populations exceeding 100 000 (Castles 1992). There was a steady increase in the proportion of the population living in urban areas between the 1920s and 1976 (from 62% to 86%), but in more recent years there has been a very small decline (Castles 1992). (The current national population is more than 17 million, growing at 1.5 per cent per annum (Castles 1992), so that between 2000 and 2005 the total population is likely to exceed 20 million).

Although the area directly affected by urbanization is small compared with the size of the continent it could be argued that nearly all human impacts on the environment are a consequence of urban development. The maintenance of urban society requires input from every component of the economy. More specifically, feeding the urban population accounts for a considerable proportion of agricultural output, while the economic value of the whole agricultural sector, although currently declining, is still a vital element in the Gross Domestic Product and thus supports, in part, urban society. Provision of water requires facilities which may be situated well beyond the urban limits, provision of raw materials for urbanization and industry requires quarries and mines, (and the exports from the mining industry provide a substantial part of the financial underpinning of urban society), meeting demands for timber extends the influence of

[1]School of Biological Science, University of New South Wales, New South Wales, Australia 2052

Pages 55–75 *in* CONSERVING BIODIVERSITY: THREATS AND SOLUTIONS, ed by R. A. Bradstock, T. D. Auld, D. A. Keith, R. T. Kingsford, D. Lunney and D. P. Sivertsen. Surrey Beatty & Sons. 1995.

Australian urbanization beyond the continent to the forests of South-east Asia, the Pacific and North America and the plantations of New Zealand. The waste products of society and industry place stresses on the environment, locally and globally.

Urban ecology has attracted considerable attention as a topic for study (see Nix 1972; Bornkamm *et al.* 1982). It is well established that urban areas have modified climates (Fitzpatrick and Armstrong 1972; Watson 1986), altered patterns of drainage and evaporation (Bell 1972; Warner 1986; Johnston 1986), and nutrient and energy cycling (Aston *et al.* 1972; Kalma *et al.* 1972). The ecology of surviving natural areas, and of artificial habitats in urban areas have been the subject of many studies (see, *inter alia* Fitter 1946; Sukopp and Werner 1982, 1987; Barker and Graf 1989; Bornkamm *et al.* 1982; Benson and Howell 1990a, b).

The impacts of urbanization on both wildlife and human health have been extensively studied overseas, but rather less in Australia. However, it is difficult to find data on the impacts of urbanization on biodiversity. In this account, attention is drawn to the multiplicity of ways in which urbanization and transport may have impacts on the environment. Where possible specific examples are given, but in many cases we can only speculate as to impacts.

From this base an attempt is made to assess the consequent impacts on biodiversity. It is necessary in so doing to recognize the breadth of the concept of biodiversity which is far more than simply species richness. However, the complexity of the concept makes it difficult both to document changes and to assess the significance of changes to particular components of biodiversity.

The focus of attention is largely, but not exclusively, Sydney. Most of the discussion is of the direct impacts of urbanization. A variety of other processes, which are a consequence of the existence and needs of urban societies and which have impacts on biodiversity will be addressed by other authors in these proceedings. The all pervasive indirect effects of urbanization have had a far greater impact on biodiversity that the localized immediate impacts.

Many of the examples may seem trivial; individually this may be so but the cumulative effect on biodiversity will be made up of the sum of numerous small impacts.

The built environment

Urbanization inevitably involves habitat destruction; buildings, roads, carparks and other constructions occupy land which was formerly "natural" habitat (although sometimes with an intervening agricultural phase).

The habitats destroyed will vary with the city; in the case of Sydney they have been documented by Benson and Howell (1990a, b). Globally, some particular systems are prone to destruction during urbanization; freshwater wetlands, riparian systems and, given that many of the world's major cities are coastal ports, estuarine and coastal wetlands.

The covering of soil by hard, impermeable surfaces increases run-off, particularly in storm events. To reduce flooding and improve the efficiency of drainage there has been a long history of canalizing natural channels, replacing natural banks by stone or concrete, straightening bends, filling in low lying floodplains, replacing surface flows by underground culverts. In addition to physical alteration of the habitat structure and flow patterns, water quality is affected by pollution, both from deliberate discharges and diffuse run-off. Aquatic habitats in urban areas have thus received severe impacts and, at the local and regional scale, have suffered a loss in biodiversity. Additionally the continuity of riparian vegetation which might have formed important wildlife corridors (Naiman *et al.* 1993) has been lost.

Many wetlands in urban areas have been filled (for which the common euphemism is reclaimed — a term implying that in some sense the sites concerned had, prior to intervention, lapsed from a superior state) and suffered pollution from stormwater and other discharges. Even when no deliberate filling has occurred there will be increased sedimentation, both from the initial erosion phase following vegetation clearance (Johnston and Gerstle 1986; Johnston 1986) and chronically from the developed catchment.

The garden

Australian cities are predominantly suburban, with, by world standards, very low population densities. Only a part of the urban area is occupied by construction, much of the rest being garden.

The presence of gardens is a major reason why the direct impact of urbanization on biodiversity has been relatively small. Both native and exotic plants in gardens provide resources for native fauna (Fox 1990), and play an important part in fostering an interest in the natural world. The barrier effect of the metropolis is mitigated by the abundance of garden and street trees which provide a diffuse network of wildlife corridors permitting at least some faunal elements (particularly birds and some insects) to move through and within the urban landscape.

Gardens also have negative impacts on biodiversity. Many of the species grown in gardens are exotic. Of these some have the ability to spread into natural vegetation; many of the most environmentally serious weeds of southern Australia were originally introduced for their perceived horticultural value. Humphries *et al.* (1991) estimate that a third of all introduced species "can be expected to significantly affect terrestrial native communities". The proportion of introduced species in floras varies between states and between regions within states. Tasmania has the highest percentage (31%) and the Northern Territory (4–5%) the lowest (Humphries *et al.* 1991). Nevertheless reciting mere numbers of introduced species, without taking into account their abundance and capacity for further spread, gives a misleading impression of the problem. Although the Northern Territory has a small number of introduced species it is host to some of the most significant environmental weeds (for example *Mimosa pigra*). Cities, but also islands, tend to have high concentrations of non-indigenous species (Humphries *et al.* 1991).

The successful growth of exotic species in Australian gardens may require application of fertilizer and or pesticides. Fertilizer run-off pollutes waterways and increases soil nutrient levels in remnant bushland, providing conditions favourable for the invasion of exotic species (Clements 1983; Fox and Adamson 1986). The environmental conditions we strive to create in gardens through cultivation (i.e., regular soil disturbance) and fertilizer application are repeated in degraded natural habitats. It is not surprising therefore that many garden plants find conditions in disturbed bushland very favourable for growth. Pesticides which are some times applied frequently and at high concentrations to gardens may reduce the populations of native as well as exotic invertebrates in gardens.

Many gardens are heavily watered. This use of water contributes to the demand for water transfers from the surrounding catchments, and, through run-off contributes to the change of originally sclerophyll communities to more mesic vegetation. A major component of many gardens, and urban parks, is lawn, a close, dense sward of exotic grasses. Turf farms are a significant component of urban fringe agriculture. Top dressing of lawns with loam is frequently practised, and requires the destruction of naturally fertile riverine habitats. The use of peat, both as a soil conditioner and in potting mixes requires the destruction of wetland habitat, either in Australia or overseas.

Exotic species make an important contribution to the cultural landscape and will continue to play a major role in the urban ecosystem. However, over recent decades there has been an increasing trend (or fashion) towards use of Australian species in both gardens and broader scale landscaping. At first glance this trend would seem to be one which would promote biodiversity conservation — it reduces the use of exotics, with the attendant risks of weed invasion, it should involve less use of fertilizers and water (although not necessarily in the case of popular rainforest species) it provides greater resources for a wider range of native fauna, and many provide an *ex situ* means of conserving rare or endangered species. Unfortunately enthusiasm for native gardens needs to be tempered with caution. "Native" is rarely restricted to "locally indigenous", and in those few cases where "locally indigenous" is specified, the genetic stock may be introduced from elsewhere in a species' range. A number of Australian species have already become serious environmental weeds outside their natural geographical range (for example, *Acacia baileyana, Pittosporum undulatum*) (see Groves 1986; Carr *et al.* 1992). The extensive planting of non-local genotypes of indigenous species in landscaping and rehabilitation threatens pre-existing patterns of genetic diversity (and is hence a direct threat to one of the components of biodiversity). Gene flow may threaten the integrity of whole species; hybridization of *Grevillea* spp. following pollen transport from garden plantings has been reported (Carr *et al.* 1992). (This phenomenon is not restricted to plants — hybridization between the Mallard *Anas platyrhynchos*, often introduced to ponds in parks and ornamental gardens, and the Native Black Duck *Anas superciliosa* is of common occurrence).

Although most suppliers of native plants for horticulture operate within both the letter and spirit of the law, there is a continuing illegal trade in wild collected species, including orchids and tree ferns. Collection involves disturbance to habitats and reduction of the wild population. The extent to which illegal collection of plants reduces biodiversity is unknown, but it must be identified as one of the many threatening processes consequent on urbanization.

There is also a substantial trade in flowers, and other plant material collected in the wild. While exports are subject to control, and legal collection for sale within Australia requires state regulations to be met, the impacts of harvesting are not well known. In the case of Christmas Bells *Blandfordia* spp. it is frequently alleged that fire regimes in some wet heathlands have been manipulated to promote harvesting, possibly to the detriment of other species.

The fashion for native plants was associated with the use, at least in Sydney, of bush rock as a conspicuous element in landscape design. Very little of this rock was legally obtained, and much came from within national parks. Collection and removal of bush rock involves extensive disturbance and also the direct loss of an important habitat for herpetofauna, invertebrates and cryptogamic plants (bryophytes and lichens). At the present time the demand for bush rock appears to be less than it was 10 years ago but the impacts of removal on the bush are still significant.

Feral animals

A consequence of urbanization has been an increase in the number of carnivores in bushland, principally cats and dogs, although even close to the centre of Sydney the fox is a significant predator.

Feral cats are a significant threat to biodiversity across the whole of Australia (ANPWS 1992). In urban and near urban areas, it is difficult to distinguish between feral and domestic cats. As domestic cats are not dependent on the bush for survival the number of part-time predators operating in small patches of remnant habitat may be far greater than would be sustainable by the natural food web. The pyramid of numbers is "top heavy", and the impact on the native fauna is substantial. It is estimated that domestic cats are responsible for the death of 75 million native animals each year (a figure which pales into significance beside the total of 12 000 million estimated annual kill of native animals by cats nationally, ANPWS 1992).

Dogs are responsible for far fewer deaths of native fauna than cats (Walraven 1993), although their contribution to mortality adds to the pressures on some species (such as urban koalas — Smith and Smith 1990). The effects of disturbance caused by dogs have not been documented. Dogs being exercised on the Botany Bay foreshores can frequently be observed putting flocks of wading birds to flight. Is the loss in feed time resulting from this sufficient to cause a reduction in the condition of the birds?

Domestic pets are important contributors to the increased level of nutrients (particularly phosphorus, Bonner and Agnew 1983) in runoff into urban bushland and thus they promote the invasion of these communities by exotic plant species.

The effluent society

Cities are the great sources of pollutants. Many pollutants know no boundaries; cities and their industries are major producers of greenhouse gases, ozone reducing gases and a vast range of chemicals which once released into the environment can enter the food chain and be dispersed around the globe. Data on the production and dispersal of pollutants in New South Wales have recently been published by the Environmental Protection Authority (1993).

Pollutants by definition, can be detected in the environment, to that extent we can be certain they cause environmental change. Many pollutants are toxic, at varying concentrations, to a wide range of organisms. Very polluted waterways support only extremely depauperate communities (Mellanby 1969) while industrial air pollution has denuded the landscape of vegetation as at Queenstown in Tasmania (Anderson 1978). At the local scale, therefore, there is clear evidence for pollution being responsible for considerable declines in biodiversity. What is less certain is whether, when placed in a larger context, these local events represent significant impacts on biodiversity, and also whether the widespread extent of low concentrations of pollutants are biologically significant. (The detection of pesticide residues in Antarctic organisms is important evidence that we must expect pollutants to be widely dispersed, and is thus a warning to consider the potential for impacts at a global scale. As yet there is little evidence that at the concentrations found, there has been or is likely to be any biological consequence. However, in the case of the increase in greenhouse gases there are plausible grounds for arguing that while it is not possible to detect (at present) any effects on biodiversity there is strong likelihood of very considerable and diverse impacts in the future (Peters and Lovejoy 1992)). At the continental scale acid rain is imputed as the cause of environmental changes to forest and aquatic ecosystems in both North America and Europe, changes which are likely to have had consequences for biodiversity (Abrahamsen 1984; Kinsman 1984).

The air quality of Australian cities has been a matter of public concern for several decades. Concentrations of pollutants, regarded as damaging, or potentially damaging, to human health are frequently recorded above levels set as acceptable standards (Castles 1992). Little is known, however, about any possible impacts on the biota as a whole. It is plausible to suggest that the organisms most likely to be affected by air pollutants will be those terrestrial cryptogamic plants which absorb water across their entire surface. The adverse impacts of air pollution on lichens (Hawksworth and Rose 1976) and bryophytes (Richardson 1981) are well established from overseas studies, but background information on both group of organisms in Australia is still sparse, despite the recent publication of flora accounts, although Rogers (1992) suggests that air pollution is responsible for the absence of a number of lichens from urban areas.

The state of aquatic environments has also been a cause of public and political concern in recent years, both in freshwaters and, particularly in Sydney, the sea. Again a major focus of concern is perceived threats to human health and considerably less attention has been given to effects on the biota. Heavily polluted waterways clearly suffer a reduction in biodiversity, but what are the impacts of lower, but far more widespread concentrations of pollutants? As well as chemical pollutants, temperature changes caused by cooling water discharges may also affect biodiversity (King 1988). Is the decline in many species of frog, which appears to be a global phenomenon, at least in part related to pollution?

Australian cities dispose of substantial volumes of garbage to landfill (Castles 1992), and current sites have very limited remaining capacity. The area directly affected by landfill is very small. There is a possibility of toxic leachates continuing to flow from tips long after filling has ceased, but the impacts of such leachates on biodiversity have not been documented.

One consequence of garbage disposal practices has been the promotion of a substantial population increase in Silver Gulls *Larus novaehollandiae* (Smith and Carlile 1992). The expansion of Silver Gull numbers has been accompanied by declines in other species utilizing the same breeding sites. Silver Gulls are also aggressive predators, partly responsible for the poor success in most years of Little Tern breeding in Botany Bay. The increased numbers of Sacred Ibis *Threskiornis aethiopica* in Sydney are supported by their habit of scavenging at litter bins.

Littering is aesthetically unpleasing and can also have adverse effects on wildlife and the environment. This is particularly the case in the marine environment where seabirds entangled in fishing line or strangled in plastic six pack harnesses are not uncommonly found on strandlines. Cumulatively the number of animals, terrestrial and marine, killed by litter must be very large.

The city at play

With increased affluence, leisure activities which were once the privilege of very few are now available to a larger proportion of the population. The impact of recreational activities must be included amongst the impacts of urbanization. In this review only a few of the range of recreational activities will be considered, to illustrate how providing for leisure activities extends both the range and geographical distribution of human impacts on the environment.

Golf

Golf is an activity which is increasing in popularity and which has a large requirement for land, not just in Australia but globally. Golf course development is currently the fastest growing property sector in the world (Pleuramon 1992). In Europe the average area of a golf course is 64 ha, but in Thailand, where golf is a recent innovation and courses are associated with a complex of developments the average area is between 160–320 ha (Pleuramon 1992). Concern over the environmental impact of golf courses has prompted the establishment of an international organization, the Global Network for Anti Golfcourse Action, based in Malaysia, which declared 29th April, 1993 as "World No Golf Day".

Golf courses may incorporate fragments of natural habitats in roughs and these may support samples of local biota and function as part of the network of wildlife corridors through urban areas.

The negative impacts of golf course developments include the destruction of far more habitat than is preserved, and degradation of surrounding habitat from run-off of fertilizers and pesticides. Maintenance of golf courses frequently requires considerable irrigation, imposing extra demands on water resources. Chemical use may result in contamination of both surface and groundwaters (Robinson and Snyder 1991). A very wide range of chemicals may be used for pest control (Sudo and Kunimatsu 1992) but data on usage in Australian courses are not readily available. While measures can be taken in the design of new golf courses to minimize environmental impacts, it can be difficult to modify existing facilities (Robinson and Snyder 1991). Additionally many new golf courses, both in Australia and overseas, are planned as a key element of a much larger tourism/recreational development and thus golf is an essential component in the creation of new urban areas (Pleumaron 1992).

Horse riding

The horse played a vital role in the European colonization of Australia. Horses are still important for the pastoral industry and feral horses have a significant degrading effect on large areas of inland Australia but they are no longer the providers of motive power nor a primary means of transport. The majority of horses today are used for recreation. (The number of horses in, and near, urban areas is difficult to determine. Data recorded by the Australian Bureau of Statistics refer only to horses on agricultural establishments so that many horses stabled in urban areas would not be included in the figures).

The riding of horses in bushland poses a number of threats to biodiversity. Viable seeds of exotic species may be deposited in faeces, while intensive use of trails can initiate erosion, damaging both the communities through which trails pass and the wetlands and aquatic communities which receive the eroded sediment and nutrients.

The maintenance of a large population of horses strongly influences the pattern of land use on the urban fringe. As urban development continues to move outwards there may be pressure to create new horse paddocks in bushland in order to meet the growing demand for riding.

Boating

Provision of facilities for boating — boat ramps, wharfs and marinas has resulted in the destruction of coastal habitats. The disposal of sewage and garbage, and the use of antifouling paints, has been responsible for environmental degradation.

Anchoring in sensitive habitats, such as sea grass beds and coral reefs, may cause localized damage which can persist for many years. River banks have been damaged by boat wash.

Hunting and gathering

For European Australians the hunting of terrestrial fauna has rarely been essential for survival. The fur trade was of considerable commercial value, but much hunting of both mammals and birds has been primarily a recreational activity. Recreational hunting, in which many urban dwellers participate, causes controversy but has probably not had a major impact on biodiversity.

The situation in the marine environment is very different. Sealing and whaling were economically important activities which contributed to the early prosperity of Sydney and subsequently other settlements, but which resulted in major reductions of target species. Commercial fishing, as elsewhere in the world, has had a major impact on the populations of some species (although Kearney (this volume) suggests relatively little long-term impact on biodiversity).

Recreational fishing is one of the major leisure activities of urban Australians. For estuarine fisheries the impact of recreational anglers may be comparable (or even exceed) that of commercial fishers. Considerable pressure has also been placed on intertidal biota, collected both for direct consumption and as bait. Collecting has resulted in major reductions in populations and physical damage to habitats (Keough *et al.* 1993; Lynch and Prokop 1993). Recognizing the threat to maintenance of intertidal ecosystems from recreational harvesting New South Wales Fisheries has recently introduced prohibition on harvesting methods which cause habitat damage and has imposed strict bag limits (Lynch and Prokop 1993).

Visiting the bush

Many city dwellers feel the need to escape from the urban world and "commune with nature". National parks play a major part in satisfying this need and, historically, their public recreational role has been a major and politically acceptable justification for dedication. Meeting the increasing and diverse recreational demands is a difficult task and Oscar Wilde summed up the heart of the problem a century ago in the *Ballad of Reading Goal* — "Yet each man kills the thing he loves".

In New South Wales, data on visitor numbers to national parks and nature reserves are mostly estimates but do serve to illustrate the much greater visitor pressure on near urban parks compared with those in more remote localities. In 1990/1991 estimated visitors to near urban parks were 2 million to Blue Mountains, 2.4 million to Ku-ring-gai, 1 million to Lane Cove SRA, 1.8 million to Royal National Park, 2 million to Botany Bay and 3 million to Sydney Harbour. In contrast Sturt National Park received 4 200 visitors (Kosciusko, however, received 3 million visitors (all data from Ramsay 1992)).

National parks have an essential part to play in biodiversity conservation but it is important to recognize that national parks are expected to fulfil a recreational need and conserve cultural heritage (both Aboriginal and European) as well as conserving biodiversity. Meeting these various obligations creates conflicts of priorities. Nevertheless it is important to maintain the goodwill and support of as many as possible of the user groups in order to ensure a constituency favourable to the "core business" of conservation agencies.

The urban fringe

The fringes of metropolitan cities are characterized by a zone of small subdivisions, used as hobby farms or for keeping horses. This pattern of development also occurs inland from many of the new coastal developments (RAC 1992).

The area occupied by these rural subdivisions is large, although they contribute little to agricultural production (Scott 1992). In the case of Hobart "the area given over to rural subdivision exceeds the total area that is currently used for more intensive forms of housing, yet the subdivisions accommodate less than 5 per cent of the total urban population" (RAC 1992).

Rural subdivisions are frequently (although not exclusively) created from larger agricultural properties. The immediate fringes of the large cities support important horticultural enterprises, which require close proximity to markets for economic viability. Agricultural land on urban fringes is often of high quality — a rarity in the Australian context (Scott 1992). Loss of such land from productive farming has economic consequences. In the context of impacts on biodiversity, its significance is that there may be pressure to replace the lost production, either now or in the future, by clearing new areas or by heavy use of fertilizer and pesticides on remaining agricultural land. Concern has also been expressed (Scott 1992) that some hobby farms are poorly managed and become havens for weeds and feral animals, thus threatening both viable farms and bushland.

The role of fire, and the impact of changes in fire regimes on biodiversity are addressed elsewhere in these proceedings. It is on the urban fringe that there may be very strong pressures to implement fire hazard management regimes, which while designed to protect life and property may have adverse impacts on biodiversity.

The collection of firewood for burning in urban fire places may cause a substantial reduction in the litter layer in urban fringe bushland (as well as resulting in a contribution to the pollution load of urban air). In 1989 it was estimated that nationally over 4 million tonnes of fuelwood were burnt in households, about a third of this total being consumed in urban areas. The amount of particulate matter released into the air in urban areas from combustion of wood is about 7 000 tonnes per annum (Forestry Technical Services and University of Tasmania 1989). About a quarter of all households in Australia use wood burning appliances at some time. Fuelwood is available commercially from a variety of outlets but a high proportion is collected by the end consumer. In Melbourne householders collect 59 per cent of their fuelwood requirements, 40 per cent of this within 5 km of home (Forestry Technical Services and University of Tasmania 1989). (I have not been able to locate reliable detailed statistics of fuelwood consumption in Sydney). Over 80 per cent of the wood collected for Melbourne was dead. The removal of standing dead and fallen timber from the forests and woodlands in the urban fringe removes habitat for both vertebrate and invertebrate fauna. A diversity of fauna is transported into urban areas in fuelwood (Forestry Technical Services and University of Tasmania 1989), so that collection and use of fuelwood may be responsible for changing natural distribution patterns.

Beyond the immediate urban fringe large areas are set aside for water storage and catchment protection.

Water storage involves loss of habitat from flooding and the creation of an ecosystem previously unknown in Australia — large, deep, permanent lakes. The construction of dams profoundly alters the downstream riparian environment through changing the patterns of flow and reducing the natural variability of the system, and also imposes barriers to the migration of fish (Lake 1980; Walker 1980).

Flood mitigation works alter the character of the instream environment and also affect adjacent wetlands and flood plains (SPCC 1978), while weirs (which are often a component of flood mitigation works) affect fish migration (Harris 1984); in New South Wales few major waterways do not have at least one obstruction along their length.

The protection of catchments, however, has protected large areas of natural vegetation in what may be regarded as *de facto* national parks, which play an important role in biodiversity conservation.

The moving city

Although the population of Australia's capital cities continues to grow, the proportion of the total population living in those cities has undergone a very slight fall in recent years (Castles 1992). Over the past two decades there has been a major expansion of non-metropolitan urban centres; as documented by the RAC (1992) this growth has been almost entirely on the coast. In New South Wales centres such as Shoalhaven, Gosford–Wyong, Port Macquarie, Coffs Harbour and the Richmond–Tweed region have greatly increased in permanent population, and extent of development, since 1970. The bulk of the population increase has been from migration from the metropolitan cities, rather than natural increase (RAC 1992).

The coast has long been Australians' favourite destination for recreation. Many centres have long attracted a seasonal, transient population of holiday makers. The summer influx of visitors to camping and caravan sites created a great demand on utilities such as water and sewerage services, which were frequently unequal to the task. Localized pollution may have caused some impacts on biodiversity, as might the sudden increase in pressure on intertidal and marine resources, but the long non-holiday season presumably permitted a degree of recovery. (Although the coast has been the principal location of seasonal cities, during the winter the snowfields are host to a large transient population. Alpine habitats are of very limited extent in Australia so that conflicts of use are inevitable. The impacts on alpine biodiversity resulting from recreational use are not well documented).

The recent trend for permanent residence on the coast has created a demand for all the normal services associated with urban living and has imposed all the impacts on the natural environment previously experienced in and around the capital cities.

Many of these new coastal urban settlements are low density (the Gold Coast being a conspicuous exception), so that they have a great demand for land, met by the clearing of natural habitat. Canal estates, a form of development previously unheard of, has permitted the expansion of urbanization into estuarine wetlands. The life style expectations of the incomers resulted in the provision of golf courses and other recreational facilities, including marinas. The provision of services for these rapidly expanding populations extends their influence both inland and, in the case of sewage, out to sea.

While many of those moving out of the metropolitan areas have been retirees, substantial numbers of younger people have also settled in coastal towns (RAC 1992). Populations in these new centres are likely to continue to grow strongly, both from continuing migration and from natural growth.

There has also been an expansion of tourism development along the coast. In the case of major resorts, the requirement for a large contiguous area of land has sometimes led to pressure to develop land with severe environmental constraints, but which in consequence could be obtained by the developer relatively cheaply. Many of the more controversial development proposals for the New South Wales coast advanced in the mid 1980s have not yet eventuated, because of difficulties in obtaining finance and/or approvals. When the state of the economy improves there is likely to be renewed pressure for a number of these developments to proceed. Development of non-coastal areas for tourism may also have impacts on biodiversity (such as have occurred at Uluru (Reid *et al.* 1993)) unless very great care is taken in planning and construction.

The early expansion of the non-metropolitan coastal centres did not occur at a time when environmental protection was given high priority. In consequence all the problems of habitat loss, erosion and pollution experienced in older centres were revisited. However, as the population grew, so did the number of environmentally concerned citizens. The communities are now often sharply polarized into pro and anti-development camps and some of the most fiercely fought battles over development in New South Wales have been in the expanding coastal towns (for example, Caves Beach, Dunbogan Canal Estate, Bonville, Iron Gates, Ocean Shores and the currently continuing debate over the Club Med proposal at Byron Bay). Council decisions now come under the very critical scrutiny of environmentally aware citizens; while this may not be sufficient to halt development it does mean that much greater attention is now given to environmental matters by both planners and developers.

The urbanization of the coastline has extended the range of communities and ecosystems affected by development. As with the earlier development of the metropolitan centres it is possible to document the areas of habitat lost, populations of both plant and animal species reduced and degradation from pollution. While local extinctions have occurred it is again difficult to demonstrate total extinctions as a direct result of coastal urbanization. However, the cumulative incremental impact on ecosystems within the newly urbanized area is likely to have considerable impacts on biodiversity, and the areas of natural environment indirectly affected by increased visitation or by works necessary to support the new centres are very large.

Many coastal areas undergoing tourism or urban development are situated in, or near, wetlands. Proximity to wetlands creates problems of conflicts between humans and insects (particularly mosquitoes and sandflies), leading to demands on local councils for controls (Dale 1993). It is important to recognize that these insects may be extremely irritating and, in some cases, through the transmission of a number of viruses, a real threat to human health. Various control measures, including habitat manipulation and insecticide use, are applied, but in many cases there is little information on the impacts on non-target species. Conflict could be avoided through more appropriate planning (but in most areas this opportunity has already been lost).

TRANSPORT

Both the provision of facilities for transport and the process of transportation itself have impacts on biodiversity.

The period since European colonization of Australia has been one of rapid changes in transport technology, and the capacity for transport to have impacts on the environment has increased in concert with these changes. The role of different forms of transport in the transport of organisms has been analysed in greater detail by Wace (1985).

One of the transport revolutions which had a considerable impact on the English environment, the canal age, was not repeated in Australia. However, from the 1850s until the early years of the present century, the Murray-Darling River system was an essential component of the national transport system. The paddle steamers had a considerable impact on the environment; river red gum forests were heavily logged to fuel the boilers and riverbeds were desnagged to improve navigation (Davis 1978).

Sea

Historically the first human settlers of Australia, both Aboriginal and European, arrived by sea. At the present time, sea transport is not a major mode of mass personal transport although the bulk of visible trade, both imports and exports travels by sea. Through facilitating the introduction, both deliberately and accidentally, of so many species to Australia (Wace 1985), shipping has had a major impact on biodiversity. Despite greater awareness of the dangers presented by the introduction of pests and diseases and the activities of the quarantine service, the risk of unplanned introductions remains a major threat.

Shipping has numerous other impacts on biodiversity.

The provision of port facilities inevitably involves the alienation of coastal habitat. In all major ports there has been reclamation of foreshore habitat for the construction of wharves, warehouses and other facilities. As the size of ships grew dredging was required, frequently on a continuing basis, resulting in loss of seagrass beds and considerable disturbance to benthic communities. The effects of dredging and port development may be experienced over wide areas. For example there is a strong case for arguing that redirection of wave energy following the development of Port Botany in the 1970s was a factor initiating the accelerated erosion of Towra Point (a site which is now a Nature Reserve listed as a Wetland of International Significance under the Ramsar Convention) (Patterson Britton 1992).

The provision of the raw materials for the construction of port facilities extended the impacts on biodiversity further afield. Wharves and warehouses were prodigious consumers of timber. Wharf piles were almost always Turpentine *Syncarpia glomulifera*, chosen because of their resistance to marine borers. The impacts on forest structure of the continued removal of a particular size cohort of a single species have not been recorded. Construction of breakwaters required large amounts of stone, extracted from quarries which might have been some distance removed from the port.

It has long been necessary to coat ships' bottoms with some form of antifouling. Antifouling reduces the growth of fouling organisms which would otherwise greatly increase the energy required to move vessels through the water. Viewed from an environmental perspective use of antifouling therefore has positive benefits in reducing fuel consumption. Unfortunately there are also adverse impacts. Antifouling in current use is effective because it is toxic to potential fouling organisms. This toxicity may extend from a ship into the surrounding waters and thus poses a threat to biodiversity. The introduction of a new generation of paints in which the active principle was tributyl tin resulted in considerable impacts on estuarine organisms in many parts of the world (Wilson *et al.* 1993; Scammell 1990). A range of regulatory measures have been introduced to limit these impacts, but it is important to note that the prohibition on the use of tributyl tin in New South Wales does not apply to vessels longer than 25 m.

One of Australia's major imports, in terms of volume but not monetary value, is water. Specialized ships involved in the transport of particular commodities only carry cargo in one direction. However, when not carrying cargo it is necessary for safety reasons for ships to be ballasted. Thus, bulk ore carriers and woodchip carriers reach Australian ports with large volumes of "foreign" seawater. Such water may contain many organisms and the introduction of species to Australian waters through ballast water has been documented (Hutchings *et al.* 1987). Introduced species are a threat to the biodiversity of coastal waters. The recent arrival in Tasmanian waters of the Northern Pacific Starfish *Asterias amurensis* is a cause of considerable concern (Anon. 1993). In the case of toxic algae, a number of which may have been imported to Australia, there are also considerable immediate economic implications (Hallegraeff 1993).

The risk presented by ballast water introductions has been recognized world wide. However, while a number of possible solutions to the problem have been suggested, implementation may be both difficult and expensive, and as such may be opposed by ship owners.

Historically, shipwrecks have been a major cause of biologically damaging introductions, most notably rats (Moors *et al.* 1989). Today, the major concern is of the threat of pollution, most particularly from either fuel oil, or oil cargo; the risk of chemical spills from other cargo should not, however, be ignored. The wreck of a large tanker (such as the Torrey Canyon, Amoco Cadiz, Exxon Valdez or the Braer) has a dramatic impact on the environment although the volume of oil released into the environment from wrecks is small compared with the chronic releases from routine shipping operations and industry. Much of the public concern relates to readily visible oil on shorelines, and little is known about the long-term impacts of oil on the biologically rich surface layer of the sea or on seabeds. It is difficult to determine whether or not oil spills have had a major persistent impact on biodiversity, but clearly oil pollution remains a potential threat to a wide range of communities and organisms. Despite advances in navigational technology the number of shipping accidents remain high.

As the recent report *Ships of Shame* (HRSCTCI 1992) makes clear, poor standards of construction, maintenance and crew training are major contributors to maritime accidents. Although Australia has been relatively fortunate in not having had a massive spill in a biologically sensitive region, this good fortune arises more by chance than by design and the risk of a major accident occurring remains a very real one.

It is unlikely that a major new commercial port will be built on the New South Wales coast in the predictable future. Modification and expansion of existing facilities will occur, as will maintenance dredging. Major developments will be subject to environmental assessment but whether these procedures will result in a significant reduction in impacts on biodiversity is unclear. The existing knowledge of even apparently well studied regions such as Botany Bay is inadequate to provide a basis for predicting and assessing impacts on estuarine ecosystems (McGuiness 1988).

Continuing vigilance will be required to limit accidental importation of organisms, either in cargo or ballast water, and regulatory authorities will need to examine the possible impacts of proposals for the introduction of new formulations for antifouling.

Air

In the late 20th century air travel is the major mode of mass long distance transport.

The speed of air travel enables the rapid transport of organisms between continents. Despite considerable publicity on inbound flights, and the presence of quarantine officers at terminals it is certain that there is continuing illegal entry of organisms, both by deliberate intent and accidentally. It is impossible to quantify this and to assess its impact on biodiversity, but it is easy to construct scenarios by which particular organisms could be introduced and have a devastating impact. Air travel also provides a ready means by which biota can be illegally exported from Australia. Most publicity is given to cases of the export of birds and reptiles. It is again difficult to assess the impacts of this illegal trade but in the case of rare and endangered species their collection (and the disturbance to habitat thus entailed) could be of considerable significance. Australia is a signatory to the Convention on International Trade in Endangered Species (CITES) so that the legitimate export of biological material is strictly regulated.

The construction and operation of airports may have major impacts on the environment. Unlike port facilities, which in many cases were developed in parallel with urbanization, the demand for airports came long after the initial phase of urban development. Airports in most cities thus tend to be on the outskirts of the urban area, both for operational reasons and to avoid conflicts with the urban population. Airport construction has therefore often involved loss of agricultural land or natural ecosystems.

The proximity of Kingsford Smith airport to densely populated suburbs is relatively unusual, and the expansion of the airport has created public and political controversy because of the impacts of increased noise in residential areas. The expansion of the airport into Botany Bay (both extension of the original north-south runway, and the current construction of the second north-south runway) has implications for biodiversity conservation. Runway construction involves altering the configuration of the Bay, and substantial dredging, with impacts on benthic communities, sea grass beds, fisheries, bird habitats (Kinhill 1990) and the destruction of some small, but scientifically interesting saltmarsh and mangrove communities (Mitchell and Adam 1989). There are also concerns that there may be changes in the wave climate on Towra Point; while these may be slight in energy terms they may nevertheless be significant in their geomorphological consequences.

The loss of wading bird habitat on the northern foreshores of Botany Bay represents a considerable impact on biodiversity, the significance of which is enhanced in the light of Australia's proclaimed commitment to shore bird conservation evidenced by it being one of the first signatories to the Ramsar Convention and by the two bilateral agreements JAMBA (Japan Australia Migratory Birds Agreement) and CAMBA (China Australia Migratory Birds Agreement).

The Environmental Impact Statement (Kinhill 1990) and subsequent Botany Bay Environmental Management Plan (FAC 1992) propose a range of measures to mitigate the impacts of construction. These include re-establishment of seagrasses and the "enhancement" of wader and Little Tern habitat elsewhere in the Bay.

Safety of aircraft and passengers is of paramount importance during airport operations. A major problem world wide is that of bird strikes. This results in measures to deter bird usage of areas on and around airports. In the case of Kingsford Smith the most significant hazard is that posed by Silver Gulls *Larus novaehollandiae* (Kinhill 1990; FAC 1992) and investigations are underway to develop an effective deterrent. Habitat modification in the vicinity of the airport to reduce bird strike risk (for example proposals to fill wetlands associated with the east-west runway approach lights in Barton Park, Arncliffe) could reduce habitat availability for some bird species and destroy locally significant stands of vegetation.

Away from the immediate vicinity of airports the direct impacts of air transport on biodiversity are minimal. Installation and operation of navigation aids may involve localized site clearance and provision of access roads. These may be visually obtrusive and have adverse impacts on a range of conservation values of areas but construction is subject to stringent environmental planning and should not have adverse impacts on biodiversity beyond those associated with any access roads through natural vegetation.

Land

Road and rail corridors occupy a surprisingly large proportion of the land surface, a much larger area than that officially regarded as urban. (In 1980 the proportion of Australia which was urban was less than 0.1% while the area of transport corridor was about 2.5%). "Road" covers a wide range of situations from a six-lane freeway, with wide central reservation and fenced margins, to rarely used unsealed tracks, which are barely more than two parallel wheel ruts. Also included are travelling stock routes, which may be the narrow verges of trafficked roads or wide (up to 2 km) swathes of semi natural vegetation (Hibberd and Soutberg 1991).

Roads and railways provide essential transport and communication links both within and between urban areas. There is a synergistic relationship between urbanization and land transport; once a route has been established between urban centres there will be a tendency for other development to occur on or adjacent to it.

The impacts of roads on biodiversity have been reviewed by Andrews (1990) and Bennett (1991), and this account does not attempt to reiterate all the points from those studies. The rail network is much smaller than the road system (Castles 1992), and in general terms the impacts of rail are similar to those of the roads (since the replacement of steam trains by diesel and electric power the importance of the railways as a source of bushfire ignition will have declined). The Very High Speed Train proposal would have the potential for substantial environmental impacts, but at present appears unlikely to proceed.

Roads have substantial impacts on biodiversity as do the vehicles which travel on them.

Road vehicles act as vectors for the transport of organisms. Wace (1977) showed from a study of car washeries in Canberra, that large numbers of viable plant propagules could be carried on motor vehicles. The car is a far more effective dispersal agent than the trouser turn ups of botanists but the large majority of transported seed is probably deposited in sites unsuitable for subsequent establishment. While it is difficult to demonstrate with certainty that the spread of exotics into the bush was aided by motor vehicles there is a strong circumstantial case. The spread of wind dispersed seed along roadsides may be encouraged by the turbulence generated by heavy traffic. Vehicles can also carry pathogenic organisms. The range of pathogens which might be transported is very large; the best documented example, and one which clearly has an impact on biodiversity is the root infecting fungus *Phytophthora cinnamomi* in Western Australia (Carbon 1982).

The accidental transport of animals by vehicles is less well documented, but a wide range of fauna could be carried long distances in agricultural or forest products.

We live in an era unprecedented in history for the ease of personal mobility. An important consequence of the high incidence of individual ownership of cars is that many people have ready access to natural areas. This has both positive and negative impacts on biodiversity conservation.

On the positive side is the existence of a large body of "friends of the environment" whose support is essential to maintain the political will to develop and implement biodiversity conservation strategies. The road system also provides the major means of access to the bush for research. In consequence, much of what is known about the distribution of organisms comes from specimens collected on, or adjacent to, roads. On the negative side the sheer pressure of visitor numbers may be a threat to biodiversity, and the provision of services and facilities for visitors consumes a considerable proportion of the conservation budget. Additionally roads provide access to those unsympathetic to conservation; arsonists, bush rock stealers, poachers and vandals.

One of the most obvious impacts of motor vehicles on the environment is as killing machines. Road kills are a familiar sight on and beside roads around the world. The total number of animals killed is large (Andrews 1990; Bennett 1991). For reptiles and amphibia Ehmann and Cogger (1985) suggest that more than five million individuals are killed on Australian roads every year. The number of invertebrates killed must be immense.

Little is known about the importance of road kills on the "target" populations, although Bennett (1991) suggests that at least for widespread and common species, the effects are not significant. For small and isolated populations, however, road kills can be a major cause of mortality. Bennett (1991) discusses a number of species of concern to conservationists for which road deaths may be a significant factor causing population decline. In the Australian context road kills have been one of the important factors leading to the decline of urban Koalas (see for example Smith and Smith 1990; Phillips 1990), and, in the rainforests of north-east Queensland, the Cassowary (Crome and Moore 1990). Road kills provide a food resource for carrion eaters, which may therefore become regular frequenters of roads (and hence themselves become increasing vulnerable to collision with vehicles).

One of the most frequent grounds for objection to major road proposals is that they will be a source of noise disturbance to near-by human populations. Little is known, however, about the impact of vehicle noise on wildlife. Anecdotal evidence indicates that some species become habituated to road noise, but more research is required.

Road vehicles are major sources of pollutants. Globally they are important contributors to the increased Greenhouse effect, which potentially has major impacts on biodiversity, and to "acid rain".

In terms of more local impacts, emissions of lead, other heavy metals, and hydrocarbons have been the subject of concern. Lead has been added to petrol, as an "anti-knock" agent, although since 1986 all new petrol driven cars sold in Australia have had to be run on lead free petrol. The reduction in the use of lead in petrol acknowledged the adverse affect of lead on human health. Numerous studies world wide have demonstrated the deposition of lead close to roads (reviewed in Bennett 1991), nevertheless there is no evidence that there are significant impacts on biodiversity from this pollution (Bennett 1991).

As well as wide spread, chronic pollution from vehicles there is the potential for acute localized pollution following accidents. Spills of fuel or chemicals may have impacts on roadside communities. (If washed off the road surface it is likely that contaminants will enter nearby waterways or wetlands). Although such accidents are regrettably of regular occurrence the biological consequences have not been studied.

As well as providing a means of access to arsonists, vehicles can be a direct ignition source of bushfires. Hot exhaust gases can ignite dry grass, and fires following accidents can spread into bushland.

In considering the impacts of vehicles, as distinct from the road system, off road vehicles, and particularly their recreational use, need to be examined.

Off road vehicles have the same impacts on the road as conventional vehicles. When used irresponsibly off road they may cause considerable habitat disturbance (SPCC 1979). In wetlands wheel tracks may interfere with natural hydrology, and the effects of even a single passage may persist for many years (Etherington 1983). Wheel tracks may also remain for very long periods in arid zone communities. Vehicle damage to sand dune vegetation may initiate blow outs, while disturbance to vegetation on steep slopes may result in extensive erosion. Vehicles on beaches may destroy nests of birds such as Little Terns.

The construction of roads represents a major disturbance to the environment, although the degree of disturbance will depend on the size of the road, the nature of its construction and the topography and geology of the region through which the road is built.

Historically construction of both highway and railway bridges created a demand for large sized construction timber, while up until recently the continuing need for railway sleepers has been met from a limited number of tree species (particularly ironbark eucalypts).

Road construction, particularly for major highways, may involve large areas of soil without any vegetation cover and thus vulnerable to erosion. Soil eroded from road sites may cause sedimentation in waterways and wetlands elsewhere in the catchment (Wright and Bamford 1986), with the potential to cause permanent changes to these ecosystems.

Roads may cause permanent alteration to catchment hydrology as drainage patterns alter — pre-existing drainage lines may be filled or diverted, and new ones created to cope with run-off from the pavement surface. Channelling of creeks through culverts may create a barrier to the movement of aquatic species (Bennett 1991). Wetland systems may be hydrologically fragmented when road construction requires filling to create causeways, as has occurred widely in both estuarine and freshwater wetlands.

Road construction frequently involves extensive cut and fill with the transport of materials both within and between catchments. In some cases additional fill is required, involving habitat disturbance at sites away from the road corridor. If the material used for fill is chemically different from the substrate at the emplacement site it may be responsible for altering the environment and thus affecting biodiversity. This is of particular concern in nutrient poor sclerophyll vegetation where run-off and dust from the road may increase the nutrient content (particularly phosphorus) in adjacent vegetation. Although the sensitivity of sclerophyll vegetation to increased phosphorus is well known (see Adam *et al.* 1989 and the references there in) firm evidence that road construction materials have been responsible for altering biodiversity adjacent to roads is lacking, and this is a topic which requires investigation. (Similar concerns are raised by the use of ballast on railways.)

Movement of fill may also result in the transport of organisms, both weeds and pathogens (for example *Phytophthora cinnamoni*, as illustrated on the cover of Saunders and Hobbs 1991). While tenders may specify that construction material be weed free this is difficult to guarantee.

The effects of roads on biodiversity vary according to the surroundings. In catchments which still have an extensive cover of natural vegetation roads are an intrusive element in the landscape, and their impacts on the landscape are negative. Within an agricultural landscape road corridors may support the last remaining remnants of natural vegetation (Bennett 1990, 1991; Breckwoldt 1986, 1990).

Roads act as barriers to the movement of wildlife (even when not protected by fences). The inhibitory effect of roads varies considerably between species (reviewed in Andrews 1990 and Bennett 1991), but for some species even a rarely used single track may be a barrier. Thus even in what appears superficially to be a continuous area of habitat, roads may fragment the population of some species into non-interacting subgroups, with possibly adverse long term genetic consequences. The siting of service roads and fire trails in national parks and state forests requires careful planning to minimize impacts.

Roadside verges may provide habitat for weed species and feral animals (such as rabbits), which may spread both longitudinally, extend their range, and laterally away from the road into adjacent natural vegetation. Weeds frequently become established on disturbed ground during and immediately after construction, and spread may be encouraged by roadside management practices. There is increasing recognition by management authorites of the need to control weeds, and improved management strategies are being developed (Vic Roads 1990).

While roads are important agents in the fragmentation of habitat, road margins support many examples of communities and species not represented in conservation reserves and additionally provide networks of habitat corridors through the landscape (see *inter alia* Andrews 1990; Bennett 1990, 1991; Breckwoldt 1986, 1990; Vic Roads 1990; Wallbrink 1992). The road system "samples" nearly all the environments in Australia and roadside vegetation probably contains a greater diversity of communities (albeit often in a degraded state) than are found within conservation reserves. Road corridors are thus important for biodiversity conservation, although in many cases a considerable management effort will be required in order to maintain conservation values. In many instances, while the upper canopy remains, understoreys have been massively changed — through mowing, grazing and changed fire regimes. Not only has much of the understorey biota been lost but regeneration of the overstorey is no longer possible. While road management authorities are now more sympathetic to the conservation value of road corridors (e.g., Vic Roads 1990) there has been little assessment of the conservation values of most road corridors and comparatively minor disturbance may, in ignorance, result in the loss of significant populations of rare species. The importance of road corridors for conservation occurs chiefly in rural areas but significant species may be found along urban roads (e.g., *Allocasuarina diminuta* in the central reservation of Anzac Parade, Kingsford).

Many stands of vegetation of conservation significance in urban areas occur on roads of the future. The survival of bushland along the proposed routes of, for example, the F5 and F6, is largely due to these areas being unavailable for other forms of development. These road corridors were designated many years ago, when the importance of the bushland was both unknown and unappreciated, and when much of the surrounding area was undeveloped. With increased development and loss of bushland, road corridors become increasingly important. Implementation of the long planned roads then becomes a matter arousing considerable controversy.

Many of the attributes of roads are shared by other utility easements through the landscape (telephone, power, water, gas). Where these easements share a common route with a road or railway, the number of separate easements is reduced, but installation, maintenance and operation of the utilities may reduce the habitat value of the road corridor. Where the routes do not follow roads, provision for vehicular access for maintenance purposes is required; these maintenance roads are frequently used by off road vehicles to gain access to the bush.

IMPACTS ON BIODIVERSITY

What is biodiversity?

The concept of biodiversity, as recognized in the international convention on biodiversity developed at the 1992 UNCED conference in Rio de Janeiro has three components:

- genetic variation within species
- variation between species (i.e., the number of species)
- the diversity of habitats and/or ecosystems.

This concept is broader than the general public understanding of the term, which concentrates on the species level component.

One of the difficulties in determining whether or not some particular process has an impact on biodiversity is that we have no single measure of biodiversity.

What would be an appropriate measure of genetic variation within a species? If we had measures for individual species how should they be combined to provide an index of within species variation for the biota of a single habitat or region?

For species we can enumerate those present in a particular region to document species richness or, to take into account the relative abundance of species, calculate the various measures of diversity which have been proposed (Magurran 1988). Such exercises are subject to taxonomic constraints. Rarely, if ever, has the total species complement of an area been documented simultaneously. What are normally presented are data for particular groups — vascular plants, birds, mammals etc. Nevertheless a complete species inventory is a theoretical possibility, even if, because of the lack of a basic taxonomy for many groups, it is not a goal which could be attained in the foreseeable future.

In the case of the community component of biodiversity there is not yet consensus as to the appropriate level for analysis. At a very broad scale level of ecosystem recognition there would be general agreement as to the existence of such entities as forest, grassland and wetland, and in a given landscape the relative areas of these units could be measured and these data combined in some way to produce an index. However, if finer scale units were to be adopted there would be considerable debate as to the nature and number of units which would be appropriate. The lack of generally accepted taxonomies for habitats and ecosystems is regarded as a major impediment to the development and implementation of a biodiversity conservation strategy (Orians 1993). The problems of developing community taxonomies are not merely scientific but are also political. As Orians (1993) points out, the coarser the classification the fewer the units and the fewer areas which will be required to conserve a representative sample.

Even if there was agreement on measurement of the components of biodiversity, how would these be combined to produce an overall index? Should all components be given equal weight or is one more "important" than the others?

I do not see any useful way of developing a quantitative measure of biodiversity. Therefore we cannot say that urbanization in the Sydney region has resulted in an x per cent decline in biodiversity, rather we must talk both in more general terms and with reference to specific examples.

Our inability to quantify biodiversity does not reduce the value of the concept. It is important to recognize that concerns for biodiversity go beyond a focus on species and that a biodiversity conservation strategy will need to be far more than an endangered species strategy, even though measures to conserve rare and endangered species will be an important component of any biodiversity conservation strategy.

Intraspecific variation

Genetic variation within species is the basis for continuing evolution. It is also the basis for the ecological versatility of species.

Widespread species are likely to exhibit considerable genetic variation in both physiological and anatomical characters. Some genetic variants may be extremely limited in distribution. For example the most widespread saltmarsh grass in Australia, *Sporobolus virginicus,* is cytologically very variable (Smith-White 1988), and several of the cytotypes are geographically widespread. However, one, a viviparous pentaploid, is restricted to a single roadside site at Greenough (Western Australia) and its long-term survival is threatened by carparking (Smith-White and Adam 1988). Species which are rare may also be genetically variable (Potts 1989).

"Extreme" habitats are characterized by assemblages of specialized ecotypes, for example saltmarsh (Adam 1990) or headlands (Auld and Morrison 1992).

Most genetic variants are not accorded any taxonomic status and thus tend not be recognized in conservation policies, and rare ecotypes are not protected by rare and endangered species legislation unless they are members of species which are in totality rare and endangered.

It is probable that urbanization has resulted in the loss of many geographically restricted ecotypes, mostly before they were recognized. In a few cases the loss can be documented; *Amperea xiphoclada* var. *pedicellata* was known only from Double Bay and is now believed to be extinct (Henderson 1992).

The importance of conserving intraspecific variation within species has important consequences for the management of urban bushland remnants (Benson 1993). In rehabilitation programmes involving replanting, care should be taken to avoid (where possible) introduction of genetic stock foreign to the site.

Species diversity

At the level of species diversity it is still difficult to construct other than an anecdotal account of changes which may be attributed to urbanization. For many taxonomic groups (particularly invertebrates) knowledge of both taxonomy and species' distributions is so imperfect that little can be said about changes to species' status, other than for those perceived as pests.

The best recorded groups are birds and vascular plants.

For birds in the Sydney region Hoskin (1991) records considerable changes in both the composition and relative abundance of species between 1958 and 1991. In the case of additions to the list, some, for example amongst the waders and seabirds, may reflect a greater number of observers but others are genuine extensions of ranges. Changes in the relative abundance of species may reflect the changed urban environment *per se*, or be in response to changes elsewhere in Australia. For example, the increase in numbers of Galahs *Cacatua roseicapilla* and Sulphur Crested Cockatoos *Cacatua galerita* in Sydney is thought to represent the establishment of breeding populations in urban parks following an influx of individuals from the inland during times of drought (Fox 1990).

The decline in small native birds (for example the Superb Blue Wren *Malurus cyaneus*, in many parts of Sydney may reflect a combination of factors, including decline in native vegetation and predation by cats and the native Pied Currawongs *Strepera graculina*. Numbers of Pied Currawongs have increased considerably, possibly as a consequence of the greater availability of fleshy fruit, both in gardens and as exotics in bushland (Recher 1972) (species such as Privet *Ligustrum* spp., Lantana *Lantana camara*, Blackberry *Rubus fruticosus* and Bitou Bush *Chrysanthemoides monilifera*). Not only do Pied Currawongs benefit from this introduced food source but by acting as dispersal agents they may facilitate the spread of the exotic plants further into the bush. There is currently a debate as to the desirability of culling *Strepera*, in order to reduce predation of the small native birds (Bowers 1993).

Urbanization has resulted in a substantial loss of plant habitat and the extinction of some species. Leigh, Boden and Briggs (1984) list seven species believed to have become extinct nationally as a result of urbanization and industrial development; one of these, *Haloragodendron lucasii* was subsequently rediscovered growing in the Sydney region (Orchard 1990), thus changing from a subject of regret to a problem to be addressed. Benson and McDougall (1991) suggest that *Hypsela sessiliflora*, endemic to the western Sydney region is now extinct. Many more species have suffered substantial reductions in both range and abundance within urban areas, and a number of species have suffered local extinctions. Nevertheless the number of such losses is remarkably small compared to the size of the flora. However, this should not engender a false sense of optimism. Many plant species are long lived and while mature individuals' may survive substantial disturbance, opportunities for natural regeneration may have been reduced so that, at some time in the future, more species may be lost despite protection of remnant populations.

It is generally assumed that native mammals have suffered considerable losses as a result of urbanization; from loss and degradation of habitat, from predation by introduced cats, dogs and foxes, and competition from rats and house mice. This is undoubtedly the case but nevertheless the number of losses may not be as large as supposed; the recent rediscovery of the Quoll *Dasyurus maculatus* in Garigal National Park suggests that a systematic mammal survey of urban bushland may be profitable exercise.

Despite considerable local extinction no mammals appear to have become extinct nationally as a result of urbanization. Some native mammals are able to maintain large populations, even within the most densely settled urban areas. The Brush Tailed Possum remains common in parks and gardens even invading roof spaces, and eats many horticultural plants as well as natives. In Sydney the Grey-headed Fruit Bat *Pteropus poliocephalus* is a familiar sight in the night sky across Sydney and maintains a large breeding colony in remnant vegetation in Gordon (Pallin 1993).

Camp followers

Urbanization in the 19th and 20th centuries has created a set of environments which are basically similar world wide — buildings and their immediate surrounds — and reflecting this similarity there is a suite of species which is cosmopolitan or nearly so. These include the Domestic Pigeon *Columba livia*, House Sparrow *Passer domesticus*, Starling *Sturnus vulgaris*, the Black and Brown Rats *Rattus rattus* and *R. norvegicus* and the household cockroaches, the Garden Snail *Helix aspera* and the Earthworm *Lumbricus*. In suburbia, these and other introduced species, compete with native species but in the heart of the concrete jungle provide us with a glimpse of a non-human world which but for their presence would otherwise be totally absent.

"Good and bad" biodiversity

The flora of Australia as a whole contains about between 10–15 per cent introduced species (Humphries *et al.* 1991) and in the Sydney region exotics number more than 20 per cent of the flora. The number of plant species which have become regionally extinct in the last 200 years is relatively small. At the species level, therefore, it could be argued that for vascular plants urbanization has resulted in an increase rather than a decrease in diversity. If the total number of species grown in gardens, but which have not (yet) become established aliens in bushland, were also considered then the increase in diversity is very large. If all the associated insects and micro-organisms were considered then the total increase in species richness within urban areas would be enormous.

However, introduced species are not normally considered as contributing to biodiversity. Indeed competition with introduced species is regarded as one of the major threats to biodiversity. At the local scale the threat becomes obvious; when a patch of coastal dune heathland is replaced by a monoculture of *Chrysanthemoides monilifera* the biodiversity of that particular stand is reduced, even though at a wider scale all the displaced species are still extant. The local loss may involve significant, but unknown, reduction in infraspecific genetic diversity, and it is easy to envisage that native species with very restricted distributions could be driven to extinction by aliens. In the case of an exotic such as *Chrysanthemoides*, with the potential to invade native vegetation over a very wide geographic range, then there is the potential to affect adversely even currently common native species.

Perhaps more importantly introduced species have an impact on the third level of biodiversity, the community. The addition of species to existing communities changes the nature of the community, in terms of composition, structure and in many cases function. For example, the invasion of mesomorphic species into previously sclerophyll communities, although possibly initiated through changes to the physical environment following disturbance, will reinforce these changes by altering patterns of mineral cycling, water use and fire behaviour. If these environmental changes are of sufficient magnitude then the possibility, through management, of restoring the previous community may be lost.

Community diversity

Although ecosystem and/or communities are included as a component of biodiversity how are they to be defined and measured? If they are defined at ecosystem level then the number of units is relatively small, and the appropriate units could be approximately synonymous with the structural vegetation formations of the Specht scheme (Specht 1981). However, these units would encompass considerable internal variation. If it were felt more appropriate to recognize more narrowly defined units then there would be considerable debate as to the appropriate criteria to be used.

Benson and Howell (1990a) in their study of Sydney's vegetation recognize vegetation types defined by a combination of structure, floristics, habitat characteristics and geography (for example Eastern Suburbs *Banksia* scrub). Such a typology is appropriate for a regional account, but we lack a broader scale (state or national) framework which defines communities in a comparable level of detail, so that determining whether particular communities are unique to the Sydney region is difficult.

Even if there were agreement as to the definition and diagnosis of communities, how are they to be incorporated in a measure of biodiversity? Is a simple enumeration of communities suitable — in which case the urbanization of the Sydney region has had little effect. Most, if not all the vegetation types which we can reasonably assume to have been present in 1788 can still be found today, albeit in many cases in extremely small fragments. However, if the relative area of vegetation types is regarded as contributing to biodiversity then urbanization has clearly had a massive and deleterious impact at the regional scale.

The history of loss of vegetation in Sydney has been documented by Benson and Howell (1990a, b). The geology and topography of the region has meant that development has had greater impact on some communities than others. Rugged sites on sandstone were unsuitable for both agriculture and building, and substantial vegetated areas still survive. The more fertile shale and alluvial sites, with gentle topography were cleared for agriculture early in the colonial period and only later (particularly post 1945) were engulfed in suburbia (Benson and Howell 1990a, b). Particularly on sandstone many areas were developed for suburbia directly from bushland, a marked contrast with Europe where the major 19th and 20th century expansion of urbanization has been at the expense of agricultural land.

At the regional level the loss of the whole range of Sydney's vegetation types has been massive, but in a broader state context some losses are more significant then others.

Despite considerable losses, much sandstone vegetation remains and very extensive areas are conserved in the ring of sandstone national parks around Sydney. Coastal sand dune and sand sheet vegetation has been substantially reduced in the urban area, but similar (but probably not identical) communities are sampled in national parks and nature reserves elsewhere on the coast. The communities on the shales of western Sydney have suffered massive reduction in area, and similar communities are either very rare or absent outside the Sydney region. The loss of these communities is probably the most important in terms of impact on biodiversity.

Communities on fertile alluvium were cleared early for agriculture, not only in what is now the metropolitan area, but along all the coastal river valleys. The loss of riparian and permanent freshwater wetland communities has been significant throughout the state, and the remaining stands in the metropolitan area are significant for biodiversity conservation.

Coastal intertidal wetlands (saltmarsh, mangrove, seagrasses) have suffered badly from urban development and industrial development throughout the world. The decline of seagrasses and saltmarsh in the Sydney region is well documented (Underwood and Hutchings 1987; Clarke and Benson 1988). The situation in the case of mangroves is slightly different, as some mangrove stands have been obliterated (Thorogood 1985) while others have grown considerably (McLoughlin 1985; Dunstan 1990). This is probably in response to sedimentation as a result of erosion during the course of catchment urbanization.

Although substantially reduced in area the remaining coastal intertidal wetland stands are in good condition and are of high conservation significance. The nature of the physical environment, with highly saline anoxic soils limits the number of species which can occupy these habitats. Seagrass beds and mangroves are amongst the few ecosystems in Australia which have not suffered invasion by introduced plants, and the number of invaders of saltmarsh is small (although significant changes to some saltmarshes in the Sydney region have resulted from invasion by *Juncus acutus*, and the spread of *Spartina anglica* in the lower intertidal has changed the structure, and possibly long-term successional processes, in marshes in Victoria and Tasmania). Around the world a number of sites of outstanding importance for estuarine biota (including migratory birds) survive in close proximity to major urban and industrial complexes. Two out of the current three wetlands of international significance listed under the Ramsar Convention for New South Wales are in urbanized estuaries — Towra Point in Botany Bay and Kooragang Island in the Hunter estuary.

Conclusions

Any discussion of the effects of urbanization on biodiversity based on a consideration of Sydney will, in general terms, be applicable to other urban areas.

Conclusions as to the significance of changes to biodiversity will be dependent upon what is meant by "significant" and the spatial context of the discussion. "Significant" will clearly have different meanings to different people. The extinction of a single species, even if it could be unequivocally ascribed to urbanization, could be regarded as significant given the irreplaceability of species and the importance placed upon maintenance of biodiversity through such instruments as the Rio convention. On the other hand, if it could be shown that the species was not a keystone species, that it represented only a tiny percentage of total biodiversity and that the total benefits to humanity outweighed the loss of a single species, then some would regard the loss as insignificant. The choice between the two conclusions inevitably remains a subjective one. Interpretations will also depend on whether losses and declines are considered in a global or local context.

It is reasonable to assume that urbanization has resulted in a loss of genetic diversity within species. The amount of loss is impossible to quantify and will vary between species.

At the species level, urbanization has not been a major factor in total extinction. However, at the regional scale, urbanization has been responsible for local extinctions, and major reductions in population size of many species. On the other hand urban areas sustain large populations of introduced species, which interact with native species in many ways, some adverse to the maintenance of biodiversity. Cities have been and are likely to remain foci for introduction of potentially environmentally damaging species.

The high species diversity of cities is largely a feature of the suburbs and urban parks; if all the urban area were developed to the extent of the Central Business District then the impacts of urbanization on all levels of biodiversity would, at the regional level, have been much greater. At the species level urban areas may support regionally high levels of diversity — although for Australian cities this effect may not be as great as for some cities in the northern hemisphere, surrounded as they are by the near biological deserts of very intensive agriculture.

At the community level, urbanization has destroyed large areas of vegetation and has been responsible for the degradation of the semi-natural areas remaining within urban areas. The very combination of climate, soils and landscape which made some sites desirable for urban development also supported biotic communities of limited distribution. The loss of these communities (for example those of fertile soils in western Sydney) is thus of national as well as local significance. The concentration of urbanization in the coastal zone has resulted in significant losses of some widespread coastal habitats, particularly coastal wetlands.

THE FUTURE

The importance of urban bushland

Cities contain many areas of bushland (see Benson and Howell 1990a). Additionally sites not normally regarded as bushland, such as cemeteries (Kirkpatrick *et al.* 1988) may support native species. Urban bushland is fragmented, is often a linear feature with a high edge to area ratio, subject to invasion by weeds and feral animals, and frequently with a physical environment very different from the pre-urban state. Management of urban bushland is time consuming, labour intensive and expensive. The management of urban bushland can be justified on the grounds that the bushland is highly valued for its contribution to the quality of city life, because of its cultural heritage significance as a living museum of the pre-European settlement environment and, to a more limited extent, because of the recreational opportunities it affords. Should it also be justified on the grounds of its contribution to biodiversity conservation?

There are a number of grounds for arguing that maintaining urban bushland is important for conserving all three components of biodiversity. Genetically distinct forms of common species may be found in urban bushland (see for examples Auld and Morrison 1992). It is clearly impossible to describe all variants, let alone maintain them *ex situ.* Urban bushland can continue to support genetic diversity within species. Individual species within urban bush may be regionally or nationally (e.g., *Haloragodendron lucasii*, *Grevillea caleyi*) rare; conservation of sites with particular rare species is thus an essential part of a biodiversity programme. The communities within urban bush may be rare regionally or absolutely (e.g., shale communities of western Sydney) and be absent, or only poorly represented, in conservation reserves.

Importantly, urban bushland is seen every day by many people and it is close to schools and thus can be a venue for ecological and environmental education. Urban bushland is where the public at large can gain practical experience of biodiversity; management of urban bushland to conserve biodiversity may assist in raising awareness of, and concern for, biodiversity conservation in general.

The future

The population of Australia will continue to grow, from a combination of natural increase and immigration. There is little probability of growth outside urban areas and the trend of continuing growth of capital cities and proportionately greater increases in other coastal urban centres is unlikely to change. There has been a continuing failure of politicians and society at large to address limitations to population growth in Australia (Taylor 1940; Birrell *et al.* 1984), and even today there is a great reluctance to even contemplate debate on the issue.

The continued growth of the urban population will impose demands on the environment and will threaten to reduce biodiversity. The impacts of continuing urbanization can be mitigated, but not eradicated, through planning, implementation of environmental protection measures and through reduction in per capita use of resources (i.e., reduction in water consumption, greater recycling of paper, glass and metals).

Existing urban bushland survived by default. Although some stands may have considerable conservation value, this was rarely the reason for preservation. With the development of new urban areas there is an opportunity to identify sites of value in advance and to plan a reserve system and wildlife corridors which will maximize the potential retention of biodiversity. Controls on development can reduce degrading impacts on bushland — through management of stormwater, erosion control measures, restrictions on pet ownership and controls on the plants used in landscaping and gardens. New approaches to the control and disposal of storm and other waste water will permit the retention of riparian and wetland habitats and may require the development of artificial wetlands. Subdivisions could be planned so as to prevent direct contact between suburban blocks and bushland, reducing the risk of weed invasion and the demand for bushfire hazard reduction. Retention of habitat, use of wetlands for waste water treatment, and modified layouts will however tend to increase the area required for urban expansion, unless the area of individual allotments is reduced. Urban consolidation within existing developed areas will reduce the need for expansion, but not the demand for resources from the surrounding region, and may result in a loss of habitat value within suburbia.

There is increasing recognition of the importance of roadside (Bennett 1990; Breckwoldt 1986, 1990; Vic Roads 1990) and streamside (DEC 1990) habitats and the development of management guidelines which recognize conservation values. Nevertheless the continuing use of existing roads, and the construction of new roads will inevitably have impacts on the environment, including reducing biodiversity.

ACKNOWLEDGEMENTS

I would like to thank Debbie Stevenson for her assistance in tracking down information.

REFERENCES

Abrahamsen, G., 1984. Effects of acidic deposition on forest soil and vegetation. *Phil. Trans. Roy. Soc. Lond.* B **305:** 369–82.

Adam, P., 1990. *Saltmarsh ecology.* Cambridge University Press: Cambridge.

Adam, P., Stricker, P. and Anderson, D., 1989. Species-richness and soil phosphorus in plant communities in coastal New South Wales. *Aust. J. Ecol.* **14:** 189–98.

Anderson, J. W., 1978. *Sulphur in biology.* Edward Arnold: London.

Andrews, A., 1990. Fragmentation of habitat by roads and utility corridors: a review. *Aust. Zool.* **26:** 130–41.

Anon., 1993. Intruder wanted — dead or alive. *Aust. Fish.* **52**(12): 5.

Aston, A. R., Millington, R. J. and Kalma, J. D., 1972. Nutrients in the Sydney area. *Proc. Ecol. Soc. Aust.* **7:** 177–91.

Auld, T. D. and Morrison, D. A., 1992. Genetic determination of erect and prostrate growth habit in five shrubs from windswept headlands in the Sydney region. *Aust. J. Bot.* **40:** 1–11.

Australian Bureau of Statistics, 1986a. *Australian standard geographical classification geographic code list. 1986 Census of population and housing supplement.* Australian Bureau of Statistics: Canberra.

Australian Bureau of Statistics, 1986b. *The 1986 Census Dictionary.* Australian Bureau of Statistics: Canberra.

Australian National Parks and Wildlife Service, 1992. *Feral Cats.* ANPWS: Canberra.

Barker, G. and Graf, A., 1989. *Principals for nature conservation in towns and cities.* Nature Conservancy Council: Peterborough.

Bell, F. C., 1972. The acquisition, consumption and elimination of water by the Sydney urban system. *Proc. Ecol. Soc. Aust.* **7:** 161–76.

Bennett, A. F., 1990. *Habitat corridors. Their role in wildlife management and conservation.* Dept. of Conservation and Environment: Melbourne.

Bennett, A. F., 1991. Roads roadsides and wildlife conservation: a review. Pp. 99–117 in *Nature conservation 2: the role of corridors* ed by D. A. Saunders and R. J. Hobbs. Surrey Beatty & Sons: Chipping Norton.

Benson, D. H., 1993. Bush regeneration and generic integrity — the practicalities. Pp. 60–64 in *Urban bushland management — into the melting pot (ponderings, processes and problems)* ed by J. L. Howard and J. Rawling. Ku-ring-gai Municipal Council and Urban Bushland Management: Sydney.

Benson, D. H. and Howell, J., 1990a. *Taken for granted. The bushland of Sydney and its suburbs.* Kangaroo Press: Kenthurst.

Benson, D. H. and Howell, J., 1990b. Sydney's vegetation 1788–1988: utilization, degradation and rehabilitation. *Proc. Ecol. Soc. Aust.* **16:** 115–27.

Benson, D. H. and McDougall, L., 1991. *Rare bushland plants of Western Sydney.* Royal Botanic Gardens: Sydney.

Birrell, R., Hill, D. and Nevill, J. (eds), 1984. *Populate and perish? The stresses of population growth in Australia.* Fontana/ACF: Melbourne.

Bonner, C. and Agnew, A. D. Q., 1983. Soil phosphorus as an indicator of canine faecal pollution in urban recreation areas. *Environ. Pollut.* (B) **6:** 145–56.

Bornkamm, R., Lee, J. A. and Seaward, M. R. D. (eds), 1982. *Urban ecology.* Blackwell: Oxford.

Bowers, P., 1993. Stone the currawongs. They're taking over. *Sydney Morning Herald,* 15th May, 1993.

Breckwoldt, R., 1986. *The last stand. Managing Australia's remnant forests and woodlands.* Australian Government Publishing Service: Canberra.

Breckwoldt, R., 1990. *Living Corridors. Conservation and management of roadside vegetation.* Greening Australia: Canberra.

Carbon, B., 1982. Competition for scarce resources: the Western Australian dilemma. Pp. 222–23 in *Quarry Australia. Managing the nation's resources* ed by R. Birrell, D. Hill and J. Stanley. Oxford University Press: Melbourne.

Carr, G. W., Yugovic, J. V. and Robinson, K. E., 1992. *Environmental weed invasions in Victoria. Conservation and management implications.* Dept. of Conservation and Environment and Ecological Horticulture Pty. Ltd.: Melbourne.

Castles, I., 1992. *Australia's environment. Issues and Facts.* Australian Bureau of Statistics. Canberra.

Clarke, P. and Benson, D., 1988. The natural vegetation of Homebush Bay — two hundred years of change. *Wetlands (Australia)* **8:** 3–15.

Clements, A. M., 1983. Suburban development and the resultant changes in the vegetation of the bushland of the northern Sydney region. *Aust. J. Ecol.* **8:** 307–19.

Crome, F. H. J. and Moore, L. A., 1990. Cassowaries in north-eastern Queensland: report of a survey and a review and assessment of their status and conservation and management needs. *Aust. Wildl. Res.* **17:** 369–85.

Dale, P., 1993. Australian wetlands and mosquito control — contain the pest and sustain the environment? *Wetlands (Australia)* **12**(2): 1–12.

Davis, P. S., 1978. *Man and the Murray.* UNSW Press: Sydney.

Department of Conservation and Environment, 1990. *Environmental guidelines for river management works for the standing committee on rivers and catchments.* Dept. of Conservation and Environment: Melbourne.

Dunstan, D. J., 1990. Some early environmental problems and guidelines in New South Wales estuaries. *Wetlands (Australia)* **9:** 1–6.

Ehmann, H. and Cogger, H., 1985. Australia's endangered herpetofauna: a review of criteria and policies. Pp. 435–47 in *Biology of Australasian frogs and reptiles* ed by G. Grigg, R. Shine and H. Ehmann. Surrey Beatty & Sons: Chipping Norton.

Emery, M., 1986. *Promoting nature in cities and towns. A practical guide.* Croom Helm: London.

Environment Protection Authority, 1993. *New South Wales; State of the environment 1993.* EPA: Sydney.

Etherington, J. R., 1983. *Wetland ecology.* Edward Arnold: London.

Federal Airports Corporation, 1992. *Botany Bay Environmental Management Plan — Stage 1 (Draft).* Federal Airports Corporation: Sydney.

Fitter, R. S. R., 1946. *London's natural history.* Collins: London.

Fitzpatrick, E. A. and Armstrong, J., 1972. The bioclimatic setting. *Proc. Ecol. Soc. Aust.* **7:** 7–25.

Forestry Technical Services Pty Ltd and University of Tasmania, 1989. *Fuelwood use and supply in Australia. NERDDP Consultancy 28.* Department of Primary Industries and Energy: Canberra.

Fox, M. D., 1990. Introductions of native and introduced species in new habitats. *Proc. Ecol. Soc. Aust.* **16:** 141–47.

Fox, M. D. and Adamson, D., 1986. The ecology of invasions. Pp. 235–55 in *A natural legacy* ed by H. F. Recher, D. Lunnery and I. Dunn. 2nd ed. Pergamon: Sydney.

Groves, R. M., 1986. Plant invasions of Australia: an overview. Pp. 137–49 in *Ecology of biological invasions. An Australia perspective* ed by R. M. Groves and J. J. Burdon. Australian Academy of Science: Canberra.

Hallegraeff, G. M., 1993. On the global increase of harmful algal blooms. *Wetlands (Australia)* **12:** 2–15.

Harris, J. H., 1984. Impoundment of coastal drainages of south-eastern Australia, and a review of its relevance to fish migrations. *Aust. Zool.* **21:** 235–50.

Hawksworth, D. L. and Rose, F., 1976. *Lichens as pollution monitors.* Edward Arnold: London.

Henderson, R. J. F., 1992. Studies in Euphorbiaceae A.L. Juss., sens lat. I. A revision of *Amperea* Adr. Jus. (Acalyphoideae Ascherson, Ampereae Muell. Arg.). *Aust. Syst. Bot.* **5:** 1–28.

Hibberd, J. K. and Soutberg, T. L., 1991. Roadside reserve condition 1977–89 in the Southern Tablelands of New South Wales. Pp. 177–86 in *Nature conservation 2: the role of corridors* ed by D. A. Saunders and R. J. Hobbs. Surrey Beatty & Sons: Chipping Norton.

Hoskin, E. S., 1991. *Birds of Sydney.* Surrey Beatty & Sons: Chipping Norton.

House of Representatives Standing Committee on Transport, Communications and Infrastructure, 1992. *Ships of Shame.* Australian Government Publishing Service: Canberra.

Humphries, S. E., Groves, R. H. and Mitchell, D. S., 1991. Plant invasions of Australian ecosystems. A status review and management directions. *Kowari* **2:** 1–134.

Hutching, P. A., van der Velde, J. T. and Keable, S. J., 1987. Guidelines for the conduct of surveys for detecting introductions of non-indigenous marine species by ballast water and other vectors and a review of marine introductions to Australia. *Occ. Rep. Aust. Mus.* **3:** 1–147.

Johnston, D., 1986. Urbanization and Total Catchment Management. *J. Soil Conserv. NSW* **42:** 22–24.

Johnston, D. and Gerstle, B., 1986. The urbanization of Narrabeen Lagoon catchment — a case study. *J. Soil Conserv. NSW* **42:** 18–21.

Kalma, J. D., Aston, A. R. and Millington, R. J., 1972. Energy use in the Sydney area. *Proc. Ecol. Soc. Aust.* **7:** 125–42.

Keough, M. J., Quinn, G. P. and King, A., 1993. Correlations between human collecting and intertidal mollusc populations on rocky shores. *Conserv. Biol.* **7:** 378–90.

King, R. J., 1988. Thermal pollution and algal growth in the Tuggerah Lakes. *Wetlands (Australia)* **8:** 30–32.

Kinhill Engineers Pty. Ltd., 1990. *Proposed third runway, Sydney (Kingsford Smith) Airport. Draft Environmental Impact Statement.* Kinhill and Federal Airports Corporation: Sydney.

Kinsman, D. J. J., 1984. Ecological effects of deposited S and N compounds: effects on aquatic biota. *Phil. Trans. Roy. Soc. Lond.* B **305:** 479–85.

Kirkpatrick, J. Gilfedder, L. and Fensham, R., 1988. *City parks and cemeteries. Tasmania's remnant grasslands and grassy woodlands.* Tasmanian Conservation Trust: Hobart.

Lake, P. S., 1980. Conservation. Pp. 163–73 in *An ecological basis for water resources management* ed by W. D. Williams. ANU Press: Canberra.

Leigh, J., Boden, R. and Briggs, J., 1984. Extinct and endangered plants of Australia. MacMillan Australia: Melbourne.

Lynch, P. and Prokop, F., 1993. Intertidal invertebrates — regulations. Fishnote DF/28, NSW Fisheries: Sydney.

Magurran, A. E., 1988. *Ecological diversity and its measurement.* Croom Helm: London.

McGuiness, K., 1988. *The ecology of Botany Bay and the effects of man's activities: a critical synthesis.* Institute of Marine Ecology: Sydney University.

McLoughlin, L., 1985. *The Middle Lane Cove River: a history and a future.* CEUS Monograph 1, Macquarie University: Sydney.

Mellanby, K., 1967. *Pesticides and Pollution.* Collins: London.

Mitchell, M. L. and Adam, P., 1989. The relationship between mangrove and saltmarsh communities in the Sydney region. *Wetlands (Australia)* **8:** 37–46.

Moors, P. J., Atkinson, I. A. E. and Sherley, G. H., 1989. *Prohibited immigrants. The rat threat to island conservation.* World Wide Fund for Nature — N.Z.: Wellington.

Naiman, R. J., Dicampes, H. and Pollock, M., 1993. The role of riparian corridors in maintaining regional biodiversity. *Ecol. App.* **3:** 209–12.

Nix, H. A. (ed), 1972. The city as a life system? *Proc. Ecol. Soc. Aust.* **7:** 1–279.

Orchard, A. E., 1990. Rediscovery of *Haloragodendron lucasii* (Haloragaceae). *Telopea* **3:** 593–96.

Orians, G. H., 1993. Endangered at what level? *Ecol. App.* **3:** 206–08.

Pallin, N., 1993. Restoration of a habitat — the Ku-ring-gai flying fox colony. Pp. 5–9 in *Urban bushland management into the melting pot (ponderings, processes and problems)* ed by J. L. Howard and J. Rawling. Ku-ring-gai Municipal Council and Urban Bushland Management: Sydney.

Patterson Britton, 1992. *Towra Point erosion. Investigation of management options.* Report to NSW and ANPWS.

Peters, R. L. and Lovejoy, T. E. (eds), 1992. *Global warming and biological diversity.* Yale University Press: New Haven.

Phillips, B., 1990. *Koalas. The little Australian we'd all hate to lose.* Australian Government Publishing Service: Canberra.

Pleumarom, A., 1992. Course and effect. Golf tourism in Thailand. *The Ecologist* **22:** 104–10.

Potts, B. M., 1989. *Population variation and conservation status of a rare Tasmanian endemic,* Eucalyptus cordata. Tasmanian Forest Research Council Inc.: Hobart.

Ramsay, A., 1992. Collection and collation of visitor use data: 1992 update. NPWS: Sydney.

Recher, H. F., 1972. The vertebrate fauna of Sydney. *Proc. Ecol. Soc. Aust.* **7:** 79–87.

Reid, J. R. W., Kerle, J. A. and Morton, S. R., 1993. Uluru Fauna. The distribution and abundance of vertebrate fauna of Uluru (Ayers Rock-Mount Olga) National Park, NT. *Kowari* **4:** 1–152.

Resource Assessment Commission, 1992. *Coastal Zone Inquiry. Draft report.* Australian Government Publishing Service: Canberra.

Richardson, D. H. S., 1981. *The biology of mosses.* Blackwell: Oxford.

Robinson, J. H. and Snyder, H. S., 1991. Golf course development concerns in coastal zone management. Pp. 431–43 in *Coastal Zone '91.* American Society of Civil Engineers: New York.

Rogers, R. W., 1992. Lichen ecology and biogeography. *Flora Aust.* **54:** 30–42.

Saunders, D. A. and Hobbs, R. J. (eds), 1991. *Nature conservation 2: the role of corridors.* Surrey Beatty & Sons: Chipping Norton.

Scammell, M., 1990. Tri-butyl tin information summary: September, 1988. *Wetlands (Australia)* **9:** 36–39.

Scott, D., 1992. *1991 State of the Environment Report. Agriculture and Victoria's environment. Resource report.* Office of the Commissioner for the Environment: Melbourne.

Smith, G. C. and Carlile, N., 1992. Habitat use by Silver Gulls *Larus novaehollandiae* in the Sydney-Wollongong region, New South Wales. *Wetlands (Australia)* **11:** 33–45.

Smith, P. and Smith, J., 1990. Decline of the urban koala *(Phascolarctos cinereus)* population in Warringah Shire, Sydney. *Aust. Zool.* **26:** 109–29.

Smith-White, A. R., 1988. *Sporobolus virginicus* (L.) Kunth in coastal Australia: the reproductive behaviour and the distribution of morphological types and chromosome races. *Aust. J. Bot.* **36:** 23–39.

Smith-White, A. R. and Adam, P., 1988. An unusual form of the saltmarsh grass *Sporobolus virginicus* (L.) Kunth. *WA Nat.* **17:** 118–20.

Specht, R. L., 1981. Foliage projective cover and standing biomass. Pp. 10–21 in *Vegetation classification in Australia* ed by A. N. Gillison and D. J. Anderson. ANU Press: Canberra.

State Pollution Control Commission, 1978. *Environmental impact assessment of Flood-mitigation works and dams.* SPCC: Sydney.

State Pollution Control Commission, 1979. *Off-road vehicles inquiry.* SPCC: Sydney.

Sudu, M. and Kunimatsu, T., 1992. Characteristics of pesticides runoff from golf links. *Water Sci. Tech.* **25:** 95–98.

Sukopp, H. and Werner, P., 1982. *Nature in cities.* European Committee for the Conservation of Nature Resources: Strasbourg.

Sukopp, H. and Werner, P., 1987. *Development of flora and fauna in urban areas.* European Committee for the Conservation of Nature and Natural Resources: Strasbourg.

Taylor, G., 1940. *Australia, A study of warm environments and their effect on British settlement.* Methuen: London.

Thorogood, C. A., 1985. Changes in the distribution of mangroves in the Port Jackson-Parramatta River estuary from 1930 to 1985. *Wetlands (Australia)* **5:** 91–96.

Underwood, A. J. and Hutchings P. A., 1987. *Australia's seashores.* Collins: Sydney.

Vic Roads, 1990. *Roadside management guide.* General Report/90-6 Vic Roads: Melbourne.

Wace, N. M., 1977. Assessment of dispersal of plant species — the car-borne flora in Canberra. *Proc. Ecol. Soc. Aust.* **10:** 167–86.

Wace, N. M., 1985. Australia — the isolated continent. Pp. 3–22 in *Pest and parasites as migrants. An Australian perspective* ed by A. Gibbs and R. Meischke. Australian Academy of Science: Canberra.

Walker, K. F., 1980. The downstream influence of Lake Hume on the River Murray. Pp. 182–91 in *An ecological basis for water resources management* ed by W. D. Williams. ANU Press: Canberra.

Wallbrink S. J. (ed), 1992. *Field guide to Victoria's native grasslands.* National Trust: Melbourne.

Walraven, E., 1993. Surviving in the suburbs, urban wildlife issues. Pp. 32–34 in *Proceedings of the 1992 urban bushland management conference. Urban bushland management — into the melting pot (ponderings, processes and problems)* ed by J. L. Howard and J. Rawling. Ku-ring-gai Municipal Council and Urban Bushland Management: Sydney.

Warner, G. B., 1986. Hydrogeomorphology of Sydney: a case study of urban impact. Pp. 42–60 in *The changing face of Sydney* ed by D. Dragovich. Geographical Society of NSW: Sydney.

Watson, I. D., (1986) Urban climate and pollution. Pp. 25–41 in *The changing face of Sydney* ed by D. Dragovich. Geographical Society of NSW: Sydney.

Wilson, S. P., Ahsanullah, M. and Thompson, G. P., 1993. Imposex in neogastropods: an indicator of tributyltin contamination in eastern Australia. *Mar. Pollut. Bull.* **26:** 44–48.

Wright, B. J. and Bamford, P. B., 1986. Impact of roads on catchments. *J. Soil Conserv. NSW* **42:** 63–67.

Mining in relation to habitat loss and reconstruction

D. WOODSIDE[1] and D. O'NEILL[2]

The long-term impact of mining activities on the environment is more presumed than it is understood or measured. There is little doubt that large scale open-cut mining and the development of infrastructure such as roads and support facilities cause change to, and sometimes complete removal of, existing habitats. There exist examples of mining scars which continue to mar the landscape over many decades and influence public perception about the impact of mining. However, the scale of initial disturbance to the landscape and the rate of recovery of habitat have not been well summarized to date.

Natural habitat recovery processes depend on the type and scale of the disturbance, the time available for recovery and proximity to natural "seeding" habitats. More proactive habitat rehabilitation involves interventionist technology and measures to overcome the limitations of the natural "re-establishment" of flora and fauna. This process can result in relatively rapid reconstruction of desired habitats such as wetlands, forests or pasture. Prescribed levels of proactive rehabilitation have been required of mining activities in Australia for several decades.

A major impediment to determining the success of the rehabilitation efforts rests with the lack of standardization of biological criteria for measuring habitat loss, habitat modification and subsequent recovery. Under these circumstances the impact and the effectiveness of post-mining rehabilitation seem rarely to satisfy the increasing expectations of scientists and the community.

This paper reviews the scale of mining activity in Australia, with a particular focus on New South Wales, the adequacy of expectations for habitat reconstruction and recommends a more collaborative approach between ecologists and mining professionals in the future. The outcomes of this improved process would enable the evolution of practical measures of habitat health, practical guidelines for habitat reconstruction or rehabitation, new knowledge of habitat dynamics, better models for impact assessment and mitigation and improved technology.

INTRODUCTION

MANY Australians view mining negatively. Some of our most prominent environmental organizations are committed to the view that mining in any form, at any scale, is to be stopped. For them, the losses outweigh the gains. One has to ask why mining engenders such an emotional response in so many. Why is it the "bogey man" of Australian conservationists (Zarsky 1990)? For many Australians mining is associated with the negative images of landscape destruction, exploitation of "the commons", wealth, power and political influence. The viewpoint, whether right or wrong, must be acknowledged, if for no other reason than that they exist in a community that has a growing awareness, anxiety and voice concerning environmental degradation of every kind. The world summit on the environment (UNCED) held in Rio in June 1992 and the existence of Agenda 21 and the Biodiversity Convention are testimony to that new level of environmental awareness and global expectation for high environmental standards.

To keep pace with the increasing expectations of the community concerning sound environmental management, environmental professionals in all sectors of society have an obligation to rapidly develop a rational, objective and common suite of criteria for assessing the environmental performance of industries such as mining. Environmental professionals also have the obligation to assist in the transfer of new environmental knowledge or low impact technologies between different land using sectors.

In this paper we have taken up the challenge of addressing some of the more critical issues relating to the impact of mining on the environment and particularly to that of habitat loss and its rehabilitation. We have taken the view that it is timely to review such industries from the point of view of threats and solutions they bring to bear on biodiversity conservation. We review the scale of mining activities in Australia, the nature of impacts on the environment, adequacy of regulation to limit impact and ensure rehabilitation and the adequacy of the criteria used to assess impact or its amelioration. We draw from this review a number of constructive recommendations for the future which will enable a more rational and less emotive debate about mining activities as well as a more co-operative approach to mining among a wider group of stakeholders, including environmental professionals and members of affected communities. Such an approach can only improve the standards of environmental management overall.

We have not conducted an exhaustive review of all aspects of mining and the environment in Australia although this needs to be done. Many of the statistics we found to be most relevant to

[1]Zoological Parks Board of New South Wales, P.O. Box 20, Mosman, New South Wales, Australia 2088.
[2]Resource Futures Pty Ltd, 8 Olive Street, Paddington, New South Wales, Australia 2021.

Pages 76–93 *in* CONSERVING BIODIVERSITY: THREATS AND SOLUTIONS, ed by R. A. Bradstock, T. D. Auld, D. A. Keith, R. T. Kingsford, D. Lunney and D. P. Sivertsen. Surrey Beatty & Sons. 1995.

our review were not easily extracted from the files of all the states. Instead we have chosen to focus primarily on the mining activities and regulations of New South Wales for which we had most information, while occasionally drawing from examples elsewhere in Australia.

LINGERING IMAGES OF THE PAST

Like most of the activities of the early white settlers, mining was pursued with minimal regard for the environment or the impact on future generations. Industry of the day was limited by its technology and understanding of the environment and society supported the view that Australia was to be "tamed" at all costs.

Most Australians have seen pictures of the bare mountains surrounding Queenstown in Tasmania and are aware that the degradation is a legacy of mining. The naked landscape, ironically maintained in its degraded state by the local council in the interests of tourism, resulted from the removal of trees to provide timber props for the mine and fuel for the smelters which in turn added another level of environmental degradation and through air water and soil pollution. Natural regeneration of the forest became impossible after the high rainfall of western Tasmania took its toll of the topsoil and seed banks through erosion and acidification. Thus even when more enlightened land management practices had developed in the mining industry, it was too late for rehabilitation of the forests at Queenstown.

Adding to the ecological woe of that region of Australia are the biologically dead King River and parts of Macquarie Harbour into which it empties, the results of almost a century of abuse. The river, the "commons", was used as a conduit for mine tailings laden with heavy metals from Mount Lyell copper/gold mine near Queenstown. Similarly, the Captains Flat mine south-east of Canberra has been a long standing study in regional heavy metal pollution of land and in water in New South Wales and the Australian Capital Territory.

Sadly, there are many historical examples of mining practices in Australia which have left dramatic scars on the landscape. With a little creativity and enterprise some of these scars have become tourist attractions and the lessons of mismanagement are often forgotten. Examples include several of the old gold mines and a black coal site in Victoria, gemstone operations in the New England district of New South Wales, the infamous alluvial gold areas in northern Queensland, opal mining in South Australia and New South Wales and parts of the Eastern Goldfields of Western Australia, to name a few.

There are also examples of "healed" scars where the land is now used for other purposes with the mining episodes apparently left far behind. For example, photographs taken of parts of the Victorian goldfields in the 1850s show large tracks of stripped, treeless valleys and hillsides. Today most of these areas are either urban developments, tourist attractions, farmland, or naturally regenerated forests and some have even recovered to the point of being incorporated into state forests and conservation reserves. A review of 215 mineral sands operations on the east coast of Australia shows that 30 of these are now included in National Parks or Nature Reserves and 61 sites are used for recreation (Allen *et al.* 1990). Few visitors to the Scenic Railway near Katoomba in the Blue Mountains west of Sydney would today associate the apparently lush surrounds with the levelled landscape which existed at the turn of the century as mining proceeded for oil shale.

Several important questions result from the exposure of skeletons of our industrial past. The first and most important question is whether we have learnt enough from these mistakes to never repeat them again. What objective criteria can we establish that guide our future planning in this regard? Does the mining industry of the 1990s perform well enough to be released in the publics mind from bondage to this legacy of destruction? Should it now be held responsible only for its current operations? If so, then whose responsibility is it to clean up the past mistakes of this, or any, industry?

The terminology used throughout this manuscript requires some clarification as there is considerable debate surrounding appropriate use of the term "rehabilitation". We have continued to use this term to reflect the broadest possible concept of re-establishing a functional ecosystem whether or not the ecological elements replicate the conditions before disturbance. This tends to be the term most commonly used by the Australian mining industry to describe every type of revegetation operation from simplistic plantations to complex ecosystems.

It might be argued that the term better used in this case is habitat "reconstruction" and that the mining industry would benefit from distinguishing between terms such as restoration, reclamation, rehabilitation, as decreasing levels of precision in the goal to replace the habitat that once occupied the site. Reconstruction is offered as a compromise term by the USA Academy of Science (1974), the Society for Ecological Restoration and adopted by Allen (1988) in a volume on The Reconstruction of Disturbed Arid Lands — An Ecological Approach.

". . . restoration implies that the exact conditions before mining will be replicated afterwards. Reclamation was defined to mean that the site will be similar in ecological functioning after disturbance, and will be habitable for similar but not necessarily the same organism. Rehabilitation means that the land will be made useful, but with a different land use and usually different species. Since all three revegetation goals especially the former two, are represented in this book, clearly none of these three terms alone would be appropriate. Reconstruction is used as a compromise term; the dictionary definition is to build an image of what something was in its original form."

In the future the mining industry might consider the advantages of partitioning the process into "rehabilitation", "reclamation" and "restoration" as increasing levels of the process and goals of re-establishing vegetation following mining.

THE NATURE AND EXTENT OF MINING IN AUSTRALIA TODAY

The term mining has many meanings to the public and the wide range of definitions of mining is partly at fault for the poor public understanding of this industry. Strictly speaking, mining includes only those operations directly involved in the extraction of mineral ores for commercial gain. It does not technically include exploration or evaluation activities. However, the public perception of mining includes exploration, extraction of organic (coal) or inorganic matter (sand, metallic minerals or gem stones) and all related activities such as support infrastructure, transport of mined material, and primary processing, particularly smelting. Some people also lump petroleum production with mineral and coal operations under the heading of mining.

In this paper, we concentrate primarily on operations involved in extraction of coal and minerals. We also include land and habitat rehabilitation activities as part of the mining process as it is part of the legal obligations of each operator and must be factored into the mining operation and budget.

For our purposes we have excluded any reference to petroleum exploration or extraction. We have also limited our reference to exploration in part because the area of significant land disturbance is small compared to the large area under exploration title at any given time. The issues relating to the "linear" impact patterns of exploration seismic lines and the wide range of low impact technologies available to exploration groups deserves special attention but some general comments can be made here.

Exploration activities in Australia are extensive but exploration is not viewed as a land use by law as is mining. Instead exploration is a temporary occupation for the purposes of gathering information with only a small proportion (one in a thousand) of exploration opportunities perhaps leading to mining operations (Fig. 1).

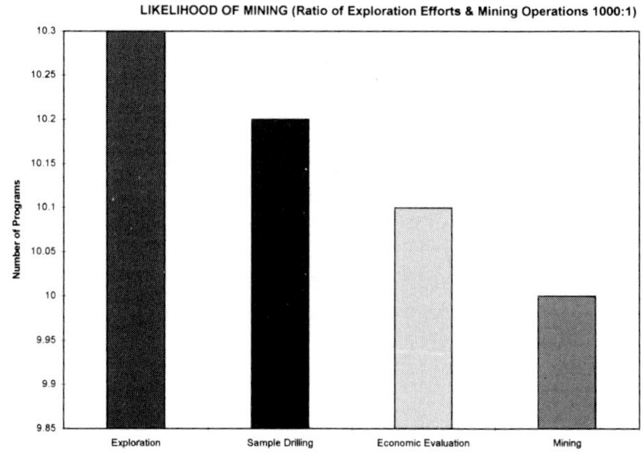

Fig. 1. Likelihood of mining operations commencing after exploration and assessment processes.

However, several aspects of exploration can cause significant local impact. These include the cutting of tracks and roads, seismic lines, the excavation of exploratory pits, intensive drilling and the cutting of trenches. Exploration is typically a punctuated event, less intense and less disturbing than the full mining operation. Exploration technology has changed greatly over the past 40 years with ever greater emphasis on remote sensing and relatively unintrusive data acquisition methods during the initial phases. During later more disruptive phases low impact techniques are increasingly used (for example, compaction of seismic lines rather than use of bulldozers, limited disturbance of vegetation at drill sites, rehabilitation of access tracks, use of helicopter access, use of tyred not tracked equipment in sensitive areas). The impact of exploration is not to be overlooked and there are both negative and positive benefits.

Mining operations in Australia make a major contribution to the Australian economy of the 1990s. Mining and minerals processing generated over $25 billion in revenue in 1991/92 and are ranked as the most important contributors to Australian exports with 52 per cent of Australia's merchandise exports in 1992 (including energy exports). The total direct employment in exploration, mining and smelting/refining in Australia is over 73 000 (AMIC 1992, 1993a).

Mining operations throughout Australia disturb a surprisingly small amount of land, about 0.02 per cent of the total Australian land mass, or about 70 per cent of the area of the Australian Capital Territory. By comparison

Table 1. Land taken up for mining in Australia.

State	Area of leases[1] for mining (km²)	Mining leases as % of state	Estimated area of disturbance[2] (km²)	Disturbed area as % of state	Disturbed area as % of mining leases	Estimated area rehabilitated (km²)
WA	14 032	0.56	510	0.02	3.6	unavailable
Qld	11 430	0.66	725	0.04	6.3	25+
NSW	3 854[3]	0.48	200	0.02	5.0	35+
NT	1 420	0.10	70	<0.01	5.0	unavailable
SA	774	0.08	35	<0.01	4.5	1
Tas.	841[3,4]	1.23	67[4]	0.10	8.0	unavailable
Vic.	560[5]	0.25	47[6]	0.02	8.4	3
Australia	32 911	0.43	1 654	0.02	5.0	

[1]Taken as the area of current production tenements for coal, metallic minerals and some extractive minerals or construction materials. The data does not include mining or extraction activity undertaken under legislation other than Mining Acts, for example, within some categories of land or certain types of construction materials, except where stated. Data currency covers mid-1992 to early 1993.
[2]Based on estimates made by state mines departments and projections where data unavailable.
[3]Includes underground leases with little or no surface expression.
[4]Includes estimates of areas for private stone leases.
[5]Does not include areas held privately for brown coal mines and estimates SECV holding.
[6]Includes areas disturbed by SECV brown coal mining.

approximately 70 per cent of Australia is used for agriculture and forestry (AUSLIG 1990; Hobbs and Hopkins 1990; other papers in this volume). Coal dominates in New South Wales with a mix of underground and open cut mines while in Queensland it is large but mainly with open cut operations (Table 1). This has implications for the ratio of surface disturbance to area of granted lease. On the other hand, in some locations near surface underground mines may have surface impacts through subsidence. Tasmania has a high proportion of underground mines while Western Australia has many large open cut mines and relatively few underground mines. In Victoria, Australias largest mines, the open cut operations of the State Electricity Commission in the Latrobe Valley, do not operate under mining legislation because they are in effect private mines operated under special legislation. Their impacts are therefore not measured in official statistics collated by the Department of Energy and Minerals. The estimate of area rehabilitated is an annual figure and is based on discussions with regulatory officials.

The land currently occupied for mining activities was reviewed using data available on the total area granted as mine leases and the estimated area of disturbance (Table 1). Area of disturbance was determined following discussions with state departments as that affected by the core mining operation as opposed to the entire lease. The area of land disturbed is minute compared to the total area granted for exploration in each state (Table 2). Areas of rehabilitation were estimated following discussions with mines inspectors and with officials responsible for environmental performance of operations. This is a conservative estimate of rehabilitation activity. Critical information of this kind has

Table 2. Area affected by exploration activity in Australia.

State	Area of state (km²)	Area of granted exploration titles (km²)[1]	Area of titles as % of state
NSW	801 400	~80 000 (26 897 sb)[2,3]	10.0
Qld	1 727 200	~150 900 (47 158 sb)	8.7
NT	1 346 200	7 183 (2 232 blocks)[2]	0.5
WA	2 525 500	185 662	7.3
SA	984 400	200 000	20.3
Vic.	227 600	31 942	14.0
Tas.	68 400	5 750	8.4
Australia	7 686 000	661 437	8.6

[1]Currency of data varies from mid-1992 to mid-1993.
[2]A "sub-block" or a "block" is a unit of area bounded by a minute of latitude and longitude and its size varies according to location on the earth's surface. In New South Wales each unit is approximately 3 km²; in Queensland and the Northern Territory about 3.2 km².
[3]This figure is slightly inflated due to "double counting" of overlapping titles granted over the same area for different suites of minerals.

not been systematically summarized by the relevant state authorities or the mining industry to date and remains fragmented. Some jurisdictions have commenced its collection and recording but there is no standardization of the information.

Surprisingly, a typical mining operation in New South Wales is still "a one-person show" which is producing gravel, sand, stone, opal or sapphires. The majority of the 1 200 mining operations examined in New South Wales employ fewer than six staff (Fig. 2). Large operations are rarer and reflect the geological scarcity of major mineral deposits. This suggests that while the public image of mining is that of a big operation, the typical mine is operated by one to five people. Such a profile affects the

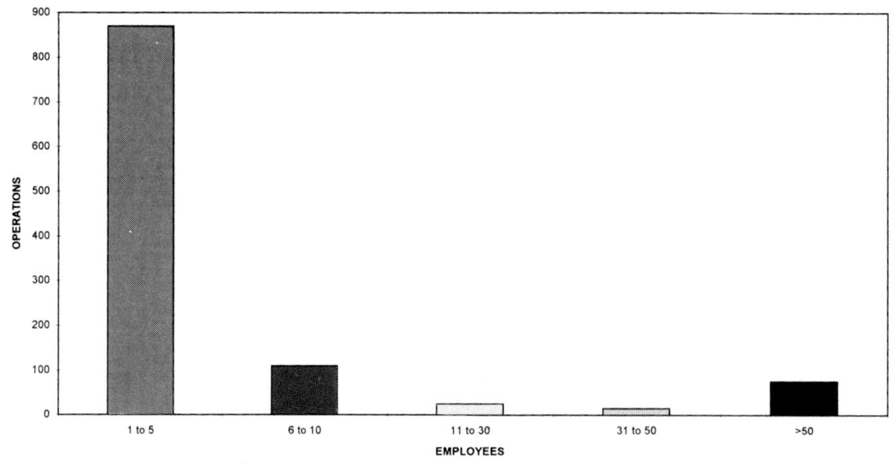

Fig. 2. Classification of mining operations in New South Wales by levels of employment.

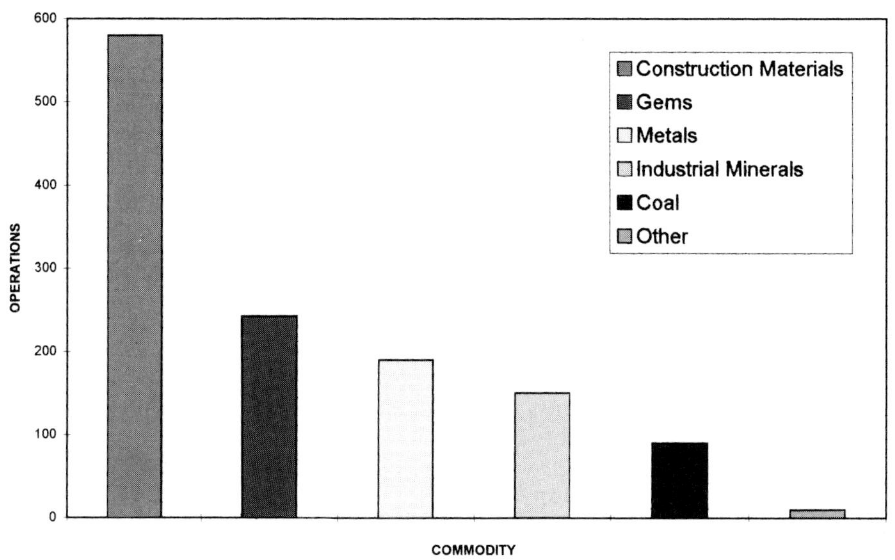

Fig. 3. Classification of mining operations in New South Wales by commodity groups.

design of the environmental controls and policing by government. It is an industry far more similar in profile to bakeries or tour operators in New South Wales than to the standard perception of mining! The number and dispersed nature of the mine operations needs to be considered in addressing future regulation of mining for biodiversity conservation, education or co-ordination of efforts.

Most of the small operators in New South Wales serve local markets through extraction of construction materials (Fig. 3). In such a highly urbanized country as Australia the demand for such operations remains very high and constant. Quarries are established for road gravel, sandstone, brick clays, construction sand and limestone for cement and so on. These operations involve the removal of large quantities of raw materials and transport to nearby locations and result in changes in local topography.

The discovery of coal in 1797 marked the start of mining in New South Wales. In addition, early settlers, etc., quarried rock and limestone used in building. However, gold was the first mineral to be selectively mined after its discovery in 1851. Gold mining was relatively widespread through the state by the late 1800s, though smaller in scale than in Victoria or Western Australia by then. Estimates of the area impacted by mining during the last century are difficult to make. The New South Wales Department of Mineral Resources estimates that there may exist in New South Wales between 100 000 and 250 000 abandoned minesites, ranging from many small individual pits and shafts to a few larger derelict plant sites, with each site relatively small in area.

At present mining in New South Wales is dominated economically by coal although this is not reflected in the type of data presented in Figure 3. These coal operations occur broadly in five areas — the Southern Coalfield centred on Wollongong and immediately inland, the Hunter and Newcastle Coalfields along the Hunter River Valley, the Western Coalfield centred on Lithgow and the Gunnedah Coalfield. Metalliferous mining is dominated by the base metals (lead/zinc) mines at Broken Hill, Cobar, Elura and Woodlawn with gold at Temora, Blayney, Lyndhurst, Cobar and Parkes and gold/copper at Parkes and Girilambone. Titanium minerals are mined at Tomago/Viney Creek, near Myall Lakes and, shortly, just north of Forster/Taree at Saltwater. Quarrying is widespread.

THE MEANING OF HABITAT LOSS OR MODIFICATION

Mining operations can affect the land, water and air that surrounds them. Whether or not these effects lead to habitat loss or modification requires definition of these terms and some means of measuring the change.

There are essential questions to ask that would enable us to understand habitat loss or modification in Australia and to place disturbance due to mining in the context of other changes to the Australian landscape.

1. What do we mean by habitat loss or modification?
2. Can impact on habitat be classified in a standardized way and what are the essential criteria?
3. How does mining as a form of disturbance compare to other forms of natural or human-induced change?
4. What is an acceptable level of change?
5. How resilient is the Australian landscape and its ecosystems to change?
6. What is the appropriate time frame for assessing recovery of habitat with respect to flora and fauna?
7. What is "natural" habitat in Australia — do benchmarks exist?
8. Is there a place for reconstructed habitats and if so what are the biodiversity objectives in reclamation?
9. How can we optimize the value of reconstructed habitat for specific conservation goals?

If these questions are asked of most environmental professionals today their answers would differ widely. The entire field of impact assessment, prediction and amelioration based on ecological science is in its infancy. Recent symposia present collections of ideas on the role of remnant native vegetation and its characterization and resilience to disturbance (Saunders *et al.* 1987), the role of corridors and their rehabilitation (Saunders and Hobbs 1991) and the reconstruction of fragmented ecosystems (Saunders *et al.* 1993). These ideas form the basis of what will become an essential science associated with all forms of land-use. However, the lack of current guidelines and the changing "goalposts" of ecologically sound resource management, cause great frustration to pragmatic industry faced with the task of staying viable in a changing world market while trying to meet ever changing expectations at home.

As Australia and its neighbours adopt goals and ideals of sustainable development, then environmental professionals must be able to objectively assess the potential impacts of industry and determine when the impacts are within the acceptable tolerance of the ecosystem. Without the ability to assess the environmental risks of a development it is difficult to address trade-offs between short-term income objectives and long-term environmental objectives (van Pelt 1993). This leads to subjective judgements on whether or not to proceed.

There is a need, therefore, for standardizing the ecological terms used and devising some common system for characterizing habitats and the nature of impact. Some ecologists have begun the hard task of addressing some of these questions and barriers and should be acknowledged.

For example, habitat (specifically wildlife habitat), might be defined as "the resources (such as food, refuge, and space) that maintain natural life history processes of a species of plant or animal" (G. Pyke, pers. comm.). "Natural habitat" is described by some ecologists as that occurring in Australia before European settlement while others describe it as that which is totally unaffected by humans. For examples of "natural" habitat of this kind we would have to return to the conditions of the Pleistocene era where the Australian landscape was unaltered by either Aboriginal or European inhabitants. Using these criteria there is no "natural" habitat remaining in Australia (Taylor 1990). Our efforts to declare benchmarks are therefore futile and the establishment of standards for the rehabilitation of habitat have no firm basis if we seek only to restore natural habitats of the past.

One approach might be the acceptance of the broadest possible concept of "habitat". It would be defined as either naturally evolved habitat *in situ,* deliberately constructed or rehabilitated habitat or that which has been altered by humans in some way. This definition of habitat places humanity within nature and allows that all habitats or ecosystems in Australia, past or present are equally natural. Using this definition an unacceptable degree of habitat modification may be described as changes which interfere with the life history processes of any native species of plant or animal living at the site of impact.

A second approach might be to use the classification of Australian ecosystems (for our purposes read "ecosystem" as "habitat") as presented by Taylor (1990) and to some extent by Bridgewater (1990). Australian ecosystems would be classified as either "natural" or human-dominated. Natural systems support native vegetation (accepting that any habitat is naturally dynamic), while human-dominated ecosystems support either cultivated, adventive (i.e., weedy) or relic (i.e., degraded native) vegetation. Bridgewater (1990) refers to the latter two as synthetic vegetation. Using this definition and classification, habitat loss might be concerned only with the unacceptable levels of change occurring in native vegetation. This could ignore the impacts on urban or rural habitats which may support native fauna even if the vegetation is no longer strictly native.

As Taylor (1990) points out, with or without human interference, ecosystems are highly dynamic. They may experience naturally occurring stochastic events such as fire, landslides, seismic, volcanic events or deterministic events such as autogenic succession, seasonality or long term climate change. Large scale or severe events, such as floods or hurricanes occasionally punctuate the pattern and regardless of their severity and impact are deemed natural. Against this background it is difficult to define unacceptable limits of disturbance induced by human activities such as mining. One might ask how the disturbance of one small minesite, however intense, is ranked against the sudden and large scale destruction of flash floods or the slower changes brought on by geological events.

Armstrong (1993) argues the case for giving priority during reconstruction of habitats to re-establishing ecosystem function over species and genetic diversity. He suggests that it is valid to make selective use of exotic species to help stabilize ecosystems immediately after disturbance and makes a number of practical suggestions for use for the planning of "synthetic" habitats. These ideas are also explored by Hobbs and Mooney (1993) and Recher (1993). This highlights the argument that exists among ecologists and the problems that can arise in setting standards for the reconstruction of vegetation communities by an industry.

Hobbs and Hopkins (1990) have provided a classification for impact based on the type and degree of disruption to natural process and the type of remedial actions required to restore native vegetation. Only human induced impact over the past 200 years are included in this classification primarily because the arrival of Europeans is seen as an "apocalyptic event for Australian ecosystems" (Adamson and Fox 1982). The impact of mining by settlers in Australia is viewed as being one of the most intense forms of impact including some minor alterations to wording. These categories are as follows:

1. *Removal.* Complete removal of native vegetation and removal or disturbance to soil and disruption of ecological processes: urban settlement, transport, industrial development and extractive industries. Native vegetation and all natural processes are completely disrupted except in small patches. Natural vegetation will not return without complete reconstruction of the ecosystem, including soil replacement.
2. *Replacement.* Removal of native vegetation and replacement with intensively managed systems: agriculture, horticulture and plantation forestry. This involves partial disruption of processes. Native vegetation (or components of it) can return after cessation of intensive management, although this may require management to restore natural processes.
3. *Utilization.* Exploitation of native vegetation, with some consequent degree of modification: forestry (in native forests), pastoralism and recreation in natural areas. This corresponds to a partial removal of the natural vegetation and a partial disruption of processes.
4. *Conservation.* Maintenance of natural vegetation for conservation and scientific purposes with minimum deliberate modification of natural processes. Lands designed as national parks and nature reserves fall into this category. In addition there are substantial areas of vacant or uncommitted government-owned land and Aboriginal lands. Designation for nature conservation does not preclude past or inadvertent modification. There are no direct or deliberate impacts on the native vegetation as a result of this land-use, but there may be indirect impacts from land-use activities on adjacent or nearby lands.

Application to mining

The determination of the acceptable levels of modification by industries such as mining might be developed around the classification of Hobbs and Hopkins (1990). Perhaps what is deemed as an acceptable modification will depend on the nature and degree of intervention required for reconstruction and the time to completion or recovery. It might therefore be acceptable to disturb the landscape if there is some certainty of habitat recovery or sufficient technology to ensure recovery within a given time frame and to an acceptable standard. This may be a conditional application of the "Precautionary Principle" as described in Agenda 21.

Acceptance of permanent modification of the habitat would be dependent on the nature of land use intended post-mining. If for example the intended use is the support of particular species such as the koala, certain features of the forest can be reconstructed to target the needs of that species (Esser 1992). There is in this case an acceptable type of habitat enhancement or change. It is also possible, through careful rehabilitation of mined lands to increase the productivity and return it to other preferred land uses. Examples of this include the replacement of monoculture forest at Viney Creek with native species, the reforestation of jarrah areas mined for bauxite in Western Australia but with reduced infestation by Phytophthora, and the improved productivity of agricultural land south of Perth after mining for titanium-bearing mineral sands.

Given the wide range of land use options that may follow mining, it is important that the role of constructed habitat in the broader Australian landscape be addressed. Habitat construction can go far beyond the development of pasture lands, forests for timber production or the return of complex native bushland for a nature reserve. In some cases wetlands may be constructed where they have not been but could serve a useful conservation role or specialized habitat can be constructed to enable the survival or reintroduction of endangered species. While some examples of cooperation between the mining industry and wildlife authorities exist (e.g., Healesville Sanctuary in Victoria, wetlands in Western Australia, koala corridors near Hawkes Nest in New South Wales) these are exceptions. The potential value of constructed habitats needs to be more widely applied as does the technology available within industry to undertake this work.

THE KNOWN IMPACTS OF MINING

Information pertaining to mining as a cause of habitat loss is virtually non-existent. We can surmise that some of the mining practices of the past caused patches of habitat to be lost. However there is no evidence that any species of plant or animal has gone extinct in Australia because of mining. Nor is there any evidence of any loss of associations of plants or major habitat types due to mining. In Australia most vertebrate extinctions have come about through a combination of pressures that persistently degrade, fragment or remove habitat (Recher and Lim 1990). The high extinction rates have occurred in spite of the fact that animal and plant communities in Mediterranean ecosystems such as Australia tend to be more resilient than others (Fox and Fox 1986). However, mining was seen as a kind of disturbance from which recovery was slowest.

Only recently have mining companies begun to undertake habitat studies and thorough faunal studies (vertebrates and non-vertebrates) before mining operations commence so that comparisons can be made with post-mining monitoring of habitat recovery. Advance studies are also being used to plan restoration programmes, to ensure adequate seed collection and to protect the local gene pool.

Mining does cause changes in habitat. Some changes are short term others are long term. The severity of these changes depends on the geographic location and on the nature of the mining operation (open pit or underground), the chemical reactivity of the ore being extracted and the degree of processing carried out on site and the resilience of the type of disturbance (Fox and Fox 1986). The impact may be in the form of acid mine drainage, heavy metal seepage (which in the case of uranium introduces radioactivity), dust or noise. A variety of strategies is used to minimize these impacts, including preventive, containment and other operational and engineered solutions.

In his review of mining and the environment, Webb (1990) drew some useful generalizations about the nature of mining and its impacts on the environment. He found that there are seven major categories of impact. We have added to his list:

1. Land Surface Impacts — The area of land disturbance ranges from a few hectares for small open pit mines or underground mines) to thousands of hectares (some strip mining). Vegetation is cleared, topsoil is removed (and often stored for replacement) and rock excavated. Nutrients are redistributed, soil profile is altered, hydrography can be changed and some soils compacted (e.g., Fox 1982; Fox and Fox 1978; Reddell and Milne 1992).
2. Land Subsidence — Underground mining for coal can lead to the eventual collapse or shift in the ground above the mined seam. Planning for such subsidence is a major issue for the coal industry in the Lower Hunter Valley, New South Wales.
3. Flora and Fauna — Surface vegetation is often removed or destroyed during mining. The duration of this disturbance is dependent on the nature and scale of the mining operation, the life span of the mine, the type of habitat that was destroyed and the isolation/proximity to "intact" habitat (Fox 1990).
4. Local Topography — Some mining operations result in large losses of substrate or the dislocation of substrate from its original position. Depending on the rehabilitation programme prescribed by the relevant government authority, the land forms may be reshaped after mining so that new hills may appear or hills may be levelled.

5. Atmospheric Emissions — Airborne chemicals have caused significant habitat losses in the past (for example Mt Lyell in Tasmania), though more so in countries other than Australia. Strict regulation of emissions, vigilant monitoring programmes and better planning for impact have reduced their effects. Point source anthropogenic sulphur dioxide (SO_2) emissions in Australia are dominated by mining-related smelting and roasting at Kalgoorlie (Western Australia) and Mt Isa (Queensland), with power generation at several sites elsewhere also contributing but at a much lower level. Williams *et al.* (1981) concluded that dry deposition and atmospheric oxidation were of approximately equal importance in removal of atmospheric SO_2 resulting from smelting at Mt Isa. The Kalgoorlie region is expected to become shortly the worlds largest point source emitter of SO_2. However the Australian Environment Council (1989) acknowledged that, at worst, the potential for acid rain impacts was only moderate, given the moderate to low acidity of Australian soils combined with relatively low rainfall over most of the country.

6. Noise and Vibration — The blasting of rock and the operation of large machinery and trucks cause the ground to vibrate, constant noise and occasional occurrence of blast shock waves. Careful planning, alternative design in equipment, tight regulations and monitoring reduce these impacts.

7. Water Pollution — Acid run-off, top soil erosion, heavy metal run-off, siltation, salination and contamination are all possible. Sampling of stream water for contaminants, changes in flow rate, temperature, silt etc are tightly controlled as is the monitoring of ground water levels and quality. Micro and macro invertebrates are used as indicators of impacts. Water is often recycled on the mine site and purified through constructed wetlands or mechanical treatment. Placement and engineering control of waste rock and low grade ore is critical if water seepage through it is to be avoided. Some mines such as Woodlawn near Goulburn and the Ranger uranium mine operate on essentially a zero release basis, with controlled release only of collected rainwater under special circumstances. Shortly after mining ended in 1971 surveys of the East Finnis River in Northern Territory was polluted to such an extent with heavy metals, acidification and radiation that the river was virtually dead. Recent surveys of the same river indicate that macro-invertebrates in the waterways continue to be less common downstream of the mining activities.

8. Radioactivity — Inconspicuous forms of pollution such as radioactivity can be the most difficult to manage and one of the most dangerous over a long time period. Radioactivity is generally associated with certain forms of heavy metal pollution and the impact of the two agents is difficult to separate.

9. Constructed and Rehabilitated Habitats — Post mining activities generally involve some proactive programme to reconstruct the landscape and the vegetation. Little if anything is done to re-establish the fauna (invertebrates or vertebrates) on land or in the water. The time frame for reconstructed or rehabilitated habitats to become "indistinguishable" from nearby bushland (if that is the target land use) is unknown. Some differences can still exist after 20 years, depending on the nature of the survey.

10. Catalyst for General Change — Mining often opens remote areas to a range of new influences. These can range from exploration tracks serving as corridors for weeds and feral animals species to the impacts of peripheral support facilities for the mine. Transport systems, townships, processing plants and local agriculture can all change or come into being as a result of a mine. Depending on the size and duration of the mining operation these changes may be irreversible so that peripheral facilities and communities continue long after the mine. Likewise the rehabilitation of a mine site to a land use different from that before mining begins (forest rehabilitated to grazing land, grazing land rehabilitated to a wetland, or grazing land made more productive or converted to recreation or conservation uses) a sequence of environmental modifications — not all negative in their impact.

11. Aesthetic Impacts — While the most important impacts on a habitat are those which affect its productivity (particularly numbers 1–7 above) or ability to support natural communities and ecological process, there are also important impacts on the aesthetics. From a public perspective this may be the most important change as it can lead either to care or abuse of habitat after mining. Lack of sensitivity by mining companies in dealing with aesthetic impacts of their operations is possibly the single most important source of public antipathy towards mining.

The persistence of habitat disturbance resulting from mining can be summarized as the interaction between at least six important factors: the nature of disturbance to topsoil, the level of disturbance or removal of vegetation, the size of the area affected, proximity to natural "seeding" habitats and refuges of fauna, the time available for recovery from this land use, and the degree of reconstruction required for habitat recovery.

Detailed studies of mining impacts on the environment and the technological developments to ameliorate the impact have been presented in publications of the Australian Mining Industry Council workshops on this topic held annually. The most recent workshop held in 1992 includes presentations on the Commonwealth Environmental Protection Agency and the mining industry (McPhail 1992), endangered species (Wamsley 1992), biodiversity (Bridgewater 1992), koala habitat reconstruction (Esser 1992) and various new technologies for impact reduction or monitoring. Information is also exchanged through the International Society for Restoration Ecology in its journal *Restoration Ecology*, other international journals such as the *Journal of Applied Ecology, Ecological Applications and Environmental and Geochemical Health,* and the *Habitat Restoration Society.* Much of the most valuable data is contained in the grey literature, reports and Environmental Impact Statements associated with the industry. Listings of available literature of all kinds are available through the bibliographic and directory service of the Australian Mineral Foundation and HERA (AHC). Current environmental research relating to mining, published and or unpublished was surveyed in 1991–92 (Woodside *et al.* 1992).

While the information on impacts and their reduction has been openly available for a long time, it was only recently that environmental scientists and managers from other sectors have taken the opportunity to participate in or scrutinize the work of the industry. Buckley (1989) reviewed the precision of resource project impacts as predicted in the EIA process in the context of a national environmental audit but as yet there is no systematic process at state level, let alone federally, to close the information loop between impact prediction and observed effects. There is great need for inter-sectoral co-operation in defining impacts, reducing them, repairing damage done (see summaries of 12 workshops relating to this issue in *Enviromine Proceedings;* Epps *et al.* 1992) and in systematically linking internal project monitoring as well as regulatory monitoring to development of predictive models.

In an attempt to work with the industry to develop standards for minesite reconstruction, to monitor direction and progress of ecosystem recovery after mining, the CSIRO Minesite Rehabilitation Research Programme studied the bauxite minesite at Gove, Northern Territory. At this operation, Nabalco Pty Ltd has implemented continuous effort toward restoration since the early 1970s. The programme provides a valuable long-term field experiment on the ways in which ecosystem process and directions are affected. Criteria used to compare progress of rehabilitation compared to surrounding sites included those which enabled assessment of relative productivity of the area, qualitative similarity, redistribution of nutrients and the ability to support native fauna (Reddell *et al.* 1992):

1. floristics and vegetation attributes

    species composition, numbers and dominance
    litter standing crop and nutrient content
    root distribution
    mycorhizal colonization.

2. soil biological characteristics

    microbial biomass
    soil animals.

3. soil properties (chemistry, structure, hydrology)

    descriptive soil features

        soil fertility (organic C, mineral and total N, pH, available and organic P, cations and trace elements)

        hydraulic properties (infiltration, bulk density, water holding capacity)

        mineralogy of whole sample and clay fractions.

The only fauna surveyed in this survey were soil invertebrates such as worms, termites and ant fauna. These animals are regarded as being critical to the flow of nutrients in northern Australia, to the redistribution and germination of seed and the fertility of soil (Reddell and Spain 1991).

Sand mining in coastal New South Wales presented the opportunity to other ecologists to observe the processes of recolonization of an area by plants and animals, particularly ants, lizards and mammals (Fox 1982; Fox and Fox 1978, 1982, 1984; Fox and Twigg 1991; Twigg and Fox 1991; Twigg *et al.* 1989). The results provide some insight into the question of how long it may take for the community of the disturbed area to recover to the point indistinguishable from its neighbouring habitat. Fox (1992) summarizes much of this information. The small mammals experience a slow continuous succession of species (Fox 1982; Fox and Fox 1978), the lizards over a period of approximately 20 years with a greater lag time before reinvasion (Twigg and Fox 1991) and ants show an early invasion but very slow shift toward the original community structure over a long period (Fox and Fox 1982; Fox 1990). It would appear the modification of the habitat in the form of faunal communities, vegetation structure and soil compaction continue for at least two to three decades. Assessments of post-mine impact and rehabilitation should therefore last for at least this period of time.

Two major issues arise from this work. Firstly, there is a need for standardization of the features used to measure the progress of success of habitat rehabilitation programmes. While some details may vary depending on the ecosystem being assessed, there should be sufficient commonality to enable comparison of rehabilitation across sites and to establish realistic objectives for most programmes. Secondly, the modern minesites (and some older ones) represent opportunities to study ecosystem process and recovery across a wide range of habitats with varying habitat features and varying degrees of impact and post mine interference.

The positive impacts of mining should also be accounted for when assessing overall threats and solutions in the management of biodiversity. Several of these have been suggested already. It is clear that mining brings with it all the community and economic benefits which are argued out in many forums. With respect to biodiversity issues and particularly habitat modification the benefits include new and long-term information about fixed geographical sites, listings of fauna, flora and geophysical features during exploration and mining which is entered into the national information bases held by NRIC (the National Resource Information Centre) and ERIN (the Environmental Resource Information Network), access to areas for study and often access to basic facilities, access to resources (financial, human and technological), access to a network of comparative research sites with geographical precision, access to long-term monitoring data enabling the study of global change and atmospheric pollution, new technology available for use in the rehabilitation of land degraded by agriculture, urbanization or natural disasters or for the enhancement of habitat for conservation of endangered species.

Opportunities exist, using mining activities and their rehabilitation, to gain information on habitat modification criteria, recovery and resilience of Australian ecosystems, succession and the potential for accelerating long-term natural processes. Given the infrastructural support provided by the mining industry for its exploration activities in remote areas, it should prove possible to develop a co-operative approach, even on a commercial basis, to include biologists on their teams and, given the observation powers of geologists, to have them undertake mapping and sampling in support of environmental objectives.

It is useful to look at the long-term sequential use of land following mining to assess both impact and benefits. A suitable summary exists for the use of land following mineral sands mining along Australia's east coast (Allen *et al.* 1990). Since 1934, mining for titanium and zirconium-bearing mineral sands occurred over 1 700 km of Australias east coast from Gladstone (Queensland) to Port Kembla (New South Wales). A survey of 109 localities, 215 mining operations showed that: 36 sites are now included in eight National Parks and seven Nature Reserves and 61 sites are used for recreation. Eight sites continue to be used for extractive purposes. The various uses to which this land has been subjected since mining include (from most to least common) recreation, urban, vacant Crown Land, conservation, agriculture, tourism, infrastructure and extraction, Aboriginal Land grants, forestry, defence and water supply. The conclusion was reached that maximum benefit is gained by planning the mining industry activities in conjunction with long term and sequential land use strategies of the region or relevant catchment.

ADEQUACY OF REGULATIONS FOR MINING

Legislation is the primary tool used for improving environmental practices and awareness in Australia. The mining industry is subject to regulation from an increasing number of laws in each state and territory. For example, Coronation Hill Joint Venture (CHJV) had to consider compliance with requirements of eight Federal and 23 Northern Territory Acts (CHJV 1990) in their original planning process. The state mining departments and environmental protection authorities set conditions on title and provide licenses to the mining industry with regard to rehabilitation and pollution for each of its activities and stages. Adherence is reviewed by regular inspection and through the establishment of periodic environmental management plans which are reviewed and endorsed by the regulatory authorities. The conditions of these licenses and approvals are responsive to the input of other agencies such as Conservation and Land Management (CALM) and water authorities particularly on issues pertaining to soil conservation and total catchment management. Where the Commonwealths jurisdiction is involved, such as foreign investment or export approvals, the Commonwealths environmental impact procedures can be triggered and subsequent approvals are contingent on meeting its standards. The most recent example of this was in relation to renewal of woodchip export approvals and the need to satisfy the requirements of the *Commonwealths Endangered Species Protection Act.*

Mining is one of the most regulated productive activities in Australia. From the industry's perspective it is over-regulated and the extent of regulation is growing. This view is not shared by everyone. If regulation is deemed the only form of control over the impact of mining on the environment, then as they now stand many regulations are questionable from both the perspective of technical adequacy and the perspective of appropriateness and effectiveness.

Where a prescriptive approach is imposed on land management and rehabilitation of the mining lease, the standards and regulations used may lag behind current knowledge of acceptable standards. This reflects some inflexibility in relation to decisions about appropriate end use and the need for community input and regional planning. Furthermore, undue haste to apply a prescriptive approach may have its environmental costs, such as the case in the 1950s of the New South Wales Soil Conservation Service dictating the use of the exotic bitou bush in dune rehabilitation and stabilization. Across Australia the perscriptions vary enormously with virtually no procedural standardization between States or regulatory jurisdictions and resulting in varying degrees of enforcement.

Fortunately, a growing number of new Australian regulations are not prescriptive and the interpretation of operational environmental management and rehabilitation procedures and standards is a mix of quantitative objectives and qualitative guidelines set out in regulations and negotiated between the operator and regulator, as well as voluntary practices developed by the operator in the interests of sound management.

There are "no prizes" (incentives) in the regulatory system for achieving "best environmental practice". In the future it is possible that economic instruments will be used to encourage more effective self-regulation, and creative solutions required to achieve desired environmental goals. Queensland, for example, in its EMOS (Environmental Management Overview Strategy) methodology, effectively provides discounts for the required level of security deposits if the regulator is satisfied with the operational and rehabilitation performance of the mining company. These discounts work in a similar way to no-claim bonuses on car insurance policies (Queensland Department of Minerals and Energy 1992).

There are other more important reasons for larger companies to be seen to "do the right thing". Major mining companies are in the business for the long haul and recognize that sound environmental performance is crucial to developing the community confidence required to gain approval for future projects. The major mining companies, in particular, are moving to improve their communication with local communities as part of ongoing project management. Previously comments from the community about project impacts were contributed through the EIA process or resulting public inquiries and were largely reactive to proposals submitted by project proponents.

In the preparation of any material pertaining to a mining activity, the legal obligation rests with the mining company (and its directors and staff) to present material that is technically accurate and truthful. No such obligation exists for the public or other bodies that wish to speak out against mining. Claims of habitat loss can be made by any member of the public with or without sound ecological information to back it up. Our legal system protects the rights of citizens to speak their minds and to demand answers to questions. Planning legislation in New South Wales provides for extensive rights to third parties to object to projects and to trigger public inquiries. Due diligence in pursuit of regulated environmental goals is the duty of the directors and staff of each mining operation. Failure to comply constitutes a crime and the director(s) and company officers are held accountable.

Why do so many environmentalists continue to feel that mining is inadequately controlled and that the results are not good enough? Are they separating the issues relating to small operators and large companies? Perhaps the problem rests as much with the regulatory system as with the level of compliance. Performance bonds are held against each operation and are not returned to the operator until rehabilitation processes have reached a satisfactory standard after mining. Perhaps the standards of rehabilitation are not seen as good enough, or perhaps low impact technology is not used frequently enough in the early extraction processes or the mining operator is not held responsible for the land rehabilitation process for long enough. Perhaps the problem lies in the early stages of planning and the need for greater community input to determining the land use in the area before and after mining.

All of these complaints may be valid given that there are no common standards for mine site rehabilitation except those established voluntarily by AMIC (1993b), no established criteria for monitoring progress of rehabilitation, and no requirement for the industry to monitor anything other than noise and pollution of the air and water. There is no requirement to monitor environmental variables beyond the boundary of the lease, but many companies do. Monitoring of the recovery of vegetation and faunal communities beyond the period of reconstructions prescribed by the licence is entirely voluntary. Such monitoring is done by companies which have made agreements with the local communities to do so or to improve their knowledge of the processes resulting in reclamation and restoration.

It should be noted that there is no systematic ecological monitoring by regulatory agencies in New South Wales following mining. Where ecological problems are anticipated, there may selective monitoring of water quality in relation to sedimentation and heavy metal contamination downstream from former mine sites. This monitoring does not extend to flora or faunal monitoring.

There are very few mines for which there exists a complete picture before and after analysis of the impact of mining. Where this has been done, such as at Olympic Dam in South Australia, the research is voluntary and collaborative.

For the most part the standard of care taken by the industry is that insisted upon by the community, not just the law. For this process to be valuable in terms of the ultimate environmental (and particularly biodiversity) outcomes the community must be well-informed and the mining companies should be actively encouraged to go beyond expectations to reach "best practice".

In the future, regulatory processes and objectives should become standardized where pollution threats are significant. The mining industry argues that this may require national standards that are not blanket numerical standards but are set in recognition of the diversity of the types of Australian habitats (Epps *et al.* 1992; AMIC 1993b). In addition the regulatory authorities and industry associations must seek to encourage self-regulation to a greater extent, encouraging good mine operators to pull others into line with higher voluntary standards and policies. The planning process must also be improved. By ensuring that all the interested and learned bodies are adequately consulted in the planning, evaluation and monitoring stages of mining and post-mining, the general standard of performance can only improve.

OUTSTANDING PROBLEMS OF THE 1990S

The outstanding environmental problems facing the mining industry include:

1. Development of objective and standardized criteria for the assessment of mining impact and reconstruction of habitat. In the process of striving to develop and improve these criteria everyone would benefit (see earlier discussion) as would the management of habitat.
2. Recognition of best environmental practices with respect to low impact technology and environmental repair and the transfer of the industry's habitat rehabilitation skills to other sectors. This would enable other landscapes and habitats to benefit from appropriate intervention and technology.
3. Some means of dealing with the conflicts between mining and other land uses, other than use of court rooms and formal inquiries, thus allowing more resources to go toward constructive environmental goals. Mining in national parks remains a source of great conflict and is likely to broaden as the issues of sequential and multiple land use are addressed more thoroughly.
4. Recognition of mining as part of a multiple or sequential land use strategy for a region or catchment rather than treating mining as an endpoint or as an isolated "island" in a sea of other land uses. This would enable better long-term environmental management and more effective use of resources.
5. The integration of mine site monitoring for water pollution, noise, atmospheric pollution and habitat rehabilitation into the network of scientific knowledge used by environmental professionals in other sectors.
6. Land access for exploration. There is a great deal of potential knowledge that can be gained about the Australian environment through exploration or in collaboration with exploration. The continual development of lower and lower impact methods for geological survey should be encouraged but is only useful if access is available.
7. Greater involvement of the community. The planning and evaluation of the mining operation with community participation ensures that the non-monetary benefits including some conservation benefits are fully supported and delivered.
8. Strategic approach by industry to minesite restoration and technology transfer. Currently vegetation reconstruction technology and knowledge is available only on a case by case, minesite by minesite basis. An opportunity exists for the industry to draw on a wide geographical area and a large number of Australian ecosystems to extract principles of habitat restoration and to use that knowledge strategically across all future mine operations or to transfer the technology to other sectors.

Most of the issues summarized above have been developed in the body of this paper. The outstanding issue is that of community involvement. This issue cannot be overlooked since the community position is one of fear for environmental degradation, awareness of the principal environmental issues and considerable influence over the decision to grant or withhold approval for a mining operation.

The experience of the mineral sands industry (Allen *et al.* 1990) shows that mine sites are readily absorbed by the community for other land uses, the most frequent use being that of recreation and conservation (38% of past mines). As a strategy to ensure the long-term protection of the habitat, whatever its use, it is important to involve the community in the design and implementation of programmes.

ACCEPTING THE VOICE AND INPUT OF THE COMMUNITY

In recent years the evolving detail of the environmental impact assessment process and changing community expectations about the social and environmental performance of resource projects have conspired to strengthen the voices of local people in influencing project planning and outcomes. There are now several well documented examples of community liaison processes undertaken to facilitate resource project approvals. Not all of these however continue to function beyond the start-up phase.

With mines having operational lives spanning a range of 3 to 25 or more years, it is sometimes difficult to envision end uses for the mine area during the start up phase. A mine's life is often extended beyond initial expectations through new discoveries, extension underground of open cuts and through changed economics or technology. Initial expectations of likely end use may require modification as a result of these events. Over time it is also possible that communities will change their preference for end use.

Those who were part of Broken Hill's start in the last century probably did not envision the greening of the city commenced during the 1930s; those who greened the city probably did not envision the tourist potential for the 1990s in retaining some old workings in their unrehabilitated state. No community or mine management can reasonably be expected to anticipate future trends beyond a few years at a time. In the worst of all worlds the anonymous regulator may decide end use because this has been prescribed in legislation without due consideration of the regional planning needs or growing need for habitat improvement programmes.

The message is clear. Local communities and relevant conservation agencies need to have an ongoing interaction with mine operations, not just to resolve day to day issues such as dust and noise control or truck movements, but to be an intimate part of the planning and decision processes which will determine desired end uses following closure. Sometimes these uses may have an ecological objective. The experience of titanium mineral sands mining in New South Wales suggests that recreational and ecological objectives may be combined satisfactorily.

Getting to this point will require innovative ways to get past historic barriers. From the corporate perspective it may require the courage to risk "letting go" and to develop listening skills as well as empathy for community needs and aspirations. These are not always expressed in economic terms. From the community perspective it may require a sense of pragmatism to identify the common ground where both qualitative or economic goals of the industry can be achieved as well as improvements in community services. In both cases it will be necessary to invest much time and energy from the earliest possible stages in a new project and to develop the persistence to see the relationship through to mine closure and beyond. Ironically, the mining industry displays these very characteristics in its exploration efforts. It needs to extend them to other areas of its operations.

CONCLUSIONS AND RECOMMENDATIONS

In reviewing an industry from the point of view of its threats and solutions in biodiversity conservation we are breaking new ground. It is relatively easy to substantiate arguments about the impact of industry or any human activity on natural ecosystems but we are not in the habit of looking to the same developers for solutions in conservation. With a more co-operative approach we can encourage new innovations in low impact technology to reduce initial impact, we can restrict the time and spatial aspects of development to assist in progressive restoration of impacted sites, we can improve the standard of habitat rehabilitation and devise technology and criteria useful in other sectors. Cairns (1993) looks at a number of social and management aspects of ecosystem restoration. He presents a sound case for a more strategic approach, involving all sectors, to ecosystem restoration with the benefit of replenishing our "national and global ecosystem capital".

It is important that ecologists and regulators foster dynamic relationships with responsible industry operators and gain all the information possible about their impact on broader environment features such as vegetation and fauna through joint research. Such co-operation encourages environmental investment by industry not now required by law. The rehabilitation of minesites and the reduction of impact in general presents ecologists with an opportunity to put principle and theories to the test. Precise location of sites of disturbance (each minesite) and the precise date of that disturbance is a treasury of opportunity for spatial and temporal studies.

Our recommendations for future directions are summarized in Figure 4. The ideas presented in this figure address, to a large extent, the outstanding environmental issues faced by Australian mining industry and summarize many of the recommendations made throughout this paper.

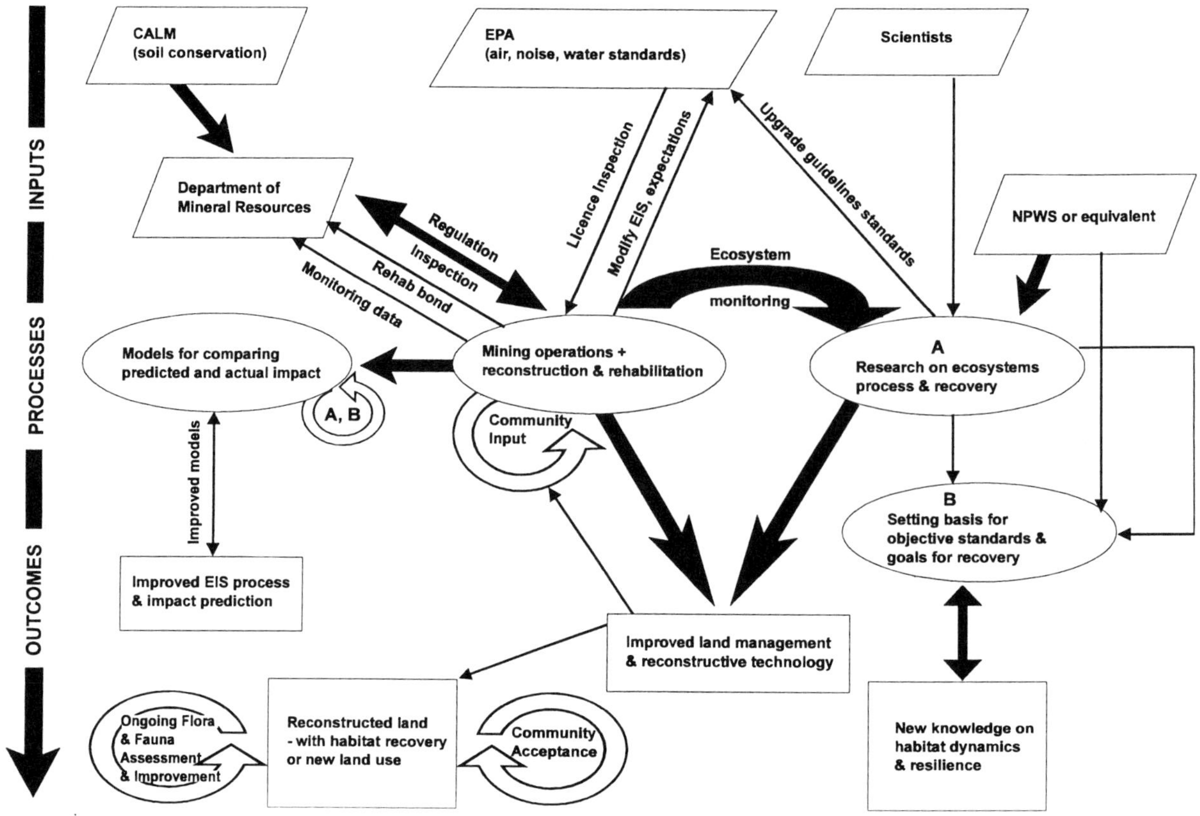

Fig. 4. Recommendations for relationships and information flow between the mining industry and other agencies or sectors with an indication of improved outcomes that would result. The need for stronger links (dialogue) are indicated by bold lines, dotted paragon boxes indicate inputs, ovals indicate processes and single lined squares indicate outcomes. Several improved information feedback loops are shown by circular arrows.

Our goal was to improve the environmental performance and output of the mining industry in Australia and to encourage the acceptance of its positive contributions to environmental management overall. This could only be done by looking at all those who officially and unofficially regulate mining and those who should be able to contribute to its performance with regard to the environment. We are particularly concerned with minimizing the impact on the habitat within the mining lease, repairing damage done to the habitat and gaining knowledge and technology that can be used elsewhere for environmental or specifically conservation purposes. To achieve this there must be a collaborative network of professionals. Information sharing and joint research must be undertaken without the backdrop of controversy such as that over Coronation Hill (Braithwaite and Woinarsky 1990). A proactive and positive approach by the mining industry to be involved in all aspects of the conservation planning process in this regard is summarized in a discussion paper put out by the Australian Mining Industry Council entitled Managing to Conserve Biodiversity (1994).

We relied on three basic principles while designing the relationships and outcomes of Figure 4 each aimed at developing a better regulatory and scientific environment for mining:

1. That the plan builds on existing relationships and mechanisms for control of information flow especially that relating to environmental variables.
2. That the new relationships lead to an enhanced capacity to manage and understand Australian habitats and predict impacts.
3. That all the players benefit.

In Figure 4 existing relationships and programmes are indicated by think double lines while the new relationships and programmes are indicated by thin single lines. The implementation of each of these suggestions is obvious and simple. Most begin with the need to build bridges for information flow and collaboration.

In summary we suggest that:

1. Ecologists be officially involved in the development of standards and guidelines for rehabilitation and that these grow out of collaborative programmes assessing ecosystem recovery process and resilience.

2. Targeted research be conducted on certain aspects of mining impacts. In particular long-term research should be established on the comparison between predicted impact (stated in EIS) and actually impacts followed by experimental research to improve the models used to predict the nature and degrees of impact on habitat or other environmental features. This should improve the EIS process and make regulatory supervision easier.
3. Environmental monitoring conducted by mining be more effectively pooled and integrated into a body of knowledge on Australian habitats, global change, etc., and that where possible this information be extended to include more direct ecological features such as vegetation and fauna.
4. Monitoring data from rehabilitation programmes should flow directly to a number of agencies such as CALM, NPWS or perhaps CSIRO which can compare results with other forms of environmental disturbance.
5. Better co-ordination between EPA (or equivalent), Land Information Systems and mining departments to ensure that critical statistics are reviewed annually across the state and nation (e.g., land currently held in mining leases, land surface disturbed annually in various categories, land rehabilitated to various land use categories).
6. The community should be more involved in planning and evaluation processes, acknowledging their environmental concerns and the need for their support for mining and subsequent land use, particularly conservation.
7. Greater integration of mining and post-mining restoration, reclamation and rehabilitation in the land use planning of a region, particularly looking at the roles of constructed habitats in regional conservation efforts.
8. A more strategic approach to rehabilitation of minesites and other land by the mining industry.
9. Rehabilitation technology and knowledge be the focus of a large and well funded effort involving environmental professionals from all the relevant sectors. This process has begun with the recent establishment of a Co-operative Research Centre for habitat rehabilitation involving CSIRO, a university and mining industry).
10. More formal efforts be made to share habitat rehabilitation skill and knowledge with other sectors and to export Australian skills.

There are a number of improvements and outcomes from the approach described above and in Figure 4. These include:

1. Better impact prediction models and more dynamic models.
2. Scientific/ecological input to design of mine programmes and habitat criteria.
3. Knowledge and technology gains especially regarding habitat dynamic, resilience and restoration.
4. More dynamic, less prescriptive approach to land restoration and to sequential land use.
5. More collaboration and better process with less legislation.
6. Community involvement at all levels and community benefit such as better regional planning for long-term land use and habitat conservation and management.

We would conclude by noting that mining cannot proceed without impacts on the habitat that surrounds it. The impact on the biodiversity of the region and on the overall quality of the habitat can be reduced through a suite of appropriate planning measures and through rehabilitation during and after mining. None of the efforts of the industry will meet the expectations of Australians if these are not spelled out clearly and objectively assessed. The goalposts may continue to shift as knowledge advances but implementation of our recommendations should reduce the speed and unpredictability of these shifts. This outcome would provide greater comfort and certainty of process and results to miners, ecologists and the community alike.

The way forward is to work with the industry in improving environmental standards and to make use of other ecological benefits and technology that can flow through more dynamic and co-operative research and management programmes.

ACKNOWLEDGEMENTS

We would like to thank the mineral and mines departments, or equivalent, in each state for sorting through files and data to answer our large information needs. We are grateful to Marilyn and Barry Fox for supplying large amounts of information about their research at such short notice and Nicole Lacey for locating some of the unpublished reports and case studies. We thank Graham Pyke for commenting on many earlier versions of this manuscript. We are mindful of the benefits of cross sectoral collaboration, particularly those fostered through Enviromine Australia in 1992, and feel encouraged to continue to explore common ground for biodiversity conservation between the two sectors we represent.

REFERENCES

Adamson, D. A. and Fox, M. D., 1982. Change in Australasian vegetation since European settlement. Pp. 109–46 in *A history of Australasian vegetation* ed by J. M. B. Smith. McGraw–Hill: Sydney.

Allen, E. B., 1988. Introduction. Pp. 1–4 in *The reconstruction of disturbed arid lands — an ecological approach* ed by E. B. Allen. American Association for the Advancement of Science Selected Symposium 109. Westview P~press. Inc.: Boulder, Colorado.

Allen, N. T., Brooks, D. R. and Jesson, J. G., 1990. Multiple land use and mineral sands mining on Australias east coast. Pp. 159–94 in *Australian Mining Industry Council Environmental Workshop 1990 Proceedings,* Volume 2. Wollongong, October 1990.

Armstrong, J., 1993. Restoration of function or diversity? Pp. 209–14 in *Nature conservation 3: reconstruction of fragmented ecosystems* ed by D. A. Saunders, R. J. Hobbs and P. R. Ehrlich. Surrey Beatty & Sons: Chipping Norton.

AUSLIG, 1990. *Atlas of Australian resources.* Third Series, Volume 6; Vegetation. Commonwealth Government Printer: Canberra.

Australian Environment Council, 1989. Acid rain in Australia: a national assessment. Report Number 25.

AMIC (Australian Mining Industry Council), 1992. Minerals Industry Survey Report.

AMIC (Australian Mining Industry Council), 1993a. Report on Mining Issues — Impediments to Trade.

AMIC (Australian Mining Industry Council), 1993b. Position Paper, National environmental standards — to suit and protect a diverse country.

AMIC (Australian Mining Industry Council), 1994. Managing to Conserve Biodiversity. (The Proposed National Reserve System — A Mining Industry Perspective). AMIC position paper.

Braithwaite, R. W. and Woinarsky, J. C. Z., 1990. Coronation Hill, Kakadu Stage III: assessing the conservation value. *Aust. Biol.* **3:** 3–13.

Bridgewater, P. B., 1990. The role of synthetic vegetation in present and future landscapes of Australia. *Proc. Ecol. Soc. Aust.* **16:** 129–34.

Bridgewater, P. B., 1992. Biodiversity — the Reality and its Management. Pp. 9–19 in *Seventeenth annual environmental workshop.* Australian Mining Industry Council: Canberra.

Buckley, R. C., 1989. Precision in Environmental Impact Prediction: First National Environmental Audit, Australia. CRES: Australian National University, Canberra.

Cairns, J. Jr., 1993. Ecological restoration: Replenishing our national and global capital. Pp. 193–208 in *Nature conservation 3: reconstruction of fragmented ecosystems* ed by D. A. Saunders, R. J. Hobbs and P. R. Ehrlich. Surrey Beatty & Sons: Chipping Norton.

Coronation Hill Joint Venture (CHJV), 1990. Coronation Hill. Coronation Hill Joint Venture submission to Resources Assessment Commission Kakadu Conservation Zone Inquiry. 44 pp.

Epps, J. E., Woodside, D. P. and O'Neill, D. R. (eds), 1992. *Enviromine Proceedings Volumes 1–3. Strategic Planning for Conservation Workshop Series.* Zoological Parks Board of New South Wales: Sydney.

Esser, D., 1992. Mining for an environmental future. Pp. 84–87 in *Seventeenth Annual Environmental Workshop.* Australian Mining Industry Council: Canberra.

Fox, B. J., 1982. The influence of disturbance (fire and mining) on ant and small mammal species diversity in Australian heathland. Pp. 213–19 in *Dynamics and management of Mediterranean-type ecosystems* ed by C. E. Conrad and W. C. Oechel. Gen. Teh. Rep. PSW-58, Pacific Southwest Forest and Range Experiment Station, Forest Service, US Department of Agriculture: Berkeley, California.

Fox, B. J., 1990. Two Hundred years of disturbance: how has it aided our understanding of succession in Australia? *Proc. Ecol. Soc. Aust.* **16:** 521–29.

Fox, B. J. and Fox, M. D., 1978. Recolonization of coastal heath by *Psuedomys novaehollandiae* following sand mining. *Aust. J. Ecol.* **3:** 447–65.

Fox, B. J. and Fox, M. D., 1984. Small mammal recolonization of open forest following sand mining. *Aust. J. Ecol.* **9:** 241–52.

Fox, B. J. and Fox, M. D., 1986. Resilience of animal and plant communities to human disturbance. Pp. 39–64 in *Resilience in Mediterranean-type ecosystems* ed by B. Dell, A. J. M. Hopkins and B. B. Lamont. Dr. W. Junk Publishers: Netherlands.

Fox, B. J. and Twigg, L. E., 1991. Experimental transplants of mice (*Pseudonyms* and *Mus*) on to early stages of post-mining regeneration in open forest. *Aust. J. Ecol.* **16:** 281–87.

Fox, M. D. and Fox, B. J., 1982. Evidence for interspecific competition influencing ant species diversity in a regenerating heathland. Pp. 99–110 in *Ant-plant interactions in Australia* ed by R. C. Buckley. Dr. W. Junk Publishers: The Hague.

Hobbs, R. J. and Hopkins, A. J. M., 1990. From frontier to fragments: European impact on Australias vegetation. *Proc. Ecol. Aust.* **16:** 93–114.

Hobbs, R. J. and Mooney, H. A., 1993. Restoration ecology and invasions. Pp. 127–33 in *Nature conservation 3: reconstruction of fragmented ecosystems* ed by D. A. Saunders, R. J. Hobbs and P. R. Ehrich. Surrey Beatty & Sons: Chipping Norton.

McPhail, I., 1992. Keynote Address. Pp. v in *Seventeenth annual environmental workshop.* Australian Mining Industry Council: Canberra.

National Academy of Sciences, 1974. *Rehabilition potential of western coal lands.* Ballinger Publishing Company: Cambridge, Massachusetts.

Queensland Department of Minerals and Energy, 1992. *Environmental Management for Mining in Queensland.*

Recher, H. F., 1993. The loss of biodiversity and landscape restoration: conservation, management, survival. An Australian perspective. Pp. 141–51 in *Nature conservation 3: reconstruction of fragmented ecosystems* ed by D. A. Saunders, R. J. Hobbs and P. R. Ehrich. Surrey Beatty & Sons: Chipping Norton.

Recher, H. F. and Lim, L., 1990. A review of current ideas of the extinction, conservation and management of Australias terrestrial vertebrate fauna. *Proc. Ecol. Soc. Aust.* **16:** 287–301.

Reddell, P. and Milnes, A. R., 1992. Mycorrhizas and other specialized nutrient-acquisition strategies: their occurrence in woodland plants from Kakadu and their role in rehabilitation of waste rock dumps. *Aust. J. Bot.* **40:** 223–42.

Reddell, P. and Spain, A. V., 1991. Earthworms as vectors of viable propagules of mycorrhizal fungi. *Soil Biol. and Biochem.* **23:** 767–74.

Reddell, P., Spain, A. V., Milnes, A. R., Hopkins, M., Hignett, C. T., Joyce, S. and Playfair, L. A., 1992. Bauxite mining and Walyamirri. The return of the living environment 3: Indicators of ecosystem recovery. Pp. 115–27 in *Seventh annual environmental workshop, Yeppoon.* Australian Mining Industry Council: Canberra.

Saunders, D. A., Arnold, G. W., Burbridge, A. A. and Hopkins, A. J. M. (eds), 1987. *Nature conservation: the role of remnants of native vegetation.* Surrey Beatty & Sons: Chipping Norton.

Saunders, D. A. and Hobbs, R. J. (eds), 1991. *Nature conservation 2: the role of corridors.* Surrey Beatty & Sons: Chipping Norton.

Saunders, D. A., Hobbs, R. J. and Ehrich, P. R. (eds), 1993. *Nature conservation 3: reconstruction of fragmented ecosystems.* Surrey Beatty & Sons: Chipping Norton.

Taylor, S. G., 1990. Naturalness: the concept and its application to Australian ecosystems. *Proc. Ecol. Soc. Aust.* **16:** 411–18.

Twigg, L. E. and Fox, B. J., 1991. Recolonization of regenerating open forest by terrestrial lizards following sand mining. *Aust. J. Ecol.* **16:** 137–48.

Twigg, L. E., Fox, B. J. and Jia, L., 1989. The modified primary succession following sand mining: A validation of the use of chronosequence analysis. *Aust. J. Ecol.* **14:** 441–47.

van Pelt, M. J. F., 1993. Ecologically sustainable development and project appraisal in developing countries. *Ecol. Econ.* **7:** 19–42.

Wamsley, J., 1992. The Business of Endangered Species. Pp. 20–26 in *Seventeenth annual environmental workshop.* Australian Mining Industry Council: Canberra.

Webb, G. J. W., 1990. Mining and environmental management — some current trends. Report to the Australian Mining Industry Council.

Williams, D. J., Carras, J. N., Milne, J. W. and Heggie, A. C., 1981. The oxidation and long-range transport of sulphur dioxide in a remote region. *Atmos. Environ.* **15:** 2255–62.

Woodside, D. P., Brett, A., Chapple, R. and Oxley, G., 1992. A catalogue of environmental information resources and a directory of environmental research relevant to the minerals and energy industries. Volume 4 in *Enviromine Australia proceedings* ed by J. E. Epps, D. P. Woodside and D. R. O'Neill. Zoological Parks Board of New South Wales: Sydney.

Zarsky, L., 1990. A sustainable future for Australia. Pp. 25–65 in *Our common future.* Aust. edition. Oxford University Press: Melbourne.

The responses of some selected arthropods and the frog *Crinia signifera* to habitat fragmentation

C. R. MARGULES,[1] K. F. DAVIES,[2] J. A. MEYERS[1] and G. A. MILKOVITS[1]

The habitat fragmentation paradigm, a widely accepted generalization, goes like this. Fragmentation creates sub-populations of species, isolates those sub-populations from one another and disrupts ecological processes thereby increasing the risk of extinction. Smaller populations are more likely to suffer local extinction and more isolated fragments are less likely to be recolonized if local extinction does occur.

However, not all species conform. An important task for ecologists is to identify those which do and are therefore vulnerable to extinction in fragmented landscapes, so that management resources can be directed most effectively. As part of a long-term field experiment at Wog Wog, New South Wales, this study reports on the selected taxa *Cercophonius squama* (Scorpionida), an amphipod, Family Tallitridae, a millipede, Family Spirobolellidae, eight carabid beetles and the frog *Crinia signifera.*

Preliminary results yielded a variety of responses ranging from increases in abundance through no response at all to dramatic decreases in abundance and local extinction and recolonization. While it is still not possible to generalize the responses of species to habitat fragmentation, it is clear that different species respond differently and therefore management should be directed at populations of selected vulnerable species.

INTRODUCTION

AGRICULTURAL development in Australia has wrought profound ecological changes. One of the most conspicuous is the fragmentation of natural habitat as vegetation is cleared to make way for crops and pasture. The resulting landscape contains isolated remnants of natural habitat varying in size and distance from one another. Most are found on soils or in parts of the landscape less suitable for agriculture. Species occupying these remnants form sub-populations of previously more widespread populations. Some may be meta-populations, sets of sub-populations which exchange individuals, while others are completely isolated, forming what amounts to new much smaller discrete populations.

There can be no doubt that habitat fragmentation has caused a loss of biological diversity and if it were to cease, the rate of loss would be slowed. Over many regions, however, habitat fragmentation is now a reality. In these areas, conservation of the regional biota depends on retaining the remnants and managing them to minimize the risk of extinction (Saunders *et al.* 1991). Concomitantly, resources for conservation management are limited and should therefore be directed to where they can make the greatest contribution to sustaining regional biological diversity. Exactly where that is, is a central problem in conservation biology and there are no clear answers.

Generalizations from theory have not helped. The equilibrium theory of island biogeography (MacArthur and Wilson 1967), which predicts that the number of species on islands is a function of island area and distance to a source of colonizing individuals, was used by, for example, Diamond (1975) and Wilson and Willis (1975) to propose a set of reserve "design rules" aimed at maximizing the number of species that could be encompassed and maintained on reserves. They argued for large reserves over small ones; for single large reserves over groups of small ones of more or less the same total area; for reserves being close together rather than far apart; for them to be connected if possible; and for them to be compact — more or less circular — rather than long and thin. These last two are not based on the equilibrium theory as such, but on implications drawn from it that connection will facilitate recolonization if it is needed and that the effective area of a habitat remnant is diminished by the length of its edge because edges are different in character from the central core area.

Arguments for and against these rules have been prominent in the literature ever since (see e.g., Simberloff and Abele 1976; Diamond 1976; Higgs 1981; Higgs and Usher 1980; Kitchener *et al.* 1980a,b; Kitchener *et al.* 1982; Margules *et al.* 1982). Haila (1986) has made the most thoughtful contribution, suggesting that the theory should be seen as a research agenda that draws attention to the dynamic nature of isolated biotas rather than as a mechanistic explanation (see also Main 1961 and Main 1987).

Part of the problem is the broad scope of the theory and its predictions of species richness based on observational data from islands varying

[1]CSIRO Division of Wildlife and Ecology, P.O. Box 84, Lyneham, Australian Capital Territory, Australia 2602.
[2]Division of Botany and Zoology, School of Life Sciences, Australian National University, Canberra, Australian Capital Territory, Australia 0200.
Pages 94–103 *in* CONSERVING BIODIVERSITY: THREATS AND SOLUTIONS, ed by R. A. Bradstock, T. D. Auld, D. A. Keith, R. T. Kingsford, D. Lunney and D. P. Sivertsen. Surrey Beatty & Sons. 1995.

in size by orders of magnitude. Generalizing to fragmented landscapes is probably not appropriate because in trying to maintain regional biotas, managers need to know which species are in which remnants and why they are there, not just how many there are. They also need to know whether or not it is possible to generalize about the responses of species to habitat fragmentation and whether or not those responses can be categorized using taxonomic or ecological relationships.

The equilibrium theory makes no predictions about the responses of particular species to habitat fragmentation, and habitat alteration, except for a reduction in island size, is not dealt with at all. Also, the reality is that habitat fragmentation is not a process that can be controlled. The option to choose, say, between a single large remnant or group of smaller remnants of the same total area rarely, if ever, exists.

Thus, management resources should not necessarily go to large remnants rather than small ones or groups of small ones. It depends on which species are present in which remnants. Neither should management be distracted by an academic debate on reserve design when there is no hope of implementation even if there is resolution. Positive attempts to manage biological diversity in fragmented landscapes will have to target vulnerable species.

This study reports on a forest field experiment begun in 1985 at Wog Wog, New South Wales. It was established in response to the research agenda set by the equilibrium theory of island biogeography. Its objectives were twofold. First, to test a prediction of the theory, that smaller habitat remnants retain fewer species than larger remnants, and second, to identify those species and their ecological characteristics most vulnerable to habitat fragmentation.

The responses detected so far reveal complex interactions between the variables time and spatial heterogeneity, and the experimental components, remnant size, edges and treatment or control.

THE WOG WOG HABITAT FRAGMENTATION EXPERIMENT AND METHODS

Experimental design

The experimental design is described in detail by Margules (1992). Briefly, 18 plots consisting of three sizes replicated six times were established in *Eucalyptus* forest in the summer of 1984/85. The sizes were 0.25 ha, 0.875 ha and 3.062 ha. Thus, the larger ones were 3.5 times the size of the smaller ones. Four replicates (12 forest fragments) were retained when the surrounding forest was cleared to establish a *Pinus radiata* plantation in 1987. Two replicates (six plots) remained as controls in adjacent continuous forest. Figure 1 is a map of the experiment.

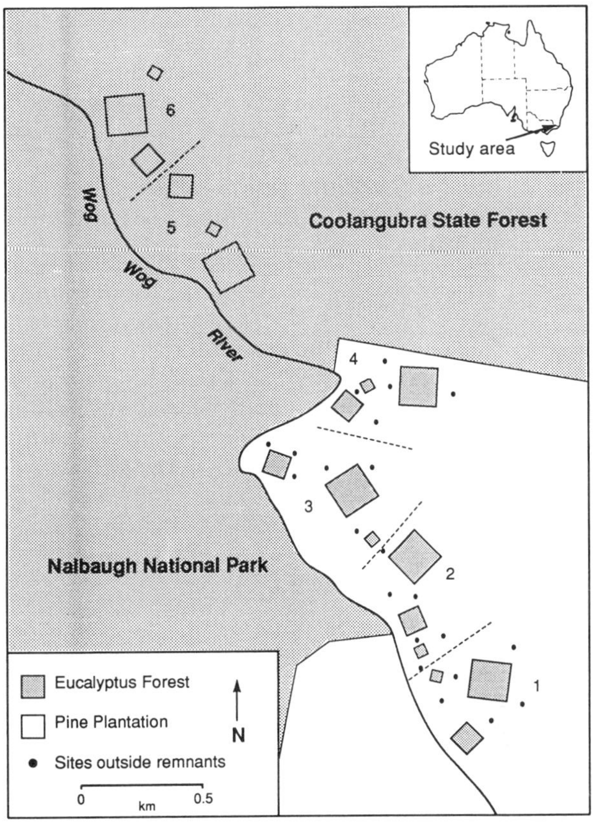

Fig. 1. A map of the experimental site showing its general location and the forest remnants and control plots in continuous forest. The dots represent the approximate locations of a pair of monitoring sites (a slope site and a drainage line site) established outside the remnants after the fragmentation treatment was completed.

Monitoring was stratified into slopes and drainage lines and inner and outer zones, and was replicated at the lowest stratum. Thus, in each fragment or plot in continuous forest there were two outer slope monitoring sites, two inner slope sites, two outer drainage line sites and two inner drainage line sites for a total of 144 permanent monitoring sites. Following the experimental treatment, a further 44 sites were established between the fragments (Fig. 1).

Ground dwelling arthropods and very small vertebrates were chosen for the study for two reasons. First, they can be sampled in a more consistent way than other groups such as flying arthropods, which may be prone to fluctuations throughout the day due to temperature changes, wind condition, etc. Second, they are easier to sample than, say, canopy insects or arboreal vertebrates. The animals were collected from a pair of permanent pitfall traps at each monitoring site, opened for seven days once each season; four times each year.

Monitoring began in February 1985. The experimental treatment took place over nine months between April and December 1987. This consisted of extracting suitable pulpwood and leaving the 12 forest fragments in a "sea" of stumps and crowns and scattered saplings. In the summer of 1988 this residue was stacked and burnt. In the winter of 1988 the pine plantation was established.

Preliminary results for the amphipod, Family Tallitridae, the scorpion *Cercophonius squama,* eight carabid beetles, a millipede Family Spirobolellidae and the frog *Crinia signifera,* are described below. More detailed analyses of the amphipod and scorpion responses can be found in Margules *et al.* (1994) where details of procedures for data analysis and statistical modelling can also be found.

RESULTS

General trends

Abundances are variable both in space and over time. Figures 2 and 3 show this variation for the amphipod and the scorpion *Cercophonius squama* (Fig. 2), and the millipede (Fig. 3) measured from monitoring sites in the controls, so there is no effect of experimental treatment. Figure 4 shows a range of responses to the experimental treatment inferred from an examination of the beetle data over four years. Species with similar pre-treatment temporal abundance patterns appear to respond differently after treatment and the same species may respond differently in different seasons. The degree of variability shown in Figures 2, 3 and 4 poses difficult problems for the analysis of experimental results and highlights the crucial role of replication and control. Spatial and temporal variability have to be accounted for otherwise natural variation and treatment effects would be confused and it would not be possible to distinguish responses due to one or the other.

Detailed responses of selected taxa

The amphipod, Family Tallitridae

Margules and Meyers (in press) and Margules *et al.* (1994), have shown that the response of the amphipod is a dramatic decrease in overall abundance and a contraction in the forest fragments to a few most favourable sites. Figure 5 shows the trends in amphipod abundance over eight years, three pre-treatment and five post-treatment. There are two peaks in both the fragments and controls, one just prior to the treatment and one beginning two years after and continuing until year eight. Abundances were similar in both fragments and controls before the treatment but there are much higher numbers in the controls after treatment.

Table 1 is a summary of amphipod abundance in year seven, the year with greatest numbers, showing that most sites in the controls had high or very high numbers but that high numbers in the fragments were restricted to only six sites. Examination of these six revealed that five of them are drainage line sites in the inner zone of the largest fragments. The sixth is a drainage line site from the outer zone of a small fragment. All of them were classified by Austin and Nicholls (1988) as grassy mature *Leptospermum* thickets. These thickets probably provide a unique micro-climate, which constitutes the most favourable habitat and therefore provides a refuge in the fragmented landscape represented by this experiment.

The scorpion Cercophonius squama

In contrast to the amphipod, there is no detectable response from the scorpion *C. squama,* five years after fragmentation. Figure 6 shows histograms of the coefficients of a statistical model (multiple Poisson regression, residual mean deviance = 1.63) to estimate abundances in different sized remnants, in different replicates, in treatments and controls before and after treatment (from Margules *et al.* 1994). The pattern is the same both before and after fragmentation. Even though estimates were marginally higher after treatment, they were higher in the controls as well.

The frog Crinia signifera

The response of the frog *C. signifera,* is different again. This species is uncommon on the experimental site, but is most likely to be found in drainage lines in summer. Thus, for analysis, the data were grouped into two-year periods and only records from inner drainage line monitoring sites in summer were used. Figure 7 summarizes a statistical model fitted to these data (multiple Poisson regression, residual mean deviance = 0.44) to estimate the abundance of *C. signifera* in different sized remnants, in treatments and controls, before and after treatment. Abundance fluctuated in the controls over the eight summers but in the fragments this species disappeared completely for the four summers following fragmentation, re-appearing again in the two years following that. *C. signifera* became extinct following fragmentation, but managed to recolonize the fragments after four years. *C. signifera* is commonly found in disturbed habitat so this recolonization was not unexpected.

Eight beetles of the Family Carabidae

Davies (1993) analysed the data for eight Carabid beetles over eight years, three pre-treatment and five post-treatment. She found that three increased in abundance or probability

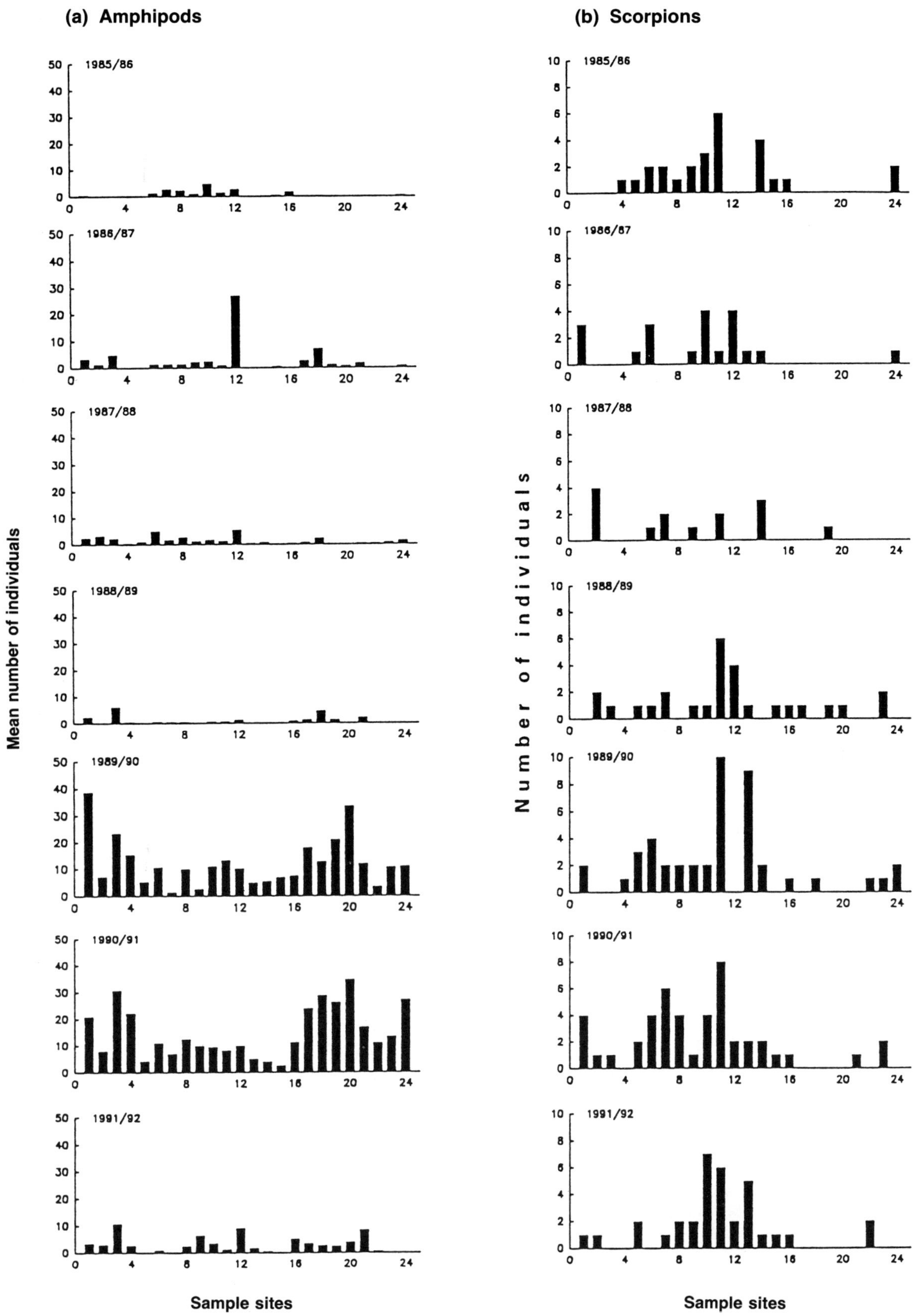

Fig. 2. Histograms of the abundance of the amphipod *Arcitalitrus sylvaticus*, and the scorpion *Cercophonius squama*, from the 24 monitoring sites on slopes in the controls over seven years. Note that the scales on the y-axis are different.

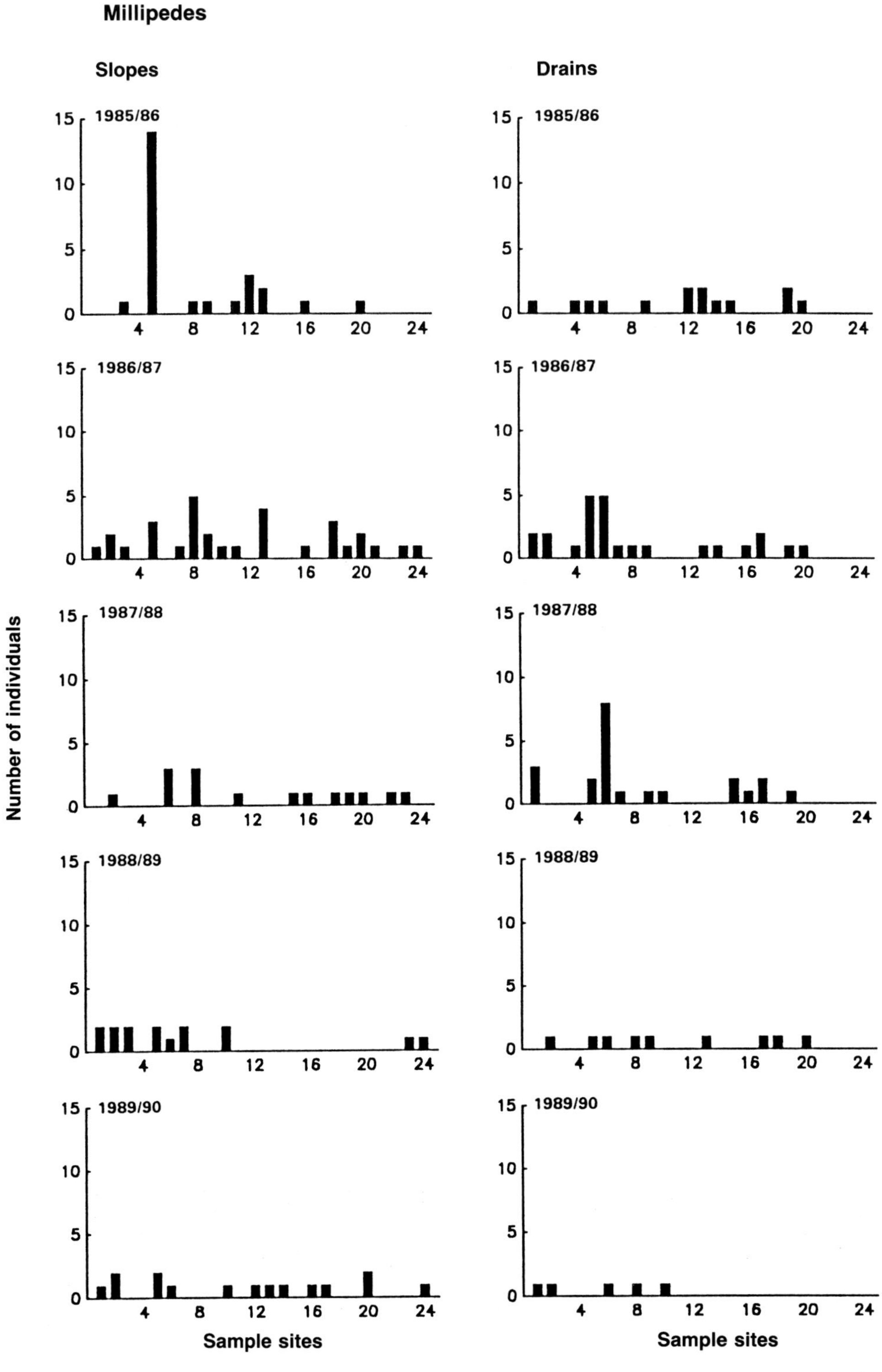

Fig. 3. Histograms of the abundance of a millipede, Family Spirobolellidae, from the 24 monitoring sites on slopes and in drainage lines from the controls over five years.

Table 1. The number of monitoring sites with large (>100) or small (<10) numbers of amphipods in year seven, October 1990 to July 1991, in the remnants and controls.

Number of individuals	Remnants No. sites	Remnants %	Controls No. sites	Controls %
≥100	6	12.5	9	37.5
>10 <100	20	41.5	15	62.5
≤10	22	46	0	0

of occurrence, three decreased in abundance or probability of occurrence, and two showed no response at all. Table 2 is a summary. Interpretation of these results is constrained by the very limited knowledge available of the biology and ecology of each species. However, the contrasting responses are somewhat surprising as these beetles share some distinctive life history characteristics, being flightless terrestrial

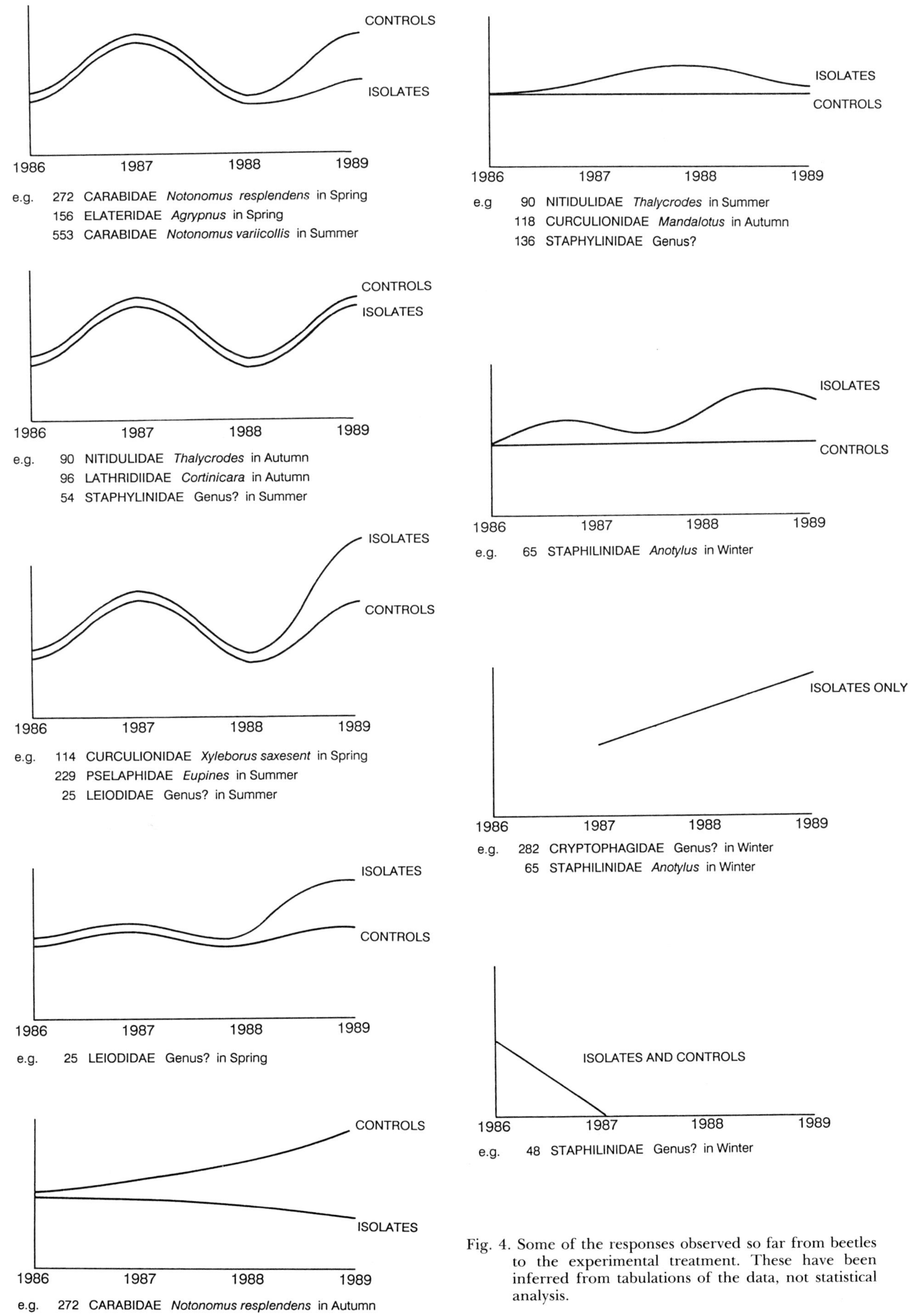

Fig. 4. Some of the responses observed so far from beetles to the experimental treatment. These have been inferred from tabulations of the data, not statistical analysis.

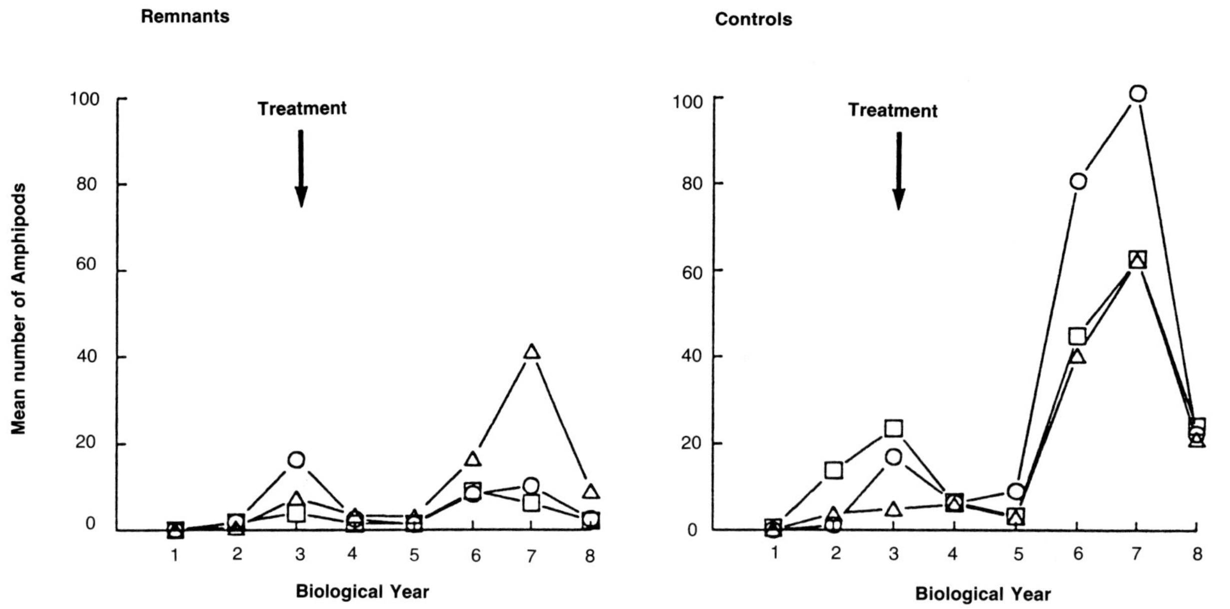

Fig. 5. Mean numbers of the amphipod, Family Tallitridae, in each remnant size class (or control plot in continuous forest) in each year. Circle = small, square = medium, triangle = large.

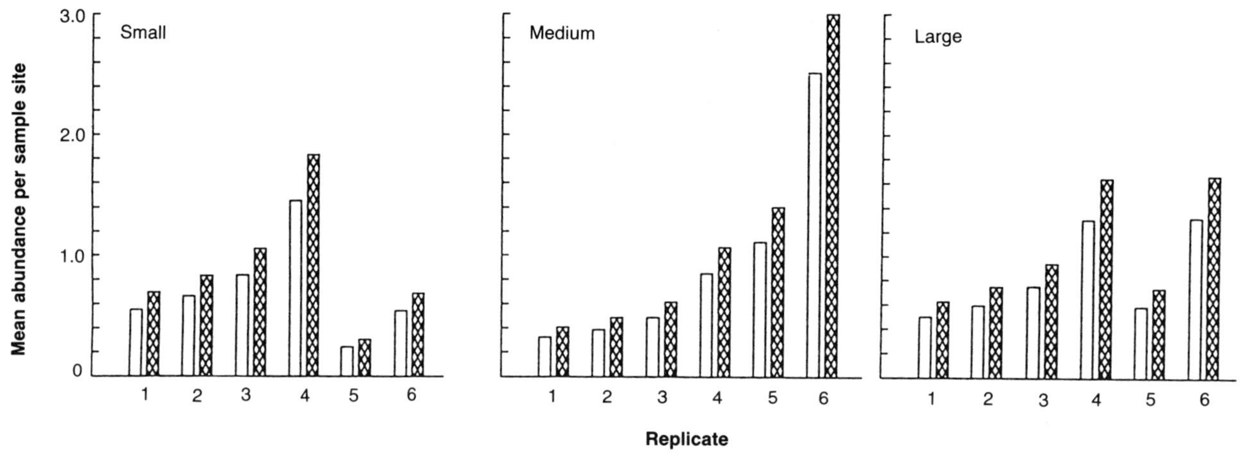

Fig. 6. Histograms of the estimated abundance of the scorpion *Cercophonius squama,* on slopes in each replicate from the coefficients of a statistical model relating observed abundance to the variables, remnant size, replicate and treatment or control. Blank bars are before treatment and cross-hatched bars are after treatment. Replicates 1, 2, 3 and 4 are the fragments and 5 and 6, the controls.

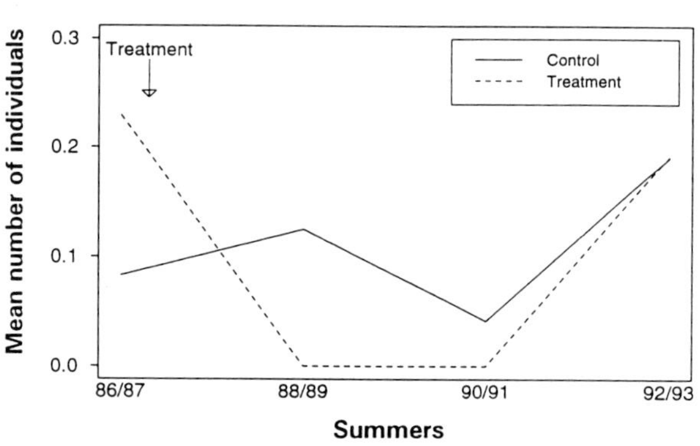

Fig. 7. Estimated mean numbers of the frog *Crinia signifera,* in inner drainage lines in summer from coefficients of a statistical model relating observed abundance to the variables remnant size, treatment or control and time.

Table 2. The responses of eight Carabid beetles in the forest fragments five years after fragmentation.

Species	Response		
	Increase	Decrease	No change
Notonomus resplendens		+	
Notonomus variicollis	+		
Notonomus minimus		+	
Notonomus metallicus		+	
Eurylychnus blagravei	+		
Promecoderus sp.			+
Carenum sp.	+		
Helluo sp.			+

predators that live for some years (Moore *et al.* 1987). Further, four of them belong to the same genus, yet one has increased whereas the other three have decreased.

DISCUSSION

The species from this habitat fragmentation experiment examined in detail so far exhibit contrasting responses. Some increase in abundance or probability of occurrence, some decrease and yet others show no response. The frog *C. signifera*, appears to have gone extinct on the fragments but recolonized them after four years.

Intuitively, the amphipod and the frog might be expected to suffer from habitat fragmentation. Both require a moist environment to avoid desiccation. Frogs lose water through a permeable skin when atmospheric humidity is lower than internal body humidity. They can only survive for a few hours in a dry atmosphere (Tyler 1976). They breed in slow flowing creeks or pools and the tadpoles are aquatic. Amphipods are crustaceans that have only recently colonized land. They live in and feed on decaying litter of the forest floor. Densities of some species have been correlated with litter thickness (Duncan 1969). They are also susceptible to drowning and so must avoid local flooding (Friend and Richardson 1986). A physical consequence of habitat fragmentation is an increase in fluxes of wind, water and solar radiation. Thus, while more water may enter a fragment via run-off when it does rain, it will also dry up quickly, and be drier between rains than it was previously.

With hindsight, the response of the scorpion *C. squama*, is also understandable given what is known of its biology and ecology. It is a nocturnal, sedentary, territorial predator living at least two or three years and probably more (G. T. Smith, pers. comm.). It has a relatively hard exoskeleton and a burrowing habit which should make it less susceptible to desiccation.

An understanding of the responses of the Carabid beetles is more problematic. It could be expected that such biologically similar animals would respond in similar ways. The fact that they do not suggests an ecological, rather than a biological, explanation and there is even less known of their ecology than their biology. It is conceivable, for example, that the niches of the four *Notonomus* spp. are narrow and similar and defined by competition. If the physical effects of habitat fragmentation constrain, for any reason, the competitive ability of the three species in Table 2 that are decreasing, the one species increasing may be expanding its niche and therefore its abundance at the expense of the other three. This is speculation and an adequate explanation still awaits a detailed study of these species. The lesson is that the same responses cannot be assumed from taxonomically and even ecologically similar species. This may have significant implications for population viability analyses if a correspondence between population dynamic processes and taxonomic relationship is assumed.

Population viability analyses will also have to take into account the reality that habitat fragments are not simply small versions of the previous continuous habitat. Continuous habitat is heterogeneous anyway and fragmentation isolates pieces that are to some extent different from each other, some dramatically and some more subtly. There appear to be complex interactions between the physical effects of habitat fragmentation, ecological relationships such as competition and predation, and the environmental conditions pertaining at the time of fragmentation. Observed responses may be highly sensitive to initial conditions and may therefore be different if those conditions were to be different. Similarly, the initial response may not be the long-term response. Just as initial extinction may be followed by re-colonization, initial increases in abundance could be followed by longer term decreases.

These various responses preclude generalizations at this stage except the common sense ones that some species are more susceptible to habitat fragmentation than others, and that some actually benefit. Usher (1987), Robinson *et al.* (1992), Sisk and Margules (1993), Simberloff (1993) and Margules and Meyers (in press) have stressed the point that management of biological diversity in fragmented landscapes should be concerned with the fate of populations and sub-populations, rather than counts of the number of species (Margules *et al.* 1994) or theoretical reserve design rules.

However, this does not mean that generalizations will never be possible. Attention in the Wog Wog experiment is focusing more now on categorizing species responses than testing predictions from the equilibrium theory of island biogeography. This is being done in an effort to identify similarities based on ecological roles; that is, "functional response groups". If responses can be categorized in terms of ecological function, then predictions of the responses of other species in other locations can be made and tested.

Saunders *et al.* (1991) emphasized the likely consequences of changes in micro-climate within habitat remnants. Habitat remnants are more exposed to the drying agents wind and solar radiation. They are also more likely to suffer flooding when it rains because run-off in the catchment above them has been increased by clearing. Remnants are, in turn, both drier and wetter than continuous forest. The reduction in both range and abundance of amphipods in this experiment to date suggests a response to increased local climatic fluctuations of this kind. Amphipods require a moist environment because they are susceptible to desiccation, but they are also susceptible to drowning. One clear management implication is this. If possible, ameliorate the increased local climatic fluctuations experienced by remnants. Two possible management strategies might be to maintain buffer zones, perhaps including new plantings of indigenous species, and to maintain the undergrowth as well as the overstorey.

Management resources are always limited and there is a tendency to overlook the importance of monitoring, which can be expensive, time-consuming and labour-intensive. But if this experiment had lasted only six years, we might have concluded that the frog *Crinia signifera,* had become extinct. Similarly, if there were no controls and no pre-treatment records, we might conclude from Figure 6 that the scorpion *Cercophonius squama,* favoured small remnants because, on average, there are more of them in the small remnants than in the medium or large remnants. However, since the pattern is identical before and after treatment and numbers are spatially variable in the controls it would appear that there is no effect of fragmentation so far. Effective management of landscapes with fragmented habitat must include the careful design and implementation of field experiments and programmes to monitor the consequences of management practices.

ACKNOWLEDGEMENTS

We are grateful to John Lawrence for identifying the beetles and maintaining the beetle collection, to Geoff Baker for maintaining the millipede collection, to Graham Smith for identifying the scorpion and providing information on its ecology, and to Fiona Gell and Natalie Barnett for laboratory and field assistance. Barbara Main and an anonymous referee both greatly improved the manuscript with their constructive criticism.

REFERENCES

Austin, M. P. and Nicholls, A. O., 1988. Species associations within herbaceous vegetation in an Australian Eucalypt forest. Pp. 95–114 in *Diversity and pattern in plant communities* ed by H. J. During, M. J. A. Werger and J. H. Willems. SPB Academic Publishing: The Hague.

Davies, K. F., 1993. The effects of habitat fragmentation on Carabid Beetles (Coleoptera: Carabidae) at Wog Wog, south-eastern New South Wales. Unpublished Honours Thesis, Division of Botany and Zoology, Australian National University.

Diamond, J. M., 1975. The island dilemma: lessons of modern biogeographic studies for the design of natural reserves. *Biol. Conserv.* **7:** 129–46.

Diamond, J. M., 1976. Island biogeography and conservation: strategy and limitations. *Science* **193:** 1027–29.

Duncan, K. W., 1969. The ecology of two species of terrestrial Amphipoda (Crustacea: Family Talitridae) living in waste grassland. *Pedobiologia* **9:** 323–41.

Friend, J. A. and Richardson, A. M. M., 1986. Biology of terrestrial amphipods. *Ann. Rev. Entomol.* **31:** 25–48.

Haila, Y., 1986. On the semiotic dimension of ecological theory: the case of island biogeography. *Biol. Philo.* **1:** 377–87.

Higgs, A. J., 1981. Island biogeography theory and nature reserve design. *J. Biogeog.* **8:** 117–24.

Higgs, A. J. and Usher, M. B., 1980. Should nature reserves be large or small? *Nature, Lond.* **285:** 568–69.

Kitchener, D. J., Chapman, A., Dell, J. and Muir, B. G., 1980. Lizard assemblage and reserve size and structure in the Western Australian wheatbelt — some implications for conservation. *Biol. Conserv.* **17:** 25–62.

Kitchener, D. J., Chapman, A., Muir, B. G. and Palmer, M., 1980. The conservation value for mammals of reserves in the Western Australian wheatbelt. *Biol. Conserv.* **18:** 179–207.

Kitchener, D. J., Dell, J., Muir, B. G. and Palmer, M., 1982. Birds in Western Australian wheatbelt reserves — implications for conservation. *Biol. Conserv.* **22:** 127–63.

MacArthur, R. H. and Wilson, E. O., 1967. *The theory of island biogeography.* Princeton University Press, Princeton: New Jersey.

Main, A. R., 1961. The occurrence of Macropodidae on islands and its climatic and ecological implications. *J. Roy. Soc. WA* **44:** 84–89.

Main, A. R., 1987. Management of remnants of native vegetation: a review of the problems and the development of an approach with reference to the wheatbelt of Western Australia. Pp. 1–13 in *Nature conservation: the role of remnants of native vegetation* ed by D. A. Saunders, G. W. Arnold, A. A. Burbidge A. J. M. and Hopkins. Surrey Beatty & Sons: Chipping Norton.

Margules, C. R., 1992. The Wog Wog habitat fragmentation experiment. *Environ. Conserv.* **19:** 316–25.

Margules, C. R. and Meyers, J. A., (in press). Biological diversity and ecosystem fragmentation: an Australian perspective. *Ekistics.*

Margules, C. R., Higgs, A. J. and Rafe, R. W., 1982. Modern biogeographic theory: are there any lessons for nature reserve design? *Biol. Conserv.* **24:** 115–28.

Margules, C. R., Milkovits, G. A. and Smith, G. T., 1994. Contrasting effects of habitat fragmentation on the scorpion *Cercophonius squama* and the amphipod. *Ecology* **75**(7): 2033–042.

Moore, B. P., Weir, T. A. and Pyke, J. E., 1987. Rhysodidae and Carabidae. *Zoological Catalogue of Australia.* Volume 4. Coleoptera: Archostemata, Myxophaga and Adephaga. Australian Government Publishing Service: Canberra. Pp. 20–320.

Robinson, G. R., Holt, R. D., Gaines, M. S., Hamburg, S. P. Johnson, M. J., Fitch, H. S. and Martinko, E. A., 1992. Diverse and contrasting effects of habitat fragmentation. *Science* **257:** 524–26.

Saunders, D. A., Hobbs, R. J. and Margules, C. R., 1991. Biological consequences of ecosystem fragmentation: a review. *Conserv. Biol.* **5:** 18–32.

Simberloff, D., 1993. Effects of fragmentation on some Florida ecosystems, and how to redress them. Pp. 179–87 in *Nature conservation 3: reconstruction of fragmented ecosystems* ed by D. A. Saunders, R. J. Hobbs and P. R. Erlich. Surrey Beatty & Sons: Chipping Norton.

Simberloff, D. and Abele, L. G., 1976. Island biogeography theory and conservation practice. *Science* **191:** 285–86.

Sisk, T. D. and Margules, C. R., 1993 Habitat edges and restoration: methods for quantifying edge effects and predicting the results of restoration efforts. Pp. 57–69 in *Nature conservation 3: reconstruction of fragmented ecosystems* ed by D. A. Saunders, R. J. Hobbs and P. R. Erlich. Surrey Beatty & Sons: Chipping Norton.

Tyler, M. J., 1976. *Frogs*. Collins: Sydney.

Usher, M. B., 1987. Effects of fragmentation on communities and populations: actions, reactions and applications to wildlife conservation. Pp. 103–21 in *Nature conservation: the role of remnants of native vegetation* ed by D. A. Saunders, G. W. Arnold, A. A. Burbidge and A. J. M. Hopkins. Surrey Beatty & Sons: Chipping Norton.

Wilson, E. O. and Willis, E. O., 1975. Applied biogeography. Pp. 522–34 in *Ecology and evolution of communities* ed by M. L. Cody and J. M. Diamond. Harvard University Press: Cambridge, Massachusetts.

Habitat reconstruction: the revegetation imperative

D. A. SAUNDERS[1] and R. J. HOBBS[1]

Widespread loss of native vegetation for a variety of reasons (agriculture, forestry, urbanization, etc.) has led to the decline and extinction of many species of native plants and animals. The full extent of these extinctions is not known, except for a few groups like the mammals. Increasing evidence now shows that other groups (e.g., birds and frogs) are following the same declining trends as the mammals.

Habitat loss is not the only factor involved in the demise of native species. Other factors include: fragmentation of remaining habitat, introduction of exotic species, and competition with native species that have invaded areas changed by European management practices. But in many areas, particularly Australia's extensive wheat-sheep zone, given the extent of the loss of native vegetation, habitat loss is the major factor and one which needs addressing rapidly if we are to halt the steady species decline.

If we examine Australia's agricultural areas where the most severe threats to biological diversity currently occur, and where the most pressing needs for habitat reconstruction exist, much of our current approach to conservation will fail unless we integrate nature conservation with agricultural production. Habitat reconstruction models must be developed within the total productive landscape. These models must address revegetation (the primary tool for habitat reconstruction) to provide habitat for a range of species, as well as addressing problems generated by agricultural production being carried out in the landscape (minimizing wind erosion, lowering water tables, etc.). Scale and context are major problems in relation to habitat reconstruction. If we develop reconstruction strategies for single species in small areas like conservation reserves they will inevitably fail unless we also address problems in the surrounding landscape and develop management strategies designed to minimize the adverse effects arising from those landscapes. We need to approach reconstruction at a scale which has ecological relevance (e.g., catchment scale), rather than at the scale of the individual farm which does not.

Conservation of Australia's biological diversity is a community responsibility; ecologists and conservation biologists need to provide the information to convince the community of their role and provide the methodology to carry it out.

HABITAT RECONSTRUCTION

HABITAT is a word in general use to describe the "the locality in which a plant or animal naturally grows or lives . . . Applied (a) to the *geographical area* over which it extends; (b) to the particular station in which a specimen is found; (c) but chiefly used to indicate the kind of locality, as sea-shore, chalk hills, or the like" *(The Shorter Oxford English Dictionary).* As pointed out by Odum (1959), its ecological meaning of the place where an organism lives, i.e., the place one would go to collect that particular organism, is the same. Ecologically, habitat may also refer to the place occupied by an entire community. In the latter case, habitat consists mainly of physical or abiotic complexes, while in the former case, habitat includes other organisms in addition to the abiotic environment.

Reconstruction is defined by *The Shorter Oxford English Dictionary* as "to construct anew . . . to restore". Accordingly, *habitat reconstruction* is the recreation of the ecological conditions required by the target organism(s) and is usually called for because of the drastic decline or disappearance of that organism (or organisms).

The scale of habitat reconstruction depends on whether it is aimed at populations of individual species or at communities. It also depends on whether it is aimed at individual remnants of native vegetation and their associated fauna (e.g., a single reserve set aside to preserve the last individuals of a species, as with the short-necked tortoise at Ellen Brook and Twin Swamps near Perth), or at whole landscapes (e.g., a collection of remnants with populations of declining species, as with those species of bird dependent on remnant native vegetation in the wheatbelt of Western Australia). Regardless of the scale, applying the dictionary definition of habitat reconstruction implies more than merely planting trees on farms or equivalent revegetation schemes. It means recreating ecologically what was there before, restoring ecological processes so that they operate at similar rates to those operating before the disruption being addressed by the reconstruction. Habitat reconstruction thus involves restoring the processes, the alteration of which, together with habitat removal, led to the decline or disappearance of the target organism(s) in the first place.

Cairns (1993) pointed out that each ecosystem is maintained by a unique sequence of climatological and biological events and that total restoration of an ecosystem is impossible. In Australia, we have a very poor understanding of what we have removed or altered. Anyone doubting this statement should realize that we have little idea of what species are present now, let alone what species were present when Europeans started settling Australia over 200 years ago. In addition, we have a very limited understanding of ecological

[1]CSIRO Division of Wildlife and Ecology, LMB 4, PO Midland, Western Australia, Australia 6056.

Pages 104–12 *in* CONSERVING BIODIVERSITY: THREATS AND SOLUTIONS, ed by R. A. Bradstock, T. D. Auld, D. A. Keith, R. T. Kingsford, D. Lunney and D. P. Sivertsen. Surrey Beatty & Sons. 1995.

processes and, as stated by Simberloff (1990), "one can hardly expect restoration of an entity only partly understood". This lack of understanding would hamper our ability to reconstruct habitats, even if we had the will to embark on a major programme of reconstruction. We are not convinced that as a Nation we do have that will at present. We see the need for all ecologists and conservation biologists to use their knowledge and influence to persuade all elements of the Australian population of the urgent need for habitat reconstruction over very large areas of this continent.

We are ecologists working in what is arguably the most seriously affected and rapidly degrading area in Australia. This is the intensively cleared wheat-sheep zone (Fig. 1) which occupies about 15 per cent of the continent (Graetz *et al.* 1992). We will use it as an example of why we need to engage in widespread habitat reconstruction, and also as an illustration of the issues that need to be confronted. We might add that the wheat-sheep zone occupies an area approaching one third of the land area of New South Wales, and is obviously one of the most dominant land uses in that state.

Fig. 1. Australia's extensively cleared wheat-sheep zone (from ABARE agricultural and grazing industries survey).

NEED FOR HABITAT RECONSTRUCTION

Australians have been extremely good at habitat modification and outright destruction of habitat. Over the past 200 years we have demonstrated this ability in every biogeographic zone on the continent (Saunders *et al.* 1990). Graetz *et al.* (1992) illustrated this point for every major landuse, using Landsat imagery in their graphic presentation of changes in landcover over the last 20 years. Some biogeographic zones have been modified directly, as in the clearing of native vegetation for agriculture, forestry plantations of exotic trees and urbanization, or the changes brought about by the grazing of introduced domestic livestock in the more than 70 per cent of Australia's land area devoted to pastoral activities (Morton 1992). Other biogeographic zones have been modified indirectly as a result of changes brought about elsewhere; for instance some desert areas now contain rabbits which were introduced into Australia at a point well away from the arid and semi-arid zones.

This destruction and modification of habitat has resulted in widespread changes in the distribution and abundance of the biota, including the extinctions of an unknown number of species. Hobbs and Hopkins (1990) point out that about three-quarters of Australia's vegetation has been modified in some way by human activity. In agricultural areas, many vegetation types have been reduced to a few percent of their original extent, with concomitant loss or endangerment of plant species (e.g., Beard and Sprenger 1984; Leigh *et al.* 1984). In some areas, clearing or modification of native vegetation is continuing.

Recher and Lim (1990) reviewed the status of Australia's terrestrial vertebrate fauna as a result of development since European settlement, and noted that the impact of European settlement on Australia's vertebrates has been greatest on mammals and birds; however, the vertebrates are the group which is most conspicuous and for which there is more data. The invertebrates are poorly described and poorly studied, and have probably been more greatly affected by the results of European settlement. The consequences of the losses of these species on ecosystem function (e.g., pedological processes or plant pollination) are also unknown and this lack of information will hamper our attempts of habitat reconstruction.

The widespread losses of medium-sized mammals have been well documented (Burbidge and McKenzie 1989; Johnson *et al.* 1989; Morton 1990; Woinarski and Braithwaite 1990). Recent information on the status of Australian birds (Ford and Howe 1980; Howe *et al.* 1981; Blakers *et al.* 1984; Saunders 1989; Saunders and Curry 1990; Recher and Serventy 1991; Robinson 1991; Reid and Fleming 1992; Saunders and Ingram 1995) indicates that birds are demonstrating the same declining trends as shown by mammals. Recher and Lim (1990) stated that they believed that declines in the abundance of reptiles and amphibians are expected to follow a similar pattern.

These losses and declines are caused by a number of factors, with habitat loss or modification being extremely important. Fragmentation of remaining habitat with some populations being isolated on patches that are too small to be viable, introductions of exotic species, competition with native species that have invaded areas changed by European management practices and disease are factors which have contributed, either singly or in combinations, to the loss of biological diversity. Given our present rate of habitat reconstruction, we agree with Cocks (1992) who summed up our present position with this depressing but accurate statement, "There is no reason to suppose that the rate of species extinction in Australia will fall".

Recher and Lim (1990) commented that it is doubtful if any continent has been altered as extensively and as rapidly as Australia, a point eloquently expressed by Lowenthal (1976); "Nowhere else has technological man lately occupied so fragile an ecosystem; nowhere else had settlers so utterly failed to identify with their new landscapes; nowhere else was man's environmental impact so patently reprehensible" (quoted in Hobbs and Hopkins 1990). The rapidity with which we have wrought these changes is shown by the fact that it took place over a period of 200 years or less, while it has taken around 10 000 years to bring about similar changes in the Middle East (Hobbs and Hopkins 1990). As a result, we are yet to gain all of the dubious benefits of these changes, and it may be decades if not centuries before we do. We **do** know that many of these changes will be detrimental and we now have the opportunity to address them while we are still fortunate enough to retain much of our unique biological diversity.

Although many people do not realize it, the most serious conservation issue facing Australia is the extreme degradation of our productive wheat-sheep areas. These occupy the semi-arid transitional zone between the high rainfall areas around the coast where most of our human population is concentrated (Cocks 1992), and the arid zone rangelands of the pastoral country (Fig. 1). The wheat-sheep zone produces a major proportion of the agricultural produce of Western Australia, South Australia, New South Wales and Queensland. It also contributes a significant amount to our export earnings and contains some of Australia's richest biological areas, as for example the Southwest Botanical Province with around 8 000 species of plant, about 75 per cent of which are endemic (Hopper 1992).

The wheat-sheep zone has experienced extensive clearing (see Cocks 1992; Fig. 2) as part of its agricultural development. In extreme cases, up to 95 per cent of the original vegetation cover has been removed (Graetz *et al.* 1992). What remains is being chipped away by further clearing or degraded by various processes (increasing soil salinity, weed invasion, nutrient enrichment, grazing, etc.). The results of our development of the wheat-sheep zone have been major changes in the hydrological balance with rises in soil salinity in susceptible areas, increased soil erosion by wind and water, soil structure decline, acidification of the soil, changed micro- and macro-climate, all of which contribute to a decline in agricultural production, and extensive loss of native flora and fauna. We have neither sustainable agricultural production nor sustainable nature conservation. If we

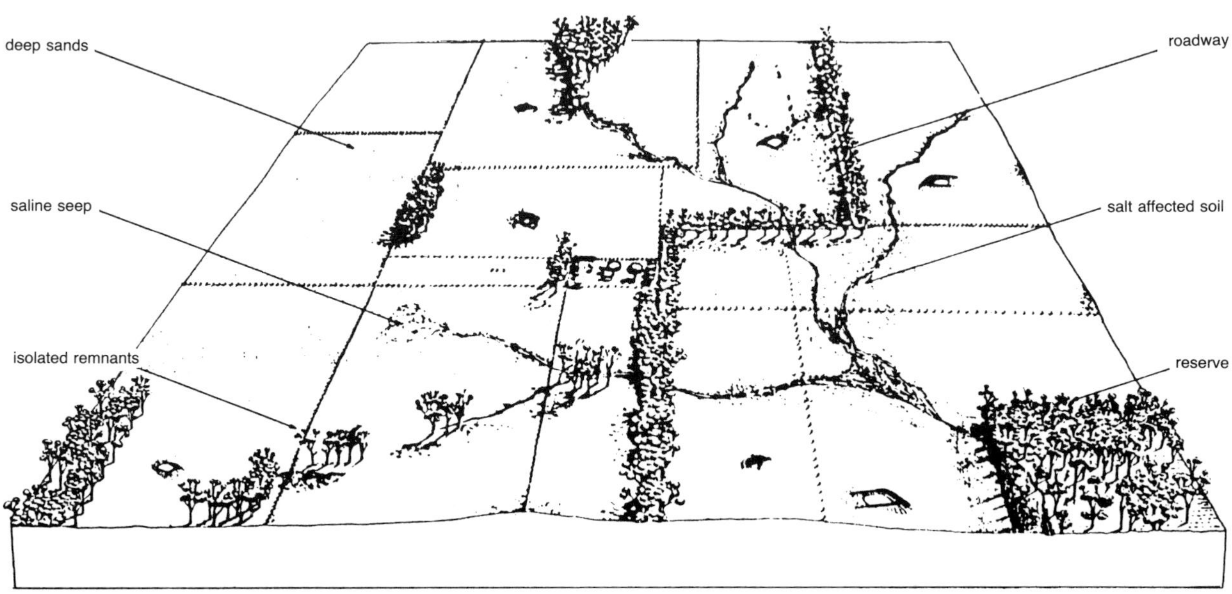

Fig. 2. Site plan and analysis showing drainage, fences, roads, vegetation and major ecological problem areas (from Lefroy *et al.* 1991 and used with permission).

continue down our present path we will end up with our agricultural heartlands as deserts from both biological and agricultural perspectives. This will have very serious ramifications for our society and our current high standard of living.

Hobbs and Saunders (1993) provide a detailed review of landscape function in one area in the central wheatbelt of Western Australia. This illustrates the increasingly alarming problems facing both agricultural production and nature conservation in the wheat-sheep zone; this example is probably typical of most of this zone, although the extent of the various problems may vary from district to district. For example, soil salinity may affect over 15 per cent of cleared land in southwestern Australia within the next 30 years and acidification of the soil may affect over 65 per cent of agricultural land with a significant decrease in productivity over the next 100 years (Nulsen 1993). Nulsen noted that if farmers do not adopt farming systems that are conservative of the soil resources, and reverse the degrading trends, the consequences will be enormous. The area is still losing species of native vertebrates (Hobbs *et al.* 1993a) and almost certainly losing species of invertebrates. It will probably continue to lose species of plants, but given the longevity of some species this process will take much longer. Whole habitat types are in danger of disappearing due to the lack of regeneration of the dominant plant species (Yates *et al.* 1994; Norton *et al.* 1995). This degrading situation will continue unless we address it as soon as possible.

Fortunately, revegetation is seen as a major tool to reconstruct these degrading landscapes in the central wheatbelt (Lefroy *et al.* 1991), and throughout the wheat-sheep zone (Hawke 1989). Ecologists and conservation biologists should use the opportunities presented by this impetus for revegetation to design strategies that not only address agricultural needs, but also address the needs for habitat reconstruction to preserve our declining flora and fauna. It is important to realize that we should work with the agricultural community in this revegetation phase and bear in mind what is possible, who will do the revegetation, and what incentives we can offer to ensure the work will be carried out. It will take more than merely demonstrating that we are mining our soils and closing the options for future generations to convince society of the needs for revegetation. Our starkly degrading landscapes are there for all to see, yet we have not even scratched the surface in our revegetation schemes. For example, Hawke (1989) called for a billion trees to be planted by the turn of the century to address Australia's land degradation problems. Note the emphasis on trees. Lefroy *et al.* (1993a) calculated that it will require the planting of 500 million to 1 billion stems (trees and shrubs) to address the hydrological balance in the wheatbelt of Western Australia alone. If we aim to do this over the next 10 years, this will require the planting of 100–200 million stems per year, allowing for 50 per cent mortality. When we realize that in 1989 Western Australian farmers planted about seven million trees for non-commercial reasons (Lefroy *et al.* 1993b), we appreciate how much we have to do to tackle the task of landscape reconstruction. The situation is similar in other states; and indeed, vegetation clearance may still out weigh revegetation in some areas. This is a social problem, not a scientific one, and we need to recognize that large shifts in community attitudes and actions are required before large-scale habitat reconstruction is even an option.

Given this current state of affairs, can conservation biologists contribute anything? As mentioned earlier, we do not have the understanding to recreate the "pristine" landscapes European settlers changed during their invasion of the continent. We must accept that we will be working with synthetic landscapes (Bridgewater 1990); that is, landscapes reconstructed with a range of vegetation types, some of which are alien. Society has dictated that vast areas of Australia are given over to production purposes and we must work within those purposes to develop practices which protect the production side **without** further compromising our native flora and fauna. However, some of these production methods must change so they do not further degrade the landscape. The onus is on ecologists and conservation biologists to provide the methodology to reconstruct ecosystems. These reconstructed ecosystems should emulate healthy, natural self-regulating systems (Cairns 1993) and be integrated within the production landscape.

We are still in the early stages of this process. There are good examples of reconstruction in mine sites which seem to emulate the structure and processes of the original ecosystem, as far as we can assess this (Hobbs 1993a). However, we have no indication as yet whether revegetation in agricultural land will enable the redevelopment of functional ecosystems, or whether fauna will actually use revegetated areas. Revegetation does not equate to reconstruction if native plants are installed, but no other ecosystem components return and ecosystem processes are not restored. While the promise is great, the problems to be overcome should not be trivialized.

HABITAT RECONSTRUCTION: WHO IS EQUIPPED TO DO IT?

Given our lack of knowledge and the enormity of the problem, what can ecologists and conservation biologists provide? In the first instance we need to examine what we do know about the way our landscapes function. Collectively throughout the scientific and farming communities there is a considerable body of relevant knowledge. This information needs to be synthesized and interpreted in a way that those who are going to carry out the reconstruction can follow and act on it. It also needs to be published in a form and place where non-scientists will see it (Ehrlich 1993). This does not mean trivializing the work, but looking at its relevance then working with others to provide the basis for constructing revegetation models which combine nature conservation with agricultural production.

This approach was followed in Western Australia in 1990 when a group of 21 people consisting of agricultural economists and scientists, botanists, farmers, hydrologists, soil scientists and zoologists spent three days together at a workshop discussing their work in the central wheatbelt. The aim was to combine their knowledge into a conceptual framework of how to reintegrate the fragmented landscape, taking into account agricultural production and nature conservation. There were three products of that workshop. One was a scientific review of the state of knowledge regarding the central wheatbelt and its problems, together with an outline of a conceptual framework to address these problems (Hobbs and Saunders 1993). This book, published by Springer-Verlag (New York), is expensive and not readily available for landcare groups and farmers who would benefit from the information; in fact one farmer not involved in the workshop stated that "it should be compulsory reading for every wheatbelt farmer". The second product distilled this knowledge into a "user friendly guide" to the landscape of the central wheatbelt with suggestions as to how to address the problems (Lefroy *et al.* 1991). This was the *Revegetation Guide to the Central Wheatbelt,* a 106 page book, which described the wheatbelt catena, the soil types illustrated in terms of the vegetation that originally grew on them, provided descriptions of 82 species of local plant available for revegetation, gave information on approaches to revegetation, paying attention to placing species on appropriate soil types, and provided contacts for specific information. The guide specified local species for revegetation on the basis that they are better suited to the environment and would be of more benefit for fauna, a point conservation biologists take to be self-evident, but which has little data to support it. In fact, site conditions can become so modified that local species will no longer establish and grow. The guide illustrated a revegetation plan that attempted to address the various problems in this degrading environment (Figs 2 and 3) and was distributed free to about 3 500 farmers in the central wheatbelt. We have been conducting a survey among farmers to establish what use has been made of the publication and whether it is influencing revegetation. Preliminary data, based on a small sample indicate that about half of the farmers surveyed have found it useful.

The third product of the workshop has been the establishment of a co-operative network involving people from a range of different disciplines and organizations. This has resulted in a number of collaborative ventures relating to landscape reconstruction. One important outcome has been that agricultural scientists are now bringing nature conservation into catchment planning as an important landscape element.

It is important to bear in mind that about 80 per cent of remnant native vegetation in the wheat-sheep zone is privately owned. This comprises a very significant proportion of what amounts to our conservation estate; the biota does not discriminate on the basis of ownership. It is up to all those committed to the preservation of our biological diversity to convince all landholders of the need to protect their remnant vegetation. It is upon that skeleton that we must build our landscape reconstruction. It is also important to realize that the scope of revegetation in the wheat-sheep zone is enormous, as mentioned earlier in this paper. Given the scope of the problem and the ownership of remnant vegetation, it is obvious that it must be a community responsibility to revegetate, with the farm being the unit of reconstruction, but set within a landscape perspective. It is the task of ecologists and conservation biologists to generate the reconstruction models and provide the guidelines to allow them to carry out the task. We are embarking on an enormous experiment, which will happen regardless of our input. The success of that experiment will depend very much on how much guidance is provided by ecologists and conservation biologists, including direct involvement in monitoring the experiment to make adjustments where necessary.

GUIDELINES FOR HABITAT RECONSTRUCTION

No remnant functions on internal processes but is driven mainly by factors arising in the surrounding matrix (Saunders *et al.* 1991; Hobbs 1993b). Therefore, the degrading processes in the surrounding matrix must be managed. Habitat reconstruction must be planned in this context, that is, at the landscape scale, not at the level of the individual patch or remnant, which is where much habitat reconstruction is carried out at present. The current landcare movement, where neighbouring farms are formed into catchment groups to tackle soil conservation problems (Goss and Chatfield 1993) provides an ideal opportunity to concentrate on ecological reconstruction. A series of steps which should

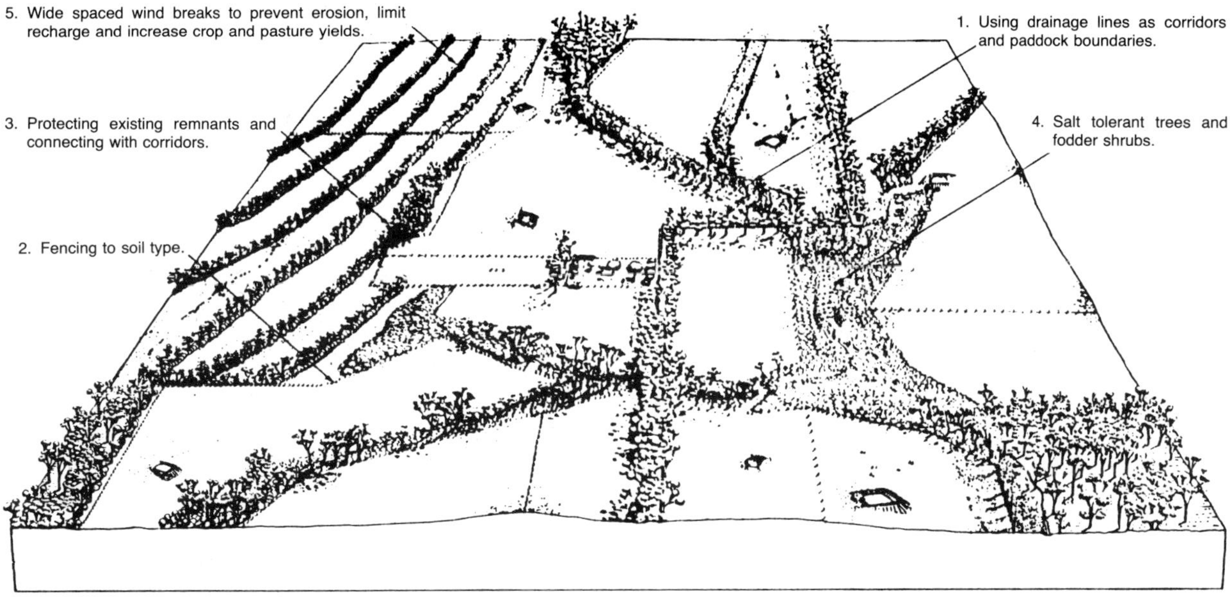

Fig. 3. A long-term revegetation plan attempting to incorporate nature conservation into the agricultural production landscape (from Lefroy *et al.* 1991 and used with permission).

be taken into account when planning habitat reconstruction are given below. These have been drawn from Saunders *et al.* (1993a and the various chapters therein) and Saunders *et al.* (1993b).

Define the problem — What is the reason for habitat reconstruction? Is it aimed at a single species, or at a community, at what scale? What are the environmental problems in the area to be reconstructed and are there production reasons directing that the area be reconstructed? Can habitat reconstruction be incorporated within revegetation plans for addressing degradation problems within the agricultural matrix?

What skeleton is available to build the reconstruction upon? — What is the distribution of remnant vegetation over the landscape and what vegetation types are involved? What animal communities are associated with these remnants? This needs to be established at the smallest unit involved in the reconstruction and this might be at the level of a remnant of native vegetation or an individual farm. However, for the reconstruction to stand any chance of success it must be planned with neighbouring units over a considerable area, be it a river catchment, shire, or some other large unit. There are now many agencies who can assist in this planning stage by providing co-ordinators trained in facilitating these plans, together with supplying computer-generated maps of soils, remnant vegetation, roads, fences and drainage lines as well as supplying aerial photographs and satellite imagery. Such planning needs to be done in association with all of the land managers in the planning unit, e.g., farmers, conservation agencies, main roads agencies, shire officials, and other Crown land managers with interests in the area (Hobbs *et al.* 1993b).

Retain and protect all remnant vegetation — Habitat reconstruction will rely on an increasing supply of seed from plants of local provenance, together with information on where in the landscape the plants grew and what animal communities were associated with them. Remnant vegetation is essential for the provision of seed and information. Accordingly, there should be no further clearing of remnant vegetation. In addition, remnant vegetation will need to be managed to prevent further degradation. Grazing by domestic livestock is a major degrading influence and stock and feral animals should be kept away from remnant vegetation. Removal of weeds may also be necessary. Retain dead or decaying vegetation, both standing and fallen, as it provides habitat for a range of species.

Establish what the landscape looked like before development and degradation — It is necessary to know what to aim at for habitat reconstruction. That can be established by examining remnant vegetation, particularly that found along linear features such as road reserves, railway reserves and fence lines. This vegetation will sample the pre-clearing vegetation associations more accurately than patches. Use these remnants to establish which vegetation grew on which soil types, and in which position in the landscape. Local knowledge among older residents and research workers can add to this information. The resulting information can then be used to plan revegetation. Existing remnants may, however, be degraded to the extent that they no longer represent a useful baseline for reconstruction. It may, in fact, be impossible to return to exactly what was there before, and it may be a waste of resources to try.

Develop goals for habitat reconstruction — Goals for reconstruction need to be developed at the smallest unit of action, be it reserve, patch or farm. These goals should be linked to goals at the catchment, landscape or regional level. To have any chance of success they must take cognisance of the predominant land use in the area and work within that landuse. For example, it is pointless attempting reconstruction in a patch which is being affected by soil salinization, unless the reconstruction addresses the rising water tables in the surrounding landscape. This will obviously involve revegetation in the area surrounding the patch. Such revegetation will not be attempted unless it is in the interests of the surrounding landholder. The priorities will differ depending on the landholder involved. However, by planning at the landscape scale, revegetation for addressing some problem in the productive landscape may be useful for habitat reconstruction and bolster the area of available habitat.

Design a habitat reconstruction plan — Working with all land managers in the area, design a habitat reconstruction plan that takes into account the various objectives of the diverse interests in the landscape. Revegetation can be used to increase the area of available habitat, to provide buffer strips around existing remnants, or to provide corridors linking otherwise isolated remnants (Hobbs 1993a). This can be done in ways compatible with other restoration goals. For example, wind breaks can be planted with a range of local species, where possible proportionally following the distribution, abundance and soil types such species occurred in before clearing. Such revegetation can provide linkages and in addition use water, reduce erosion and provide shelter (Fig. 3). Our current state of knowledge does not allow an assessment of how useful such corridors might be for fostering movement (Hobbs 1992, 1993a), but if their development allows major land conservation problems to be tackled, nothing is lost in the attempt. It is better to try things that might work than do nothing because we do not know enough: we will never know enough.

Aim at structural and patch diversity — Unfortunately many revegetation schemes specify trees and ignore the need for structural diversity. Habitat reconstruction should aim at developing all the former structural layers and provide a range of patches emulating the former uncleared landscape. For example, some nectivorous animals need a range of patches which flower at different times of year so that they have food available all year somewhere in the landscape. Some of the animals may require native vegetation to foster their movements around the landscape if they will not fly over cleared agricultural fields. These species will need vegetation linkages to allow them access to the full array of patches. These sorts of linkages may also serve as wind breaks on farms as well as hydrological pumps lowering the water table in areas subject to rising soil salinity.

Monitor and record results — It is essential to record what is being planned, what has be carried out, as well as monitoring the results of revegetation and of faunal use of revegetated areas. Monitoring is essential to allow the revegetation programme to be adjusted where necessary. Communicate these results so that others may learn from that experience.

CONCLUSION

Many areas of agricultural Australia have lost most of their native vegetation and hence most of their habitat for fauna. What remains is not sufficient to retain functioning ecosystems which will maintain viable faunal and plant populations in the long-term, even with considerable management input. Increasing the extent of areas useful for conservation is the only hope of reversing the current situation, and this involves large-scale habitat reconstruction. Extension of current revegetation programmes can form the basis for this, but this requires a radical rethink about how agricultural landscapes are planned and managed. It also requires a huge increase in revegetation/restoration activity, which can only come about via greater incentives. Given that the current level of activity is less than optimal, we need to make sure that it will have the maximum benefit. Our present level of knowledge is insufficient to provide detailed guidelines as to how to restore habitats on a large-scale basis. Nevertheless, such guidelines are urgently required, and scientists and managers need to come up with answers now. As ecologists and conservation biologists we need to extract what we can out of our current knowledge and understanding. We either fly with what we have got and hope we can refine things as we go along, or we wait until we know more but are left with much less to conserve. Which is it to be?

REFERENCES

Beard, J. S. and Sprenger, B. S., 1984. Geographical data from the vegetation survey of Western Australia. Vegetation Survey of Western Australia Occasional Paper 2. Vegmap Publication: Perth.

Blakers, M., Davies, S. J. J. F. and Reilly, P. N., 1984. *The atlas of Australian birds.* Melbourne University Press: Melbourne.

Bridgewater, P. B., 1990. The role of synthetic vegetation in present and future landscapes of Australia. *Proc. Ecol. Soc. Aust.* **16:** 129–34.

Burbidge, A. A. and McKenzie, N. L., 1989. Patterns in the modern decline of Western Australia's vertebrate fauna: causes and conservation implications. *Biol. Conserv.* **50:** 143–98.

Cairns, J., 1993. Ecological restoration: replenishing our national and global ecological capital. Pp. 193–208 in *Nature conservation 3: reconstruction of fragmented ecosystems; global and regional perspectives* ed by D. A. Saunders, R. J. Hobbs and P. R. Ehrlich. Surrey Beatty & Sons: Chipping Norton.

Cocks, D., 1992. *Use with Care.* New South Wales University Press: Kensington.

Ehrlich, P. R., 1993. Communication: how can ecologists get their message out? Pp. 295–301 in *Nature conservation 3: reconstruction of fragmented ecosystems; global and regional perspectives* ed by D. A. Saunders, R. J. Hobbs and P. R. Ehrlich. Surrey Beatty & Sons: Chipping Norton.

Ford, H. and Howe, R., 1980. The future of birds in the Mount Lofty Ranges. *SA Ornithol.* **28:** 85–89.

Graetz, D., Fisher, R. and Wilson, M., 1992. *Looking back. The changing face of the Australian continent 1972–1992.* CSIRO: Canberra.

Goss, K. F. and Chatfield, J., 1993. Landcare groups in Western Australia: the role of self help groups in restoring degraded farmland. Pp. 281–93 in *Nature conservation 3: reconstruction of fragmented ecosystems; global and regional perspectives* ed by D. A. Saunders, R. J. Hobbs and P. R. Ehrlich. Surrey Beatty & Sons: Chipping Norton.

Hawke, R. J. L., 1989. *Our country, our future.* Australian Government Publishing Service: Canberra.

Hobbs, R. J., 1992. Corridors for conservation: solution or bandwagon? *Trends Ecol. Evol.* **7:** 389–92.

Hobbs, R. J., 1993a. Can revegetation assist in the conservation of biodiversity in agricultural areas? *Pac. Conserv. Biol.* **1:** 29–38.

Hobbs, R. J., 1993b. Effects of landscape fragmentation on ecosystem processes in the Western Australian wheatbelt. *Biol. Conserv.* **64:** 193–201.

Hobbs, R. J. and Hopkins, A. J. M., 1990. From frontier to fragments: European impact on Australia's vegetation. *Proc. Ecol. Soc. Aust.* **16:** 93–114.

Hobbs, R. J. and Saunders, D. A. (eds), 1993. *Reintegrating fragmented landscapes: towards sustainable production and nature conservation.* Springer-Verlag: New York. 332 pp.

Hobbs, R. J., Saunders, D. A., Lobry de Bruyn, L. A. and Main, A. R., 1993a. Changes in biota. Pp. 65–106 in *Reintegrating fragmented landscapes: towards sustainable production and nature conservation* ed by R. J. Hobbs and D. A. Saunders. Springer-Verlag: New York.

Hobbs, R. J., Saunders, D. A. and Arnold, G. W., 1993b. Integrated landscape ecology: a Western Australian perspective. *Biol. Conserv.* **64:** 231–38.

Hopper, S. D., 1992. Patterns of plant diversity at the population and species level in south-west Australian mediterranean ecosystems. Pp. 27–46 in *Biodiversity in mediterranean ecosystems in Australia* ed by R. J. Hobbs. Surrey Beatty & Sons: Chipping Norton.

Howe, R. W., Howe, T. D and Ford, H. A., 1981. Bird distributions on small rainforest remnants in New South Wales. *Aust. Wildl. Res.* **8:** 637–51.

Johnson, K. A., Burbidge, A. A. and McKenzie, N. L., 1989. Australian Macropodoidea: status, causes of decline and future research and management. Pp. 641–57 in *Kangaroos, wallabies and rat-kangaroos* ed by G. Grigg, P. Jarman and I. Hume. Surrey Beatty & Sons: Chipping Norton.

Lefroy, E. C., Hobbs, R. J. and Atkins, L. J., 1991. *Revegetation guide to the central wheatbelt.* Greening Western Australia, Western Australian Department of Agriculture and CSIRO: Perth.

Lefroy, E. C., Hobbs, R. J. and Scheltema, M., 1993a. Reconciling agriculture and nature conservation: toward a restoration strategy for the Western Australian wheatbelt. Pp. 243–57 in *Nature conservation 3: reconstruction of fragmented ecosystems; global and regional perspectives* ed by D. A. Saunders, R. J. Hobbs and P. R. Ehrlich. Surrey Beatty & Sons: Chipping Norton.

Lefroy, E. C., Salerian, J. and Hobbs, R. J., 1993b. Integrating economic and ecological considerations: a theoretical framework. Pp. 209–44 in *Reintegrating fragmented landscapes: towards sustainable production and nature conservation* ed by R. J. Hobbs and D. A. Saunders. Springer-Verlag: New York.

Leigh, J., Boden, R. and Briggs, J., 1984. *Extinct and endangered plants of Australia.* MacMillan: Melbourne.

Lowenthal, D., 1976. Perceiving the Australian environment: a summary and commentary. Pp. 357–65 in *Man and landscape in Australia. Towards an ecological vision* ed by G. Seddon and M. Davis. Australian Government Publishing Service: Canberra.

Morton, S. R., 1990. The impact of European settlement on the vertebrate animals of arid Australia: a conceptual model. *Proc. Ecol. Soc. Aust.* **16:** 201–13.

Morton, S. R. (ed), 1992. Wildlife and conservation in the rangelands. *Rangel. J.* **14.**

Norton, D. N., Hobbs, R. J. and Atkins, L., 1995. Fragmentation, disturbance and plant distribution: mistletoes in woodland remnants in the Western Australian wheatbelt. *Conserv. Biol.* **9** (in press).

Nulsen, R. A., 1993. Changes in soil properties. Pp. 107–45 in *Reintegrating fragmented landscapes: towards sustainable production and nature conservation* ed by R. J. Hobbs and D. A. Saunders. Springer-Verlag: New York.

Odum, E. P., 1959. *Fundamentals of ecology.* W. B. Saunders Company: Philadelphia.

Recher, H. F. and Lim, L., 1990. A review of current ideas of the extinction, conservation and management of Australia's terrestrial vertebrate fauna. *Proc. Ecol. Soc. Aust.* **16:** 287–301.

Recher, H. F. and Serventy, D. L., 1991. Long term changes in the relative abundances of birds in Kings Park, Perth, Western Australia. *Conserv. Biol.* **5:** 90–102.

Reid, J. and Fleming, M., 1992. The conservation status of birds in arid Australia. *Rangel. J.* **14:** 65–91.

Robinson, D., 1991. Threatened birds in Victoria: their distributions, ecology and future. *Vict. Nat.* **108:** 67–77.

Saunders, D. A., 1989. Changes in the avifauna of a region, district and remnant as a result of fragmentation of native vegetation: the wheatbelt of Western Australia. A case study. *Biol. Conserv.* **50:** 99–135.

Saunders, D. A. and Curry, P. J., 1990. The impact of agricultural and pastoral industries on birds in the southern half of Western Australia: past, present and future. *Proc. Ecol. Soc. Aust.* **16:** 303–21.

Saunders, D. A. and Ingram, J. A., 1995. *Birds of southwestern Australia. An atlas of changes in the distribution and abundance of the wheatbelt avifauna.* Surrey Beatty & Sons: Chipping Norton.

Saunders, D. A., Hobbs, R. J. and Ehrlich, P. R., 1993a. *Nature conservation 3: reconstruction of fragmented ecosystems; global and regional perspectives.* Surrey Beatty & Sons: Chipping Norton. 326 pp.

Saunders, D. A., Hobbs, R. J. and Ehrlich, P. R., 1993b. *Repairing a damaged world.* Surrey Beatty & Sons: Chipping Norton. 16 pp.

Saunders, D. A., Hobbs, R. J. and Margules, C. R., 1991. Ecological consequences of ecosystem fragmentation: a review. *Conserv. Biol.* **5:** 18–32.

Saunders, D. A., Hopkins, A. J. M. and How, R. A., 1990. Australian ecosystems: 200 years of utilization, degradation and reconstruction. *Proc. Ecol. Soc. Aust.* **16,** 602 pp.

Simberloff, D., 1990. Reconstructing the ambiguous: can island ecosystems be restored? Pp. 37–51 in *Ecological restoration of New Zealand islands* ed by D. R. Towns, C. H. Daugherty and I. A. E. Atkinson. Conservation Sciences Publication No. 2. Department of Conservation: Wellington.

Woinarski, J. C. Z. and Braithwaite, R. W., 1990. Conservation foci for Australian birds and mammals. *Search* **21:** 65–68.

Yates, C. J., Hobbs, R. J. and Bell, R. W., 1994. Landscape-scale disturbances and regeneration of semi-arid woodlands of South-western Australia. *Pac. Conserv. Biol.* **1:** 214–21.

Degradation and Pollution of Water Resources

PLATE 7 *(left)*. The Macquarie Marshes, on the Macquarie River, are a wetland in western New South Wales, comprising an important habitat for waterbirds, as recognized internationally under the Ramsar Convention. Recent debate over the future of the marshes has occurred because of the supply of water for irrigation of cotton and other crops. *(Photograph by D. Eastburn.)*

PLATE 8 *(right)*. Burrendong Dam is the largest storage on the Macquarie River with a capacity of one million megalitres. Usage and regulation of stored water for irrigation has reduced the size of the Macquarie Marshes by at least 40–50 per cent over the period 1944–93. *(Photograph by Richard Kingsford.)*

PLATE 9 *(left)*. Large off-river storage with flooded trees at Cubbie Station in southern Queensland. Water is harvested from irregular floods and stored for cotton irrigation.The resulting change to water regimes threatens the future of wetlands downstream such as Narran Lake which is home to one of the largest colonies of Ibis *Threskiornis* spp. in Australia. *(Photograph by Richard Kingsford.)*

Water resources and biodiversity: past, present and future problems and solutions

P. CULLEN[1] and P. S. LAKE[1]

Australia's natural waterways have been vastly degraded by the agricultural development of the countryside. Many of these flowing waters may be restored if land uses and land management can be improved to reduce the erosion and flow of nutrients that have caused such devastation. While the impacts of dams can probably not be removed, there are ways to reduce these impediments to fish movements and to release water to allow natural ecosystems to function.

We need to improve considerably our knowledge base if we are to better manage these systems. We need to ensure that communities and decision makers understand the connections between land use and water quality and make sure the real environmental impacts of major decisions are identified as part of the decision making process.

INTRODUCTION

THE biodiversity movement in Australia has so far paid little attention to freshwater ecosystems. Compared with our terrestrial systems we have a very poor knowledge of the aquatic species that exist except for fish, and consequently we have a poor understanding of the loss of biodiversity.

The Australian landscape has been severely impacted by the adoption from Europe of often inappropriate land management practices (Mott and Bridgewater 1992). Present Australian attitudes to the land can be traced back to early explorers. Three key elements of our attitude to the bush were developed in this period, and still persist today.

The development imperative

From their arrival, the prime concern of the First Fleeters was survival. The main threat was famine, so it was essential to clear land, sow crops and husband stock. Exploration was encouraged as a means of opening agricultural land to provide food for the new colony.

This need to clear and develop land became deeply embedded in the Australian "frontier ethic". It was how we marked our territory; it was how we earned our right to control the land. As Tighe (1984) points out this development ethos turned into a belief in development for its own sake. The clearing of native forests at Heytesbury in Victoria in the mid sixties to produce what was recognized even then as uneconomic dairy farms is an example of an unsustainable activity supported by tax relief. Davidson (1969) would argue that much of our irrigation development, especially the Ord Dam is in the same mould. Grand, symbolic and ignorant gestures to development have caused a futile loss of natural resources, the loss of large amounts of public money, and considerable misery for many of the supposed beneficiaries. The main beneficiaries were the professionals who designed and built the monuments, and the politicians who supported them. Fearnside (1989) called these giant water projects "pharonic projects" in that like the pyramids of ancient Egypt these public works demand the effort of an entire society to complete but bring virtually no economic returns.

Ignorance is no impediment to action

To the early explorers much of what they saw was the reverse of their expectations. Mountains had rounded tops rather than the scraggy rocky cliffs of Europe. Australian rivers did not conform to the expectations of large rivers draining the country and providing navigation from the sea. As Sturt, quoted by Carter (1987), points out when describing Australian rivers:

"Falling rapidly from the mountains in which they originate into a level and depressed country . . . they fail before they reach the coast and exhaust themselves in marshes."

Cook's bountiful description of Botany Bay and its surrounds ("some of the finest meadows in the world") led to it becoming a cause in England, ending in the arrival of the First Fleet. Yet within six months of arrival, Tench, quoted by Carter (1987), wrote "Of which the natural meadows which Mr Cook mentions near Botany Bay, we can give no account." A year later Tench wrote (Sept 1789):

"Had not the nautical part of Mr Cook's description . . . been so accurately laid down,

[1]Co-operative Research Centre for Freshwater Ecology, University of Canberra and Monash University.

Pages 115–25 *in* CONSERVING BIODIVERSITY: THREATS AND SOLUTIONS, ed by R. A. Bradstock, T. D. Auld, D. A. Keith, R. T. Kingsford, D. Lunney and D. P. Sivertsen. Surrey Beatty & Sons. 1995.

there would exist the utmost reason to believe that those who described the contiguous country, had never seen it."

Seasonal variability was something the new-comers could expect, however the annual variability in Australian rainfall was outside their experiences. This misunderstanding of the Australian climate led to a plethora of poor decisions of land settlement made with no understanding of land capability. These were not just early mistakes, but continue to the present. Clearing of forest to provide farmland, the extension of irrigation into saline areas, proposals to farm the Little Desert in Victoria and so on, show a preparedness to chance one's luck without understanding.

The early explorers had no basis for understanding the landscapes they found; modern exploiters of the land do not have this excuse. Plenty is known about the causes of land degradation for instance, but the exploiters choose to ignore it.

Heroes capture public goods for private gain

Australians have long made heroes of the bush battler who overcomes drought, bushfire, cane toads, rabbits and so on to survive and prosper from what is seen as an inhospitable bush. Crocodile Dundee encapsulates the cool, fearless and competent bush image many Australians dream of as they drive their 4WDs to work on the freeway.

These images have their origins in the squatters who followed on the heels of the explorers to take possession of huge tracts of the good land. Cotton farmers blithely spray organic pesticides around communities because "unless they do, they would go broke." Private gain seems often dependent on public loss.

RIVERS AND THEIR CATCHMENTS

The emphasis in this paper will be on flowing waters, since these are the dominant water bodies in Australia. We do have some salt lakes, subalpine lakes and wetlands of significance, and some of these are damaged. We recognize the connection between catchments, floodplains and wetlands as they control river health, but in this paper we concentrate on flowing waters. Most large standing waters in Australia are impoundments of major rivers. Wetlands in the Murray Darling basin have been recently reviewed by Pressey (1990) who identified considerable human impact on these systems.

The essential coupling

Since we know that the health of a river is a direct reflection of the health of the catchment, it should come as no surprise that many of our rivers are thought to be in a very poor condition.

Cyanobacterial blooms throughout the Murray-Darling system and elsewhere have alerted the public to what limnologists have long been saying. That we have been degrading our water resources in ways that will prevent them sustaining natural aquatic ecosystems, and the simplified systems that are now replacing the natural ones will be of much less use to humans.

When rain runs off land it transports sediment and a variety of other contaminants to the receiving waters. An aquatic ecosystem is tightly coupled with its catchment. Land use decisions, and the quality of land management are critical determinants of water quality.

When considering these non-point sources it is important to appreciate that they operate mainly when rain is falling and runoff is being generated. The rain and runoff provide the energy that detaches and transports particles.

Forested catchments generally produce the best quality water. As land is cleared for grazing and agriculture, the transport of sediment, nutrients and organic matter to the waterway increases. The "peakiness" of runoff events also increases as large volumes of water runoff in a shorter time scale than for well vegetated catchments. This has hydraulic effects on the river channel.

Special features of Australian water bodies

Australian water bodies reflect the unusual features of low flow, high variability of flow and unpredictability of both precipitation and river flow. Many of them flow only intermittently. McMahon (1982) has demonstrated that as a group, Australian streams are significantly more variable than world and world arid zone streams.

The rivers drain an old weathered landscape with naturally low levels of geologically derived nutrients such as phosphorus. They tend to be more turbid due to the erodable soils, and many are also coloured. Many of them are noticeably saline, especially under low flow conditions.

Leaf fall in Australian sclerophyll forests is predominantly in summer, and it seems that leaf material enters the river also in summer when water temperatures are high and flow is low. This is in marked contrast to overseas studies where deciduous vegetation dominates.

BIODIVERSITY IN AUSTRALIAN FLOWING WATERS

The poverty of our knowledge base

If the health of rivers is characterized by species composition and by biodiversity, then the paucity of taxonomic work in large parts of Australia becomes a major impediment. Very little work has been done over large parts of the continent. Northern and Central New South Wales, Central Queensland, most of central Australia, and most of Western Australia, other than the south-west corner, are notably depauperate in taxonomic knowledge.

Lake (1987) presented a table (see Table 1) assessing the state of knowledge of various components of Australian stream ecosystems.

Observations on biodiversity of Australian rivers

Despite these knowledge gaps, we do have reasonable information from some areas, and various workers have established some key observations.

a. Invertebrate diversity in both temporary and permanent streams is high by international standards (Lake *et al.* 1985).
b. Fish diversity is low but not unexpectedly so in view of catchment sizes and discharge (Bishop and Forbes 1991).
c. Latitudinal gradients in invertebrate biodiversity do exist but are not marked. (Lake *et al.* 1994).
d. There are areas of high species richness (hot spots) and these may be a priority area for conservation management once identified (e.g., Kosciusko plateau).
e. Australian freshwaters have a high degree of endemism for many faunal groups. For example all 196 species of Plecoptera, all 84 described species of Ephemeroptera, all 10 species of extant anaspidacean syncarid crustaceans, 94 out of the 96 species of freshwater crayfish as compared with 130 out of 183 species of freshwater fish.

What have we lost; what are we losing?

There is as yet no comprehensive assessment of the conservation status of Australian freshwater biota. This is a major challenge awaiting

Table 1. State of knowledge of components of stream ecosystems. Updated from Lake (1987).

	Nil–Poor	Rudimentary*	Moderate–Good*
Bacteria			
Taxonomy and distribution		×	
Ecology	×		
Aquatic fungi			
Taxonomy and distribution		×	
Ecology	×		
Micro-algae			
Taxonomy		×	
Ecology	×		
Macro-algae			
Taxonomy		×	
Ecology	×		
Macrophytes			
Taxonomy and distribution			×
Ecology		×	
Micro-invertebrates			
Taxonomy and distribution	×		
Ecology	×		
Macro-invertebrates			
Taxonomy and distribution		×	
Ecology		×	
Fish			
Taxonomy and distribution			×
Ecology		×	
Amphibia			
Taxonomy and distribution			×
Ecology		×	
Freshwater-associated reptiles			
Taxonomy and distribution			×
Ecology		×	
Freshwater-associated birds			
Taxonomy and distribution			×
Ecology		×	
Freshwater-associated mammals			
Taxonomy and distribution			×
ecology		×	

*Rudimentary means there are some studies but there are still major gaps in knowledge. Moderate means there are many studies available but the geographic coverage is still incomplete.

the limnological community. The fragmented information available suggests the following.

a. Of the freshwater fish one is extinct in the wild, six are regarded as endangered and six as vulnerable.
b. Hill and Michaelis (1988) identify 10 species of aquatic insect that are regarded as threatened.
c. Horwitz (1990) has identified 57 species of freshwater Crustacea regarded as threatened in Australia. One, *Austrogammarus australis,* is known to now be extinct.
d. Geddes (1990) reports the loss of River Murray crayfish from the Murray downstream of Mildura since the 1950s.

THE CONCEPT OF RIVER HEALTH

There are no clear and agreed criteria for what constitutes health of a river ecosystem. Hart (1992) has discussed the characteristics of possible indicators of river health. He concludes that commonly used physico-chemical indicators are not sufficiently sensitive to environmental change, and that biological indicators must be developed.

This debate has a long way to go. Some would argue "health" is characterized by process measures such as rates of nutrient cycling and primary productivity, while others would argue these do not matter in comparison with species composition. "River Health" would seem to be a relative rather than absolute concept.

Even in this situation where we have no agreement on appropriate indicators, and very little information, we do have assessments of river condition that paint an alarming picture. Much of this work has been done in Victoria where the knowledge base is substantially better than other mainland states. Hart (1992) reports various studies that indicate one-third of Victoria's 64 000 km of streams are in poor or very poor condition. The most severely degraded are in agricultural areas, again confirming the link between catchment quality and river quality.

Rapport *et al.* (1985) identify typical responses of ecosystems to stress in that diversity is reduced, the downstream transportation of nutrients is increased and the ratio of production to biomass is increased.

The new ANZECC Australian Water Quality Guidelines make an important step forward by recommending national biological indicators of water quality alongside the more traditional chemico-physical approaches. This step recognizes that the important outcomes of water quality from an ecological perspective are biological and direct measurement is better than using chemistry as a surrogate.

THE FACTORS CAUSING DESTRUCTION OF AQUATIC ECOSYSTEMS

It is often not possible to identify a single factor that leads to the extinction of a species, due to both lack of data and the fact that often multiple factors are important. Allan and Flecker (1993) propose six factors as being of critical importance in the loss of river biodiversity:

- habitat loss and degradation
- spread of exotic species
- overexploitation
- secondary extinctions
- chemical and organic pollution
- climate change

This framework will be used to identify the major stresses on river ecosystems in Australia.

Habitat loss and degradation

Habitat loss is obvious from the large engineering works that convert riverine systems to lacustrine systems. It is less obvious from the cumulative impacts of small decisions affecting drainage, "river improvement", land use and so on.

The physical impacts of dams

The extensive development of the Murray-Darling basin is based on capturing water in storages and having it available to irrigate plants in dry periods. In the southern part of the basin this is through large storages operated by the States. In the northern part of the basin it is based on pumping from rivers during high flow and holding the water on-farm until it is needed.

Dams introduce a physical barrier that restricts fish movement. Many Australian dams have been constructed with inadequate outlet structures which means that only bottom waters can be released. These waters are frequently much colder than the normal river flow, and frequently are anoxic and high in iron, manganese and hydrogen sulphide. The ecological impacts of these release waters have been well documented. Immediately downstream of these dams it is normal to have a dramatic loss of invertebrate species, with recovery with increasing distance from the dam (Lake and Marchant 1990).

Reservoirs also act as a trap for sediment and nutrients. Some of these stay in the water column and cause impacts through algal growth. Others settle to the bed of the lake, and may later become available to stimulate algal growth in the storage or downstream after release. Many New South Wales storages are now eutrophic and support major blooms of toxic algae over summer. Many have had to be closed for recreation and for domestic supply. The algae in storages may seed downstream rivers and stimulate algal problems throughout the system. These problems include toxicity to humans, stock and wildlife (Cullen 1986).

The alteration of flow regime

Walker (1985) has discussed the alteration and simplification of flow regimes following construction of dams on a drainage basin basis. Cadwallader and Lawrence (1990) discuss these flow variations in terms of reduced flows, altered seasonality of flows, reduced frequency of flood-flows and consequent altered river levels, and increased rate of fall of river levels and discuss the impact of each on fish communities. The naturally low summer flow condition of the Murray River has been transformed into a bankfull condition in summer to maximize the carriage of irrigation waters. The winter flows are now lower due to the capture of water by the dams. The dams have also prevented small and moderate floods in the system from inundating floodplains (Cadwallader and Lawrence 1990). This has had significant impacts on river red gum, Murray cod and other native fish. The most marked reduction in natural flows is below dams or major offtake structures.

The ill-conceived "river improvement" movement

"River Improvement" has been advocated to control bank erosion, minimize flooding, improve the discharge capacity of river channels and remove obstacles to river traffic. This has largely taken the form of "desnagging" by removing fallen trees from rivers, removing bank vegetation likely to fall into the river and sometimes straightening the river channel itself. These activities may well make the river more hydraulically efficient but have had serious impacts on fish by reducing cover and potential spawning sites for species like river blackfish (Cadwallader and Lawrence 1990).

Hortle and Lake (1983) found significantly higher species richness, total fish biomass, fish density and fish numbers in unchannelized sites than those that had been "improved". The "improvements" had reduced snags, reduced areas of slack water and reduced the length of bank with natural riparian vegetation.

The impacts of irrigation

Irrigation has been a significant force in shaping rural Australia, especially within the Murray-Darling basin. It has enabled a dramatically increased agricultural production from a low and unreliable rainfall. This has supported thousands of people on the land and supported many thriving rural communities. It may have helped "drought proof" Australia and ensure a supply of food to urban Australia. Irrigation does make a contribution to our balance of payments.

The environmental impacts of irrigation are of course not confined to an irrigator's property. Water must be stored, transported to the farm, applied and removed. Each step involves impacts.

Drainage waters include surface runoff from irrigated lands, which includes excess irrigation water, and rainfall that is more than the soil's ability to take up water. It also includes sub-surface drainage water from structures designed to lower water tables. Both sorts of water may contain significant contaminants. Drainage waters may be high in nutrients and salt.

Salinity

Close (1990) points out that native species are adapted to widely fluctuating natural salinities in large rivers in Australia. Under low flow summer conditions it is common for highly saline groundwater to enter the river. Close suggests that the range of salinities in the regulated rivers is less than in the previous unregulated condition. Hart *et al.* (1991) have reviewed the impact of saline waters on Australian stream and wetland biota. They identify a lack of data on the sensitivity of freshwater plants and animals to salinity increase. They suggest that direct adverse biological effects are likely to occur in rivers and wetlands once salinity exceeds 1 000 mg l^{-1}.

Increased salinity has changed the nature of many wetlands. For example the explorer Oxley named the Great Cumbungi Swamp, terminating the Lachlan River, presumably because of the presence of cumbungi. This has now been replaced by the more salt tolerant *Phragmites australis* (Sainty and Jacobs 1990).

The impacts of land use change

The development of agriculture in many catchments has had impacts on sediment and nutrients entering rivers. Nutrients have stimulated algal blooms causing considerable concern, while sediment and increased turbidity is implicated in the loss of aquatic plants from many river systems (Sainty and Jacobs 1990). Siltation of streams has been implicated in the decline of fish such as Macquarie perch and Australian bass, and in the decline of invertebrate populations.

The destruction of the riparian zones

The riparian zone is an essential component of a stream system. Riparian zone vegetation shades the water affecting both light and temperature, and harbours insect populations that may be food for fish. The leaf fall is important for some of the invertebrate population in the stream. The vegetation limits bank erosion and the movement of sediment and nutrients into the river channel from surrounding lands. These riparian areas have special conservation significance in their own right as vegetation communities and as habitats and corridors for other species.

While far sighted legislators in the last century ensured the riparian zone was in public ownership, decisions to allow farmers to lease these lands has meant we have not conserved the riparian zones in our agricultural areas. This is one of the causes of stream degradation through bank erosion and other processes.

Spread of exotic species

Australia has a long and disastrous history of the deliberate and accidental introduction of species from elsewhere which have displaced native species and led to widescale ecosystem destruction. Earlier this century we had a very active "fish acclimatization" movement dedicated to introducing fish to improve angling as a recreational activity. Some of this was due to the improved angling provided by the new species; some of it seemed a homesickness for Europe and the need to introduce familiar European species to the Australian landscape.

Barmuta *et al.* (1992) point to the lack of substantial scientific work to back up the anecdotal claims of the impacts of these introduced species on native species. Nevertheless the anecdotal evidence is strong and there is overseas evidence to demonstrate the impact of introduced fish on native populations (Allan and Flecker 1993). The mechanisms include competition for food, competition for habitat, predation, hybridization and the introduction of disease. It is unclear which of these have been the main mechanisms involved in the impacts on native fish of the introduction of salmonids. However, there is evidence from Australian studies that trout can reduce populations of small native fish (such as galaxiids) and invertebrates by predation (Fletcher 1986).

The introduction of willows and blackberries have dramatically altered riparian vegetation and stream morphology. Introduced plants such as *Elodea, Eichhornia* and *Salvinia* have in many areas replaced native aquatic plants, and in places caused major nuisance. There has been considerable effort to develop chemical controls for these plants.

There is also much anecdotal evidence that the introduction of European Carp has led to the destruction of native aquatic plants and native fisheries and given rise to the increased turbidity in billabongs and rivers. There is no agreement, however, on the ecological impact of carp.

The introduction of Nile Perch, *Lates niloticus*, was suggested by the Queensland Government in 1980. After a detailed consideration of its biology, especially its temperature tolerance and role as an opportunistic predator, it was recommended it would be an unwise introduction and the project was abandoned in 1986 (Barlow and Lisle 1987).

Townsend and Winterbourn (1992) discuss a New Zealand case study of a decision to introduce an exotic fish for an aquaculture development. New Zealand, unlike Australia, requires a formal environmental impact assessment of such a proposed introduction, but this was not sufficient to provide the information needed to make a rational decision. Townsend and Winterbourn propose a four step procedure:

a. environmental impact assessment;
b. publicity;
c. independent scientific review;
d. implementation.

In this New Zealand case, extensive publicity following the environmental impact assessment lead to interest groups actively searching for additional information. Much of the information that finally came to light had not been published but came from seeking information from experienced biologists in other countries that had knowledge of the proposed introduction. Deliberate introductions to Australia will hopefully be subjected to similar scrutiny.

Australia has no effective mechanisms to prevent the importation of aquarium fish which often escape and establish feral populations. These provide a significant hazard to native fish populations. The Weather Loach *Misgurnus anguillicaudatus* is now widespread following multiple releases, initially in Ginninderra Creek near Canberra following its importation as an aquarium fish. It is now established in the Murray River, Ovens River, Yarra River and the Maribynong River. It has the potential to become a major problem. We have no effective mechanisms to try and eradicate such populations when they establish.

Overexploitation

Overfishing

Cadwallader and Lawrence (1990) identify overfishing as a problem in the Murray-Darling, although the impact is obscured by other stresses. Various angling restrictions have been introduced to protect the endangered trout cod. Geddes (1990) discusses the impacts of overfishing on both yabby and River Murray crayfish populations. Commercial yabby fishing in Lake

Alexandrina from 1970–75 is thought to have removed 2 million mature yabbies per year. Yabby populations have not as yet returned to the high populations of the 1970s, although habitat destruction and other factors are important as well as the overfishing. Fishing for crayfish removes the largest, older animals that may be 20–50 years old. A continual decline in the size of River Murray Crayfish has been observed in New South Wales since the 1960s and the average weight of the catch was 200 g per animal. Since they do not breed until attaining a weight of about 250 g the future looks bleak despite more restrictive fishing regulations recently introduced.

Overfishing and habitat have had a very damaging impact on populations of the Giant Freshwater Crayfish *Astacopsis gouldi* in Tasmania. This is the largest freshwater arthropod in the world and has now been given protection in a series of rivers.

Excessive extraction of water

Irrigation water is trapped in storages in winter and released to provide water in summer. This has meant reduced water availability to recharge wetland areas in winter and consequent reduction in the viability of some wetland species. The reduction of medium sized floods on regulated rivers has various effects and conservationists argue for the maintenance of such flood regimes.

There is now growing pressure to provide water for the maintenance of wetlands and fisheries (Viessman 1990). This is hotly argued by irrigation interests who assert that environmental flows are just a waste of water, especially when the well-being of irrigators is at stake during a major drought. This is an interesting public policy issue on rationing of a scarce resource made available from a public investment. It would be normal in justifying such public investment nowadays to attempt to spread the benefits over a wide range of stakeholders. Providing water for one use does introduce the opportunity cost from the alternative uses of the water.

The algal blooms on the Darling River in recent years have focused attention on the problem of how much water should be taken from a river for irrigation or other purposes.

Secondary extinctions

Secondary extinctions occur when the removal of one species has a cascading effect throughout the species assemblages in that area, and other species unaffected by the primary stress, also become extinct (Allan and Flecker 1993). While there are documented studies overseas of this effect, there appear to be no documented cases in Australia, although it is likely the mechanism would be important in Australian waters. These mechanism may become important for freshwater mussels. The mussel larvae (glochidia) are parasitic on fish for part of the life cycle. Not all fish are suitable as hosts. Native fish are suitable, but some introduced species like carp are not (Walker 1990). As native fish populations decline in some areas this secondary effect may become apparent.

Chemical and organic pollution

Pollution from the discharge of sewage effluents and from urban and rural runoff has been well documented, and is a cause of the algal blooms in many inland waters. The subtle impacts of even low levels of effluent discharges have been discussed by Cullen and Norris (1989).

Herbicides are widely used in agriculture and in maintaining channels and drains within irrigation areas. There is no systematic long-term monitoring of these agricultural chemicals in Australia. There is not even agreement about where to monitor these substances — in the water, in sediment or in the biota. Analyses are expensive. Sampling is tricky since the movement of chemicals is often as pulses with water flow rather than as continuous inputs. Biological assessment would seem more useful in these situations, yet the agencies concerned rarely have this capability (Cullen 1990).

These agricultural chemicals have impacts on the biota in receiving waters and some of the few studies available have been discussed by Bowmer (1990). Periodic fish kills in the cotton growing areas are generally blamed on the cotton industry.

A number of Australian streams have a long-term impact from previous mining activities. The Molonglo River near Canberra is a well documented example where contamination by heavy metals has persisted despite mining having ceased in 1962. Various restoration attempts have reduced some sources of contamination but there has been little discernible change in the distribution and abundance of invertebrates (Norris 1986).

Climate change

Any effects of global climate change on Australian aquatic biota will be almost impossible to detect given the variety of stresses on the system and the lack of any long-term biological monitoring sites in Australia. A small increase in temperature might give some species that are near the limits of their temperature range some difficulties. Close (1988) has examined some of the implications of climate change on the River Murray.

REVERSING THE TRENDS

Rivers are unusual ecosystems in that there is a movement of water mass and energy down the river. Consequently, impacted areas can often recover once the cause of degradation is removed since propagules arrive from upstream and can recolonize the impacted area. This, of course, assumes there are no barriers to movement, such as dams, and that there are sources for such colonization in upstream areas. As Allan and Flecker (1993) point out the potential for recovery of damaged river ecosystems is considerable. Prerequisites for recovery are however that we have sufficient knowledge to guide managers and that the value of rivers for conservation, science, recreation and aesthetics are recognized as significant values alongside the values of supplying water and cheap waste disposal. There is evidence that the community is now demanding that water managers maintain these wider community values, brought about partly by the toxic cyanobacterial blooms along 1 000 km of the Darling River.

Our knowledge base

Inventories of river health

Lake and Marchant (1990) and Hart (1992) have both called for a National Rivers programme to undertake reconnaissance surveys of Australian rivers. Some studies of this nature have been undertaken already in Victoria which provide a base to develop appropriate techniques. Mitchell (1990) provides important information on 868 sites in Victoria. Physical variables used included streambed composition, flow velocity and water depth. Vegetation features included cover of bank, underwater vegetation, cover for fish and amount of organic debris. Each feature was scored according to its suitability for aquatic life and an overall "environmental rating" developed.

Conservation status of selected species

Many of the gaps in our knowledge have been identified earlier in this paper. Taxonomic work has not been a high fashion area in science in recent years but if we cannot identify the species in our aquatic ecosystems one wonders how we can effectively manage them. We need to go beyond taxonomic description and increase the number of species for which we have good distributional information that can be correlated with environmental attributes.

Ecological understanding

It is clear from Lake (1987) that our ecological understanding of Australian rivers and other aquatic ecosystems is rudimentary. Research on energetics, nutrient dynamics and community structure and function of representative rivers is essential if we are to develop useful predictive models.

Science is about prediction and understanding. We have crude predictive models in ecology. They will not be improved unless we have long runs of data to calibrate and validate the models or explanatory theories in a variety of environments. These models are fundamental to improving environmental prediction (Cullen 1990). Since the models deal with ecological processes that vary over long periods, and over large spatial areas, it is apparent that sustained ecological research is important if we are to improve our predictive capability. The methods used in ecological research must consider the scale of the phenomena being studied (Swanson and Sparks 1990). Most important questions in ecology deal with predicting ecosystem response to stress or change, testing the correctness of ecological concepts and prediction of future conditions.

If long-term environmental studies are to be a useful investment in our research portfolio it is essential we carefully select key hypotheses that are to be tested in each study. Mindless collection of data, while keeping young scientists in work, will hardly advance environmental science in Australia. We have many examples in management agencies of extensive inventories of natural resource information being collected at great cost, and never being used to guide any decisions or to test any ecological theory. Commonly when one tries to use these general purpose inventories one finds the wrong attributes have been measured, the wrong class intervals chosen and the wrong scales adopted. State-of-the-art experimental design must be used to address a specific problem. Science rarely advances by hoping that something will emerge from a morass of randomly collected data. The same lesson can be learned from water quality monitoring where extensive data sets have been collected, but the data are often flawed because of changes in sites and analytical methods.

Three types of long-term environmental study are envisaged:

a. *Long-term observational studies* need to be established in unimpacted ecosystems to establish baselines for assessing change. The Australian Science and Technology Council (ASTEC) (1990) proposes a series of biomes which could be the basis for selecting sites. These sites need a long-term commitment, and need to ensure the collection of a comprehensive data set that characterizes the structure and function of the ecosystem. It may be such sites will be located in national parks as providing an appropriate management structure for such sites. The scientific direction of such sites needs to include a variety of disciplines and agencies rather than be restricted to the prime management agencies mindsets.

b. *Long-term experimental sites* need to be established to identify the consequences of our land management strategies in forest, agricultural, rangeland and aquatic environments. The hydrologic consequences of logging may be seen over a period of 50–100 years; the ecological impacts on wildlife over an even more uncertain time frame. The chances of identifying anything useful in three year studies on such issues is negligible given the temporal and spatial scale of the phenomena concerned. The Hubbard Brook Experimental Forest in New Hampshire in the USA is very important in showing the flow of materials through ecosystems at the catchment scale in response to clear felling (Borman and Likens 1979). Unexpectedly, this long-term experimental programme was also critical in proving the occurrence of acid rain in North America.

c. *Long-term monitoring sites* should provide the core of any environmental monitoring programme, with short term stations being used as needed. We can also expect a return on investment in analysing existing data sets coming from monitoring programmes. Many of these programmes are poorly designed and are poorly managed. Clarification of objectives, professional study design and competent implementation may not cost much more but would yield a much better return than now.

In each of these three types of long-term study it is essential to provide continuity in data collection, professional quality control and periodic scientific review of findings, guiding hypotheses and future directions. Long term does not mean static, and new techniques of data collection and new insights into ecosystem behaviour will require continuing scientific direction of a high standard.

It is also important that we review the need for the data. These sorts of long-term experimental studies and data collection exercises do need to provide a pay-off for the resources invested. The pay-off will be in better understanding of how ecosystems work, and better predictive models to guide decision makers.

It is clear for these scientific reasons, as well as for the other reasons of conserving the genetic material of Australian aquatic ecosystems, that we need to establish a series of National Aquatic Reserves. Some aquatic systems receive adequate protection within National Parks, but often these agencies have neither the skill, the interest or the jurisdiction to manage to maintain aquatic ecosystems. There are many situations where appropriate type systems are not in the existing Parks systems. Much work needs to be done on this proposal, since ideally the aquatic reserves would include floodplains and intact riparian vegetation, as well as be subject to a natural range of flows. It is already too late to find such sites throughout much of the Murray-Darling basin.

Strategies for restoration

Restoration generally requires removal or reduction in the stressor causing the degradation. There seems little point in just treating symptoms when it is possible to cure the disease. It is interesting however to see how cyanobacterial blooms have led to much research on toxin identification and water treatment processes when we know the causes of blue green blooms is mismanagement of nutrients on a catchment basis. In many situations we have enough knowledge to proceed to field and demonstration trials to fine tune better management approaches. There is much folk lore in Australian resource management and if effective techniques are to be demonstrated they need to be embedded in a strong ecological context and subject to professional monitoring and evaluation.

Environmental flows

There are knowledge gaps in determining appropriate environmental flows that must be maintained in rivers if we wish to maintain functional aquatic ecosystems. Managers, limnologists and the public need to interact to negotiate the objectives for such interventions. Do we seek to maintain flows to maintain fish numbers, to maintain fish breeding, to prevent deoxygenation, to flush nutrients and contaminants or to inhibit the development of algal biomass? Environmental flows might be required to allow periodic flooding to maintain riparian vegetation (the River Red Gums of the Barmah forest area for example) or to connect billabongs and rivers to provide a flush of food to allow fish breeding.

The purpose of the flow needs to be clearly articulated, and this will allow assessment of the required amount and timing of flow required. It may even be that some wetlands can cope without water in a dry year if they are given periodic flooding in years when water is not so scarce.

There are many social problems in delivering these environmental flows. Irrigators and other water users believe their uses of the water should have a higher priority for any available water. The irrigators in New South Wales have been arguing for even further alienation of public waters to give them better "security of supply". This is a recurring example of public goods being used for private gain that has so characterized much of Australia's development.

Total catchment management

The quality of water in a river is an ideal performance indicator for the health of a catchment. If the water quality is poor, so is the level of land management within the catchment. Australia has developed some exciting community-based approaches to managing land and water, and it is within this total catchment context that improvements will be made. The land uses that are allowed in a catchment, and the skills that land managers demonstrate determine the erosion and other contaminants that enter the waterways. It is important that biologists, fishermen and conservationists involve themselves in these community processes to ensure the interests of aquatic ecosystems are at least advocated in these processes.

Regional water quality management requires a consideration of point and non-point inputs to waterways, and a consideration of the likely change these inputs will have on the receiving ecosystem. Such planning should be undertaken for high, medium and low flow conditions. The impacts of sewage discharge and the impacts of inadequately treated wastes from intensive animal industries are critical inputs to catchment-based water quality planning.

SUMMARY AND CONCLUSION

Australia's natural waterways have been seriously degraded by the agricultural development of the countryside. Many of these flowing waters may be restored if land uses and land management can be improved to reduce the erosion and flow of nutrients that have caused such devastation. While the impacts of dams can probably not be removed, there are ways to reduce impediment to fish movements and to release water to allow natural ecosystems to function.

We need to considerably improve our knowledge base if we are to better manage these systems. We need to ensure communities and decision makers understand the connections between land use and water quality and make sure the real environmental impacts of major decisions are identified as part of the decision making process.

REFERENCES

Allan, J. D. and Flecker, A. S., 1993. Biodiversity conservation in running waters. *BioScience* **43:** 32–43.

Australian and New Zealand Environment and Conservation Council, 1992. Australian water quality guidelines for fresh and marine waters.

Australian Science and Technology Council, 1990. *Environmental research in Australia: the issues.* Australian Government Publishing Service: Canberra. 87 pp.

Barlow, C. G. and Lisle, A., 1987. Biology of the Nile Perch *Lates niloticus* (Pisces: Centropomidae) with reference to its proposed role as a sports fish. *Biol. Conserv.* **39:** 269–98.

Barmuta, L. A., Marchant, R. and Lake, P. S., 1992. Degradation of Australian streams and progress towards conservation and management in Victoria. Pp. 65–79 in *River conservation and management* ed by P. Boon, G. Petts and P. Calow. John Wiley and Sons: Chichester.

Bishop, K. A. and Forbes, M. A., 1991. The freshwater fishes of northern Australia. Pp. 79–107 in *Monsoonal Australia: landscape ecology and man in the Northern Lowland* ed by C. D. Haynes, M. G. Ridpath and M. A. J. Williams. A. A. Balkema: Rotterdam.

Borman, F. H. and Likens, G. E., 1979. *Pattern and process in a forested ecosystem.* Springer-Verlag: New York.

Bowmer, K. H., 1990. Herbicides in water: economics or environment? CSIRO Div. Water Res. Tech Memo 90/4.

Cadwallader, P. L. and Lawrence, B., 1990. Fish. Pp. 317–35 in *The Murray* ed by N. Mackay and D. Eastburn. Murray-Darling Basin Commission: Canberra.

Carter, P., 1987. *The road to Botany Bay. An essay in spatial history.* Faber and Faber: London.

Close, A., 1988. Potential impact of greenhouse effect on the water resources of the River Murray. Pp. 312–23 in *Greenhouse: planning for climate change* ed by G. I. Pearman. CSIRO and L. J. Brill.

Close, A., 1990. River salinity. Pp. 127–44 in *The Murray* ed by N. Mackay and D. Eastburn. Murray-Darling Basin Commission: Canberra.

Cullen, P., 1986. Managing nutrients in aquatic systems: the eutrophication problem. Pp. 539–54 in *Limnology in Australia* ed by P. De Dekker and W. D. Williams. CSIRO/Dr W. Junk: Melbourne/Dordrecht.

Cullen, P., 1990. Biomonitoring and environmental management. *Environ. Monit. and Assessment* **14:** 107–14.

Cullen, P. and Norris, R., 1989. Significance of lakes and rivers in the Australian mainland alps. In *The scientific significance of the Australian Alps* ed by R. Good. Aust. Alps National Parks Liaison Ctee and Aust. Acad. of Science: Canberra.

Davidson, B., 1969. *Australia, wet or dry?* Melbourne University Press.

Fearnside, P. M., 1989. Brazil's Balbina Dam: environment versus the legacy of the pharaohs in Amazonia. *Environ. Manage.* **13:** 401–23.

Fletcher, A. R., 1986. Effects of introduced fish in Australia. Pp. 231–38 in *Limnology in Australia* ed by P. DeDekker and W. D. Williams. CSIRO/Dr W. Junk: Melbourne/Dordrecht.

Geddes, M., 1990. Crayfish. Pp. 303–37 in *The Murray* ed by N. Mackay and D. Eastburn. Murray-Darling Basin Commission: Canberra.

Hart, B. T., 1992. Ecological conditions of Australia's rivers. *Search* **23:** 33–37.

Hart, B. T., Bailey, P., Edwards, R., Hortle, K., James, K., McMahon, A., Meredith, C. and Swadling, K., 1991. A review of the salt sensitivity of the Australian freshwater biota. *Hydrobiologia* **210:** 105–44.

Hill, L. and Michaelis, F. B., 1988. Conservation of insects and related wildlife. Occ. Pap. 13. Australian National Parks and Wildlife Service: Canberra.

Hortle, K. and Lake, P. S., 1983. Fish of channelized and unchannelized sections of the Bunyip River, Victoria. *Aust. J. Mar. Freshwat. Res.* **34:** 441–50.

Horwitz, P., 1990. The conservation status of Australian freshwater Crustacea. Rpt 14. Australian National Parks and Wildlife Service: Canberra.

Lake, P. S., 1987. Stream ecosystem research in Australia: problems and strategies. National Water Research Seminar. Dept Res. and Energy: Canberra.

Lake, P. S. and Marchant, R., 1990. Australian upland streams: ecological degradation and possible restoration. *Proc. Ecol. Soc. Aust.* **16:** 79–91.

Lake P. S., Barmuta, L. A., Boulton A. J., Campbell, I. C. and St Clair, R., 1985. Australian streams and northern hemisphere stream ecology: comparisons and problems. *Proc. Ecol. Soc. Aust.* **14:** 61–82.

Lake, P. S., Schreiber, E. S. G., Milne, B. J. and Pearson, R. G., 1994. Species richness in streams; patterns over time, with stream size and latitude. *Verhandlungen der Internationalen Vereinigung fur theoretische und angewandte Limnologie* **25:** 1822–836.

McMahon, T. A., 1982. World hydrology: does Australia fit? In *Hydrology and water resources symposium 1982.* The Institution of Engineers: Australia. 7 pp.

Mitchell, P., 1990. The environmental condition of Victorian streams. Dept of Water Resources, Vic.: Melbourne.

Mott, J. J. and Bridgewater, P. B., 1992. Biodiversity conservation and ecologically sustainable development. *Search* **23:** 284–87.

Norris, R. H., 1986. Mine waste pollution of the Molonglo River, NSW and ACT: effectiveness of remedial works at Captains Flat mining area. *Aust. J. Mar. Freshwat. Res.* **37:** 147–57.

Pressey, R. L., 1990. Wetlands. Pp. 167–81 in *The Murray* ed by N. Mackay and D. Eastburn. Murray-Darling Basin Commission: Canberra.

Rapport, D. J., Regier, H. A. and Hutchinson, T. C., 1985. Ecosystem behaviour under stress. *Amer. Naturalist* **125:** 617–40.

Sainty, G. R. and Jacobs, S. W. L., 1990. Water plants. Pp. 265–73 in *The Murray* ed by N. Mackay and D. Eastburn. Murray-Darling Basin Commission: Canberra.

Swanson, F. J. and Sparks, R. E., 1990. Long-term ecological research and the invisible place: the local to global spatial scales of the long term ecological research programme. *BioScience* **40:** 502–08.

Tighe, P. G., 1984. The Australian developmental ethos and the past, present and future of Australian rainforests. In *Proceedings of a workshop on the past, present and future of Australian rainforests* ed by G. L. Werren, and A. P. Kershaw. Monash University: Melbourne.

Townsend, C. R. and Winterbourn, M. J., 1992. Assessment of the environmental risk posed by an exotic fish: the proposed introduction of Channel Catfish *Ictalurus punctatus* to New Zealand. *Cons. Biol.* **6:** 273–82.

Viessman, W., 1990. Water management issues for the nineties. *Wat. Res. Bull.* **26:** 883–91.

Walker, K. F., 1985. A review of the ecological effects of river regulation. *Hydrobiologia* **125:** 111–29.

Walker, K. F., 1990. Mussels. Pp. 309–14 in *The Murray* ed by N. Mackay and D. Eastburn. Murray-Darling Basin Commission: Canberra.

The impact of salinization and sedimentation on aquatic biota

L. METZELING,[1] T. DOEG[2] and W. O'CONNOR[2]

Salinization and sedimentation are major threats to aquatic biota. Although there are aquatic animal and plant species in Australia able to tolerate high salinities, many groups are restricted to freshwaters and will be adversely affected by even small increases in salinity. Macroinvertebrates, macrophytes and algae are the groups considered to be most at risk from salinization, with impacts expected to occur with an increase of as little as 1 g/L.

Sedimentation is caused by a variety of activities and occurs in three phases: suspended, deposited and hyporheic. Five species of fish and many species of invertebrate have been conclusively found to be adversely affected by sedimentation; many other species with similar requirements will also be threatened. Increased sediment to streams due to human activities is listed as a potentially threatening process under the *Flora and Fauna Guarantee Act* 1988 of Victoria.

INTRODUCTION

SALINIZATION and sedimentation are major environmental threats to aquatic species and ecosystems. In Victoria, salinization is considered the greatest single threat facing the environment (Salt Force 1988) and increased sediment to streams due to human activity is listed as a potentially threatening process under the *Flora and Fauna Guarantee Act* 1988. Both problems are widespread and potentially threaten many species and ecosystems (Lake and Marchant 1990; Williams 1987). Despite this, there is considerable scientific uncertainty as to what degree of increase over natural levels will impact on individual species, communities or populations. While some of the recent studies described below fill some of these information gaps, considerably more work is still needed.

SALINIZATION

Salinization has been defined as the process which leads to an increase in concentration of dissolved salts, or the result of this process (Williams 1987). It is a natural process, most evident in endorheic drainages such as the Lake Eyre basin and the Wimmera River in western Victoria. These systems typically contain salt lakes within their catchments or at the terminus of the river. Periods of high salinity have occurred many times in Australia's past, and have been well documented for the past 500 000 years for the Murray Basin (Bowler 1988). These past episodes of salinization have occurred in response to climatic fluctuations with resulting salinities sometimes exceeding those recorded in recent times.

The current episode of salinization, however, is not due to natural events but began with the introduction of European technology (Macumber 1991). There are two major causes. The first is the extensive clearing of deep-rooted native vegetation from the highlands and plains, replacing it with shallow-rooted crops and pasture. The second is the introduction of large-scale irrigation schemes in some parts of the country, particularly within the Murray-Darling Basin (Macumber and Dyson 1988). Both of these result in greater infiltration of surface water into the groundwater and the gradual filling of aquifers. This raises the water table to or near the surface, bringing accumulated salts, and causes the salinization of soils and water-bodies. The salinity problem is fundamentally a groundwater problem, and long-term control will depend on the success of reducing ground-water pressures (Macumber 1991).

Some species of the Australian aquatic biota are tolerant of very high salinities, particularly some invertebrates and algae (Bayley 1972; Bayley and Williams 1973; Williams 1981; Timms 1981, 1983; Greenway 1986). This is of little relevance to the current salinization problem, which can occur when freshwaters of less than 0.5 g/L are increased to 1–3 g/L and perhaps up to 10 g/L (Hart *et al.* 1990).

Most of the effort addressing the salinity problem has been directed towards the economic and social impacts (Peck *et al.* 1983) while the environmental impacts, including those from control strategies, have received scant attention (Williams 1987). Major reviews by Hart *et al.* (1990, 1991) considered the possible impact of salinization to streams and wetlands, and summarized what was then known on the salt sensitivity of Australian aquatic biota.

[1]Environment Protection Authority, Victoria, 477 Collins Street, Melbourne, Victoria, Australia 3000.
[2]Department of Conservation and Natural Resources, Victoria, 123 Brown Street, Heidelberg, Victoria, Australia 3084.

Pages 126–36 *in* CONSERVING BIODIVERSITY: THREATS AND SOLUTIONS, ed by R. A. Bradstock, T. D. Auld, D. A. Keith, R. T. Kingsford, D. Lunney and D. P. Sivertsen. Surrey Beatty & Sons. 1995.

Hart *et al.* (1990, 1991) concluded that freshwater aquatic ecosystems are likely to show adverse effects to an increase in salinity of as little as 1 g/L. The effects may be direct or indirect and could affect all functional components of an aquatic ecosystem (Lake 1987). Sub-lethal effects may include alterations to behaviour, ecosystem processes such as metabolism, production and nutrient cycling or spiralling, and the structure and functioning of lowland aquatic communities, the communities most at risk from salinization (Hart *et al.* 1990). Most sensitivity information relied on distributional data rather than specifically designed studies such as laboratory toxicity tests or in situ manipulations of salinity. The review highlighted significant information gaps regarding the response of the biota to a small increase in salinity.

The following discussion covers only biota which are largely restricted to aquatic environments, and excludes biota associated with (e.g., riparian vegetation) or living on rather than in the water (e.g., waterfowl). It begins with the findings of Hart *et al.* (1990, 1991).

Effects of salinization on aquatic biota

Mammals, reptiles and amphibians

Three species of mammals inhabit aquatic environments in Australia. Of these, the two species of water rat are unlikely to be adversely affected by salinization as they can live in coastal environments and may survive out of water if conditions become harsh. The platypus is also thought not to be at risk (Hart *et al.* 1991) but may be indirectly effected by habitat degradation and an altered food supply.

Most reptiles are also considered to be safe from salinization (Hart *et al.* 1991). The only doubt concerns freshwater turtles but one species possesses a salt gland, which enables excess salts to be excreted. The other freshwater turtles may also possess such a gland, and thus the group may not be at risk.

Nothing is known about the salinity tolerance of Australian amphibians. The most sensitive stages may be the egg masses and tadpoles (Hart *et al.* 1991).

Fish

Adults of most fish species found in the Murray-Darling Basin are thought to be able to tolerate up to 10 g/L (Anderson 1991; Hart *et al.* 1991). The Southern Pygmy Perch *Nannoperca australis* is also known to tolerate only salinities up to about 10 g/L (Chessman 1971). Such a salinity may not, however, result in viable reproduction and growth of larvae and young fish.

Bacher and Garnham (in press) examined the salinity tolerance of four native and two exotic fish species (Table 1). They measured the osmolality of the blood in response to different salinities and from this determined the no observable effect level (NOEL). Physical stress was usually observed at higher concentrations than was a change in osmolality. Tolerances varied greatly, the most sensitive being the native species River Blackfish *Gadopsis marmoratus* followed by Australian Grayling *Prototroctes maraena*. Bacher and O'Brien (1989) demonstrated how the salinity tolerance of *P. maraena* varied with life stage, the NOELs for eggs and larvae being 7 and >30 g/L respectively, whereas adults had a NOEL of 4.9 g/L (Table 1). These tolerances can be interpreted relative to the life history of this species as the eggs are laid in fresh water, the larvae migrate down to estuaries and the adults move back into fresh waters before breeding. The adults appear to lose their ability to osmoregulate their internal fluids in saline waters (Bacher and Garnham, in press).

Table 1. Salinity tolerance of six species of fish determined from laboratory tests. Adapted from Bacher and Garnham (in press).

Species	NOEL (g/L)[a]
Salmo trutta (Brown Trout)	>24
Perca fluviatilis (Redfin)	9.8
Gadopsis marmoratus (River Blackfish)	<2
Pseudophritis urvilli (Tupong)	17
Prototroctes mareana (Australian Grayling)	4.9
Galaxias maculatus (Common Galaxias)	17.3

[a]NOEL — No observable effect level.

It can be difficult relating data derived from laboratory tests to natural situations as concentrations tolerated in the laboratory are often far greater than encountered in nature. To determine standards or criteria, it is common to use an application factor of ×0.1 for laboratory data for non-persistent toxicants. If this approach is adopted for salinity, then concentrations should not exceed 0.7 g/L in streams where grayling are breeding or 0.5 g/L in order not to effect adult populations. Such concentrations seem to be quite low and would be exceeded during low flow periods in many streams. Further consideration needs to be given to what, if any, is the approriate application factor for salinity.

Macrophytes and algae

Freshwater macrophytes and algae are sensitive to small increases in salinity (Hart *et al.* 1990, 1991). Field observations (Brock 1981; Brock and Lane 1983) have suggested that 4 g/L is the upper limit tolerated by many common and widespread species of macrophyte. Above this limit, freshwater species are replaced by halophytic species, some of which can tolerate very high salinities.

James and Hart (1993) examined the effect of salinity increases on four common species of freshwater macrophyte in glasshouse trials. Response varied greatly between species, and was also dependent on the concentration of salt and duration of submersion in saline water. *Potamogeton tricarinatus* was the most sensitive species followed by *Myriophyllum crispatum,* with significant mortalities occurring at 5 and 6 g/L respectively. Neither *Eleocharis acuta* nor *Trichglochin procera* suffered any mortalities from 7 g/L during the 72 days of study. All species showed delayed adverse effects in salinities greater than 1 g/L, including reduced growth rate, reduced numbers of leaves or shoots, development of necrotic regions, and death of apical tips. In the short term, no response was apparent, even at high salinities; but even low increases in salinity produced adverse effects over the longer time period. Three species of *Potamogeton* in the Lower Rhine River also had reduced biomass and growth rates associated with higher salinities (Brink and Velde 1993).

A prolonged increase in the salinity of a wetland or stream, even if the increase is small, will result in an altered macrophyte community with the possible replacement of sensitive species with halophytic species.

There is little direct information on the salinity tolerance of algae, although some species are tolerant of high levels of salinity. Distributional data suggests that some species will be sensitive to small increases in salinity. Hart *et al.* (1991) considered an increase of about 1 g/L will have adverse effects on many species of algae.

Invertebrates

Some species of invertebrate are tolerant of very high salinities whereas many others are more restricted in their distributions being only found in fresh waters. Based largely on distributional information, Hart *et al.* (1990, 1991) concluded that many species of invertebrate from most major taxonomic groups would be adversely affected by an increase in salinity of about 1 g/L.

Bacher and Garnham (in press) determined the salinity tolerance for two common species of invertebrate, *Daphnia carinata* and *Paratya australiensis* in laboratory trials. The LC50 and NOEL for *D. carinata* were, respectively, 2.5 and 1 g/L. The equivalent concentrations for *P. australiensis* were 4.5 and <2 g/L. These species were more sensitive to an increase in salinity than any of the six species of fish they studied (Table 1). Interestingly, Williams (1984) found *P. australiensis* able to tolerate much higher salinities, with a calculated LC50 of 25.8 g/L and a NOEL of 16.6 g/L. The difference in these results may reflect differences in the salinity of the environments from which the animals were collected. Bacher and Garnham's animals were from the Yarra River, Victoria, which rarely exceeds 0.2 g/L, whereas the animals used by Williams came from the Torrens River, South Australia, which had a salinity of 0.75 g/L at the time of collection and is often higher. These two populations had adapted to their local salinity, suggesting that, given time, a population inhabiting fresh waters may become tolerant of more saline conditions. This may not be a general response of invertebrates, as *P. australiensis* is widespread and can clearly tolerate a broad range of salinities.

Laboratory tests on common and widespread stream invertebrates (Metzeling, unpubl. data) have shown some to be sensitive to small increases in salinity. Three species of Trichoptera *Cheumatopsyche modica, Cheumatopsyche* sp. 1 and *Asmicridea edwardsii* showed increased mortality above 1.2–2.4 g/L. Adult elmid beetles were more tolerant with increased mortality above 3.6 g/L whereas the hemipteran *Micronecta gracilis* was unaffected by salinities up to 6 g/L. These data support the conclusion of Hart *et al.* (1990, 1991) regarding an increase in salinity of 1 g/L having adverse affects on aquatic biota.

Laboratory studies may be linked to field observations by *in situ* manipulations of salinity. Such work is underway in a stream and a wetland in central Victoria (Paul Bailey, Monash University, pers. comm.). Salinities in isolated sections within these two ecosystems have been increased and the response of the fauna will be compared with that in control sections.

Some ecological studies in recent years have examined macroinvertebrate community structure in streams of different salinities. Studies of two very saline streams in western Victoria (salinities ranged from 5–30 g/L; Mitchell and Richards 1992) and the Blackwood River in Western Australia (Williams *et al.* 1991) concluded tentatively that the invertebrate fauna of Australian streams may be tolerant to higher levels of salinity than previously expected. These streams were in areas with a long history of salinity, thus suggesting that the fauna had adapted to high salinity, or may represent the salt tolerant remains of a once more diverse fauna.

Metzeling (1993) quantitatively examined invertebrate communities at nine sites on six streams where salinities ranged from 0.05 to 1.2 g/L (and the more saline streams were historically higher, up to 2 g/L) and found no relationship between salinity and the simple measures of species number or faunal abundance. The sites showed clear differences in community structure in multivariate analyses which were strongly related to salinity. The less common or "rare" taxa were the main difference between the communities. Common or abundant taxa did not distinguish between low and high salinity sites. This may be due to wide tolerance by the common taxa of environmental factors including salinity. The "rare" taxa may have narrower environmental requirements and may therefore be more susceptible to salinization. Although the combined contribution of these "rare" taxa to the overall invertebrate abundance is usually less than 10 per cent, their contribution to the biodiversity of the ecosystem is large, usually making up more than two-thirds of all species found. The loss of rare species may reduce the capacity of the community to respond to severe catastrophes such as extreme drought or toxic pollution (Patrick 1988a).

Although there are significant information gaps, many groups of invertebrates are probably sensitive to salinization (Table 2). The extent of the impact and the ability of the fauna to adapt to increased concentrations of salt remains uncertain. The loss of sensitive species, even though they may in time be replaced by some salt tolerant forms, will result in an overall loss of biodiversity.

Table 2. Freshwater invertebrate groups considered by Hart *et al.* (1991) most sensitive to salinization.

	Lowland rivers	Wetlands
Cnidaria	*Hydra* sp.	*Hydra* sp.
Platyhelminthes	*Dugesia* sp.	
	Turbellarians	Turbellarians
Annelida	Naididae	Spionidae
	Lumbricidae	Naididae
		Lumbricidae
Crustacea		Cladocera
		Ostracoda
	Parastacidae	Parastacidae
Insecta	Ephemeroptera	Ephemeroptera
	Plecoptera	
	Odonata	Odonata
		Heteroptera
		—*Anisops*
		—*Naucoris*
		—*Notonecta*
		—*Rantra*
		—*Laccotrephes*
	Trichoptera	Trichoptera
	—Leptoceridae	—Leptoceridae
	—Hydroptilidae	—Hydroptilidae
	Diptera	Diptera
	—Tipulidae	—Tipulidae
	—Simuliidae	
Mollusca	Gastropoda	Gastropoda
	—Ancylidae	—Ancylidae
	—Planorbidae	—Planorbidae
	—Lymnaeidae	—Lymnaeidae
	—Physidae	
	Bivalvia	Bivalvia
	—Sphaeriidae	—Sphaeriidae

Saline pools

A saline pool is a pool within the watercourse of a river where the salinity of the bottom waters is higher, often considerably higher, than the surface waters. They are believed to be formed by the intrusion of saline groundwater directly into the riverbed or via saline seeps along the banks (Anderson and Morison 1989).

Salinity differences between surface and bottom waters may be small (<1g/L) or large (greater than seawater, i.e., >35g/L), but both result in stable stratifications within the pool. High flows can remove stratifications but they can re-establish within one to two months (Anderson and Morison 1989). Due to lack of contact with the surface, the bottom waters quickly become low in dissolved oxygen or anoxic and can also become sinks for nutrients. In the Wimmera River, Victoria, the bottom waters of saline pools have had concentrations of total phosphorus of up to 6.2 mg/L and total Kjeldahl nitrogen of up to 11 mg/L, two to three orders of magnitude greater than surface water concentrations (Metzeling, in press).

Although salinity levels may be prohibitively high in these pools, the most immediate limiting factor for the fauna is the lack of oxygen. Surveys of benthic invertebrates have found a very poor fauna in these pools, with only small numbers of a few taxa (Metzeling, in press). It is unlikely that fish would inhabit such pools or would visit them only briefly. As the stratifications often occur within 1 to 2 m of the surface, the habitat rendered unsuitable for most forms of life represents a significantly large proportion of the total habitat within the river. The increased incidence of saline pools is another way in which salinization is impacting on the aquatic biota.

Salinization will cause more widespread occurrence of these pools. There are saline pools in many lowland Victorian rivers with intermittent or low flow regimes (Anderson and Morison 1989; McGuckin *et al.* 1991) and also in south-west Western Australia (Morrissy 1979).

Salinity control strategies

Controlling the salinity problem is a formidable task. Doing nothing may be the only economically viable course of action in some areas. Various strategies attempt to control the salinity problem. These include reafforestation of recharge areas, establishment of deep-rooted pasture, landforming and improved irrigation techniques reducing water use, groundwater pumping and improved drainage to remove groundwater before it reaches the soil surface (Macumber 1991). These control options may ultimately benefit the environment even though implemented in terms of economic costs and benefits.

The environmental consequences of control strategies have received scant attention. For example, sub-surface drainage has been recommended in many of the salinity management plans proposed in Victoria but deliberate discharge of any saline waste water is likely to result in the degradation of receiving streams and wetlands. Great care must be taken to ensure that efforts to arrest an environmental problem in one area do not cause it in another.

SEDIMENTATION

Sedimentation is currently affecting many rivers and streams throughout Australia causing ecological degradation in Australian upland (Lake and Marchant 1990) and lowland streams (Mitchell 1990). While the following discussion focuses on Victorian examples, the principles outlined are generally applicable to all areas of Australia.

Although erosional processes are essentially a natural event, the input of sediment is generally low in streams draining relatively undisturbed catchments (Olive and Rieger 1987). It is well established that these inputs have been greatly accelerated by human intervention and interference (Ministry for Conservation, Victoria 1983). A variety of land uses have been shown, or are believed to substantially increase the accession of sediment to streams (Mitchell 1990). While the deleterious effects of this increase in sediment on the biological resources of streams has been recognized for some time (e.g., Cordone and Kelly 1961; Chutter 1969), subsequent data suggests that sediment may be a more serious environmental threat than previously thought.

Changes in catchment-wide land-use, such as land clearance for agriculture, and grazing or cropping beside the river channel have increased erosion and sedimentation of streams and estuaries (OCE 1988; DCFL 1987). Uncontrolled grazing and clearing of streamside frontages has reduced the protective vegetation cover, resulting in stream bank erosion and higher than normal stream turbidity and sedimentation (Patrick 1988b). For example, it is estimated that the Avon River in Victoria and its tributaries have washed away around 72 million m^3 of soil (DWR 1989). While the major concern of this has been economic, it should be realized that this soil has entered the stream system as additional sediment.

Forestry activities have increased sediment accession to streams (e.g., Olive and Rieger 1987; Campbell and Doeg 1989; Papworth *et al.* 1990), even where they have been carefully controlled (MMBW 1980). For example, the export of sediment in a small catchment near Healesville, Victoria, increased from 40 kg/ha/year to 90 kg/ha/year immediately after logging (MMBW 1980). Of 27 forestry studies in Australia that investigated water quality, all but one showed some increase in sediment levels in affected streams, even though the monitoring techniques in these studies would tend to underestimate the increase (Doeg and Koehn 1990). All streams where forestry occurs are potentially affected by sedimentation.

The prime cause of this increase in sediment input following forestry comes from roading operations (Anderson 1971). In the example above, MMBW (1980) identified road crossings as the main source of the observed increase. According to DWR (1989), "Upland forestry areas are riddled with unsealed roads and unimproved tracks" and ". . . the potential for erosion and subsequent deterioration of water quality is considerable where drainage is poor, where tracks are located on slopes and where they are close to the stream" (p. 24). Increases in the frequency of landslips following forestry can also contribute elevated levels of sediment to streams (Doeg and Koehn 1990).

Dam construction has been the major source of sediment in some streams (EPA 1980; Blyth *et al.* 1984; Davey *et al.* 1987; West *et al.* 1984). For example, the mean suspended solids concentration in the Mitta Mitta River, Victoria, (April 1976–October 1977) below the Dartmouth Dam construction site was 23.1 mg/L, compared with 10.1 mg/L above the dam (West *et al.* 1984). Maximum concentrations (based on daily samples) were 205 mg/L below the dam compared with 133 mg/L above the dam site. Large releases for irrigation purposes (e.g., Doeg 1984) can also lead to increased bank erosion and sedimentation.

Following the completion of the dam, the barrier acts as a sediment trap, reducing the effectiveness of the structure. For example, between 1916 and 1975, sedimentation from poor land use upstream reduced the capacity of the Melton Reservoir in Victoria by 29 per cent (approximately 6 million m^3) (DWR 1989). To maintain the use of the barrier, this sediment has to be removed, and the cleaning of weirs (using inappropriate methods) has resulted in increases of sediment in some streams (e.g., 9 500 mg/L suspended solids 4 km downstream of a weir; Doeg, McKenzie-Smith and Metzeling, unpubl. data).

In-stream activities such as mining and gravel extraction, off-stream mining, quarries, run-off from construction sites, tourist or urban developments and industrial discharges may also increase sedimentation (Hellawell 1986; DWR 1989).

The effect of sedimentation is not restricted to the immediate site of the disturbance.

Sediment introduced into a stream will move downstream in cycles of suspension and deposition, until it is stopped by some barrier (e.g., a dam or a lake) or until it reaches the estuary or the sea.

Effects of sedimentation on aquatic biota

Most aquatic species may be affected by sedimentation. While many terrestrial species have the ability to move out of areas where environmental disturbance has occurred (and to recolonize after the disturbance has finished), aquatic species are generally confined to the river or stream and cannot avoid the threat. Only few taxa may be able to escape sedimentation events by migration from the disturbed zones.

Sediment in a stream system can be suspended (particles in the water column), deposited (particles lying on the stream-bed) or hyporheic (particles within the matrix of the stream bottom). While these phases are distinct, particles move between the phases depending on changing physical characteristics in the stream. Adverse effects on the biological resources of the streams may depend on the particular phase of any increased sediment. However, it is often impossible to distinguish which phase is responsible for observed effects on freshwater biota. All systems affected by sedimentation usually display elevated levels of all three phases of sedimentation mentioned above at some stage during a normal hydrograph (or between sampling efforts). Hence, unless the sediment is particularly fine (and stays in suspension) or coarse (and deposits rapidly), or specific experimental techniques are used (Doeg and Milledge 1991), it often cannot be concluded whether observed changes are due to suspended, deposited or hyporheic sediment.

It is important to realize that suspended sediment should be viewed like other toxicants where exposure time and concentration together determine the effect on aquatic communities (Hellawell 1986). Newcombe and MacDonald (1991) analysed 70 studies (mostly from the USA) that investigated the effect of suspended sediment on fish and aquatic macroinvertebrates, concluding that the effects of low levels of sediment sustained over a long period of time were the same as high levels of sediment over shorter periods, as for many other toxicants and pollutants.

Mammals, birds, reptiles and amphibians

None of the three species of mammals that inhabit freshwater environments are likely to be directly affected by sediment levels. Platypus have wide environmental tolerances and are often reported in the water during periods of high turbidity. Both species of Water Rats have the ability to avoid turbid waters and survive on non-aquatic food sources. However, reductions in macroinvertebrate food supplies due to sedimentation (see later) may have deleterious effects on local Platypus populations.

Birds and reptiles that live outside aquatic systems and rely heavily on aquatic food resources (e.g., Kingfishers, Water Dragons) may also not be affected directly by sedimentation, but may be indirectly affected by impacts to food supplies (e.g., frogs or fish).

Frogs in rivers and streams may be adversely affected by sedimentation at certain stages of their life cycle. While adults may avoid turbid systems until they clear, eggs laid in sticky, jelly-like clumps in streamside pools or backwaters (e.g., Hero *et al.* 1991) may be affected by coatings of adhered sediment, reducing oxygen exchange. Tadpoles restricted to the stream channel are likely to display similar reactions to increased sediment as fish (see later).

While there are concerns regarding the impact of sedimentation on higher vertebrates, there is generally insufficient data relating increased sedimentation to higher vertebrate species in Australia.

Fish

Lloyd (1987) reviewed the available data on the effects of turbidity on North American and European salmonids, with reported effects including direct fatality, reduced growth rates, reduced feeding, altered diet, increased stress, increased incidence of disease, altered behaviour and displacement. Other potential or reported effects include a reduction in oxygen uptake by the coating or clogging of gills by fine particles and a reduction in feeding efficiency for visual predators (DWR 1989). There is no reason to suspect that native Australian fish species would not display these effects, although the levels of suspended sediment that may induce them is generally unknown.

Deposited sediment also has direct adverse effects on fish populations. The smothering of the substrate and wood debris, and the filling of pools and scour holes reduces the usable habitat areas. This affects sites for deposition of eggs, and as habitat, shelter and rearing areas by juvenile fish and small species (Cadwallader and Backhouse 1983; Koehn and O'Connor 1990a; Mitchell 1990). The coating of eggs already in place reduces gas exchange which can prevent egg development (DWR 1989).

Native fish feed mainly on benthic invertebrates, algae and plants or other fish. Thus factors affecting invertebrates and other food sources will indirectly affect fish (Jackson 1986).

Sedimentation has been identified as a major cause of loss of fish habitat and breeding grounds (OCE 1988), a factor threatening native freshwater fish (Koehn and O'Connor 1990b; Jackson *et al.* 1993) and a cause for the decline in the range and abundance of native fish species (Cadwallader 1978). Few species of native Australian fish have been directly proven to be affected by sediment, although there are legitimate concerns for their survival, given the behaviour of sediment in streams and the life histories of many species of fish. Some native species occurring in Victoria for which data exist are listed below.

The serious decline in the range and abundance of Macquarie Perch *Macquaria australasica* has been particularly attributed to siltation in streams (Cadwallader 1981) as they lodge their eggs among stones and gravel (Llewellyn and MacDonald 1980). In laboratory tests, significantly higher mortalities have been recorded for eggs covered in silt (Koehn and O'Connor, unpubl. data).

Similarly, the range and abundance of Freshwater Blackfish *Gadopsis marmoratus* has undergone a significant reduction in Victoria (Koehn and Morison 1990). This species requires clean sites for the deposition of eggs and is particularly susceptible to smothering by sediment (Blyth and Jackson 1985). If potential spawning sites are covered with sediment, spawning may not be successful (Koehn and O'Connor 1990a). Tunbridge (Department of Conservation and Natural Resources, pers. comm.) blamed increased sedimentation in areas of the Thomson River downstream of dam construction for preventing the spawning of Blackfish. A reduction in the population of Blackfish (up to 93%) has been observed below a weir cleaning site on Armstrong Creek, Victoria (Doeg and Koehn 1994). In laboratory tests, significantly higher mortalities have been recorded for eggs covered in silt, and larvae in increasing concentrations of suspended sediment (Koehn and O'Connor, unpubl. data). Dead and dying adults of *G. marmoratus,* as well as adults of Australian Smelt *Retropinna semoni* and larvae of Short-headed Lamprey *Mordacia mordax,* observed in the Thomson River, were attributed to high suspended solid concentrations of at least 190–200 mg/L caused by dam construction (Tunbridge, op. cit., pers. comm.).

In laboratory tests, significantly higher mortalities have been recorded for juvenile Common Galaxias *Galaxias maculatus* in increasing concentrations of suspended sediment (Koehn and O'Connor, unpubl. data). Fertilized eggs of Broad-finned Galaxias *Galaxias brevipinnis* and Spotted Galaxias *Galaxias truttaceus* covered in silt in laboratory tests showed 100 per cent mortality, while no mortalities occurred for eggs without silt (Koehn and O'Connor, unpubl. data).

Specific experimental data are not available for other native fish which are known to deposit eggs on or amongst the stream bed. The eggs would, however, be liable to smothering from increased sediment deposition and must be considered threatened by increased sedimentation (Table 3).

Macrophytes and algae

Freshwater flora may be affected by sediment in a number of ways, including smothering by deposited sediment. But probably the most severe effect will be the reduction of light levels due to the optical effects of suspended material, thus adversely affecting photosynthetic activity (Lloyd 1987; DWR 1989) and growth rates of plants (Tanner *et al.* 1993). In New Zealand, mean increases in turbidity of as low as 9 mg/L (or 9 NTU) were found to reduce algal biomass by more than 40 per cent (Davies-Colley *et al.* 1992). It should be noted that these low increases were continuous as a result of leakage of turbid material from tailings dams associated with off-stream mining, and may have been sustained for a number of years. If such data are applicable in Australian conditions, even low levels of sediment may have serious consequences for freshwater flora (and animals higher in the food chain).

Macroinvertebrates

Suspended sediments may affect aquatic macroinvertebrates through similar mechanisms as for fish (mortality, reduced oxygen uptake, physiological and behavioral changes). There are insufficient data available in Australia relating suspended sediment levels to the subsequent effect on aquatic macroinvertebrate biota. However, in New Zealand, Quinn *et al.* (1992) suggest that aquatic macroinvertebrates there are more sensitive to suspended sediment than equivalent communities in Alaska, which leads to the possibility that Australian fauna may also be more sensitive (and hence that suspended sediment standards of water quality derived solely from data from overseas studies may be inappropriate in local conditions).

Many aquatic invertebrate species spend large parts of their life-cycle (particularly during the larval phase) on the river-bed or on woody debris (Williams 1980), and some species, such as Simuliidae (Diptera) need clean surfaces for attachment (Williams 1980). Smothering by sediment reduces habitat diversity and suitability, a key factor in determining and maintaining the diversity of macroinvertebrate

Table 3. Native Victorian fish species which lay eggs on or amongst the stream-bed and would be liable to smothering due to increased sediment deposition.

Species	Conservation status[a]
Geotria australis (Pouched Lamprey)	Potentially threatened
Galaxias oldidus (Mountain Galaxias)	Indeterminate
Galaxias brevipinnis (Climbing Galaxias)	Potentially threatened
Galaxias rostratus (Flat-headed Galaxias)	Indeterminate
Retropinna semoni (Australian Smelt)	Common/Widespread
Prototroctes maraena (Australian Grayling)	Vulnerable
Tandanus tandanus (Freshwater Catfish)	Vulnerable
Craterocephalus stercusmuscarum (Freshwater Hardyhead)	Indeterminate
Nannoperca australis (Southern Pigmy Perch)	Common/widespread
Gobiomorphus coxii (Cox's Gudgeon)	Indeterminate
Maccullochella peelii (Murray Cod)	Vulnerable
Maccullochella macquariensis (Trout Cod)	Endangered
Arenigobius bifrenatus (Bridled Goby)	Common/Widespread

[a]Conservation status in Victoria (Koehn and Morison 1990).

communities in streams and rivers. Grazing animals (e.g., many mayfly nymphs) may be adversely effected by the smothering of their algal food source. Deposited sediment may also smother leaves on the bottom of the stream, interrupting decomposition and reducing habitat for specialized leaf pack fauna (Bunn 1988).

Interstitial spaces serve as important habitats for many macroinvertebrate species (Hynes 1970). Some use them as habitat over the entire life cycle while others use them during a particular part of their life cycle (e.g., Campbell 1980). The hyporheic zone is also used as refugia during times of stress (Coleman and Hynes 1970; Hynes 1974; Marchant 1988). Hyporheic sediment reduces the suitability of habitats within the stream bed.

The prime Victorian studies of the effect of additional sediment input on flora and fauna have come from studies of dam construction (Blyth *et al.* 1984; Chessman *et al.* 1987; Doeg *et al.* 1987; Marchant 1989). Some particular examples of macoinvertebrate taxa which have been shown to be affected in one or more of these studies, are given in Table 4. Supporting evidence has been derived for a number of these taxa, including comparisons of abundances between sandy and rocky streams (e.g., Metzeling *et al.* 1984) and the reaction to artificial sediment increases (Doeg and Milledge 1991) and other studies.

Several species have been consistently affected in these studies, including *Austrophlebioides pusillus* (Leptophlebiidae, Ephemeroptera), *Notalina* spp. (*N. fulva* and *N. bifaria;* Leptoceridae, Trichoptera), *Austrolimnius* spp. (Elmidae, Coleoptera), *Agapetus* spp. (Glossosomatidae, Trichoptera) and *Helicopsyche* spp. (Helicopsychidae, Trichoptera). The first three of these have been shown to substantially increase drift rates as a response to relatively low levels of suspended sediment (Doeg and Milledge 1991). *Austrophlebioides pusillus* was also less common in sandy streams (Metzeling *et al.* 1984), suggesting an intolerance for fine particles. This genus was found to be adversely affected by sedimentation from the collapse of a forestry road in New South Wales, increasing drift rates with increasing turbidity (Richardson 1985).

The studies cited above have shown that sedimentation has a serious effect on aquatic macroinvertebrate communities as a whole by reducing the diversity (total number of species) of the community, or substantially changing the composition of the community.

Strategies to reduce sediment input to streams

It is a truism that the valley rules the stream. Controlling and reducing the amount of sediment entering streams requires changes to the way that catchments are managed. Given the unidirectional nature of the flow in streams, any disturbance at a single point in a catchment may have a deleterious effect at points far distant downstream. On a broad scale, the trend towards total or integrated catchment management is a positive step in addressing this problem.

At a smaller scale there are many ways that sediment input to streams can be reduced. These include fencing off stream frontages from grazing animals and putting local streamside erosion control works in place. Establishing or maintening a well vegetated riparian zone or buffer strip of appropriate width will reduce overland runoff and stabilize stream banks. Alternative agricultural practices can be adopted which maintain soil structure and result in less erosion. Other land uses near rivers or streams, including quarrying and construction, should adopt practices that reduce erosion. Roads could be better sited and sealed (bitumen or crushed rock) to prevent runoff, and the adoption of a

Table 4. Aquatic macroinvertebrate taxa whose abundance was decreased by sedimentation during dam construction.

Order	Species	Ref[a]
Plecoptera	*Austroheptura picta*	2
	Stenoperla australis	2
	Trinotoperla nivata	1
	T. yeoi	1
	Dinotoperla serricauda	1
	D. brevipennis	4
	Riekoperla rugosa	2
	R. tuberculata	2
	Austrocercella marianne	2
	Newmanoperla thoreyi	3
Ephemeroptera	*Baetis* spp.	3
	Coloburiscoides spp.	1,2
	Austrophlebioides pusillus	1,2,4
Trichoptera	*Tamasia variegata*	2
	Notalina spp.	1,2,3
	Asmicridea edwardsii	2,3
	Cheumatopsyche sp. 3	2,3
	Smicrophylax sp.	2,3
	Ethochorema sp. 2	2
	Apsilochorema obliqua	2
	Agapetus spp.	1,2,3,4
	Helicopsyche spp.	1,2,4
	Chimarra sp. 1	2
	Hydroptilidae sp. 5	1
	Costora delora	3
	Orthotrichia? bishopi	3
Coleoptera	Ptylodactilidae	2
	Austrolimnius spp.	1,2,3
	Simsonia spp.	2,3
	Scirtidae	2
	Sclerocyphon spp.	1,3
Diptera	*Polypedilum* sp. 16E	1
	Eukiefferiella spp.	1,2
	Ablabesmyia spp.	3
	Thienemaniella spp.	2,3
	Riethia spp.	2,3
	Procladius sp. 66E	1
	Aphrotiniinae sp. 18E	1
	Paraheptagyia sp. 23E	1
	Stempellina sp. 54E	2
	Harnischia sp.	2
	Rheotanytarsus sp. 202	3
	Tipulidae sp. A	3
	Tipulidae sp. 2	1
	Athericidae sp. 2	2

[a]1 — Blyth *et al.* 1984; 2 — Doeg *et al.* 1987; 3 — Chessman *et al.* 1987; 4 — Marchant 1989.

scientifically determined forestry code of practice should aim to limit erosion.

There are many river management practices that could be refined to reduce the amount of sediment generated as a byproduct. Alternative methods for weir cleaning operations could be adopted and greater care taken during dam construction (recently built major dams in Victoria — Dartmouth, Thomson and Blue Rock — sequentially showed less impact due to "greater care"). The prevention of in-stream mining activities such as gravel extraction and eductor dredging would reduce sedimentation. Also, the prevention of inappropriate "river improvements works" such as snag removal would decrease the energy of the stream and hence lead to less erosion.

CONCLUSION

Salinization and sedimentation are widespread and threaten a wide variety of aquatic species, communities and populations. Both require concerted effort from all levels of government and from the community. Some action is taking place, such as the implementation of salinity management plans, the push towards total catchment management and the development of forestry codes of practice, but there is still much to be done. Although there is still considerable scientific uncertainty as to the levels of salinization and sedimentation that produce impacts on the aquatic biota, this should not prevent implementation of action to prevent the potential loss of many species due to one or both of these processes.

REFERENCES

Anderson, H. W., 1971. Relative contributions of sediment from source areas, and transport processes. Pp. 55–63 in *Proceedings of a symposium on forest land uses and the stream environment* ed by J. T. Krygier and J. D. Hall. Oregon State University: Corvallis, Oregon.

Anderson, J. R., 1991. The implications of salinity, and salinity management initiatives, on fish and fish habitat in the Kerang Lakes Management Area. Arthur Rylah Institute for Environmental Research, Technical Report Series No. 103. Department of Conservation and Environment. ISSN 0810 5774.

Anderson, J. R. and Morison, A. K., 1989. Environmental flow studies for the Wimmera River, Victoria — Part C: Water quality and the effects of an experimental release of water. Arthur Rylah Insistute for Environmental Research, Technical Report Series No. 75. Department of Conservation and Environment. ISSN 0810 5774.

Bacher, J. and Garnham, J. (in press). The effect of salinity on several freshwater species of southern Victoria. Environment Protection Authority Scientific Report Series 92/003. Environment Protection Authority: Melbourne.

Bacher, J. and O'Brien, T., 1989. Salinity tolerance of the eggs and larvae of the Australian Grayling *Prototroctes maraena* Gunther (Salmoniformes: Prototroctidae). *Aust. J. Mar. Freshw. Res.* **40:** 227–30.

Bayley, I. A. E., 1972. Salinity tolerance and osmotic behaviour of animals in athalassic saline and marine hypersaline waters. *Ann. Rev. Ecol. Sys.* **3:** 233–68.

Bayley, I. A. E. and Williams, W. D., 1973. *Inland waters and their ecology*. Longman: Australia.

Blyth, J. D. and Jackson, P. J., 1985. The aquatic habitat and fauna of East Gippsland, Victoria. *Bull. Aust. Soc. Limnol.* **10:** 89–109.

Blyth, J. D., Doeg, T. J. and St Clair, R. M., 1984. Response of the macroinvertebrate fauna of the Mitta Mitta River, Victoria, to the construction and operation of Dartmouth Dam. 1: Construction and initial filling period. *Occas. Pap. Mus. Vic.* **1:** 83–100.

Brink, F. W. B. van den and Velde, G. van der., 1993. Growth and morphology of four freshwater macrophytes under the impact of the raised salinity level of the Lower Rhine. *Aquatic Botany* **45:** 285–97.

Bowler, J., 1988. Environmental and salinity history of the Murray Basin in the last 500 000 years. Department of Primary Industry and Energy, Bureau of Mineral Resources, Geology and Geophysics, Record 1988/7, pp. 19–21.

Brock, M., 1981. The ecology of halophytes in the south-east of South Australia. *Hydrobiol.* **81:** 23–32.

Brock, M. and Lane, J. A. K., 1983. The aquatic macrophyte flora of saline wetlands in Western Australia in relation to salinity and permanence. *Hydrobiol.* **105:** 63–76.

Bunn, S. W., 1988. Processing of leaf litter in two northern jarrah streams, Western Australia: II. The role of macroinvertebrates and the influence of soluble polyphenols and inorganic sediment. *Hydrobiol.* **162:** 211–23.

Cadwallader, P. L., 1978. Some causes of the decline in range and abundance of native fish in the Murray-Darling River system. *Proc. Roy. Soc. Vic.* **90:** 211–24.

Cadwallader, P. L., 1981. Past and present distributions and translocations of Macquarie Perch *Macquaria australasica* (Pisces: Perchichthyidae), with particular reference to Victoria. *Proc. Roy. Soc. Vic.* **93:** 23–30.

Cadwallader, P. L. and Backhouse, G. N., 1983. *A guide to the freshwater fish of Victoria.* Government Printer: Melbourne.

Campbell, I. C., 1980. Diurnal variations in the activity of *Mirawara purpurea* Riek (Ephemeroptera: Siphlonuridae) in the Aberfeldy River, Victoria, Australia. Pp. 297–308 in *Advances in ephemeroptera biology* ed by J. F. Flannagan and K. E. Marshall. Plenum: New York.

Campbell, I. C. and Doeg, T. J., 1989. Impact of timber harvesting and production on streams: a review. *Aust. J. Mar. Freshw. Res.* **40:** 519–39.

Chessman, B. C., 1971. Studies of Victorian salt lake fishes, with special reference to *Galaxias maculatus* (Jenyns). 1842 (Pisces : Salmoniformes : Galaxiidae). B.Sc. (Hons) thesis, Monash University, Clayton Victoria.

Chessman, B. C., Robinson, D. P. and Hortle, K. G., 1987. Changes in the riffle macroinvertebrate fauna of the Tanjil River, southeastern Australia, during construction of Blue Rock Dam. *Regulated Rivers* **1:** 317–29.

Chutter, F. M., 1969. The effects of silt and sand on the invertebrate fauna of streams and rivers. *Hydrobiol.* **34:** 57–76.

Coleman, M. S. and Hynes, H. B. N., 1970. The vertical distribution of the invertebrate fauna in the bed of a stream. *Limnol. and Oceanog.* **15:** 31–40.

Cordone, A. J. and Kelly, D. W., 1961. The influences of inorganic sediment on the aquatic life of streams. *Calif. Fish and Game* **47:** 189–228.

Davey, G. W., Doeg, T. J. and Blyth, J. D., 1987. Changes in benthic sediment in the Thomson River, Southeastern Australia, during construction of the Thomson Dam. *Reg. Riv.* **1:** 71–84.

Davies-Colley, R. J., Hickey, C. W., Quinn, J. M. and Ryan, P. A., 1992. Effects of clay discharges on streams. 1. Optical properties and epilithon. *Hydrobiol.* **148:** 215–34.

DCFL, 1987. Better Rivers and Catchments. Department of Conservation Forests and Lands, Government Printer: Melbourne.

Doeg, T. J., 1984. Response of the macroinvertebrate fauna of the Mitta Mitta River, Victoria, to construction and operation of Dartmouth Dam. 2. Irrigation release. *Occas. Pap. Mus. Vic.* **1:** 101–28.

Doeg, T. J. and Koehn J. D., 1990. A review of Australian studies on the effects of forestry practices on aquatic values. Silvicultural Systems Project Technical Report Series No. 5. Department of Conservation and Environment: Melbourne.

Doeg, T. and Koehn, J., 1994. Effects of draining and desilting a small weir on downstream fish and macroinvertebrates. *Reg. Riv.* **9:** 263–77.

Doeg, T. J. and Milledge, G., 1991. Effect of experimentally increasing concentrations of suspended sediment on macroinvertebrate drift. *Aus. J. Mar. Freshw. Res.* **42:** 519–26.

Doeg, T. J., Davey, G. W. and Blyth, J. D., 1987. Response of the aquatic macroinvertebrate community to dam construction on the Thomson River, southeastern Australia. *Regulated Rivers* **1:** 195–209.

DWR, 1989. Water Victoria: An Environmental Handbook. Department of Water Resources: Melbourne.

EPA, 1980. Observations of sedimentation in the Mitta Mitta River during construction of the Dartmouth Dam. Publication No. 113. Environment Protection Authority: Melbourne.

Greenway, P., 1986. Salt and water balance in the fauna of Australian athalassic waters. Pp. 193–206 in *Limnology in Australia* ed by P. De Deckker and W. D. Williams. CSIRO/Junk Publication.

Hart, B. T., Bailey, P., Edwards, R., Hortle, K., James, K., McMahon, A., Meredith, C. and Swadling, K., 1990. Effects of salinity on river, stream and wetland ecosystems in Victoria, Australia. *Water Research* **24:** 1103–17.

Hart, B. T., Bailey, P., Edwards, R., Hortle, K., James, K., McMahon, A., Meredith, C. and Swadling, K., 1991. A review of the salt sensitivity of the Australian freshwater biota. *Hydrobiol.* **210:** 105–44.

Hellawell, J. M., 1986. *Biological indicators of freshwater pollution and environmental management.* Elsevier Applied Science Publishers: Essex.

Hero, J.-M., Littlejohn, M. and Marantelli, G., 1991. *Frogwatch. Field guide to Victorian frogs.* Department of Conservation and Environment: Melbourne.

Hynes, H. B. N., 1970. *The ecology of running water.* Liverpool University Press: Liverpool.

Hynes, H. B. N., 1974. Further studies on the distribution of stream animals within substratum. *Limnol. Oceanog.* **19:** 92–99.

Jackson, P. D., 1986. Fish habitat for river management works, in river management foremans workshop: workshop proceedings. Department of Water Resources, Melbourne.

Jackson, P. D., Koehn, J. D. and Wager, R., 1993. Australia's threatened fishes 1992 listing — Australian Society for Fish Biology. Appendix 1 in *Sustainable fisheries through sustaining fish habitat* ed by D. A. Hancock. Australian Society for Fish Biology Workshop, Victor Harbour, South Australia, 12–13 August, Bureau of Resources Sciences Proceedings, Australian Government Publishing Service: Canberra.

James, K. R. and Hart, B. T., 1993. Effect of salinity on four freshwater macrophytes. *Aust. J. Mar. Freshw. Res.* **44:** 769–77.

Koehn, J. and Morison, A. K., 1990. A review of the conservation status of native freshwater fish in Victoria. *Vic. Nat.* **107:** 13–25.

Koehn, J. and O'Connor, W., 1990a. Biological Information for the Management of Native Freshwater Fishes in Victoria. Department of Conservation and Environment, Government Printer: Melbourne.

Koehn, J. and O'Connor, W., 1990b. Threats to Victorian native freshwater fish. *Vic. Nat.* **107:** 5–12.

Lake, P. S., 1987. Stream ecosystem research in Australia: problems and strategies. Pp. 49–80 in *National water research seminar discussion papers.* Department of Resources and Energy: Canberra.

Lake, P. S. and Marchant, R., 1990. Australian upland streams: ecological degradation and possible restoration. *Proc. Ecol. Soc. Aust.* **16:** 79–91.

Llewellyn, L. C. and MacDonald, M. C., 1980. Family Percichthyidae. Australian freshwater basses and cods. Pp. 142–59 in *Freshwater fishes of south-eastern Australia.* ed by R. M. McDowall. Reed: Sydney.

Lloyd, D. S., 1987. Turbidity as a water quality standard for salmonid habitats in Alaska. *N. Amer. J. Fish. Man.* **7:** 34–45.

Macumber, P. G., 1991. Interaction between groundwater and surface water systems in Northern Victoria. Department of Conservation and Environment: Melbourne.

Macumber, P. G. and Dyson, P. R., 1988. The salinity problem in southern Australia — an overview. Department of Primary Industry and Energy, Bureau of Mineral Resources, Geology and Geophysics, Record 1988/7. Pp. 121–26.

Marchant, R., 1988. Vertical distribution of benthic invertebrates in the bed of the Thomson River, Victoria. *Aust. J. Mar. Freshw. Res.* **39:** 775–84.

Marchant, R., 1989. Changes in the benthic invertebrate communities of the Thomson River, southeastern Australia, after dam construction. *Reg. Riv.* **4:** 71–89.

McGuckin, J., Anderson, J. A. and Gasior, R. J., 1991. Salt affected rivers in Victoria. Arthur Rylah Insistute for Environmental Research, Technical Report Series No. 118. Department of Conservation and Environment. ISSN 0810 5774.

Metzeling, L., 1993. Benthic macroinvertebrate community structure in streams of different salinity. *Aust. J. Mar. Freshw. Res.* **44:** 335–51.

Metzeling, L. (in press). Invertebrates, algae and water quality of saline pools in the Wimmera River. Environment Protection Authority, Victoria, SRS 92/002.

Metzeling, L., Graesser, A., Suter, P. and Marchant, R., 1984. The distribution of aquatic macroinvertebrates in the upper catchment of the LaTrobe River, Victoria. *Occas. Pap. Mus. Vic.* **1:** 1–62.

Ministry for Conservation 1983. The State of the Rivers. Government Printer: Melbourne.

Mitchell, B. and Richards, K., 1992. Macroinvertebrate communities in two salt affected tributaries of the Hopkins River, Victoria. *Intern. J. Salt Lake Res.* **1:** 81–102.

Mitchell, P. A., 1990. The environmental condition of Victorian streams. Department of Water Resources: Melbourne.

MMBW, 1980. Summary of technical conclusions to 1979. Water Supply Catchment Hydrology Research Report MMBW-W-0012. Melbourne Metropolitan Board of Works: Melbourne, Victoria.

Morrissy, N. M., 1979. Inland (non-estuarine) halocline formation in a Western Australian river. *Aust. J. Mar. Freshw. Res.* **30:** 343–53.

Newcombe, C. P. and MacDonald, D. D., 1991. Effects of suspended sediments on aquatic ecosystems. N. Amer. *J. Fish. Man.* **11:** 72–82.

OCE, 1988. State of the Environment Report. Office of the Commissioner for the Environment, Government Printer: Melbourne.

Olive, L. J. and Rieger, W. A., 1987. Eden catchment project: sediment transport and catchment disturbance, 1977–83. Monograph Series No. 1. Department of Geography and Oceanography, University College, Australian Defence Force Academy: Canberra, Australian Capital Territory.

Papworth, M. P., Hartland, R. and Lucas, A., 1990. Logging Alpine Ash in the East Kiewa River Catchment. Part I: Effects on Stream Sediment Levels. Department of Conservation and Environment: Melbourne.

Patrick, R., 1988a. Importance of diversity in the functioning and structure of riverine communities. *Limnol. Oceanog.* **33:** 1304–307.

Patrick, R., 1988b. Catchment Management, in Workshop on Native Fish Management, Canberra. Murray-Darling Basin Commission: Canberra.

Peck, A. J., Thomas, J. F. and Williamson, D. R., 1983. Water 2000: Consultants Report No. 8. Salinity issues. Effects of man on salinity in Australia. Department of Resources and Energy, Australian Government Publishing Service: Canberra.

Quinn, J. M., Davies-Colley, R. J., Hickey, C. W., Vickers, M. L. and Ryan, P. A., 1992. Effects of clay discharges on streams. 2. Benthic invertebrates. *Hydrobiol.* **248:** 235–47.

Richardson, B. A., 1985. The impact of forest road construction on the benthic invertebrate and fish fauna of a coastal stream in southern New South Wales. *Bull. Aust. Soc. Limnol.* **10:** 65–88.

Salt Force 1988. Salt action: joint action; Victorian Government salinity strategy.

Tanner, C. C., Clayton, J. S. and Wells, R. D. S., 1993. Effects of suspended solids on the establishment and growth of *Egeria densa. Aquatic Botany* **45:** 299–310.

Timms, B. V., 1981. Animal communities in three Victorian lakes of different salinity. *Hydrobiol.* **81:** 181–93.

Timms, B. V., 1983. A study of benthic communities in some shallow saline streams of western Victoria, Australia. *Hydrobiol.* **105:** 165–77.

West, G., Blyth, J. D., Balkau, F. and Slater, S., 1984. Chemical and physical features of construction sediment in the Mitta Mitta River, Victoria. *Occas. Pap. Mus. Vic.* **1:** 129–38.

Williams, W. D., 1980. *Australian freshwater life.* MacMillan Books: Melbourne.

Williams, W. D., 1981. The limnology of saline lakes in Western Victoria. A review of some recent studies. *Hydrobiol.* **82:** 233–59.

Williams, W. D., 1984. Salinity as a water quality criterion and determinant in Australia. Australian Water Resources Council Research Project 80/121.

Williams, W. D., 1987. Salinization of rivers and streams: an important environmental hazard. *Ambio* **16:** 180–85.

Williams, W. D., Taffe, R. G. and Boulton, A. J., 1991. Longitudinal distribution of macroinvertebrates in two rivers subject to salinization. *Hydrobiol.* **210:** 151–60.

Water pollution, biodiversity and the effectiveness of water quality criteria

R. T. BUCKNEY[1]

The effects of water pollution on species composition and species richness have long been known and used in the assessment of pollution levels. A number of examples are presented briefly to illustrate the effects of industrial, agricultural and urban contamination of waters. The strengths and weaknesses of water quality criteria, as applied in Australia, in preserving species richness in fresh waters are considered. Many laboratory-derived data, including toxicological and physiological data (from which many of the criteria are derived) have not usually been directed at examining potential impacts on biodiversity, yet current trends in their derivation suggest that they could be so used. The results point to a need for a closer examination of water quality criteria with respect to biodiversity issues. There are particular problems where a number of pollutants occur together; the criterion may be met for each component individually, but there can be clear reductions in species richness caused by pollutant mixtures. Even in the case of single pollutants, receiving waters which meet quality criteria may have a significant proportion of species seriously disadvantaged by the pollutant.

INTRODUCTION

POLLUTION of waters has long been recognized as a phenomenon which affects species diversity and species richness. Indeed, if it did not have such effects, it probably would not be regarded as pollution.

Despite a widespread understanding that pollution often results in a decrease of species richness, it is difficult to assess whether biodiversity or any of its components (genetic variability, species richness or community diversity) has been affected by water pollution, certainly in Australia. This difficulty stems principally from a lack of knowledge. Lake (1982) addressed many of these issues in relation to Australian upland streams.

The aim of this paper is to describe briefly the effects of water pollution on species richness, to broadly assess the likelihood of regional extinctions of species and to evaluate the effectiveness of applying water quality criteria in protecting biodiversity.

Consideration is largely restricted to fresh waters in Australia. This does not imply that estuarine or coastal waters are less important than fresh ones; indeed, the concentration of population along the coast has meant that the longest affected and most severely polluted waters often are the estuarine and coastal ones.

TYPES OF POLLUTANTS AND THEIR EFFECTS

Analysis of the effects of pollutants is not straightforward. In the field, pollutants rarely occur singly and the possibility of interaction between them makes precise definition of the effects of single pollutants difficult. Another complication stems from the confounding effects of the physical alteration of waters subject to pollution; many water bodies receiving pollutants have been altered by damming, channelling or diversion to serve the needs of the polluting population, so flow regimes are different to those in the natural state. A further factor is the alteration of the catchment which often accompanies development.

These effects are particularly evident in urban areas, where many streams have been transformed into concrete drains, the catchment has a high percentage of impervious surfaces and water flow typically changes very rapidly (to levels capable of scouring a stream bed) in response to rainfall. Definitive factorial experiments to separate the effects of habitat alteration and changed flow regimes from those of pollution *per se* are extremely rare, as are studies which incorporate full BACI (Before-After-Control-Impact) design (Stewart-Oaten *et al.* 1986) or a derivative of it (see Underwood 1993) to assess the impact of a development on a water body.

Attempts to eliminate complicating factors, as in laboratory experiments, suffer no less serious drawbacks. Single species often are studied in isolation and typically without the variations in temperature, light intensity and water quality expected in a natural system. Moreover, natural habitats are absent in most glass tank experiments and the responses of many species could well be expected to be aberrant in laboratory tests; indeed, the range of species actually studied in the laboratory is very narrow. The use of laboratory-derived data is dealt with later.

Despite the inherent complications of field studies, there have been sufficient examples of pollution by single substances, or relatively few of them together, for a few generalizations to be made. Some such generalizations follow.

---

[1]Centre for Environmental Toxicology, Department of Environmental Biology and Horticulture, University of Technology, Sydney. P.O. Box 123, Broadway, New South Wales, Australia 2007.

Pages 137–43 *in* CONSERVING BIODIVERSITY: THREATS AND SOLUTIONS, ed by R. A. Bradstock, T. D. Auld, D. A. Keith, R. T. Kingsford, D. Lunney and D. P. Sivertsen. Surrey Beatty & Sons. 1995.

1. The principal pollutants

(a) Industrial

Under this heading, I have included mining and manufacturing industries. The pollutants capable of being released from such ventures are very varied. The common ones include metals, fuels and lubricants, organic chemicals and a variety of solid materials. Metals are probably the most studied of these and are dealt with below.

(b) Agricultural

The principal agricultural pollutants of waters are pesticides, nutrients, herbicides and solid materials, mostly derived from eroded soils. Norris (1991) reviewed these and other effects of agriculture on waters in Australia.

(c) Urban

Urban pollutants form perhaps the most complex mixture of substances contaminating water bodies. Contaminants include all those already mentioned. Sewage must be added as a significant component in urban streams, since leakage and overflow of sewerage can occur. Sewage is a complex mixture of organic matter, organic solids, nutrients and, in some places, industrial wastes. Petrochemicals, their combustion products and other motor vehicle wastes are also important. In urban streams, water quality varies enormously in a very short time during storm events (Cordery 1977). Urban waters, as noted above, probably experience the greatest modification of flow regime of all streams so far studied in Australia (see Arthington *et al.* 1982).

2. Effects of some pollutants

The list of possible water pollutants is far too long to attempt a comprehensive treatment here. In the following sections, an effort is made to summarize the main effects and to point to some useful references.

(a) Metals

Dissolved metals often have been released through bacterial oxidation of sulfide minerals in mine and industry waste dumps. Typically, such processes yield waters of very low pH and high dissolved metal content and have proved capable of polluting waters long after the operation has been abandoned. Long-studied examples include the Molonglo River, New South Wales (Weatherley *et al.* 1967; Norris 1986), and the South Esk River, Tasmania (Tyler and Buckney 1973; Thorpe and Lake 1973). Non-mining industrial processes, such as smelting and electro-plating, are also capable of generating significant metal loadings to water bodies.

Macroinvertebrate distribution and abundance has been the most studied biological response to the metal pollution and typically, a marked reduction of species richness is found. For example, Weatherley *et al.* (1967) demonstrated a reduction in species richness from more than 40 species per sample (from more than 10 orders) to less than five species (two orders) at sample locations with less than 1 and more than 30 mg zinc per litre respectively. Recent evidence from Sunny Corner, New South Wales (Chapman *et al.* 1993; Napier 1993) shows similar trends, which are also evident in the riparian plant communities (Higgs 1991).

(b) Pesticides

Pesticides may be found in waters draining agricultural land. In Australia, the use of rivers for irrigation ensures that the potential for contamination of water by pesticides is quite high. Bowmer (1987) has reported on the levels of some pesticides in irrigation water and sediments. Scribner *et al.* (1987) have reported on the levels of pesticides and other substances in freshwater aquatic organisms. Concerns about such substances are warranted even when levels in the environment are very low, since they have a potential to bioaccumulate and (perhaps) to concentrate along food chains.

The potential for non-target species to be affected by pesticides has long been recognized. Recent studies on the insecticide endosulfan, as used in the cotton-growing areas of northern New South Wales (Napier 1993; Bacher *et al.* 1992), did not detect an effect of pesticide residues on the biodiversity of aquatic organisms in some receiving waters, despite levels in fish being sufficiently high to cause some subcellular alterations and, presumably, metabolic and behavioural changes (Nowak 1992). The possibility of the sediments storing pesticide, and hence becoming a long-term source of contamination, was recognized during these studies. Work on sediments are likely to help overcome some of the difficulties encountered in studying pesticide contamination events, which tend to be sudden and short-lived relative to the frequency of sampling programmes.

(c) Nutrients

The phenomenon of eutrophication, in which nutrient enrichment results in algal blooms, with concomitant changes in the oxygen status and species composition and biodiversity of water bodies has been a very intensively studied area of freshwater ecology. Norris (1991), however, pointed to the relative scarcity of detailed studies in Australia. Chapman and Simmons (1990) described the effects of nutrient enrichment and oxygen depletion from sewage discharges on the biomass and number of taxa of periphyton in some high-altitude streams in the Mt Koskiusko area.

(d) Solids

The input of solid materials to streams and other water bodies occurs principally during storm flows when cleared or urbanized land occurs in the catchment. Solid material, if remaining in suspension, can affect light penetration and hence the productivity of algae and aquatic macrophytes; it can also affect the behaviour of animals which rely on sight for food capture. Settling solids can modify the substrate and destroy habitats of aquatic organisms. Particularly heavy solids loads can smother some organisms. Richardson (1985) described the effects of sediment input to a stream over a nine-month period, and identified changes in faunal composition and in the drift rate of macroinvertebrates.

WHAT EFFECT HAS POLLUTION HAD ON BIODIVERSITY?

As pointed out above (see also Lake 1982), current information is insufficient to provide a definitive answer to this question.

The deficiency in the knowledge base has at least three components. The first is a shortage of ecological studies on many of Australia's water bodies, so there is little confidence that even most of the species present are known, let alone their conservation status adequately understood.

The second is the relatively poor state of the taxonomy. Hence, it is often impossible to know whether what one has in one's sample is the same thing that someone else found there several years ago. This is less of a problem for plants and large animals than for small animals. Of 36 wetlands of high conservation significance listed for Tasmania (Kirkpatrick and Tyler 1988), 24 are regarded as significant for their plant communities or species, while eight are significant for their water birds or fish. Jackson (1993) assessed the conservation status of those Australian fish which are fully or predominantly found in fresh waters. He estimated that 24 of 217 species are threatened with extinction, most from habitat modification or interaction with introduced species. One is threatened through a deterioration of water quality.

The third component is the high hydrological variability of water in Australia, which confounds analysis of the effects of pollution.

Most freshwater organisms are small and poorly known, so whether any have become extinct would be impossible to determine. Even given a heightened awareness of the importance of conserving biodiversity, the effort needed to detect an extinction (or an imminent one) for such organisms is probably prohibitive. It is therefore important to ensure that pollution control strategies adequately protect all the life in the water bodies.

The concept of indicator species has been widely applied in evaluations of the quality of waters. For example, among the macroinvertebrates, the Ephemeroptera, molluscs and Coleoptera are regarded as metal-sensitive. The Chironomidae are regarded as tolerant to metals (Chapman *et al.* 1993). Hynes (1960) described a general model of changes in stream community composition downstream of an input of organic matter (such as a sewage inflow); tubificid worms and chironomids are generally tolerant of the low oxygen conditions found in such situations.

The fact that similar pollution events have had sufficiently similar effects in a number of different areas to allow the classification of aquatic taxa as generally "sensitive" or "tolerant" has an important ramification in the context of effects of pollution on biodiversity. It is likely to be the same (or related) taxa which disappear from aquatic systems wherever water pollution occurs.

One consequence of this is that there almost certainly have been many extinctions of species from significant sections of many water bodies. In the large urban centres, this could have amounted to regional extinctions of several species. Moreover, the surviving aquatic communities in such areas are likely to be very similar in geographically dispersed areas, since they would tend to be dominated by tolerant species. A comparison of the urban streams of Australia's largest cities might be expected to show high levels of similarity in the aquatic fauna and flora; the same could be expected of other pollution types.

As stated above, the taxonomic knowledge of many groups of aquatic organisms is rudimentary, so the significance of the inferred extinctions would be difficult to assess.

WATER QUALITY CRITERIA AND THE PROTECTION OF SPECIES

Water quality criteria are numbers derived from the known responses of organisms to contaminants in the water. They allow prediction of the effect (or probability of an effect) of a contaminant on a water user; in the present context, the water users are plants, animals, algae, etc. which live in the water. Water quality criteria are used by a number of regulatory agencies to aid in the setting of emission standards for effluents.

Several approaches have been developed to generate water quality criteria (ANZECC 1992). For many pollutants, the approach has been to derive "safe" concentrations based on toxicity test results.

The derivation of "safe" concentrations

Toxicity tests consist of experimental exposures of organisms to known concentrations of toxicants for defined periods of time and the calculation of the concentration which will have a defined effect on the "average" (actually "median") organism in the test population in a defined time. The most commonly-used numbers come from tests where the defined effect is death; such tests are known as acute lethal tests and the important numbers to come from them are known as *median lethal concentrations,* usually abbreviated LC_{50}. The most common LC_{50} values relate to four-day (96-hour) exposures, but other test durations are frequently used. A variety of other tests are used, many of which have, as the defined effect, subtle, sub-lethal changes in the performance of the test organisms; these are often referred to as chronic tests.

The system adopted in Canada for the generation of "safe" concentrations (CCREM 1991) relies strongly on chronic test data. The Australian guidelines (ANZECC 1992) tend to follow the same system.

A common practice in deriving "safe" concentrations has been to take LC_{50} values (or, preferably, chronic test data) and multiply by a safety (or application) factor (most often in the range 0.1 to 0.01) to derive the desired number. In deriving "safe" levels for toxicants where only LC_{50} data were available, ANZECC (1992) used an application factor of 0.05 for materials that are not persistent or likely to accumulate and 0.01 for those that are.

Other approaches in deriving "safe" levels have been described by Bongaerts *et al.* (1990). These include a range of algorithms which assess the concentration which will protect a defined percentage of species (see Stephan *et al.* 1985; Kooijman 1987; Wagner and Lokke 1991). A typical target is the protection of 95 per cent of the species, which in the present context represents a discounting of the value of 5 per cent of the species. Clearly, one of the dangers of such an approach is that if 5 per cent of species have truly been discounted, it is likely that some of them will disappear from the pool of species; then, a value which will protect 95 per cent of the remaining species is less protective than the original value.

These techniques are statistically-based and assume certain properties of the data. They incorporate extrapolations from real data to the edges of the distributions of the numbers. Uncertainty levels become very high when this is attempted, and the resulting numbers vary greatly depending on the number of species which have been studied. One interesting result is that the lowest safe concentrations (i.e., the safest) come from data sets with the smallest number of species and the highest from large data sets. The increase in the safe concentration as the data base is improved is not unlimited, however. Wagner and Lokke (1991) provide data which suggest that the "safe" level might increase by a factor of less than ten when the number of test species is increased from eleven to (the unrealizable) infinity.

An evaluation of the effectiveness of water quality criteria in conserving fish species

The use of application factors and the statistical derivation of water quality criteria can yield similar values, so it is important to assess whether the criteria currently used in Australia protect all species, or only a proportion of them. A brief analysis for freshwater fish species follows. Fish were chosen because there is a very high frequency of similar (96-hour LC_{50}) tests in the published literature; data for other groups of organisms are derived from much more diverse experimental protocols.

Among other things, the following analysis highlights some of the difficulties encountered when water quality criteria are derived from toxicological data.

Table 1 compares the water quality criteria for some dissolved metals, as recommended by ANZECC (1992), with the inferred concentration which would be lethal to 5 per cent, 2 per cent and 1 per cent of fish in four days. The toxicity data here come from a compilation of toxicity test results (Skidmore and Firth 1983). Where more than one LC_{50} value was recorded for the same species, the highest value has been used. The data in each case are consistent with a log-normal distribution, so the derivation of the concentration which is lethal for a particular proportion of species is a relatively simple matter of finding the lower 90, 96 and 98 per cent confidence limits of the distributions. The water quality criteria are given as ranges, the higher value being for "hard" water.

Table 1. Recommended water quality criteria (ANZECC 1992) for four metals, the average LC_{50} and the metal concentrations estimated to be lethal to 5 per cent, 2 per cent and 1 per cent of fish species. See text for details. All numbers are in micrograms per litre.

Metal	Recommended water quality criterion	Mean LC_{50}	Concentration lethal to 5%	2% of species	1%
Copper	2– 5	1 010	50	10	6
Zinc	5–50	6 190	1 040	340	222
Cadmium	0.2– 2	930	52	8	4
Lead	1– 5	1 170	148	72	39

For copper and cadmium, the recommended criterion is of the same order of magnitude as the inferred concentration lethal to 1 per cent of species (Table 1). This is cause for concern, especially given the stochastic nature of the latter. For the freshwater fish species in Australia (Jackson 1993), perhaps up to two species would be close to death in water which meets the recommended criterion for copper and up to a further two for cadmium.

For the four metals copper, zinc, cadmium and lead, chronic effects in fish have been recorded at very low concentrations (10, 26, 0.2 and 4 ppb respectively: Skidmore and Firth 1983); the last three numbers clearly fall within the recommended criterion ranges. Hence, waters which meet the water quality criteria may well have a number of fish species suffering low-level effects of metal exposure, perhaps to the extent that behaviour or physiology is detrimentally affected and prospects of survival diminished.

If macroinvertebrates or plants are included, a somewhat different picture might emerge. Whether that picture would be more or less comforting is impossible to tell. Further analysis should be conducted, if only to evaluate the effectiveness of the criteria in meeting their objective of protecting "all forms of aquatic life in all life stages from an indefinite exposure to a chemical" (Chapman *et al.* 1993). The toxicity data set for some chemicals is probably too small for realistic analysis; however, in such cases derived criteria tend to be very conservative, as explained above.

There are other reasons for suspecting that a fuller picture would show that the water quality criteria are more conservative than suggested by the analysis presented here. For example, there is almost certainly a concentration below which no actual effect of a chemical could occur; the above analysis does not allow for such a threshold concentration. Additionally, some zero level of toxicity might be expected for some substances, since (in the case of the metals), some are essential for life. Both of these influences would tend to ensure that the toxicity distribution curve is truncated at a concentration greater than the zero concentration implied in the analysis. Consequently, a concentration calculated to be toxic to 1 per cent of species (as in Table 1) may well be much lower than any real toxic concentration for any species. And, as indicated above, additional information would be likely to amplify this difference.

A further possibility is that, although the lethal toxicity data used here are consistent with a log-normal distribution, the real distribution of the data is somewhat different; the test for compliance with a log-normal distribution is incapable of identifying any but huge deviations from the model distribution. Another distribution might better describe the data.

Other approaches to deriving "safe" concentrations

Two of the serious shortcomings of toxicity tests (and, therefore, of any numbers derived from them), is that they are conducted in simplified laboratory conditions, using a single toxicant. In nature, nothing is that simple.

The simplification of test conditions — tests are conducted in glass tanks, usually at constant temperature — almost certainly has meant that the data refer only to species which can survive exceptional habitat deprivation. Those incapable of tolerating such conditions would probably rate as "sensitive" species in any test, being subjected to abnormal, perhaps threatening, conditions throughout the test. Such species are unlikely to survive general laboratory protocols and hence are under-represented in the toxicity test data bases. Yet if water quality criteria are to be confidently regarded as protecting all species, some of the "sensitive" species should feature in the toxicity data base. The study of these may require special conditions which acknowledge habitat requirements and even interspecific interactions. These studies must include controlled mesocosm and macrocosm investigations, of which field studies are the ultimate examples. Proper studies of such systems are costly, a feature that probably works against their wholesale adoption.

ANZECC (1992) emphasized the need to develop and introduce biological indicators for assessing water quality conditions, particularly for broad scale monitoring of ecosystem condition. The biota respond simultaneously to all changes in water quality, so marked changes in biotic condition and composition can be expected to signal deterioration of water quality when the chemical nature of the pollutants is unknown or unpredictable.

Where toxicants occur in combination, water quality criteria may prove ineffective in protecting all aquatic life. Chapman *et al.* (1993) described the results of studies on a metal-contaminated stream system in New South Wales. They found detectable differences between the polluted stream macroinvertebrate community and those of adjacent unpolluted streams, even where the metal concentrations had fallen below the recommended water quality criteria of ANZECC (1992). They suggested that this may have been due to interactive effects of the metals and asserted the need for site-specific assessments of water quality criteria. ANZECC (1992) also recognized the need for regional and site-specific guidelines.

Water quality criteria cannot be used to regulate all examples of water pollution. Diffuse source pollution, such as runoff from developed land, is virtually immune from any regulatory application of the criteria. In these cases, the criteria may be used as indicative "levels of concern" (ANZECC 1992) about the substances and changes in land use practices and drainage characteristics evaluated by determining the extent to which they assist the receiving water to comply with the criteria.

SUMMARY

Pollution of water causes local reductions in species richness and results in domination of aquatic communities by tolerant species. There have probably been species extinctions on regional scales, but the state of ecological and taxonomic knowledge, certainly in Australia, is at present inadequate to allow a confident evaluation of its seriousness.

Water quality criteria are possibly better than they need to be to protect all aquatic life from imminent death, but they may not provide adequate protection from sublethal influences likely to cause long-term deterioration and gradual population decline. Their usefulness is limited to cases where point-source discharges represent the major input of pollutants. Moreover, they may fail to predict impacts where several contaminants occur together. Closer evaluation of their effectiveness in protecting all aquatic life, particularly those sensitive species which do not presently contribute to the toxicological data base, is needed. Without such continuous evaluation, including detailed field validation studies on receiving waters, the use of water quality criteria by regulatory agencies might well constitute a threat to biodiversity, rather than a solution.

ACKNOWLEDGEMENTS

The author is grateful to John Chapman and Therese Manning for useful discussion and assistance in the preparation of this paper. Barry Hart, Roger Croome, Richard Kingsford and Moreno Julli provided constructive comments on the manuscript.

REFERENCES

Arthington, A. H., Conrick, D. L., Connell, D. W. and Outridge, P. M., 1982. The ecology of a polluted urban creek. Australian Water Resources Council Technical Paper No. 68. Canberra.

Australian and New Zealand Environment and Conservation Council (ANZECC), 1992. Australian water quality guidelines for fresh and marine waters. ANZECC: Canberra.

Bacher, G. J., Chapman, J. C., Lim, R. P., Napier, G. M., Nowak, B., Sunderam, R. M. and Thompson, G. B., 1992. The impact of agriculture on inland temperate waters: field validation. *Aust. Biol.* **5:** 196–202.

Bongaerts, Kuyper and Huiswaard, Consulting Engineers, 1990. Ecotoxicological effects assessment: extrapolation from single species toxicity data. The Hague.

Bowmer, K. H., 1987. Residues of Dalapon and tCA in sediments and irrigation water. *Pesticide Science* **18:** 1–13.

Canadian Council of Resource and Environment Ministers (CCREM), 1991. Canadian water quality guidelines. CCREM, Inland Water Directorate. Environment Canada: Ottawa.

Chapman, J. C. and Simmons, B. L., 1990. The effects of sewage on alpine streams in Koskiusko National Park, New South Wales. *Environ. Mon. Assess.* **14:** 275–95.

Chapman, J. C., Napier, G. M., Sunderam, R. I. M. and Wilson, S. P., 1993. The contribution of ecotoxicological research to environmental protection. *Aust. Biol.* **6:** 72–81.

Cordery, I., 1977. Quality characteristics of urban storm water in Sydney, Australia. *Water Resources Research* **13:** 197–202.

Higgs, J. R., 1991. Effect of metalliferous mine pollution on aquatic macrophytes in the Daylight-Williwa Creek system, New South Wales. Honours thesis, Department of Applied Biology, University of Technology: Sydney.

Hynes, H. B. N., 1960. *The biology of polluted waters.* Liverpool University Press: Liverpool.

Jackson, P. D., 1993. The conservation status of Australian freshwater fishes: the fish action plan. Pp. 61–66 in *Water. An endangered national resource* ed by A. H. Arthington. Australian Institute of Biology, Fortitude Valley: Queensland.

Kirkpatrick, J. B. and Tyler, P. A., 1988. Tasmanian wetlands and their conservation. Pp. 1–16 in *The conservation of Australian wetlands* ed by A. J. McComb and P. S. Lake. Surrey Beatty & Sons: Chipping Norton.

Kooijman, S. A. L. M., 1987. A safety factor for LC50 values allowing for differences in sensitivity among species. *Water Research* **21:** 269–76.

Lake, P. S., 1982. The 1981 Jolly Award Address *Ecology of the macroinvertebrates of Australian upland streams — a review of current knowledge. Aust. Soc. Limnol. Bull.* **8:** 1–15.

Napier, G. M., 1993. Field validation of laboratory toxicity data. Ph.D. Thesis, Macquarie University: North Ryde.

Norris, R. H., 1986. Mine waste pollution of the Molonglo River, New South Wales and the Australian Capital Territory: effectiveness of remedial works at Captains Flat mining area. *Aust. J. Mar. Freshw. Res.* **37:** 147–57.

Norris, R. H., 1991. Agriculture and modified temperate environments. *Aust. Biol.* **4:** 220–27.

Nowak, B., 1992. Histological changes in gills induced by residues of endosulfan. *Aquatic Toxicol.* **23:** 65–84.

Richardson, B. A., 1985. The impact of forest road construction on the benthic invertebrate and fish fauna of a coastal stream in southern New South Wales. *Aust. Soc. Limnol. Bull.* **10:** 65–88.

Scribner, E. A., Kaetl, A., Moodie, E. G., Williams, R. J. and Young, J. W., 1987. Organochlorine pesticide and polychlorinated biphenyl (PCB) residues in fish and other aquatic organisms in New South Wales. Part 1 — Fresh water. Miscellaneous Bulletin 4, Division of Fisheries, Department of Agriculture: New South Wales.

Skidmore, J. F. and Firth, I. C., 1983. Acute sensitivity of selected Australian freshwater animals to copper and zinc. Australian Water Resources Council Technical Paper No. 81. Department of Resources and Energy: Canberra.

Stephan, C. E., Mount, D. I., Hansen, D. J., Gentile, J. H., Chapman, G. A. and Brungs, W. A., 1985. Guidelines for deriving numerical National water quality criteria for the protection of aquatic organisms and their uses. US-EPA, PB85-227048. US Dept. Commerce: NTIS.

Stewart-Oaten, A., Murdoch, W. W. and Parker, K. R., 1986. Environmental impact assessment: "pseudoreplication" in time? *Ecology* **67:** 929–40.

Tyler, P. A. and Buckney, R. T., 1973. Pollution of a Tasmanian river by mine effluents. I. Chemical evidence. *Internat. Rev. der gesamten Hydrobiol.* **58:** 873–83.

Thorpe, V. and Lake, P. S., 1973. Pollution of a Tasmanian river by mine effluents. II. Distribution of macroinvertebrates. *Internat. Rev. der gesamten Hydrobiol.* **58:** 855–92.

Underwood, A. J., 1993. The mechanics of spatially replicated sampling programmes to detect environmental impacts in a variable world. *Aust. J. Ecol.* **18:** 99–116.

Wagner, C. and Lokke, H., 1991. Estimation of ecotoxicological protection levels from NOEC toxicity data. *Water Res.* **25:** 1237–242.

Weatherley, A. H., Beevers, J. R. and Lake, P. S., 1967. The ecology of a zinc-polluted river. Pp. 252–78 in *Australian inland waters and their fauna — eleven studies* ed by A. H. Weatherley. ANU Press: Canberra.

Ecological effects of river management in New South Wales

R. T. KINGSFORD[1]

Most large rivers in New South Wales are managed with structures that alter their flow (dams, weirs, levee banks and pumps). By 1995, there will be 144 large (capacity exceeding 1 000 megalitres) water storages in New South Wales. Most (67%) were built after the mid 1950s. They are generally located along the coast or in the headwaters of the Murray-Darling river system. These dams and weirs have reduced flooding in many floodplains. Some wetlands that provide habitat for plants and animals are no longer flooded or not flooded for as long as they used to be. Some natural wetlands are also used as permanent storages, killing many of the plant species and affecting waterbirds. Three major river systems in northwestern New South Wales, have not been seriously affected by river management: the Warrego, Paroo and Bulloo Rivers. They supply particularly important wetland habitats for waterbirds. Their future is threatened by water harvesting, principally for cotton. On other severely compromised river systems, attempts should be made to limit and mitigate the impacts of river management. Our knowledge about the effects of river management remain primitive but sufficient to clearly demonstrate detrimental impacts on aquatic flora and fauna.

INTRODUCTION

RIVERS have not usually featured in the biodiversity debate even though they supply some of the most threatened species and ecosystems in the world (Turner 1991; Allan and Flecker 1993). Vulnerability stems mainly from the utility of rivers to humans. Around the world a few strong interests have usually managed rivers (Mellquist 1992) and Australia is no exception. Most large rivers in New South Wales are now managed. Levee banks, drains and channel works mitigate flooding. Dams and weirs store and regulate water for navigation, drinking, industry, cooling power stations, hydro electricity and irrigation while pumps take water out of and deposit waste into rivers. Many populations of aquatic invertebrates, plants, fish and waterbirds in Australia are in decline; the same may be true of frogs and bats. River management is one of the main culprits (Walker 1985; Lake 1980; Lake and Marchant 1990; Hart 1992; Sheldon and Walker 1993; Walker and Thoms 1993).

The aims of this paper are: (1) to present geographical and historical data on river management in New South Wales; (2) to give an overview of the effects of river management on biodiversity; and (3) to recommend management options to ensure that we minimize or reverse losses in biodiversity.

STATUS OF RIVER MANAGEMENT

History

Early settlers transported goods up and down rivers. They managed the rivers accordingly, clearing obstructions (Williams 1974) and building weirs when rivers became too low (Helman and Estella 1983). River management for agriculture began in the 1850s when canals and dams were built along the Murrumbidgee River. This sometimes led to the sabotaging of dams as squatters defended their rights to water (Buxton 1974). Along the Murray River, the 1870s heralded irrigation (Buxton 1974; Myers 1974; Close 1990) which now uses 75 per cent of surface water in New South Wales (WRC 1993). *The Barren Jack Dam and Murrumbidgee Canals Construction Act* passed in 1906 provided the impetus for irrigation. The New South Wales Government began building Burrinjuck Dam in 1912 for the Murrumbidgee Irrigation Area (MIA) but lost money on the investment (Davidson 1974). New South Wales now has an irrigation industry that employs more than 20 000 people and produces about \$1.4 billion each year, about 20–30 per cent of agricultural production (WRC 1993).

Towns and cities also need water (15% of total water consumption in New South Wales), particularly Sydney, which uses 65 per cent of all urban water (WRC 1993). Prospect Dam was built in the late 19th century (Appendix) while various weirs ponded water for towns and farms around Sydney (Warner 1984). One of Sydney's early supplies of water, the Tank Stream, stored water in tanks to overcome problems of supply during low flows (WCIC 1971). Managing water for irrigation, flood mitigation, power generation and domestic needs were, and are, the driving forces for river management.

An index of the extent of river management is the number of large (>1 000 megalitres = ML) dams or weirs in the state (Appendix). There will be 144 large water storages in New South Wales by 1995 (Fig. 1). Most large storages

[1]National Parks and Wildlife Service, P.O. Box 1967, Hurstville, New South Wales, Australia 2220.

Pages 144–61 *in* CONSERVING BIODIVERSITY: THREATS AND SOLUTIONS, ed by R. A. Bradstock, T. D. Auld, D. A. Keith, R. T. Kingsford, D. Lunney and D. P. Sivertsen. Surrey Beatty & Sons. 1995.

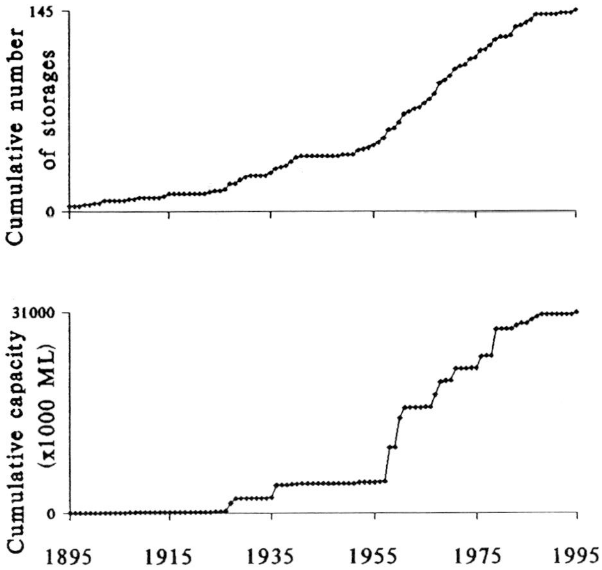

Fig. 1. Establishment of large storages (>1 000 ML) within New South Wales, 1882–1995. Enlargements of >1 000 ML are defined as new storages (see Appendix). (a) Cumulative frequencies of storages; (b) Cumulative storage capacity.

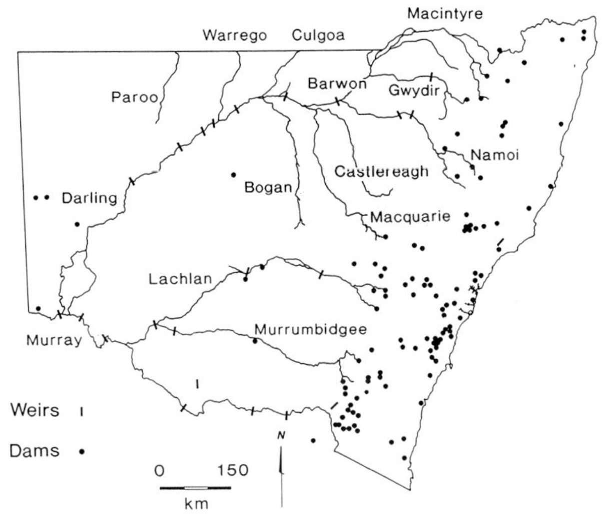

Fig. 2. Distribution of main rivers and locations of large (>1 000 ML) dams (filled in circles) and weirs (dashes) in New South Wales (see Appendix).

in New South Wales before the 1930s were built for urban and industrial supply and irrigation along the Murray and Murrumbidgee Rivers (Appendix). Most dams and weirs in New South Wales (67%), holding about 85 per cent of stored water, were built after the mid 1950s in New South Wales (Fig. 1). These included most dams of the Snowy Mountain Scheme, dams to supply water for irrigation on the lower Darling, Macquarie, Namoi, Gwydir and Border Rivers (Dumaresq, Macintyre) catchments as well as expansion of Sydney's water supply. Accelerated development of water resources within 25 years was the plan for the sixties and seventies at a cost, in 1971, of $723 million (WCIC 1971, p. 209).

The locations of large storages reflect the purposes of river management and topography of land (Fig. 2). Most are along the coast or in the headwaters of the Murray-Darling system (Fig. 2). Storages east of the Great Dividing Range principally supply urban, industrial and coal-fired power generation needs. Dams of the Snowy Mountains Scheme store water for electricity generation. Other storages within the Murray-Darling Basin are mainly for irrigation (Fig. 2). There are more storages for public supply (urban, industrial, stock and domestic) than for other purposes (Fig. 3a). Irrigation is second in the number of storages but stores more water than any other use (Fig. 3b). Most large storages serve more than one purpose. There are also thousands, possibly hundreds of thousands, of farm dams within New South Wales (Timms 1980). Most appear to

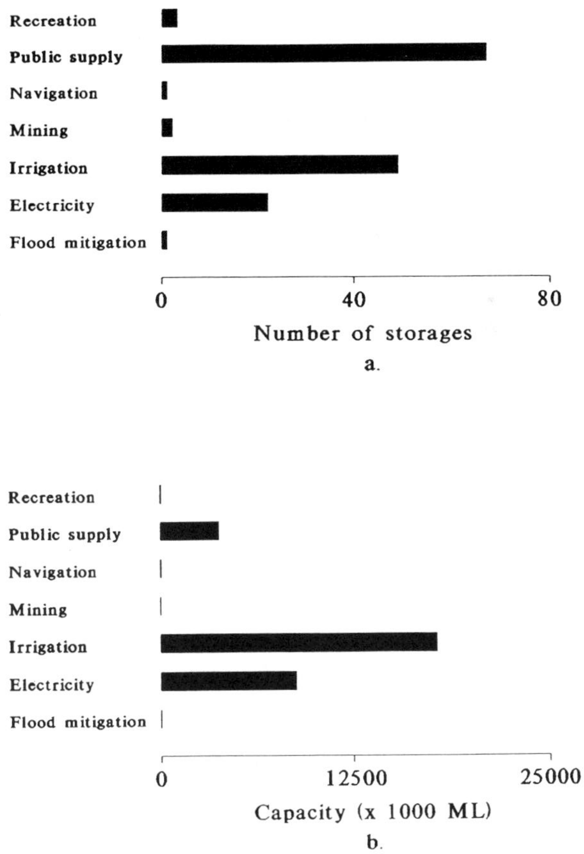

Fig. 3. Purposes for which large storages (>1 000 ML) were built, 1882–1995. Only the primary purpose is recorded for storages which fulfil more than one purpose (see Appendix). Enlargements of >1 000 ML are defined as new storages: (a) frequency versus use; (b) storage capacities versus use.

have been established in the 1960s and 1970s (see Kingsford 1992).

The index of large storages does not recognize effects of small storages; flood mitigation works; water use from unregulated rivers or; water use from unregulated flow on regulated rivers. Water managers define regulated rivers as those where water is released from upstream dams to meet specific demand (DWR 1992a; Wettin *et al.* 1994). Water not committed (mostly for indirect or direct human use) in unregulated or regulated rivers is defined as "surplus" or unregulated water and this can be pumped from rivers when allowed by DWR (Wettin *et al.* 1994). This happens particularly in the mid-north-west of the state. Water is pumped into large off-river storages (Boyd 1992) and water use often exceeds storage capacity because of refilling (Wettin *et al.* 1994). Volumes held in off-river storages are large (e.g., 105 000 ML — Border Rivers; 282 000 ML — Gwydir River; 98 000 ML — Namoi River; 41 000 ML — Macquarie River; 182 000 ML — Barwon/Darling Rivers: K. Boyd, pers. comm., from DWR 1992 data). The Queensland Government has recently approved a 100 000 ML off-river storage, close to the border, which covers part of the Culgoa River floodplain.

Future

Increased demand for water in Australia will be driven by increasing population and agricultural and industrial development (Brown *et al.* 1983). Welcome Reef Dam on the Shoalhaven River, with a storage capacity of 2.7×10^6 ML, could be completed by 2005 to augment Sydney's water supply. Other towns with growing populations, particularly along the north coast (e.g., Coffs Harbour, Lismore) need water. At least double and perhaps three times the storages built to hold water to cool coal-fired power stations (see Appendix) may need to be constructed next century to satisfy increased energy consumption (WRC 1986).

Demands for water for irrigation remain strong in New South Wales and may increase as the profitability margin between livestock and irrigated production rises. The eight times enlargement of Pindari Dam, scheduled for completion in 1995, will expand irrigation along the Macintyre River. Another large dam may be established in the Border Rivers region in the next 15–20 years. Although demand for water remains strong in other parts of the Murray-Darling Basin (Close 1990), there is little prospect for further large headwater storages in the Basin (MDBMC 1987, p. 74).

The likely increased demand for irrigation water may be met by transferring existing allocations of water ("sleeper licences") and pumping from rivers into off-river storages. More water could be harvested in all river systems (Wettin *et al.* 1994). Increased demand by irrigators for water can be satisfied through improving efficiency of current water allocations and appropriate costing of water. Consumption in urban communities may be reduced by education or an appropriate pricing structure.

BIOLOGICAL EFFECTS OF RIVER MANAGEMENT

River management has reduced the amount of flooding on many rivers, increased flooding along parts of others and changed the river environment, while subsequent agricultural developments have resulted in loss of wildlife from developed areas.

Reduced flooding

There are usually more plants and animals, and more species, on the floodplains of Australian rivers than in the channels (Boon *et al.* 1990; Shiel 1990; Roberts and Ludwig 1991; Boulton and Lloyd 1991; Walker *et al.* 1991; Walker *et al.* 1992). Not just water but its spatial and temporal variation sustains high biodiversity on floodplains (Brock 1986; Klimas 1988; Boon *et al.* 1990; Roberts and Ludwig 1991; Welcomme 1992; Allan and Flecker 1993) — the flood pulse concept (Junk *et al.* 1989). Australian wetland plants and animals are generally adapted to highly variable flooding of our rivers. Where there is no flooding, aquatic biodiversity is reduced. Storages (and pumps) reduce the extent of flooding downstream (Fig. 4). On parts of some floodplains in New South Wales, water no longer flows on to the floodplain (e.g., Tandou Lake on Darling River, North coast wetlands). River management has reduced the frequency and extent of flooding of many rivers in New South Wales (Lloyd *et al.* 1994). There is less water for floodplain wetlands (Hillman 1986; Walker 1986; Barmuta *et al.* 1992). Eighty-seven per cent of mean natural flows in the Murray-Darling Basin (excluding Queensland) are now diverted (Thomson 1992); although some water returns to wetlands after being used for irrigation. Flows to the lower Murray River are about 44 per cent of what they were before river management (Walker and Thoms 1993) and only about 36 per cent at the mouth (Close 1990). The Barmah Millewa forest now floods effectively about 35 per cent of years but it used to naturally flood about 80 per cent of years (Close 1990). The frequency and extent of flooding on the Macquarie River is less than it used to be (Kingsford and Thomas 1995). Water diversions from the Darling River (upstream of Menindee Lakes) and the Murrumbidgee River increased substantially between the early 1960s and 1990/91 (respectively 1 per cent to

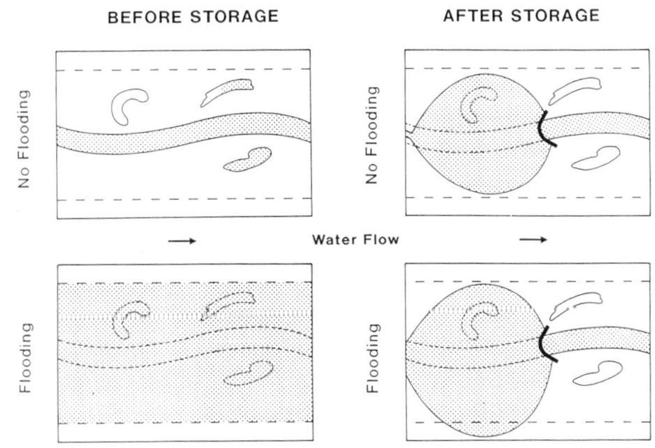

Fig. 4. Theoretical effects of building a storage on downstream flooding of a river and associated floodplain, including three floodplain wetlands. Dashed lines signify the floodplain and dotted area is flooded area.

40 per cent, and 25 per cent to 93 per cent of mean natural flow per year) (Thomson 1992). Based on landholders' claims, Copeton Dam and water harvesting reduced wetland area along the Gwydir River from 47 000 ha to 8 400 ha, (Keyte 1992; WRC 1993). A great deal of water is pumped from the Gwydir River, Darling River and Border Rivers (Wettin *et al.* 1994) and has dramatically impacted on downstream hydrology (Meyer 1992). On coastal rivers, flood gates, drains and levees have reduced wetland area (e.g., 96 per cent wetland area affected by drainage in lower Macleay floodplain; Pressey 1989).

Flooding is essential for plant species on floodplains. In arid Australia Red Gum forests *Eucalyptus camaldulensis* cannot survive on rainfall alone. Fewer and shorter floods have reduced growth and health of Red Gums and Black Box Eucalypts *E. largiflorens* on the Murray River (Bren 1988; Smith and Smith 1991). Riparian vegetation along the Darling River (Thomson 1992) and in the Macquarie Marshes is lost (Brander 1987) or degraded (Johnson 1992; Wettin *et al.* 1994).

When flooding is less frequent, vegetation that is less reliant on flooding replaces more aquatic vegetation. Along the Murray River, Red Gums have replaced Moira Grass plains *Pseudoraphis spinescens* (Ward 1992; Bren 1992). Dry grassland is replacing Red Gum and Coolabah *E. coolabah* woodland in the Macquarie Marshes (Johnson 1992) and Black Box may eventually take over Red Gum forests along parts of the Murray River (Bren 1988). Dryland pastures appear to have replaced wetland plants in north coast wetlands (Briggs 1977). Lippia *Phyla nodiflora,* Blown Grass *Agrostis avenacea,* Black Roly-poly *Bassia quinquecuspis* and annual burrs and thistles have replaced wetland vegetation along the Gingham Watercourse (McCosker and Duggin 1992) and in the lower Gwydir River (Wettin, pers. comm.) where flooding is 75 per cent less than before river management (Wettin *et al.* 1994).

Many animals rely on flooding which appears to release nutrients from the vegetation and litter (Briggs and Maher 1983; Maher 1984; Briggs *et al.* 1985). The water fern *Azolla* spp. fixes nitrogen and increases nutrient availability (Boon *et al.* 1990). High densities of invertebrates exploit newly flooded areas (Maher and Carpenter 1984; Crome 1986; Crome and Carpenter 1988) by moving from permanent water or rivers, emerging from a desiccated state, and hatching from eggs or cysts left in the soil from when the floodplain dried. Within natural limits, frequency of flooding is positively related to invertebrate abundance and biomass (Boulton and Lloyd 1992).

Reductions in floodplain habitats have also affected fish, amphibians, reptiles, waterbirds and mammals that consume aquatic plants and invertebrates. Many native fish species spawn and mature on the floodplain (Lloyd *et al.* 1991). This habitat can be lost (e.g., Yarrahapinni levee-Macleay Valley; LMP 1980). Declines of Golden Perch *Macquaria ambigua,* Silver Perch *Bidyanus bidyanus* and Murray Cod *Maccullochella peeli* populations in the Murray River are attributed to loss of spawning and nursery habitat (Geddes and Puckridge 1989; O'Connor 1989; Cadwallader and Lawrence 1990).

The problems for waterbirds are similar: loss of feeding and breeding areas. By 1969, flood mitigation had reduced or destroyed about 60 per cent of good waterfowl habitat on the north coast of New South Wales (Goodrick 1970). Loss of coastal wetlands, including Seven Oaks in the Macleay Valley (LMP 1980), continued (Pressey and Middleton 1982) until 1985 when State Environmental Planning Policy No. 14 of the *Environmental Planning and Assessment (EP and A) Act* 1979 required planning approval for flood mitigation works. Coastal flood mitigation particularly affected waterbird species with a coastal distribution like Black-necked Storks *Xenorhynchus asiaticus,* Comb-crested Jacana *Irediparra gallinacea* (Braithwaite 1975; Briggs 1977) but also others like Hardhead *Aythya australis* (Frith 1982, p. 257). Channelling of water in the Gwydir River has reduced feeding and breeding habitat for waterbirds (McCosker and Duggin 1992). High numbers of invertebrates in intermittently flooded wetlands are a driving force for waterbird breeding in semi-arid Australia (Maher and Carpenter 1984; Crome 1986) and loss of this habitat along inland rivers has led to reduced waterfowl (Anatidae) breeding (Briggs 1990). Reduced flooding has probably detrimentally affected microchiropteran bats (e.g., Large-footed

Mouse-eared Bat *Myotus adversus*, Greater Long-eared Bat *Nyctophilus timoriensis*) which feed and roost on the floodplain (G. Richards, pers. comm.).

Prolonged flooding

Paradoxically too much flooding is also detrimental to many plant and animal species. Effects of increased flooding are generally more obvious than those of decreased flooding. Dead trees or lignum in a wetland are the most obvious sign of too much flooding.

Dams and weirs permanently flood areas, sometimes natural wetlands (Fig. 4). The animals have to move or die with the dryland plants. The effect of permanent flooding on wetland communities is more gradual but equally detrimental as many aquatic plants in many wetlands in Australia are adapted to variable flooding and drying. Such plants need this hydrological regime for viable seed germination and root zone aeration (Gopal 1986; Klimas 1988; Roberts and Ludwig 1991). Too much flooding may lead to reduced growth and survival (Brock and Casanova 1991).

Weirs have made almost 35 per cent (38 000 ha) of the wetlands along the floodplain of the Murray River permanently full (Pressey 1986; Pressey 1990). Many Red Gums along the Murray River are now physiologically stressed or dead (Dexter *et al.* 1986; Bren 1990; Smith and Smith 1991). Flooding of Menindee Lakes (Darling River) killed many Red Gums (Somerville 1988), Black Box trees and extensive Lignum areas (WCIC 1971, cf photo p. 85). Since 1983 annual aerial survey counts of Lake Menindee have not exceeded 4 000 waterbirds, most of which are fish-eating birds (Kingsford, unpubl. data). Similar sized lakes on the Paroo River which flood and dry naturally support much larger populations of waterbirds (Kingsford 1994). Most (>90%) Black Box trees in a temporary swamp in Kinchega National Park died when flooded for 18 months after a levee was breached in 1988 (T. Auld, unpubl. data). Subsequent flooding killed regenerating and adult trees. Prolonged flooding has stressed or killed Red Gums in the Macquarie Marshes, near artificial channels (Wettin *et al.* 1994; Johnson 1992) and has destroyed vegetation of many swamps in the Murrumbidgee River system (Red Gum, Black Box trees, sedges, Nardoo, Canegrass and Lignum) (DWR 1991; Shepheard 1992; Thornton and Briggs 1994).

The changed hydrological regime from temporary to permanent flooding in many wetlands has affected invertebrates adapted to variable flooding. Fewer plankton species survive in these new habitats which are dominated by species more adapted to lakes and standing water (Shiel 1990; Bennison *et al.* 1991). The lower Murray River has changed from a flowing river to a series of cascading pools and with it the range of the River Mussel *Alathyria jacksoni*, adapted to flowing waters, has decreased while the Floodplain Mussel *Velesunio ambiguus*, which is able to exploit this new habitat, has extended its range (Walker 1990). Water storages support fewer benthic invertebrate species than rivers (Blyth 1980; O'Brien *et al.* 1983), possibly because they are deeper. Continuous flooding may also reduce invertebrate abundance (Maher 1984; Maher and Carpenter 1984).

Reduced populations of plants and invertebrates may have adversely affected some species of waterbirds (Briggs 1993). Long-term losses of structural vegetation for waterbird nesting, like Lignum, canopies of trees and tree holes, are the most obvious effect. Even cormorants and ibis, that benefit in the short term as dead trees and Lignum provide nesting and roosting sites (DWR 1991; Briggs 1994), will eventually lose habitat when these plants decay.

Creation of artificial wetlands and permanent flooding of natural wetlands has been thought to counteract negative effects of river management on biodiversity (LMP 1979; Braithwaite 1980; Pigram 1986, p.255), but this is unlikely. Initially, most storages provide good habitat for fish and waterbirds as organic matter and nutrients are released and invertebrate populations build up after flooding. With extended inundation (>2 years) biodiversity decreases (Cadwallader 1978; Maher 1990; Briggs *et al.*, in press). Introduced fish species may become dominant in reservoirs (Crome 1980; Cadwallader 1978). Permanently flooded wetlands along the Murrumbidgee River (Briggs *et al.*, in press) and the Darling River (Kingsford, unpubl. data) support waterbird communities dominated by fish-eating waterbirds. These have replaced duck and wading bird communities. A few artificial wetlands may benefit a range of waterbirds (e.g., in the Northern Territory and arid Australia (Braithwaite 1980), and artificial evaporation pans also provide important habitat for migratory waders (Smith 1991). Permanently flooded wetlands also act as refuges and may increase survival of waterbirds during droughts (LMP 1979; Braithwaite 1980; Briggs 1990; Briggs and Lawler 1991; Briggs 1993). Apart from the impact on aquatic flora and fauna, permanent flooding of wetlands may have a negative impact on dryland fauna. Dry wetlands may provide refuge habitat for dryland fauna (and flora) during dry periods (Briggs 1992).

Some wetland biota have benefited from permanent flooding. Reedbeds *Phragmites australis, Typha domingensis, Juncus ingens* and introduced Willows *Salix babylonica* flourish along the Murray River (Walker 1985; Sainty and Jacobs 1990; Smith and Smith 1990; Walker *et al.* 1991). Cumbungi *Typha* sp. has increased on the Murrumbidgee River (DWR 1991; Shepheard 1992) and in the Macquarie Marshes (Wettin *et al.* 1994). These plants may provide spawning grounds for introduced fish (Cadwallader 1978). Introduced plants (Water Hyacinth *Eichornia crassipes,* Salvinia) have infested permanent channels of wetlands (e.g., Tuckean Swamp, Hexham Swamp, Gingham Watercourse; see Crome 1980; Pressey and Middleton 1982; McCosker and Duggin 1992). Fish-eating birds such as cormorants and Australian pelicans have probably benefited from increased flooding because this has allowed permanent fish populations (often introduced species) to establish. Maned Duck *Chenonetta jubata* have benefited from the increased number of farm dams (Kingsford 1992) and other storages.

Changes to river environment

Barriers

Walls of storages have contributed to declines in native fish populations (Barmuta *et al.* 1992) because they restrict migration (see Reynolds 1983). Some fish species are now locally extinct on coastal rivers (Harris 1984a), because only about 9 per cent of about 300 dams, weirs and causeways on coastal rivers in New South Wales allow fish passage (Harris 1984b). Tallowa Dam on the Shoalhaven River precipitated local extinction of the endangered Australian Grayling *Prototroctes maraena* (Bell *et al.* 1980) and possibly Bass *Macquaria novemaculeata* upstream (O'Brien *et al.* 1983, p. 28). There are only 22 fishways (mostly inefficient) over the entire Murray-Darling Basin to help 33 fish species pass more than 150 dams, weirs and causeways (Mallen-Cooper 1989). Boat locks in regulating weirs also allow some fish movement (Walker 1983). Golden and Silver Perch, once common upstream of Yarrowonga Weir on the Murray River, are now extinct in this stretch of the river (Walker *et al.* 1978). Walls of weirs and dams may also have more subtle effects. Aquatic vegetation upstream of weirs tends to be correlated to disturbance (see Walker *et al.* 1991). Diversity and abundance of invertebrates are generally reduced below weirs and dams possibly because organic material is restricted from flowing downstream (Bravard *et al.* 1986; Marchant 1989). Water managers now recognize the problem and there are now new fish ladders in the Murray-Darling Basin.

"Improvements"

River management procedures, anthropocentrically termed "improvements", include the removal of obstructing vegetation (snags or woody debris) from river channels and floodplains and dredging and channelling of the river bed. Between Lake Hume and Yarrawonga Weir on the Murray River, 24 000 snags were removed over a decade (MDBMC 1987, p. 154). Improvements were done for efficient navigation, recreation, flood mitigation and to increase flow rates to supply water for agriculture. Detrimental effects of desnagging and channelling often occur elsewhere (Warner 1984) and last a long time (Lloyd *et al.* 1994).

Desnagging and channelling detrimentally affect many biota (Pressey and Harris 1988; Lloyd and Walker 1989; Koehn and O'Connor 1990; Lloyd *et al.* 1991). Increased flow reduces numbers of benthic invertebrates (Borchardt 1993). Snags increase heterogeneity of river environments (Walker and Lloyd 1987); such heterogeneity is correlated with biodiversity. Algae, bacteria, and fungi colonize snags and provide food for invertebrates (Walker *et al.* 1992; Lloyd *et al.* 1994). Snags increase densities of fish and crustaceans by providing refuges from predation (Borchardt 1993; Everett and Ruiz 1993). Fish such as Murray Cod, Blackfish *Gadopsis marmoratus* and Gudgeons *Hypseleotris* sp., *Philypnodon* sp. rely on snags for spawning (Lloyd and Walker 1989; Cadwallader and Lawrence 1990; Lloyd *et al.* 1994). Platypus, Water Rats, Long-necked Tortoises, Water Dragons, cormorants, ducks and pelicans which rest on or feed around snags are probably also negatively affected by the removal of snags (Briggs 1977; Walker and Lloyd 1987). River improvement is generally not practised in New South Wales now (see Lloyd *et al.* 1994) but the legacy remains.

Fluctuations in water level

Water levels often change rapidly with delivery of water for irrigation or power generation, particularly below dams, weirs or around irrigation pumps. Before 1925, the Murray River had sustained rises or falls in water level but water level now fluctuates rapidly below weirs (Walker *et al.* 1991). Flood duration is less on the Murray River (Walker and Thoms 1993). Irrigation pumps for cotton reduced a flow of 500 ML/day to zero in three days (e.g., Barwon River, December 1992, K. Boyd, pers. comm. from DWR data). Few animals or plants survive in such environments (O'Brien *et al.* 1983, p. 30). More insidious effects may occur downstream. Plant, invertebrate and fish populations may decrease if floods are not sustained because they have insufficient time to complete life cycles (Lloyd *et al.* 1994). Rapidly dropping water levels may strand eggs and young of fish that lay eggs in shallow water (Walker 1983; Geddes and Puckridge 1989; Cadwallader and Lawrence 1990). Murray Cod and Catfish may abandon nests and eggs if subjected to severe water level fluctuations (Lloyd *et al.* 1991). Water is sometimes completely cut off from wetlands. Large numbers of several species of fish, including 312 Australian Grayling died when water stopped flowing below Tallowa Dam (Bishop and Bell 1978). In the late 1960s a rapid drop in the water level of Lake Menindee resulted in the death of many young black swans when parents deserted their young (J. Eveleigh, pers. comm.). Similarly a rapid drop in water level caused ibis to desert eggs and young in a colony on the Lachlan River system (Pressey and Harris 1988).

Timing of flooding and flow

River management has altered the seasonality of flooding in some rivers because of the need for irrigation water during summer. Flooding has shifted from spring-early summer to a late summer pattern at the top end of the Murray River (Close 1990) and red gums are replacing grasslands (Bren 1992). Changes in flooding season may also affect invertebrates that have breeding cycles correlated with season (O'Brien *et al.* 1983, p. 30). The ideal time for native fish to breed is in the spring or summer when physiological requirements for breeding are met (Lloyd *et al.* 1991) but along the Murray River, flows at this time may not be long or extensive enough for successful breeding (Lloyd *et al.* 1991). Floods outside this period are of little use to breeding native fish. There may be no floodplain habitat available of the right type even for juveniles of fish who rely on stimuli other than flooding to spawn (Geddes and Puckridge 1989; Lloyd *et al.* 1991). The flow-on effects from invertebrates to waterbirds are also probably negative because increasing warmth in spring stimulates production of many waterbird foods (Briggs 1990). Dexter (1978) proposed that spring flooding with receding water in late spring and early summer was beneficial for waterbird breeding along the Murray River.

Temperature

Water temperature in the Murray River is often lower than normal and often low in oxygen because it comes from the bottom of storages (Lake 1980; Walker 1985; Lloyd *et al.* 1991). Low temperatures have reduced diversity and populations in the macroinvertebrate community (Doeg 1984; Bennison *et al.* 1991). This can be avoided if water comes from the surface of a dam (Marchant 1989). Hypolimnetic releases produce a downstream benthic fauna adapted to cold but impoverished in species (Lake and Marchant 1990). Changed temperatures have been a major cause of local extinctions of Golden Perch, Catfish *Tandanus tandanus,* and possibly other fish species (Walker 1980; Walker 1983). Reduced temperature may also affect survival by reducing swimming performance (Harris 1984a). Decreased temperatures may favour exotic species like trout (Salmonidae) and Canadian Pondweed *Elodea* sp. (Lake 1980; Lloyd *et al.* 1994).

Erosion, sedimentation and turbidity

Sediment transport has changed with river management and altered flows (Walker 1985). The regulated parts of the Murray River transport most sediment during irrigation periods (October–March) (Thoms and Walker 1992). Highly variable water levels and high water energy promote erosion of river banks immediately downstream of weirs and dams (O'Brien *et al.* 1983, p. 182; Walker *et al.* 1991; Thoms and Walker 1992; Walker and Thoms 1993). High sedimentation rates and increased turbidity also occur during dam construction. These reduce the number and abundance of most invertebrate species although favouring a few others (Blyth 1980; Doeg 1984; Doeg *et al.* 1987; Marchant 1989). Where erosion is minimized, biodiversity of invertebrate populations is better maintained (Chessman *et al.* 1987). Sediments fill deep water holes (Bower 1980), and may destroy downstream habitat where fish and reptiles feed and shelter. Sediments along the Murray River were coarse sands before regulation but are now fine silts and clays (Walker and Thoms 1993). Mobility of sediments in the Thredbo River could be threatening Platypus populations more than any other factor, by disturbing their prey and filling holes (D. Goldney, pers. comm.). Sediment may also smother eggs of fish that spawn in gravel (Cadwallader 1978; Lloyd *et al.* 1991).

Salinity

Conversion of saltwater wetlands to freshwater wetlands (and the other way around) fundamentally affects biota. Wetlands become less estuarine when floodgates on coastal wetlands restrict tidal exchange (Pressey and Harris 1988). Mangroves die (Middleton *et al.* 1985; Briggs 1977; Pressey and Harris 1988) and freshwater plants replace salt-adapted plants (Briggs 1977, 1983). Fish species which rely on the saline estuarine habitat are also affected. Similarly freshwater biota die when salt water invades freshwater wetlands along drainage lines (e.g., Nambucca floodplain, Lake Innes; Pressey and Middleton 1982) or when floodplain wetlands become salt inception or evaporation basins (L. Lloyd, pers. comm.).

Indirect effects

Irrigation is generally the driving force behind river management but river management also often leads to increased irrigation. There is now irrigation along all major rivers in the Murray-Darling Basin. Irrigation development has had as great an impact on biodiversity at the local scale as other intensive agriculture or urban development; it removes natural dryland and floodplain vegetation along with the animals. Even when aquatic plants can survive irrigation development, as in rice fields, overall densities of species are less and exotic species are more abundant than in natural habitats (McIntyre *et al.* 1988).

In 1984/85, about 600 000 ha of New South Wales were irrigated, excluding pasture irrigation (WRC 1986). Irrigation in the Riverina has converted 456 000 ha of dryland pasture into intensively managed farmland (Leigh and Noble 1972) and destroyed many small swamps during land-forming (Churchward and Flint 1956). After the storages on the Namoi River were built (see Appendix), large areas of dryland grazing were converted into intensively cultivated agricultural land over a period of 20 years (WCIC 1971; Pigram 1986, p. 176). The change in land use is stark (see Graetz *et al.* 1992). In the last 10 years, irrigation developments for cotton have increased. The area under cotton production, excluding the Macintyre River system, has increased from about 63 000 ha in 1981/82 to 170 000 ha in 1991/92 in New South Wales (CYB 1992) with relatively higher increases in the Darling (Bourke), Macquarie and Macintyre River valleys (Fig. 5); a small percentage of the Namoi and Gwydir valley production is dryland cotton.

River management can indirectly promote irrigation development. With less flooding, floodplains revert to drylands and cultivation can proceed (e.g., Gwydir River, see McCosker and Duggin 1992). Legislation can indirectly promote irrigation development by not recognizing natural flooding as a legitimate use for water. An irrigation licence for flooding of native wetland plants may be refused because grasses are not sown or improved (*Water Act* 1912: Section 13C(1)). All irrigation licences stipulate that water not be used for flooding "unsown" or "unimproved" pasture.

HALTING AND REVERSING BIODIVERSITY LOSS: SOLUTIONS

Establishment of conservation reserves alone will not sustain biodiversity of rivers; we must devise management strategies for rivers that incorporate conservation as well as river use (Welcomme 1992). This means active management of whole river systems (Arthington *et al.* 1991). Management must hinge on identification and understanding of the processes that cause loss of biodiversity. Four options provide a framework for halting decline in biodiversity in rivers of New South Wales (after Boon 1992): conservation (= preservation); limitation and mitigation; restoration and; dereliction. We need economic, legislative, educational and social methods for achieving biodiversity conservation (MDBMC 1987, p. 165).

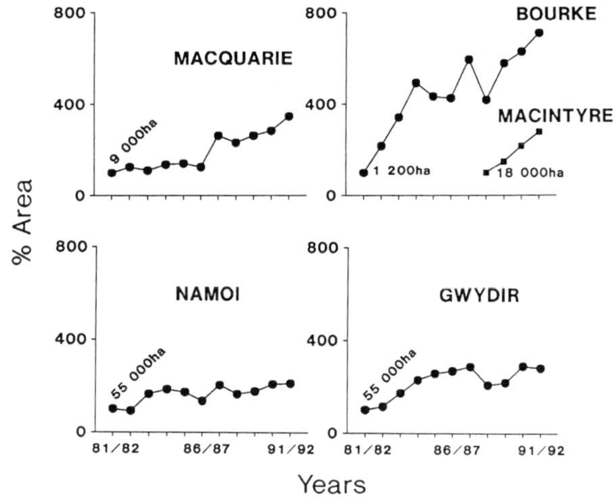

Fig. 5. Change in area under cotton production in New South Wales, 1981–92. Data were from Cotton Year Book (1993) and were produced by adding up areas under production in the Macquarie, Darling (Bourke), Macintyre, Namoi and Gwydir River valleys.

Conservation

Some rivers, concentrated in north western New South Wales, have little river management: the Warrego (some dams), Paroo and Bulloo and other rivers east of the Narran River (see Fig. 2). Water is extracted for domestic, stock and town water needs. There are plans to extract water from the Warrego River near Cunnamulla and water is being extracted from the Narran and Culgoa Rivers near Dirranbandi in Queensland for irrigated cotton. Water in Narran River ends up in Narran lake and the other floodplain wetlands. Narran Lake is an extremely important wetland for waterbird conservation (Magrath 1991; Brooker 1993; Smith 1993). Similar wetlands further west, supplied by these rivers and local catchments, hold some of the highest concentrations of waterbirds in Australia (Braithwaite *et al.* 1986; Maher 1991; Maher and Braithwaite 1992; Kingsford and Porter 1994; Kingsford 1994) as well as supporting other significant fauna and flora (Green 1992; Timms 1993). Many wetlands within this area are wetlands of national and international importance and carry fauna for which we have international responsibilities (see Michaelis and O'Brien 1988).

There should be a moratorium on water harvesting for irrigation, until developers rigorously demonstrate that developments are ecologically sustainable. This moratorium should include other smaller rivers or creeks that fill high value or significant wetlands in the state (e.g., Bancannia Creek; Cuttaburra channels; Talyawalka-anabranch of the Darling

River; Bootra and Yancannia creek; lower Lachlan River wetlands; Castlereagh Marshes). There may be other rivers with little management elsewhere in the state where development should be discouraged because wetlands are significant (see ANCA 1993).

Many of the rivers in the north-west of the state originate in Queensland. The moratorium must involve co-operation with Queensland, possibly through the Murray-Darling Basin Ministerial Council, and designations of floodplains of high environmental value under *Water Resources, Environmental Planning and Assessment Acts* and *Total Catchment Management Acts.* Extension of Border Rivers Commission offers another avenue for protection. River management has had a detrimental effect on current land use in the area (Marshall 1993) as it has in other parts of New South Wales and the world (Turner 1991; McCosker and Duggin 1992) so this would probably be welcomed by pastoralists. Similarly, New South Wales should recognize the importance of providing beneficial floods to South Australia.

Limitation and mitigation

Many of our managed rivers have significant biota, albeit reduced in number and distribution. Many are adapted to highly variable flow regimes. To sustain biodiversity, we must try to limit and mitigate processes identified as deleterious (see above) and attempt to mimic natural variability of rivers. It may not be possible or socially responsible to restore natural river regimes but we can gradually reduce the amount of extraction where current levels are undesirable. To achieve this, most unregulated flows and a little of regulated flow of river systems should be allocated for environmental use. There should be a ceiling on extractions for existing developments set at current levels and requirements to increase efficiency so that environmental improvement can occur. This could be done by designating floodplains (see below). On fully committed river systems there must be an appropriate basic flow that provides water or protects some of the natural flow for the river ecosystem (Lake and Marchant 1990; Arthington *et al.* 1991). DWR has established this on some regulated and unregulated rivers (Wetting, pers. comm.). Basic flow regimes should include sufficiently prolonged and frequent small to medium floods (as well as unpreventable large floods) to flush pools, and restore and maintain ecosystems. New developments that use water must demonstrate net benefit after assessing ecological and social consequences (see Statzner and Sperling 1993), particularly in a regional context. Development of water resources must be subject to review which includes rigorous research on effects of development after the impact as well as before (see Underwood and Fairweather 1993). Such research will provide future developers with an objective basis for sustainable use and regulatory bodies with the data to make decisions about sustainability.

Current legislation provides some review of environmental impacts of water resource developments. Proposed or existing works for irrigation, water conservation, water supply, drainage or prevention of flooding must be licensed by DWR under Part II of the *Water Act* (1912). Where these works include excavation, building or construction of permanent structures the DWR, as the determining authority under Part V, Section 110 of the *EP and A Act*, can require an Environmental Impact Statement if the environmental impact is considered significant (see Part V, Sect. 112 of *EP and A Act* and Part VII Sect. 56, EP and A Regulations). Developers may need to consider the impact on endangered fauna as required by the *Endangered Fauna (Interim Protection) Act* 1991 (see Sect. 4A) which may require a Fauna Impact Statement. Even if these procedures are followed, approval of a licence may be favoured. Irrigation works become legal as soon as a licence application is registered (Sect. 10, Div. 3 of *Water Act*). Inevitably there is a time lag before approval or refusal and the economic loss caused by refusal may be particularly high. DWR has proposed legislation to address this problem. Local Government can also insist on assessment of environmental impact by designating water resource projects as developments under the *EP and A Act* (Sect. 29). Then when the development is applied for, it must be accompanied by an Environmental Impact Statement which is widely publicized. Regional and Local Environmental Plans provide another opportunity for environmental assessment at the local scale. The opportunities exist for increased involvement of local councils with reductions in state control of water for irrigation (Wettin *et al.* 1994). If local decisions on water developments are made without due regard to downstream effects that may be outside local government control or interest then water use may increase.

Floodplains can be declared under Part VIII of the *Water Act.* If this is done, DWR must regulate earthworks, levees, access roads, irrigation channels and storages that could affect the distribution of floodwaters. Environmental approval is granted on the basis of significant impact (see discussion above). It is an offence to construct works in a designated floodplain without approval. The Macquarie Marshes is a designated floodplain where such works have been built without approval. Designation of a floodplain may provide a means to regulate water harvesting. Part VII should apply to all floodplains to prevent developments occurring without environmental assessment. Although not yet used, Section 61 of the *National Parks and Wildlife Act* 1974, allows for protection of rivers within National Parks or Nature Reserves by declaring them as wild and scenic rivers.

Local impacts of river improvements, rapid water level fluctuations, effects of barriers and temperature can be managed. DWR maintain water to breeding ibis on Booligal Swamp off the Lachlan river and have other environmental contingency plans. Better designed fish ladders are proposed in the Murray-Darling Basin on weirs (Lloyd *et al.* 1991; DWR 1992b) through a fishways programme (DWR and Fisheries). Every dam built by DWR since 1980 has variable offtake outlets which release water from the upper surface and avoid problems with water temperature and oxygen depletion. These can also be added to older dams (e.g., Glenbawn Dam). There should be no desnagging of rivers unless there is a clear local need (Lloyd and Walker 1989; Cadwallader and Lawrence 1990). All options for overcoming environmental problems should be considered (e.g., modification or abolition of weirs instead of establishing expensive fish ladders).

Young (1992, p. 62) advocates a mixed approach to resource use. Government provides the legal, regulatory framework and policies for sustainable use and intervenes where necessary while markets efficiently produce goods and services from a resource which is appropriately costed for future impacts.

Restoration

River systems can sometimes be restored with minimal assistance (Allan and Flecker 1993). We can restore whole or parts of rivers and the wetlands that they supply either passively through management of catchments or actively by repairing local problems (Lake and Marchant 1990). Total catchment management will achieve some objectives of passive restoration (e.g., Westernport Bay, see Barmuta *et al.* 1992).

Active restoration efforts should aim to restore the physical environment by constructing small wetlands and establishing and maintaining buffer strips of vegetation (Petersen and Petersen 1992). These limit effects of erosion and nutrient input. Restoration attempts along rivers in Australia began with revegetation, often with exotic species such as willows, and "improving" (see above) the river environment (Barmuta *et al.* 1992).

Manipulating hydrology of the river can also effect recovery of parts of the floodplain. Implementing more sophisticated techniques of licensing may help to mimic natural flows. Flooding may unintentionally create high value wetland habitat for waterbirds (e.g., Lowbidgee district, Kingsford *et al.* 1992). Sometimes restoration can be simple. A new regulator on Nearie Lake, on the Darling Anabranch, will allow natural fluctuations in water levels which have been held artificially high (Briggs and Townsend 1993). Other sites along the Murray River in South Australia have also benefited from restoration of hydrological regimes (Jensen *et al.* 1994). There are preliminary guidelines for restoring and managing wetlands for fauna and flora according to their natural hydrological regime (Briggs 1988), although more details are needed for specific biota.

It may be possible to re-establish fauna that are locally extinct (e.g., Murray Crayfish in Murray River, Mackay 1990). Wetlands that provide habitat for waterbirds and aquatic plants and invertebrates can also be created (e.g., Shortlands Wetland centre, Newcastle and Tullakool evaporation pans). Restoration is sometimes costly and it should not be considered as an alternative to conservation (see above). For example, four options to provide water for the Barmah-Millewa red gum forest cost about $16.5 million on average (MDBC 1992).

Dereliction

Restoration will be impossible or seriously limited on some rivers (e.g., Cooks River). We can use these rivers for other purposes. While ensuring that they are not treated as waste dumps, we should acknowledge their depauperate biological status and concentrate resources for limiting and mitigating biodiversity loss in the river systems that will most benefit. This may also help to draw community attention to deleterious anthropogenic effects on river systems.

CONCLUSIONS

River management for irrigation has been costly, economically (Davidson 1974; Young 1992, p. 55) and environmentally (Walker 1985). We need to manage our rivers more sustainably. This means acquiring a better scientific understanding of river systems and the effects of river management (Walker 1985; Lake and Marchant 1990; Hart 1992; Barmuta *et al.* 1992; Cullen and Lake, this volume). Regular and repeatable long-term surveys of fauna and flora can provide the necessary information to determine the status of populations of animals and plants. Conversions of wetlands to agricultural land or for other human uses have largely proceeded because of information failure (Turner 1991). Rigorous design of studies and reporting of results in peer-reviewed scientific journals should be encouraged if we are to produce reliable knowledge (Boon 1992). With better information on needs and effects of river management on biota, environmental demands could conceivably be mounted on river systems already exploited to capacity. We should learn from management decisions and discard management based on untested dogma and emotional appeal for modelling and experimental design if we are to make the right decisions (Walters 1993).

Until the river ecosystem is acknowledged as a legitimate user of water with concomitant rights to water use and variability, biodiversity will continue to decline. Flooding is essential for conservation and its worth needs to be acknowledged and understood by the community. Managing authorities of rivers recognize the importance of providing water for the environment (MDBMC 1987, p. 165; Hart 1992; Wettin *et al.* 1994) but there are relatively few actual concessions to the environment. This problem is not unique to New South Wales (see Mellquist 1992). Currently fish and wildlife only receive 0.7 per cent of all water allocations in New South Wales (WRC 1993) which includes an annual environmental allocation of 50 000 ML for the Macquarie Marshes (WRC and NPWS 1986); indirectly the environment obviously receives more. There is a formal agreement between states to provide water for waterbird breeding in the red gum wetlands of Barmah-Millewa, along the Murray River (MDBC 1992). Wetlands are managed for conservation in the Lowbidgee district (DWR 1991). DWR is also preparing various policies to provide water for environmental use (Wettin *et al.* 1994). The Murray Darling Basin Commission provides minimum releases for water quality, environmental maintenance and navigation but generally water for the environment is only released on an *ad hoc* basis (Fitzpatrick 1991).

There is little legislation guaranteeing water rights for the environment (MDBMC 1987, p. 187). Even the oft-touted environmental allocation of water for the Macquarie Marshes has no legal basis. Even when there is uncommitted water and an agreed environmental allocation, actual provision of water may take a long time (e.g., 12 years for Dartmouth Dam; see Fitzpatrick 1991). This problem is not unique to water use. Development planning seldom successfully integrates the environment (Slocombe 1993) probably because of conceptual and information problems (Hohl and Tisdell 1993). Society perhaps does not consider such environmental problems sufficiently critical to embark on the drastic solutions needed (Fairweather 1993). More likely, society does not know how critical the problems are.

Policies for rivers primarily revolve around their high economic importance. There will be incentives for conservation if inputs for water developments and use are not subsidized by governments (see Young 1992, p. 56). Managers of water resources must stop the worst impacts, limit further deterioration and try to restore already damaged systems (Welcomme 1992). Communities, politicians and other policy-makers must continue to be educated about the past, present and future environmental costs of river management. The challenge is to determine a cost for water which recognizes sustainability, particularly avoidance of environmental degradation.

ACKNOWLEDGEMENTS

Views expressed in this paper are those of the author and not necessarily those of the National Parks and Wildlife Service (NSW). Sue Briggs helped develop the outline and provided useful comments and information for the paper. I thank René Ferster Levy for collecting, collating and checking data on water storages. I thank Tim Cooney for collecting data for some of the weirs. Michael Hood and Julianne Smart clarified legislative points. I thank John Porter and Rachael Thomas for helping with the drawings. The paper benefitted from comments by Kate Boyd, Terry Hillman, Lance Lloyd and Paul Wettin.

APPENDIX

Names, dates of completion, storage capacities, locations and original purposes of storages >1 000 megalitres built in New South Wales, up to 1995. Enlargements are listed, giving only the increased storage capacity. Data mainly compiled from WCIC (1971), DSC (1992). [a]Some data for weirs were not available: Berembed (3 100 ML — Murrumbidgee River), Dubbo (1929); Gin Gin (1896 — Macquarie River), Marebone (1975 — Macquarie River), Narromine (Macquarie River), Nyngan (Bogan River), Warren (1896 — Macquarie River), Yanko (1928 — Murrumbidgee River). [b]×1 000 megalitres (ML).

Storage name[a]	Date	Capacity[b]	Location	Purpose
Three Mile Dam	1882	1.8	Three Mile Creek	Mining
Prospect Reservoir	1888	50.2	Prospect Creek	Public supply
Manly Dam	1892	2.0	Curl Curl Creek	Flood mitigation
Stephens Creek Dam	1892	17.5	Stephens Creek	Public supply
Junction Reef Dam	1898	7.3	Belubula River	Mining
Cobar Dam	1900	1.8	Off-stream	Recreation
Lake Cargellico	1902	36.3	Lachlan River	Irrigation
Booberoi Weir	1902	2.9	Lachlan River	Irrigation
Cataract Dam	1907	94.3	Cataract River	Public supply
Stephens Creek Dam	1909	6.3	Stephens Creek	Public supply
Umberumberka Dam	1914	8.2	Umberumberka Creek	Public supply
Cotter Dam	1915	5.0	Cotter River	Public supply
Upper Cordeaux Dam	1915	1.2	Cordeaux River	Public supply
Chichester Dam	1923	16.9	Chichester River	Public supply
Torrumbarry Weir	1924	38.0	Murray River	Irrigation
Cordeaux Dam	1926	93.6	Cordeaux River	Public supply
Mildura Weir	1927	37.0	Murray River	Irrigation
Avon Dam	1927	214.6	Avon River	Public supply
Burrinjuck Dam	1927	951.9	Murrumbidgee River	Irrigation
Yass Dam	1927	1.1	Yass River	Public supply
Lake Victoria	1928	680.3	Murray River	Irrigation
Puddledock Creek Dam	1928	1.7	Puddledock Creek	Public supply
Wentworth Weir	1929	47.0	Murray River	Navigation
Sooley Dam	1930	4.5	Sooley Creek	Public supply
Tenterfield Creek Dam	1930	1.3	Tenterfield Creek	Public supply
Spring Creek Dam	1931	2.7	Spring Creek	Public supply
Nepean Dam	1935	81.4	Nepean River	Public supply
Stevens Weir	1935	9.0	Edward River	Irrigation
Hume Dam	1936	1 520.0	Murray River	Irrigation
Winburndale Dam	1936	1.9	Macquarie River	Public supply
Wyangala Dam	1936	374.7	Lachlan River	Irrigation
Euston Weir	1937	38.0	Murray River	Irrigation
Cascade 3 Dam	1938	1.7	Cascade Creek	Public supply
Inverell Dam	1939	1.5	Macintyre River	Public supply
Lake Endeavour	1939	2.4	Billabong Creek	Public supply
Yarrawonga Weir	1939	120.0	Murray River	Irrigation
Jemalong Weir	1940	1.6	Lachlan River	Irrigation
Maude Weir	1940	5.0	Murrumbidgee River	Irrigation
Redbank Weir	1940	6.0	Murrumbidgee River	Irrigation
Woronora Dam	1941	71.8	Woronora River	Public supply
Oberon Dam	1949	9.0	Fish River	Public supply
Lake Brewster	1952	153.0	Lachlan River	Irrigation
Lake Brewster Weir	1952	7.0	Lachlan River	Irrigation
Rocky Creek Dam	1952	13.6	Richmond River	Public supply
Rylstone Dam	1953	3.3	Cudgegong River	Public supply
Lake Rowlands	1954	4.7	Coombing River	Public supply
Guthega Dam	1955	1.6	Snowy River	Electricity
Walgett Weir	1955	4.0	Darling River	Public supply
Burrinjuck Dam	1956	80.5	Murrumbidgee River	Irrigation
Dungowan Dam	1956	5.9	Namoi River	Public supply
Ben Chifley Dam	1957	16.0	Campbells River	Public supply
Coeypolly 2 Dam	1957	5.4	Namoi River	Public supply
Oberon Dam	1957	36.0	Fish River	Public supply
Cochrane Dam	1958	3.1	Bega River	Electricity
Lake Eucumbene	1958	4 799.1	Snowy River	Electricity
Glenbawn Dam	1958	361.4	Hunter River	Irrigation
Tumut Pond	1958	52.8	Tumut River	Electricity
Wilcannia Weir	1958	4.0	Darling River	Public supply
Gogeldrei Weir	1958	7.4	Murrumbidgee River	Irrigation
Bundanoon Dam	1959	2.0	Bundanoon Creek	Public supply
Lake Keepit	1960	425.6	Namoi River	Irrigation
Menindee Lakes scheme	1960	1 794.0	Darling River	Irrigation
Tantangara Reservoir	1960	254.1	Murrumbidgee River	Electricity
Warragamba Dam	1960	2 091.8	Warragamba River	Public supply
Bendora Dam	1961	10.7	Cotter River	Public supply
Deep Creek Dam	1961	4.5	Moruya River	Public supply

Appendix — continued

Storage name[a]	Date	Capacity[b]	Location	Purpose
Hume Dam	1961	1 520.0	Murray River	Irrigation
Tooma Reservoir	1961	28.1	Tooma River	Electricity
Tumut 2 Pond	1961	2.7	Tumut River	Electricity
Upper Mooney Reservoir	1961	4.6	Mooney Mooney Creek	Public supply
Grahamstown Dam	1962	7.3	Off-stream	Public supply
Suma Park Dam	1962	18.1	Summerhill Creek	Public supply
Mardi Reservoir	1963	7.3	Off-stream	Public supply
Scrivener Dam	1963	27.7	Molongolo River	Recreation
Medway Reservoir	1964	1.3	Medway rivulet	Public supply
Island Bend Reservoir	1965	3.0	Snowy River	Electricity
Khancoban Reservoir	1965	21.5	Swampy Plains	Electricity
Rangers Valley Dam	1965	5.0	Cam Creek	Irrigation
Geehi Reservoir	1966	21.1	Geehi River	Electricity
Narranbulla Reservoir	1966	1.5	Off-stream	Irrigation
Weir 19a Darling River	1966	3.7	Darling River	Public supply
Bootawa Reservoir	1967	2.3	Off-stream	Public supply
Burrendong Dam	1967	1 190.1	Macquarie River	Irrigation
Jindabyne Reservoir	1967	689.8	Snowy River	Electricity
Seaham Weir	1967	16.1	Williams River	Public supply
Blowering Reservoir	1968	1 631.4	Tumut River	Electricity
Corin Dam	1968	75.0	Cotter River	Public supply
Jounama Reservoir	1968	43.8	Tumut River	Electricity
Liddel Reservoir	1968	148.1	Gardiners Creek	Electricity
Malpas Dam	1968	13.0	Gara River	Public supply
Murray 2 Reservoir	1968	2.3	Khancoban Creek	Electricity
Porters Creek Dam	1968	2.5	Clyde River	Public supply
South Creek Dam	1968	1.5	Hawkesbury River	Irrigation
Grahamstown Dam	1969	152.6	Off-stream	Public supply
Pindari Dam	1969	37.0	Severn River	Irrigation
Carcoar Dam	1970	35.8	Belubula River	Irrigation
Liddel Reservoir	1970	4.6	Tinkers Creek	Electricity
Brewarrina Weir	1970	1.5	Darling River	Public supply
Danjerra Creek Dam	1971	7.7	Shoalhaven River	Public supply
Lostock Dam	1971	19.8	Paterson River	Irrigation
Talbingo Reservoir	1971	950.5	Tumut River	Electricity
Toonumbar Dam	1971	20.2	Iron Pot Creek	Irrigation
Wyangala Dam	1971	842.3	Lachlan River	Irrigation
Tilpa Weir	1972	3.0	Darling River	Public supply
Tareelaroi Weir	1972	2.4	Gwydir River	Irrigation
Bendeela Pond	1973	1.2	Off-stream	Public supply
Glenquarry Cut Dam	1974	34.5	Off-stream	Public supply
Wingecarribee Dam	1974	34.5	Wingecarribee River	Public supply
Weir 21 Darling River	1974	4.0	Darling River	Public supply
Gunidgera Weir	1974	2.0	Namoi River	Irrigation
Ginninderra Dam	1975	3.7	Ginninderra River	Recreation
Brogo Dam	1976	9.8	Brogo River	Irrigation
Copeton Reservoir	1976	1 364.0	Gwydir River	Irrigation
Fitzroy Falls Dam	1976	22.2	Yarrunga Creek	Public supply
Glenlyon Dam	1976	260.0	Pike Creek	Irrigation
Tallowa Dam	1976	110.2	Shoalhaven River	Public supply
Googong Dam	1977	125.0	Queanbeyan River	Public supply
Gunidgera Weir	1978	1.9	Namoi River	Irrigation
Mollee Weir	1978	4.0	Namoi River	Irrigation
Wallerawang Reservoir	1978	4.3	Coxs River	Electricity
Chaffey Dam	1979	62.0	Peel River	Irrigation
Dartmouth Dam	1979	4 057.0	Mitta Mitta River	Irrigation
Karangi Dam	1979	2.2	Off Clarence River	Public supply
Pejar Dam	1979	9.0	Wollondilly River	Public supply
Port Macquarie Dam	1980	1.5	Off-stream	Public supply
Tombullen Lake	1980	11.3	Off-stream	Irrigation
Clarrie Hall Dam	1982	16.0	Doon Doon Creek	Public supply
Bamarang Dam	1983	3.8	Off-stream	Public supply
Broombee Dam	1983	1.7	Saumarez Creek	Irrigation
Glennie Creek Dam	1983	284.0	Hunter River	Irrigation
Lyell Dam	1983	26.0	Coxs River	Electricity
Mangrove Creek Dam	1983	176.0	Mangrove Creek	Public supply
Weir 20a Darling River	1983	7.3	Darling River	Public supply
Windamere Dam	1984	353.0	Cudgegong River	Irrigation
Chichester Dam	1985	21.5	Chichester River	Public supply
Duncans Creek Dam	1985	1.5	Duncans Creek	Irrigation
Glenbawn Dam	1986	510.0	Hunter River	Irrigation
Hovell Weir	1986	2.0	Mannus Creek	Irrigation

Appendix — continued

Storage name[a]	Date	Capacity[b]	Location	Purpose
Molong Creek Dam	1987	1.0	Molong Creek	Public supply
Plashett Dam	1987	70.0	Saltwater Creek	Electricity
Split Rock Dam	1987	372.0	Manilla River	Irrigation
Yellow Pinch Dam	1987	3.0	Yellow pinch Creek	Public supply
Thompsons Creek Dam	1992	27.5	Thompsons Creek	Electricity
Pindari Dam	1995	259.0	Severn River	Irrigation
Lyell Dam	1995	7.5	Coxs River	Electricity

REFERENCES

Allan, J. D. and Flecker, A. S., 1993. Biodiversity conservation in running waters. *BioScience* **43:** 32–43.

ANCA (Australian Nature Conservation Agency), 1993. A directory of important wetlands in Australia. Australian Nature Conservation Agency: Canberra.

Arthington, A. H., Bunn, S. E., Pusey, B. J. and Bluhdorn, D. R., 1991. Development of an holistic approach for assessing environmental flow requirements of riverine ecosystems. Pp. 69–76 in *Water allocation for the environment* ed by J. J. Pigram and B. P. Hooper. Proceedings International Seminar and Workshop, Centre for Water Policy Research, University of New England.

Barmuta, L. A., Marchant, R. and Lake, P. S., 1992. Degradation of Australian streams and progress towards conservation and management in Victoria. Pp. 65–79 in *River conservation and management* ed by P. J. Boon, P. Calow and G. E. Petts. John Wiley and Sons Ltd: Chichester.

Bell, J. D., Berra, T. M., Jackson, P. D., Last, P. R. and Sloane, R. D., 1980. Recent records of Australian grayling *Prototroctes maraena* Gunther (Pisces: Prototroctidae) with notes on its distribution. *Aust. Zool.* **20:** 419–31.

Bennison, G., Sullivan, C. and Suter, P., 1991. Macroinvertebrate and phytoplankton communities of the River Murray. Pp. 231–39 in *Conservation and management of the River Murray system — making conservation count.* Proceedings of the third Fenner Conference on Environment. Canberra, September 1989, South Australian Department of Environment and Planning with Australian Academy of Science.

Bishop, K. A. and Bell, J. D., 1978. Observations on the fish fauna below Tallowa Dam (Shoalhaven River, New South Wales) during river flow stoppages. *Aust. J. Mar. Freshw. Res.* **29:** 543–49.

Blyth, J., 1980. Environmental impact of reservoir construction: the Dartmouth Dam invertebrate survey: a case history. Pp. 174–81 in *An ecological basis for water resource management* ed by W. D. Williams. Australian National University Press: Canberra.

Boon, P. J., 1992. Essential elements in the case for river conservation. Pp. 11–33 in *River conservation and management* ed by P. J. Boon, P. Calow and G. E. Petts. John Wiley and Sons Ltd: Chichester.

Boon, P. I. and Shiel, R. J., 1990. Grazing on bacteria by zooplankton in Australian billabongs. *Aust. J. Mar. Freshw. Res.* **41:** 247–57.

Boon, P., Frankenberg, J., Hillman, T., Oliver, R. and Shiel, R., 1990. Billabongs. Pp. 183–98 in *The Murray* ed by N. Mackay and D. Eastburn. Murray Darling Basin Commission: Canberra.

Borchardt, D., 1993. Effects of flow and refugia on drift loss of benthic macroinvertebrates — implications for habitat restoration in lowland Streams. *Freshw. Biol.* **29:** 221–27.

Boulton, A. J. and Lloyd, L. N., 1991. Aquatic macroinvertebrate assemblages in floodplain habitats of the lower River Murray. *Reg. Riv.* **6:** 183–201.

Boulton, A. J. and Lloyd, L. N., 1992. Flooding frequency and invertebrate emergence from dry floodplain sediments of the River Murray, Australia. *Regul. Riv.* **7:** 137–51.

Bower, C., 1980. Is the West Darling becoming a desert? *Nat. Parks J. NSW* **24:** 10–12.

Boyd, K., 1992. Inland rivers need restoration (not more irrigation). *Nat. Parks J. NSW* **36:** 11–15.

Braithwaite, L. W., 1975. Managing waterfowl in Australia. *Proc. Ecol. Soc. Aust.* **8:** 107–28.

Braithwaite, L. W., 1980. Waterfowl resources and their management. Pp. 287–94 in *An ecological basis for water resource management* ed by W. D. Williams. Australian National University Press: Canberra.

Braithwaite, L. W., Maher, M., Briggs, S. V. and Parker, B. S., 1986. An aerial survey of three game species of waterfowl (Family Anatidae) in eastern Australia. *Aust. Wildl. Res.* **13:** 213–23.

Brander, D., 1987. Environmental changes in the southern Macquarie Marshes: 1934 to 1987. Unpublished Honours Thesis, University of New South Wales, Sydney. 109 pp.

Bravard, J. P., Amoros, C. and Pautou, G., 1986. Impact of civil engineering works on the successions of communities in a fluvial system. *Oikos* **47:** 92–111.

Bren, L. J., 1988. Effects of river regulation on flooding of a riparian red gum forest on flooding of a riparian red gum forest on the river Murray, Australia. *Regul. Riv.* **2:** 65–77.

Bren, L., 1990. Red gum forests. Pp. 231–42 in *The Murray* ed by N. Mackay and D. Eastburn. Murray Darling Basin Commission: Canberra.

Bren, L. J., 1992. Tree invasion of an intermittent wetland in relation to changes in the flooding frequency of the River Murray, Australia. *Aust. J. Ecol.* **17:** 395–408.

Briggs, S. V., 1977. Flood mitigation. *Nat. Parks J. NSW* **21:** 5–8.

Briggs, S. V., 1988. Guidelines for management of inland wetlands in southern New South Wales. *Wetlands* (Australia) **8:** 1–2.

Briggs, S. V., 1990. Waterbirds. Pp. 337–44 in *The Murray* ed by N. Mackay and D. Eastburn. Murray Darling Basin Commission: Canberra.

Briggs, S. V., 1992. Wetlands as drylands: a conservation perspective. *Bull. Ecol. Soc. Aust.* **22:** 109–10.

Briggs, S. V., 1994. The future of waterbirds in western New South Wales. Pp. 149–54 in *The future of fauna of western New South Wales* ed by D. Lunney, S. Hand, P. Reed and D. Butcher. Royal Zoological Society of New South Wales: Mosman.

Briggs, S. V. and Lawler, W. G., 1991. Management of Murray-Darling wetlands for waterbirds. Pp. 146–57 in *Conservation and management of the River Murray system — making conservation count.* Proceedings of the third Fenner Conference on Environment. Canberra, September 1989, South Australian Department of Environment and Planning with Australian Academy of Science.

Briggs, S. V. and Maher, M. T., 1983. Litter fall and leaf decomposition in a River Red Gum *(Eucalyptus camaldulensis)* swamp. *Aust. J. Bot.* **31:** 307–16.

Briggs, S. V. and Townsend, G., 1993. Restoration and management of Nearie Lake. Unpubl. Report to Murray-Darling Basin Commission (Natural Resource Management Strategy). National Parks and Wildlife Service (NPWS): Buronga.

Briggs, S. V., Hodgson, P. F. and Ewin, P., (in press). Changes in populations of waterbirds on a wetland following water storage. *Wetlands* (Australia).

Briggs, S. V., Maher, M. T. and Tongway, D. S., 1985. Dry matter and nutrient loss from decomposing *Vallisneria spiralis* L. *Aquat. Bot.* **22:** 387–92.

Brock, M. A., 1986. Adaptations to fluctuations rather than to extremes of environmental parameters. Pp. 131–40 in *Limnology in Australia* ed by P. De Deckker and W. D. Williams. CSIRO/W. Junk: Melbourne/Dordrecht.

Brock, M. A. and Casanova, M. T., 1991. Vegetative variation of *Myriophyllum variifolium* in permanent and temporary wetlands. *Aust. J. Bot.* **39:** 487–96.

Brooker, M. G., 1993. Aerial counts of waterbirds on Narran Lake, New South Wales. *Aust. Bird Watcher* **15:** 13–18.

Brown, J. A. H., Harrison, R. D. and Jacobson, G., 1983. Water demand and availability with reference to particular regions. Water 2000: Consultants Report No. 12. Australian Government Publishing Service: Canberra.

Buxton, G. L., 1974. The middle reaches in the nineteenth century. Pp. 109–23 in *The Murray waters* ed by H. J. Frith and G. Sawer. Angus and Robertson: Sydney.

Cadwallader, P. L., 1978. Some causes of the decline in range and abundance of native fish in the Murray-Darling system. *Proc. Roy. Soc. Vic.* **90:** 211–24.

Cadwallader, P. and Lawrence, B., 1990. Fish. Pp. 317–35 in *The Murray* ed by N. Mackay and D. Eastburn. Murray Darling Basin Commission: Canberra.

Chessman, B. C., Robinson, D. P. and Hortle, K. G., 1987. Changes in the riffle macroinvertebrate fauna of the Tanjil River, southeastern Australia, during construction of Blue Rock Dam. *Regul. Riv.* **1:** 317–29.

Churchward, H. M. and Flint, S. R., 1956. Jenargo extension of the Berriquin Irrigation District, New South Wales. CSIRO Soils and Land use Series, No. 18, CSIRO, Melbourne.

Close A., 1990. The impact of man on the natural flow regime. Pp. 61–74 in *The Murray* ed by N. Mackay and D. Eastburn. Murray Darling Basin Commission: Canberra.

CYB (Cotton Year Book), 1992. Toowoomba, Queensland: Australian Cottongrower.

Crome, F. H. J., 1986. Australian waterfowl do not necessarily breed on a rising water level. *Aus. Wildl.* **13:** 461–80.

Crome, F. H. J. and Carpenter, S. M., 1988. Plankton community cycling and recovery after drought — dynamics in a basin on a flood plain. *Hydrobiol.* **164:** 193–211.

Crome, R. L., 1980. Lake Hume. Pp. 305–10 in *An ecological basis for water resource management* ed by W. D. Williams. Australian National University Press: Canberra.

Cullen, P. and Lake, P. S. (this volume). Water resources and biodiversity: past, present and future problems and solutions. In *Conserving biodiversity: threats and solutions* ed by R. A. Bradstock, T. D. Auld, D. A. Keith, R. T. Kingsford, D. Lunney and D. P. Sivertsen. Surrey Beatty & Sons: Chipping Norton.

Davidson, B. R., 1974. Irrigation economics. Pp. 193–212 in *The Murray waters* ed by H. J. Frith and G. Sawer. Angus and Robertson: Sydney.

Dexter, B. D., Rose, H. J. and Davies, N., 1986. River regulation and associated forest management problems in the River Murray red gum forests. *Aust. For.* **49:** 16–27.

Doeg, T. J., 1984. Response of the macroinvertebrate fauna of the Mitta Mitta River, Victoria, to the construction and operation of the Dartmouth Dam: 2. Irrigation release. *Occas. Pap. Mus. Vic.* **1:** 101–28.

Doeg, T. J., Davey, G. W. and Blyth, J. D., 1987. Response of the macroinvertebrate communities to the dam construction on the Thomson River, southeastern Australia. *Regul. Riv.* **1:** 195–209.

DSC (Dam Safety Committee of NSW), 1992. Annual report 1991/92. Dam Safety Committee of New South Wales.

DWR (Department of Water Resources), 1991. Lowbidgee management plan: stage two: land and water management 1991–96. 43 pp.

DWR (Department of Water Resources), 1992a. Management of unregulated flow in regulated systems. NSW Department of Water Resources. 14 pp.

DWR (Department of Water Resources), 1992b. Interim unregulated flow management plan for the north-west, NSW Department of Water Resources. 8 pp.

Everett, R. A. and Ruiz, G. M., 1993. Coarse woody debris as a refuge from predation in aquatic communities — an experimental test. *Oecologia* **93:** 475–86.

Fairweather, P. J., 1993. Links between ecology and ecophilosophy, ethics and the requirements of environmental management. *Aust. J. Ecol.* **18:** 3–19.

Fitzpatrick, C., 1991. Normal effects of water quality, water regulation and salinity management on conservation values. Pp. 49–59 in *Conservation and management of the River Murray system — making conservation count.* Proceedings of the third Fenner Conference on Environment. Canberra, September 1989, South Australian Department of Environment and Planning with Australian Academy of Science.

Frith, H. J.,1982. *Waterfowl in Australia — 2nd edition.* Angus and Robertson: Sydney.

Geddes, M. C. and Puckridge, J. T., 1989. Survival and growth of larval and juvenile native fish: the importance of the floodplain. Pp. 101–16 in *Proceedings of workshop on native fish management, Canberra 16–17 June.* Murray-Darling Basin Commission: Canberra.

Goodrick, G., 1970. A survey of wetlands of coastal New South Wales. CSIRO Division of Wildlife Research Technical Memorandum No. 5.

Gopal, B., 1986. Vegetation dynamics in temporary and shallow freshwater habitats. *Aquat. Bot.* **23:** 391–96.

Graetz, R. D., Fisher, R. and Wilson, M., 1992. Looking back. The changing face of the Australian continent, 1972–1992. CSIRO: Australia.

Green, D. L., 1992. Survey of wetlands of the Warrego river. Department of Water Resources, Technical Services Division. 32 pp.

Harris, J. H., 1984a. A survey of fishways in streams of coastal southeastern Australia. *Aust. Zool.* **21:** 219–33.

Harris, J. H., 1984b. Impoundment of coastal drainages of southeastern Australia, and a review of its relevance to fish migrations. *Aust. Zool.* **21:** 235–50.

Hart, B. T., 1992. Ecological condition of Australian rivers. *Search* **3:** 33–37.

Helman, P. and Estella, P., 1983. The conservation status of riverine ecosystems in western New South Wales. Report to State Pollution. 59 pp.

Hillman, T. J., 1986. Billabongs. Pp. 457–70 in *Limnology in Australia* ed by P. De Deckker and W. D. Williams. CSIRO/W. Junk: Melbourne/Dordrecht.

Hohl, A. and Tisdell, C. A., 1993. How useful are environmental safety standards in economics — the example of safe minimum standards for protection of species. *Biodiver. Cons.* **2:** 168–81.

Jensen, A., Lloyd, L. and Bennett, M., 1994. Rehabilitation techniques for wetlands in the Murray-Darling Basin. Pp. 53–60 in *Murray-Darling Basin floodplain wetlands management: Proceedings of the floodplain wetlands management workshop, Albury, New South Wales, 20–22 October 1992* ed by A. J. Sharley and C. A. Huggan. Murray-Darling Basin Commission: Canberra.

Johnson, B., 1992. The Macquarie Marshes: overdue for environmental restoration. *Nat. Parks J. NSW* **36:** 9–12.

Junk, W. J., Bayley, P. B. and Sparks, R. E., 1989. The flood-pulse concept in river-floodplain systems. Proceedings International Large Rivers Symposium (LARS) ed by D. P. Dodge. *Can. Spec. Publ. Fish. Aquat. Scien.* **106:** 110–27.

Keyte, P., 1992. Lower Gwydir wetland — preliminary identification of resource management issues — Discussion paper. Lower Gwydir wetland management plan steering Committee. NSW Department of Water Resources. 16 pp.

Kingsford, R. T., 1992. Maned ducks and farm dams: a success story. *Emu* **92:** 163–69.

Kingsford, R. T., 1994. Occurrence of high concentrations of waterbirds in arid Australia. *J. Arid Environ.* **29:** 421–25.

Kingsford, R. T. and Porter, J. L., 1994. Waterbirds on an adjacent freshwater and salt lake in arid Australia. *Biol. Cons.* **69:** 219–28.

Kingsford, R. T., Porter, J. L. and Ferster Levy, R., 1992. An aerial survey of wetland birds in eastern Australia — October 1991. New South Wales National Parks and Wildlife Service Occasional Paper 12. 40 pp.

Kingsford, R. T. and Thomas, R. F., 1995. The Macquarie Marshes in arid Australia and their waterbirds: a 50-year history of decline. *Environ. Manage.* **19** (in press).

Klimas, C. V., 1988. River regulation effects on floodplain hydrology and ecology. Pp. 40–49 in *The ecology and management of wetlands Vol 1: Ecology of wetlands* ed by D. D. Hook, W. H. McKee Jr, H. K. Smith, J. Gregor, V. G. Burrell Jr, M. R. DeVoe, R. E. Sojka, S. Gilbert, R. Banks, L. H. Stolzy, C. Brooks, T. D. Matthews and T. H. Shear. Croom Helm: London and Sydney.

Koehn, J. D. and O'Connor, W. G., 1990. Biological information for management of native freshwater fish in Victoria. Government Printer: Melbourne.

Lake, P. S., 1980. Conservation. Pp. 163–73 in *An ecological basis for water resource management* ed by W. D. Williams. Australian National University Press: Canberra.

Lake, P. S. and Marchant, R., 1990. Australian upland streams: ecological degradation and possible restoration. *Proc. Ecol. Soc. Aust.* **16:** 79–91.

Leigh, J. H. and Noble, J. C., 1972. Riverine plain of New South Wales: its pastoral and irrigation development. CSIRO Division of Plant Industry, Canberra.

Lloyd, L. N. and Walker, K. F., 1989. Management of snags (woody debris) and river and floodplain vegetation Pp. 99–100 in *Proceedings of workshop on native fish management, Canberra 16–17 June.* Murray-Darling Basin Commission: Canberra.

Lloyd, L., Puckridge, J. and Walker, K., 1991. The significance of fish populations in the Murray-Darling system and their requirements for survival. Pp. 86–99 in *Conservation and management of the River Murray system — making conservation count.* Proceedings of the Third Fenner Conference on Environment. Canberra, September 1989, South Australian Department of Environment and Planning with Australian Academy of Science.

Lloyd, L., Atkins, B., Boon, P., Jacobs, T. and Roberts, J., 1994. Natural processes within Murray-Darling floodplain wetlands. Pp. 6–23 in *Murray-Darling Basin floodplain wetlands management: Proceedings of the floodplain wetlands management workshop, Albury, NSW, 20–22 October 1992* ed by A. J. Sharley and C. A. Huggan. Murray-Darling Basin Commission: Canberra.

LMP (Laurie, Montgomery and Pettit Pty Ltd), 1979. Benefits of large water storages in New South Wales. Water Resources Commission: Sydney.

LMP (Laurie, Montgomery and Pettit Pty Ltd), 1980. Macleay Valley. New South Wales Coastal Rivers Flood Plain Management Study.

Mackay, N., 1990. Understanding the Murray. Pp. ix–xix in *The Murray* ed by N. Mackay and D. Eastburn. Murray Darling Basin Commission: Canberra.

Magrath, M. J. L., 1991. Waterbird breeding in the Narran wetlands. Technical Services Division, New South Wales Department of Water Resources.

Maher, M., 1984. Benthic studies of waterfowl breeding habitat in southwestern New South Wales. I. The fauna. *Aust. J. Mar. Freshw. Res.* **35:** 85–96.

Maher, M. T., 1991. An inland perspective on the conservation of Australian waterbirds. Ph.D. University of New England. 172 pp.

Maher, M. T. and Braithwaite, L. W., 1992. Patterns of waterbird use in wetlands of the Paroo: a river system of inland Australia. *Rangelands J.* **14:** 128–42.

Maher, M. and Carpenter, S. M., 1984. Benthic studies of waterfowl breeding habitat in South-Western New South Wales. II. Chironomid populations. *Aust. J. Mar. Freshw. Res.* **35:** 97–110.

Maher, P. N., 1990. Bird survey of the Lachlan/Murrumbidgee confluence wetlands. NSW National Parks and Wildlife Service. Unpubl. Report. 153 pp.

Mallen-Cooper, M., 1989. Fish passage in the Murray-Darling Basin. Pp. 123–36 in *Proceedings of workshop on native fish management, Canberra 16–17 June.* Murray-Darling Basin Commission: Canberra.

Marchant, R., 1989. Changes in the benthic invertebrate communities of the Thomson River, southeastern Australia, after dam construction. *Regul. Riv.* **4:** 71–89.

Marshall, A., 1993. The disappearing Darling. *The Land* (May 27). Pp. 12–13.

McCosker, R. O. and Duggin, J. A., 1992. Gingham watercourse resource management issues Gwydir River Basin, Moree NSW. Unpubl. Report. 20 pp.

McIntyre, S., Ladiges, P. Y. and Adams, G., 1988. Plant species-richness and invasion by exotics in relation to disturbance of wetland communities on the Riverine Plain, New South Wales. *Aust. J. Ecol.* **13:** 361–73.

MDBC (Murray-Darling Basin Commission), 1992. Watering the Barmah-Millewa red gum forest. Issues Paper. 23 pp.

MDBMC (Murray-Darling Basin Ministerial Council), 1987. Murray-Darling Basin Environmental Resources Study. State Pollution Control Commission, Sydney. 426 pp.

Mellquist, P., 1992. River management-Objectives and applications. Pp. 1–8 in *River conservation and management* ed by P. J. Boon, P. Calow and G. E. Petts. John Wiley and Sons Ltd: Chichester.

Meyer, W. S., 1992. Sustainability of land and water resources use for Australia's irrigated agriculture. CSIRO Division of Water Resources, Seeking solutions, Series No. 8. 57 pp.

Michaelis, F. B. and O'Brien, M., 1988. Preservation of Australia's wetlands: A Commonwealth approach Pp. 167–77 in *Conservation of Australian Wetlands* ed by A. J. McComb and P. S. Lake. Surrey Beatty & Sons: Chipping Norton.

Middleton, M. J., Rimmer, M. J. and Williams, R. J., 1985. Structural flood mitigation works and estuarine management in New South Wales — case study of the Macleay River. *Coastal Zone Manage. J.* **13:** 1–23.

Myers, L. F., 1974. Irrigation development: reasons and consequences. Pp. 160–75 in *The Murray waters* ed by H. J. Frith and G. Sawer. Angus and Robertson: Sydney.

O'Brien, W. T., McGregor, A. M. and Crawshaw, B. W., 1983. Water 2000: Consultants Report No. 9. In-stream uses of water in Australia. Australian Government Publishing Service: Canberra.

O'Connor, P. F., 1989. Fisheries management in inland New South Wales. Pp. 19–24 in *Proceedings of workshop on native fish management, Canberra 16–17 June.* Murray-Darling Basin Commission: Canberra.

Petersen, R. C., Jr. and Petersen, L. B.-M., 1992. A building block model for stream restoration: Pp. 293–309 in *River conservation and management* ed by P. J. Boon, P. Calow and G. E. Petts. John Wiley and Sons Ltd: Chichester.

Pigram, J. J., 1986. *Issues in the management of Australia's water resources.* Longman Cheshire: Melbourne.

Pressey, R. L., 1986. Wetlands of the River Murray below Lake Hume. River Murray Commission Environmental Report 86/1. 134 pp.

Pressey, R. L., 1989. Wetlands of the Lower Macleay floodplain, northern coastal New South Wales. *Proc. Linn. Soc. NSW* **111:** 157–68.

Pressey, R. L., 1990. Wetlands. Pp. 167–81 in *The Murray* ed by N. Mackay and D. Eastburn. Murray Darling Basin Commission: Canberra.

Pressey, R. L. and Harris, J. H., 1988. Wetlands of New South Wales. Pp. 35–57 in *Conservation of Australian Wetlands* ed by A. J. McComb and P. S. Lake. Surrey Beatty & Sons: Chipping Norton.

Pressey, R. L. and Middleton, M. J., 1982. Impacts of flood mitigation works on coastal wetlands in New South Wales. *Wetlands* (Australia) **2:** 27–44.

Reynolds, L. F., 1983. Migration patterns of five fish species in the Murray-Darling river system. *Aust. J. Mar. Freshw. Res.* **33:** 663–70.

Roberts, J. and Ludwig, J. A., 1991. Riparian vegetation along current-exposure gradients in floodplain wetlands of the River Murray, Australia. *J. Ecol.* **79:** 117–27.

Sainty, G. and Jacobs, S., 1990. Waterplants. Pp. 265–73 in *The Murray* ed by N. Mackay and D. Eastburn. Murray Darling Basin Commission: Canberra.

Sheldon, F. and Walker, K. F., 1993. Pipelines as a refuge for freshwater snails. *Regul. Riv.: Res. Manage.* **8:** 295–99.

Shepheard, M., 1992. Flooding frequency and *Eucalyptus largiflorens* (Black box) wetland health in the Wah Wah District. Technical Report 92/14, Resource Assessment Group, Murrumbidgee Region, Department of Water Resources.

Shiel, R., 1990. Zooplankton. Pp. 275–84 in *The Murray* ed by N. Mackay and D. Eastburn. Murray Darling Basin Commission: Canberra.

Slocombe, D. S., 1993. Environmental Planning, Ecosystem Science, and Ecosystem Approaches for Integrating Environment and Development. *Environ. Manage.* **17:** 289–303.

Smith, J., 1993. A report on the vertebrate fauna of the Narran River floodplain in New South Wales. New South Wales National Parks and Wildlife Service, unpubl. report.

Smith, P., 1991. The biology and management of waders (Suborder Charadrii) in New South Wales. New South Wales National Parks and Wildlife Service, Species management report number 9, 167 pp.

Smith, P. and Smith, J., 1990. River Murray riparian vegetation study. Margules and Partners Pty Ltd, P. and J. Smith Ecological Consultants, Department of Conservation Forests and Lands. 187 pp.

Smith, P. and Smith, J., 1991. The significance of riparian vegetation in the Murray system and the requirements for its survival. Pp. 61–67 in *Conservation and management of the River Murray system — making conservation count.* Proceedings of the third Fenner Conference on Environment. Canberra, September 1989, South Australian Department of Environment and Planning with Australian Academy of Science.

Somerville, J., 1988. Conservation of River Red Gums. *Nat. Parks J. NSW* **32:** 31–33.

Statzner, B. and Sperling, F., 1993. Potential Contribution of System-Specific Knowledge (SSK) to Stream-Management Decisions — Ecological and Economic Aspects. *Freshw. Biol.* **29:** 313–42.

Thomson, C., 1992. The impact of river regulation on the natural flows of the Murray-Darling Basin. Murray-Darling Basin Technical Report 92/5.1, Murray-Darling Basin Commission, Canberra, 11 pp.

Thornton, S. A. and Briggs, S. V., 1994. A survey of hydrological changes to wetlands of the Murrumbidgee river. *Wetlands* (Australia) **13:** 1–13.

Thoms, M. C. and Walker, K. F., 1992. Sediment transport in a regulated semi-arid river: the River Murray, Australia. Pp. 239–50 in *Aquatic ecosystems in semi-arid regions: implications for resource management* ed by R. D. Robarts and M. L. Bothwell. WHRI Symposium Series 7, Environment Canada: Regina.

Timms, B. V., 1980. Farm dams. Pp. 345–59 in *An ecological basis for water resource management* ed by W. D. Williams. Australian National University Press: Canberra.

Timms, B. V., 1993. Saline lakes of the Paroo, inland New South Wales, Australia. *Hydrobiol.* **267:** 269–89.

Turner, K., 1991. Economics and wetland management. *Ambio* **20:** 59–63.

Underwood, A. J. and Fairweather, P. J., 1993. Introduction: international workshop on solutions to environmental problems at the Institute of Marine Ecology, University of Sydney, 21–24 April. *Aust. J. Ecol.* **18:** 1.

Walker, K. F., 1980. The downstream influence of Lake Hume on the River Murray. Pp. 182–91 in *An ecological basis for water resource management* ed by W. D. Williams. Australian National University Press: Canberra.

Walker, K. F., 1983. The Murray is a floodplain river. *Sth Aust. Nat.* **58:** 29–33.

Walker, K. F., 1985. A review of the ecological effects of river regulation in Australia. *Hydrobiol.* **125:** 111–29.

Walker, K. F., 1986. The Murray-Darling river system. Pp. 631–59 in *The ecology of river systems* ed by B. R. Davies and K. F. Walker. W. Junk: Dordrecht.

Walker, K. F., 1990. Mussels. Pp. 309–14 in *The Murray* ed N. Mackay and D. Eastburn. Murray Darling Basin Commission: Canberra.

Walker, K. F. and Lloyd, L. N., 1987. Fallen wood in the River Murray. *Water News* **(June):** 15–17.

Walker, K. F. and Thoms, M. C., 1993. Environmental effects of flow regulation on the Lower River Murray, Australia. *Regul. Riv.* **8:** 103–19.

Walker, K. F., Hillman, T. J. and Williams, W. D., 1978. Effects of impoundments on rivers: an australian case study. *Verh. Int. Ver. Theor. Limnol.* **20:** 1695–701.

Walker, K., Thoms, M. and Sheldon, F., 1991. The significance of the river-edge boundary for conservation of the River Murray. Pp. 69–84 in *Conservation and management of the River Murray system — making conservation count.* Proceedings of the third Fenner Conference on Environment. Canberra, September 1989, South Australian Department of Environment and Planning with Australian Academy of Science.

Walker, K. F., Thoms, M. C. and Sheldon, F., 1992. Effects of Weirs on the littoral environment of the River Murray, South Australia. Pp. 271–91 in *River conservation and management* ed by P. J. Boon, P. Calow and G. E. Petts. John Wiley and Sons Ltd: Chichester.

Walters, C. J., 1993. Dynamic models and large scale field experiments in environmental impact assessment and management. *Aust. J. Ecol.* **18:** 53–61.

Ward, K. A., 1992. Investigation of the flood requirements of the Moira Grass plains in Barmah Forest, Victoria. Integrated watering strategy Report, Department of Conservation and Environment, Victoria. 64 pp.

Warner, R. F., 1984. Man's impacts on Australian drainage systems. *Australian Geog.* **16:** 133–41.

WCIC (Water Conservation and Irrigation Commission), 1971. Water Resources of New South Wales. Government Printer: Sydney.

Welcomme, R. L., 1992. River conservation — future prospects Pp. 453–62 in *River conservation and management* ed by P. J. Boon, P. Calow and G. E. Petts. John Wiley and Sons Ltd: Chichester.

Wettin, P., Jacobs, T., Johnston, P. and Ryan, N., 1994. Water sharing and floodplain wetlands in the Murray-Darling Basin. Pp. 24–41 in *Murray-Darling Basin floodplain wetlands management: Proceedings of the floodplain wetlands management workshop, Albury, NSW, 20–22 October 1992* ed by A. J. Sharley and C. A. Huggan. Murray-Darling Basin Commission: Canberra.

Williams, M., 1974. The lower reaches. Pp. 140–59 in *The Murray waters* ed by H. J. Frith and G. Sawer. Angus and Robertson: Sydney.

WRC (Water Resources Commission, NSW), 1986. Water and the natural resources of New South Wales. Water Resources Commission, NSW. 112 pp.

WRC (Water Resources Council, NSW), 1993. Our water. 51 pp.

WRC and NPWS (Water Resources Commission and, NSW National Parks and Wildlife Service), 1986. Water management plan for the Macquarie Marshes. NSW Water Resources Commission and National Parks and Wildlife Service.

Young, M. D., 1992. Sustainable investment and resource use. Man and the Biosphere Series, Vol 9 ed by J. N. R. Jeffers. CSIRO, Australia, UNESCO, Parthenon Publishing Group: Melbourne.

Managing water resources: seeking unity in the interests of diversity

P. CRABB[1]

Administrative boundaries and jurisdictions are unknown to the natural world but they are critical to its management. Nowhere is this more evident than in inter-jurisdictional river basins. Drawing on long international experience, certain principles can be identified for their management, principles that are equally applicable to the management of river basins within federations, such as the Murray-Darling Basin. A history of mismanagement compounded by inter-jurisdictional issues provides the background for a remarkably innovative management structure. The Murray-Darling Basin Agreement, signed in 1987 and revised in 1992, established political, bureaucratic and community structures to facilitate the management of the entire basin. In addition, it provided a process for the Basin's management. The success of the Agreement is seen in such measures as the Natural Resources Management Strategy and the Salinity and Drainage Strategy. Whilst these measures tackle the Basin's urgent immediate problems, the Agreement's continuing success is dependant on its ability to deal with new problems as they emerge, such as the outbreaks of blue-green algae. The Murray-Darling Basin Agreement goes further than any other comparable agreement in meeting the principles for the managing of inter-jurisdictional river basins. Only with such an Agreement is it possible to successfully manage the Basin and ensure the maintenance of its biodiversity.

INTRODUCTION

We live with and within the many administrative boundaries and jurisdictions we create. Such creations are unknown to our flora and fauna. They are also unknown by our other natural and environmental resources, including water. The natural world establishes its own boundaries without any consultation with or consideration for human beings. It is rare for there to be any coincidence between natural and human-imposed boundaries. Yet we are all being made increasingly and acutely aware that it is the natural boundaries and the natural systems that matter. They are the ones that are essential to the survival of environmental and ecological integrity, to the conservation of biodiversity, people included. They are also essential in economic terms, for we cannot separate the economy and the environment.

Administrative confusion and inadequacies, especially the many boundaries that separate administrations, are probably as much a threat to biodiversity as bulldozers, pollution and feral animals! So many agencies are involved in the management of any one of our natural and environmental resources. Nowhere is this more evident than in our waterways and aquatic environments. More often than not, the lack of co-ordination and co-operation between agencies is a major contributor to the pollution and degradation of aquatic environments. There is frequently a lack of agreed objectives; as a result, there is little substantive action.

Water and aquatic environments are essential to the conservation of biodiversity. "Many of the benefits Australians obtain from aquatic systems can only be obtained if the ecosystems themselves are protected from degradation" (ANZECC 1992, p. 2–1/2). How do we preserve such environments? The focus of this chapter is one critical part of any comprehensive answer. Nothing is more important than the administrative framework, yet it is frequently overlooked and the issues are rarely addressed. For example, other than giving a key place to total catchment management, issues of administration are not considered in the CEPA report, *Towards Healthier Rivers* (CEPA 1992); inadequate administration is not seen as one of "the ills affecting our rivers". A similar comment can be made with regard to the recently released Australian water quality guidelines (ANZECC 1992). For all their very positive value, if there is any mention of administration in such documents, it is brief and platitudinous, suggesting that it must remain in the "too hard basket" (see, for example, ADPC 1990, pp. 16–17). Is this an area we would rather not discuss, especially where it involves federal-state issues? Some years ago, former Prime Minister John Gorton observed that the states cannot set national policy, but they can frustrate it. This continues to be evident in so many ways, though by no means all of the responsibility should be levelled at the states!

The focus of this paper is the administrative framework within which water and land resources management take place, paying particular attention to inter-jurisdictional issues. Consideration of the international situation provides a setting for a detailed consideration of the management of Australia's key inter-jurisdictional resource, the Murray-Darling Basin.

[1]Department of Geography and Oceanography, University College, University of New South Wales; Australian Defence Force Academy, Canberra, Australian Capital Territory, Australia 2600.

Pages 162–73 *in* CONSERVING BIODIVERSITY: THREATS AND SOLUTIONS, ed by R. A. Bradstock, T. D. Auld, D. A. Keith, R. T. Kingsford, D. Lunney and D. P. Sivertsen. Surrey Beatty & Sons. 1995.

ADMINISTRATIVE BOUNDARIES AND THEIR ASSOCIATED PROBLEMS

There are at least three levels at which administrative issues can be considered: international, intra-national (especially within federal nations), and intra-governmental or interdepartmental.

International

Taking account only of surface water resources and leaving aside river basins that have been linked by canals (of which there are a number in Europe), United Nations studies list 240 international river basins (UNFAO 1978). Most span only two or three countries, but some involve many more, such as the Nile (9 countries), Rhine (9), Niger (10), and Danube (13).

There is a long history of negotiations and a vast number of agreements relating to international river basins. A United Nations study has listed "treaties and other international legal instruments dealing with the various aspects of non-maritime water resources management. In addition to bilateral and multilateral treaties, unilateral declarations bearing on the legal status of international waters and multilateral declarations by governments, national legislative texts providing for measures having international implications, as well as arbitral awards and international court cases" have also been included (UNFAO 1978, p. xi; 1984). Though not everything included is directly relevant here, over 3 700 items are listed, spanning the period from the year 805 to 1983. It is possible to go back even earlier than 805, to antiquity. Rivers and water resources were critical to the earliest civilizations, for both transportation and agriculture, in the Nile, Tigris-Euphrates and Indus River basins (Teclaff 1967, 1985).

In the gradual evolution of relationships between nations concerning shared water resources, there are two aspects that merit particular attention. Firstly, there is the resource. Initial concern was with the major streams of river systems, rivers that either crossed from one state to another or formed boundaries between states, along with lakes that were crossed by an international frontier of two or more countries. In the first half of this century, concern gradually moved from the single river to at least the implicit consideration of the river basin. In the late 1950s this was extended to the total river basin. As was agreed at the 1958 New York conference of the International Law Association, "A system of rivers and lakes in a drainage basin should be treated as an integrated whole (and not piecemeal)" (Teclaff 1985, p. 426).

The second point is the uses to which the water and waterways are put. Expanding the focus from the main streams of rivers to total river basins has been largely a consequence of the changing and increasing uses of water resources. The earliest and for long the major use of rivers was for navigation. So, the concern was with navigable rivers and the freedom of navigation. Initially, these were matters for the riparian states, but in the nineteenth century, a number of treaties extended such principles beyond the riparian states, giving recognition to the freedom of navigation on a number of "international rivers". Gradually, other uses of international water resources entered the picture, such as fishing, irrigation, hydro-electric power, urban and industrial uses. There were also other matters relating to water resources that required agreements between nations, such as the maintenance of stream flow, deepening of river beds, construction of bridges and dams, flood control, and the maintenance of water quality.

With time, there was not only this broadening of concerns. Attention began to focus on management rather than simply development or exploitation. For example, a 1957 report to the United Nations stated: "It is now widely recognized that individual water projects — whether single or multipurpose — cannot as a rule be undertaken with optimum benefit for the people affected before there is at least the broad outlines of a plan for the entire drainage area" (Teclaff 1985, p. 427). With increasing water pollution, concerns for wildlife and their habitats, and aesthetic considerations, such statements become even more valid. The broadening issues go well beyond a simple concern for the water in a river to the total complex of a basin's resources and especially the land resources. As a result, the problems have multiplied in both number and complexity.

There is little argument that each river basin is unique, with its own geography and particular problems. Cano (1985) stated that "the golden rule is that there does not exist any golden rule. It is necessary to adopt an ad hoc solution for each river basin". It is hardly surprising that after considering over two hundred bilateral and multilateral treaties and agreements relating to shared watercourses, Lewis (1988, p. iv) concluded that each was "a specific reflection of the geography and hydrology of the river basin in question, the historical pattern of water use in each of the co-basin States, the history of relations between the States and the internal water resource policies of each State". Nonetheless, there are issues common to most if not all inter-jurisdictional river basins. Further, such common issues are generally the major ones. They include, for example, the apportionment of resources between the states concerned, the use of the water, land and other resources, the maintenance of water quality, and decision-making about the resources and their management. Given the similarity of issues and experiences, it is hardly surprising that certain common responses can be recognized in the many agreements. As a result, it is possible to identify a number of basic principles for the management of international river basins (Table 1).

Table 1. Principles for the management of inter-jurisdictional river basins.

- the river basin is the basic hydrological management unit; acceptance of this principle has implications for virtually all of the other principles;
- no state can claim exclusive sovereignty to the waters of an inter-jurisdictional river basin;
- reasonable and equitable participation in the control of the resources;
- reasonable and equitable apportionment of the water resources of the basin;
- protection and non-abuse of the basin's resourcers; and
- acceptance of the inter-relationships of natural resources, especially water and land and surface and groundwater.

Over time, these principles have gained increasing acceptance. At least some of them have also acquired status in international law. After all, the various treaties themselves are set within the context of the principles of international law. Further, "Collectively, the treaties themselves, become over time, evidence of customary practice in international law and the evolution of international water resources and environmental law reflects this customary practice, as well as the consensus of progressive thought on the subject by legal and water resource experts" (Lewis 1988, iv). However, quite apart from their international legal status, they now carry considerable moral weight.

Next to acceptance of such fundamental principles, the most important and difficult issue in the management of inter-jurisdictional river basins is the creation of appropriate and acceptable institutional arrangements, i.e., the means by which the principles and inter-governmental agreements are put into practice. With the growing number, complexity and severity of the problems that have to be confronted, not least in the area of water quality, this is becoming increasingly difficult. Further, many issues are going beyond water to embrace other resources, especially land. At the same time, "states' rights" limit the extent to which states will concede any of their jurisdictional or constitutional rights to an extra-state authority.

In spite of the difficulties, there are successful agreements and institutional arrangements. Without doubt, the best example is the International Joint Commission, established by the Boundary Waters Treaty of 1909 between Canada and the United States. Even where countries do not always get along particularly well, successful agreements have been reached, such as for the Senegal River in West Africa. There are also parts of the world where unless agreement is reached over shared water resources, trouble is inevitable, in particular the Middle East, where a future war could well be over water rather than oil (Mitchell and Downey 1993).

There have also been international agreements that conserve biodiversity, especially waterbirds, such as the *North American Waterfowl Management Plan* (Canadian Wildlife Service 1986). With the continuing and increasing pressures on habitats, especially wetlands and grasslands, from actual destruction, pollution, chemicals, predators, and disease, habitat preservation is of pre-eminent importance in the agreement. Australia is a signatory to the 1971 Convention on Wetlands of International Importance (the Ramsar Convention), with 27 Australian wetlands included in the list, including the Barmah Forest, the Kerang Wetlands, and the Macquarie Marshes Nature Reserves. Australia has also signed agreements with Japan and China for the Protection of Migratory Birds and their Habitats, which impose obligations with regard to certain species and their habitats (Michaelis and O'Brien 1988). Given the continued degradation and destruction of habitats, especially waterways and aquatic environments, the question has to be asked whether or not countries such as Australia and Canada are meeting their international obligations under the treaties they have signed.

Intra-national

Inter-jurisdictional issues within federations, countries such as Australia, Canada, the United States, South Africa, Malaysia, Switzerland, and Germany, have much in common with those at the international level. The particular nature of a federation is dependent on the terms of the country's constitution, but in general, the situation of a river basin in a federation is very similar to that of an international river basin. This is certainly the case in those countries where the responsibility for and ownership of water, land and other natural resources rests essentially with the state level of government. There are some countries, such as South Africa, where responsibility for water resources rests with the federal level of government. This may eliminate some inter-jurisdictional problems, but it is unlikely that it would eliminate all of them, as there are bound to be water-related issues that come within the state level of jurisdiction.

In most federations, the member states are in many respects sovereign states. Thus the basic principles listed in Table 1 are just as relevant to intra-national river basins within federations as to international ones. Equally applicable are the observations made with respect to the institutional arrangements for the management of water and other resources. In at least some situations, the principles of international law are also relevant. For example, in the United States, the Supreme Court rules on disputes between states over water resources and in doing so, it regards the states as sovereign entities and applies what it regards as the relevant international law (Lewis 1988, p. iv).

Intra-governmental or inter-departmental

Jurisdictional boundaries do not only exist between nations and within federations. They are also evident within states, at the inter-departmental level, where administrative fragmentation affects so many issues, not least the management of river basins. Government departments usually have narrow and compartmentalized views and mandates that bring both benefits and costs. "From an organizational perspective, clear demarcation lines must be drawn not only to ensure efficiency, but also in the interests of accountability. . . . However, in addition to splitting problems into manageable pieces, decomposition splinters the wider perspective" (Barrie 1992, p. 2). The Brundtland report regarded this as a situation in need of attention (WCED 1987, p. 310–313). Some years earlier, in a review of New South Wales Government administration, Wilenski came to a similar conclusion. Though referring particularly to statutory authorities, his words have wider relevance:

> "Like generals using the strategies that would have been effective in a previous war, many such government organizations, bound by the Acts which created them, are geared to deal best with the problems that were most relevant to a previous generation. Furthermore, as each pursues the objectives laid down in its legislation — and often pursues these very successfully — problems of co-ordination arise, sometimes resulting in considerable conflict, duplication and waste, while new problems, which cut across (or fall between) organizational boundaries, are often overlooked" (Wilenski 1977, p. 4).

Paterson (1978) was even more direct: "The specialized public works authorities are often long on technical competence, but short on adaptability."

Bureaucracies have certainly changed since Wilenski and Paterson made their comments in the late 1970s, but there is still room to improve departmental structures and their political management. Conflicts still occur between departments. The NSW National Parks and Wildlife Service and Department of Water Resources (or rather some of its clients) argue over the allocation of water for the Macquarie Marshes, as has been acknowledged (NSWWRC and NPWS 1986; Johnson 1992; Millington 1991). Opinions differ over total catchment management between the Department of Water Resources and the former Soil Conservation Service (see NSWDWR 1990). There is the need for greater liaison between departments. This involves much more than changing structures; it means changing people and their attitudes. In the words of the Brundtland report, "The real world of interlocked economic and ecological systems will not change; the policies and institutions concerned must" (WCED 1987, p. 9). The institutions and the links between them must reflect the issues that are being addressed.

An overview

There can be no doubt about the relevance of the international lessons for resource management to federal nations and their individual states. The various administrative arrangements outlined above have been put in place essentially because of the inherent conflict between a natural hydrological unit or system and a political/administrative one and the non-coincidence of their boundaries, i.e., a river basin or catchment and a territorial or jurisdictional unit. Such problems exist within and between all state jurisdictions, even within so-called unitary states, as it is rare for there not to be a local level of government. Thus, for example, the introduction of measures at the catchment level in Victoria were frustrated by a lack of co-operation by local governments (Clark 1984 p. 6). However, in an inter-jurisdictional situation, at whatever level this occurs, divided jurisdiction compounds the problems of managing river basins. The problems are by no means confined to water, but extend to other natural resources, such as mineral deposits (see Saunders 1986) and the atmosphere, with acid rain being a particularly good illustration. Whatever level of government is involved, far more powerful arrangements for total or integrated catchment management are required.

THE MURRAY-DARLING BASIN: A SUCCESS STORY

In his statement on the environment, the Prime Minister stated that "Nowhere is the link between the Australian environment, the Australian economy and Australian culture better described" than in the Murray-Darling Basin (Keating 1992, p. 3). There is no better demonstration of the consequences of inadequate institutional arrangements and no better illustration of the opportunities for change than in the Murray-Darling Basin. Further, there is no better example of an inter-jurisdictional water resource (Fig. 1). This is not the place for an extended outline of the Basin's vast diversity of environments and its overall geography. This has been done elsewhere (see Crabb 1988; McKay and Eastburn 1990; MDBC 1991–93). Similarly, accounts have been provided of earlier endeavours in the joint management of some of the Basin's resources, such as through the River Murray Waters Agreement and the River Murray Commission (see Clark 1982, 1983).

Given the constraints within which it had to operate, the River Murray Commission was a major achievement, especially when considered from an historical point of view. However, through no fault of its own, it was unable to cope with the issues of the 1970s and early 1980s. The continued environmental deterioration in the Murray Valley and the larger Murray-Darling Basin, together with the realization that the existing management arrangements were unable to tackle the problems, had a significant outcome in late 1985, when senior ministers responsible for water, land and the environment in the Australian, New South Wales, Victorian and South Australian Governments met in Adelaide. Other factors certainly contributed to this meeting, including changing attitudes to water management overseas, the changes in water management within the various Australian jurisdictions, certain key politicians and bureaucrats, and four governments of the same political persuasion. The next two years probably saw more inter and intra-government contacts in the interests of the Murray-Darling Basin than had ever occurred before. These intensive and extensive ministerial and bureaucratic discussions had their outcome in the Murray-Darling Basin Agreement, signed by the four governments on October 30, 1987.

The Murray-Darling Basin Agreement (MDBA) involved a significant relaxation of concern by the states with "states' rights". This only happens in exceptional circumstances. It came about because of the severity of the problems that had to be tackled and the realization that they could only be tackled by co-operation. The continued co-operation is evident in the completely revised Agreement, signed by all governments involved with the Basin in June 1992. The 1987 Agreement was an amendment to the River Murray Waters Agreement, but the 1992 one was a totally new document which replaced

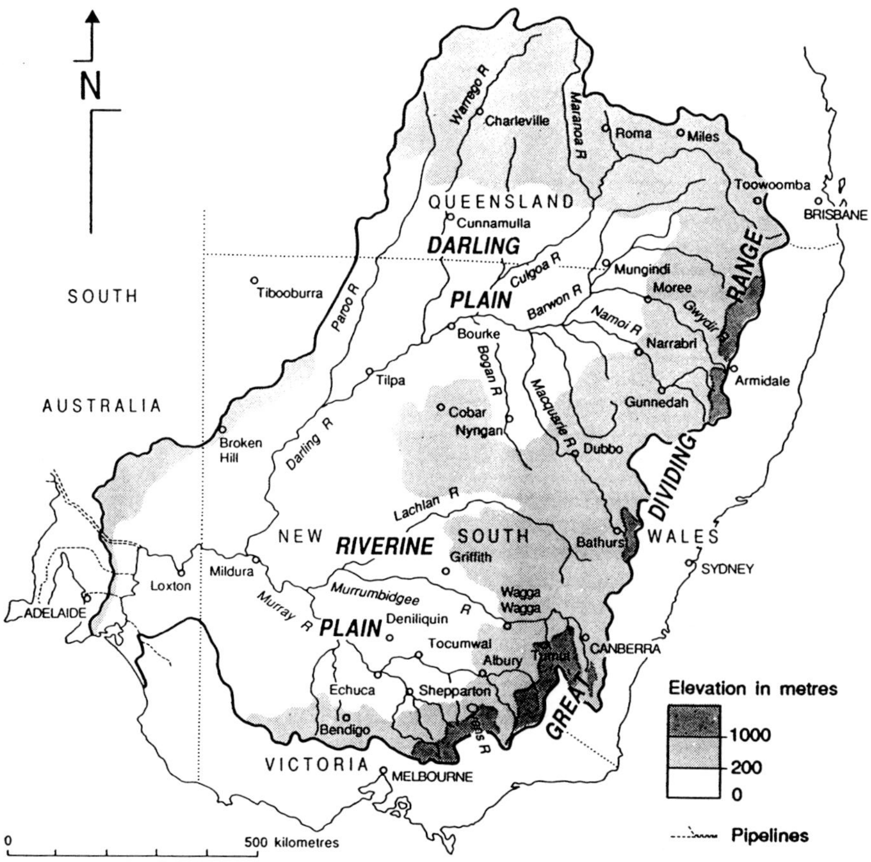

Fig. 1. The Murray-Darling Basin.

the River Murray Waters Agreement and its many amendments. It has yet to be ratified by the various parliaments, but since July 1, 1992, the governments have been acting in accordance with the new Agreement.

Although Queensland accounts for some 25 per cent of the Basin (and 15 per cent of Queensland is within the Basin), it was not a party to the initial Agreement. This was largely for political reasons, but such non-participation is not atypical of up-stream states. However, Queensland is a signatory to the revised Agreement, though on somewhat different terms to the other states, essentially because it is not a party to the water sharing part of the Agreement.

The prime objective of the Murray-Darling Basin Agreement is "to promote and co-ordinate effective planning and management for the equitable, efficient and sustainable use of the water, land and other environmental resources of the Murray-Darling Basin" (Murray-Darling Basin Agreement 1992, Cl. 1). To achieve this objective, new political, bureaucratic and community participation institutions were formed, namely the Murray-Darling Basin Ministerial Council (MDBMC), the Murray-Darling Basin Commission (MDBC), and the MDBMC Community Advisory Committee (CAC) (Fig. 2). The MDBMC consists of three ministers from each of the signatory governments, representing the portfolios of water, land and environmental resources. The Council's prime function is "generally to consider and determine major policy issues of common interest to the contracting governments concerning effective planning and management for

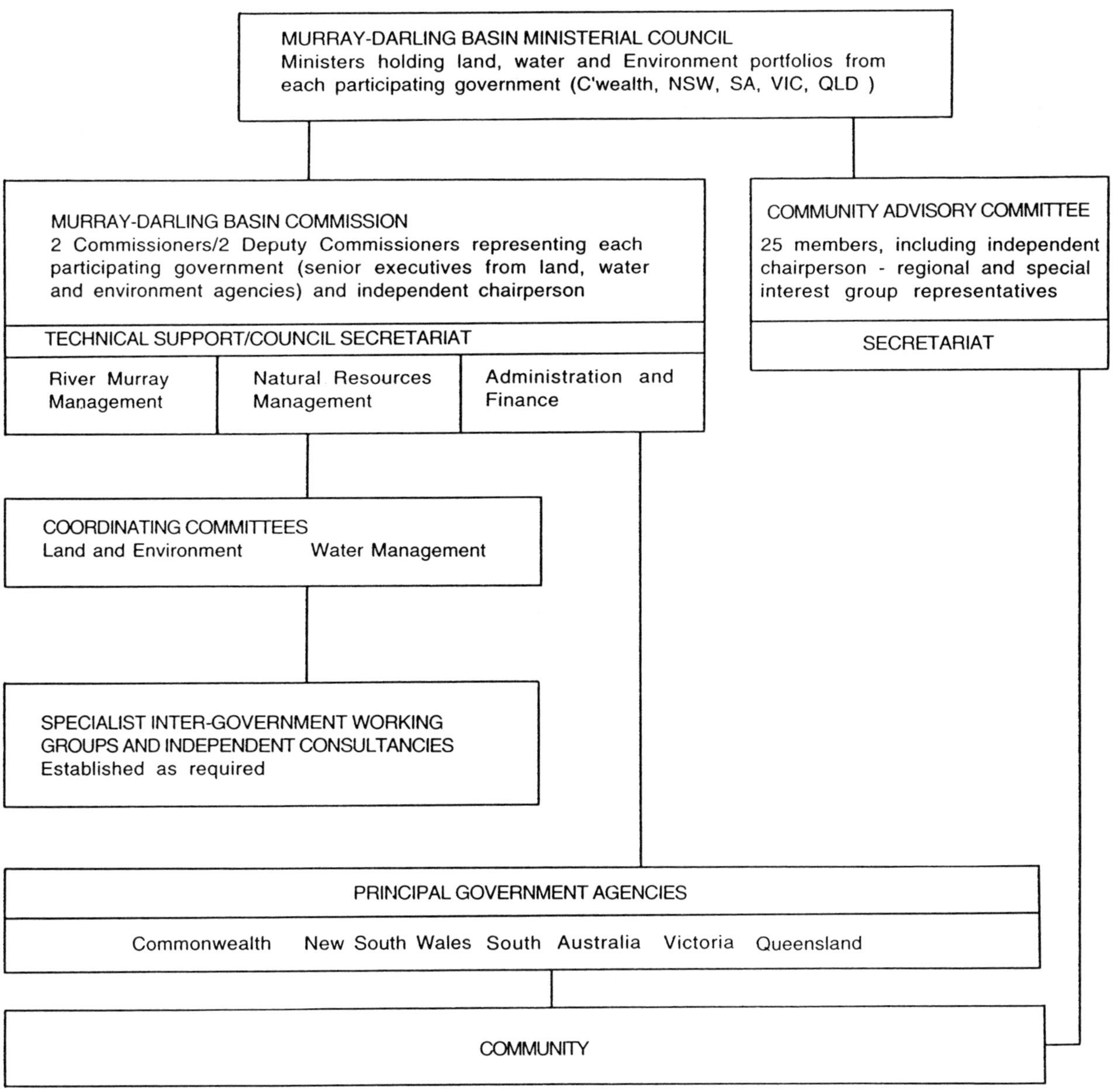

Fig. 2. The Murray-Darling Basin management structure (MDBC 1992).

the equitable, efficient and sustainable use of the water, land and other environmental resources of the Murray-Darling Basin" (Murray-Darling Basin Agreement 1992, Cl. 9(a)). Being a political forum, the Council is able to make decisions in a way that a bureaucratic one cannot, and, in terms of the Agreement, its potential powers are considerable. The MDBC replaced the former River Murray Commission and took over all the latter's responsibilities with respect to the regulation of the main stream of the Murray. In addition, however, the Commission has a much broader task, namely "to advise the Ministerial Council in relation to the planning, development and management of the water, land and other environmental resources within the Murray-Darling Basin" (Murray-Darling Basin Agreement 1992, Cl. 17.(1)(a)). This is not a discretionary power but a mandatory one, in effect covering all of the Basin's natural resources. There are two components to the Commission. The Commission comprises two senior bureaucrats from each of the signatory governments (representing water, land and environmental resources) and an independent President. The Office of the Commission is the technical executive, which carries out the various functions of the Commission, as well as acting as a secretariat for the Ministerial Council. The third institution created by the MDBA is the CAC, a large group representing the many regional and special interest groups within the Basin. It is an acknowledgment of the fact that if any progress is to be made in tackling the Basin's problems, "grass roots" or community involvement and support are essential. With direct access to the Council, it has the difficult task of trying to provide a two-way means of communication between the people who live and work in the Basin and the Council. It reports directly to the Ministerial Council, but its role is limited to an advisory one.

In less than ten years, inter-jurisdictional arrangements have gone from a very narrow focus on sharing the water in the main stream of the Murray between three states to a comprehensive concern with the management of water, land and other environmental resources of the entire hydrologic Basin.

THE ACHIEVEMENTS OF THE AGREEMENT

The 1987 MDBA and its 1992 revision are remarkable achievements in the inter-jurisdictional management of land and water resources. Such a view is valid in an Australian context and an international one. It is perhaps the largest exercise in total catchment management ever attempted anywhere in the world.

In the short history of the Agreement, there have been major achievements which can be considered in terms of "substance" and "process".

Substance

Thus far, the new management structure has produced three significant agreements for tackling the Basin's critical resource and environmental problems. These are of critical importance to biodiversity conservation.

(a) The Salinity and Drainage Strategy

Land salinization and waterlogging and river salinity are increasingly critical problems that are imposing huge costs on the whole community. Further, without intervention, these costs are likely to double in less than 25 years, quite apart from the impact on the natural environment. The problems provided much of the stimulus for the establishment of the existing management structures. The Salinity and Drainage Strategy (SDS), approved in 1989, is a co-ordinated plan to tackle these problems (MDBMC 1989). It has three objectives:

- to improve water quality in the River Murray (and the South Australian portion in particular) for all users;
- to control existing land degradation, prevent further land degradation and, where possible, rehabilitate land resources, especially in the Murray and Murrumbidgee valley irrigation areas; and
- conserve the natural environments of these valleys.

A critical goal is to maintain River Murray salinity, as measured at Morgan, below 800EC for 95 per cent of the time for a 20-year period from 1990. To achieve this, salt interception schemes are being constructed in the Riverland and Sunraysia districts to reduce the flow of saline groundwater to the Murray. The first of these at Woolpunda has been completed and those at Waikerie and Mallee Cliffs are under construction. They will reduce Murray water salinity at Morgan by 140EC (Fig. 3). Of this sum, 30EC will be to compensate for "natural deterioration" in river salinity. In recognition of their contributions to the cost of the salt interception works, New South Wales and Victoria can each increase river salinity by 15EC. These "salinity credits", a form of pollution sharing, will enable the states to drain at least some of their high value irrigation areas to the Murray. However, given that the credits are very small in relation to the total demand for salt disposal, land management and drainage schemes for the irrigation areas are also an essential element of the Strategy. Such schemes are being put into operation in the Benerembah and Berriquin districts of New South Wales and the Shepparton, Kerang and Sunraysia districts of Victoria. The overall outcome of the Strategy is that there will be a net 80EC reduction in salinity at Morgan.

The increasing size of the dryland salinity problem (MDBC 1993) and its impact on stream salinity will almost certainly require future reconsideration and adjustment of the Salinity and Drainage Strategy to ensure that its objectives are achieved.

(b) The Natural Resources Management Strategy

Because it dealt with the Basin's most difficult and contentious issues, the Salinity and Drainage Strategy came first. However, it has to be seen as part of the Natural Resources Management Strategy (NRMS), which has been described as the "cornerstone" of the Basin's management (MDBMC 1990). The NRMS is not a document with a solution to every resource problem. Rather, it is "the framework for co-operative and co-ordinated community and government action to address the Basin's natural resource management problems on a long-term, integrated basis" (MDBMC 1990, p. 1). The identified problems and the aims of the NRMS contain no surprises (Table 2). Critical to the achievement of these aims is community involvement and commitment, based on the philosophy that the people who use the resources are the ones best able to identify the problems and the solutions. With government assistance and co-operation, "Communities of Common Concern" (groups of people covering all kinds of situations) identify local natural resource problems, prepare plans to tackle them, and then implement the plans. Examples of such community-based projects in New South Wales include revegetation in the Yass Valley, woody weed control at Ford's Bridge, north-west of Bourke, and the establishment of a catchment management plan for the Belubula River.

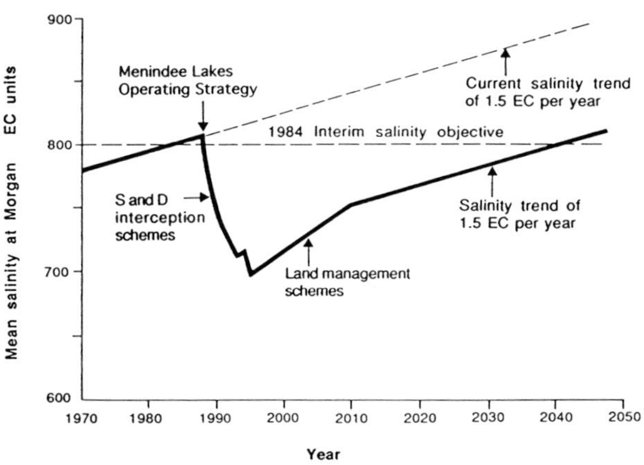

Fig. 3. Based on salinity as measured at Morgan, the graph shows the current salinity trend for Murray water, the impact of the groundwater interception schemes and land management schemes of the Salinity and Drainage Strategy, and the likely future trend (EWS 1990).

Co-operation is not only necessary between governments and the community at the local level; it is also essential between the various levels of government at regional and Basin-wide levels. Because of their large and inter-state nature, certain issues are seen as government responsibilities, namely vegetation, groundwater and aquatic and riverine environments, issues that are clearly of relevance to biodiversity conservation. Among the interstate projects are a native fish management plan for the Murray-Darling river system (a recognition of the serious

Table 2. Problems identified by and aims of the Murray-Darling Basin Natural Resources Management Strategy (MDBMC 1990).

Problems	Rising salinity levels in soils and streams
	Deteriorating quality of water supplies
	Land degradation (e.g., soil erosion and acidification)
	Decline and loss of native vegetation
	Loss of native habitats
	Over-commitment of, and competing demands for, water supplies
	Cultural losses (e.g., Aboriginal heritage sites)
Aims	Prevent further degradation of natural resources
	Restore degraded resources
	Promote sustainable user practices
	Ensure appropriate resource use planning and management
	Ensure a long-term viable economic future for those dependent on the Basin
	Minimize adverse effects of resource use
	Ensure community and government co-operation
	Ensure self-maintaining populations of native species
	Preserve cultural heritage
	Conserve recreational values

decline in the numbers and distribution of native species, largely due to the regulation of the Basin's waterways), which includes the construction of fish ladders on a number of the water regulatory structures (Lawrence 1991). Other interstate projects include mallee and other vegetation management (MDBC 1991); mapping of groundwater over a large area of the Basin; investigations of nutrient pollution and blue-green algae (Gutteridge Haskins and Davey 1992); co-operative projects to manage wetlands and forests along the Murray (Sharley and Huggan 1994); and research into the importance for the breeding of waterfowl of ephemeral lakes in western New South Wales. Finally, the Commission has the important task of ensuring that the solution of a problem in one place does not result in a new problem somewhere else.

(c) Water sharing

A third important achievement of the new management structure is a new method of sharing River Murray water, agreed to in 1990 after many years of negotiation. "Continuous accounting" provides a more equitable system of water sharing and a greater level of security for the individual states. Like the SDS, this forms part of the new Agreement. It also makes possible the trading of water entitlements between states (which could be the first step towards a Basin-wide market for water) and of saline water emissions.

Process

The three agreements reached thus far can be seen as the initial substantive achievements of the new management structure. They also indicate what is perhaps an even more important achievement, namely that the MDBA has established a process for decision making and reaching agreement on inter-jurisdictional management issues. This could be even more important in the future with increasing pressures on the Basin's limited natural resources.

There can be no doubt that the Murray-Darling Basin Agreement has made an important contribution to arresting further decline of biodiversity within the Basin. At its 1993 meeting, the Ministerial Council approved a draft Algal Management Strategy for the Basin and, on a more specific issue, agreed in principle to a permanent annual allocation of 100 GL of water for the Barmah-Millewa Forest. The Council also directed the Commission to prepare a Nature Conservation Strategy for the Basin which will consider:

- development of an environmental water allocation strategy;
- development and implementation of a sustainable rivers programme;
- development and implementation of a pest animal and plant management programme for the Basin;
- protection of biodiversity;
- protection of remnant vegetation;
- development and maintenance of a sound database of the environmental resources of the Basin (MDBMC 1993).

Quite apart from the importance of such a measure to the conservation of the Basin's biodiversity, the Council's decisions provide further illustrations of the way in which the Murray-Darling Basin Agreement is continuing to provide a process for the resolution of difficult and critical issues.

SOME CONTINUING ISSUES

There is no doubt about the unique and innovative nature of the Murray-Darling Basin Agreement, both from a national and international perspective. However, this does not mean that it is free of difficulties (see Crabb 1991) or that there are no more issues for the Ministerial Council and Commission to address. There are current issues of particular relevance to the conservation of biodiversity. Some will provide a major test for the MDB management process, because they highlight the fine dividing line between state and inter-state Basin-wide issues.

One major issue is the increasing abstraction of water along the upper Darling and its tributaries (Marshall 1993). This is occurring particularly through the "harvesting" of flood flows (which are frequently regarded as "surplus" water) and the use of "sleeper licences". The overall reduced flows in the rivers seriously impact on downstream water users as well as on the rivers themselves. There is a loss of "beneficial flooding", with particular impacts on graziers and wetlands, such as Narran Lake (a nature reserve which is particularly important for waterbirds (Brooker 1993)) and other areas along the intersecting streams that cross from Queensland into New South Wales. In the context of these issues, cotton growing is of particular concern. It is an industry of major economic importance, but one that has a voracious appetite for water and floodplain land, points that are highlighted by the controversial developments and plans on "Cubbie Station". Located in southern Queensland where the Balonne River divides into the Culgoa, Birrie and Narran Rivers, the developments — including storage facilities for some 180 000 megalitres of water — have the potential to affect a vast area of southern Queensland and northern New South Wales (Fig. 4). Fortunately, further work has been put on hold for three years, which, among other things, will allow the needed hydrological studies to be undertaken before any firm decisions are made regarding further abstractions from these rivers. Quite apart from the impacts of droughts, the implications of such developments for the rivers and their aquatic ecosystems are surely self-evident. They are irreplaceable and destroying them is "tantamount to smashing up the natural plumbing" (Alexandra and Eyre 1993; see also McCosker and Duggan 1993).

Without in any way diminishing the importance of such issues as outlined above, there are others of greater significance, especially in terms of biodiversity and the long term. For example, groundwater levels continue to rise over 80 per cent of the Murray hydro-geological basin (which covers much of the southern part of the MDB), with serious implications for all land uses from dryland salinity (Brown 1989; Walker *et al.* 1993). It is in this context that measures to preserve areas of dryland vegetation are important, such as through the government purchase of a number of large properties between Mildura and Renmark.

If only one issue of concern could be nominated, for the Basin as a whole and especially the conservation of its biodiversity, then it would have to be the state of the Basin's rivers, in terms of the quantity and quality of their flows and their riverine environments. Critical to any improvement is total or integrated catchment management and its application at all levels. This has been stressed in the *Towards Healthier Rivers* statement (CEPA 1992) and in the Prime Minister's *Statement on the Environment* (Keating 1992, p. 10). Nowhere are such observations more relevant than in the Murray-Darling Basin. To address this problem, many decisions may well be painful, such as

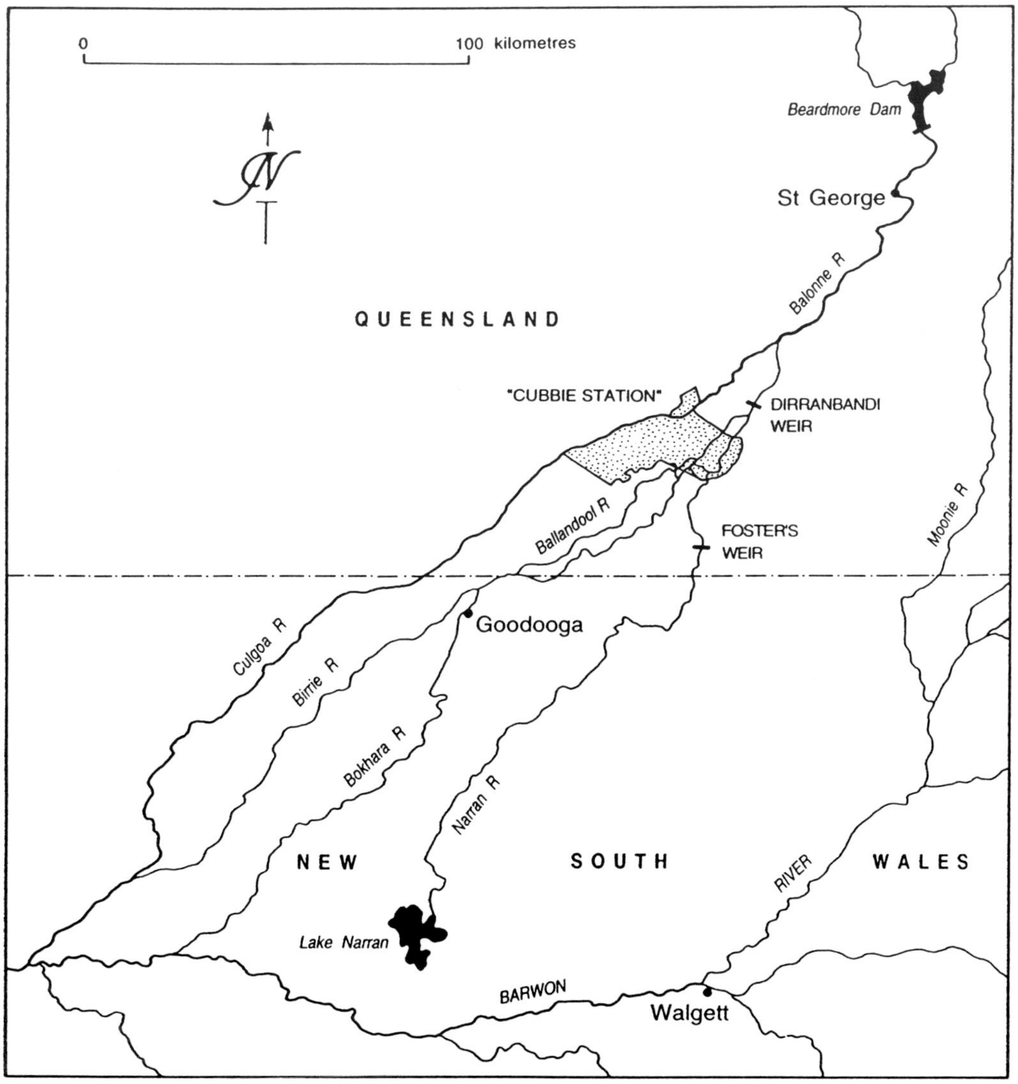

Fig. 4. The location of "Cubbie Station", southern Queensland, and the related river systems.

increased allocations of water for in-stream purposes, changes to agricultural landuse practices, and the off-stream disposal of all forms of urban, industrial and agricultural effluent (Walker 1985). There may well have to be significant re-allocations of water in many parts of the Basin. As Johnson (1992) has concluded:

> "It has become obvious that traditional water uses, without regard for environmental requirements, have seriously degraded the rivers and river environments of the Murray-Darling Basin. Determining the best way to allocate further water for environmental purposes and to manage these environmental allocations most efficiently is a matter of urgency."

The acknowledgment of this fact and the allocation of water specifically for the needs of the rivers mark a significant change in management practices by a number of authorities, one that is of particular significance for biodiversity conservation (DCEV 1991; NSWWRC 1993).

CONCLUSION

In his *Statement on the Environment,* the Prime Minister stated that "We are determined to restore the Murray-Darling to its former stature" (Keating 1992, p. 3). That is not possible; as Heraclitus observed, "You can't step into the same river twice." More realistically, in 1987, Senator Gareth Evans said, "It is essential, if we are to improve the management of the Basin, that all governments get their collective act together." To a significant degree, this has been achieved. It is a classic example of the recognition of the need for a co-operative approach to resource and environmental management problems. Critical to the establishment of the MDBA was political will, and this is essential for its continuance, along with co-operation, trust, and the people of the Basin owning the problems and the solutions. Governments are, after all, ultimately responsible for the success or failure of the institutions they create

The Murray-Darling Basin Agreement goes further than any other comparable agreement in meeting the general principles detailed earlier for the management of inter-jurisdictional river basins (Table 1). It is recognized as a model for other river basins. Much has been achieved in a very short period of time in terms of the Basin's overall management. But there are also difficult and painful issues ahead, and they will have to be dealt with sooner rather than later. In many respects, they focus on the Basin's environment and, if nothing else, recognize that what is good for the environment is good for the economy — though the reverse may not be the case. For example, as in the western United States, "environmental non-consumptive, in-stream uses of water are now found to have economic values that compete with the traditional consumptive uses of water" (Moore and Willey 1991). It is evident that total or integrated catchment management at all levels is being given primacy by the Murray-Darling Basin Ministerial Council and Commission and among the issues under consideration are possible severe limitations on future extractions of water from the Basin's rivers and the rewriting the NRMS to strengthen its conservation and environmental objectives.

However, concern for the environment must go well beyond economic matters. At stake is the very well-being of the Basin and the maintenance of its total biodiversity, including human beings and their diverse activities. Only within the unified framework provided by the Murray-Darling Basin Agreement — probably the world's largest exercise in total catchment management — is there any prospect of the necessary decisions being made and the objectives being achieved.

REFERENCES

ADPC (Australia, Department of Prime Minister and Cabinet), 1990. Ecologically sustainable development: a Commonwealth discussion paper. Australian Government Publishing Service: Canberra.

Alexandra, J. and Eyre, D., 1993. Restoring our rivers. *Habitat Aust.* **21**(1): 22–31.

ANZECC, 1992. Australian water quality guidelines for fresh and marine waters. Australian and New Zealand Environment and Conservation Council: Canberra.

Barrie, D., 1992. Environmental protection in federal states: interjurisdictional co-operation in Canada and Australia. Federalism Research Centre, Australian National University: Canberra.

Brooker, M. G., 1993. Aerial counts of waterbirds on Narran Lake, New South Wales. *Aust. Bird Watcher* **15:** 13–18.

Brown, C. M. (ed.), 1989. The Murray, special issue of *BMR J. Aust. Geol. Geoph.* **11:** 127–395.

Canadian Wildlife Service, 1986. North American Waterfowl Management Plan: a strategy for co-operation. Environment Canada/Supply and Services Canada: Ottawa.

Cano, G. J., 1985. Legal and administrative tools for river basin development. Pp. 189–99 in *Strategies for river basin management: environmental integration of land and water in a river basin* ed by J. Lundqvist, U. Lohm and M. Falkenmark. Reidel Publishing: Dordrecht.

CEPA, 1992. Towards healthier rivers: the ills afflicting our rivers and how we might remedy them. Commonwealth Environmental Protection Agency: Canberra.

Clark, S. D., 1982. The River Murray Waters Agreement: down the drain or up the creek? *Civil Engin. Trans.* **24:** 201–08.

Clark, S. D., 1983. Inter-governmental quangos: the River Murray Commission. Pp. 154–72 in *Quangos: the Austlian experience* ed by G. R. Curnow and C. A. Saunders. Hale & Iremonger: Sydney.

Clark, S. D., 1984. Inter-governmental co-operation in salinity control: a discussion paper for the Salinity Committee of the Victorian Parliament. Government Printer: Melbourne.

Crabb, P., 1988. Managing the Murray-Darling Basin. *Aust. Geog.* **19:** 64–88.

Crabb, P., 1991. Resolving conflicts in the Murray-Darling Basin. Pp. 147–59 in *Negotiating water: conflict resolution in Australian water management* ed by W. J. Handmer, A. H. J. Dorcey and D. I. Smith. Centre for Resource and Environmental Studies, Australian National University: Canberra.

DCEV, 1991. Water Victoria: the next 100 years. Department of Conservation and Environment Victoria: Melbourne.

EWS, 1990. Waikerie Salt Interception Scheme: public environment report. Engineering and Water Supply Department: Adelaide.

Gutteridge Haskins and Davey, 1992. An investigation of nutrient pollution in the Murray-Darling River System. Murray-Darling Basin Commission: Canberra.

Johnson, B., 1992. The Macquarie Marshes: overdue for environmental restoration. *Nat. Parks J.* **36**(4)**:** 9–12.

Keating, P., 1992. Statement on the environment: Australia's environment: a natural asset. Australian Government Publishing Service: Canberra.

Lawrence, B. W., 1991. Fish management plan. Murray-Darling Basin Commission: Canberra.

Lewis, G. D., 1988. The development of international water resources law and principles to guide the management and use of international watercourse systems. Water Planning and Management Branch, Environment Canada: Regina.

Marshall, A., 1993. The disappearing Darling. *The Land, May 27,* 12–13, 83.

McCosker, R. O. and Duggin, J. A., 1993. Gingham Watercourse Wetland Management Options: a discussion paper. Department of Ecosystem Management, University of New England: Armidale.

McKay, N. and Eastburn, D. (eds), 1990. *The Murray.* Murray-Darling Basin Commission: Canberra.

MDBC, 1991. Mallee Vegetation Management in the Murray hydrogeological Basin. Murray-Darling Basin Commission: Canberra.

MDBC, 1991–93. Videos produced to support the Natural Resources Management Strategy. Murray-Darling Basin Commission: Canberra.

MDBC, 1992. Murray-Darling Basin Commission 1992 Annual Report. Murray-Darling Basin Commission: Canberra.

MDBC, 1993: Dryland salinity management in the Murray-Darling Basin. Murray-Darling Basin Commission: Canberra.

MDBMC, 1989. Murray-Darling Basin salinity and drainage strategy. Murray-Darling Basin Ministerial Council: Canberra.

MDBMC, 1990. Murray-Darling Basin Natural Resources Management Strategy. Murray-Darling Basin Ministerial Council: Canberra.

MDBMC, 1993. Official Communique Meeting No. 12, 25 June 1993, Melbourne, Victoria. Murray-Darling Basin Ministerial Council: Canberra.

Michaelis, F. B. and O'Brien, M., 1988. Preservation of Australia's wetlands: a Commonwealth approach. Pp. 167–77 in *The conservation of Australian wetlands* ed by A. J. McComb and P. S. Lake. Surrey Beatty & Sons: Chipping Norton.

Millington, P., 1991. The water market — a NSW case study. The New Environmentalism Conference.

Mitchell, B. and Downey, T. J. (eds), 1993. Water in the Middle East. *Water Inter.* **18:** 1–68.

Moore, D. and Willey, Z., 1991. Water in the American West: institutional evolution and environmental restoration in the 21st century. *Uni. Col. Law Rev.* **62:** 775–825.

NSWDWR, 1990. Draft New South Wales State Rivers Policy. Department of Water Resources: Sydney.

NSWWRC, 1993. Our water: a review of the current status of the water resources of New South Wales and the key issues relevant to their future development. New South Wales Water Resources Council: Sydney.

NSWWRC and NPWS, 1986 Water Management Plan for the Macquarie Marshes 1986. Water Resources Commission and National Parks and Wildlife Service: Sydney.

Paterson, J., 1978. Urban management: the Australian context. Pp. 7–14 in *Urban management processes* ed by P. F. Ryan. Australian National Committee for UNESCO/Australian Government Publishing Service: Canberra.

Saunders, J. O. (ed.), 1986. *Managing natural resources in a federal state.* Carswell Co.: Toronto.

Sharley, A. J. and Huggan, C. A. (eds), 1994. Murray-Darling Basin floodplain wetlands management: proceedings of the floodplain wetlands management workshop, Albury, NSW, 20–22 October 1992. Murray-Darling Basin Commission: Canberra.

Teclaff, L. A., 1967. *The river basin in history and law.* Martinus Nijhoff: The Hague.

Teclaff, L. A., 1985. *Water law in historical perspective.* William S. Hein: Buffalo.

UNFAO, 1978. Systematic Index of International Water Resources Treaties, Declarations, Acts and Cases by Basin. Legislative Study No. 15. UN Food and Agricultural Organization: Rome.

UNFAO, 1984. Systematic Index of International Water Resources Treaties, Declarations, Acts and Cases, by Basin, Volume II. Legislative Study No. 34. UN Food and Agriculture Organization: Rome.

Walker, J., Bullen, F. and Williams, B. G., 1993. Ecohydrological changes in the Murray-Darling Basin. I. The number of trees cleared over two centuries. *J. Appl. Ecol.* **30:** 265–73.

Walker, K. F., 1985. A review of the ecological effects of river regulation in Australia. *Hydrobiol.* **125:** 111–29.

WCED (World Commission on Environment and Development), 1987. Our common future. Oxford University Press: Oxford.

Wilenski, P., 1977. *Directions for change.* Government Printer: Sydney.

PLATE 10 *(above)*. Examples of stream invertebrates considered to be good indicators of water quality. Many of these are expected to be intolerant of salinization and sedimentation. *(Photograph courtesy of Museum of Victoria.)*

PLATE 11 *(left)*. Bank stabilization works, although intended to protect streams, can result in significant short-term inputs of sediment. *(Photograph by Leon Metzling.)*

Weeds and Feral Animals

PLATE 12 *(top left)*. While baiting provides an effective method of killing foxes *Vulpes vulpes*, such programmes are expensive, highlighting the need for careful evaluation of the cost effectiveness of control measures in terms of biodiversity conservation. *(Photograph by David Priddel.)*

PLATE 13 *(above)*. Rabbits are inhibiting the regeneration of many long-lived plants in arid Australia. Here a young vegetative root sucker of the rare plant *Acacia carnei* has been ringbarked by rabbits. *(Photograph courtesy of NSW NP&WS.)*

PLATE 14 *(centre left)*. Feral cats pose a threat to many native birds, reptiles and mammals. *(Photograph by Tony Auld.)*

PLATE 15 *(bottom left)*. Bitou Bush *Chrysanthemoides monilifera* from South Africa now dominates many vegatation communities in coastal southeastern Australia. *(Photograph by Kathie Atkinson.)*

Conserving biodiversity: impact and management of exotic organisms

M. D. FOX[1]

The biota of Australia has been affected by the arrival of many new species of plants, animals and micro-organisms particularly over the past two centuries. Although some components of the original biota were robust and withstood competition from the introduced species, other components were particularly vulnerable and have been driven to extinction or endangerment. The present situation is one of flux between a mix of native and introduced species in a landscape of continuing disturbance. In different taxa, and at different trophic levels, there are species that are particularly troublesome, i.e., they exhibit rapid and uncontrolled growth at the expense of native species. The preponderance of such weed and pest species is in ecosystems most subject to disturbance.

The long-term retention of native biodiversity will depend on our abilities to effect changes to both the population growth of the pest and the continued disturbance of the habitat. Management will necessarily involve measures aimed both at inhibiting the pest species and at minimizing disturbance. This paper reviews the history of introductions in Australia, and discusses the current problems and future threats posed by invasive species to the retention of our biodiversity.

AUSTRALIAN BIOLOGICAL DIVERSITY

BIOLOGICAL diversity is defined by the Biological Diversity Advisory Committee (BDAC) as "the variety of all life forms — the different plants, animals and micro-organisms, the genes they contain and the ecosystems of which they form part" (BDAC 1992). The ecological and perhaps genetic basis of such a definition is clear, however there may well be problems in interpreting such a broad and diffuse spectrum, especially when it comes to the more rigid requirements of the law (see for example Bradsen 1992). For the purposes of discussing biological introductions, most interest has focused on the species level of biodiversity, on the impact of introduced species on the status of resident species and, to a lesser extent, on the impact of introduced species on ecosystem functioning.

In a review of the components of global environmental change, Vitousek (1992) lists both loss of biological diversity and biological invasions as constituting sufficiently widespread phenomena to be global. Biological invasions now primarily result from human activity breaking down biogeographical barriers to species dispersal (D'Antonio and Vitousek 1992). Certainly in Australia the recent history of invasions is a rich research field (e.g., Groves and Burdon 1986; di Castri 1990) and much research money is focused on biological control of pest species.

Australia has presented some fascinating cases of successful introductions leading to pest proportions. This partly is accounted for by the previous isolation of the continent. The Australian biota developed largely over the >30 m.y. period of relative isolation after its separation from Antarctica before encountering the Asian Plate. Some parts of both the flora and fauna pre-date that separation and form the Gondwanan element of the biota. A small even more ancient part, including the Cycads, relates to earlier stages of the continent's history. The typical Australian element, the scleromorphic plant taxa and the macropodid marsupials for example, developed once the island continent reached latitudes where aridity became more pronounced. Once the leading northern edge of the continent reached tropical latitudes, there was a series of invasions of species from the island archipelago to its north. One example was the arrival of murid rodents before 5 m.y. BP; these have radiated here and are known as the "old endemics". A more recent invasion of rats about 1 m.y. BP has evolved into a small number (6) of species of native rats known as the "new endemics".

With the rise and fall of the world's sea level with the successive Pleistocene ice-ages, landbridges arose between some of the islands to the north of Australia and such bridges facilitated the exchange of other species. One such species was *Homo sapiens* which arrived in northern Australia at least 40 000 y. BP. There is no evidence of the intentional introduction of any species by these people, except for the Dingo *Canis familiaris dingo* about 8 000 y. BP.

BIOLOGICAL INTRODUCTIONS

The millennia rolled by with dramatic environmental changes building and submerging land bridges, with the "last" ice-age peaking about 18 000 y. BP and sea level reaching its

[1]School of Geography, University of New South Wales, Sydney, New South Wales, Australia 2052.

Pages 177–83 *in* CONSERVING BIODIVERSITY: THREATS AND SOLUTIONS, ed by R. A. Bradstock, T. D. Auld, D. A. Keith, R. T. Kingsford, D. Lunney and D. P. Sivertsen. Surrey Beatty & Sons. 1995.

present stage at about the time of the arrival of the Dingo. Since then the burning practices of the Aboriginal people continued to impact on the rest of the biota. Perhaps 500 years ago a new force to change our biota was to reach the Australian shores. This was the European seafarer and his retinue of ship-borne biota. It is possible that cats were first introduced four or five centuries ago from shipwrecks on the western coastline. Species were intentionally released by sailors. For example, 220 years ago Captain Cook released pigs near Cooktown in Queensland, 20 years before the arrival of the first settlers. The fear of shipwreck fostered the notion of "seeding" islands with species that could sustain the marooned mariner — for this reason goats were released on many Pacific islands and rabbits released on Macquarie Island.

The decision by the British Government to make the newly claimed southern continent a colonial outpost of the Empire impacted most upon the rate of arrival of new species. The first arrivals were eight hundred convicts from the over-crowded prisons of England, four companies of marines, and 400 sailors. The other live cargo on the First Fleet included hens, sheep, goats, pigs, greyhound dogs, turkeys, ducks, rabbits, horses and cattle, together with cuttings of fruit trees and seed of many crop plants contaminated as it would be with the seed of agrestal weeds (Adamson and Fox 1982; Fox and Adamson 1986). Successive arrivals brought more and more individuals of the same species, and added more species expected to thrive in the new land. The rest, as they say, is history.

The settlement of Australia happily coincided with a flowering of interest in the natural world, indeed the colony and Victorian naturalism had a similar birthday (Barber 1980), and so exploratory parties commonly included naturalists and artists to capture the new-found biota on paper or canvas. This meant that much of the flora and fauna was documented before the impacts of introduced species could affect communities to any great degree.

Unfortunately, another cultural development at the time of settlement was the rise of acclimatization societies. These (still extant) societies aimed to introduce "innoxious" species to Australia. An indication of their level of activity is that some 60 species of mammals were introduced to Australia between 1860 and 1880, most of which failed to naturalize. Often the introductions were for sport such as fresh-water fish and deer. However, many were "amenity" additions such as the introduction of house sparrows, blackberry, Indian mynah, privet for hedges, etc. Such multiple introductions forced a rapid naturalization process. For example, the introduction of sparrows from repeated imports of hundreds of individuals, gave the founding population high genetic diversity and increased its chances of survival.

The process of introduction is an ongoing one. As well as the intentional, assisted passage of some species, the accidental introductions continue apace. Recent marine introductions have come from the illegal emptying of bilge water from international ships in coastal waters (Jones 1991). Of eight species of marine fish introduced to Australian waters, four were probably introduced with ballast water (Pollard 1990). Timber containing borers from Tanzania was recently detected in New Zealand on a ship that had already passed through Australian ports.

THE NEW MIX

The end result, although one that is still in flux, is a contemporary biota that comprises a mix of the original, i.e., pre-European biota and a set of new species from all over the world. Many introductions fail to thrive, but those that are introduced to a compatible climate are often successful and may become troublesome due to their extravagant population growth (Fox and Adamson 1986). Particularly those species emanating from the Mediterranean Basin found a congenial environment in southern Australia (Fox 1990a).

The size of the new component is relatively large for some taxa and is not proportional to the size of the endemic biota (Table 1). For example, with a pre-European mammalian biota comprising 288 species, there are now an additional 20–30 species of placental mammals. Myers (1986) lists 47 mammalian species introduced into Australia, however 16 died out and nine are restricted to local small populations. Some of these are from families already well represented in the biota, such as the rodent family. However, other families have been introduced. Importantly, the ungulates are now well represented where previously they were unknown. Generally specialist species failed to establish, while generalists thrived (Myers 1986).

Other vertebrate groups were richer, and the new component is quite small. The Australian reptile and amphibian fauna is particularly rich numbering 700 species, and the naturalized species number only three, two small geckos and the Marine Toad *Bufo marinus* (Cogger 1975). The very few instances of naturalizations of amphibians and reptiles in Australia has been observed elsewhere. For example of over 50 introductions only five naturalized species have been added to the 124 native species of reptiles and amphibians in California (Jennings 1983). Similarly the Australian birds comprise 850 species, however, partly because of the acclimatization societies, they are now joined by

Table 1. Components of the Australian biota with the estimated species richness, recent (European) extinctions, number of threatened species and number of naturalized species.

Taxonomic group	Estimated richness	Recent extinctions	Number currently threatened	Number of naturalized species
Vascular plants	20 000	100	3 232[h]	1 500–2 000[c]
Non-vascular	28 000 (poss. 50 000)	unknown	unknown	unknown
Vertebrates				
Birds	850	10[g]	11[g] (103)[b]	32 (96 introduced)[d]
Marsupials	146	10[g]	20[g]	nil
Monotremes	2	nil	nil	nil
Placentals	140	10[g]	7[g]	25[e]
Reptiles	700[m]	nil	6[g] (155)[b]	2
Amphibians	(above)	nil	7[g] (41)[b]	1
Fish — fresh	150[j]	nil	79[g]	Fresh 21 spp.[i]
— marine	3 600[k]	nil	5	Maring 8 spp.[f]
Invertebrates	107 571[l]			
Insects	54 000	unknown	unknown	unknown
Molluscs	10 000s	unknown	unknown	unknown

[a]CSIRO 1990; [b]Kennedy 1990; [c]Humphries, Groves and Mitchell 1992; [d]Long 1981; [e]O'Brien 1989; [f]Pollard 1990; [g]ANZECC 1991; [h]Briggs and Leigh 1988; [i]Arthington and Mitchell 1986; [j]Harris 1984; [k]Castles 1992; [l]Williams 1990; [m]Cogger 1975.

an additional 64 species (Newsome and Noble 1986). The Australian marine fish fauna is large (3 600 species), and the introductions (8 species) relatively small (Pollard and Hutchings 1990). The freshwater fish fauna is much smaller (150 species) but has relatively more introduced species (21 species) (Arthington and Mitchell 1986). Some of these are particularly noxious, altering the quality of the habitat and displacing native species. Di Castri (1991) has commented on the vulnerability of endemic fish faunas of inland rivers to the intentionally introduced sport fish.

With as many as 107 571 species in the major groups (Williams 1990), the invertebrates far outnumber the combined richnesses of plants and vertebrates. Yet, estimates of invertebrate species richness are poor because of poorly described and under-collected groups. Similarly poorly known is the estimated extent of the introduced invertebrate assemblage. Some well known invertebrate groups are also well represented in the naturalized fauna, for example the land molluscs are particularly rich. For groups of invertebrates that are still relatively unknown the status of the introductions is confused.

What then is the impact of these new species to the original biodiversity? The new arrivals comprise predators and competitors and almost invariably the interspecific interactions of indigenous and introduced species is negative, to the detriment of the indigenous species (Fox 1990b). The spectacular demise of middle-sized mammals in the semi-arid and arid zones is a case in point. Apart from suggestions that this was a particularly "vulnerable" group in a variable and unreliable habitat (Morton 1990), the introduction of herbivorous competitors in the form of millions of sheep and rabbits also played its part, together with the introduction of new predators in the guise of foxes, dogs and cats.

VULNERABLE TAXA

In some instances there were taxa that proved to be vulnerable to extinction or endangerment with the changes wrought by the new settlers. Of all records, the mammalian fauna is well documented and relatively small. Of the marsupial mammals, the family Potoroidae (Potoroos, Bettongs and Rat-kangaroos) has been hard hit. Of nine species two are extinct and another two only survive on islands. This group of small marsupials was probably in decline prior to European settlement (Strahan 1982). However, their rapid reduction in range was the result of vegetation clearance and the introduction of predators and competitors.

Other small-sized marsupials were also vulnerable. Of the nine bandicoots (Family Peramelidae) two species are extinct and one persists on islands only. The family Macropodidae contains some small kangaroos and of the five hare-wallabies two are extinct and two others rare. Two families have been particularly affected, the Thylacomyidae had only two species and one, the Lesser Bilby, is probably extinct. The Thylacinidae was a mono-specific family and with the death of the last Tasmanian Tiger, the family also died out. In all cases of mammalian extinctions and the equally alarming high proportion of endangered species, land-use changes including the introduction of mammalian predators and competitors are most often listed as the causal factors.

The other 10 mammalian extinctions were from the native rodents, again small sized mammals vulnerable to loss of cover and to predation and competition. Of the introduced mammal fauna three species were from the order Rodentia, the house mouse, ship rat and black rat. This was one case where introductions were from an order already represented in the Australian fauna. The introduced cat, dog and fox (Order Carnivora) join the dingo. The other mammalian introductions are from three orders not previously found here: the Lagomorpha (rabbit and hare), the Perissodactyla (horse and donkey) and the Artiodactyla (pig, camel, deer, water buffalo and goat). The spectacular success of the European rabbit in Australia was not experienced in North America which already had several closely related lagomorph species and their predators and diseases (Redhead *et al.* 1991). The latter two orders are hard-hoofed browsers and grazers, a mammalian trophic type not previously found here.

By and large, the mammalian biota has suffered proportionally the greatest disruption. The 10 extinct marsupials (from a total of c. 140 species) are matched by a further 65 species that are currently threatened. Similarly the 10 extinct rodents (from a total of c. 140 species also) are matched by 54 threatened species (Kennedy 1990). Importantly, all were small-sized. The introduced mammalian fauna comprises all placental mammals which are split between some small rodents, the rabbit, fox, cat and dog and the much larger browsers.

INTRODUCTIONS AFFECTING ECOSYSTEM FUNCTIONING

Researchers in South Africa have coined the term "transformer" species for invasive plant species that become dominant and transform aspects of ecosystem functioning. Transformer species "transform habitats or landscapes, i.e., change the character, condition, form or nature of a natural ecosystem over a substantial area" (Wells *et al.* 1986). Vitousek (1992) suggested three main ways in which such species are effective: those that alter resource levels such as nitrogen-fixing plants, those that alter trophic structure particularly by adding a new predator, and those that alter the disturbance regime as some introduced grasses can alter the fire regime.

Some recent plant introductions in Australia are clearly causing changes in ecosystem functioning. Athel Pine or Salt-Cedar *Tamarix aphylla*, introduced from Eurasia as a shade tree, has spread along the Finke River system in central Australia, changing siltation patterns, soil salinity, wildlife habitats and displacing the native River Red Gum *Eucalyptus camaldulensis*. It has the potential to spread throughout arid Australian river systems (Griffin *et al.* 1989; Humphries *et al.* 1991). Congeners of this invasive tree have also spread over the arid zone of North America with severe impacts on flora, wildlife and fluvial geomorphology (Loope *et al.* 1988), it also lowers the water table, sucking water holes dry, with disastrous effects on wildlife (Johnson 1986). Tamarix is also a nitrogen-fixer and alters nutrient cycling and status of soils.

Prickly Acacia *Acacia nilotica*, a highly invasive small tree from the Middle East, is spreading uncontrolled across the indigenous grasslands (*Astrebla* spp.) in central Queensland, northern New South Wales and is in the Barkly Tablelands of the Northern Territory. When canopy cover of the tree exceeds 50 per cent it excludes the perennial native grasses which are replaced by annual grasses and bare soil. It is considered a threat to this biome continent-wide (Humphries *et al.* 1991).

Mesquite *Prosopis* spp. is a seriously invasive tree that has a very deep and extensive root system and competes strongly with native species for available water. At present it is restricted to parts of Queensland, western New South Wales and Western Australia, but could spread through much of inland Australia (Humphries *et al.* 1991). In the United States where it is native, Mesquite has expanded its range to cover over 28 million hectares. Mesquite is also invasive of arid communities in southern Africa (Brown and Gubb 1986).

These three plant examples illustrate the landscape transforming ability of single species. They are all also current invasions, still expanding and affecting ecosystems. Some of the vertebrate species such as the European rabbit, first naturalized over a century ago, have probably equally impacted at the broad landscape level but have now reached their limits of distribution.

A CONTINUING PROBLEM

Vitousek (1988) has considered the special case of oceanic islands and species invasions. Australia can be viewed as the biggest oceanic island, and what has been observed on islands elsewhere may occur here, but more slowly. Such a prognosis is a worrying one. For example, before the arrival of the Polynesians the islands of Hawaii supported a vascular flora of almost 1 800 species (almost all of which were endemic). More than 4 600 species of introduced vascular plants are now known to occur in the islands, of which at least 700 species are naturalized, a 40 per cent exotic component. The corollary has been the extinction of 200 (11%)

endemic species and endangerment of another 800 (44%) (Vitousek 1988). The current situation in the Australian flora is not as extreme, but of a vascular flora of possibly 20 000 species (Barlow 1981), there have been 1 400 (6%) naturalized species in the past two centuries and possibly 100 species (4%) have become extinct in the same time-frame. Many more species of plants have been introduced, as garden plants, crops, etc., but are not as yet naturalized in Australia. One alarming prospect is that with postulated climate change, some of these will be triggered to become invasive.

THE CONTINUING WAVE

Invasions continue. In June 1993 news items appeared featuring "starfish eating their way to local waters" (*Sydney Morning Herald* 25.6.93). The species in question is the Pacific Seastar *Asterias amurensis,* currently introduced to Tasmanian waters but with unconfirmed sightings in Victoria. Its normal prey are bivalves, so its northward spread could endanger the oyster industry. As simplified by Fox and Adamson (1986), this invader has the basic biological advantage of all invading species, "it has no known predator" (L. Turner, Tasmanian Museum). This was an accidental introduction.

Some introductions are intentional attempts at the biological control of species already established in pest proportions. This source of new species is significant, in 1989–90 for example, one leaf-eating beetle, two parasitic wasps (plus one that failed to establish), two parasitic nematodes, a sap-sucking bug, a stem-boring weevil, a leaf-eating weevil, and two strains of a bacterium were released from quarantine (Castles 1992). Although selected to be very host-specific, these are nonetheless introduced species.

MANAGEMENT OPTIONS

To turn the tide of invasions is perhaps a futile goal at the continental scale, however, for those areas that we have set aside for conservation purposes and other areas that we wish to retain in a "natural" condition, all or some combination of the following steps are necessary:

1. Remove the exotic species.

This may seem an unrealistic goal but it is achievable depending on degree of infestation and funds available for control. Ecologically desirable results may be achieved with reduced population of the pest species.

This can be achieved by a variety of means.

Physical control is exemplified by the bush management practices now in operation in much urban bushland. Warren ripping and shooting rabbits are also examples of physical agents of control.

Chemical control involves the use of herbicides and pesticides. The application of chemicals is costly; it has been estimated that the total costs of pesticides each year is about 3 per cent of the value of the agricultural produce (Anderson 1986). The ecological cost of mis-use of chemicals in the environment is substantial.

Biological control involves the introduction of a natural predator, parasite or disease organism, strict selection and quarantine procedures are necessary. Another form of biological control involves the use of pathogenic nematodes which invade pest insects and destroy them.

Reproductive control may employ pheromones to attract pests to sterile mates, hormones are being investigated to interfere in the reproductive cycle of some pests.

Plant breeding which incorporates resistance genes into local pasture and crop plants has had remarkable success in combating introduced diseases on crop plants.

2. Reintroduction of native species.

This is necessary if a dominant transformer species is removed by one or a combination of control measures. For example, coastal dunes cleared of Bitou Bush *Chrysanthemoides monilifera,* need to be stabilized and revegetated with native dune species. An accelerated revegetation with native grass such as *Spinifex sericeus* and nursery stock of shrub species such as *Banksia* spp. and *Leptospermum* spp. helps to re-establish a viable natural community.

3. Maintenance of natural disturbance regimes.

"In general terms, the principal management policy which minimizes invasion of conservation areas is to minimize disturbance of all kinds. Maintenance of a high level of species richness (one of the main conservation objectives) will also increase the community's resistance to invasion. However, there is an interaction between these objectives: the maintenance of high species richness is often achieved as a result of a degree of disturbance that creates habitat mosaics (Fox 1984). A fine balance is then required between the degree and type of disturbance required to maintain a high value for native species richness, and that which may result in successful invasion by introduced species." (Fox and Fox 1986: p. 66).

4. Minimize exogenous disturbance.

Even aggressively successful invasive species may not breach the integrity of undisturbed natural communities. However, where natural communities have been disturbed the outcome is very different. Disturbance leads to the creation of spare resources and underutilized resources

(Fox and Fox 1986). In such situations competition for these resources is biased in favour of the introduced species.

One pervasive form of disturbance is the fragmentation of previously extensive habitat. Such fragmentation leads to the creation of a mosaic of islands of habitat, each of which has a long perimeter across which disturbances can penetrate the remnant. The long-term fate of habitat fragments is the loss of native species and their replacement by exotic species (Soulé 1990; Fox 1990b, 1992).

5. Vigilance

This is not a static situation, changes are still working through the system and new problem species manifest themselves from time to time. It is necessary to consider the ecological requirements of potential pest species first with respect to introduction to a homologous climate. However, with predicted climate change innocuous species already here may prove to be biological timebombs awaiting a trigger. The second key factor is the usually high reproductive success of pest species (Fox and Adamson 1986). This is manifested in many "r" characteristics leading to high fecundity. However, this is enhanced by the absence of normal population control from diseases and predators. These two positive pressures, one for prolific growth and the other for an absence of the usual checks leads to rapid and uncontrolled growth.

Continued vigilance is necessary to prevent the establishment of potentially invasive species from being made readily available. For example, advertisements in the national press for a fast-growing small tree "Powton Sapphire Dragon" *Paulownia kawakami* seems a dubious addition to our horticultural array of plants. The tree is currently propagated from tissue culture only, however, the advertisements for it claim it will reach 8 m in three years and flower profusely. Such rapid vegetative growth and early maturation is very "weedy" in character.

6. Education

The fairly recent community recognition of the value of our natural biological diversity needs to be coupled to a repugnance for the invasive non-natural part of the biota. Public education through various media could assist in both controlling unwanted species and accepting control measures. The widespread acceptance of community initiatives such as Greening Australia and the Landcare programmes is an indication of both interest in and acceptance of the natural landscape. More formal educational avenues are also a growth area, such as the National Trust's bush regeneration programmes.

THE FUTURE

Especially in our conservation estate we must strive for the retention of biological diversity. However, even there we may have to accept mixes of native and exotic species (Bridgewater 1990; Fox 1990b). Future decades will see a reduction or dilution of the uniqueness of the Australian biota. From a starting point, only two centuries ago, of an essentially indigenous biota, a sizeable part has been lost or is in the process of being lost, as landscapes are modified by a range of human agencies. These modified or disturbed landscapes accommodate the naturalized species that now form a part of many Australian ecosystems.

ACKNOWLEDGEMENTS

I thank Tony Auld for considerable patience and an anonymous referee for very constructive criticism which improved the manuscript.

REFERENCES

Adamson, D. A. and Fox, M. D., 1982. Change in Australasian vegetation since European settlement. Pp. 109–46 in *A history of Australasian vegetation* ed by J. M. B. Smith. McGraw-Hill Scientific Publishers: Sydney.

Anderson, T. (ed), 1986. *Weeds pests and plant diseases.* CSIRO: Canberra.

ANZECC, 1991. *Australian and New Zealand environment and conservation council list of endangered vertebrate fauna.* ANPWS: Canberra.

Arthington, A. H. and Mitchell, D. S., 1986. Aquatic invading species. Pp. 34–56 in *Ecology of biological invasions: an Australian perspective* ed by R. H. Groves and J. J. Burdon. Australian Academy of Science: Canberra.

Barber, L., 1980. *The heyday of natural history.* Jonathan Cape: London.

Barlow, B. A., 1981. The Australian flora: its origin and evolution. Pp. 25–76 in *Flora of Australia vol. 1.* Australian Government Publishing Service: Canberra.

BDAC, 1992. *A national strategy for the conservation of Australia's biological diversity.* Biological Diversity Advisory Committee, Department of the Arts, Sport, the Environment and Territories: Canberra.

Bradsen, J., 1992. Biodiversity legislation: species, vegetation, habitat. *Environ. Plan. Law J.* 175–80.

Bridgewater, P. B., 1990. The role of synthetic vegetation in present and future landscapes of Australia. Pp. 129–34 in *Australian ecosystems — 200 years of utilization, degradation and reconstruction* ed by D. A. Saunders, A. J. M. Hopkins and R. A. How. Ecological Society of Australia.

Briggs, J. D. and Leigh, J. H., 1988. *Rare and threatened Australian plants.* Special publication 14. Australian National Parks and Wildlife Service: Canberra.

Brown, C. J. and Gubb, A. A., 1986. Invasive alien organisms in the Namib desert, Upper Karoo and the arid and semi-arid savannas of western southern Africa. Pp. 93–108 in *The ecology and management of biological invasions in southern Africa* ed by I. A. W. Macdonald, F. J. Kruger and A. A. Ferrar. Oxford University Press: Cape Town.

Castles, I. (ed), 1992. *Australia's environment: issues and facts.* Australian Bureau of Statistics: Canberra.

Cogger, H. G., 1975. *Reptiles and amphibians of Australia.* A. H. and A. W. Reed: Sydney.

Commonwealth Scientific and Industrial research Organization (CSIRO), 1990. *Australia's environment and its natural resources, an outlook.* CSIRO: Canberra.

D'Antonio, C. M. and Vitousek, P. M., 1992. Biological invasion by exotic grasses, the grass/fire cycle and global change. *Ann. Rev. Ecol. System.* **23:** 63–88.

di Castri, F., 1990. On invading species and invaded ecosystems: the interplay of historical chance and biological necessity. Pp. 3–16 in *Biological invasions in Europe and the Mediterranean Basin* ed by F. di Castri, A. J. Hansen and M. Debussche. M. Kluwer Academic Publishers: Dordrecht.

di Castri, F., 1991. The biogeography of mediterranean animal invasions. Pp. 439–52 in *Biogeography of Mediterranean invasions* ed by R. H. Groves and F. di Castri. Cambridge University Press: Cambridge.

Fox, B. J., 1984. Small scale patchiness and its influence on our perception of animal species' habitat requirements. Pp. 162–78 in *Survey methods for nature conservation: volume 1* ed by K. Myers, C. R. Margules and I. Musto. CSIRO: Canberra.

Fox, M. D., 1990a. Mediterranean weeds: exchanges of invasive plants between the five Mediterranean regions of the world. Pp. 179–200 in *Biological invasions in Europe and the Mediterranean Basin* ed by F. di Castri, A. J. Hansen and M. Debussche. Kluwer Academic Publishers: Dortrecht.

Fox, M. D., 1990b. Interactions of native and introduced species in new habitats. Pp. 141–47 in *Australian ecosystems: 200 years of utilization, degradation and reconstruction* ed by D. A. Saunders, A. J. M. Hopkins and J. D. Majer. *Proc. Ecol. Soc. Aust.* 16.

Fox, M. D., 1992. The green archipelago — urban bushland islands. Pp. 23–26 in *Urban bushland management* ed by J. Howard and J. Rawlings. Ku-ring-gai Council.

Fox, M. D. and D. Adamson, 1986. The ecology of invasions. Pp. 235–55 in *A natural legacy* (2nd ed) ed by H. Recher, D. Lunney and I. Dunn. Pergamon Press: Sydney.

Fox, M. D. and Fox, B. J., 1986. The susceptibility of natural communities to invasion. Pp. 57–66 in *The ecology of biological invasions* ed by R. H. Groves and J. Burdon. Australian Academy of Science: Canberra.

Griffin, G. F., Stafford-Smith, D. M., Morton, S. R., Allan, G. E., Masters, K. A. and Preece, N., 1989. Status and implications of the invasion of Tamarisk *(Tamarix aphylla)* on the Finke River, Northern Territory, Australia. *J. Environ. Manag.* **29:** 297–315.

Groves, R. H. and Burdon, J. J. (eds), 1986. *Ecology of biological invasions: an Australian perspective.* Australian Academy of Science: Canberra.

Harris, J., 1984. Zoogeography of the Australian freshwater fish fauna. Pp. 211–23 in *Vertebrate zoogeography and evolution in Australasia* ed by M. Archer and G. Clayton. Hesperian Press: Carlisle.

Humphries, S. E., Groves, R. H. and Mitchell, D. S., 1992. Plant invasions of Australian ecosystems, a status review and management directions. In *Plant invasions: the incidence of environmental weeds in Australia, Kowari, 2.* ANPWS: Canberra.

Humphries, S. E., Groves, R. H., Mitchell, D. S., Hallegraeff, G. M. and Clark, J., 1991. *Plant invasions of Australian ecosystems.* Australian National Parks and Wildlife Service: Canberra 134 pp.

Jennings, M. R., 1983. An annotated check list of the amphibians and reptiles of California. *California Fish and Game* **69:** 151–71.

Johnson, S., 1986. Alien plants drain western waters. *Nat. Conserv. News* **36:** 24–25.

Jones, M. M., 1991. Marine organisms transported in ballast water, a review of the Australian scientific position. Australian Bureau of Rural Resources, Bulletin No. 7. Australian Government Publishing Service: Canberra.

Kennedy, M. (ed), 1990. *Australia's endangered species: the extinction dilemma.* Simon Schuster: Brookvale.

Long, J. L., 1981. *Introduced birds of the world.* Reed Pty Ltd: Sydney.

Loope, L. L., Sanchez, P. G., Tarr, P. W., Loope, W. L. and Anderson, R. L., 1988. Biological invasions of arid land nature reserves. *Biol. Cons.* **44:** 95–118.

Morton, S. R., 1990. The impact of European settlement on the vertebrate animals of arid Australia: a conceptual model. Pp. 201–13 in *Australian ecosystems: 200 years of utilization, degradation and reconstruction* ed by D. A. Saunders, A. J. M. Hopkins and J. D. Majer. *Proc. Ecol. Soc. Aust.* **16.**

Myers, K., 1986. Introduced vertebrates in Australia, with emphasis on the mammals. Pp. 120–36 in *The ecology of biological invasions* ed by R. H. Groves and J. Burdon. Australian Academy of Science: Canberra.

Newsome, A. E. and Noble, I. R., 1986. Ecological and physiological characters of invading species. Pp. 1–20 in *The ecology of biological invasions* ed by R. H. Groves and J. Burdon. Australian Academy of Science: Canberra.

O'Brien, P. H., 1990. Managing Australian wildlife. *Search* **21**(1): 24–27.

Pollard, D. A. (ed), 1990. Introduced and translocated fishes and their ecological effects. Australian Society for Fish Biology Workshop, Proceedings No. 8, BRR. Australian Government Publishing Service: Canberra.

Pollard, D. A. and Hutchings, P. A., 1990. A review of exotic marine organisms introduced to the Australian region. I. *Fishers Asian Fisheries Sci.* **3:** 205–21.

Redhead, T. D., Singleton, G. R., Myers, K. and Coman, B. J., 1991. Mammals introduced to southern Australia. Pp. 293–308 in *Biogeography of Mediterranean invasions* ed by R. H. Groves and F. di Castri. Cambridge University Press: Cambridge.

Soulé, M. E., 1990. The onslaught of alien species, and other challenges in the coming decades. *Conserv. Biol.* **4:** 233–39.

Strahan, R. (ed), 1982. *Complete book of Australian mammals.* Angus and Robertson: Sydney.

Strahan, R. (ed), 1988. Preface. Pp. xi–xii in *Complete book of Australian mammals.* Cornstalk Publishing: Australia.

Vitousek, P. M., 1988. Diversity and biological invasions of oceanic islands. Pp. 181–89 in *Biodiversity* ed by E. O. Wilson. National Academy Press: Washington.

Vitousek, P. M., 1992. Global environmental change: an introduction. *Ann. Rev. Ecol. System.* **23:** 1–14.

Wells, M. J., Poynton, R. J., Balsinhas, A. A., Musil, K. J., Joffe, H., van Hoepen, E. and Abbott, S. K., 1986. The history of introduction of invasive alien plants to southern Africa. Pp. 21–35 in *The ecology and management of biological invasions in southern Africa* ed by I. A. W. Macdonald, F. J. Kruger and A. A. Ferrar. Oxford University Press: Cape Town.

Williams, G., 1990. Invertebrate conservation. Pp. 135–43 in *Australia's endangered species* ed by M. Kennedy. Simon Schuster: Australia.

The threat of environmental weeds to biodiversity in Australia: a search for solutions

R. J. ADAIR[1]

Environmental weeds are alien plants that invade natural ecosystems. Many pose a serious threat to the maintenance and enhancement of biodiversity and may also degrade cultural, economic and recreational values of native vegetation. Managing the impact of environmental weed invasions requires a combination of preventative and ameliorative approaches, both at the local and national level. Reforms to Australia's quarantine legislation are urgently required to provide stricter control on the importation of plant material and its evaluation for weed potential. In cases where more stringent quarantine legislation may cause a conflict of interest, clearly defined procedures are needed to facilitate decisions which are in the national interest. More flexible and uniform state noxious weed legislation's to regulate the movement, propagation and sale of selected environmental weed species are required to assist the control of established species. The early detection of new weed invasions and the development of rapid response plans are vital for successful eradication programmes. Factors governing the invasion of environmental weeds in Australian ecosystems are not well understood but disturbance appears to be an important precurser for many species. Identifying and managing disturbance regimes that induce weed invasions can be a useful tool for reducing the impact of some species. Other managment options for environmental weeds include: biological control for large scale or inaccessible infestations; chemical control where low levels of non-target damage can be achieved; fire for susceptible species, and mechanical control (slashing, hand pulling) for small or specific infestations. The reduction of environmental weed infestations to levels that are ecologically acceptable will generally require the integration of a range of control methods, particularly classical biological control, and the planned restoration of treated sites with appropriate native vegetation. Detailed technical information on control techniques is available for some environmental weeds, but most lack specific control data. The development and implementation of environmental weed control in Australia cannot advance without recognition of the problem within federal, state and local governments and, importantly, the general community. An increased awareness of the ecological, economic, and social consequences of environmental weed invasions is required before remedial and preventative action can be effectively undertaken. Encouraging the support and participation of the broader community in environmental weed control is paramount to reducing the impact of these weeds now and in the future.

INTRODUCTION

WEED invasions in natural ecosystems, have until recently, received little attention, although their impact on the structure, function and composition of native plant and animal communities has probably been profound (Humphries *et al.* 1991; Carr *et al.* 1992; Carr 1993). Weeds of native vegetation are generally referred to as "environmental weeds", a term originally used by Amor and Stevens (1975) and defined by Holzner and Numata (1982) and Carr (1993). Recent analyses of environmental weed invasions on off-shore islands (Keighery 1993), in Tasmania (Young 1993) and northern Australia (Swarbrick 1993; Thompson 1993; Cowie and Werner 1993) represent an increasing concern of the ecological impact of this group of weeds.

Few, if any, Australian ecosystems appear to be immune to invasion by environmental weeds (Groves 1991; Fox and Fox 1986; Carr 1988). In southern Australia, multiple species invasions are the predominant pattern and are a reflection of habitat fragmentation from intense development, steep climatic and topographic gradients (Humphries *et al.* 1991) and high levels of exposure to invasive propagules. In northern Australia, large-scale invasions by single species reflect more gradual climatic gradients, extensive rather than intensive land-use operations (Humphries *et al.* 1991) and a more restricted exposure to potentially invasive species because of low human population densities.

Quantitative studies on the impact of environmental weeds provide a sound basis for the ranking of invasive taxa and facilitates resource allocation into the development and implementation of appropriate management strategies. Such studies are rare, however. Evaluation of the impact of the Giant Sensitive Plant *Mimosa pigra* on wetlands in the Northern Territory (Braithwaite *et al.* 1989) has facilitated a biological control programme for this weed as have studies on Boneseed and Bitou Bush *Chrysanthemoides monilifera* (Paterson and Volframs 1976; Weiss and Noble 1984a, 1984b; Weiss 1986) and English Broom *Cytisus scoparius* (Waterhouse 1986), weeds of coastal and highland areas of eastern Australia. Similarly, impact studies on Athel *Tamarisk aphylla* (Griffin *et al.* 1989) have identified the potential damage this species may cause to inland river systems. Ecological studies on the pattern and process of invasion by the woody weed Sweet Pittosporum *Pittosporum undulatum* (Gleadow 1982; Gleadow and Ashton 1981; Gleadow and Rowan 1982; Gleadow *et al.* 1983) and the annual grass, Quaking Grass *Briza maxima* (Rayner 1989) have provided important

[1]Department of Conservation and Natural Resources, Keith Turnbull Research Institute, P.O. Box 48, Frankston, Victoria, Australia 3199.

Pages 184–201 *in* CONSERVING BIODIVERSITY: THREATS AND SOLUTIONS, ed by R. A. Bradstock, T. D. Auld, D. A. Keith, R. T. Kingsford, D. Lunney and D. P. Sivertsen. Surrey Beatty & Sons. 1995.

clues for the development of control techniques. However, for most environmental weeds, even basic information such as distribution, abundance or control options are non-existent.

While there have been no documented cases of national species extinctions attributed solely to the invasion of weeds, Leigh and Briggs (1992) list weeds as a major cause in the extinction of four plant species: *Gentiana baeuerlenii* (Gentianaceae), *Hydatella leptogyne* (Hydatellaceae), *Hypsela sessiliflora* (Lobeliaceae), *Trachymene scapigera* (Apiaceae). The fragmentation and disintegration of native vegetation by environmental weed invasions has lead to range contractions, reduced abundance and a decline in the diversity of native biota. In many areas species extinctions have occurred at the local and regional levels (e.g., Kirkpatrick 1974). Several authors have demonstrated a negative correlation between weed invasion and native species diversity (McIntyre 1993; Braithwaite *et al.* 1989) or the cover of an ecological analogue (Weiss and Noble 1984a). The inevitable expansion of the current weed flora, and the arrival of new environmental weeds in the future, are likely to increase the rate of species extinction through direct competition and disruption to essential ecosystem functions and structure (fire, nutrient cycling, hydrological cycles, geomorphological processes). Leigh and Briggs (1992) list 57 plant species that are nationally endangered by competition with exotic plants, while in Victoria 166 native plant taxa are listed as threatened by environmental weeds (Carr *et al.* 1992). Vegetation formations are also at risk. The vigorous Madagascan liana, Rubber Vine *Cryptostegia grandiflora* threatens the species-rich native vine thickets, gallery forests and dry rainforest in the monsoonal belt of northern Australia (Humphries *et al.* 1991); while Prickly Acacia *Acacia nilotica* and Buffel Grass *Cenchrus ciliaris* threaten plant and animal communities in central Australia (Griffin 1993; Humphries *et al.* 1991). In southern Australia, particularly in dry coastal communities, the South African geophyte, Bridal Creeper *Myrsiphyllum asparagoides* smothers understorey vegetation and may lead to substantial decline in biodiversity.

Environmental weeds can have a range of effects on the survival prospects of vertebrate fauna by altering the availability of food, nest sites, cover and protection from predators (Brown *et al.* 1991; Loyn and French 1991). On tidal mud flats in south-east Australia, the maritime grasses, Cord Grass *Spartina* spp. threaten internationally significant wading bird communities (Lane 1992) due to siltation and eventual displacement of critical feeding areas. Similarly, the long-term survival prospects of the Greater Stick-nest Rat may be under threat by the invasion of Ice Plant *Mesembryanthemum crystallinum*, a South African annual, which forms part of the rats' diet, but may displace other plant species important as a source of food and cover (Copley 1988; Brown *et al.* 1991). In the Dandenong Ranges National Park, Melbourne, English Ivy *Hedera helix* invades the ground stratum of moist sclerophyll forest and has reduced the feeding areas of the Lyrebird and Wombat (Freshwater 1991), contributing to their decline. Weeds can be beneficial for some species of wildlife, such as, the Red-tailed Black Cockatoo *Calyptorhynchus magnificus* which has shifted its dietary and habitat preferences to feed almost exclusively on Spiny Emex *Emex australis*, a South African annual (Saunders *et al.* 1985).

Several national environmental management strategies or programmes have recognized environmental weed invasions as a threatening process to either the conservation of biological diversity (BDAC 1992; ESAC 1992; AWC 1993; DHAE 1984) or the maintenance of ecologically sustainable development (ESDSC 1992; COA 1991) and, in some cases, have made policy recommendations addressing biological invasions. Humphries *et al.* (1991) and Carr *et al.* (1992) provide comprehensive and practical proposals for alleviating the threat to nature conservation posed by environmental weeds. The aim of this review is to examine current approaches used for the management or prevention of plant invasions in native vegetation and, where possible, suggest where improved techniques could be adopted.

QUARANTINE AND NOXIOUS WEED LEGISLATION

Preventing the entry of undesirable plants into Australia is the most reliable and cost effective method of weed control available. The *Quarantine Act* 1908 under proclamation 86P prohibits or restricts the entry of 66 species (Schedule 1) and 19 genera (Schedule 2) considered to be a threat to Australian interests and, together with the *Seed Act* 1982, are the only functional national legislation regulating the importation of potential weeds into Australia. Seventy-seven per cent of the species listed under proclamation 86P are already naturalized in Australia.

Under the existing legislation, further additions to the weed flora of Australia are inevitable. Unlisted taxa are not subject to mandatory evaluation for their weed potential and plants known to be risks elsewhere have legally entered and established in Australia, e.g., *Kochia scoparia* ssp. *densiflora* (Dodd and Moore 1993) or could do so in the future, e.g., *Hieracium* spp. (Groves 1986). Forty-six per cent of Australia's noxious weed flora have been introduced intentionally (Panetta 1993), but this proportion is considerably higher when non-proclaimed agricultural and environmental weeds are included. In Victoria, 59 per cent of environmental weeds (346 species) have been deliberate introductions (Carr 1993) made mostly for horticultural and to a lesser extent agricultural purposes.

The *Quarantine Act* 1908, and importation protocols for plant material, have recently been reviewed to reduce the arrival of new weed taxa, particularly environmental weeds. The Department of Primary Industries through the Australian Quarantine and Inspection Service (AQIS) administer the Act and are addressing its deficiencies by proposing to increase the number of plants subject to importation evaluation to around 1 600 taxa. These are grouped into six categories: (A) Exotic plants not known to be present in Australia, but weeds elsewhere; (B) Exotic weed species present in Australia, but not included in state legislation as noxious plants; (C) Plants considered a danger to humans. Listed on the Customs schedule of prohibited plants; (D) Weedy, spiny, thorny exotic *Acacia* spp.; (E) Weeds already present in Australia and included in state and territory legislation; and (F) Cacti (W. Wright, pers. comm. 1994). Plants not present in these categories are assessed for weed potential using the Australian Weeds Committee system described by Hazard (1988), which scores plants based on their history of weediness, morphology, taxonomy, and reproductive and dispersal characteristics. However, Panetta (1993) argues that biological and ecological attributes are poor indicators of weediness due to the diversity of such systems displayed by weeds and undue emphasis can be placed on these and therefore proposes an alternative risk assessment system based on the reliability of predictive attributes. Weedy behaviour elsewhere coupled with a close bioclimatic match between a species native range and Australian conditions, indicators of weediness for plants from southern Africa (Scott and Panetta 1993), form the primary criteria for rejection for importation. Species are then assessed for the presence of noxious characteristics, weedy congeners and multiple modes of reproduction and dispersal. As nearly all Australian noxious weeds exhibit features in at least one of these groups, Panetta's (1993) system has a high detection level for this group of plants. The same system, however, is less reliable for non-noxious weeds and failed to detect 25 environmental weeds listed by Humphries *et al.* (1991) (Table 1) which lack features in each of the four groups listed above (assuming "weediness elsewhere" does not include species listed merely as naturalized). Further screening criteria would therefore enhance the sensitivity of this system to detect potential environmental weeds.

Evidence of weediness elsewhere is a key feature in weed risk assessment systems utilized or proposed for Australia. Its value as an indicator of weed potential in agricultural areas is demonstrated by Scott and Panetta (1993) and supported by the observation that numerous invasive native plants in Australia are serious environmental weeds in other countries (e.g., *Acacia longifolia, A. mearnsii, Paraserianthes lophantha, Hakea sericea, Leptospermum laevigatum* in South Africa, *Pittosporum undulatum* in Jamaica). In agricultural systems, documentation of weed invasions has long been practised and considerable information on weed distribution, even at the global scale, are available (e.g., Holm *et al.* 1979). By comparison, environmental weed census are mostly not available or poorly documented. The effectiveness of risk assessment systems that incorporate a history of weediness as an indicator of invasion potential will be considerably handicapped without the construction of a global database of environmental weeds. Such a task may fall within the charter of international conservation organizations such as UNESCO or the World Wide Fund for Nature.

Panetta (1993) identifies two important noxious characteristics relevant to the success of a number of environmental weeds: a scrambling or climbing growth form and competitiveness. Vines and scrambling plants, although not disproportionally represented in the Australian weed flora, contain taxa that can result in the rapid and complete collapse of native vegetation communities (e.g., *Cryptostegia grandiflora, Thunbergia grandiflora, Anredera cordifolia*) by over-topping the dominant canopy layer, a predominantly tropical to warm temperate phenomena, or by smothering the understorey and ground stratum, thereby preventing the regeneration of indigenous species (e.g., *Tradescantia albiflora, Hedera helix, Myrsiphyllum asparagoides*). Panetta's listing of "competitive" as a noxious characteristic recognizes a fundamental process underlying the invasion of weeds into both natural and agricultural ecosystems. Weeds are successful invaders due to their ability to compete more effectively for limited resources than more desirable species occupying the same niche or habitat. The process of invasion therefore involves the interaction of attributes inherent in the invading species with the biotic and abiotic components of the host habitat. Such interactions are complex and difficult to define and present substantial difficulties in predicting the weed status of foreign plants. Attributes that may contribute to a weeds ability to compete effectively for limited resources are listed by Baker (1965), Bazzaz (1986), Newsome and Noble (1986) and Noble (1989). In native vegetation, growth-related features that enable a plant to dominate habitat space, and reproductive strategies that enable the production of large quantities of viable and readily dispersed propagules are common to Australian environmental weeds.

Table 1. Environmental weeds in Australia (listed in Humphries *et al.* 1991) not recorded as weedy elsewhere and lacking noxious characteristics, weedy congeners, multiple modes of reproduction; and one predominate means of dispersal.

Family	Species	Common name
Agavaceae	*Phormium tenax*	New Zealand Flax
Annonaceae	*Annona glabra*	Pond Apple
Asteraceae	*Chrysanthemoides monilifera* ssp. *monilifera*	Boneseed
Balsaminaceae	*Impatients* spp.	Balsam
Brassicaceae	*Carrichtera annua*	Wards Weed
Leguminosae	*Cytisus palmensis*	Tagasaste
	Erythrina x sykesii	Coral Tree
	Trifolium spp.	Clover (annual)
Moraceae	*Ficus elastica*	Rubber Tree
Ochnaceae	*Ochna serrulata*	Mickey Mouse Bush
Oleaceae	*Olea europaea*	Olive
Onagraceae	*Oenothera glazioviana*	Evening Primrose
Orchidaceae	*Monadenia bracteata*	South African Orchid
Poaceae	*Ammophila arenaria*	Marram Grass
	Ehrharta calycina	Perennial Veldt Grass
	Hyparrhenia hirta	Tambookie Grass
	Melinis minutiflora	Molasses Grass
	Phalaris aquatica	Toowoomba Canary Grass
Palmae	*Arecastrum romanzofianum*	Cocos Palm
Polygalaceae	*Polygala myrtifolia*	Myrtle-leaf Milkwort
Rubiaceae	*Coffea arabica*	Coffee
	Coprosma repens	Tuapata
	Coprosma robusta	Karamu
Rutaceae	*Murraya exotica*	Murraya
Ulmaceae	*Celtis sinensis*	Chinese Sinensis

Key ecological attributes are listed in Table 2 and could increase the detection rate for environmental weeds if used in conjunction with the screening system suggested by Panetta (1993). Although all listed attributes are not required in one species for its success as an invader, the probability of success is likely to increase with an increase in the number of attributes present. Taxa with none of the listed key attributes are considered to have a negligible weed risk potential and therefore are considered safe for importation; taxa with 1–5 attributes have a low to moderate risk potential and require further evaluation, while those with 6–12 attributes are considered high risk plants and should be rejected for importation.

Many of the attributes contributing to weediness listed in Table 2 are considered valuable features for the planned naturalization of plants with economic or utilitarian value. Establishing broad criteria for the detection of weediness may well hamper the introduction of potentially useful, but non-invasive taxa for horticultural, agricultural or medicinal purposes, a point argued recently by Harrison (1994). In cases where the weed potential of taxa proposed for importation require evaluation, assessment under quarantine conditions would enable the documentation of essential characteristics such as life history features, behaviour under field conditions and competitive ability with relevant crop plants and ecological analogues. Logistic difficulties would occur with such an assessment process, particularly for an evaluation of the invasion potential of native vegetation where consideration of the diverse habitat types and management practices present in Australia, need to be taken into account. Given the serious impact of environmental weeds and the weakness of existing screening procedures, such an evaluation process, which in most cases would be expensive, seems justified. By placing the costs of weed risk evaluation on prospective importers, fewer importations and a more critical selection of plants by the proponent is likely to occur. However, illegal importations may also increase, a factor which AQIS would need to prepare for.

Table 2. Characteristics of plants which indicate a potential for weediness in native ecosystems.

High output of viable propagules
Short (<2 yr) development time to reproductive maturity
Seeds or other reproductive units with prolonged (>5 yr) periods of dormancy
High rate of aerial or subterranean biomass production, particularly under conditions of low light, water or nutrient availability
Dense and spreading foliage canopy
Efficient long distance (>1 km) dispersal capabilities
Presence of interspecific alleopathic properties or absence of intraspecific alleopathic properties
Presence of specialized vegetative food storage organs
Successful colonizer of disturbed or bare ground
Reproductive strategies that facilitate survival in fire prone environments
Broad distribution over a range of distinct climate types
Low susceptibility to attack by phytophagous organisms

Conflicts are likely to arise where plant introductions with high potential value to rural industries may have an undesirable impact on native ecosystems. At this stage an informed assessment of the proposed introduction needs to be considered in the context of the national interest. This process is well developed for the importation and release of biological control agents for the control of noxious plants and animals. A similar, but streamlined approach is required to evaluate and regulate the importation of plant material into Australia to reduce the acquisition of new weeds.

EARLY DETECTION AND RAPID RESPONSE PLANS

Where new weeds penetrate quarantine barriers the eradication of infestations is likely to be the most cost-effective long-term control option, but the early detection of weed immigrants is critical for the success of such programmes. Where containment of more established weed populations are required, the early detection and eradication of satellite infestations should be the highest priority (Moody and Mack 1988). Assessment, control and rehabilitation of the main infestations should then proceed as the resources become available. As not all weeds present the same degree of threat to agricultural or conservation values, the efficiency of early detection efforts may increase by focusing efforts on priority weeds. For weeds already present in Australia, empirically based impact evaluations enable priority weed lists to be readily prepared; for weeds not yet present, priority ranking based on predictive evaluation is required. Michael (1989) compiled a list of 23 serious tropical weed species present in Asia, but not known in Australia. Nine of these are now targeted for early detection in the Northern Australian Quarantine Strategy in anticipation of their arrival. Similar proactive weed detection efforts are required for weeds of importance from other biogeographic regions, such as those of South African origin listed by Scott and Panetta (1993). The shrub Heather *Calluna vulgaris* and the vine Old Man Beard *Clematis vitalba*, both serious environmental weeds in New Zealand (Timmins and Williams 1987), are presently rare in Australia, but have high weed potentials and should be targeted for eradication in temperate Australia. The widespread publication of literature aiding the identification of priority weeds would greatly enhance the prospects of early detection.

A rapid response to new or potential weed invasions increases the likelihood of achieving eradication. Early control efforts for *Acacia karroo* (Scott 1991), Ivy Gourd *Coccinea grandis* and *Kochia scoparia* (Dodd and Moore 1993) in Western Australia, Parthenium Weed *Parthenium hysterophorus* in Northern Territory and Taurian Thistle *Onopordum tauricum* in Victoria, have or are likely to result in the eradication of these weeds. In contrast, the slow and uncoordinated response of authorities to the invasion of Nodding Thistle *Carduus nutans* in New South Wales (Medd 1987) largely prevented the eradication of this species. A similar scenario seems inevitable for Pampas Grass *Cortaderia* spp. in Victoria.

Weed invasions often encompass areas under both public and private ownership. Control campaigns therefore require the support and co-ordinated integration of both government and community control efforts. Where there is a clear identification and acceptance of roles and responsibilities of all parties participating in weed eradication or containment programmes the prospects of success are greatly enhanced. While local and state government weed control guidelines are mostly well defined, stronger interstate and inter-agency co-ordination is required to develop and implement an effective national weed management programme, particularly for the destruction or containment of newly arrived weed species. The development of complementary state and commonwealth weed legislation with the ability to list environmental weeds and restrict their sale, intentional propagation or use within relevant climatic zones rather than within political boundaries (state borders) would make a significant contribution to the control of environmental weeds at a national level. The need for national co-ordination of weed management is recognized in the National Weeds Strategy (AWC 1993).

MONITORING

Monitoring naturalized alien plants is important for the early detection of population expansions. As predicting problematic weeds is difficult, all foreign plants in native vegetation, especially in nature reserves, should be subject to monitoring, particularly with the anticipated change in weed distribution and impact associated with global warming forecasts (Fox 1991; McIntyre 1990). Information on the distribution and rate of spread of weed populations may assist with the development of control priorities and resource allocation. To this effect, a number of state-based weed information systems are in operation (Cropper 1991; Fenn 1988; Lane *et al.* 1989; Schultz 1987; Swarbrick and Skarratt 1994) providing biological and management information. There is also a clear need for a national programme for monitoring weeds to ensure the rapid assimilation of weed activity data across all major weed management authorities. Ready access to a national weed database would facilitate early action against weed invasions. The integration of a national weed database with GIS (Geographic Information System) which could enable the overlaying of susceptible land-use areas, vegetation types and bioclimatic areas would enhance the development of proactive weed extension and detection efforts.

DISTURBANCE AND WEED INVASION

Numerous authors have demonstrated a relationship between disturbance and weed invasion both in Australia (Fox and Fox 1986; Amor and Piggin 1977; Hobbs 1989) and elsewhere (Crawley 1987; Hobbs and Huenneke 1992; Kruger *et al.* 1986; Rejmanek 1989; Lepart and Debussche 1991). Disturbance enhances weed invasion primarily by affecting the availability of suitable microsites and in some cases the availability of invasive propagules (Hobbs and Huenneke 1992). As almost all of Australia's most serious environmental weeds are either promoted or dispersed by habitat disturbance (Humphries *et al.* 1991), understanding the ecology of disturbance-related weed invasions may lead to the successful management of these weeds.

Various definitions of disturbance relevant to biotic communities have been proposed (Petraitis *et al.* 1989; Pickett *et al.* 1989; Rykiel 1985). Pickett and White's (1985) definition of "any relatively discrete event in time that disrupts ecosystem, community or population structure and changes resources, substrate availability, or the physical environment" is adopted here.

Fox and Fox (1986) distinguish between endogenous disturbance events which occurred through evolutionary time (e.g., fire, drought, fossorial activities by native animals) and exogenous disturbance events, which invariably are human-mediated. However, in natural systems where disturbance regimes are poorly defined or understood distinguishing between these two forms of disturbance is difficult, and moreover, may be unimportant due to similarities between the effects of natural and human induced disturbance events (Bazzaz 1983).

All vegetation formations are prone to various forms of disturbance (Fox and Fox 1986). When invasive species are available, incursion into the host community appears to occur through changes in the availability of limiting resources (Fox and Fox 1986; Hobbs 1989, 1991; Pickett and White 1985) and the disruption of species interactions resulting in reduced competition or predation (Kruger *et al.* 1986; Crawley 1986; Fox and Fox 1986).

Vegetation exposed to multiple forms of disturbance (Hobbs and Mooney 1985, 1991), or intense disturbance (Crawley 1987; Fox and Fox 1986) are most likely to be subject to weed invasion, although plant communities that are naturally subject to high levels of endogenous disturbance events may be less prone to the recruitment of alien plants (McIntyre *et al.* 1988). Riparian vegetation formations are highly susceptible to weed invasion (Humphries *et al.* 1991; Carr *et al.* 1992; Gullan 1988) due to frequent and multiple disturbance events. Similar levels of disturbance occur on the relatively fertile soils supporting lowland grasslands and grassy woodlands in the agro-pastoral belt of southern Australia where a long history of grazing by domestic stock and pastoral improvement programmes have caused soil disturbance, nutrient enrichment and vegetation depletion, rendering them prone to weed invasion. In Victoria, these vegetation formations have the highest number of naturalized taxa (Gullan 1988; Carr *et al.* 1992).

Although disturbance in native vegetation is generally recognized as a precursor to weed invasion, there are instances where disturbance causes little or no increase in weed invasion, such as fire in Western Australian heathland (Hester and Hobbs 1992) and spring fire (cf. autumn fire) in *Banksia* woodland (Hobbs and Atkins 1990). Conversely, weed invasions are reported to occur in the apparent absence of disturbance (Carr *et al.* 1992; Carr 1993) and these authors cite the invasion of Engligh Ivy *Hedera helix* in tall-open forest in the Otway Ranges (Victoria) and Himalayan Honeysuckle *Leycestaria formosa* at Mount Buffalo (Victoria) as examples. However, historical disturbance events (timber harvesting) and subtle endogenous disturbance processes may have been important for the invasion of these species.

As all Australian plant communities are subject to disturbance, formal testing of the Fox and Fox (1986) hypothesis that there is no invasion without disturbance is fraught with difficulties. Measuring the impact of exogenous disturbance regimes on weed invasion, however, is feasible and has been undertaken using an experimental approach by Hobbs and Atkins (1988). As Hobbs (1991) suggests, the more relevant question, from a practical point of view, is which disturbances are most likely to lead to weed invasions. The identification and treatment of disturbance types that clearly enhance weed invasion must be a priority in the management of native vegetation. In relation to this point, two examples have been taken where changes in resouce levels created by disturbance can result in weed invasion.

Within the closed canopy of rainforest vegetation light intensity is an important factor governing plant growth. The creation of canopy gaps and subsequent increase in light availability is an integral part of forest regeneration and occurs naturally through endogenous processes, such as tree fall. The succession from disturbance pioneers through to late phase species can however be deflected or arrested by the establishment of invasive gap colonists (Floyd 1988; Summerbell 1991) such the shrub *Lantana camara* and particularly the alien canopy vines *Anredera cordifolia, Thunbergia grandiflora, Macfadyena unguis-cati* and *Cardiospermum halicacabum*. From a weed management perspective, disturbance events that create large or permanent canopy openings (tracks, roads, clearings), thus allowing light into an otherwise light-limiting environment, should be avoided. Where revegetation of disturbed sites is required, rapid restoration of an overhead canopy using clumped mixtures of pioneer, secondary colonists and mature phase species (Kooyman 1988) which reduce light intensities will disadvantage the establishment and growth of persistent invaders. Pioneer weeds unable to grow under shade e.g., Wild Tobacco *Solanum mauritianum* can be used as nurse crop to provide critical protection to developing rainforest seedlings (Floyd 1988; Stockard 1988).

Nutrient enrichment of aquatic and terrestrial environments is often linked with the invasion of alien plants. Australian heathland and related shrublands occur on acid, nutrient-poor soils, usually sands, that are invariably deficient in phosphorus and nitrogen (Groves 1977). The low nutrient status of heathland soils can act as a biological barrier and inhibit the invasion of weeds. Increasing soil nutrient levels in heathland alters its floristic and structural composition and can lead to the invasion of herbaceous aliens (Connor and Wilson 1968; Specht 1963; Heddle and Specht 1975), particularly when combined with soil disturbance (Hobbs and Atkins 1988). In nutrient poor environments, management practices that prevent the assimilation of foreign nutrient sources may provide a powerful means of reducing the ingress of weeds that are dependent on high fertility levels for their survival or maintaining their competitive superiority over native plants.

The "ecologically-sensitive" management of disturbance in native vegetation is important for both the maintenance of biological diversity (Connell 1978) and the treatment or prevention of weed invasions. Although the link between weed invasion and disturbance has been subject to scientific debate, a greater effort is required for the production of disturbance-orientated extension material, and training and public education programmes for those that manage and use areas of native vegetation.

BIOLOGICAL CONTROL

The introduction of the cochineal insect *Dactylopius ceylonicus* (syn. *Coccus indicus*) to its natural host Drooping Prickly Pear *Opuntia vulgaris* during 1912–15 (White 1913) was the first successful case of classical biological control undertaken in Australia. Since this time, 77 weed species have been targeted for biological control (Julien 1992; Cullen, pers. comm.), the majority of which are weeds of primary importance to agriculture, waterways or are damaging to human health. Although 69 environmental weed species are subject to biological control, only 30 are regarded as having a serious impact on natural ecosystems (Table 3). Seven species (*Chrysanthemoides monilifera, Cytisus scoparius, Mimosa pigra, Myrsiphyllum asparagoides, Protasparagus* spp.) are candidates for biological control primarily for the protection of nature conservation values.

Biological control of environmental weeds has a number of advantages over alternative control techniques, notably target specificity, continuous action, long-term cost effectiveness and self-dispersal, irrespective of terrain. In infestations where alternative control techniques may cause unacceptable non-target damage, or are impractical or ineffective, biological management of environmental weeds offers, in many cases, the only possibility for control. This is particularly so for environmental weeds in remote areas such as northern Australia, where sparse human populations severely restrict the implementation of alternative control techniques. However, weedy taxa that are closely allied to plants of economic importance or to Australian native plants are generally precluded from biological control. In this respect, many weedy Poaceae are not considered suitable biological control targets, although Wapshere (1990) has identified genera, including several serious environmental weeds (*Anthoxanthum, Briza, Cenchus, Cortaderia, Ehrharta, Spartina, Vulpia*) where classical or inundative biological control methods may be suitable.

Mycoherbicide techniques (utilization of fungi) have considerable potential for weed control (Templeton 1987) and where native organisms are utilized offer control prospects for taxa that cannot be targeted for classical biological control. This technique would be constrained by application difficulties in native vegetation, difficulties also experienced with chemical control methods. However, host-specific inundative agents are likely to result in the selective suppression of the target species, an outcome that is rarely achieved by chemical control techniques in native vegetation. Developments in the use of *Colletotrichum gloeosporoides* to control the invasive shrub *Hakea sericea* in South Africa illustrate the potential for this technique (Morris 1991). Lack of sufficient host-specificity by inundative agents may cause a conflict of interest where native plants or species of commercial importance are affected. For example, White Rot disease of Onion

Table 3. Environmental weeds listed in Swarbrick and Skarratt (1994) subject to classical biological control (Adapted from Julien 1992; Parsons and Cuthbertson 1992; J. Cullen, pers. comm. 1993).

Species	Primary ecosystem targeted	Ecological impact	Agents first released	No. agents released	Impact on weed/programme status
Acacia nilotica	A/E	S	1982	2	Reduced seed production in non-grazed areas.
Ageratina adenophora *A. riparia*	A	S	1953	2	Suppression in some areas.
Ambrosia artemisiifolia	A	L	1984	4	Plant vigour and size reduced.
Baccharis halamifolia	A	M	1969	13	Significant damage in some areas.
Carduus pycnocephalus *C. tenuiflorus*	A	L	1993	1	Impact under evaluation.
Carthamus lanatus	A	L	—	—	Preliminary evaluation underway.
Chrysanthemoides monilifera	E	S	1989	4	Reduced flower and seed production in some areas.
Cirsium vulgare	A	L	1990	1	Agent not established.
Chondrilla juncea	A	L	1971	4	Effective control of narrow leaf-strain.
Cryptostegia grandiflora	A/E	S	1988	2	One agent not established, one under evaluation.
Cytisus scoparius	E/A	S	1993	2	Impact under evaluation.
Echium plantagineum	A	M	1980	3	One established. Limited impact in some areas.
Eichornia crassipes	Aq/E	S	1975	4	Effective control in many areas.
Emex spinosa *E. australis*	A	L	1974	2	No control.
Eriocereus spp. (3)	A	L	1974	4	Effective control.
Heliotropium amplexicaule	A	L	—	—	Preliminary evaluation underway.
Heliotropium europaeum	A	L	1979	2	No control.
Hypericum androsaemum	—	S	1991	1	Severe defoliation damage by the fungus *Melampsora hypericorum*, an accidental or illegal introduction.
Hypericum perforatum	A	S	1930	10	Effective control in some open areas. No control in timbered areas.
Lantana camara	A/E	S	1914	24	Effective control in limited areas.
Marrubium vulgare	A/E	M	1994	1	Agent under evaluation.
Mimosa pigra	E/A	S	1983	8	Reduced seed production and foliage damage.
Myrsiphyllum asparagoides	E	S	—	—	Potential agents under evaluation.
Onopordum acaulon	A	L	1992	—	Releases planned after establishment achieved on other *Onopordum* spp.
Opuntia spp. (7)	A	L	1914		Effective control in most areas.
Oxalis pes-caprae	A	M	—	—	Preliminary evaluation underway.
Parkinsonia aculeata	A/E	S	1989	2	Impact under evaluation.
Parthenium hysterophorus	A		1980		Damage occurring but insufficient for control.
Pistia stratiotes	Aq/E	L	1982	1	Successful control.
Protasparagus spp. (3)	E	S	—	—	Potential agents under evaluation.
Prosopis spp. (5)	A/E	S	—	—	Potential agents under evaluation.
Reseda luteola	A	L	—	—	Preliminary investigations underway.
Rubus fruticosus spp. agg. (8)	A/E	S	1991	1	Major impact on susceptible taxa.
Rumex spp. (4)	A	L	1989	1	Agent established, impact under evaluation.
Salvinia molesta	Aq/E	S	1980	2	Effective control in most areas.
Senecio jacobaea	A	L	1930	5	Effective control in some areas.
Senna obtusifolia	A	L	—	—	Survey for agents underway.
Sida acuta	A	L		1	Severe defoliation and reduced seed *S. rhombifolia* production at some release sites.
Silybum marianum	A	L	1988	1	Agent not established.
Tribulus terrestris	A	L	—	—	Preliminary evaluation underway.
Ulex europaeus	A	M	1939	1	Established, but negligible impact.

( ) Number of species in the genus targeted for control.
Ecological Impact: S = serious, M = moderate, L = low.
Primary Ecosystems targeted for biological control: A = Agricultural, E = Natural, Aq = Aquatic.

Sclerotium cepivorum has potential for the long-term control of Angled Onion *Allium triquetum*, an invasive geophyte of southern Australia, but as commercial Onion *Allium cepa* is highly susceptible to this disease (D. Lane, pers. comm.) its use as an inundative biological control agent would be restricted to sites distant from domestic and market gardens. Genetic manipulation of plant pathogens to alter their host range and level of pathogenicity offers numerous and diverse scenarios for biological control of weeds (Sands and Miller 1993) with possibilities for the selective control of weeds in native vegetation. Such technology, however, awaits development and acceptance as a suitable and ecologically safe form of weed control.

Management programmes that integrate biological control with other control techniques offer the best prospects for reducing environmental weed infestations. The ultimate criteria for the successful control of environmental weeds, regardless of the techniques used, must be measured in the level of replacement of the target infestation with other vegetation, or the level of protection provided to uninfested vegetation by a reduction in the rate of spread. Huffaker and Kennett (1959) utilizes such measures for evaluating the impact of control programmes for *Hypericum perforatum* in California. Biological control programmes that fail to reduce the fitness of weed infestations to levels below the "critical ecological threshold" that allows natural regeneration, or fails to significantly reduce dispersal rates cannot be regarded as successful. Where the target weed is replaced by undesirable vegetation with an equal or greater weed status, limited success may be claimed, but the contribution of the programme to the protection or enhancement of biological diversity must be rated as negligible. In this respect, biological control programmes for environmental weeds need be incorporated into management plans that embrace the concept of "ecological control" as defined by Groves (1989) and used to manipulate post-control vegetation succession from weedy to native vegetation. Incorporating biological control methods with existing management practices in native vegetation, particularly the use of fire, is important in this respect (Briese 1994).

Funding biological control programmes for environmental weeds with no tangible economic importance (cf. weeds of agriculture) will, by necessity, come from government sources. Biological control programmes for the environmental weeds *Chrysanthemoides monilifera, Protasparagus* spp. and *Myrsiphyllum asparagoides,* are funded mainly by the Australian and New Zealand Environment and Conservation Council (ANZECC). Financial contributions are received on a pro-rata basis from states and territories with infestations of these weeds and used to undertake foreign surveys for potential agents, host-specificity testing, and the breeding and distribution of agents approved for release. A multi-disciplinary task force of weed scientists, public land managers and entomologists with representatives from the funding sources, supervise implementation of the projects. This co-operative approach has been successful, as the financial burden is spread over a number of agencies, and provides a model for future control programmes for environmental weeds. The high initial capital and financial costs involved in the development of biological control programmes severely constrain the number of environmental weeds subject to this form of control. However, if current expectations and standards of nature conservation in Australia are to be maintained, a greater emphasis on the biological control of environmental weeds is required. To achieve this, further allocation of funds are required either by increasing expenditure, altering spending priorities within public land budgets, or by generating new revenue, perhaps by adopting the user-pays system and increasing the costs of utilizing or visiting natural ecosystems.

In addition to weeds of foreign origin, many Australian native plants are now recognized as invaders of native vegetation. Promoted for their horticultural potential, a large number of Australian native taxa are cultivated outside their natural geographic range. In Western Australia, 33 species of non-indigenous native plants are recorded as naturalized (Keighery 1991; Swarbrick and Skarratt 1994) most of which are likely to have originated from eastern Australia. In Victoria, 112 native species including 30 species from Western Australia are naturalized in native vegetation (Carr 1993). Naturalized native plants that have been introduced across previously effective biogeographic barriers such as the Nullarbor Plain are likely to have been transferred without the phytophagous fauna associated with these plants in their natural habitat. Although polyphagous or oligophagous organisms present in the extended range may attack the introduced host, the likely absence of specialist phytophages offers the possibility that these organisms could be used as biological control agents. However, the likelihood of success with this technique is more uncertain than conventional classical biological control, where foreign organisms are utilized, as the chances of attack by parasitoids or diseases from closely related taxa are higher. Phytophagous fauna surveys in the introduced and natural range of native weeds are required to establish the feasibility of this approach. High levels of host specificity are reported for organisms associated with weedy Australian genera such as seed-feeding insects associated with *Acacia* (Auld 1983; van den Berg 1980, 1982) and *Hakea*

(Neser 1968). Host specificity testing and introduction protocols, similar to those outlined for foreign organisms (Cullen and Delfosse 1985) would ensure that only beneficial introductions were made. Serious environmental weeds potentially suitable for this form of classical biological control are: Blue Bell Creeper *Sollya heterophylla*, Cape Wattle *Paraserianthes lophantha* and Cootamundra Wattle *Acacia baileyana* in eastern Australia; and Coast Tea Tree *Leptospermum laevigatum* and *Pittosporum undulatum* in Western Australia, Tasmania and South Australia. In New South Wales, the native scale insects *Austrotachardia* sp. and *Paratachardina* sp. have killed large areas of *Cassinia* spp., native weeds in some agricultural areas, and are under evaluation as potential biological control agents for these weeds (Campbell *et al.* 1994).

CHEMICAL CONTROL

Herbicide recommendations are generally available for environmental weeds that are problematic in agricultural, industrial or urban environments, e.g., Blackberry *Rubus fruticosus* agg. but few are specifically available for use in native vegetation. Herbicides play a particularly important role in the eradication or control of new weed outbreaks where hand-removal is impracticable. Selective control of environmental weeds in native vegetation is difficult to achieve as herbicides have either a broad-spectrum action or, at best, are specific to particular taxonomic groups of plants. The challenge native vegetation managers face is to select herbicide prescriptions that minimize non-target damage and reduce the extent of gap creation, a common precursor to the invasion of weeds in bushland. Non-target damage can be reduced by the choice of herbicides, season of application, method of application and concentration of active constituent. Herbicide application techniques with high levels of target accuracy such as stem injection and basal bark treatment are highly effective, relatively safe and cause minimal non-target damage, but are labour intensive and generally only suitable for small infestations of woody weeds and lianas. Aerial application techniques provide a useful method of treating environmental weeds in mono-specific stands or dense canopies develop over the upper stratum of native vegetation, such as Rubber Vine (Vitteli 1993), Giant Sensitive Plant (Miller 1988) and Bitou Bush (Toth *et al.* 1993).

In situ trials examining the impact of herbicides on target weeds as well as associated native species are particularly useful, although seldom undertaken. In coastal vegetation of eastern Australia, selective control of Bitou Bush *Chrysanthemoides monilifera* spp. *rotundata* can be achieved by winter rather than summer application of herbicides (Toth *et al.* 1993); and in the Mount Lofty Ranges Boneseed *C. m.* spp. *monilifera* was selectively controlled by dilute applications of glyphosate (Earl 1990). Recent herbicide evaluation trials for the environmental weeds *Agave* (Foley and Bolton 1990), *Cortaderia* spp. (Duckett and Wilkinson 1988), *Leycesteria formosa* (Pritchard 1993), *Ligustrum* (Madden and Swarbrick 1990), *Marrubium vulgare* (McMillan 1991), *Myrsiphyllum* (Pritchard 1991), *Delairea odorata* (Fagg 1989), *Spartina* (Pritchard 1992), *Tradescantia albiflora* (Murphy 1993) and *Thunbergia grandiflora* (van Haaren 1993) have contributed much to developing chemical control options for these species. Herbicide evaluation trials provide data that is required for the inclusion of specific weeds on the manufacturer's label. Considerably more research effort is required to produce similar information for numerous other environmental weeds of concern. The weedy Poaceae, particularly *Cenchrus, Brachiaria, Hymenachne, Pennizetum polystachion* and *Themeda quadrivalvis*, require the development of herbicide regimes that have minimal impact on indigenous grasses and other monocotyledons. Hitchmough's *et al.* (1994) work on temperate grasses indicates a potential for this approach. Herbicide manufacturers are unlikely to undertake efficacy trials on weeds where the financial return on their investment is not profitable. As most environmental weeds are of indeterminate economic value, herbicide control research will mostly have to be undertaken by government laboratories and tertiary education establishments.

Community groups make significant contributions to weed control in native vegetation, but often avoid the use of herbicides because of lack of knowledge, and concerns over personal safety and impact on the environment. In many cases, these impediments could be alleviated by the provision of basic training in the safe use of herbicides. Government or conservation agency sponsored training courses would enhance the weeding capabilities of volunteer groups and would provide a stimulus for a larger range of control programmes to be undertaken.

FIRE

Fire has been implicated as a pre-disposing factor to the invasion of native vegetation by environmental weeds (Fox and Fox 1986; Bridgewater and Blackshall 1981; Baird 1977; Trabaud 1991) and its impact depends on the frequency, intensity, seasonality, stage of vegetation development and the biologies of both native and exotic species present (Christensen and Burrows 1986; Hobbs and Atkins 1990). Baird (1977) attributes the spread of the perennial Tussock Grass *Ehrharta calycina* (South African Veld Grass) in temperate *Eucalyptus* woodland due to frequent burning and the construction of fire breaks. Preventing fire over a 15 year period resulted in a gradual decrease in the number of *Ehrharta* tussocks. In undisturbed and dense native vegetation, invasion by *E. calycina* was resisted. A similar decline occurred with the pioneer shrub, African Daisy *Senecio pterophorus* following the suppression of fire and grazing in the Mount Lofty Range, Adelaide (Groves 1989). Fire patterns may be irreversibly altered by the invasion of exotic plants leading to regimes that may promote further invasion and the suppression of indigenous plants and animals. The introduced perennial grasses, Mission Grass *Pennisetum polystachion* and Buffel Grass *Cenchrus ciliaris* alter fire regimes in northern Australian savannah woodlands and mesic areas of arid Australia respectively, by increasing fuels loads and causing more intense fires later in the season, and in the case of *C. ciliaris* more frequent fires (Humphries *et al.* 1991). Similar impacts are reported with the invasion of *E. calycina* in King's Park, Perth (Wycherley 1984).

Fire has considerable potential for the management of environmental weeds that lack the capability of post-fire coppicing. The fire-sensitive native tree Sweet Pittosporum *Pittosporum undulatum* is naturally confined to coastal and sub-coastal forests of eastern Australia. Fire suppression programmes have facilitated the invasion of *P. undulatum* in vegetation communities outside its natural range where it may substantially alter the composition and structure of native vegetation (Gleadow and Ashton 1981) and form disclimax communities dominated by *P. undulatum.* Seedlings and seed of *P. undulatum* are killed by temperatures exceeding 175°C and 120°C respectively, and low to moderately intense prescribed fire can reduce the density of *P. undulatum* juveniles and soil-stored seed (Narayan 1993). However, moderate to high intensity fire may be necessary to kill larger plants of *P. undulatum* (Gleadow and Ashton 1981) and penetrate dense infestations where fuel moisture levels are relatively high. The invasive coastal shrub Coast Tea Tree *Leptospermum laevigatum* is fire sensitive (Burrell 1981; Judd 1990) and spring burning kills plants at a time when few fruits are stored on the canopy and soil-seed loads are minimal thus providing an opportunity to minimize the rate of spread (Molnar *et al.* 1989). The woody weeds Boneseed *Chrysanthemoides monilifera* spp. *monilifera* and Cluster Pine *Pinus pinaster* are also killed by fire, but extensive seedling regeneration occurs in the post-fire period from soil-stored seed in the case of *C. monilifera* (Lane and Shaw 1978) and from woody canopy-stored fruits in the case of *P. pinaster* (Corbett 1988). Removal of seedlings before they reach sexual maturity, either by hand pulling, herbicide application or subsequent burning, offers a control option for these weeds. A combination of mechanical removal techniques and fire are practised for the control of *P. pinaster* in South Africa (Kruger 1977a) and French Island (Corbett 1991). In temperate grasslands dominated by Kangaroo Grass *Themeda triandra*, burning is recommended to enhance native species diversity and reduce the abundance of alien plants (Robertson 1985; Stuwe and Parsons 1977; Stuwe 1986; McDougall 1989); it may also be critical for the regeneration of a number of endangered plant species (Scarlett and Parsons 1982). Fire can reduce the density of the exotic grasses *Briza maxima* (McMahon 1991), *Cynosurus echinatus* and *Lolium rigidum* (Robertson 1985) and *Lolium perenne* and *Bromus hordaceus* (McDougall 1989), but may also promote a range of exotic monocotyledons, notably *Aira, Vulpia* and *Romulea rosea*, and the dicot *Trifolium subterraneum* (Lunt 1990; McDougall 1989). Despite the widely held view that the season of burning can influence exotic plant communities in *Themeda* grassland and recommendations for annual spring burning in weedy sites (Stuwe 1986; Robertson 1985), no clear evidence is available to suggest that the season of burning has a significant effect on introduced grasses and forbs. This, however is not the case for the introduced geophyte *Watsonia pyramidata* (= *W. borbonica*), a weed of southern Australia, where flowering and subsequent seed set was depressed following spring fires compared to autumn burnt or unburnt plants (Kruger 1977b).

Fire is widely used throughout Australia for the manipulation of vegetation for agricultural or forestry pursuits and fire prevention in areas close to human habitation (Gill *et al.* 1981). We are only just beginning to understand the relationships between fire and weed invasion processes. There is considerable scope for the use of fire in managing weed populations through the manipulation of intensity, frequency and seasonality. In the case of invasive perennial grasses of northern and central Australia, manipulation of fire regimes, particularly the reduction in fire frequency to break the possible self-perpetuating fire-invasion cycle, may offer the only practical means of minimizing the impact of these weeds on a broad scale. In agricultural ecosystems, fire can be used to control a number of weeds but its greatest role is to improve the efficiency of other control measures (Johnson and Purdie 1981). The challenge we face with control of environmental weeds is to develop fire regimes that fit in with the concept of ecological control (Groves 1989) so that weedy invaders are disadvantaged while associated native biota are given a competitive advantage (e.g., Briese 1994; Robertson 1985; Weiss 1983).

ECOLOGICAL CONTROL

Groves (1989) defined the term "ecological control" of weeds as "the planned use of one or several methods of control when integrated with an understanding of the dynamics of the ecosystem in which the plant occurs". This concept is familiar in the management of agro-ecosystems where biological, chemical, and mechanical techniques, crop rotation and fire may be used singly or in combination to suppress weeds while promoting more desirable species. The use of competitive pasture species in combination with biological control agents appears particularly useful for the suppression of pasture weeds (Burdon *et al.* 1981) and have been described for Skeleton Weed *Chondrilla juncea* (Groves and Williams 1975), St John's Wort *Hypericum perforatum* (Campbell and Delfosse 1984) and is recommended for *Mimosa pigra* where infestations have been treated with fire and herbicides (Miller and Lonsdale 1992). In natural areas managed for nature conservation, the options available for weed control are fewer than those in agro-ecosystems due to their complex structural and floristic composition. The re-establishment of competitive indigenous plants is an integral part of successful weed control in native vegetation. In habitat restoration programmes on abandoned pasture in South Australia and Victoria, a combination of chemical and mechanical methods were used to suppress weed growth and encourage establishment of the native grass Kangarro Grass *Themeda triandra*. Spring burning of *Themeda* hay, collected and laid 8–9 months previously, provided seed-bed conditions favourable to *Themeda* establishment. Herbicide treatments prior to burning killed competitive exotics and simultaneously provided additional fuel to carry the fire (McDougall 1989; Stafford 1991). Subsequent mowing reduced weed competition and favoured the development of a dominant canopy of *Themeda* (Stafford 1991). Classical biological control can play an important role in the integrated management of invasive weeds in natural ecosystems by reducing target plant vigour and thereby facilitating the re-establishment of native vegetation. In some cases, biological control offers the only possibility for suppression or containment as alternative techniques are either unavailable or inappropriate for large-scale infestations. For *Chrysanthemoides monilifera*, a biological programme (Scott and Adair 1990, 1994) has established the geometrid shoot-feeding moth *Comostolopsis germana* on Bitou Bush *C. monilifera* ssp. *rotundata* infestations in coastal New South Wales, resulting in reduced flower and fruit production (Holtkamp and Maguire 1993). Lowered reproductive output may slow the rate of spread of *C. m. rotundata* and reduce the annual contribution to the soil-seed bank. This action would enhance the herbicide-burn or slash-burn control strategy outlined by Weiss (1993). Although fire would be detrimental to *Comostolopsis germana* and other canopy dwelling agents, post-fire colonization is likely to be rapid if unburnt patches of *C. germana* infested Bitou Bush are left adjacent to the burnt areas. Burning *Chrysanthemoides* dominated coastal vegetation is also likely to stimulate the germination of Coastal Sallow Wattle *Acacia sophorae*, a native analogue that is displaced by *C. m. rotundata* (Weiss and Noble 1984b). Where soil-seed levels have been depleted, direct seeding of appropriate fire-promoted species prior to burning may facilitate the regeneration of native dominated vegetation and therefore reduce the susceptibility to further weed invasion. Direct seeding rather than hand planting offers the potential for large-scale revegetation, but is generally limited by difficulties in obtaining large quantities of seed. The development of seed orchards of competitive, locally indigenous plants may alleviate this problem. On the Bogong High Plains, Victoria, groomed ski slopes are sown with native snow grass seed (*Poa hothamensis* and *P. hiemata*) collected from wild stands and seed crops cultivated at low altitudes (W. Papst, pers. comm.). The availability of snow-grass seed reduces the need for exotic based revegetation mixes which may contain invasive species such as White Clover *Trifolium repens* and Brown Top Bent *Agrostis capillaris*.

In native vegetation close to urban areas, hand removal of weeds by humans, a little recognized form of biological control, followed by natural or assisted revegetation has contributed much to the conservation of plant and animal communities (Buchanan, this volume). Such activities are mostly undertaken by volunteer community groups who feel compelled to contribute on a practical level to an obvious environmental problem. Community involvement in weed control generates a sense of responsibility for public land and its management. During times of limited government funding for nature conservation, this is critical for the maintenance of biological diversity of many bushland remnants. Greater financial assistance through both private and public sponsorship for weed management activities undertaken by community based groups in schemes such as Landcare (Atkins and Molloy 1993), ParkCare (Ward 1992) and Dune Care (Smith 1992) would contribute much to weed control pursuits in native vegetation. In more remote areas where the human resource is scarce, the opportunities for community involvement in weed control programmes are more limited, but where adequate extension and encouragement is provided, community groups can play an important role in the detection and monitoring of infestations, and the eradication of small weed outbreaks. Aboriginal communities have been involved in such activities (Melville 1992) and can play a vital role in the protection of isolated areas.

SUMMARY AND CONCLUSIONS

Environmental weeds pose a serious threat to biodiversity in areas dedicated to nature conservation. Although control efforts have been directed at many of Australia's most serious invasive plants, few have been adequately contained or eradicated. As new weed introductions occur and established species continue to spread, further modifications to natural ecosystems seems inevitable. The development and implementation of environmental weed control in Australia requires recognition, both within government and the general community, of the ecological, economic and social problems they cause. Management options for environmental weeds include biological control for large inaccessible infestations, chemical control where low levels of non-target damage can be achieved, fire for susceptible species and mechanical techniques for small, accessible infestations. The reduction of environmental weed infestations to levels that are ecologically acceptable will require the integration of a range of control methods and the planned restoration of treated sites with appropriate native vegetation.

Reforms to Australia's quarantine legislation, to provide greater control over the importation of new plant material, and more flexible and uniform noxious weed legislation to regulate the movement, propagation and sale of selected environmental weed species are urgently required. In cases where more stringent legislation may cause conflict, clearly defined procedures are needed to facilitate decisions which are in the national interest. The creation of working groups to develop national environmental weed policies and priorities, and to develop plans for the eradication of newly arrived weeds would be an important step in reducing the threat of environmental weeds in the future.

Increasing funds for research into the use of biological control agents, fire, herbicides and disturbance management would enhance present weed control techniques. Developing integrated weed management methods, following the concept of ecological control (Groves 1989), is particularly important for the suppression of weeds in natural ecosystems. Community interest and participation in areas dedicated to nature conservation has considerable advantages for the preservation of biological diversity, but requires encouragement and training.

ACKNOWLEDGEMENTS

I thank E. Bruzzese, R. Hobbs, G. Pritchard, J. Scott, W. Wright (AQIS) and particularly an anonymous referee for constructive criticism of the manuscript. I also thank A. Bruzzese, M. Millar and P. Gillespie for sharing my normal work duties so that this review could be written.

REFERENCES

Amor, R. L. and Piggin, C. M., 1977. Factors influencing the establishment and success of exotic plants in Australia. *Proc. Ecol. Soc. Aust.* **10:** 15–26.

Amor, R. L. and Stevens, P. L., 1975. Spread of weeds from a roadside into sclerophyll forests at Dartmouth, Australia. *Weed Res.* **16:** 111–18.

Atkins, D. L. and Molloy, J. M., 1993. The role of Landcare groups in weed control. Pp. 454–57 in *Proceedings of the 10th Australian and 14th Asian-Pacific weed conference, Brisbane, 6–10 September 1993, Vol. 1* ed by J. T. Swarbrick, C. W. L. Henderson, R. J. Jettner, L. Streit and S. R. Walker. Weed Society of Queensland: Brisbane.

Auld, T., 1983. Seed predation in native legumes of south-eastern Australia. *Aust. J. Ecol.* **8:** 367–76.

Australian Weeds Committee (AWC), 1993. The national weeds strategy. Discussion paper and recommendations for action. Australian Weeds Committee and Plant Production Committee. 55 pp.

Baird, A. M., 1977. Regeneration after fire in King's Park, Perth, Western Australia. *J. Roy. Soc. WA* **60:** 1–22.

Baker, H. G., 1965. Characteristics and modes of origin of weeds. Pp. 147–69 in *The genetics of colonizing species* ed by H. G. Baker and G. L. Stebbins. Academic Press: New York.

Bazzaz, F. A., 1983. Characteristics of populations in relation to disturbance in natural and man-modified ecosystems. Pp. 259–75 in *Disturbance and ecosystems* ed by H. A. Mooney and M. Gordon. Springer-Verlag: Berlin.

Bazzaz, F. A., 1986. Life-history of colonizing plants: some demographic, genetic and physiological features. Pp. 96–110 in *Ecology of biological invasions of North America and Hawaii* ed by H. A. Mooney and J. A. Drake. Springer-Verlag: New York.

Biological Diversity Advisory Committee (BDAC), 1992. A national strategy for the conservation of Australia's biological diversity. Draft report. Department of the Arts, Sport, the Environment and Territories: Canberra. 38 pp.

Braithwaite, R. H., 1985. The implications for management. Kakadu Fauna Survey final report Vol. 4. CSIRO Division of Wildlife and Rangelands Research: Canberra.

Braithwaite, R. W., Lonsdale, W. M. and Estbergs, J. A., 1989. Alien vegetation and native biota in tropical Australia: the impact of *Mimosa pigra*. *Biol. Conserv.* **48:** 189–210.

Briese, D. T., 1994. Biological control of weeds and fire management in protected natural areas: Are they compatible strategies? In *Proceedings of the eighth international symposium on biological control of weeds.* (In press).

Bridgewater, P. B. and Blackshall, D. J., 1981. Dynamics of some Western Australian ligneous formations with special reference to the invasion of exotic species. *Vegetatio* **46:** 141–48.

Brown, P. R., Wallis, R. L., Simmons, D. and Adams, R., 1991. Weeds and wildlife. *Plant Prot. Quart.* **6**(3): 150–53.

Buchanan, R. (this volume). Solutions to weed problems in urban bushland. In *Conserving biodiversity: threats and solutions* ed by R. A. Bradstock, T. D. Auld, D. A. Keith, R. T. Kingsford, D. Lunney and D. P. Sivertsen. Surrey Beatty & Sons: Chipping Norton.

Burdon, J. J., Marshall, D. R. and Groves, R. H., 1981. Aspects of weed biology important to biological control. Pp. 21–30 in *Proceedings of the 5th international symposium on biological control of weeds, Brisbane, 22–29 July, 1980* ed by E. S. Delfosse. CSIRO: Canberra.

Burrell, J. P., 1981. Invasion of coastal heaths in Victoria by *Leptospermum laevigatum* (J. Gaertn.) F. Muell. *Aust. J. Bot.* **29:** 747–64.

Campbell, M. H. and Delfosse, E. S., 1984. The biology of Australian weeds. 13 *Hypericum perforatum* L. *J. Aust. Inst. Agri. Sci.* **50:** 63–73.

Campbell, M. H., Holtkamp, R. H., McCormick, L. H., Wykes, P. J., Donaldson, P. J., Gullan, P. J. and Gillespie, P. S., 1994. Biological control of the native shrubs *Cassinia* spp. using the native scale insects *Austrotachardia* sp. and *Paratachardina* sp. (Hemiptera: Keriidae) in New South Wales. *Plant Pro. Quart.* **9:** 64–68.

Carr, G. W., 1988. Perspectives on environmental weeds in Victoria. Pp. 52–59 in *Weeds on public land: an action plan for today* ed by R. G. Richardson. Weed Science Society of Victoria and the School of Environmental Science, Monash University: Melbourne.

Carr, G. W., 1993. Exotic flora of Victoria and its impact on indigenous biota. Pp. 256–98 in *Flora of Victoria, Vol. 1* ed by D. B. Foreman and N. G. Walsh. Inkata Press: Melbourne.

Carr, G. W., Yugovic, J. V. and Robinson, K. E., 1992. Environmental weed invasions in Victoria: conservation and management implications. Department of Conservation and Environment and Ecological Horticulture Pty Ltd: Melbourne.

Christensen, P. E. and Burrows, N. D., 1986. Fire: an old tool with a new use. Pp. 97–105 in *Ecology of biological invasions: an Australian perspective* ed by R. H. Groves and J. J. Burdon. Australian Academy of Science: Canberra.

Commonwealth of Australia (COA), 1991. Decade of landcare plan. Commonwealth component. Australian Government Publishing Service: Canberra. 62 pp.

Connell, J. H., 1978. Diversity in tropical rain forests and coral reefs. *Science* **199:** 1302–310.

Connor, D. J. and Wilson, G. L., 1968. Response of a coastal heath community to fertilizer application. *Aust. J. Bot.* **16:** 117–23.

Copley, P., 1988. *The stick-nest rats of Australia.* World Wildlife Fund (Australia): Adelaide.

Corbett, D. P., 1988. Aspects of the ecology of *Pinus pinaster* invasion in the French Island State Park. M. Env. Sci. Thesis, Graduate School of Environmental Science, Monash University, Melbourne, Victoria.

Corbett, D. P., 1991. Control of Cluster Pine on French Island, Victoria. *Plant Pro. Quart.* **6**(3)**:** 128.

Cowie, I. D. and Werner, P. A., 1993. Alien plant species invasive in Kakadu National Park, tropical northern Australia. *Biol. Conserv.* **63:** 127–35.

Crawley, M. J., 1986. The population biology of invaders. *Phil. Trans. R. Soc. Lond.* **314:** 711–31.

Crawley, M. J., 1987. What makes a community invasible? Pp. 429–54 in *Colonization, succession and stability* ed by A. J. Gray, M. J. Crawley and P. J. Edwards. Blackwell Scientific Publications: Oxford.

Cropper, S., 1991. The resource evaluation and monitoring system: a system for monitoring natural and cultural assets. Department of Conservation and Environment: Victoria.

Cullen, J. M. and Delfosse, E. S., 1985. *Echium plantagineum:* catalyst for conflict and change in Australia. Pp. 249–92 in *Proceedings VI international symposium on the biological control of weeds, Vancouver, 19–25 August, 1984* ed by E. Delfosse. Agriculture Canada: Ottawa.

Department of Home Affairs and Environment (DHAE), 1984. A national conservation strategy for Australia. Australian Government Publishing Service: Canberra. 24 pp.

Dodd, J. and Moore, J. H., 1993. Introduction and status of *Kochia scoparia* in Western Australia. Pp. 496–500 in *Proceedings of the 10th Australian and 14th Asian-Pacific weed conference, Vol. 1 Brisbane 6–10 September, 1993* ed by J. T. Swarbrick, C. W. L. Henderson, R. J. Jettner, L. Streit and S. R. Walker. Weed Society of Queensland: Brisbane.

Duckett, T. and Wilkinson, G., 1988. Pampas, a serious new weed threat to Tasmania's forests. Forestry Commission: Tasmania.

Earl, T. J., 1990. Boneseed control using glyphosate. *Plant Pro. Quart.* **5**(4)**:** 169–70.

Ecologically Sustainable Development Steering Committee (ESDSC), 1992. Draft national strategy for ecologically sustainable development. Australian Government Publishing Service: Canberra. 148 pp.

Endangered Species Advisory Committee (ESAC), 1992. An Australian national strategy for the conservation of Australian species and communities threatened with extinction. Australian National Parks and Wildlife Service: Canberra. 36 pp.

Fagg, P. C., 1989. Control of *Delairea odorata* (Cape Ivy) in native forest with the herbicide clopyralid. *Plant Pro. Quart.* **4**(3)**:** 107–10.

Fenn, G. R., 1988. A simple agricultural land information system. *J. Aust. Inst. Agri. Sci.* **54:** 38–41.

Floyd, A. G., 1988. Natural regeneration of rainforest in NSW. Pp. 20–25 in *Rainforest remnants: proceedings of a workshop on rainforest rehabilitation* ed by S. Phillips. NSW National Parks and Wildlife Service: Hurstville.

Foley, M. J. and Bolton, M. P., 1990. Control of Sisal Hemp (*Agave* sp.) in native bushland. Pp. 145 in *Proceedings of the 9th Australian weeds conference Adelaide, 6–10 August, 1990* ed by J. W. Heap. Council of Australian Weed Science Societies: Adelaide.

Fox, M. D., 1991. Developing control strategies for environmental weeds. *Plant Pro. Quart.* **6**(3)**:** 109–10.

Fox, M. D. and Fox, B. J., 1986. The susceptibility of natural communities to invasion. Pp. 57–66 in *Ecology of biological invasions: an Australian perspective* ed by R. H. Groves and J. J. Burdon. Australian Academy of Science: Canberra.

Freshwater, V., 1991. Control of English ivy *(Hedera helix)* in Sherbrooke forest — a practical experience. *Plant Pro. Quart.* **6**(3)**:** 127.

Gleadow, R. M., 1982. Invasion by *Pittosporum undulatum* of the forests of central Victoria. II. Dispersal, germination and establishment. *Aust. J. Bot.* **30:** 185–98.

Gleadow, R. M. and Ashton, D. H., 1981. Invasion by *Pittosporum undulatum* of the forests of central Victoria. I. Invasion patterns and plant morphology. *Aust. J. Bot.* **29:** 705–20.

Gleadow, R. M. and Rowan, K. J., 1982. Invasion of *Pittosporum undulatum* of the forests of central Victoria. III. Effects of temperature on growth and drought resistance. *Aust. J. Bot.* **30:** 347–57.

Gleadow, R. M., Rowan, K. J. and Ashton, D. H., 1983. Invasion by *Pittosporum undulatum* of the forests of central Victoria. IV. Shade tolerance. *Aust. J. Bot.* **31:** 151–60.

Gill, A. M., Groves, R. H. and Noble, I. R. (eds), 1981. *Fire and the Australian biota.* Australian Academy of Science: Canberra.

Griffin, G. F., 1993. The spread of Buffel Grass in inland Australia: Land use conflicts. Pp. 501–04 in *Proceedings of the 10th Australian and 14th Asian-Pacific weed conference, Vol. 1, Brisbane, 6–10 September, 1993* ed by J. T. Swarbrick, C. W. L. Henderson, R. J. Jettner, L. Streit and S. R. Walker. Weed Society of Queensland: Brisbane.

Griffin, G. F., Stafford Smith, D. M., Morton, S. R., Allan, G. E., Masters, K. A. and Preece, N., 1989. Status and implications of the invasion of Tamarisk *(Tamarix aphylla)* on the Finke River, Northern Territory, Australia. *J. Environ. Manage.* **29:** 297–315.

Groves, R. H., 1977. Heathland soils and their fertility status. Pp. 143–50 in *Ecosystems of the world 9B* ed by R. L. Specht. Elsevier: Amsterdam.

Groves, R. H., 1986. Plant invasions in Australia: an overview. Pp. 137–49 in *Ecology of biological invasions: an Australian perspective* ed by R. H. Groves and J. J. Burdon. Australian Academy of Science: Canberra.

Groves, R. H., 1989. Ecological control of invasive terrestrial plants. Pp. 437–61 in *Biological invasions. A global perspective* ed by J. A. Drake, H. A. Mooney, F. Di Castri, R. H. Groves, F. J. Kruger, M. Rejmanek and M. Williamson. Wiley: Chichester.

Groves, R. H., 1991. Status of environmental weed control in Australia. *Plant Pro. Quart.* **6**(3)**:** 95–97.

Groves, R. H. and Williams, J. D., 1975. Growth of skeleton weed (*Chondrilla juncea* L.) as affected by growth of subterranean clover (*Trifolium subterraneum* L.) and infection by *Puccinia chondrillina* Bubak and Syd. *Aust. J. Agri. Res.* **26:** 975–83.

Gullan, P., 1988. Weeds in Victoria: where are we? Pp. 7–17 in *Weeds on public land — an action plan for today* ed by R. J. Richardson. Weed Science Society of Victoria and the School of Environmental Science, Monash University: Melbourne.

Harrison, P. G., 1994. Potentially weedy plant introductions to Australia. *Plant Pro. Quart.* **9**(1)**:** 40.

Hazard, W. H. L., 1988. Introducing crop, pasture and ornamental species into Australia — The risk of introducing new weeds. *Aust. Plant Intro. Rev.* **19:** 19–36.

Heddle, E. M. and Specht, R. L., 1975. Dark Island heath (Ninety-Mile Plain, South Australia). VIII. The effect of fertilizers on composition and growth 1950–1972. *Aust. J. Bot.* **23:** 151–64.

Hester, A. J. and Hobbs, R. J., 1992. Influence of fire and soil nutrients on native and non-native annuals at remnant vegetation edges in the Western Australian wheatbelt. *J. Vegeta. Sci.* **3:** 101–08.

Hitchmough, J. D., Kilgour, R. A., Morgan, J. W. and Shears, I. G., 1994. Efficacy of some grass specific herbicides in controlling exotic grass seedlings in native grassy vegetation. *Plant Pro. Quart.* **9:** 28–34.

Hobbs, R. J., 1989. The nature and effects of disturbance relative to invasions. Pp. 389–405 in *Biological invasions. A global perspective* ed by J. A. Drake, H. A. Mooney, F. Di Castri, R. H. Groves, F. J. Kruger, M. Rejmanek and M. Williamson. Wiley: Chichester.

Hobbs, R. J., 1991. Disturbance a precursor to weed invasion in native vegetation. *Plant Pro. Quart.* **6**(3)**:** 99–104.

Hobbs, R. J. and Atkins, A., 1988. The effect of disturbance and nutrient addition on native and introduced annuals in the Western Australian wheatbelt. *Aust. J. Ecol.* **13:** 171–79.

Hobbs, R. J. and Atkins, L., 1990. Fire-related dynamics of a *Banksia* woodland in south-west Western Australia. *Aust. J. Bot.* **38:** 97–117.

Hobbs, R. J. and Huenneke, L. F., 1992. Disturbance, diversity, and invasion: Implications for conservation. *Conserv. Biol.* **6**(3)**:** 324–37.

Hobbs, R. J. and Mooney, H. A., 1985. Community and population dynamics of serpentine grassland annuals in relation to gopher disturbance. *Oecologia* **67:** 342–51.

Hobbs, R. J. and Mooney, H. A., 1991. Effects of rainfall variability and gopher disturbance on serpentine annual grassland dynamics in N. California. *Ecology* **72:** 59–68.

Holm, L., Pancho, J. V., Herberger, J. P. and Plucknett, D. L., 1979. *A geographical atlas of world weeds.* Wiley and Sons: New York.

Holtkamp R. and Maguire, A. J., 1993. NSW Biological Control Program. Pp. 24–25 in *Proceedings of a national workshop on* Chrysanthemoides monilifera, *Port Macquarie, 28–30 April, 1993* ed by R. H. Holtkamp. New South Wales Department of Agriculture: Tamworth.

Holzner, W. and Numata, N. (eds), 1982. *Biology and ecology of weeds.* Dr W. Junk Publishers: The Hague.

Huffaker, C. B. and Kennett, C. E., 1959. A ten-year study of vegetational changes associated with biological control of Klamath weed. *J. Range Manage.* **12:** 69–82.

Humphries, S. E., Groves, R. H. and Mitchell, D. S., 1991. Plant invasions of Australian ecosystems: a status review and management directions. *Kowari* **2:** 1–134.

Johnson, R. W. and Purdie, R. W., 1981. The role of fire in the establishment and management of agricultural systems. Pp. 497–528 in *Fire and the Australian biota* ed by A. M. Gill, R. H. Groves and I. R. Noble. Australian Academy of Science: Canberra.

Judd, T., 1990. Invasion of native plant communities by *Kunzea ambigua, Kunzea ericoides* and *Leptospermum laevigatum.* PhD. Thesis, School of Botany, University of Melbourne, Victoria.

Julien, M. H., 1992. *Biological control of weeds: a world catalogue of agents and their target weeds.* Third edition. CAB International and ACIAR: Wallingford.

Keighery, G. J., 1991. Environmental weeds of Western Australia. *Kowari* **2:** 180–88.

Keighery, G. L., 1993. Weeds of Western Australia's west coast offshore islands. Pp. 167–71 in *Proceedings of the 10th Australian and 14th Asian-Pacific weed conference, Vol. 1, Brisbane, 6–10 September, 1993* ed by J. T. Swarbrick, C. W. L. Henderson, R. J. Jettner, L. Streit and S. R. Walker. Weed Society of Queensland: Brisbane.

Kirkpatrick, J. B., 1974. Plant invasion and extinction in a suburban coastal reserve. *Aust. Geog. Studies* **12:** 107–18.

Kooyman, R. M., 1988. Rainforest regeneration, reforestation and maintenance — Recommendations for the far north coast of NSW. Pp. 74–84 in *Rainforest remnants. Proceedings of a workshop on rainforest rehabilitation* ed by S. Phillips. NSW National Parks and Wildlife Service: Hurstville.

Kruger, F. J., 1977a. Invasive woody plants in the Cape Fynbos, with special reference to the biology and control of *Pinus pinaster.* Pp. 57–74 in *Proceedings of the second national weeds conference* ed by D. P. Annecke. Cape Town: Balkema.

Kruger, F. J., 1977b. Ecology of Cape Fynbos in relation to fire. Pp. 230–44 in *Proceedings of a symposium on the environmental consequences of fire and fuel management in mediterranean ecosystems.* USDA Forest Service: Washington D.C.

Kruger, F. J., Richardson, D. M. and van Wilgen, B. W., 1986. Processes of invasion by plants. Pp. 145–55 in *The ecology and management of biological invasions in South Africa* ed by I. A. Macdonald, F. J. Kruger and A. A. Ferrar. Oxford University Press: Cape Town.

Lane, B. A., 1992. Impact of *Spartina* on international migratory waders. Pp. 14–20 in *The sea has weeds too! Proceedings of a conference on the problem of* Spartina, *May 1991* ed by G. Fraser-Quick and A. Phillips. Bird Observers Club, Marine Education Society of Australasia, Australasian Wader Studies Group, Department of Conservation and Environment and South Gippsland Conservation Society: Inverloch.

Lane, D., Yugovic, J., Murphy, G. and Backholer, J., 1989. A pest management information system in Victoria. *Plant Pro. Quart.* **4**(1): 11–13.

Lane D. W. and Shaw K., 1978. The role of fire in boneseed (*Chrysanthemoides monilifera* (L.) Norl.) control in bushland. Pp. 333–35 in *Proceedings 1st Conf. council Australian weed science society.* Weed Science Society: Melbourne.

Leigh, J. H. and Briggs, J. D., 1992. Threatened Australian Plants: Overview and Case Studies. Australian National Parks and Wildlife Service: Canberra.

Lepart, J. and Debussche, M., 1991. Invasion processes as related to succession and disturbance. Pp. 159–78 in *Biogeography of mediterranean invasions* ed by R. H. Groves and F. Di Castri. Cambridge University Press: Cambridge.

Loyn, R. H. and French, K., 1991. Birds and environmental weeds in south-eastern Australia. *Plant Pro. Quart.* **6**(3): 137–49.

Lunt, I. D., 1990. Impact of an autumn fire on a long-grazed *Themeda triandra* (Kangaroo Grass) grassland: implications for management of invaded remnant vegetation. *Vic. Nat.* **107:** 45–51.

McDougall, K. L., 1989. The re-establishment of *Themeda triandra* (Kangaroo Grass): Implications for the restoration of grassland. Arthur Rylah Institute for Environmental Research, Technical Report Series No. 89. Department of Conservation, Forests and Lands: Melbourne.

McIntyre, S., 1990. Invasion of a nation: Our role in the management of exotic plants in Australia. *Aust. Biol.* **3**(2): 65–74.

McIntyre, S., 1993. The nature of exotic invasions in herbaceous vegetation of high and low species richness: Implications for conservation and management. Pp. 466–70 in *Proceedings of the 10th Australian and 14th Asian-Pacific weed conference, Vol. 1, Brisbane, 6–10 September, 1993* ed by J. T. Swarbrick, C. W. L. Henderson, R. J. Jettner, L. Streit and S. R. Walker. Weed Society of Queensland: Brisbane.

McIntyre, S., Ladiges, P. Y. and Adams, G., 1988. Plant species-richness and invasion by exotics in relation to disturbance of wetland communities on the Riverine Plain, NSW. *Aust. J. Ecol.* **13:** 361–73.

McMahon, A. R. G., 1991. Control of annual grasses with particular reference to *Briza maxima. Plant Pro. Quart.* **6**(3): 129.

McMillan, M. G., 1991. Chemical control of Horehound. Pp. 387–91 in *Proceedings of the 9th Australian weeds conference, Adelaide, 6–10 August, 1990* ed by J. W. Heap. Council of Australian Weed Science Societies: Adelaide.

Madden, J. E. and Swarbrick, J. T., 1990. Chemical control of *Ligustrum lucidum. Plant Pro. Quart.* **5**(4): 145–47.

Melville, I., 1992. Landcare for aboriginal lands of the Northern Territory. Pp. 223–31 in *Proceedings of the 7th international soil conservation organization conference, Sydney, 1992* ed by P. G. Haskins and B. M. Murphy. CALM: Sydney.

Medd, R. W., 1987. Impact of legislative actions on the invasion of *Carduus nutans.* Pp. 290–93 in *Proceedings of the eighth Australian weeds conference, Sydney, 21–25 September 1987* ed by D. Lemerle and A. R. Leys. Weed Society of New South Wales: Sydney.

Michael, P. W., 1989. Weeds of concern to northern Australia. Report commissioned by the Australian Quarantine and Inspection Service.

Miller, I. L., 1988. Aspects of the biology and control of *Mimosa pigra* L. M. Sc. Agr. Thesis, University of Sydney, 248 pp.

Miller, I. L. and Lonsdale, W. M., 1992. Ecological management of *Mimosa pigra:* Use of fire and competitive pastures. Pp. 104–06 in *A guide to the management of* Mimosa pigra ed by K. L. S. Harley. CSIRO: Canberra.

Molnar, C. D., Fletcher, D. and Parsons, R. F., 1989. Relationships between heath and *Leptospermum laevigatum* scrub at Sandringham, Victoria. *Proc. Roy. Soc. Vic.* **101:** 77–87.

Moody, M. E. and Mack, R. N., 1988. Controlling the spread of plant invasions: The importance of nascent foci. *J. Appl. Ecol.* **25:** 1009–21.

Morris, M. J., 1991. The use of plant pathogens for biological weed control in South Africa. *Agri. Ecosys. Environ.* **37:** 239–55.

Murphy, A. R., 1993. Selective control of wandering jew with fluroxypyr foliar spray. Pp. 265–68 in *Proceedings of the 10th Australian and 14th Asian-Pacific weed conference, Vol. 1, Brisbane, 6–10 September, 1993* ed by J. T. Swarbrick, C. W. L. Henderson, R. J. Jettner, L. Streit and S. R. Walker. Weed Society of Queensland: Brisbane.

Narayan, I., 1993. Can invasions of *Pittosporum undulatum* be controlled by fire? Hons. Thesis, School of Botany, University of Melbourne, Victoria.

Neser, S., 1968. Studies on some potentially useful insect enemies of needle-bushes (*Hakea* spp.— Proteaceae). PhD Thesis, Australian National University, Canberra, 194 pp.

Newsome, A. E. and Noble, I. R., 1986. Ecological and physiological characters of invading species. Pp. 1–20 in *Ecology of biological invasions: an Australian perspective* ed by R. H. Groves and J. J. Burdon. Australian Academy of Science: Canberra.

Noble, I. R., 1989. Attributes of invaders and the invading process: Terrestrial and vascular plants. Pp. 301–10 in *Biological invasions. A global perspective* ed by J. A. Drake, H. A. Mooney, F. Di Castri, R. H. Groves, F. J. Kruger, M. Rejmanek and M. Williamson. Wiley and Sons: Chichester.

Panetta, F. D., 1993. A system of assessing proposed introductions for weed potential. *Plant Pro. Quart.* **8**(1): 10–14.

Paterson, I. G. and Volframs, A., 1976. Boneseed: a threat to native plant communities. A study on the Queen's Domain, Hobart. University of Tasmania, Environmental Studies Occasional Paper 3.

Petraitis, P. S., Latham, R. E. and Neiesenbaum, R. A., 1989. The maintenance of species diversity by disturbance. *Quart. Rev. Biol.* **64**(4): 393–418.

Pickett, S. T. A. and White, P. S., 1985. Patch Dynamics: A synthesis. Pp. 371–84 in *The ecology of natural disturbance and patch dynamics* ed by S. T. A. Pickett and P. S. White. Academic Press: Florida.

Pickett, S. T. A., Kolasa, J., Armesto, J. J. and Collins, S. L., 1989. The ecological concept of disturbance and its expression at various hierarchical levels. *Oikos* **54:** 129–36.

Pritchard, G. H., 1991. Control of bridal creeper with herbicides. *Plant Pro. Quart.* **6**(1)**:** 126.

Pritchard, G. H., 1992. Control of *Spartina* with herbicides. Pp. 23–29 in *The sea has weeds too! Proceedings of a conference on the problem of* Spartina, *May 1991* ed by G. Fraser-Quick and A. Phillips. Bird Observers Club, Marine Education Society of Australasia, Australasian Wader Studies Group, Department of Conservation and Environment and South Gippsland Conservation Society: Inverloch.

Pritchard, G. H., 1993. Control of Himalayan Honeysuckle *(Leycestaria formosa).* Pp. 142 in *Proceedings of the 10th Australian and 14th Asian-Pacific weed conference, Vol. II, Brisbane, 6–10 September, 1993* ed by J. T. Swarbrick, C. W. L. Henderson, R. J. Jettner, L. Streit and S. R. Walker. Weed Society of Queensland: Brisbane.

Rayner, G., 1989. Aspects of the biology and population ecology of the environmental weed *Briza maxima.* Hons. Thesis, Department of Botany and Zoology, Monash University, Victoria.

Rejmanek, M., 1989. Invasibility of plant communities. Pp. 369–88 in *Biological invasions. A global perspective* ed by J. A. Drake, H. A. Mooney, F. Di Castri, R. H. Groves, F. J. Kruger, M. Rejmanek and M. Williamson. Wiley: Chichester.

Robertson, D., 1985. Inter-relationships between kangaroos, fire and vegetation dynamics at Gellibrand Hill Park, Victoria. PhD Thesis, University of Melbourne.

Rykiel, E. J., 1985. Towards a definition of ecological disturbance. *Aust. J. Ecol.* **10:** 361–65.

Sands, D. C. and Miller, R. V., 1993. Evolving strategies for biological control of weeds with plant pathogens. *Pest. Sci.* **37:** 399–403.

Saunders, D. A., Rowley, I. and Smith, G. T., 1985. The effects of clearing for agriculture on the distribution of cockatoos in the south-west of Western Australia. Pp. 309–21 in *Birds of eucalypt forests and woodlands: ecology, conservation and management* ed by A. Keast, H. F. Recher, H. Ford and D. Saunders. Surrey Beatty & Sons: Chipping Norton.

Scarlett, N. H. and Parsons, R. F., 1982. Rare plants of the Victorian plains. Pp. 89–105 in *Species at risk: research in Australia* ed by R. H. Groves and W. D. L. Ride. Australian Academy of Science: Canberra.

Schultz, G. C., 1987. A computer based noxious weed recording system. Pp. 297–300 in *Proceedings of the 8th Australian weeds conference, Sydney, 21–25 September 1987* ed by D. Lemerle and A. R. Leys. Weed Society of New South Wales: Sydney.

Scott, J. K., 1991. *Acacia karroo* Hayne (Mimosaceae), a potentially serious weed in Australia. *Plant Pro. Quart.* **6**(1)**:** 16–18.

Scott, J. K. and Adair, R. J., 1990. The commencement of biological control of bitou bush and boneseed *(Chrysanthemoides monilifera).* Pp. 126–29 in *Proceedings of the 9th Australian weeds conference, Adelaide, 6–10 August 1990* ed by J. W. Heap. Crop Science Society of South Australia: Adelaide.

Scott, J. K. and Adair, R. J., 1994. Selection of biological control agents for *Chrysanthemoides monilifera,* a weed of protected natural areas in Australia. In *Proceedings of the 8th international symposium on biological control of weeds* ed by E. S. Delfosse and R. R. Scott. DSIR/CSIRO: Melbourne. (in press).

Scott, J. K. and Panetta, F. D., 1993. Predicting the Australian weed status of South African plants. *J. Biogeo.* **20**(1)**:** 87–93.

Smith, T., 1992. Community Dune Care. Active public participation in coastal land management. In *Proceedings of a coastal management conference, 4–6 November 1992.* Kiama City Council: Kiama.

Specht, R. L., 1963. Dark Island heath (Ninety-mile Plain, South Australia). VII. The effect of fertilizers on composition and growth, 1950–1960. *Aust. J. Bot.* **11:** 67–94.

Stafford, J. H., 1991. Techniques for the establishment of kangaroo grass in South Australian conservation reserves. *Plant Pro. Quart.* **6**(3)**:** 120–22.

Stockard, J., 1988. Rainforest rehabilitation at Wingham Brush. Pp. 94–100 in *Rainforest remnants. Proceedings of a workshop on rainforest rehabilitation* ed by S. Phillips. NSW National Parks and Wildlife Service: Hurstville.

Stuwe, J., 1986. An assessment of the conservation status of native grasslands on the Western plains, Victoria and sites of botanical significance. Arthur Rylah Institute for Environmental Research Technical Report Series No. 48. Department of Conservation, Forests and Lands: Melbourne.

Stuwe, J. and Parsons, R. F., 1977. *Themeda australis* grasslands on the basalt plains, Victoria: floristics and management effects. *Aust. J. Ecol.* **2:** 467–76.

Summerbell, G., 1991. Regeneration of complex notophyll vine forest (humid subtropical rainforest) in eastern Australia — a review. *Cunninghamia* **2**(3)**:** 391–410.

Swarbrick, J. T., 1993. Naturalized weeds of rainforests and associated habitats in tropical Queensland. Pp. 172–75 in *Proceedings of the 10th Australian and 14th Asian-Pacific weed conference, Vol. 1, Brisbane, 6–10 September, 1993* ed by J. T. Swarbrick, C. W. L. Henderson, R. J. Jettner, L. Streit and S. R. Walker. Weed Society of Queensland: Brisbane.

Swarbrick, J. T. and Skarratt, D. B., 1994. The bushweed 2 database of environmental weeds in Australia. University of Queensland Gatton College: Gatton.

Templeton, G. E., 1987. Mycoherbicides — achievements, developments and prospects. Pp. 489–97 in *Proceedings of the 8th Australian weeds conference, Sydney, 21–25 September 1987* ed by D. Lemerle and A. R. Leys.

Thompson, W. L., 1993. The current status of weeds in selected national parks of south east Queensland. Pp. 162–66 in *Proceedings of the 10th Australian and 14th Asian-Pacific weed conference, Vol. 1, Brisbane, 6–10 September, 1993* ed by J. T. Swarbrick, C. W. L. Henderson, R. J. Jettner, L. Streit and S. R. Walker. Weed Society of Queensland: Brisbane.

Timmins, S. M. and Williams, P. A., 1987. Characteristics of problem weeds in New Zealand's protected natural areas. Pp. 241–47 in *Nature conservation: The role of remnants of native vegetation* ed by D. A. Saunders, G. W. Arnold, A. A. Burbidge and A. J. M. Hopkins. Surrey Beatty & Sons: Chipping Norton.

Toth, J., Milham, P. and Nazer, C., 1993. Aerially applied herbicides selectively control Bitou Bush growing in communities of indigenous plants on sand dunes. Pp. 67–71 in *Proceedings of the 10th Australian and 14th Asian-Pacific weed conference, Brisbane, 6–10 September, 1993* ed by J. T. Swarbrick, C. W. L. Henderson, R. J. Jettner, L. Streit and S. R. Walker. Weed Society of Queensland: Brisbane.

Trabaud, L., 1991. Is fire an agent favouring plant invasions? Pp. 179–90 in *Biogeography of mediterranean invasions* ed by R. H. Groves and F. Di Castri. Cambridge University Press: Cambridge.

van den Berg, M. A., 1980. Natural enemies of *Acacia cyclops* A. Cunn. ex G. Don. and *Acacia saligna* (Labill.) Wendl. in Western Australia. II. Coleoptera. *Phytophylactica* **12:** 169–71.

van den Berg, M. A., 1982. Coleoptera attacking *Acacia dealbata* Link., *Acacia decurrens* Willd., *Acacia longifolia* (Andr.) Willd., *Acacia mearnsii* de Willd. and *Acacia melanoxylon* R. Br. in Australia. *Phytophylactica* **14:** 51–55.

van Haaren, P. E., 1993. The ecological impact and control of Blue Thunbergia *(Thunbergia grandiflora)*. Pp. 156 in *Proceedings of the 10th Australian and 14th Asian-Pacific weed conference, Vol. II, Brisbane, 6–10 September, 1993* ed by J. T. Swarbrick, C. W. L. Henderson, R. J. Jettner, L. Streit and S. R. Walker. Weed Society of Queensland: Brisbane.

Vitteli, J. S., 1993. Effect of aerial water volume on triclopyr/picloram applied to Rubber Vine *(Cryptostegia grandiflora)*. Pp. 103 in *Abstracts of the 10th Australian and 14th Asian-Pacific weed conference. Brisbane, 6–10 September, 1993.* Weed Society of Queensland: Brisbane.

Wapshere, A. J., 1990. Biological control of grass weeds in Australia: an appraisal. *Plant Pro. Quart.* **5**(2): 62–75.

Ward, D., 1992. ParkCare — Educating local communities for bush conservation. Pp. 98–102 in *Open to interpretation. Proceedings of a conference. Melbourne, 16–17 November, 1992.* Interpretation Australia Association: Melbourne.

Waterhouse, B. M., 1986. Broom *(Cytisus scoparius)* at Barrington Tops, New South Wales. *Aust. Geog. Studies* **26:** 239–48.

Weiss, P. W., 1983. Invasion of coastal *Acacia* communities by *Chrysanthemoides.* PhD Thesis, Australian National University.

Weiss, P. W., 1986. The biology of Australian weeds 14. *Chrysanthemoides monilifera* (L.) T. Norl. *J. Aust. Inst. Agri. Sci.* **52:** 127–34.

Weiss, P. W., 1993. Ecology of *Chrysanthemoides monilifera* in relation to control. Pp. 2–5 in *Proceedings of a national workshop on* Chrysanthemoides monilifera, *Port Macquarie, 28–30 April, 1993* ed by R. H. Holtkamp. New South Wales Department of Agriculture: Tamworth.

Weiss, P. W. and Noble, I. R., 1984a. Status of coastal dune communities invaded by *Chrysanthemoides monilifera. Aust. J. Ecol.* **9:** 93–98.

Weiss, P. W. and Noble, I. R., 1984b. Interactions between seedlings of *Chrysanthemoides monilifera* and *Acacia longifolia. Aust. J. Ecol.* **9:** 107–16.

White, J., 1913. Report of the Officer in Charge of the Prickly Pear Experimental Station, Dulacca, up to 30th June, 1913. Department of Public Lands, Queensland.

Wycherley, P., 1984. People, fire, and weeds: can the vicious spiral be broken? Pp. 11–17 in *The management of small bush areas in the Perth metropolitan region* ed by S. A. Moore. Department of Fisheries and Wildlife: Perth.

Young, K. R., 1993. Environmental weeds in Tasmania. Pp. 159–61 in *Proceedings of the 10th Australian and 14th Asian-Pacific weed conference, Vol. 1, Brisbane, 6–10 September, 1993* ed by J. T. Swarbrick, C. W. L. Henderson, R. J. Jettner, L. Streit and S. R. Walker. Weed Society of Queensland: Brisbane.

Solutions to feral animal problems: ecological and economic principles

M. BOMFORD,[1] A. NEWSOME[2] and P. O'BRIEN[1]

The objective of vertebrate pest management is cost-effective damage control. This paper examines and challenges some commonly held beliefs about the impact of feral animals, including cats *Felis catus*, foxes *Vulpes vulpes*, and rabbits *Oryctolagus cuniculus*, and the approaches that should be taken for their control.

Not enough is known about the impact of feral animals in Australia or their response to control for effective management. Killing is by far the most common technique used for control, but little is known about its effectiveness for reducing environmental damage. We need to improve our understanding of the factors that regulate pest populations and the effect that reducing pest densities has on levels of damage. This is the key to assessing whether a chosen control technique will be an effective management tool, and if so, in what circumstances it can be used for each species.

We also need to do more to prevent populations of native species becoming isolated in small "islands" of suitable habitat. Even low densities of vertebrate pests can drive small, isolated populations to extinction.

High costs and increasing community concern over moral and animal welfare issues associated with killing pests have focussed interest on fertility control and on permanent eradication. Although many chemicals are known to cause infertility in animals, the present role of fertility control in wildlife management in Australia is extremely limited. The main barrier to its use is the lack of field delivery techniques and the high costs for treatment of widespread and abundant pests. Current research on the use of anti-fertility agents passively spread via genetically engineered infectious organisms could overcome these limitations for some pest species. This is high risk research and it is too early to predict the chances of its successful development. Eradication seems an attractive approach, offering perpetual freedom from the pest. This paper presents criteria for assessing the feasibility of eradication. These are not met for any pest animals in mainland Australia.

INTRODUCTION

THIS paper challenges five orthodoxies about pest control:

1. We know enough about pest impact
2. Cats are a major problem for wildlife conservation
3. Fertility control can be used to reduce pest damage
4. Pest eradication is the best approach
5. Reserves alone can save biodiversity

We examine these commonly held beliefs, basing our analysis on the assumption that the objective of vertebrate pest management is cost-effective damage control, where cost-effectiveness includes assessments of amenity, aesthetic, moral, heritage and any other values society wants. Cost-effective damage control occurs when the cost of pest control is more than met by savings in protecting such values.

PEST IMPACT AND MANAGEMENT PRINCIPLES

Orthodoxy Number 1: "We know enough about pest impact"

We suggest not enough is known about the impact of vertebrate pests or their response to control to allow effective management. Many commonly held beliefs about the impacts of pests, especially impacts caused by predation or competition, are based on anecdotal rather than scientific evidence.

Only four potentially cost-effective techniques are currently available for vertebrate pest management in Australia:

- killing by poisoning, trapping or shooting;
- habitat manipulation to reduce its suitability for pests;
- exclusion to prevent entry; and
- biological control for rabbits using myxomatosis.

Killing is by far the most common technique used in Australia, and each year many hundreds of thousands of rabbits, pigs, foxes, goats and other pests are killed (Norton and Pech 1988; Wilson *et al.* 1992). In some instances, killing pests can protect native species. For example, killing foxes has allowed populations of Rock Wallaby *Petrogale lateralis* (Kinnear *et al.* 1988) and Chuditch *Dasyurus geoffroii* (Morris 1992) to increase.

Unfortunately, killing pests may often do little to prevent environmental damage. We cannot assume that reduced pest numbers leads to reduced environmental damage (Fig. 1). For example, rabbit densities lower than 1 ha^{-1}, an almost imperceptibly low density to a casual

[1]Bureau of Resource Sciences, P.O. Box E11, Queen Victoria Terrace, Australian Capital Territory, Australia 2600.
[2]CSIRO Division of Wildlife and Ecology, P.O. Box 84, Lyneham, Australian Capital Territory, Australia 2602.

Pages 202–09 *in* CONSERVING BIODIVERSITY: THREATS AND SOLUTIONS, ed by R. A. Bradstock, T. D. Auld, D. A. Keith, R. T. Kingsford, D. Lunney and D. P. Sivertsen. Surrey Beatty & Sons. 1995.

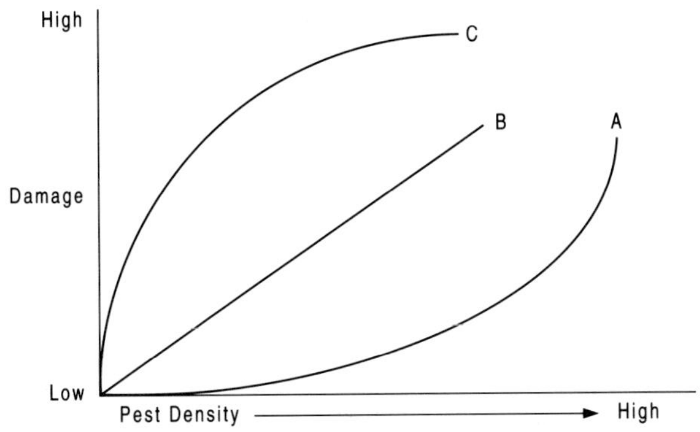

Fig. 1. Relationship between pest density and damage for three theoretical situations. A: damage only occurs at very high pest densities. B: damage is directly proportional to pest density. C: damage occurs even at very low pest densities. We do not know the relationship between pest density and damage levels for most pests.

observer, cause severe damage to *Acacia* seedlings (Lange and Graham 1983). So reducing rabbit numbers to moderate densities may do nothing to help regeneration of trees in *Acacia* woodlands, if this is the management goal. This relationship between damage and pest density is illustrated by line C in Figure 1.

The level of control which will optimize damage reduction is not known for most pests. This is because there has been hardly any good experimental work on the effects of pest control on damage levels. By good experimental work we mean controlled, replicated field experiments in which pest density is reduced and the environmental response measured. For example, it is not known whether reducing cat numbers in mainland habitats would cause native prey species to increase in abundance. This is why we challenge the second orthodoxy:

Orthodoxy Number 2: "Cats are a major problem for wildlife conservation"

It is not known whether feral cats are a major problem for conservation of native wildlife, even though feral cats are known to eat hundreds of thousands of native animals each year, including some rare species (Potter 1991). But reducing cat numbers may do little to protect these native species. This may sound like a contradiction. But by analogy one could falsely conclude that because people eat many hundreds of thousands of sheep, cattle and poultry each year, these populations are threatened, or because people kill lots of rabbits, feral pigs, and foxes each year, these pest populations are also threatened. A clear distinction needs to be made between the deaths of individual animals and the impact this may have on prey populations. While we may have sympathy for individual animals that die, it is possible to take a substantial ongoing harvest of animals from a population and not cause any decline in numbers. It is perfectly possible that cats simply take a sustained harvest of many native species without threatening their populations at all. How can the key question of whether cat predation is threatening native species populations be answered? More cat diet studies are certainly not going to help. We know what cats eat. But the effects on prey populations are not known. Cat densities need to be reduced experimentally and the effect on prey populations measured. If cats are found to be a problem we will have to determine how much cat densities need to be reduced in order to protect the prey species.

Similar analogies could be drawn for many of the environmental impacts attributed to vertebrate pests. But unless the relationships between pest densities and environmental impacts are assessed, much money may be wasted on pest control activities that will do little or nothing to protect conservation values.

To help achieve cost effective pest management, the Bureau of Resource Sciences is using expert task forces to collate information on impact and control techniques for each of Australia's major vertebrate pests. The information is being used to develop best-practice guidelines for management, which will be tested by large scale demonstration projects (Braysher 1993).

High costs, lack of effective damage control and increasing community concern over moral and animal welfare issues associated with traditional pest control techniques have focussed interest on alternative approaches, such as fertility control or permanent eradication.

A ROLE FOR FERTILITY CONTROL?

Orthodoxy Number 3: "Fertility control can be used to reduce pest damage"

Fertility control is currently an ineffective, expensive technique for reducing animal damage.

Fertility control involves artificially reducing birth rates by sterilizing animals, or preventing the implantation or development of the fertilized egg. Bomford (1990a and b) reviewed the literature on fertility control for pest management, and identified seven criteria for assessing the feasibility of using fertility control for reducing damage caused by pests. Currently available and proposed fertility control techniques are examined here to see how well they meet these criteria for Australian pests.

Availability of drug or technique to reduce fertility

Many chemicals and techniques are known to cause infertility in animals so the application of fertility control to managing pests is not restricted by a lack of suitable sterilizing agents.

An effective delivery mechanism

The lack of practical techniques to deliver sterilizing agents to an adequate proportion of the target population is a major obstacle to using fertility control to manage wild animal populations. Many tests on caged animals, or on animals in small enclosures, have relied on drugs delivered by surgical implantation, by repeated injections by hand or dart gun, or by frequent oral doses in food or drink. These options are not feasible for widespread and abundant wildlife.

The main problems are high costs, failure to treat sufficient numbers of animals, accidental treatment of non-target species, and meeting requirements for repeated dosing of individuals. Immunization techniques usually require a minimum of two injections with a break between, plus annual booster injections. Many orally active drugs require daily ingestion. Such delivery techniques are impractical for 200 000 wild horses *Equus caballus* spread over much of northern and central Australia, let alone 200 000 000 wild rabbits.

Virally vectored immunocontraception is a new approach being developed in Australia to control foxes and rabbits (Tyndale-Biscoe 1990 and in press). It uses new developments in molecular technology. It is now possible to use key proteins involved in the process of gametogenesis, fertilization and implantation in mammals to provoke an immune response which prevents fertilization and implantation. The genes which encode the key proteins can also be identified and cloned and the DNA sequence determined. Such proteins and genes can then be delivered to target pest species. Three methods are possible.

Oral baits are the simplest, with a key protein as the antigen, encapsulated in micro-spheres which are then incorporated into the bait. The second inserts the gene coding for the key protein into a non-disseminating virus using molecular engineering. The vaccinia virus has been used in this way to vaccinate foxes to prevent the spread of rabies in Europe and North America. A third, more sophisticated and economical method, is to use a disseminating recombinant virus that is specific to the target pest. Australia has such a virus for the rabbit, myxoma, but a suitable virus has yet to be found for the fox.

Only this third method, using a live, disseminating virus, could overcome the delivery problems described above, and hence the huge expense associated with techniques that require field delivery of sterilizing agents. But this is high risk research and there are many technical hurdles still to be overcome. It is too early to predict the chances of its successful development, but success would bring high benefits for conservation of biodiversity.

Reduced damage caused by animals

The focus of research in this field has been on reducing fertility. This is not enough. The goal must be to reduce pest numbers and so reduce damage caused by the pest. No studies were found that demonstrated this. A major problem with some antifertility drugs is that they disrupt hormonal control of sexual and social behaviour. When treated animals lose their social dominance, they can be replaced in the social structure by untreated fertile individuals which defeats the purpose of fertility control. Even when antifertility drugs leave animals hormonally intact, populations may not decline. Many populations have high juvenile mortality. If fertility control simply prevents the birth of young that would otherwise die and disperse without breeding, it will not lead to a population decline.

Compensatory responses can be expected in populations subjected to fertility control, which may prevent population declines. These include reduced mortality and dispersal, increased immigration, and increased birth rates in the remaining fertile individuals. Also young animals may reach sexual maturity earlier, and so produce more offspring. The extent of compensation determines whether fertility control will work and how well it will work. Unfortunately little is known about this central aspect of the effect of fertility control on wild populations.

Humane and non-toxic effects

Many fertility control drugs are humane, but some have side-effects or are toxic at high doses.

Target specificity

Unfortunately, few fertility control drugs are species specific. Genetically engineered viruses used for immunological fertility control could be target specific, though this would be difficult for feral species that are kept as livestock or domestic pets, and possibly also for closely related species, such as foxes and dingoes *Canis familiaris dingo*. There is also a slight risk that such viruses could mutate to infect non-target species, but this would not cause them to become infertile if the virus was engineered to affect only genes and proteins present in the target species.

Environmental acceptability

Fertility control drugs generally pose low environmental risks, but residues in baits or carcasses could enter the food chain and put at risk native wildlife, livestock or people.

Cost effectiveness

Cost is another major obstacle in the use of fertility control as a wildlife management technique. Fertility control using currently available technology is prohibitively expensive for widespread and abundant pests. Cost-effectiveness is the main area where fertility control is weak in comparison to currently used pest control techniques.

In summary, the availability of drugs or techniques that cause infertility in pest animals, and that are environmentally acceptable, are relatively small problems for the use of fertility control. Meeting requirements for humane, non-toxic and target-specific actions are moderate problems but in general these can probably be overcome. But delivery of drugs to wild populations, and achieving population regulation and animal damage control that is cost-effective are major problems. We conclude that currently available fertility control techniques cannot be used to manage our pest animals.

ERADICATION OF PESTS

Orthodoxy Number 4: "Pest eradication is the best approach"

Eradication of pest animals is not feasible, except under exceptional circumstances.

Permanent eradication would offer perpetual freedom from the pest, its harmful effects and from current high costs of ongoing control. Eradication would also solve some environmental problems associated with ongoing control, such as potential build up of poisons in food chains, and killing non-target species. But attempts to eradicate pest animals have not been successful.

We define "eradication" as the complete and permanent removal of all wild populations by a time-limited campaign. "Time-limited" is important in this definition:- eradication without a time limit is just ongoing control.

No widely established introduced vertebrate pest has ever been eradicated from mainland Australia or from any other continent. There have been some eradications from islands, including feral goats *Capra Hircus*, feral cats, rabbits, possums *Trichosurus vulpecula* and coypus *Myocastor coypus* (Hermes 1987; Gosling and Baker 1989; Parkes 1990; Cowan 1992).

Bomford and O'Brien (1993) identified six criteria that need to be met before eradication is technically possible and preferable to ongoing control. In developing these criteria, they drew on the experience of past eradication campaigns, both successful and unsuccessful (Gosling and Baker 1987 and 1989; Parkes 1990) and attempts to eradicate infectious human diseases around the world (Yekutiel 1980).

The following examination of these six criteria makes the reasons for the lack of success of eradication campaigns conducted against widespread pest populations apparent.

Rate of removal must exceed rate of increase at all population densities

If the removal rate does not exceed the rate of replacement at all densities, the population will not decline further and eradication will not be achieved. This seems simple and obvious; but in practice it is an extremely stringent requirement for two main reasons. First, many pest populations have high natural rates of increase. Second, as density declines, it takes progressively more time and expense to locate and remove individual animals (Fig. 2).

The most difficult part of an eradication campaign occurs just before eradication is achieved. For example, it took 1 000 hunter days to find and remove the last feral goat on Raoul Island in New Zealand, and the cost of removing the last five goats was $A20,600 each (Parkes 1990). This was in an area of only 2 943 ha. In a mainland ecosystem it would be far harder, and cost much more.

Hence removal rates due to pest control operations tend to decline at low population densities, and if the removal rate declines to less than the rate of increase, the population does not decline further. This is both a technical and a management problem. There must be a motive and resources to continue removing animals at low densities, when numbers caught and damage inflicted are extremely low.

Immigration is zero

If animals can immigrate onto the eradication area, or be released from captive populations, then eradication will be unachievable or transient. This criterion can be met on remote offshore islands, but on the mainland, imports or releases of new animals need to be prevented. Zero immigration can be achieved for regional eradication where fencing or control on the margins creates a perfect barrier. For example, the dingo-proof fence along the New South Wales border (Wilson *et al.* 1992) and the local control of foxes to prevent their predation on remnant populations of endangered animals. But these are both expensive options and both have a chance of failure.

All animals must be at risk

It is not necessary that the removal technique takes all animals, at any one attempt. But all reproductive and potentially reproductive members of the population must be susceptible to removal. For example, if trap-shyness is

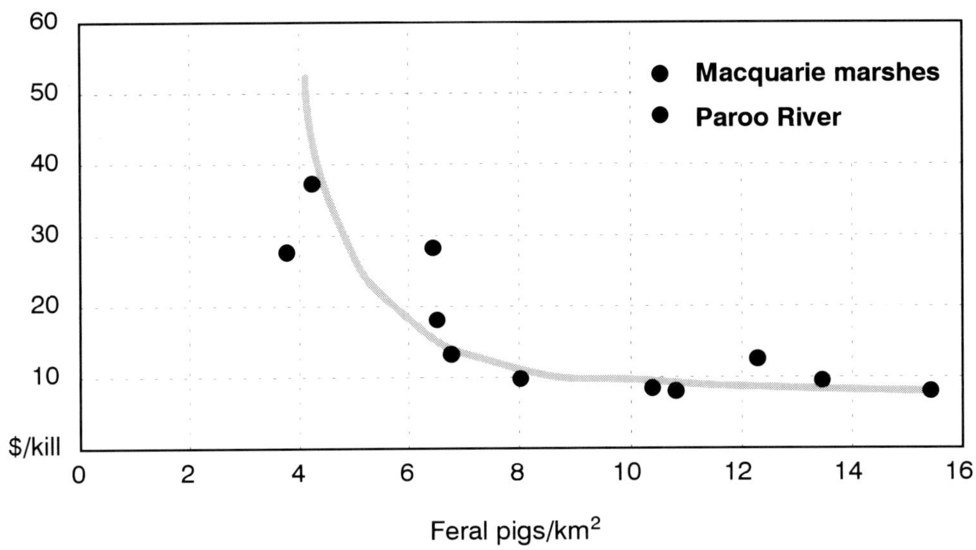

Fig. 2. Cost of removal of feral pigs *Sus scrofa* by helicopter shooting at different densities. The cost of finding and removing a single extra pig is extremely high at low densities. (From unpublished data, D. Choquenot and G. Saunders).

inherited or learnt, or if animals develop neophobia to poison baits, then a subset of the population is not at risk and will not be eradicated.

Populations can be monitored at all densities

If animals cannot be detected at low densities, there is no performance indicator for measuring success or failure. This means there will be no way of ensuring that control efforts are still causing the population to decline, and no way of determining if eradication has been achieved. This technical requirement is difficult to meet for many species.

Discounted cost-benefit analysis favours eradication over control

Intuitively one might think the value of perpetual freedom from a pest is very high. But future benefits have a lower economic value than equivalent benefits achieved immediately because of discounting. Calculating discount rates involves using the reverse equation to that used for calculating interest rates on invested money. Discounting means that benefits which occur in more than 30 years time are perceived to have little current economic value. Hence market discount rates have a major impact on the current value of control as compared to eradication. The loss of value of the benefit of eradication is determined by the discount rate and the time until the benefit is attained. Eradication requires a large initial outlay, and, if successful there are no further costs, but most of the benefits accumulate more than ten years in the future.

In Figure 3 we compare three different discount rates for a hypothetical data set on the cumulative benefit/cost ratio of eradication (the benefits of protecting resources damaged by the pest and the cost of achieving eradication). We optimistically assume the benefits of eradication are high relative to the costs. When the discount rate is set at zero (i.e., ignoring discounting altogether), eradication becomes cost effective after 28 years. Using the same data set and a 3.5 per cent discount rate, eradication becomes cost effective after 47 years. With a 10 per cent discount rate, eradication is never cost effective. There is, however, considerable debate about the selection of appropriate discount rates, particularly when non-market values, such as conservation and animal welfare, are considered.

Suitable socio-political environment

Even when technical and economic criteria can be met, social and political factors may play an overriding role in determining the prospects for successful eradication. The benefits of eradication as an alternative to ongoing control must be convincing, mainly because of the huge cost of achieving eradication and because community attitudes may oppose killing large numbers of wild animals for moral, emotional or cultural reasons (Van Vuren 1992).

These six criteria determine whether eradication is technically feasible, and preferable, to ongoing control. The possibility of failure also needs to be considered when deciding whether to opt for eradication as a management option. Eradication is a risky business, and a failed eradication attempt is very expensive.

Eradication should only be attempted when all six criteria can be met. When applied to Australian pests, national eradication is not feasible for any species. It is even doubtful if these criteria can be met for the small population of feral Ferrets *Mustela furo* near Launceston in

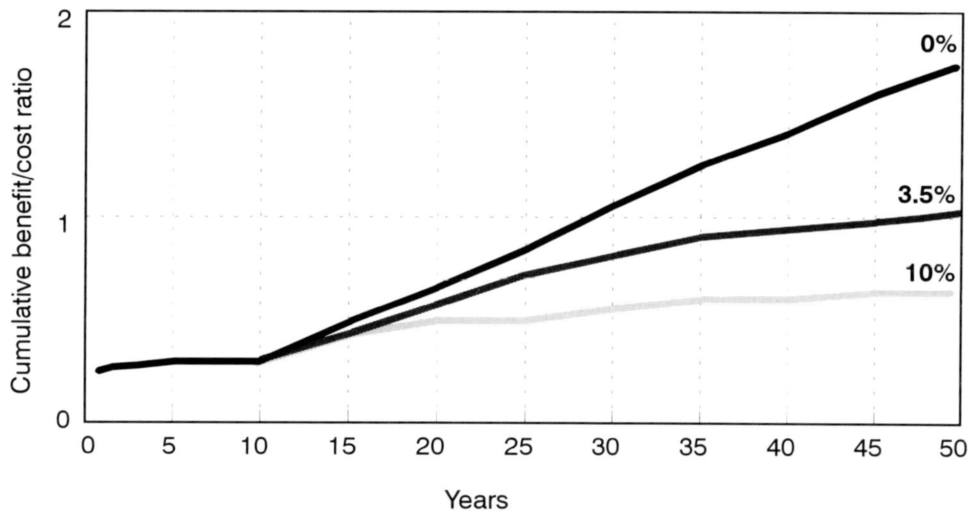

Fig. 3. The effects of three different discount rates on a hypothetical cumulative benefit-cost data set for eradication. We have assumed the cost of eradication is relatively low (\$40 $ha^{-1}year^{-1}$) compared with the benefits (\$10 $ha^{-1}year^{-1}$ in perpetuity), and that eradication is achieved after 10 years with no further costs. These are probably highly optimistic assumptions for widespread pests.

Tasmania. The main problem for species kept as pets or livestock is the high risk of re-release. Regional eradication does not appear much more feasible than national eradication, and has the additional complication of animals immigrating from the edges, increasing the costs because of the need for perpetual surveillance and margin control or fencing. Research would be required to overcome these technical problems, and use of the six criteria brings research needs into sharp focus and allows us to assess research needs for proposed eradication attempts.

PEST IMPACTS ON SMALL POPULATIONS

Finally, we want to look at one area where there is good information on pest impact. This is the devastating impact which introduced pests have on island flora and fauna. Unfortunately there are many case histories where introduced pests are known to have caused species extinctions and major declines on islands (Loope and Mueller-Dombois 1989).

There are two main reasons. First, island species are often not adapted to coping with mammalian grazers and predators. Second, island populations are small and isolated, so it does not take much to reduce them to critically low numbers. If this happens, there are no colonizers to help populations recover, as would happen for widespread mainland populations. This is, of course, why it is possible to eradicate pest populations on islands. We suggest this second cause of vulnerability of island species has great significance for the approach that should be taken to managing conservation on the mainland, in relation to the risk that mainland reserves could become equivalent to "habitat islands" for many species (Burbidge and McKenzie 1989).

Orthodoxy Number 5: "Reserves alone can protect biodiversity"

Small populations occupying isolated "islands" of mainland habitat may be extremely vulnerable to vertebrate pests.

Our pest species cannot be eradicated, yet even low densities of introduced pests may threaten populations of native species that are isolated in mainland habitat islands. This is why foxes and cats are a problem for captive breeding and release programmes. Cat predation is a problem for a very small isolated population of Eastern Barred Bandicoots *Perameles gunnii* near Hamilton, Victoria (Seebeck *et al.* 1991). In contrast, barred bandicoots are more widespread in Tasmania, where they coexist with cats and have shown no decline in numbers (Strahan 1983). The impact of predators is, in general, likely to be more severe on small, low-density populations than on widespread and abundant populations (Ricklefs 1979; Sinclair 1989).

The ability of some native species to withstand competition and predation by introduced pests may be enhanced by the maintenance of suitable habitat on land used for production, in addition to land in reserves. Unfortunately little is known about the extent to which habitat loss and modification may make native species more vulnerable to the impacts of introduced pests. This area requires more research to allow reliable assessments of the most cost effective approaches to biodiversity conservation. For some species, particularly those with reasonably widespread populations, the most cost-effective approach for conservation may be to direct funds towards off-reserve initiatives, to prevent their populations reaching such critically low numbers that the only viable conservation option is management in citadel reserves. There is a need to investigate ways to make land used for production, grazing, mining and forestry, more environmentally friendly so that species conservation and productive use can occur together on the same land.

Conservation agencies and policy makers need to put a much stronger emphasis on providing incentives, particularly economic incentives, to managers of land outside reserves to maintain habitats with conservation value. There has already been much good research in this field in Australia, for example in forests used for logging (Davey and Norton 1990), the rangelands (Stafford-Smith and Morton 1992), and in the wheat belt (Hobbs and Saunders 1993). Unless a high priority is given to this approach for both research funding and land management policy, we may fail to protect many native species from extinction.

In Australia, with so many invasive introduced species, our primary goal for conservation of biodiversity should be on minimizing further habitat fragmentation and isolation of populations, in the most cost effective way possible. Realistically, as a nation, we cannot afford to protect most of our native species in reserves run as citadels. For example, it is costly to reduce and maintain fox populations at sufficiently low numbers to allow Mallee Fowl *Leipoa ocellata* populations to recover in Yathong Nature Reserve and surrounding areas in central New South Wales. Although control of predators in reserves is now necessary to protect remnant populations of some endangered species (Newsome 1993), this is not a viable approach for all the native species which are potentially vulnerable to introduced pests.

CONCLUSIONS

Not enough is known about the impact of pests or their response to control measures. Current vertebrate pest control techniques are successful at killing animals, and they do reduce numbers; but they are expensive and there are concerns about animal welfare. We do not know enough about the impact of pests on biodiversity, or their response to different levels of control for effective management.

Fertility control is currently an ineffective, expensive technique for reducing animal damage. There are no suitable agents available for current use. If research on passive spread of immunological fertility control agents by genetically engineered organisms is successful, the potential economic and environmental advantages will be great. But a successful outcome is by no means guaranteed. Even if the research is successful, this technique is likely to be suitable only for some species, and the benefits will need to be weighed against social, ethical and ecological considerations and risks.

It is unlikely that any widespread pest animal could be eradicated on the mainland using currently available technology and it is uncertain whether regional eradication would be feasible or cost-effective. Even if the technology were up to the task, economic and social constraints and the risk of re-infestation currently make eradication an unattainable goal for most species.

Our most sensible option is to use all available information on the effect of control techniques on the dynamics of pest populations, to plan and implement cost-effective damage control. We need good information on the types and patterns of damage being inflicted by pests so we can focus and target our control efforts to get maximum damage reductions for dollars spent. The response of native animal and plant populations to feral animal control operations needs to be monitored to ensure conservation values are being protected and money is not being wasted by ineffective programmes. Even if research provides us with additional pest control tools, they will still need to be used strategically, weighing up costs against benefits.

Also, the emphasis of species conservation needs to be moved away from "citadel" reserves, and more towards integrated regional management, to prevent populations becoming increasingly fragmented and isolated.

REFERENCES

Bomford, M., 1990a. A role for fertility control in wildlife management? Bureau of Rural Resources Bulletin 7. 50 pp. Australian Government Publishing Service: Canberra.

Bomford, M., 1990b. The potential for fertility control in Australian wildlife management. Proceedings Fertility Control in Wildlife Conference: Melbourne.

Bomford, M. and O'Brien, P., 1993. Feral goat control or eradication: assessment criteria for decision making. Proceedings National Feral Goat Workshop: Dubbo, 1992.

Braysher, M., 1993. *Managing vertebrate pests: principles and strategies.* Australian Government Publishing Service: Canberra.

Burbidge, A. A. and McKenzie, N. L., 1989. Patterns in the decline of the mammal fauna of Western Australia: causes and implications for future conservation. *Biol. Conserv.* **50:** 143–98.

Cowan, P. E., 1992. The eradication of introduced Australian brushtail possums, *Trichosurus vulpecula,* from Kapiti Island, a New Zealand Nature Reserve. *Biol. Conserv.* **61:** 217–26.

Davey, S. M. and Norton, T. W., 1990. State forests and their role in wildlife conservation. *Proc. Ecol. Soc. Aust.* **16:** 323–45.

examples in Sydney are Small-leaf Privet *Ligustrum sinense* and Camphor Laurel *Cinnamomum camphora.* Large individuals have wide-spread and sometimes deep root systems and large trenches need to be dug to remove them. Glyphosate is generally used to kill large specimens of these plants.

- DICOTS WITH PRONOUNCED TAP-ROOTS. Herbs such as Catsear *Hypochaeris radicata* and the shrub Ochna *Ochna serrulata,* which reshoots from deep on the tap roots, are best dug out with a bulb trowel.

c. Machinery

Machinery is used infrequently. Piles of alien weed infested soil, heaped on the edge of bushland or along tracks can be removed. Batters along the edges of ovals or roads also frequently support an abundance of densely growing weeds. Heavy equipment has been used to scalp the propagule laden top soil and reslope the batter so that it can be planted and mulched.

SELECTION OF TECHNIQUES

Technique selection depends upon the species of weeds, the size of the plants, their reproductive state and their density. It also depends on the density of native plants present, their size and life form. Other considerations include the possibility of erosion, safety of the operator and the public, habitat preservation, public perceptions and cost.

Examples of control techniques

There is no simple answer to the question "How do I kill species x"? For example;

a. An edge with Queensland Blue Couch *Digitaria didactyla* invading the native grass *Microlaena stipoides.*

Strands of the stoloniferous Queensland Blue Couch were removed from amongst the Microlaena using a broad boning knife. At the edge, tufts of Microlaena amongst the Queensland Blue Couch were protected by milkshake cups and the dense mass of Microlaena was protected by newspaper. The remainder of the Queensland Blue Couch was then sprayed with 1:100 Glyphosate.

b. Dense germination of the highly invasive grass *Ehrharta erecta* and *Allocasuarina littoralis* after a fire.

Hand pulling the Ehrharta was rejected as the widely spreading fibrous roots dislodged the tiny She-oak seedling. Instead the area was thoroughly searched for She oak seedlings. These were covered by paper and plastic cups. The area was then sprayed with 1:100 Glyphosate.

c. A dense Blackberry *Rubus fruticosus* infestation, at least 30 years old. The infestation was sprayed four times with 1:100 Glyphosate from January to May. The whole of the area was then burnt in the following September to get rid of the dead canes and to encourage native plant germination. In the following summer any resprouting roots and rhizomes were dug out. Several rhizomes were wedged underneath and between large rocks. These were allowed to produce abundant foliage until March, when they showed signs of rooting at the cane tips. The growth was coiled up so that it occupied minimal area and was then sprayed with Tree Blackberry and Woody Weed Killer® (Triclopyr) at the recommended dose.

d. Dense *Lantana camara.* The removal of dense Lantana can pose a real conundrum for regenerators. It often provides shelter and nesting requirements for scrub-wrens and fairy wrens and its removal may cause the local extinction of these species. No solution has been found to this problem; decisions on how much Lantana to remove are normally made on the amount of other habitat and the rate of expected regrowth by native ferns and other plants.

e. Wandering Jew *Tradescantia albiflora.* No single solution has been found for the elimination of this species from a site.

- Dense growth — spray and respray repeatedly at 1:50. When the biomass is sufficiently reduced the last remaining fragments can be carefully removed manually.

- Dense growth — roll it up and rake the large pieces into a pile. This can then be composted on site (only marginally successful), composted and the surface of the pile sprayed with herbicide, or lifted out of the site and taken to a land-fill site.

  Raking often encourages native germination but many broken fragments of Wandering Jew regrow rapidly. These can be sprayed with herbicide or manually removed.

- Sparse growth — this is infrequently encountered but it is best controlled by carefully removing every fragment.

No matter what method is selected, repeated treatment will be needed before it is eliminated.

f. Cutting and Painting or injecting large stems 5 cm–30 cm diameter.

For thin stems, it is most cost effective to cut and paint but as stem size increases it becomes more effective to inject stems. This can leave a forest of dead standing stems that may be considered unsightly or unsafe by park users or managers.

Dead stems may also be left to provide perches for birds and to encourage deposition of bird dispersed fruit (especially in rainforests). On the other hand, trees may be removed so that perches are not provided and bird dispersed fruit of weeds such as Privet, Camphor Laurel and African Olive is not deposited.

Herbicide use versus manual removal

At a philosophical level this must be resolved by the park manager and the individual operator. All herbicides should be used with care so that the operator, the public and the ecosystem is not damaged. Bush regeneration in school grounds, especially when involving children, often relies solely on manual techniques. This would seem to me to be a reasonable decision. Where valuable ecosystems are under threat of collapse from rapidly growing weeds, such as vines, the benefits of careful herbicide application use far out weighs any potential damage it may do to the ecosystem.

Rubbish removal

An enormous quantity of waste biomass is created when weeds are removed. Some of the main choices are:

- removal to a land-fill site — this can be very expensive as carrying time, truck and dumping fees can be high. It does, however, remove propagules, nutrients and unwanted biomass so that the leaflitter remains sparse.
- Scatter biomass over the site — this is occasionally done when no weed propagules are present, when little native germination is expected from the soil seed bank and to prevent erosion.
- Leave plants standing dead — after trees have been injected with herbicide, leaves are shed on to the soil and the trunk and branches take years to decompose.
- Heap weeds and let them decompose — this is often used for herbaceous and small plants a long way from road access. Regrowth from plants such as Wandering Jew may need to be treated with herbicide.
- Heap weeds and burn — in Local Council areas where this is permissible, this method is being increasingly used in co-operation with the fire authorities. A mosaic of regeneration often results from this method.

A SITE EXAMPLE

This site is typical of most regeneration sites as the site history had to be obtained from local residents and was therefore vague and at times contradictory.

The entire 0.7 ha site is a western facing slope on Hawkesbury Sandstone in Sydney. It is covered in woodland to open forest dominated by *Eucalyptus haemastoma, E. gummifera* and *Angophora costata* on the upper slopes and *E. sieberi, E. piperita* and *A. costata* on the lower slopes. A trail probably constructed early this century, zig-zags down the site. A house and septic tank were constructed about 1958. Use of the septic tank stopped in the mid 1980s. A drainage easement is present along one boundary. It collects drainage from a 50 m length of sealed road. The site, except the house and possibly site 3, was burnt in wild fires in the spring of 1968.

The whole site was neglected until the summer of 1991/1992 when it was weeded. All bush was burnt by the Fire Brigade in September 1992.

All replanting in 1993 was from propagules collected on or adjacent to the site.

Site 1. Stormwater drain (Fig. 2)

Pre 1991–1992

The stormwater pipe terminated without a formal channel at the outlet. A channel had eroded for part of the length but water then fanned out widely over an area of shallow soil underlain by a sandstone shelf.

Pre-fire treatment 1991–1992

The Small and Large-leaf Privet and Mirror Bush *Coprosma repens* were cut and painted. Crofton weed was pulled out. Wandering Jew and *Erigeron karvinskianus,* present as a dense mat where the water fanned out, were sprayed several times with Glyphosate. The weeds on the edge of the fire trail, Cobblers Peg *Bidens pilosa,* Sow thistle *Sonchus oleraceus* and Fleabane *Erigeron* spp. were sprayed several times.

A small silt trap was constructed at the pipe outlet and a channel dug to the cliff edge. The channel was filled with rip-rap of sandstone. The silt trap has worked efficiently but a severe thunderstorm washed much of the sandstone rocks down below the rock shelf.

Post fire response and actions

The fire personnel failed to ignite part of the area due a low fuel load and the remainder was burnt at a very low intensity. The weeds on the fire trail continued to germinate. A few disturbance weeds also germinated along the channel. Crofton weed germinated prolifically and was pulled out as soon as it could be grasped so that it did not out-compete establishing fern gametophytes. Wandering Jew fragments continued to regrow for many months and were

Fig. 1. Native plants can be protected with plastic bags, cups and bottles before large areas of weeds are sprayed.

Fig. 2. Stormwater is controlled by a sediment trap and a channel, May 1992.

Fig. 3. Septic tank soak in the drought of 1980. Dead Crofton Weed dominates the soak.

Fig. 4. Septic tank soak, May 1992. Crofton Weed has been removed by hand. The bank of Lantana and Blackberry on the left hand side has been sprayed with Glyphosate.

Fig. 5. Septic tank soak, September 1992, one week after the fire. Scorched foliage illustrates the low intensity fire. The pipe from the septic tank is finally revealed.

Fig. 6. Septic tank soak, June 1993. The dense regrowth of the native *Imperata cylindrica* and *Hypolepis muelleri* is clearly visible. Hours of weeding were needed to stop Blackberry and Crofton Weed redominating the site.

hand removed. The most prolific native germination was *Centella asiatica* that rapidly spread over the area. Ferns established along the cut edges of the channel. Blackwattles *Callicoma serratifolia* and Gahnia *Gahnia sieberiana* were planted near the channel to see if they aided resistance to reinvasion by Privets and Crofton Weed.

Site 2. The septic tank soak (Figs 3–6)

Pre 1991–1992

The central core of the soak was dominated by Crofton Weed with scattered Blackberry. The batter of the trail above the soak was covered in dense Blackberry and the edge supported a wall of Lantana. Smaller Lantana were also present at the toe of the soak. After the mid-80s, when the septic tank was no longer used, the Crofton Weed suffered severe drought stress in dry times but still dominated the site. Blackberry cover increased in the soak from the mid-80s to 1991. Two native species Bladey Grass *Imperata cylindrica* and *Hypolepis muelleri* were always sparsely represented at the toe of the soak. After the mid 1980s these increased in cover and spread upwards onto the centre of the soak to provide a sparse cover.

Pre-fire treatment 1991–1992

The Crofton Weed was pulled out when weakened by a summer drought. Little vegetative regrowth occurred but thousands germinated. This was pulled out and the soil disturbance appeared to encourage further germination. This treatment was repeated several times to try and reduce the seed bank in the soil.

The dense wall of Lantana was initially treated by crawling into the base of the plant and cutting and painting the original stem. None of the plant died as layering was so prolific. The wall was then sprayed with 1:100 Glyphosate, which killed the plants. The Lantana at the toe of the soak were all successfully treated by cut and paint. Some smaller Blackberries were dug out but most were repeatedly treated with 1:100 and 1:50 Glyphosate over the summer of 1991–1992 and autumn of 1992.

Post-fire response and actions

The wall of Lantana and Blackberry produced a moderate to high intensity fire but the central part of the soak only experienced a low intensity burn due to the absence of fuel.

Over the following summer both the Bladey Grass and Hypolepis produced a dense cover at the toe of the soak. They slowly spread upwards onto the main body of the soak, forming a dense cover over approximately 60 per cent of the area.

At the edges of the soak, native germination was fairly prolific and included: *Eriostemon australasicus, Acacia longifolia, A. ulicifolia, Eucalyptus sieberi* and *Gahnia sieberiana.* The only native species to germinate in the centre were *Gahnia sieberiana* and a *A. longifolia.* All of these species (except *G. sieberiana*), suffered severe yellowing in the unusually overcast December and January and some died. In late February and March the yellowing was greatly reduced and the *A. longifolia* growth rates increased rapidly.

The Blackberry covered slope produced some sparse *Smilax glyciphylla* but no other native plants and very little weed. The slope remains bare.

Weed growth on the soak and immediate surrounds was dominated by dense Crofton Weed, Blackberry Nightshade *Solanum nigrum* and Ink Weed *Phytollacca octandra.* These were pulled out at approximately fortnightly intervals over the summer period so that they did not compete with native growth. This disturbance encouraged further germination of these seeds and reduced the soil seed bank. Germination decreased in early March but a torrential thunderstorm in late March resulted in substantial soil disturbance. Crofton Weed germination was again prolific, 18 months after the last flowering on the site. Most Blackberry rhizomes had been killed by prior treatment but suckering from roots was fairly common. Most were dug out. Several rhizomes were buried underneath rocks and regrowth was sprayed with Triclopyr in early March.

All weed biomass was left on site even though its removal may have reduced the amount of nutrient on site. The difficulty transporting it up a steep slope, the cost of disposal and its usefulness as erosion control material was judged to outweigh the benefits of its removal.

Several Blackwattle and Gahnia were planted at the break of slope between the trail batter and the soak in early December. These have flourished without any leaf yellowing. Before more planting was undertaken, soils from the soak and nearby were sent for analysis (Table 1). The comments on the septic tank soil included the remarks . . . "greatly elevated calcium, phosphorus and nitrate levels. This is just the sort of effect that would be seen from a pasture improvement programme, i.e., liming and superphosphate. This P level would exclude P sensitive native plants, and would greatly benefit weeds".

These results and natural regeneration indicate that Proteaceae and some members of the Epacridaceae and Myrtaceae will probably not flourish in these conditions (Handreck and Black 1991). More Blackwattles, Christmas Bush *Ceratopetalum gummiferum,* River Dog Rose *Bauera*

Table 1. Soil test results from the septic tank soak and nearby.

Test	Results Septic tank soak	Nearby soil
pH in water 1:2	4.6	4.7
pH in CaC/2 1:2	3.9	3.7
E C MS/cm 1:2	0.14	0.08
Soluble Salts	**meq.%**	
Sodium	0.05	0.21
Potassium	0.02	0.02
Calcium	0.10	0.10
Magnesium	0.08	0.28
Soluble + Exchangeable Cations	**meq.%**	
Sodium	0.15	0.35
Potassium	0.10	0.18
Calcium	2.81	0.94
Magnesium	0.82	0.86
Aluminium	0.15	0.20
Sum of Cations	4.03	2.53
Available Nutrients	**ppm**	
Phosphorus Bray No. 1	56.4	9.8
Ammonium	12.0	12.6
Nitrate	10.6	0.3
Sulphate	90.9	140.8

Adapted from results provided by Sydney Environment and Soil Laboratory.

rubioides and Black She-oak *Allocasuarina littoralis* will be planted as they tolerate periodically moist and high nutrient sites (pers. obs.). Hopefully they will increase the inertia of the system.

Site 3. Side of the property

This site measures approximately 22 × 11 m.

Pre 1991–1992

This site has been disturbed by construction of the old trail and a small shed in the late 1950s or early 1960s.

When the area was first seen in 1979 there were two large *Banksia ericifolia* present and one *Hakea teretifolia*. The *H. teretifolia* died of old age in the next few years. The site was dominated by old *Allocasuarina littoralis*, approximately half to a quarter of which died and collapsed in the next 13 years. A few heath species were present but the only clearly remembered species was *Grevillea sericea*. Over the next few years these vanished as the *Pittosporum undulatum* thickened to 50–70 per cent canopy cover. This canopy provided moist, cool conditions for the germination and establishment of a thick understorey of Small-leaf Privet, with occasional Large-leaf Privet, Camphor Laurel and Ochna.

The sewer main was laid along the fence line in 1985–86. The 2–3 metre wide clearing was mulched and planted. Most plants died under the canopy of the Allocasuarina and Pittosporum.

One corner of the site was invaded by Japanese Honeysuckle *Lonicera japonica*.

Pre fire treatment 1990–1992

In the summer of 1990–1991 the Japanese Honeysuckle was dug out and Privets, Camphor Laurels and Ochna were cut off and herbicides applied to the stump. A few large Privets and Camphor Laurel were injected with herbicide. About half a tonne of weed was removed from the site.

Over the next year *Ehrharta erecta* invaded a small section near the fence and fragments of Wandering Jew from the neighbours spread rapidly. The native grass *Microlaena stipoides* increased in cover enormously (to 5–10%), but the cover of grasses, *Paspalidium* sp., *Entolasia* sp. *Poa* sp. and *Imperata cylindrica* remained unaltered and sparse. No sclerophyllous shrubs germinated.

Prior to the fire a control line was raked along the fence line. Several small piles of timber were added to help increase the fuel load, and hence increase fire and soil temperatures. A low intensity fire burnt the whole site. A couple of the piles burnt for several hours.

Post fire response and actions

There appeared to be no link with indigenous shrub germination and the piles as germination was fairly even over the site. One of the piles of fuel obviously aided the germination of the non-local *Acacia baileyana*.

Native germination (Table 2) was unexpectedly prolific. *Acacia suaveolens* germinated quite prolifically over the whole site. *Acacia binervia*, never recorded on the property, was less dense but scattered all over the site. The eucalypts, *Lepidosperma* sp, *Platysace lanceolata* and *Micrantheum* were only represented by less than half a dozen individuals. *Gahnia, Bossiaea, Dillwynia, Pultenaea, Acacia longifolia* and *A. linifolia* were slightly more prolific, but still represented by limited number of individuals (<20 each). *Grevillea sericea* germinated in three distinct groups of 5–20 plants each. *Hakea teretifolia* and *Banksia ericifolia* failed to germinate.

Over the summer (1992–1993) following the fire the *Microlaena, Oplismenus, Paspalidium* and *Entolasia* germinated and flowered prolifically. Cover by these grasses and *Imperata* increased to approximately 60 per cent by April 1993. The Poa re-shot and flowered but germination or increase in cover was not noted.

Allocasuarina littoralis germination was prolific and continued to at least the next April. The old trees were not killed by the low intensity fire but they were considerable weakened. Several died over the next few months and piles of sawdust, as a result of borer damage, were a common feature.

Table 2. Post-fire regrowth of native species. Site 3 — Mt Ku-ring-gai.

	Germination only			No germination of regrowth		Vegetative regrowth	
Species	Remembered 13 years ago	Not present 18 months prior to fire	Present at time of fire	Present 13 years ago	Present at time of fire	Vegetative growth only	Vegetative regrowth and germination
Acacia binervia	N	√					
Acacia linifolia	N	√					
Acacia longifolia	N	√					
Acacia suaveolens	N	√					
Allocasuarina littoralis			√				
Banksia ericifolia					√		
Bossiaea scolopendria	N	√					
Dianella caerulea							√
Dillwynia retorta	N	√					
Entolasia sp.							√
Eucalyptus gummifera	N	√					
Eucalyptus haemastoma	N	√					
Gahnia sieberiana	N	√					
Grevillea sericea	Y	√					
Hakea teretifolia				√			
Hypericum sp.	N	√					
Imperata cylindrica						√	
Juncus sp.			√				
Lepidosperma laterale	N	√					
Micrantheum ericoides	N	√					
Microlaena stipoides							√
Omalanthus populifolius			√				
Oplismenus sp.			√				
Paspalidium sp.							√
Pittosporum undulatum			√				
Platysace lanceolata	?	√					
Poa sp.						√	
Pultenaea sp.	N	√					

Many *Pittosporum undulatum* shrubs were killed outright in the fire. Bark damage to those that remained alive was severe (often more than 75 per cent of the circumference killed) and borer attack was severe over the summer. Many trees died. Over-all canopy cover from the *Allocasuarina* and *Pittosporum* was reduced from approximately 50–70 per cent to <5 per cent.

Despite the fact that the fire stimulated germination, it was still considered that the ecosystem had collapsed. Two plant families prolific in the nearby healthy bush, Epacridaceae and Rutaceae, were totally absent. Myrtaceae was only represented by a couple of *Eucalypt* seedlings, Proteaceae only represented by three groups of one *Grevillea* species and Fabaceae (Faboideae) only represented very sparsely by three species.

The Proteaceae, *Banksia ericifolia, Banksia oblongifolia, Hakea sericea* and *Banksia serrata* have been replanted. In the next spring a diversity of sclerophyllous shrubs will be planted. Grass regrowth may need to be controlled.

Privet, Camphor Laurel and Ochna germination was almost nil. Pittosporum germinated in a few small clumps. Wandering Jew survived the fire and was manually removed as it regrew. Weeds of disturbed places such as *Erigeron* spp., *Sonchus oleraceus* and *Euphorbia peplus* germinated readily as there was a nearby seed source. Ink weed and Blackberry Nightshade germinated from the soil seed bank. All weeds were manually removed, except Ehrharta which flourished. This was sprayed with 1:100 Glyphosate when it could be separated from native regrowth.

Site 4. Healthy bush

The healthy bush consisted of an aging heathy understorey dominated by *Banksia ericifolia* and *Hakea teretifolia.* Little pre or post-fire treatment was given. All existing shrubs and herbs are resprouting or re-establishing from seed. No intervention, apart from the removal of an occasional weed, will be required. For example, Crofton Weed has germinated sparsely over the site. This will be low priority (in contrast to the drain site) as it is competing very poorly for resources. It will be removed before flowering.

Summary

With hindsight the lessons were;

- the site should have been treated earlier before deterioration was as serious, for financial as well as biological reasons.
- even though 0.7 ha is a very small site, it may have been better to burn the site sections at a time so that extreme weather conditions would have had less impact on regenerating native plants.

- weed control of resprouting weeds before a fire is essential as post fire control should be kept to a minimum as trampling and minute amounts of soil disturbance crush and dislodge tiny seedlings.
- control of geminating weeds by herbicide or mechanical means is painstaking for the above reasons.
- stable storm water channels are essential before a fire so that regeneration is not disturbed after the fire.
- germination of native plants was better than expected in all areas of moderate fire intensity.
- sites which carried little fuel should have had fuel transferred to them as germination of native plants was extremely poor in areas of very low intensity or where the fire failed to burn.

The future appears to be;

- replanting in areas with little native germination.
- several years of intensive and extensive weed control of species such as Crofton Weed, Fleabane and Catsear.
- healthy vigorous bushland which will support a far greater diversity of wildlife than it did pre-fire.

REFERENCES

Bradley, J., 1988. *Bringing back the bush: the Bradley method of bush regeneration.* Lansdowne Press: Sydney.

Buchanan, R. A., 1989. *Bush regeneration, recovering Australian landscapes.* TAFE, New South Wales: Sydney.

Buchanan, R. A., 1989. Pied Currawongs *(Strepera graculina):* their diet and role in weed dispersal in suburban Sydney, New South Wales. *Proc. Linn. Soc. NSW* **111:** 241–55.

Handreck, K. A. and Black, N. D., 1991. *Growing media for ornamental plants and turf.* New South Wales University Press: Sydney.

Hobbs, R. J. (ed), 1992. *Biodiversity of mediterranean ecosystems in Australia.* Surrey Beatty & Sons: Chipping Norton.

Majer, J. D., 1990. Rehabilitation of disturbed land. Long-term prospects for the recolonization of fauna. Pp. 509–19 in *Australian ecosystems: 200 years of utilization, degradation and reconstruction* ed by D. A. Saunders, A. J. M. Hopkins and R. A. How. Proceedings of the Ecological Society of Australia. 16.

Majer J. D., 1992. Maintaining and restoring diversity. Pp. 169–91 in *Biodiversity of mediterranean ecosystems in Australia* ed by R. J. Hobbs. Surrey Beatty & Sons: Chipping Norton.

Monsanto, 1985. Control of unwanted tree growth with Roundup Herbicide. Technotes, March, Vol. 1. No. 2/R.

National Trust of Australia, New South Wales, 1991. *Bush regenerators' handbook.* Sydney.

Parsons, W. T. and Cuthbertson, E. G., 1992. *Noxious weeds of Australia.* Inkata Press: Melbourne.

Rawling, J., 1991. *The bush next door.* Brisbane City Council and Greening Australia.

Weeds and feral animals: management of future threats

F. D. PANETTA[1] and J. K. SCOTT[2]

While it is obvious that the progenitors of feral animals were purposefully brought to this country, it is not widely appreciated that a large proportion of today's serious weeds were also intentionally introduced. At present there are strict protocols concerning the entry of animals into Australia, but the quarantine "net" for plants is much looser, with fewer restrictions to the entry of new species for the nursery and aquarium trades, as well as for agricultural production. Methods for identifying potential weeds are rudimentary and will remain so until the phenomenon of weediness is better understood. Accordingly, the Precautionary Principle should be employed when determining which plant species are to be introduced to Australia. Quarantine considerations aside, it can be argued that many incipient environmental weeds are already present in Australia but their potential effects have not been recognized. Early detection of new pest organisms is critical; if early action is taken, eradication may be achieved. However, campaigns to eradicate pests are expensive and commonly fail. Management of chronic, longstanding problems must utilize a variety of methods, including biological control when an effective agent(s) can be found. Since the best approach to future pest problems emphasizes prevention, educational programmes should cover all aspects of the problem, from the threats to the environment posed by commercially-driven introduction of new species, to the prompt recognition of emerging weeds.

INTRODUCTION

NUMEROUS instances are known where the impact of feral animals and weeds in Australia has led to habitat simplification and the decline of native animal and plant species (Alexiou 1983; Braithwaite *et al.* 1984; Braithwaite *et al.* 1989). The effects of pest organisms (our definition of "pest" includes both plants and animals) upon biodiversity may be direct or indirect. As an example of the latter, feral herbivores are often vectors for weed spread. Furthermore, interactive effects occur between feral animals and weeds, for example the interactions between water buffalo and *Mimosa pigra* in the Northern Territory (Braithwaite *et al.* 1989) or between Rubber Vine *Cryptostegia grandiflora* and feral pigs in the vine thickets of tropical northeastern Australia (M. Barker, pers. comm.). Levels of biodiversity in different parts of Australia reflect the impacts resulting from 200 years of naturalization and spread of exotic animals and plants. In this paper our primary emphasis is upon how new threats, in particular weed threats, can be recognized and managed. We also address briefly the future impact and management of the feral animals and weeds which already occur in Australia.

FERAL ANIMALS

We use the definition of a feral animal as a wild descendant of a domesticated animal (Wilson *et al.* 1992). Feral animals listed by Wilson *et al.* (1992) include the camel, cat, cattle, dog, donkey, goat, horse, pig, sheep and water buffalo. The only invertebrate which could qualify as a feral animal is the Honeybee *Apis mellifera*.

Continued management effort will be required to prevent the population growth and spread of most of these species. Owing to the very close affinities with domestic populations, however, it is unlikely that biological control methods will be seriously contemplated. Otherwise, it is critical that management of animal pests is developed and implemented in a flexible manner (Walters and Holling 1990), in order to cope best with the vicissitudes of climate, market fluctuations and social goals, to name a few factors.

The Vertebrate Pests Committee (VPC), a technical committee of the Standing Committee on Agriculture and Resource Management (SCARM), provides advice on the control of the import and keeping of exotic vertebrates. VPC undertakes a continuing review of vertebrate pests in Australia, including measures for the exclusion of potential pests from overseas. Animals are assigned to one of five categories, ranging from exclusion to unrestricted entry and keeping (Table 1). The procedure for recategorization (from Category 1 to Category 2 or from Category 3 to Category 4) is rigorous and an application may be rejected if opposed by any VPC member. A decision model for assessing risk in recategorizing exotic vertebrates to Categories 3 or 4 constitutes a further refinement of this procedure (Bomford 1991).

Since feral animals are the descendants of domesticated populations, new feral taxa in Australia would most likely arise from Category 4 (Table 1). This indicates that the introduction of either domesticated species new to Australia or new breeds of species already resident in Australia should be contingent upon an

[1]Alan Fletcher Research Station, Queensland Department of Lands, P.O. Box 36, Sherwood, Queensland, Australia 4075.
[2]CSIRO Division of Entomology, Private Bag P.O. Wembley, Western Australia, Australia 6014.

Pages 220–28 *in* CONSERVING BIODIVERSITY: THREATS AND SOLUTIONS, ed by R. A. Bradstock, T. D. Auld, D. A. Keith, R. T. Kingsford, D. Lunney and D. P. Sivertsen. Surrey Beatty & Sons. 1995.

Table 1. Categorization of animals by the Vertebrate Pest Committee (SCA 1991).

Category 1 *Animals, the entry and keeping of which is prohibited*

(a) Animals which are considered to have major pest potential and which should not be allowed to enter, nor be kept in state or territory. (Special consideration may be given to scientific institutions on a case by case basis).

(b) Animals not assigned to any other category.

Category 2 *Animals limited to restricted collections*

Animals believed to have a high pest potential, significant conservation value or to be highly dangerous to humans. May be introduced and/or kept only in statutory zoos or other special collections endorsed for the relevant taxa.

Category 3 *Animals permitted in other collections*

(a) Animals for which permits may be issued to zoological premises maintained primarily for the purpose of exhibition, education, entertainment or conservation which are recognized by the relevant Authority (this category applies to 'B CLASS' zoos, circuses and wildlife parks as well as statutory zoos).

(b) Animals which may be kept for private, commercial or exhibition purposes but which require the issue of an appropriate permit. This category will include animals kept in such commercial enterprises as deer or game bird farms.

Category 4 *Animals, the entry and keeping of which is unrestricted*

Animals which are recognized as domestic and/or farm animals, or other animals having no pest potential. Included in this category are animals that may have pest potential and that are kept widely in Australia.

assessment of potential pest status. While the recommendation was made by the Animal Production Committee of the SCA that an investigation of environmental impact be included in test programmes for new breeds of sheep and goats in arid or semi-arid areas (Anon. 1979), this appears not to have been adopted (Bomford 1991). Species with prospects for adoption as pets, and considered to have major pest potential (e.g., the hamster or gerbil) would be assigned to either Category 1 or 2. Thus, the current requirement for assessment of pest potential should prevent the widespread introduction of new species likely to establish feral populations, particularly if a wide margin for error is permitted in the risk assessment process (Bomford 1991).

While we have employed the strict definition of "feral", we acknowledge that many "non-feral" exotic animals are current (e.g., the wild rabbit and fox) or potential environmental pests. There is a need for improved recognition and management of the risks associated with keeping of potential pest vertebrate species in private collections as well as in major zoos. Category 1 species are being kept, and have escaped from, some zoos. Emergency plans for responding to animal escapes only take account of the risk to human safety, not the risk that escaped animals could establish wild populations (M. Bomford, pers. comm.). Many exotic bird and aquarium fish species are kept in private collections and there is a high likelihood that accidental or intentional releases from such collections will eventually lead to established wild populations. In addition, a substantial smuggling industry brings exotic species, especially birds and fish, into Australia. This obviously operates outside all of the risk assessment and management mechanisms that governments have established for environmental protection (M. Bomford, pers. comm.).

WEEDS

Naturalized species comprise about 15 per cent of the Australian flora. Of these, about 50 per cent occur within native plant communities and approximately 25 per cent are either significant environmental weeds or have the potential to be so (Humphries *et al.* 1991). Today, managers of native plant communities are confronted by many seemingly intractable weed problems. There are rarely sufficient resources to keep a single plant invader under control and yet the number of introduced plant species is increasing steadily.

The total number of species naturalized in Australia has increased linearly with time since European settlement with a rate of accretion of from four to six species per year (Specht 1981). Not every naturalized species has the potential to interfere with natural vegetation. Whatever the proportion of introduced species becoming serious environmental weeds, however, the rate of naturalization is relatively high. Unless circumstances affecting the introduction of plant species change dramatically, therefore, the potential for new invasions is essentially unlimited.

HOW ARE WEEDS INTRODUCED?

Most people are aware of the potential for accidental introduction of weeds. Weed seeds may be introduced inadvertently on the clothes, shoes or baggage of international travellers.

There is a general awareness of the necessity to import only uncontaminated plant products. Quarantine bodies have engendered a "Protect Australia" mentality in an attempt to reduce the risks of introduction of pests and diseases in general. This reflects an overemphasis, in our opinion, on accidental introduction as a means of bridging the geographical barriers surrounding Australia (see Wace 1985). It also underplays the difficulty, from a weed's point of view, of establishing a viable population from a single or possibly a few seeds. However, what proportion of the general public is aware that perhaps the majority of our most serious weeds have been introduced and spread *intentionally*?

With regard to the legislation which is enacted to enforce control of weeds, more than 220 plant introductions have been proclaimed noxious weeds in Australia (Parsons and Cuthbertson 1992). Of these at least 46 per cent were introduced intentionally, 31 per cent of the total as ornamental plants (Fig. 1). Such noxious plants include agricultural weeds as well as environmental weeds. However, an examination of the plants considered by Humphries *et al.* (1991) to be Australia's major environmental weeds (Table 2) reveals that in only one instance (*Mimosa pigra*) is there a *possibility* that introduction might have been accidental and without a history of intentional spread by man.

The mode of introduction of *Mimosa pigra* has been contentious, in relation to whether the plant was first introduced on stock illegally imported from south-east Asia or if it was introduced to the Darwin Botanic Gardens, which had an active seed introduction programme (Swarbrick 1983; Miller and Lonsdale 1987). In the latter case, it is unclear whether *M. pigra* was

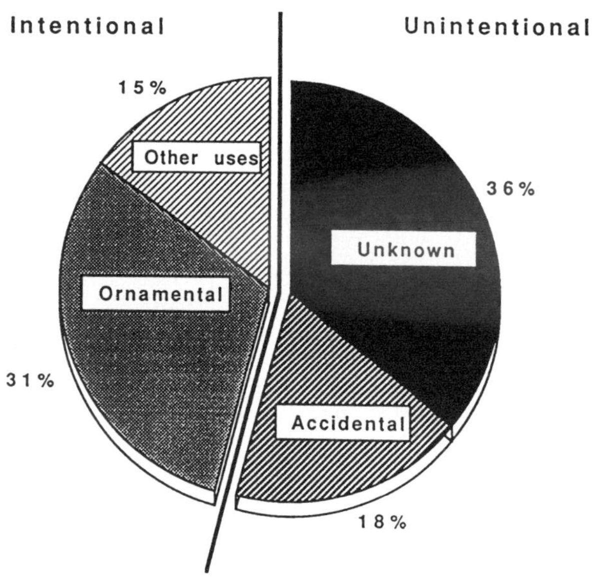

Fig. 1. Modes of introduction of Australian noxious weeds (from Panetta (1993)).

intentionally introduced to Australia or was a contaminant in other seed (Miller and Lonsdale 1987). Its early introduction and cultivation at botanic gardens in Bogor and Lae, however, is suggestive (Miller and Lonsdale 1987). An interesting contrast is provided by the example of Bitou Bush *Chrysanthemoides monilifera* ssp. *rotundata*. This plant appears to have been introduced in ships' ballast which was dumped on the banks of the Hunter River about 1908 (Parsons and Cuthbertson 1992). Because it was recognized as a vigorous invader in the Newcastle district, Bitou Bush was proclaimed a noxious weed there. However, because of its ability to stabilize drift areas in coastal habitats, Bitou Bush was subsequently *removed* from the noxious weeds list prior to widespread utilization for dune stabilization (Mort and Hewitt 1953).

Clearly any attempt to deal with the threat of intentionally introduced species must address conflicts of interest within society. The Bitou Bush example is an unusual one where different aspects of environmental protection (weed control and sand stabilization) were in conflict. Instances involving ornamental or aquarium plant species involve the opposition of commercial and environmental interests. In Victoria, 251 of the 584 weedy taxa listed by Carr *et al.* (1992) are available from commercial nurseries. Of these, 137 (54%) were regarded as serious or very serious threats to one or more vegetation formations, 110 (44%) were regarded as potential threats and only 4 (1.4%) were considered to be no threat. In most cases it can be argued retrospectively that the commercial gains made by selling such species are very small in relation to the environmental damage incurred by their introduction. Another line of cost-benefit argument might address the benefits to a small section of the community in comparison with the costs experienced by the community at large. However, the most difficult conflicts to resolve will be those involving the introduction of species for the purpose of increasing primary production.

The introduction of Prickly Acacia *Acacia nilotica* as a shade tree for sheep in western Queensland led to increased lambing percentages. Ironically, at the time of introduction (c. 1926), a warning was made that seedling recruitment was much more frequent from seeds that had been ingested by cattle, as compared with sheep (Parsons and Cuthbertson 1992). Later introductions for shade and food value preceded a change from sheep to cattle husbandry during the 1970s, which aided the subsequent massive buildup of infestations. Prickly Acacia poses a serious threat to the formerly treeless Mitchell Grass Downs, yet where it is present at relatively low densities (a stage of invasion where the benefits gained outweigh the costs incurred), some landholders oppose its removal.

Table 2. Australia's major environmental weeds (after Humphries *et al.* 1991).

Species	Habitat/community threatened	Approx. date introduced	Purpose for which introduced or used
Cryptostegia grandiflora	vine thickets, riparian	1860s	ornamental
Eichhornia crassipes	aquatic	1890s	ornamental
Myrsiphyllum asparagoides	coastal and inland woodlands and shrublands	1880s	ornamental
Thunbergia grandiflora	rainforest	1930s	ornamental
Tamarix aphylla	riparian	1930s	ornamental/windbreak
Chrysanthemoides monilifera			
ssp. *monilifera*	coastal/subcoastal non-aquatic	1850s	ornamental
ssp. *rotundata*	coastal/subcoastal non-aquatic	1910	soil stabilization
Parkinsonia aculeata	grasslands, wetlands	1890s	ornamental, shade
Prosopis spp.	riparian, grasslands	1920s	ornamental, shade, fodder
Acacia nilotica	grasslands	1890s	ornamental, shade, fodder
Mimosa pigra	wetlands and floodplains	1870–90	botanic gardens (accidental?)
Salvinia molesta	aquatic	1950s	aquarium plant
Pennisetum polystachion	wet-dry forests and woodlands	1970s	agronomic
Brachiaria mutica	wetlands, aquatic	1880s	agronomic
Glyceria maxima	wetlands, aquatic	?	agronomic
Echinochloa polystachya	wetlands, aquatic	1970s	agronomic
Hymenachne amplexicaulis	wetlands, aquatic	1970s	agronomic
Cenchrus ciliaris	mesic habitats in arid/ semi-arid regions	1950s	agonomic

Perennial grasses are a major component of the weeds originally introduced for agronomic purposes (Table 2). Of particular concern are the grasses that colonize continuously and periodically wet areas, including the ponded pasture systems in the wet-dry climatic regions of tropical Australia. Such plants, e.g. *Brachiaria mutica,* are highly invasive and have the potential to dominate wetland systems of many types.

Herein lies the crux of the problem: some of the characteristics that make a species useful in an agronomic context also predispose it to invasive behaviour. Mott (1986) stated that in many cases the attributes considered necessary for desirable pasture species parallel those forwarded by Baker (1974) as ideal characteristics for weeds. The need for low-cost, non-mechanical (e.g., aerial) methods for sowing pasture legumes and grasses selects for species that are able to establish readily and spread spontaneously. Furthermore, the requirement for only moderate palatability (Partridge and Wright 1992) ensures that plant losses owing to grazing pressure are relatively low. In the absence of grazing pressure, the spread and eventual dominance of such "weedy" species is enhanced. Even a plant such as the leguminous *Leucaena leucocephala,* which suffered high failure rates when first adopted (Lesleighter and Shelton 1986), has emerged as an environmental problem in tropical situations where grazing pressure is low, including urban and semi-rural areas, some rehabilitated mining areas and the banks of many coastal and sub-coastal watercourses (S. Csurhes, pers. comm.).

To date, there has been little debate in the scientific community concerning the pros and cons of the introduction of species to increase agricultural production. Not only have the issues of weediness and depletion of biodiversity been avoided, there has been no requirement (hence no attempt) to conduct environmental impact assessments, which detail the potential costs and benefits arising from intentional introductions. Assessing the costs incurred by introductions would not be easy; it is probably impossible to capture the value of biodiversity in the utilitarian and anthropocentric terms of economics. However, it is possible that society would be unwilling to accept certain risks, such as the endangerment of native plants and animals, regardless of projected benefits. Unfortunately, there is currently *no* reliable mechanism whereby proposed agricultural introductions may be evaluated, either for potential weediness or for their environmental impact. Establishing such mechanisms is arguably the most important goal in the management of future threats to Australian biodiversity.

CAN WEEDY SPECIES BE IDENTIFIED PRIOR TO INTRODUCTION?

Several analyses have shown that the majority of introduced species do not become pests. For example, in a study of pest invasions of the British Isles, Williamson and Brown (1986) found that approximately 10 per cent of potential invaders became established, and 10 per cent of those established became pests. Similarly, for Australian conditions, Groves (1986) suggested that 5 per cent of all plant introductions became naturalized and 1–2 per cent became weeds. Such statistics suggest that highly developed techniques will be necessary to predict the few species that will become significant weeds. Unfortunately, such techniques appear to be beyond reach. Weeds display a rich variety of biological and ecological attributes (Barrett and

Richardson 1986; Newsome and Noble 1986); there are no genetic, physiological or ecological characteristics that represent necessary and sufficient conditions for weediness (Panetta 1993). Furthermore, weediness may be as much an expression of the physical and biotic environment as of inherent plant attributes; many species have been promoted from relative insignificance in a native flora (e.g., *Myrsiphyllum asparagoides* in South Africa) to major weed status when introduced to Australia.

Panetta and Mitchell (1991a,b) have argued that the task of prediction is simplified when the only plants addressed are those which are considered already to be weeds somewhere in the world. This approach has recently been formalized as a basis for assessing proposed plant introductions for weed potential (Panetta 1993). However, the method is more likely to be useful with agricultural, as opposed to environmental, weeds. Scott and Panetta (1993) employed multiple logistic regressions to identify variables that best predicted Australian weed status of the approximately 240 naturalized species whose native distributions are confined to southern Africa. The highest level of variation explained (43%) was obtained for agricultural weeds using a single variable: weed status within native distribution. No variables were suitable predictors of environmental weeds.

The dynamic nature of climatic patterns raises an additional source of difficulty where predictions are concerned. From a quarantine point of view, plants that pose no threat under current climatic conditions could become weedy as global warming proceeds (Panetta 1992). Fox (1991) has suggested that if climate change in Australia were to lead to a southward shift of summer-dominated rainfall, pre-adapted species grown in southern gardens could serve as sources of propagules for the invasion of native vegetation.

MANAGEMENT

Under certain circumstances, eradication can be a highly cost-effective form of weed management. Unfortunately, the period during which eradication can be effected is all too brief: ". . . once a species has been in a region, State or country for several years — the time varying with the species and its characteristics — and is established even sparsely in a number of places, there is little possibility of eradication. The species will eventually occupy all of the sites to which it is adapted and for most species the chances of containment are negligible" (Moore 1975).

The rapidity with which prospects for eradication diminish varies according to species biology (Dodd 1990), eradication being favoured by delayed reproduction and the production of short-lived (or no) seeds. Eradication of an intentionally introduced plant species is unlikely to be achieved because: (1) multiple, widespread introductions are likely to have occurred; and (2) some proportion of the public favours utilization of the species. The existence of "lag phases" prior to the major range expansions exhibited by some invasive species (Groves 1986) could in certain instances give time to achieve eradication, *if* such species are recognized as threats to the environment. However, a "watching brief" approach will, in all probability, allow populations of potential invaders to pass beyond the eradicable stage. Again, we are limited by our predictive capacity.

The most favourable eradication prospects exist for species that are detected soon after accidental introduction. The importance of education and publicity in this regard are obvious. An alert was sounded some time ago for one such candidate, Siam Weed *Chromolaena odorata* (McFadyen 1989). Seeds of this species, an important weed in South African national parks (Macdonald 1983), could be inadvertently brought into international terminals at Cairns and Darwin on the clothes, shoes or baggage of travellers from south-east Asia. It is also essential that Australian workers remain abreast of the progress in overseas research on biological control; such research has the spinoff that if Siam Weed is controlled in Asia, there will be less danger of it invading tropical Australia (McFadyen 1989).

Conventional herbicide and cultivation techniques are not suitable for long-term control of weeds in areas reserved for the protection of native species. Biological control, combined with other management techniques such as fire, is often the only ecologically sensitive option available. Research directed towards the biological control of weeds that threaten biodiversity is comparatively recent in Australia when compared with research for agricultural weeds, and few projects have had the time to progress to the point of significant control. Half the weeds listed in Humphries *et al.* (1991) (Table 2) are currently targeted for biological control. One of the world's most important aquatic weeds, *Salvinia molesta,* is a spectacular example of how biological control can work. This weed initially proved difficult to control until the introduction of the weevil, *Cyrtobagous singularis,* which has provided continuous control over a large part of the weed's world distribution (Room *et al.* 1981). Another aquatic weed, *Eichhornia crassipes,* is under good control in some situations (Wright 1981; Julien 1992). To date, these successes have not been matched by the degree of control obtained for terrestrial weeds. However, projects on *Chrysanthemoides monilifera* (Adair and Scott

1989), *Mimosa pigra* (Lonsdale *et al.* 1989) and *Parkinsonia aculeata* (Julien 1992) have progressed in Australia to the point of release of biological control agents. One of the greatest challenges facing biological control research lies in the selection of effective agents. Many agents that are released either do not survive or do not provide significant control (Julien 1992). Crawley (1989) has estimated that one in six of the insects released for biocontrol purposes achieved some degree of control. Unfortunately, predicting which agents will be effective is proving as elusive as predicting which plants will become weeds. For this reason, biological control should not be assumed to be the "magic bullet" that will always solve a weed problem.

Furthermore, there is a real risk that, even should the biocontrol agent(s) effectively reduce the density and vigour of a target weed, the end result could be replacement by another weed species. The approach of integrated or "ecological" control (Groves 1991) aims to use several methods of control, considering the biology of the weed in relation to that of the native species with which it is growing. Hobbs and Huenneke (1992) assessed the importance of disturbance regimes to the diversity and invasibility of natural ecosystems. One of their conclusions was that land managers must make choices of which taxa to favour, since species vary so much in their response to disturbance. It is clear from the above that managers should be acting, where possible, to disfavour the invasive species with the greatest potential impact upon biodiversity, while favouring their natural competitors. The complexity of this task increases as each additional weed invades a community.

MOVEMENT OF PLANT SPECIES WITHIN AUSTRALIA

In addition to the threat posed to the environment by introducing new species to Australia, problems can arise from the translocation of native species within the country. That relatively few Australian plants have become weedy in native communities was considered by Gullan (1988) to be a consequence of historical, rather than ecological factors. Plants that are opportunistic and are native to ecologically similar areas would be most likely to become invaders when given a chance; this usually occurs when they are cultivated outside their normal range. Species such as *Pittosporum undulatum, Acacia longifolia* and *Leptospermum laevigatum* have invaded and achieved dominance in woodland communities near Melbourne (Gullan 1988). Similarly, the Umbrella Tree *Schefflera actinophylla*, endemic to the rainforests of northern Queensland, has invaded both rainforests and coastal woodlands in southeastern parts of the state.

Biological control may be a future option even though a plant is native to Australia. Native weeds of eastern Australia that come from floristically isolated regions such as southwestern Australia (e.g., *Sollya heterophylla* and *Paraserianthes lophantha*) could be surveyed for potential biological control agents in the region of origin. In the case of *P. lophantha* and for other Australian legumes that are weeds overseas, a number of biological control agents are already known (Julien 1992).

ROLE OF LEGISLATION AND EDUCATION IN RELATION TO WEEDS

The Federal *Quarantine Act* 1908 provides the power to control plant importations and contains schedules under Proclamation 86P listing species that are prohibited entry to Australia. The Australian Quarantine and Inspection Service (AQIS) administers this Act. At present, however, the Act is inadequate to prevent the introduction of many potential weeds, especially those which could invade native plant communities. Furthermore, under the present system, the onus rests with AQIS and its advisers to establish why material should not be imported. There is thus an urgent need to incorporate a method of assessment that can identify species that pose potential risks to the environment. Given the difficulties of predicting environmental weeds, such a method would have to operate in a highly conservative manner, according to the Precautionary Principle (Young 1993). This principle highlights the caution that is required in making decisions when there is a risk of irreversible environmental degradation. It is to be expected that mistakes will be made with any assessment system. According to the Precautionary Principle, it would be better to exclude a harmless (yet potentially beneficial) plant than to admit erroneously a harmful one.

Scott and Panetta (1993) have suggested that, given the unreliability of predictions relating to potential environmental weed status, any species intended for introduction should be grown under secure conditions in a number of environments until its weediness can be assessed. Hazard (1988) stated earlier that nursery and early palatability evaluation of pasture grass introductions should be conducted under virtual quarantine conditions. Features which should be examined include characteristics of seeds and seed banks, growth rates, competitiveness, capacity for vegetative regeneration, age at first reproduction, fecundity and attack by natural enemies (Panetta 1993). An appraisal of the plant's susceptibility to commonly used herbicides should also be made at this stage. Unfortunately, one of the most important aspects of weediness can not be investigated; intelligent guesses might be made about the effectiveness of

Table 3. Potential environmental weeds whose introduction is prohibited in Queensland under the *Rural Lands Protection Act* 1985. (Note: a few species have restricted distributions elsewhere in Australia).

Species	Habitat/community threatened	Purpose for which introduction may occur
Zizyphus spina-christi	grasslands	agronomic
Acacia spp.	grasslands	agronomic
Mikania spp.	rainforest margins	ornamental
Equisetum spp.	wetlands and riparian	ornamental
Sesbania punicea	riparian	ornamental
Clidemia hirta	rainforest margins	ornamental
Pontederia rotundifolia	wetlands	ornamental
Pithecellobium dulce	grasslands and coastal communities	ornamental or fodder plant
Rubus argutus	forests	ornamental or food plant
Eichhornia azurea	aquatic	ornamental or aquarium plant
Sagittaria pygmaea	wetlands	ornamental or aquarium plant
Limnocharis flava	wetlands, aquatic	ornamental or aquarium plant
Ludwigia peruviana	wetlands	ornamental or aquarium plant
*Trapa*spp.	aquatic	ornamental, aquarium or food plant
Myriophyllum spicatum	aquatic	aquarium
Salvinia spp.	aquatic	aquarium
Gymnocoronis spilanthoides	wetlands, aquatic	aquarium
Lagarosiphon major	aquatic	aquarium

potential dispersal agents, but the need for security must exclude an examination of dispersal *per se*. For potential environmental weeds, pains should be taken to ensure that assessments are undertaken in representative natural habitats. Otherwise the procedure will be more appropriate for the identification of plants that will be weedy in simplified, agricultural environments.

This type of procedure would undoubtably prove expensive; it has been suggested that the potential importer should bear the associated costs (Humphries *et al.* 1991). If such a screening mechanism were in place, however, the number of candidates proposed for introduction would decline, presumably to include only those from which substantial benefits could be anticipated. Furthermore, the savings gained through avoiding purposeful introductions of weedy species are bound to be greater than screening costs by some orders of magnitude (Hazard 1988). However, it must be recognized that increased restrictions upon, or costs for, plant importation may lead to a greater incidence of smuggling.

Csurhes (1991) used a scoring system devised by S. W. L. Jacobs (see Panetta 1993) to rank 788 weed species not occurring in Australia but recognized as major problems overseas. Of the 30 most highly ranked species, plants considered to be climatically pre-adapted and likely to be intentionally introduced to Queensland were identified. These were subsequently declared prohibited species under the *Rural Lands Protection Act* (Table 3). While the use of simple scoring systems for the evaluation of weediness has its drawbacks (Panetta 1993), it is at least a step in the right direction, pending the development of more refined (and expensive) methods for assessing weed potential.

Whatever system might come into place to screen potential plant importations, the diversity of organizations which are involved in plant importation will have an important bearing upon screening effectiveness. Government and the larger commercial organizations could be expected to comply with screening system requirements, but importations are currently initiated by many small nurseries as well as by individuals. It cannot be expected that the latter groups would have the resources to underwrite formal assessments of proposed importations. Nor can it be anticipated that they would comply with the system by importing solely through legitimate channels. Undoubtedly, this is an area where education will be invaluable, with regard to heightened environmental awareness of both merchants and the public.

The depletion of biodiversity by exotic species has progressed over the past 200 years, partially out of ignorance and partly owing to the predominant values espoused by society. Importers of ornamental species are probably more concerned with continuing profitability than about the ecological consequences of disseminating potential invaders. Similarly, a goal of agricultural research has been to increase production, irrespective of the risk of recruiting serious environmental weeds. There is no doubt that the general community lacks information about the ecological consequences of plant invasions. If the public were more aware of what is at stake, would their garden purchases reflect this? Would they demand a closer scrutiny of the species whose introductions are proposed for agricultural reasons? Perhaps not. In the current situation, however, a lack of awareness precludes the exercise of choice. A broad range of education activities is required, from treatment in school curricula to issues carried by the media. Demand for weedy garden plants could be reduced by the provision in each state or region of brochures illustrating plants that should not be purchased. Members of a more environmentally aware public may not only be more careful about dumping garden cuttings, and refrain from purchasing exotic garden plants, but may also notice and report newly naturalized species.

CONCLUSIONS

Although new vertebrate threats may arise as a result of accidental or intentional release and establishment of exotic fish, birds and mammals, we consider that the greater future threat to biodiversity in Australia will come from weeds, especially from the continuing introduction of plants with undesirable characteristics. Stricter conditions are needed for plant introductions, including assessments of weed potential and environmental impact as parts of the importation requirements. For accidentally introduced plants, vigilance and rapid recognition of weediness are necessary if eradication is to be a realistic objective.

The future of Australia's natural communities depends, essentially, upon the values of her human population. Given the record of this species, there is every chance that priorities will remain largely economically driven. However, economic considerations needn't necessarily be pitted against environmental concerns. For example, the nursery trade could explore the commercial potential of a diverse array of native species, in addition to many exotic taxa which are proven to be non-invasive. Given a certain degree of flexibility from all parties involved, the necessary compromises should be achievable. Otherwise, the very real threat is that losses of biodiversity over the next 200 years will exceed those that have occurred during the past two centuries.

ACKNOWLEDGEMENTS

We thank Mary Bomford, Steve Csurhes, Stella Humphries, Mark Lonsdale, James Ridsdill-Smith, Joe Scanlan and Paul Yeoh for discussion and comments on drafts of the manuscript.

REFERENCES

Adair, R. J. and Scott, J. K., 1989. The life-history and host specificity of *Comostolopsis germana* Prout (Lepidoptera: Geometridae), a biological control agent of *Chrysanthemoides monilifera* (Compositae). *Bull. Ent. Res.* **79:** 649–57.

Alexiou, P. N., 1983. Effect of feral pigs *(Sus scrofa)* on sub-alpine vegetation at Smokers Gap, Australian Capital Territory. *Proc. Ecol. Aust.* **12:** 135–42.

Anon., 1979. Exotic breeds of sheep and goats for import into Australia. Report of the Expert Panel appointed by the Animal Production Committee of Standing Committee on Agriculture. SCA Technical Report Series No. 1: Canberra.

Baker, H. G., 1974. The evolution of weeds. *Ann. Rev. Ecol. Syst.* **5:** 1–24.

Barrett, S. C. H. and Richardson, B. J., 1986. Genetic attributes of invading species. Pp. 21–33 in *Ecology of biological invasions: an Australian perspective* ed by R. H. Groves and J. J. Burdon. Australian Academy of Science: Canberra.

Bomford, M., 1991. Importing and keeping exotic vertebrates in Australia. Criteria for the assessment of risk. Bureau of Rural Resources Bulletin No. 12, Australian Government Publishing Service: Canberra.

Braithwaite, R. W., Dudzinski, M. L., Ridpath, M. G. and Parker, B. S., 1984. The impact of water buffalo on the monsoon forest ecosystem in Kakadu National Park. *Aust. J. Ecol.* **9:** 309–22.

Braithwaite, R. W., Lonsdale, W. M. and Estbergs, J. A., 1989. Alien vegetation and native biota in tropical Australia: the impact of *Mimosa pigra*. *Biol. Conserv.* **48:** 189–210.

Carr, G. W., Yugovic, J. V. and Robinson, K. E., 1992. *Environmental weed invasions in Victoria: conservation and management implications.* Department of Conservation and Environment, and Ecological Horticulture Pty Ltd: Melbourne.

Crawley, M. J., 1989. The success and failure of weed biocontrol using insects. *Biocon. News Inform.* **10:** 213–24.

Csurhes, S. M., 1991. Review of 86P quarantine legislation. Plant species of which importation represents a threat to the Australian economy and environment. Queensland Dept Lands: Brisbane (under contract to Australian Quarantine and Inspection Service, DPIE: Canberra).

Dodd, J., 1990. The role of ecological studies in assessing weed eradication programmes. Pp. 416–26 in *Proceedings of the 9th Australian weeds conference* ed by J. W. Heap. Adelaide.

Fox, M. D., 1991. Developing control strategies for environmental weeds. *Plant Prot. Quart.* **6:** 109–10.

Groves, R. H., 1986. Plant introductions in Australia: an overview. Pp. 137–49 in *Ecology of biological invasions: an Australian perspective* ed by R. H. Groves and J. J. Burdon. Australian Academy of Science: Canberra.

Groves, R. H., 1991. Status of environmental weed control in Australia. *Plant Prot. Quart.* **6:** 95–98.

Gullan, P., 1988. Weeds in Victoria: where are we? Pp. 42–45 in *Weeds on public land* ed by R. G. Richardson. Weed Science Society of Victoria: Parkville.

Hazard, W. H. L., 1988. Introducing crop, pasture and ornamental species into Australia — the risk of introducing new weeds. *Aust. Plant Intro. Rev.* **19:** 19–36.

Hobbs, R. J. and Huenneke, L. F., 1992. Disturbance, diversity, and invasion: Implications for conservation. *Conserv. Biol.* **6:** 324–37.

Humphries, S. E., Groves, R. H. and Mitchell, D. S., 1991. Plant invasions of Australian ecosystems. A status review and management directions. *Kowari* **2.**

Julien, M. H., 1992. *Biological control of weeds: a world catalogue of agents and their target weeds.* 3rd ed. C.A.B. International: Wallingford.

Lesleighter, L. C. and Shelton, H. M., 1986. Adoption of the shrub legume *Leucaena leucocephala* in central and south-east Queensland. *Trop. Grassl.* **20:** 97–106.

Lonsdale, W. M., Miller, I. L. and Forno, I. W., 1989. The biology of Australian weeds. 20. *Mimosa pigra* L. *Plant Prot. Quart.* **4:** 119–31.

McFadyen, R. E. C., 1989. Siam weed: a new threat to Australia's north. *Plant Prot. Quart.* **4:** 3–7.

Macdonald, I. A. W., 1983. Alien trees, shrubs and creepers invading indigenous vegetation in the Hluhluwe-Umfolozi Game Reserve Complex in Natal. *Bothalia* **14:** 949–59.

Miller, I. L. and Lonsdale, W. M., 1987. Early records of *Mimosa pigra* in the Northern Territory. *Plant Prot. Quart.* **2:** 140–42.

Moore, R. M., 1975. An ecologist's concept of a noxious weed: plant outlaw? *J. Aust. Inst. Agric. Sci.* **41:** 119–21.

Mort, G. W. and Hewitt, B. R., 1953. Vegetation survey of the marine sand drifts of New South Wales. Some remarks on useful stabilizing species. *J. Soil Conserv. NSW* **9:** 59–69.

Mott, J. J., 1986. Planned invasions of Australian tropical savannas. Pp. 89–96 in *Ecology of biological invasions: an Australian perspective* ed by R. H. Groves and J. J. Burdon. Australian Academy of Science: Canberra.

Newsome, A. E. and Noble, I. R., 1986. Ecological and physiological characters of invading species. Pp. 1–20 in *Ecology of biological invasions: an Australian perspective* ed by R. H. Groves and J. J. Burdon. Australian Academy of Science: Canberra.

Panetta, F. D., 1992. The role of prediction in determining the quarantine significance of weeds. Pp. 381–83 in *Proc. 1st int. weed control conf., Vol. 2* ed by R. G. Richardson. Weed Science Society of Victoria: Melbourne.

Panetta, F. D., 1993. A system of assessing proposed plant introductions for weed potential. *Plant Prot. Quart.* **8:** 10–14.

Panetta, F. D. and Mitchell, N. D., 1991a. Homoclime analysis and the prediction of weediness. *Weed Res.* **31:** 273–84.

Panetta, F. D. and Mitchell, N. D., 1991b. Bioclimatic prediction of the potential distributions of some weed species prohibited entry to New Zealand. *N.Z. J. Agric. Res.* **34:** 341–50.

Parsons, W. T. and Cuthbertson, E. G., 1992. *Noxious weeds of Australia.* Inkata Press: Melbourne.

Partridge, I. J. and Wright, J. W., 1992. The value of round-leafed cassia (*Cassia rotundifolia* cv. Wynn) in a native pasture grazed with steers in south-east Queensland. *Trop. Grassl.* **26:** 263–69.

Room, P. M., Harley, K. L. S., Forno, I. W. and Sands, D. P. A., 1981. Successful biological control of the floating weed salvinia. *Nature* **294:** 78–80.

Scott, J. K. and Panetta, F. D., 1993. Predicting the Australian weed status of southern African plants. *J. Biogeog.* **20:** 87–93.

Specht, R. L., 1981. Major vegetation formations in Australia. Pp. 165–297 in *Ecological biogeography of Australia* ed by A. Keast. Dr W. Junk: The Hague.

Standing Committee on Agriculture, 1991. Guidelines for the control of exotic vertebrate animals. Report of a working party of the Vertebrate Pests Committee of the Standing Committee on Agriculture, SCA Technical Report 37, CSIRO Australia, 20pp.

Swarbrick, J. T., 1983. The introduction of *Mimosa pigra* into northern Australia. *Aust. Weeds* **2:** 124.

Wace, N. M., 1985. Australia — the isolated continent. Pp. 3–22 in *Pests and parasites as migrants; an Australian perspective* ed by A. J. Gibbs and H. R. C. Meischke. Australian Academy of Science: Canberra.

Walters, C. J. and Holling, C. S., 1990. Large-scale management experiments and learning by doing. *Ecology* **71:** 2060–68.

Williamson, M. H. and Brown, K. C., 1986. The analysis and modelling of British invasions. *Phil. Trans. Roy. Soc. Lond., Ser. B,* **314:** 505–22.

Wilson, G., Dexter, N., O'Brien, P. and Bomford, M., 1992. Pest animals in Australia. a survey of introduced wild mammals. Bureau of Rural Resources and Kangaroo Press: Kenthurst, New South Wales.

Wright, A. D., 1981. Biological control of water hyacinth in Australia. Pp. 529–35 in *Proceedings of the V international symposium on biological control of weeds* ed by E. S. Delfosse. CSIRO: Melbourne.

Young, M. D., 1993. For our children's children: Some practical implications of inter-generational equity, the precautionary principle, maintenance of natural capital, and the discount rate. Working Document 93/5. CSIRO Division of Wildlife and Ecology: Canberra.

Commercial Use of Native Biota

PLATE 16 *(above)*. The harvesting of native forests, particularly the practice of clear-felling as shown here in Mumbulla State Forest in southeastern New South Wales, remains as one of the most controversial conservation issues. At issue is the compatability of a range of forest management practices with the conservation of forest biodiversity, including ecosystem services such as yield of water to users downstream of commercially exploited forests. *(Photograph by Dan Lunney.)*

PLATE 17 *(below)*. Illegal trade in wildlife, such as the Galah *Cacatua roseicapilla* shown here, promotes demand and inflicts unknown pressures on both common and rare species of animals. *(Photograph courtesy of Australian Customs Service.)*

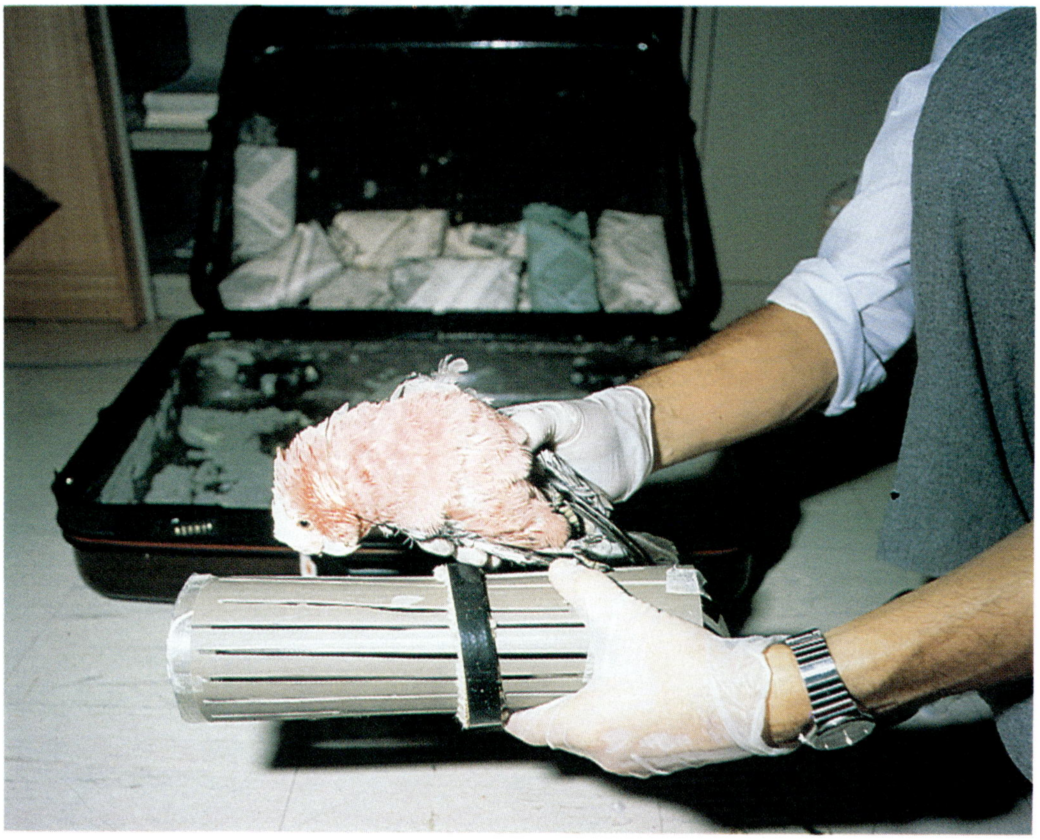

Resource harvesting and biodiversity: the contribution of ecologists to policy decisions

I. R. NOBLE[1]

With the increasing importance of biodiversity issues in policy formulation, ecologists are being called upon to make specific and practical suggestions as to how biodiversity might best be maintained. Most ecologists are reluctant to be drawn too far into this process. I use an example from the Resources Assessment Commission Inquiry into forest resources to contrast the preparedness of the economic community to use less than perfect models and ecological community's reluctance to do so. It is important that ecologists consider what our best understanding of ecological systems can contribute to policy advice. I outline the contribution that is being made by basic theory about efficient reserve selection and about coup selection practices ("new forestry"). I then ask what insights ecological theory can provide into the question of whether harvesting intensively, so that we can extract our needs from the smallest possible area, or harvesting at some less intensive rate over a larger area will best maintain biodiversity. I conclude that the theory marginally favours intensive use of a smaller area but that there is relatively little difference between the two options.

INTRODUCTION

THE importance of biodiversity is becoming apparent to even the most exploitative of resource users. They may not agree that there is a great need to sustain the diversity of the earth's biotic resources; they almost certainly think that they should not be conserved in their backyard. However, they know that they have to consider it. Biodiversity has also entered the perception of all levels of our administrative and business world. The impacts of the Biodiversity Convention alone galvanizes the attention of even the most "economically rational" of our administrative elite.

This means that there is greater pressure on ecologists to make specific suggestions about how biodiversity might best be maintained. These suggestions have to be clear and practical. They may not be what we, as ecologists, see as the ideal solution, but one group of society rarely achieves all its ideals. We may not be certain whether it is the optimal solution, but we have to be prepared to make the best decision possible with the information available at the time, defend it and implement it.

Ecologists, when confronted with issues about the long-term conservation of biodiversity, must stop wringing their hands and saying "we do not know enough". Too often our response has been to say "Well, I could answer that question if only I had a bit more data" and rather than presenting the inquirer with a well formulated best opinion, we present him or her with a request for a research grant. Already some conservation organizations are by-passing scientific advice and are simply going-for-broke on anything they can preserve.

In this paper I look at the problems confronting ecologists when asked to provide advice on how to cope with the often incompatible demands of harvesting and the conservation of biodiversity. First I look back at the input of ecologists into the Resources Assessment Commission Inquiry into forest resources (RAC 1992) and contrast this with that of economists. I then look ahead and discuss some simple but effective models that might provide the basis for greater ecological input in the future.

RESOURCES ASSESSMENT COMMISSION INQUIRY INTO FOREST RESOURCES

Ecological inputs

What can we learn from the Resources Assessment Commission Inquiry into forest resources? The Inquiry documented Australia's sad history in failing to implement sustainable management of our forest resources. It noted the early clearing of major forest and woodland systems for agriculture and the wilful exploitation of particular timber species or forest communities beginning with the arrival of European settlers and continuing to our own generation. It described the attempts by individual foresters to see our forest resource better managed and conserved prior to World War II. It also documented the blatant overcutting of our forests ordered by many of our governments in the post war reconstruction period. Unfortunately this edict was not thrown out for several decades. During this period some Forestry Commissions seem to have forgotten the by then well established principles of scientific forestry. To some the future was *Pinus radiata* and more *Pinus radiata*. To some individuals it still is.

[1]Ecosystem Dynamics Group, Research School of Biological Sciences, Australian National University, Canberra, Australian Capital Territory, Australia 0200.
Pages 231–39 *in* CONSERVING BIODIVERSITY: THREATS AND SOLUTIONS, ed by R. A. Bradstock, T. D. Auld, D. A. Keith, R. T. Kingsford, D. Lunney and D. P. Sivertsen. Surrey Beatty & Sons. 1995.

The Inquiry also documented the recent moves towards sustained yield management and how this move is hindered by the legacy of over-cutting and by the paucity of useful yield data. But when it came to commenting directly on the maintenance of the biological diversity of our forest and woodland systems the Inquiry was forced to come to the "wimpish" conclusion that "we do not know enough". A number of factors that might impinge on biodiversity could be listed and a large number of case histories could be documented, but it was not possible to state explicit, quantifiable rules for the better management of biological diversity.

Coming into the RAC Inquiry in its last phase I started asking the question of conservation agencies and groups, "What constitutes adequate conservation". Often I was met with responses that ranged from blank amazement to stumbling discussions that were difficult for even an ecologist to follow let alone operationalize. I appreciate that just how much importance society is prepared to place upon conservation is not a decision for ecologists alone but for society as a whole. But as the professionals in this area we should be in the position to provide firmer advice than we do. Instead we have 1 per cent, 5 per cent, and 50 per cent rules of what area to conserve, 500 and 5 000 minimum populations, the SLOSS debate (Single Large Or Several Small reserves) and a few "rules-of-thumb". Best management of our biotic resources will always be a matter of debate within our profession. There will always be a need for more research and refinement of any rules of operation. But we do need to have some firm rules that are the best advice that we can provide at present.

Economic inputs

In contrast to the response of the ecologists is that of the economists. On the issue of the conservation of old-growth forest the Inquiry commissioned an investigation of the economic value of the old-growth resource in the National Estate in southeastern Australia (Streeting and Hamilton 1991). In this investigation it was assumed that all forests on the register of the National Estate are worthy of protection. The value foregone in sawlogs and pulplogs from the National Estate was estimated along with the impacts of the loss of jobs, disruption of local communities, the value of other options in the use of old-growth (mainly tourism) and the result expressed in dollar terms. The conclusion was that the net value of logging the old-growth forest was $11M. Sensitivity analyses showed that this value could vary between about $3M to $20M if extreme sets of parameter values were used. The $11M value can be interpreted as asking the residents of New South Wales and Victoria to make a one-off payment of $1.55 per person. If a compensation package for the people affected by such a decision of about the same scale as that adopted for Fraser Island was negotiated then the net cost would be $43M requiring a one-off payment of $6.05.

Independently, and before the results of the economic analysis were known, two methods of estimating the willingness of this same population to pay to have the old-growth forest conserved were conducted. One was based on the Travel Cost method. In this approach a survey was conducted to find how many people visited National Estate forests in the south-east. The average expenditure of each visitor was estimated and this led to the conclusion that the total annual expenditure on visiting National Estate areas in the south-east was about $3.6M per year. The next stage of the analysis involved estimating the benefit of the forest use itself over and above the financial expenditure incurred. This is estimated by asking respondents whether they would visit if specified entrance fees were charged. The outcome was that the value placed on the National Estate areas was about $950 000 per year or about $8.90 per visitor.

The second method of estimating willingness to pay was a Contingent Valuation Analysis. In this approach a sample of the population was asked how much they were willing to pay to preserve the National Estate forests. There are many rules about how samples are selected, how questions are phrased and how the results are interpreted, but I will not discuss them here. The result was that the willingness to pay was an average of about $22 per head. This is about three times the estimate from the Travel Cost method and greatly exceeds the estimate of the net cost to the residents of New South Wales and Victoria of ceasing logging in the National Estate which was $1.55 or $6.05 depending on whether a compensation package was provided.

Krutilla and Fisher (1985) have argued that whereas the future value of most goods are discounted (e.g., I will accept a dollar now but I will want $1.05 plus inflation if I have to wait a year for it), most environmental services and preservation benefits will actually increase in value into the future. This is because it is not possible to create more high quality conservation areas and there is an increasing consumer demand for them. Taking these considerations into account the economic analysis showed that if the adult population of Victoria and New South Wales are prepared to forego more than 7 cents per year, increasing by 3.5 per cent each year to 50 cents in 2050, then society will be better off by preserving the National Estate forests rather than logging them.

The Travel Cost method and Contingent Valuation estimate different values. Travel Cost estimates the preparedness to make direct expenditure on visiting the forest (i.e., recreation value, etc.); Contingent Valuation includes the value placed on the forest by individuals who may never visit it; i.e., it includes what economists call the existence value. Previous comparisons of these two methods have shown about a three fold difference between the two estimates. Both estimation techniques have their limitations. Travel cost is based on past usage for a limited number of purposes. Contingent Valuation Analysis looks more to the present and future but is based on the response to hypothetical questions about the willingness to pay.

The methods are far from ideal. However, they can be used to compare the values that society places on alternative options. Direct use value placed on the forest by visitors is a significant proportion of the estimated value derived from logging it. The willingness to pay as estimated by the Contingent Valuation Analysis is well above the estimated economic values of logging the forest.

The methods by which the economic analysis of the effects of a cessation of logging were criticized by some economists; ideological factions seemed to be involved here. The Contingent Valuation Analysis and Travel Cost Analysis have been criticized as weak tools with many problems in their implementation and interpretation. Note, for example, that the question asked in the Contingent Valuation Analysis was "how much are you prepared to pay to protect the National Estate from logging". But state forests, including the National Estate, are already owned by the public and thus the question should have been "how much compensation would we want to allow them to be logged". Would changing the question have affected the outcome? Would the same willingness to pay have been evident if the respondents had also been asked about willingness to pay for the rehabilitation of land degradation in New South Wales and pollution of our water ways in the same questionnaire?

Nevertheless, most accept that they are the best tools available and considerable effort is being expended on improving them. For the first time, they are giving decision makers rough estimates of the relative values of intangible and tangible goods and services.

QUANTITATIVE ECOLOGICAL ADVICE

Does ecology have similar methods that might be used to provide policy advice? I believe that the answer is "yes", but ecologists have been reluctant to promote their use. This reluctance arises for a number of reasons. Ecologists, like most others in the natural sciences, tend to be perfectionists and sometimes impose inappropriate standards of precision, accuracy and verification for ecological models. Many ecologists will argue that we rarely have enough data to develop a valid model. This is true, but models are merely analytical and extrapolative tools. The decision will be made on the same currently available data no matter what the method of analysis and extrapolation. In policy related matters we must consider the alternative sources of advice. A model is useful if it provides better advice than the alternative method that might be used. Natural scientists also have often had a fixation with finding the optimal solution to a problem. But in many policy matters this is often not feasible and unnecessary. In many decisions the real options are limited and all a model has to do is provide advice as to which of those limited options is the better.

We must balance these positive comments about the utility of models with the cautions from some of the ecological community who argue that many of these quantitative assessments in resources policy are exercises in "crackpot rigour" (Erhlich and Daily 1993). That is, where detailed mathematical analysis is applied to a problem that is intractable due to gross uncertainties. Another caveat is that we are involved in "suboptimization", where great effort is expended to find the best way of doing something that we should never have contemplated doing in the first place.

In the rest of this paper I will discuss some attempts to provide better advice based on basic ecological research. I will deal mainly with the issue of adequate conservation in a forest estate that is also used for harvesting for wood products. A series of government, NGO (Non-government Organizations) and international inquiries and working groups have formulated similar conclusions as to the best approach to managing the use of our biotic resources. Essentially it is to:

- Set up a comprehensive land use planning system.
- Place some of your land in areas with conservation priority to ensure adequate conservation.
- Reduce human activity and impacts in sensitive areas where the biotic resource plays a role in water catchment erosion control and so on.
- Devise "codes of practice" to govern the use of the remaining biotic resource and in many cases pay cognisance to the off-reserve maintenance of biological diversity.
- Establish an independent agency to monitor the effectiveness of the land allocation and compliance with the codes.
- Allow the market to determine the use of the remaining biotic resource.

I will discuss briefly three attempts to provide quantitative ecological advice to planners and policy formulators. Each is relevant to the task of establishing a comprehensive reserve network while at the same time supporting this by appropriately designed off-reserve management to conserve biodiversity. They are the reserve selection methods developed by collaboration between groups of Australian scientists based in conservation agencies, CSIRO and universities (summary by Pressey *et al.* 1993); the analysis of coupe selection patterns in North America by Franklin and Forman (1987), and a new model investigating the trade off between the intensity of harvesting and the area harvested.

Reserve selection

There has always been a discrepancy between the theory and practice of reserve selection. Ideally we should select reserves so that they protect as many species, communities and environments as possible. In practice their selection has been much more *ad hoc* but many conservation agencies are becoming much more proactive in their planning and selection of areas for conservation. A series of tools for systematic reserve selection have been developed independently by several groups in Australia, Britain and South Africa (Pressey *et al.* 1993). These use iterative methods normally based on the criterion of selecting a set of reserves that includes the greatest number of species (or communities) in the smallest number of reserved units. When applied to a large number of potential land units the computational demands increase rapidly and efficient search algorithms have to be developed. If more complex criteria of success are required (e.g., multiple representation of species and communities in different reserves, building the new reserve system from existing holdings, factors taking issues such as management, tourism, etc., into account) the feasibility of reaching an "optimal" solution becomes even more difficult.

Nevertheless, this small research community has gone a long way towards developing basic principles for reserve selection. These principles include efficiency, where the number or area of reserves required to achieve a particular representation (or a more complex criterion) is kept as small as possible. Another is irreplaceability, which measures the extent to which options are lost if a site is lost from the potential reserve system. These two principles guide the design of algorithms that place importance on selecting reserves that are most complementary to the existing system. Most algorithms are based on iterative searches guided by a set of simple rules. The efficiency of these rule sets is improving and will continue to do so as more ideas from artificial intelligence and some of the newer genetic algorithms are invoked.

We must be careful to remember that we are seeking practical tools for the real world. We should not be too concerned whether the algorithms will deliver the optimal solution. In practice the selection criteria will change, packets of land will be altered by human activities or natural events and thus the selection process will need to be dynamic. Flexibility is the third principle recognized by Pressey *et al.* (1993). Usually there will be many combinations that achieve the same level of selection criteria and it is important that an algorithm find as many of these as possible and present them to the planners so that they are able to apply the more empirical criteria that exist in real world planning.

My point here is that, despite the caveats expressed by the researchers in this area that they have a long way to go, there are practical tools that will assist land-use planners very effectively. We must expect the power and accessibility of these tools to increase rapidly over the next few years. As ecologists we should be forthright in promoting their use.

Coupe selection patterns

If we look at the "Codes of Practice" devised by the various state forest agencies, we will find that in general their recommendations on coupe siting and design favour the wide spread dispersal of small coupes throughout the forest. Intervening forested areas are to be left unharvested as long as possible. Emphasis is placed on harvesting small coupes within the limitations of the economics of timber felling, roading and topography. These codes are based on a response to many years of pressure to move away from large areas of logged forest stretching from horizon to horizon.

In some places these recommendations have found an extreme form in a checker-board approach to coupe planning. In this approach beta diversity (ie the diversity of species along environmental gradients or across a series of habitats, Noss 1983) is assumed to be high. Many wildlife texts teach that wildlife habitat is maximized by this approach. However, concern has been expressed that this approach may not be the best either ecologically or economically (Franklin 1993). Among the other factors to be taken into account are the possibility of blowdown along exposed edges; changes in animal populations because of the advantages given to edge specialists, and the vulnerability to the spread of weeds and feral animals.

Franklin and Forman (1987) tackled the questions: "Should we cut small patches and disperse them as much as possible? Should we start at one side and clear-cut progressively to the other? Or should we use some other cutting strategy?" They analysed the amount of edge exposed by various logging patterns and the amount of interior forest remaining. They found that a checkerboard pattern of cutting created a greater amount of forest edge than other approaches such as cutting patterns radiating from one or more nuclei, strip cutting or a progressive cut starting at one edge and proceeding to the other. It also tended to result in smaller patches of remnant forest, a result that was accentuated if the edge effects of forest clearing were taken into account.

They concluded: "We suggest a two-point guide for managers of natural landscapes based on this landscape analysis and on other ecological considerations. First, reduce the emphasis on dispersing small clearcut patches through the forest landscape. The fragmentation does not enhance many resource values. Approaches featuring progressive or clustered cuts from scattered nuclei should be considered. . . . Second, identify and reserve large patches of primeval forest in the landscape for maintenance of interior species and amenity values. It is urgent because many current cutting programmes are rapidly reducing the size of available patches".

Harvest intensity model

As a third example I present a simple model that tackles the question "in order to maintain the highest possible local diversity, should we harvest intensively so that we can extract our needs from the smallest possible area, or should we harvest at some less intensive rate but use a larger area". Appendix 1 gives the full mathematical background to the arguments and can be read as an alternative to the summary presented here.

Imagine that we have an area of forest, isolated from other forests, from which a particular yield is to be obtained in a relatively short period. We have the options of achieving the required yield of either harvesting part of the area at full intensity leaving the rest untouched, or harvesting more of the area at lesser intensity. What is the best choice in terms of conserving local diversity?

Basic ecological principles and data tell us that within the forest diversity falls slowly as the area harvested increases until only a small area remains undisturbed. Then diversity falls more rapidly as the rest of the area is harvested (Fig. 1). The curve is similar to the species-area model of McArthur and Wilson (1967) for isolated patches surrounded by a sea of completely inhospitable areas.

It can be argued that harvesting (e.g., for wood products in a forest) will not cause the local extinction of all species even when the whole

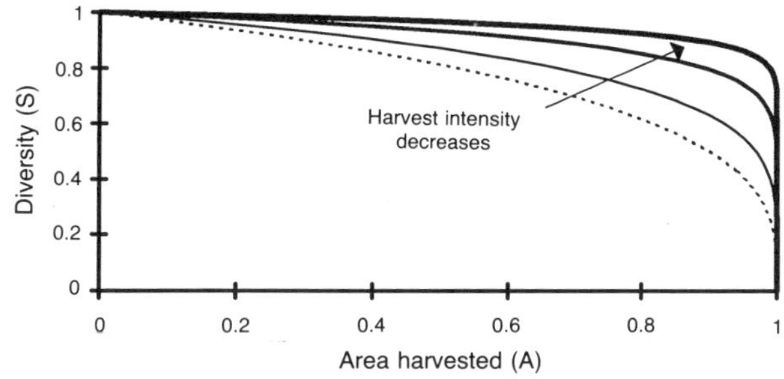

Fig. 1. The relationship between the local diversity remaining and the area harvested as the harvest intensity changes. Based on the McArthur and Wilson (1967) function for diversity/area relationships: $S = (1-A)^z$ where S is local diversity, A is area harvested and z is a measure of harvest intensity.

area is used. Some species will survive on the modified site and others will invade. However, here we consider only those species that are at risk due to the harvesting process and the resultant environment. The next question to ask is what is the effect of reducing the harvest intensity on the curve presented in Figure 1? Clearly diversity should be higher in all cases with reduced harvest intensity resulting in a diversity-area relationship is shown in Figure 2.

The usual management goal will be to maintain maximum local diversity while achieving a particular yield and some simple mathematics shows that if a reduction in harvest intensity leads to a smaller proportional reduction in the loss of local diversity then it is always best to harvest the smallest area at the maximum intensity. In some situations maximum biodiversity can be conserved by harvesting a larger area at less than full intensity. This occurs mainly when the total yield being sought from the area is small (ca. 50%). When the yield to be harvested from the forest is high (80 to 90%) it appears to be best to harvest most of the area at full intensity and to leave the remainder unharvested.

The nature of the relationship between harvest intensity and species loss will be an extremely difficult value to measure since we have neither areas for the multiple experiments nor the time

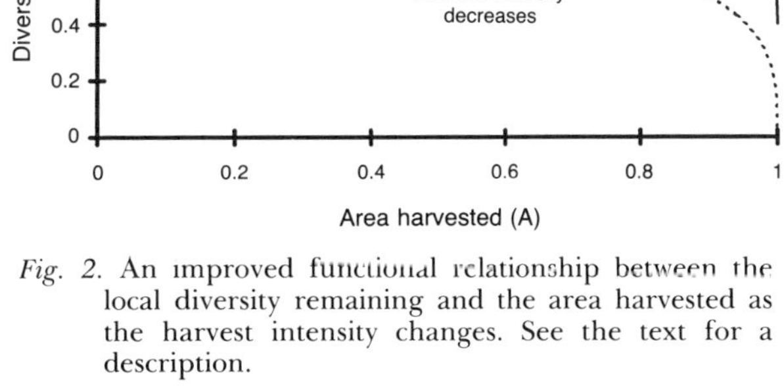

Fig. 2. An improved functional relationship between the local diversity remaining and the area harvested as the harvest intensity changes. See the text for a description.

for local extinctions to play themselves out. Rough estimates of the shape of the curve may be possible in some cases, for example by estimating the loss of wood harvest by leaving habitat trees. It is also possible to obtain estimates of the shape of the diversity-area curve shown in Figure 2 by examining diversity in small fragments of forest. The mathematical formulation of the model can provide some specific questions that can be answered with field data. In the mathematical explanation of the model in Appendix 1 an example is given which leads the decision of how to harvest dependent on the answer to the question "will spreading the harvesting over the whole area but leaving 10 per cent of the trees (actually timber yield) as habitat trees retain more than 50 per cent of the vulnerable species?".

But closer examination of the model shows that the reductions in the loss of local diversity by adopting the optimum harvesting strategy rather than alternative approaches are small. It is likely that the result that can be achieved by adopting a reduced harvest intensity over a wider area rather than full intensity over the smallest area are <10 per cent reduction in the loss of local diversity. I suspect that when other factors such as aesthetic values and the intangible benefits of retaining areas relatively unaffected by human activities, the decision will usually be made to harvest intensively the smallest possible area. Circumstances where lower intensity harvest is justified are rare and lead to only small gains.

These conclusions should be qualified. Recall that the model is a simple one that deals with only those species at risk due to harvesting. It would be improved by better biological data to underpin the model, although the conclusions seem to be robust for a wide range of estimates of biological parameters. A bigger shortcoming is that it deals with patches that are isolated from other patches and does not deal with changes in species population in time or staggered harvesting schedules. To deal with these issues, more detailed meta-population models are needed. Possingham (this volume) describes some examples. But for the near future, meta-population models will remain case specific and heavily dependent on the quality of the biology and data incorporated in them. Meta-population modelling is an area where the potential to model management scenarios has outstripped the availability of appropriate biological inputs.

CONCLUSIONS

From the above studies we can conclude that, even though attempts to provide quantitative ecological advice in policy matters related to reserve selection and the design of harvesting processes are still in their infancy, there are some general principles emerging. First, there are techniques available that will help in reducing the ad hoc nature of reserve acquisition which will result in better conservation and lower costs in the long run. Secondly, that in harvesting we should reduce the emphasis on dispersing small clearcut patches through the forest landscape and favour progressive or clustered cuts from scattered nuclei. Within harvested patches there are situations where a reduced harvest intensity over a wider area might lead to a smaller loss of local diversity, but these are relatively rare and the gains made are small. Thus, it may be better to harvest at full intensity so reducing the total area affected by harvesting as much as possible.

Put simply, these models suggest that there should be forests for conservation and there should be forests for wood production and we have techniques that can help in making the decisions of how to allocate land between the two. These conclusions are controversial in that they appear to place less emphasis on concepts of multiple use forests, and they should be qualified that they are preliminary answers based on simple models. But are the current planning models better?

ACKNOWLEDGEMENTS

I thank Alison Saunders for her comments and assistance in preparing this paper.

APPENDIX 1

The text in this appendix parallels that in the section "Harvest intensity model" but presents the full mathematical description of the model. It can be read in place of the text in the main body of the paper.

Imagine that we have an area of forest, isolated from other forests from which a particular yield (Y) is to be obtained in a relatively short period. We have the options either to harvest part of the area at full intensity leaving the rest untouched, or to harvest more of the area at lesser intensity. What is the best choice in terms of conserving local diversity?

If we harvest (H) at the highest intensity the area harvested is A = Y/H. I suggest that the relationship between the area harvested in this way and species diversity (S) remaining in the forest is as shown in Figure 1. That is, diversity falls slowly as the area harvested increases until only a small area remains undisturbed. Then diversity falls more rapidly as the rest of the area is harvested. The curve is similar to the species-area model of McArthur and Wilson (1967) and I have used their equation, $S = (1-A)^z$, as the basis of the model although I recognize that their equation referred to the diversity found in isolated patches surrounded by a sea of completely inhospitable areas. Various summaries of species-area data suggest that $z < 0.7$ and often about 0.2 to 0.4.

It can be argued that harvesting (e.g., for wood products in a forest) will not cause the local extinction of all species even when the whole area is used. Some species will survive on the modified site and others will invade. However, here we deal with only those species that are at risk due to the harvesting process and the resultant environment. I will also come back to the added complexities that harvesting could be staggered in time and that harvested sites are not totally isolated from other forested areas.

What would be the effect of reducing the harvest intensity on the curve presented in Figure 1? Clearly diversity should be higher in all cases with reduced harvest intensity which is equivalent to reducing the value of z (Fig. 1). But this model is too simplistic since it implies that local diversity will fall to zero in situations where the total area is harvested even if the intensity of the harvesting is undetectably small. So I have selected a slightly more complex model based on the loss of species (i.e., 1–S). As harvest intensity decreases so the number of species lost will decrease. Assume that this loss is a fixed proportion over all areas for a given harvest intensity, i.e.,

$$\text{species loss} = (1-S)\, f(H)$$

The resultant diversity area relationship is shown in Figure 2. The only thing that remains is to define the shape of the f(H) function. I have used $f(H) = H^x$ which can take a wide range of relationships (Fig. 3) including the linear form, f(H) = H when x = 1.0. Thus the full equation is:

$$S = 1-(1-(1-A)^z)\, H^x$$

which results in a relationship between local diversity, harvest intensity and area as shown in Figure 4. The usual management goal will be to maintain maximum local diversity while achieving a particular yield, i.e., Y = HA or H = Y/A. Thus the goal is to maximize

$$S = 1-(1-(1-A)^z)\,(Y/A)^x$$

where

$$dS/dA = z(1-A)^{z-1}(Y/A)^x-((1-(1-A)^z)x(Y/A)^x)/A$$
$$= (Y/A)^x\,[z(1-A)^{z-1}-(x(1-(1-A)^z))/A]$$

which is a maximum when

$$x = (1-(1-z)\,A)/((1-A)^{1-z})$$

For any feasible value of A (i.e., $0.0 \leq A < 1.0$), the optimal solution can be found only for x > 1.0 (because x > 1 for all A and z). Thus, if the shape of the f(H) function is such that $x \leq 1$ then the best course will always be to harvest at maximum intensity. That is, if a reduction in harvest intensity leads to a smaller proportional reduction in the loss of local diversity then it is always best to harvest the smallest area at the maximum intensity. If x > 1 then maximum biodiversity can sometimes be conserved by harvesting a larger area at less than full intensity. For a given z, the optimum area to be harvested (and thus the harvest intensity) is independent of the yield and is a function only of x as shown in Figure 5. For example, if x = 1.5 the optimum area is 0.65. Thus for yields of less than 0.65, 65 per cent of the area should be harvested and the harvest intensity reduced to give the required yield. For higher yields the full harvest intensity should be used.

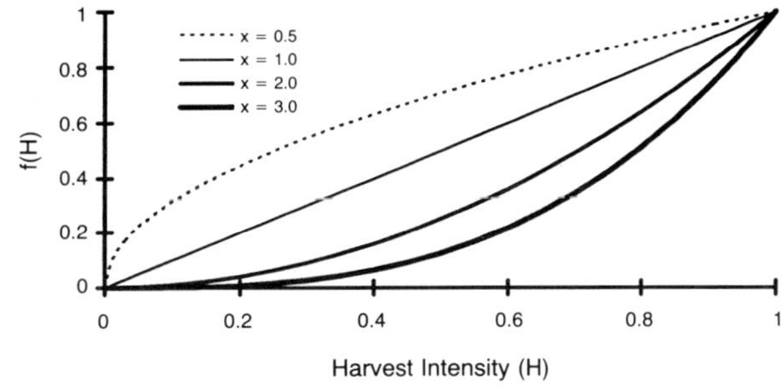

Fig. 3. The relationship between the increase in species loss, F(H), and the harvest intensity for $F(H) = H^x$. High values of x result in small losses of local biodiversity for lower harvest intensities which seems to be most likely in forested systems.

These results appear at first sight to give further weight to arguments in support of multiple use since they imply that in some circumstances more local diversity will be conserved by adopting lighter harvesting over wider areas. However, this is not necessarily the case and the results must be treated with some caution. If the required yields are high (e.g., Y = 0.9 which means that 90% of the resource is to be harvested) then x must be about 3 before there is any benefit in reducing harvest intensity and increasing the area affected. An x = 3 implies that for a 20 per cent reduction in harvest intensity a 50 per cent reduction in the loss of local species will occur.

The nature of the relationship between harvest intensity and species loss will be an extremely difficult value to measure since we have neither areas for the multiple experiments nor the time for local extinctions to play themselves out. Rough estimates of the shape of the curve may be possible in some cases, for example by estimating the loss of wood harvest by leaving habitat trees. It is also possible to obtain estimates of the shape of the diversity-area curve shown in Figure 2 by examining diversity in small fragments of forest.

As an example, consider the case where there is 100 ha of a particular forest type that is available for harvesting. Via whatever planning processes that apply to the area it is agreed that 90 per cent of the wood resource is to be harvested on that site. The question then is whether it is better to harvest 90 per cent of the area fully leaving 10 per cent in a reserve, or to harvest 100 per cent of the area but leaving some trees (10% of them) as habitat trees, or possibly some compromise between the two extremes. With z = 0.3, leaving 10 per cent in reserve would maintain 50 per cent of the vulnerable species whereas leaving no reserve would protect none of the vulnerable species locally. Thus the question to be answered by ecologists is whether leaving 10 per cent of the trees as habitat trees and harvesting the whole area will retain more than 50 per cent of the vulnerable species.

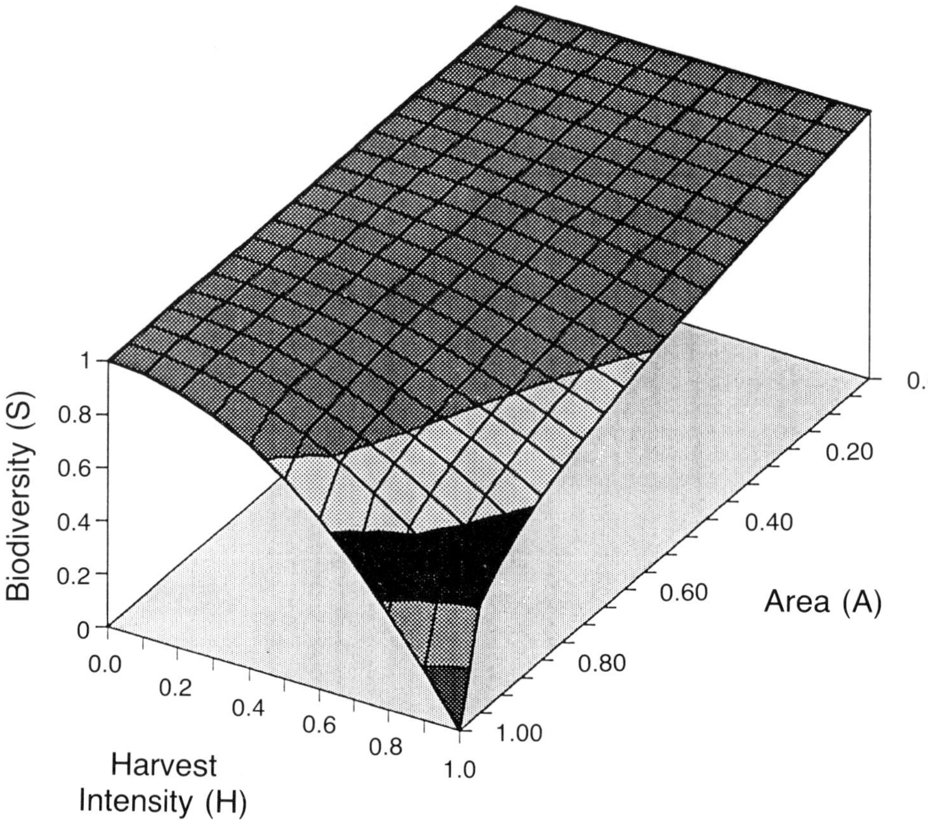

Fig. 4. The relationship between local biodiversity retained, the area harvested and the harvest intensity based on the function $S = 1-(1-(1-A)^z)H^x$ where S is local diversity, A is area harvested, H is harvest intensity and in this graph z = 0.3 and x = 1.5.

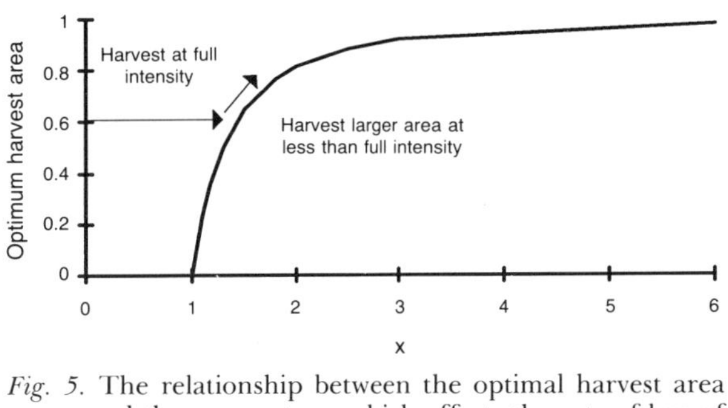

Fig. 5. The relationship between the optimal harvest area and the parameter x which affects the rate of loss of local diversity as harvest area increases. For all cases where x < 1.0 the optimal tactic is to harvest at full intensity whatever the required yield. For x > 1.0 it is sometimes better to harvest a larger area at less than full intensity. For example, if the site is to yield 0.6 of its resource then for x < 1.3 the optimal tactic is to harvest 0.6 of the area at full intensity. For x > 1.3 the tactic is to harvest a larger area, as shown by the curve, at a lesser intensity; e.g., for x = 2.0, harvest 0.8 of the area at 0.6/0.8 = 75 per cent of full harvest intensity.

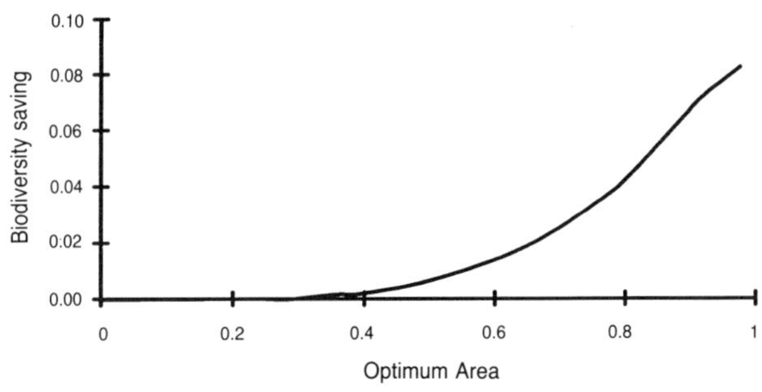

Fig. 6. The reduction in the loss of local biodiversity as a result of taking the optimal balance between harvest area and intensity compared with harvesting at full intensity [for z = 0.3]. In the example in Figure 5 with x = 2.0, adopting the optimal tactic for harvesting 0.6 of the resource by harvesting 0.8 of the area at 75 per cent intensity rather than harvesting 0.6 of the area at 100 per cent intensity will reduce the loss of biodiversity by 0.04, i.e., 4 per cent of the non-harvested situation.

But there may be a more general rule that can be used. Closer examination of the model shows that the reductions in the loss of local diversity are small (Fig. 6). With z = 0.3 (a mid-range value from island biogeography theory), the best savings that can be achieved by adopting a reduced harvest intensity over a wider area rather than full intensity over the smallest area are <10 per cent reduction in the loss of local diversity. I suspect that when other factors such as aesthetic values and the intangible benefits of retaining areas relatively unaffected by human activities, the decision will usually be made to harvest intensively the smallest possible area. Circumstances where lower intensity harvest is justified are rare and lead to only small gains.

These conclusions should be qualified. Recall that the model is a simple one that deals with only those species at risk due to harvesting. It would be improved by better biological data to underpin the model, although the conclusions seem to be robust for a wide range of estimates of biological parameters. A bigger shortcoming is that it deals with patches that are isolated from other patches and does not deal with changes in species population in time or staggered harvesting schedules. To deal with these issues, more detailed meta-population models are needed. Possingham (this volume) describes some examples. But for the near future, meta-population models will remain case specific and heavily dependent on the quality of the biology and data incorporated in them. Meta-population modelling is an area where the potential to model management scenarios has outstripped the availability of appropriate biological inputs.

REFERENCES

Erlich, P. R. and Daily, G. C., 1993. Population extinction and saving biodiversity. *Ambio* **22:** 64–68.

Franklin, J. F., 1993. Scientific basis for new perspectives in forests and streams. Pp. 25–72 in *Watershed management: balancing sustainability and environmental change* ed by R. J. Naiman. Springer Verlag.

Franklin, J. F. and Forman, R. T. T., 1987. Creating landscape patterns by forest cutting: ecological consequences and principles. *Landscape Ecology* **1:** 5–18.

Krutilla, J. and Fisher, A., 1985. The economics of natural environments. Resources for the Future: Washington, D.C.

MacArthur, R. H. and Wilson, E. O., 1967. *The theory of island biogeography.* Princeton University Press: Princeton, New Jersey.

Noss, R. F., 1983. A regional landscape approach to maintain diversity. *BioScience* **33:** 700–06.

Possingham, H. (this volume). The practical applications of population viability analysis for conservation planning. In *Conserving biodiversity: threats and solutions* ed by R. A. Bradstock, T. D. Auld, D. A. Keith, R. T. Kingsford, D. Lunney and D. P. Sivertsen. Surrey Beatty & Sons: Chipping Norton.

Pressey, R. L., Humphries, C. J., Margules, C. R., Vane-Wright, R. I. and Williams, P. H., 1993. Beyond opportunism: Key principles for systematic reserve selection. *Trends Ecol. Evol.* **8:** 124–28.

Resources Assessment Commission (RAC), 1992. Forest and Timber Inquiry: Final Report. Australian Government Publishing Service: Canberra.

Streeting, M. and Hamilton, C., 1991. An economic analysis of the forests of south-eastern Australia. Resource Assessment Commission, Australian Government Publishing Service: Canberra.

Sustainable forestry — the urgency to make the myth a reality

T. W. NORTON[1] and J. B. KIRKPATRICK[2]

Forest biodiversity conservation in the State of New South Wales from a national and international perspective, and scientific understanding of the responses of forest-dependent biota to modification and fragmentation arising from forest harvesting operations are discussed. While tropical forests have received much attention, the destruction and modification of temperate forests is of equal concern. These ecosystems have had a larger part of their area lost to clearance and their removal, modification and degradation continues in many regions at a rate rivalling tropical forests. Currently, some 2 billion ha of temperate forest remain, and most temperate forest ecosystems are poorly reserved. The conservation status of temperate eucalypt forests in Australia is little different from that globally despite the fact that this continent was only recently occupied by Europeans. Approximately half of the continent's forests have been cleared or severely modified since 1788 and rates of deforestation remain high. The forest conservation reserve system is not currently representative of the full range of biodiversity. Many impacts related to forestry are significantly detrimental since they can (i) destroy and modify complex forest landscapes and ecosystems; (ii) destroy and modify the natural environmental heterogeneity of forest ecosystems; (iii) destroy, prevent or hinder ecological processes that are the basis for species' persistence and evolution; (iv) destroy or significantly modify the habitat of species; and (v) destroy individual organisms and significantly modify populations and assemblages of species. We argue that many forestry impacts on biodiversity may be irreversible or effectively irreversible, and the harvesting of many old growth eucalypt forest ecosystems in New South Wales does not appear to be ecologically sustainable. Essential components of a move towards sustainability include (i) the establishment of a comprehensive conservation reserve network centred on old growth forest ecosystems on productive sites; (ii) enhanced off-reserve codes of forest practice on both public and private lands; (iii) an ecologically-conservative approach to forest management; and (iv) the development of a systematic ecological monitoring programme across the forest estate.

INTRODUCTION

RECENT inquiries into the management of Australia's native forests have concluded that there is a real need for enhanced measures to protect forest biodiversity (e.g., Ecologically Sustainable Development Working Group on Forest Use — Commonwealth of Australia 1991; Resource Assessment Commission Inquiry into Forest and Timber Resources — RAC 1992). It has consistently been recommended that forest biodiversity conservation be enhanced by upgrading the conservation reserve system, with particular emphasis on the reservation of old growth forest and wilderness, and improving off-reserve management.

In this paper we (i) discuss aspects of forest biodiversity conservation in the State of New South Wales from a national and international perspective; (ii) briefly review scientific understanding of the responses of forest-dependent biota to modification and fragmentation arising from forest harvesting; and (iii) outline our view of some major needs and priorities for reform in forestry research, planning and management. Forestry harvesting operations are defined generally here as the process of logging eucalypt forest for pulpwood, sawlogs etc., and the suite of activities such as roading, prescribed fire and silviculture that are associated with harvesting. Biodiversity is the variety of life — the different plants, animals and micro-organisms, the genes they contain and the ecosystems of which they form a part (Commonwealth of Australia 1992a).

THE GLOBAL EXTENT OF TEMPERATE FOREST ECOSYSTEMS

Given a broad definition of temperate forests that includes various types of boreal forest in the Northern Hemisphere, approximately 2 billion hectares of these ecosystems (mainly closed forest) occur globally (WWF 1992). This represents more than half of the planet's remaining forest cover. The majority of this forest estate, by area, is found in Russia (41%) and in North America (32%). However, the nature of this forest cover, the biodiversity complement of the forests, and the use of forests by humans can vary significantly within and between geographic regions.

Temperate forests have experienced a major reduction in areal extent due to human activities such as clearing for fuel wood, agriculture and urban expansion. The extent of remaining old growth forest varies from less than 1 per cent in western Europe to about 25 per cent in New Zealand (WWF 1992; Norton and Mitchell 1994). As a general rule, virtually all remaining temperate forest ecosystems are poorly protected in conservation reserves (WWF 1992).

[1]Centre for Resource and Environmental Studies, Institute of Advance Studies, The Australian National University, Canberra, Australian Capital Territory, Australia 0200.
[2]Department of Geography and Environmental Studies, University of Tasmania, GPO Box 252 C, Hobart, Tasmania, Australia 7001.

Pages 240–48 *in* CONSERVING BIODIVERSITY: THREATS AND SOLUTIONS, ed by R. A. Bradstock, T. D. Auld, D. A. Keith, R. T. Kingsford, D. Lunney and D. P. Sivertsen. Surrey Beatty & Sons. 1995.

CONSERVATION STATUS OF TEMPERATE FORESTS IN AUSTRALIA

Analysis of the present conservation status of temperate forests in Australia requires consideration of the historical changes and impacts on the forests, irrespective of their sources (Norton and May 1994). Three aspects are important.

Clearance of old growth forest ecosystems

Approximately half of Australia's forests have been cleared or severely modified since 1788 (RAC 1992). Patterns of forest clearance and modification have not been consistent across forest ecosystems, including eucalypt forests, and have typically been heavily biased to relatively more accessible sites (often lowlands) of higher productivity (often underlain by fertile soils). Consequently, relatively large areas of some forest types have been removed or significantly modified while other forest ecosystems have suffered less impact from human exploitation (Braithwaite *et al.* 1993). The current national rate of deforestation is in the range of 0.22–0.27 per cent a year (Young *et al.* 1990; RAC 1992).

Forest ecosystem modification

No comprehensive and reliable estimates of the extent of remaining forest that has been modified, or remains un-modified, are available. In the absence of these, the RAC (1992) calculated the extent of unlogged forest — which can be used as a crude approximation of old growth forest ecosystems. On average, nationally, it was estimated that roughly half of all remaining eucalypt forest in conservation reserves is unlogged, and about 20 per cent of all remaining eucalypt forest is unlogged (RAC 1991). For three eucalypt forest groups, southeastern dry forest and woodland, northeastern coastal eucalypt forest, and river red gum forest, less than 10 per cent of the total remaining forest is unlogged (RAC 1992). The extent of unlogged eucalypt forest on private land is largely unknown.

Unrepresentative system of forest conservation reserves

Some nine million hectares of forested land occurs in conservation reserves in Australia (RAC 1991). A coarse assessment of the representation of different forest groups in these reserves was undertaken by the RAC in their Forest Resource Survey. Many forest groups are poorly represented within reserves in the states and territories. For example, combined, the southeastern dry eucalypt forest, central coastal eucalypt forest and northeastern eucalypt forest occupy about 12.5 million hectares, of which some 10 per cent (1.32 million ha) occur in conservation reserves. Compared to other forest groups, the RAC (1992) considered these forests to be under-represented in conservation reserves.

It should be noted that the percentages given above are based on existing forest cover rather than pre-European forest cover. If the original area of forest types is used as the basis of calculation, then the estimates would typically be considerably smaller.

CONSERVATION STATUS OF OLD GROWTH FORESTS IN NEW SOUTH WALES

Forest clearance and modification

Forest clearance and modification is extensive in many parts of New South Wales. In the north-east region, for example, a substantial percentage of the major vegetation associations have been cleared (Fig. 1). Since 1986, deforestation rates have been about 20 000 hectares per year in the Eastern and Central Divisions, and about 80 000 hectares per year in the Western Division (RAC 1991).

Logging and modification of old growth eucalypt forests occurs widely across the State on public and private lands (Braithwaite *et al.* 1993; Norton and May 1994). For example, in south-east New South Wales some 500 000 tonnes of woodpulp and 65 000 cubic metres of sawlog can be harvested annually in the Eden Management Area (EMA). This activity is primarily reliant on the logging of old growth forest ecosystems. In 1991, it was estimated that the old growth forests available for logging in the EMA would be fully cut out by about 2012 AD if the 59 000 cubic metres per annum sawlog quota was maintained (Streeting and Hamilton 1992). Since that time the sawlog quota has increased.

Adequacy of reserve system

The forest conservation reserve system in New South Wales is not currently representative of the full range of forest biodiversity occurring in the state (Fig. 2; RAC 1991; Norton and May 1994). In north-east New South Wales, for example, only 17.9 per cent of the "dry/cool" environmental units on flatter terrain with fertile soils remain uncleared and only 0.8 per cent of these are reserved. Of the "dry/warm" environmental units on flatter terrain with fertile soils, 61.8 per cent remain uncleared and only 0.1 per cent are reserved (Love 1993; Table 1).

Wet old growth forest ecosystems in the north-east of the state are also very poorly reserved: 61.8 per cent of the "wet/cool" environmental units on flatter terrain with fertile soils remain uncleared but only 5.4 per cent of these are reserved; 45.9 per cent of the "wet/warm" environmental units on flatter terrain with fertile soils remain but only 0.9 per cent are reserved (Love 1993; Table 1).

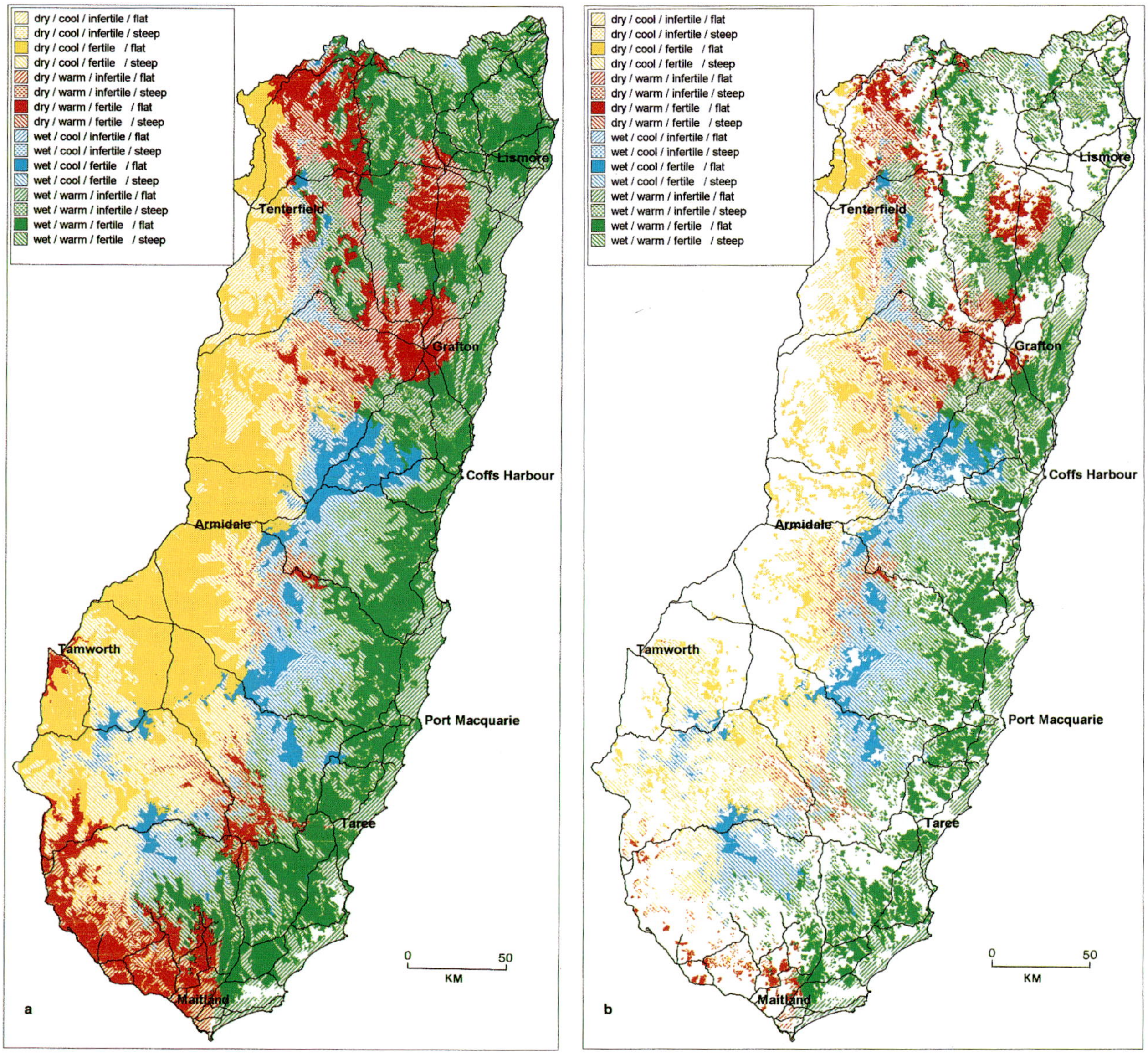

Fig. 1. A real extent of 16 broad environmental units (a) prior to clearing and (b) remaining uncleared in north-east New South Wales. Environmental units represent the 16 combinations of dry and wet forest associations on different slopes and substrates shown in Table 1; percentage of units remaining and reserved is given in Table 1; adapted from Love (1993).

The reservation of productive old growth forests and their old growth dependent flora and fauna is just as poor in south-east New South Wales. For example, in the Eden Management Area 70 per cent of the forests and woodlands on crown lands with a slope <20 degrees are in State Forests and are subjected to roading, logging, prescribed fire, and the incursion of exotic predators. (M. J. Saxon, NSW National Parks and Wildlife Service, pers. comm.). Virtually all of the remaining old growth forests that are underlain by fertile soils occur in State Forest or on private land (T. W. Norton, unpubl. data). These factors contribute significantly to serious conflicts over the protection of species such as the Koala, *Phascolarctos cinereus* and old growth forest ecosystems in the south-east forests (Norton and Saxon 1993; Norton 1994).

Braithwaite *et al.* (1993) have analysed the conservation status of broad forest associations in the Batemans Bay forestry region on the south coast of New South Wales. This region represents an area of 3 500 km^2 and was assessed in relation to land tenure, site productivity (i.e., flats and lower slopes, particularly if underlain by fertile soils), accessibility, the distribution of 31 forest tree associations, and the extent of disturbance from logging and clearing activities. Braithwaite *et al.* (1993) reported that 19.3 per cent of the area occurs in national parks. Tree associations such as *Eucalyptus maculata* — *E. globoidea* and *E. saligna* — rainforest that characterize the most productive sites and, on average, the most accessible land are: (i) poorly represented in reserves, (ii) well represented on freehold land and in state forests, and (iii) most

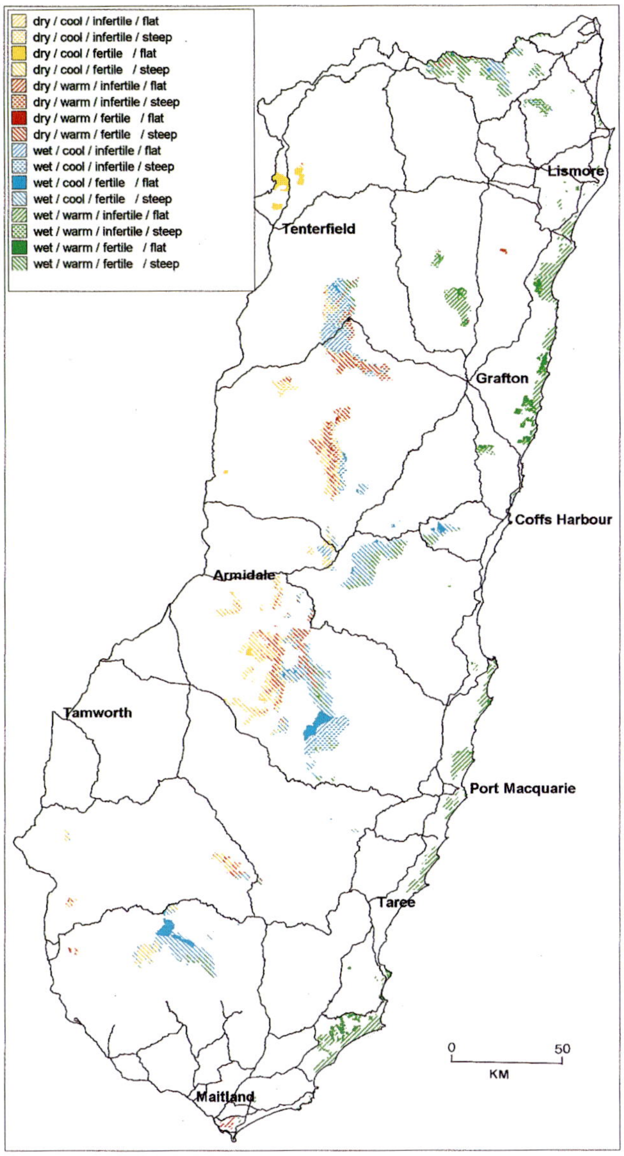

Fig. 2. Representation of 16 broad environmental units in national parks and nature reserves in north-east New South Wales; see Table 1 for percentages; adapted from Love (1993).

extensively disturbed by logging and clearing. Three tree associations occur on the most productive sites. Of these associations, less than 7 per cent of each are reserved, and less than 9 per cent remains in an undisturbed or minimally disturbed condition (Braithwaite *et al.* 1993).

BIOLOGICAL SIGNIFICANCE OF THE REMAINING OLD GROWTH EUCALYPT FORESTS IN NEW SOUTH WALES

A number of the most biologically rich old growth eucalypt forest ecosystems remaining on the Australian continent occur in New South Wales (Benwell 1993; Keith and Sanders 1990; Common and Norton 1992; Mosley and Costin 1992). A number of these forests exhibit exceptional biological values since they support diverse assemblages of species (e.g., arboreal marsupials in the Dorrigo Forest Management Area) and exceptionally high abundances of a variety of fauna (S. Ferrier, NSW National Parks and Wildlife Service, pers. comm.). The population density of the Greater Glider, *Petauroides volans* in Chaelundi State Forest, for example, is unsurpassed elsewhere and very high population densities of this animal and other arboreal marsupials occur in a number of old growth forests in the north-east of the state (T. W. Norton, unpubl. data).

The old growth forests of the state also have very high wilderness values and other attributes of value to the people of New South Wales (Mosley and Costin 1992). Many of the forests in the north east are located in a regional climatic overlap zone which becomes sub-tropical to the north. Because of the particularly wide range of climate, geology and terrain in the region, many of the forests and associated vegetation support a range of sub-tropical and temperate biota. The region has several endemic plants, for example, and the highest number of rare plants in the state (Benwell 1993).

Old growth forest ecosystems in state forests in northern and southern New South Wales provide essential food and shelter resources and high quality habitat for a large number of plants and animals. These include many species that are listed on Schedule 12 of the *Endangered Fauna (Interim Protection) Act* 1991. In north-east New South Wales alone, the high quality habitat of over 100 species of amphibians, reptiles, birds and mammals is located in old growth forests (see Table 2). Of these vertebrates, 27 species — over one quarter — are listed on Schedule 12.

As a general characterization, it is the diversity of the dominant structural features of old growth forest, and their relative stability over time compared to highly disturbed and logged forests, that are essential in supporting critical food and shelter resources and habitat for old growth dependent biota (Table 3). Old growth forest ecosystems on the more productive sites are especially important because they often exhibit a different fauna compared to old growth forest on poorer sites (Braithwaite 1984; Davey and Norton 1990; Scotts 1991). It is often on the more productive sites that the greatest diversity of species and assemblages occurs, and where the population densities of many species can reach a maximum. This aspect is illustrated in Figure 3 using several species of arboreal marsupial — all species are most abundant in forests older than at least 150 years of age (Davey and Norton 1990).

Table 1. Original, remaining and reserved area (km²) of broad environmental units in north-east New South Wales. Percentages represent proportion of area of forest remaining or reserved versus original extent; infertile = low nutrient rating, fertile = medium to high nutrient rating; adapted from Love (1993).

Rainfall/Temperature	Substrate/slope	Original area	Remaining area	Reserved area
DRY (<1 100 mm per annum)				
	infertile flat <	1 837	716 (38.0%)	43 (2.3%)
	infertile steep	801	677 (84.5%)	120 (15.0%)
Cool (<16° annual mean)				
	fertile flat	11 541	2 075 (17.9%)	91 (0.8%)
	fertile steep	8 284	5 951 (71.8%)	766 (9.3%)
	infertile flat	3 783	963 (25.5%)	6 (0.2%)
	infertile steep	730	505 (69.2%)	101 (13.8%)
Warm (>16° annual mean)				
	fertile flat	6 011	1 477 (24.6%)	6 (0.1%)
	fertile steep	3 695	2 468 (66.8%)	512 (13.9%)
	infertile flat	344	314 (91.3%)	210 (61.0%)
	infertile steep	652	639 (98.0%)	317 (48.5%)
WET (>1 100 mm per annum) Cool				
	fertile flat	2 823	1 746 (61.8%)	152 (5.4%)
	fertile steep	5 525	5 079 (91.9%)	1 033 (18.7%)
	infertile flat	8 022	3 791 (47.3%)	921 (11.5%)
	infertile steep	1 140	942 (82.6%)	57 (4.9%)
Warm				
	fertile flat	12 629	5 801 (45.9%)	119 (0.9%)
	fertile steep	7 875	6 080 (77.2%)	475 (6.0%)
		67 781		

Table 2. Old growth forest ecosystems in north-east New South Wales provide high quality habitat for a wide range of biota including the amphibians, reptiles, birds and mammals listed below (adapted from A. M. Gilmore, Southern Cross University, pers. comm.).

Amphibians:

Assa darlingtoni
Lechriodus fletcherii
Mixophyes balbus
Mixophyes fleayi
Mixophyes iteratus
Litoria subglandulosa
Litoria pearsoniana

Reptiles

Diplodactylus vittatus
Varanus varius
Coeranoscincus reticulatus
Ophioscincus reticulatus
Orphioscincus truncatus
Anomalopus verrauxii
Cryptoblepharus virgatus
Egernia mcpheei
Leiolopisma zia
Eulamprus murrayi
Eulamprus martini
Eulamprus tenuis
Ramphotyphlops nigrescens
Cacophis krefftii
Hoplocephalus bitorquatus
Hoplocephalus stephensii
Vermicella annulata

Birds:

Grey Goshawk
Superb Fruit-dove
Rose-crowned Fruit-Dove
Wompoo Fruit-Dove
Topknot Pigeon
Red-tailed Black-Cockatoo
Yellow-tailed Black-Cockatoo
Rainbow Lorikeet
Scaly-breasted Lorikeet
Musk Lorikeet
Little Lorikeet
Double-eyed Fig-Parrot
Australian King Parrot
Crimson Rosella
Channel-billed Cuckoo
Powerful Owl
Barking Owl
Sooty Owl
Marbled Frogmouth
Forest Kingfisher
Sacred Kingfisher
Collared Kingfisher
Dollarbird
Noisy Pitta
Alberts Lyrebird
Superb Lyrebird
Rufous Scrub-bird
Tree Martin
Yellow-eyed Cuckoo-Shrike
Russet-tailed Thrush
Bassian Thrust
Pale-yellow Robin
Spectacled Monarch
Rufous Fantail
Red-browed Treecreeper
Regent Honeyeater
Mangrove Honeyeater
Black-chinned Honeyeater
Painted Honeyeater
Eastern Spinebill
Scarlet Honeyeater
Mistletoebird
Spotted Pardalote
Striated Pardalote
Figbird
Regent Bowerbird
Paradise Riflebird

Mammals:

Spotted-tailed Quoll
Brush-tailed Phascogale
Dusky Antechinus
Long-nosed Bandicoot
Koala
Greater Glider
Yellow-bellied Glider
Squirrel Glider
Eastern Pygmy Possum
Feather-tailed Glider
Red-legged Pademelon
Black-striped Wallaby
Grey-headed Flying Fox
Little Red Flying Fox
Black Flying Fox
Queensland Tube-nosed Bat
Queensland Blossom Bat
Eastern Horseshoe Bat
White-striped Mastiff-bat
Queensland Long-eared Bat
Gould's Long-eared Bat
Little Long-eared Bat
Chalinobus dwyeri
Gould's Wattled Bat
Chocolate Wattled Bat
Hoary Bat
Large-footed Mouse-eared Bat
Greater Broad-nosed Bat
Eastern Broad-nosed Bat
Golden-tipped Bat
Great Pipistrelle
Eptesicus pumilis
Eptesicus regulus
Eptesicus vulturnus
Eptesicus darlingtoni
Hastings River Mouse

Table 3. Examples of some of the essential foraging, nesting, roosting, basking and perching resources provided to fauna in old growth forest ecosystems; adapted from Scotts (1991).

Foraging Resources:

- canopy gaps on which species such as aerial feeding insectivorous birds and forest bats depend.
- multiple layer canopies which provide the dense canopy required by many insectivorous birds.
- loose and decorticating bark which harbour an invertebrate food resource utilized by insectivorous birds (e.g., treecreepers) and gliding possums.
- understorey and ground layer patchiness which may result from the changing characteristics of mature trees as they age. The characteristics of the canopy itself influences the patchiness of the ground layer which is important to animals such as potoroos and bandicoots.
- the level of nectar and other plant exudates.
- the importance of floristic diversity which can ensure the supply of food across seasons.
- moist litter layer which produces a microclimate important to frogs and other litter feeders.
- a dead wood medium, which various species of arthropod find suitable as habitat. These species, in turn, are a food resource for insectivorous birds and mammals.

Nesting and Roosting Resources:

- hollows — the number, range, shape and context of hollows are critical for a wide variety of species.
- dead branches — certain birds may nest only on horizontally oriented forks or vertically oriented forks.
- multiple strata — many birds have very specific requirements for nesting which are only found in structually diverse forest. Old growth forest may also support optimum densities of nesting materials.

Basking and Perching Resources:

- dead stags and dead branches provide perching sites for raptors, including endangered species.
- dead tops, stags and logs are used for basking by reptiles.

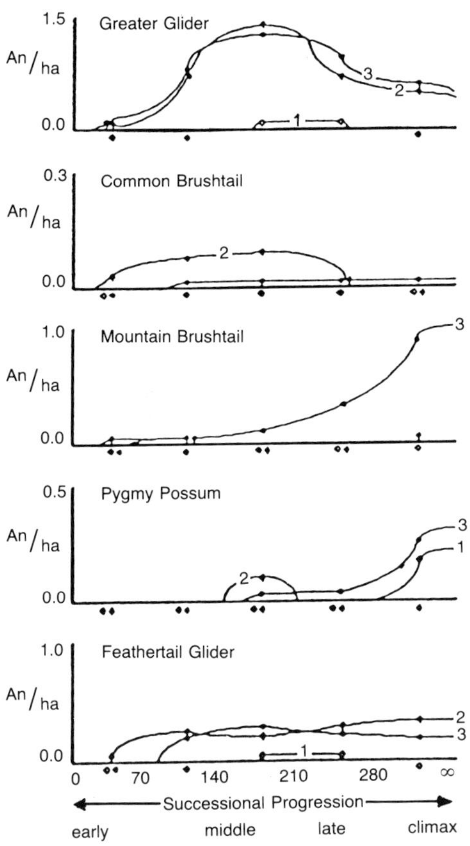

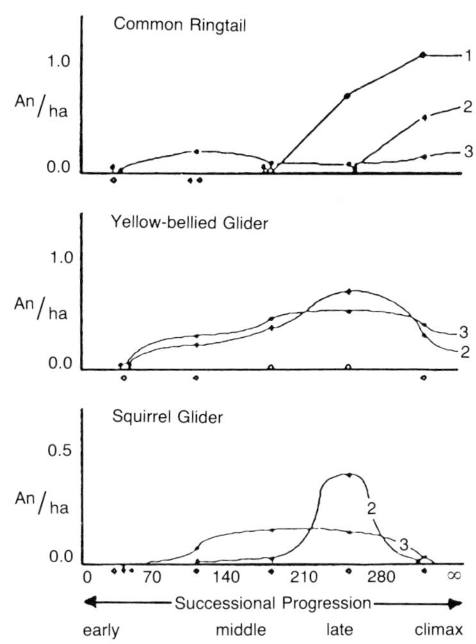

Fig. 3. Estimates of population density of arboreal marsupials along a vegetation succession, 0 to in excess of 280 years of age; adapted from Davey and Norton (1990).

IMPACTS ON BIODIVERSITY RESULTING FROM FORESTRY OPERATIONS

There are many aspects to forestry harvesting other than simply logging. For example, roads have to be built, sites have to be regenerated, poison baiting may need to be undertaken to help protect seedlings from native herbivores, and prescribed burning is employed for various tasks. Many of these aspects cannot be avoided if harvesting is to be undertaken, although there may be much scope for minimizing their impacts on biodiversity. The impacts on biota resulting from harvesting are not

Table 4. Impacts of forestry harvesting on forest biodiversity, adapted from Norton and May (1994). Note that forestry harvesting includes logging and associated roading, and the use of prescribed fire which may occur at various spatial and temporal scales.

Impacts on biodiversity resulting from forestry harvesting:

- may have positive, negative or indeterminate affects on the distribution, abundance, likelihood of persistence and evolution of taxa and other components of biodiversity.
- many are significantly detrimental since they can:
  1. destroy and modify complex forest landscapes and ecosystems.
  2. destroy and modify the natural environmental heterogeneity of forest ecosystems.
  3. destroy, prevent or hinder ecological processes that are the basis for species' persistence and evolution.
  4. destroy or significantly modify the habitat of species.
  5. destroy individual organisms and significantly modify populations and assemblages of species.
  6. result in loss of genetic diversity and may increase the likelihood of extinction.
- particularly threaten the continuation of various ecological and biological processes, genetic diversity, fauna — including Schedule 12 species in New South Wales, rare and endangered taxa, riparian environments and aquatic systems, landscape diversity, wilderness values.
- many may be irreversible or effectively irreversible given the legacies of past forest-use, cumulative and synergistic effects, and the incompatibility between the natural rates of turnover of species compared to the time horizons of intensive forestry.

point-based but normally will extend spatially well beyond the forest location that is actually harvested.

The ecological significance of impacts on eucalypt forest biodiversity that result from harvesting depends on factors such as: (i) the proportion of the total forest ecosystem or habitat, irrespective of land tenure, that is to support harvesting; (ii) the landscape pattern of impact — that is, the spatial configuration (e.g., clumped versus dispersed) of all of the forests supporting harvesting, including areas logged for roading and subjected to prescribed burning; (iii) the nature of impacts arising from other human activities in the same forest ecosystems or habitat, again irrespective of land tenure; (iv) the frequency of the harvesting relative to the natural rate of turnover of the target forest ecosystem or habitat; (v) the duration and frequency of significant ecological impacts arising from other human activities in the forest ecosystem or habitat targeted for integrated harvesting; and (vi) the resilience of the target forest ecosystems or habitat, and their constituent biota, to climate change and other global change (Norton and May 1994). Because of these factors, the conservation of biodiversity in a forest region requires a landscape approach to forest planning and management if the cumulative impacts on biota resulting from various human activities are to be minimized.

THREATS TO EUCALYPT FOREST BIODIVERSITY IN NEW SOUTH WALES

Norton and May (1994) reviewed a number of the key impacts on biodiversity resulting from forestry harvesting (Table 4) and detailed the components of old growth eucalypt forest biodiversity that they considered are likely to be under most threat. These components include: (i) biological and ecological processes which are significantly altered by forest fragmentation and modification; (ii) the genetic diversity of plants and animals; (iii) forest fauna including Schedule 12 species, fishes and invertebrates; (iv) rare and endangered species and assemblages; (v) riparian environments and aquatic systems; and (vi) ecosystem and landscape diversity and connectivity. The persistence of many other components of forest biodiversity, such as non-vascular plants and micro-organisms, is also of great concern (e.g., Maser and Trappe 1984; Kirkpatrick *et al.* 1991; Commonwealth of Australia 1992b; Franklin 1992; Brown 1994).

In general, we consider that a number of old growth forest ecosystems and/or their constituent biota, which occur on the more productive sites, are under significant threat of extinction from logging and clearing in many parts of the state.

TOWARDS SUSTAINABLE FORESTRY

The new National Forest Policy (Commonwealth of Australia 1992b) presents a framework for a move towards ecologically sustainable forestry. The policy outlines an approach based on a dedicated system of conservation reserves that are to be representative, comprehensive and adequate. The effectiveness of the reserve system is to be complemented by improved off-reserve management. In the policy it is stated that:

> "The governments have agreed to a strategy designed to conserve and manage areas of old-growth forests and wilderness as part of the reserve system . . . The governments' agreed approach to conserving and managing old-growth forests and forested wilderness has five basic elements: first, agreed criteria for old-growth forests and wilderness will be determined through the working group process already described. Second, using these criteria, the relevant state agencies will, as a matter of high priority, undertake

assessment of forests for conservation values, including old-growth values, and of forested land for wilderness values. Third, until the assessments are completed, forest management agencies will avoid activities that may significantly affect those areas of old-growth forest or wilderness that are likely to have conservation value. Fourth, forested wilderness areas will be protected by means of reserves developed in the broader context of protecting the wilderness values of all lands. For old-growth forest, the nature conservation systems will be the primary means of protection . . . The governments agree that, conditional on satisfactory agreement on criteria by the Commonwealth and the states, comprehensive, adequate and representative reservation to protect old-growth forests and wilderness values will be in place by the end of 1995 . . . Fifth, the relevant management agencies will develop management plans to appropriately protect old-growth forest and wilderness values."

As a signatory, the Government of New South Wales is required to ensure that until the forest assessments are completed, forest management agencies avoid activities that may significantly affect those areas of old-growth forest or wilderness that are likely to have conservation value. The protection of currently unreserved forest ecosystems with known or likely conservation value regionally is important as they provide the major options for reservation.

If enhanced off-reserve management is to be achieved then a number of changes to the design and implementation of forestry prescriptions and codes of practice in forest used for wood production are required. In future, we suggest that off-reserve forestry prescriptions intended to enhance biodiversity conservation be based on guiding ecological principles (Shepherd *et al.* 1992), the precautionary principle (Commonwealth of Australia 1992b) and be amenable to testing (Murphy and Noon 1991). Guiding ecological principles include the need to minimize forest fragmentation and degradation, the need for adequate forest conservation planning prior to road building, and the need for additional and special forestry prescriptions in areas identified as having high conservation values (Shepherd *et al.* 1992). Additional prescriptions include increased retention levels of habitat trees and trees with the potential to become habitat trees within a coupe, wider buffer zones along streams, and the exclusion of prescribed fire from specified areas.

CONCLUSIONS

The government and land management agencies of the State of New South Wales have statutory responsibilities to protect and conserve the full range of biological diversity occurring in the state's native forests, particularly the old growth forests on productive sites. Since many of these forests and their biota are of national and international significance, the need for commitment, leadership, innovation and vision by the state government to ensure the long-term viability of forest biodiversity is critical.

The essential components of a move towards sustainability will be (i) the establishment of a scientifically-defendable conservation reserve network focussed on old growth forests ecosystems on productive sites; (ii) markedly enhanced off-reserve codes of forest practice on both public and private lands that are based on scientific principles; (iii) an ecologically-conservative approach to forest allocation and management; and (iv) the development of a systematic ecological monitoring programme across the forest estate. These needs are outlined in the National Forest Policy Statement (Commonwealth of Australia 1992b). In our view, the rate of implementation of the above needs will be one critical measure of the extent to which the New South Wales and Commonwealth governments are committed to the adequate protection and conservation of old growth forest ecosystems and wilderness in New South Wales. We believe that there is an urgent need to put forestry on a sustainable basis in Australia, and particularly in New South Wales, if all components of forest biodiversity are to have a reasonable probability of persistence in the long term.

REFERENCES

Benwell, A., 1993. The biological significance of the forests of north eastern New South Wales. Internal report, New South Wales. National Parks and Wildlife Service: Sydney.

Braithwaite, L. W., 1984. On identifying important habitat characteristics and planning a conservation strategy for arboreal marsupials within the Eden Woodpulp Concession Area. Pp. 501–08 in *Possums and gliders* ed by A. P. Smith and I. D. Hume. Surrey Beatty & Sons: Chipping Norton.

Braithwaite, L. W., Belbin, L., Ive, J. R. and Austin, M. P., 1993. Land use allocation and biological conservation in the Batemans Bay forests of New South Wales. *Aust. For.* **56:** 4–21.

Brown, M. J., 1994. Global change and the monitoring of biodiversity in Tasmania's cool temperate forests. Pp. 107–16 in *Ecology and sustainability of southern temperate ecosystems* ed by T. W. Norton and S. R. Dovers. CSIRO: Melbourne.

Commonwealth of Australia, 1991. Ecologically Sustainable Development Working Groups Final Report — Forest Use. Australian Government Publishing Service: Canberra.

Commonwealth of Australia, 1992a. A National Strategy for the Conservation of Australia's Biological Diversity. BDAC, Australian Government Publishing Service: Canberra.

Commonwealth of Australia, 1992b. National Forest Policy Statement. Commonwealth of Australia: Perth.

Common, M. S. and Norton, T. W., 1992. Biodiversity: its conservation in Australia. *Ambio* **21:** 258–65.

Davey, S. M. and Norton, T. W., 1990. State forests in Australia and their role in wildlife conservation. *Proc. Ecol. Soc. Aust.* **16:** 323–45.

Franklin, J. F., 1992. Scientific basis for new perspectives in forests and streams. Pp. 25–27 in *Watershed management: balancing sustainability and environmental change* ed by R. J. Naiman. Springer-Verlag: New York.

Keith, D. A. and Sanders, J. M., 1990. Vegetation of the Eden Region, South East Ausstralia: Species composition, diversity and structure. *J. Veg. Sci.* **1:** 203–32.

Kirkpatrick, J. B., Meredith, C. M., Norton, T. W., Plumwood, V. and Fenshaw, R., 1991. The Ecological Future of Australia's Forests. Australian Conservation Foundation: Melbourne.

Love, L. A., 1993. The need for a regional approach to forest conservation planning in north east New South Wales. New South Wales National Parks and Wildlife Service, Grafton office, mimeo.

Maser, C. and Trappe, J. M., 1984. The Seen and Unseen World of the Fallen Tree. USDA Pacific Northwest Forest Experiment Station GTR PNW-164.

Mosley, G. and Costin, A., 1992. World heritage values and their protection in far south east New South Wales. Report to the Earth Foundation Australia: Sydney.

Murphy, D. D. and Noon, B. D., 1991. Coping with uncertainty in wildlife biology. *J. Wild. Manage.* **55**(4): 773–82.

Norton, T. W., 1994. Koala management in south-east Australia and the potential roles for reintroduction biology. In *Reintroduction biology of Australian and New Zealand fauna* ed by M. Serena. Surrey Beatty & Sons: Chipping Norton, in press.

Norton, T. W. and May, S. A., 1994. Towards sustainable forestry in Australian temperate eucalypt forests: ecological impacts and priorities for conservation, research and management. Pp. 10–30 in *Ecology and sustainability of southern temperate ecosystems* ed by T. W. Norton and S. R. Dovers. CSIRO: Melbourne.

Norton, T. W. and Mitchell, N. D., 1994. Towards the sustainable management of southern temperate ecosystems: lessons from Australia and New Zealand. *Pac. Conserv. Biol.* in press.

Norton, T. W. and Saxon, M. J., 1993. Predicted potential habitat of the Koala, *Phascolarctos cinereus* in south-east New South Wales and recommendations for field survey. Consultancy report to New South Wales National Parks and Wildlife Service, Queanbeyan, New South Wales.

RAC, 1991. Forest and Timber Inquiry Draft Report Volume 1. Resource Assessment Commission: Canberra.

RAC, 1992. Forest and Timber Inquiry Final Report Volume 1. Resource Assessment Commission: Canberra.

Scotts, D., 1991. Old growth forests: their ecological characteristics and value to forest-dependent fauna of south-east Australia. Pp. 147–60 in *Conservation of Australia's forest fauna* ed by D. Lunney. Royal Zoological Society of New South Wales: Mosman.

Shepherd, T. G., Saxon, M. J., Lindenmayer, D. B., Norton, T. W. and Possingham, H. P., 1992. A proposed management strategy for the Nalbaugh Special Prescription Area based on guiding ecological principles. South East Forest Series No. 2, New South Wales National Parks and Wildlife Service, Queanbeyan.

Streeting, M. and Hamilton, C., 1992. An economic analysis of the forests of south-eastern Australia. Resource Assessment Commission Research Paper No. 5.

WWF, 1992. The conservation status of temperate forests globally. WWF: Gland, Switzerland.

Young, M. D., Cocks, K. D. and Humphries, S. E., 1990. Australia's Environment and its Natural Resources — An Outlook. CSIRO Institute of Natural Resources and the Environment: Canberra.

How does grazing of native pastures affect their biodiversity?

M. H. FRIEDEL[1] and C. D. JAMES[1]

While biodiversity of native pastures can be addressed at the genetic level, we restrict our considerations to those levels which are amenable to management, that is, from single species through to whole landscapes. At large spatial scales, where landscape-level processes are important, biodiversity can be expressed in terms of life forms, structural characteristics and mosaic diversity. At smaller scales, biodiversity is reflected in identity, abundance, distribution and dominance of species or taxa, and demographic processes become significant. Variation through time, on many scales, compounds spatial diversity.

The major impact of grazing is through the redistribution of resources within and between habitat patches, and recovery does not necessarily follow on from removing grazing pressure. Grazing can affect biota directly or indirectly through reducing the productive potential of the soil. Managing for biodiversity depends on an understanding of patterns and processes at a variety of spatial and temporal scales. Managers can get feedback on their actions, and adjust them accordingly, by monitoring changes in key attributes of the different scales.

Conservation of biodiversity is a regional goal, not a requirement of individual grazing enterprises, but it cannot be achieved within the existing system of reserves. Integrating conservation with sustainable grazing of native pastures is essential. Grazing can be consistent with conserving biodiversity, provided basic biological processes are maintained. As a complementary strategy to broad-scale sustainable grazing, the existing reserve system could be supplemented, but the process would require socio-political changes.

INTRODUCTION

NATIVE pastures take many forms. In New South Wales, livestock graze in the alpine and subalpine country of the south-east and in the arid and semi-arid regions of the inland (Fig. 1). Inhospitable terrain or capricious climate make these areas unattractive to more intensive agriculture. In these rangelands, grazing livestock (sheep and cattle) have been introduced for commercial production, but otherwise the landscape is altered little compared to the intensive modifications imposed by agriculture or forestry. The temperate tablelands and slopes of the Great Divide are sometimes classed as rangelands too, but they have been much modified by tree removal, introduction of exotic pastures and weeds, addition of fertilizers and by consequent high stocking levels. Their ecology has been changed to such an extent that we cannot equate them with other native pastures and so we do not consider them further.

The alpine and subalpine rangelands are confined to the area above about 1 500 m, the winter snow line. They cover a very small area — about 5 000 km^2 in New South Wales and Victoria together — and grazing production is decreasing (Wilson 1990). Semi-arid and arid rangelands occur where annual rainfall is below 500 mm, and cover most of the western two-thirds of New South Wales. Sheep are the predominant domestic grazers of this region although some beef cattle are run as well (Atlas of Australian Resources 1980). Non-domestic grazers include kangaroos, feral goats and rabbits. We will limit our discussion to the semi-arid rangelands because of their large size and the amount of information available on grazing. For a comprehensive bibliography of the alpine and subalpine rangelands, we refer readers to Williams (1990).

We believe that biodiversity of native pastures should be considered at several levels. For instance, within the rangelands there are distinctly different types of vegetation, caused by regional variation in soils, topography and climate. In the Western Division of New South Wales (Fig. 1), large tracts of semi-arid woodland cover much of the west and north-west, giving way to bluebush and saltbush in the chenopod shrublands of the south and west, and mallee in the far south-west (Atlas of Australian Resources 1990). Of these, the semi-arid woodlands and chenopod shrublands are the major grazing lands of western New South Wales. The mallee accounts for a small area of the state, and grazing activities are relatively limited.

Within each major vegetation type there is further diversity at smaller spatial scales. For example, Harrington *et al.* (1984a) discern five main floristic alliances within the semi-arid woodlands, characterized variously by Poplar Box *Eucalyptus populnaea*, Cypress Pine *Callitris glaucophylla*, Mulga *Acacia aneura* and other dominants. At a finer spatial scale it is possible

[1]National Rangelands Programme, CSIRO Division of Wildlife and Ecology, P.O. Box 2111, Alice Springs, Northern Territory, Australia 0871.
Pages 249–59 *in* CONSERVING BIODIVERSITY: THREATS AND SOLUTIONS, ed by R. A. Bradstock, T. D. Auld, D. A. Keith, R. T. Kingsford, D. Lunney and D. P. Sivertsen. Surrey Beatty & Sons. 1995.

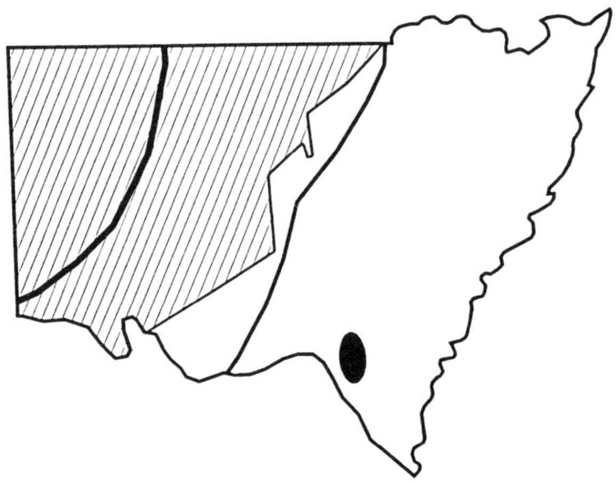

Fig. 1. Location of semi-arid (less than 500 mm annual rainfall) ______ and arid (less than 250 mm annual rainfall) ______ boundaries of the rangelands, the alpine and subalpine rangelands ■ and the Western Division of New South Wales ▨.

to see individual watersheds within the mulga woodlands and, on a component slope of that watershed, patterns of mulga groves and treeless intergroves (Tongway and Ludwig 1990). Individual plants become apparent on closer inspection of an intergrove.

Vegetation patterns such as these result in variation in the diversity of plants and animals at different spatial scales. In addition, major influences like rainfall and grazing act at different time scales, so that there is variation in the distribution of species through time. Both spatial and temporal variability in the distribution and abundance of species in native pastures are important for understanding the impact of any disturbance on biodiversity.

Biodiversity can best be described as "the variety of life and its processes" (West 1993), and so it cannot be expressed at a single scale in time or space (Noss 1990). Four levels of organization are commonly recognized in the biodiversity literature: landscape-region, community-ecosystem, species, and genetic (e.g., Noss 1990; West 1993). The different levels are not necessarily nested at different spatial scales. The landscape-region level can be defined in explicit spatial dimensions, for example, from 10^2 to 10^6 km^2. However, other levels of biodiversity are best defined by ecological dimensions. For example, niches of species are defined by various interactions of soil, climate, elevation and aspect, as well as by biotic interactions. These ecological dimensions are not necessarily encompassed by the geographic area that is chosen to define a region. In other words, landscapes and regions contain representative parts (more rarely, whole units) of a variety of communities or ecosystems, which in turn contain fractions of the geographic ranges of species, and fractions of the genetic diversity of a species.

It follows that the impact of grazing on biodiversity cannot be interpreted solely at a paddock scale; instead it needs to be integrated over regional scales so that the spatial variation in the distribution of biota, and the spatial variation in management, can be integrated. In this paper, we will focus on the species, community and landscape scales because we believe that these are the levels amenable to grazing and land management practices, that is, scales at which management actions can lead to predictable outcomes. Undoubtedly, management can have an impact at the genetic level too but the likely outcomes are obscure.

In this paper, we discuss the way in which livestock and other mammalian grazers use landscapes, and how that might affect "the variety of life and its processes" at different scales of time as well as space. We review the information available on the effect of grazing on biotic diversity, and discuss whether grazing brings about deleterious changes, whether these are reversible, and what land managers can do to maintain or enhance biodiversity.

PRINCIPLES AND PROCESSES OF CHANGE

Resource distribution

In the semi-arid woodlands, perennial grasses are the basis of a sustainable grazing system. These grasses require relatively high quality edaphic habitat, that is, nutrient-rich soils, good supplies of water (input and storage in the ground), and soils resistant to erosion (D. J. Tongway, pers. comm.). In a country renowned for its soils of poor nutrient quality and unreliable rainfall (Stafford Smith and Morton 1990), how do these grasses persist? The answer is that they grow in situations with relatively high levels of these resources. The redistribution of water and nutrients at a variety of scales from a few metres upwards results in patches of habitat of different quality.

In a semi-arid landscape with a degree of relief, upslope areas shed water, nutrients and sediment down slope. If the gradient is shallow, the movement of resources might be by sheet flow, ending in shallow depressions and floodouts. On steeper slopes, flow may become channelled into tributary drainage lines and move further away from the source. Either way, there are some areas that are enriched with water and nutrients relative to others, and these areas support different suites of species. At a landscape scale, the enriched areas often carry a higher proportion of perennial plants and more large woody species than the upslopes, because the supply of resources for establishment and growth are more reliable. Thus, there is diversity in life forms, structure and patchiness, depending on the size of catchments, resulting from differences in water, nutrient and energy flows.

Similar processes operate at the community scale. For example, the patterning of vegetation and soil in the semi-arid mulga woodlands into groves and intergroves is associated with differences in runon-runoff dynamics (Tongway and Ludwig 1990). The intergrove and grove pattern has three distinct vegetation components: *Eragrostis eriopoda* savannah on runoff slopes, *Monochather paradoxa* savannah at the bottom of each runoff slope, and a grove of mulga on each runon zone. The lowest intergrove-grove series abuts a dendritic drainage line with fairly dense mulga. Rainfall is shed most readily from the *E. eriopoda* zone, while the mulga grove retains much of the runon from smaller rains and the soils are consequently much wetter (Greene 1992). Only substantial rainfall produces runoff sufficient to flow into the drainage lines.

Although there is little published literature on the distribution and abundance of animals in relation to the patterning of soil and vegetation across the landscape, it is intuitively and demonstrably obvious that such correlations exist. For example, kangaroos are more abundant on floodouts where there is a better quality and higher cover of perennial grasses compared with runoff zones; small mammal, reptile, and invertebrate species respond to the grove-intergrove patterning of mulga woodlands (C. D. James, unpubl. data). However, a lack of specific knowledge limits our understanding of how grazing-induced changes in the processes and patterns of plant associations affect the distribution and abundance of fauna.

Grazing and resource distribution

When animals graze too heavily, many of these processes are disrupted. To begin with, overuse of the mulga woodlands eliminates the perennial grasses. With fewer or no plants to obstruct flow and stabilize the soil, runoff will increase. Trampling will release fragments of soil, and stock pads will channel water and sediment through or around the groves, into drainage lines and, eventually, out of the system (D. J. Tongway, pers. comm.). Once the groves becomes starved of resources from upslope, regeneration will fail and the vegetation will gradually die out. Resources within the groves will no longer be retained and, it can be argued, the slopes will become increasingly homogeneous as the groves are lost. Diversity at the scale of the original groving pattern is lost, and deposition and plant production will intensify somewhere in the larger drainage system. The scale of spatial biodiversity is now different and the temporal scale changes too. For example, light rain may produce no plant growth on the slopes where once the perennial grasses would have responded. Alternatively, the variety and abundance of ephemeral plants may increase briefly after rain, in place of perennials.

This description may suggest that the effect of grazing can be generalized across whole landscapes but the impact is by no means uniform, even in a relatively homogeneous environment. Utilization will be heavy around watering points, diminishing with distance. In semi-arid saltbush country, sheep graze intensively within 1.5 km of water but when conditions are very dry they graze up to 5 km away. Cattle range further, within 3 km of water when forage supply is good, and more than twice that distance when it is poor (Wilson and Harrington 1984).

Factors other than water supply also affect the distribution of grazing impacts. Spatial variation in the intensity of grazing is also created because animals seek out patches of preferred vegetation (J. Landsberg, pers. comm.). At the landscape scale, in the order of square kilometres, grazing stock may encounter plains and hills and associated contrasting pastures within a single paddock. Not all the pasture types will be equally preferred, so that the effective stocking rate on one may be very high, while that on another may be only a fraction of the whole-paddock rate (Lynch *et al.* 1992). The grazing impact on a particular pasture type will depend on the mix of other pasture types in the same paddock. As well, smaller scale features like the presence of shade trees or upslope areas for night-time camps, and the direction of prevailing winds will also influence grazing patterns. Thus, grazing is spatially heterogeneous in response to distance from water, edaphic heterogeneity, microclimate and the natural patterning of vegetation.

CHANGES IN THE SEMI-ARID RANGELANDS

There have been substantial changes in the biodiversity of the semi-arid rangelands since the arrival of Europeans. Here we describe the major changes and their causes, and whether there is potential for changes to be reversed.

By far the greatest change in the flora of the semi-arid woodlands has been a shift from an open grassy state to one dominated by unpalatable shrubs. Prior to the introduction of livestock, the vegetation contained high quality ephemerals, medium quality perennials and palatable browse, and the potential carrying capacity for livestock was higher and more reliable than it is now (Harrington *et al.* 1984a). As previously explained, degradation in these woodlands is characterized by reduced cover of perennial grasses, and a change in the scale and functioning of the habitat patches. Without the competition provided by vigorous perennial grasses, seedlings of woody species establish readily. Once the trees and shrubs are mature, re-establishment of perennial grasses is minimized through competition for scarce resources, and woody species become increasingly dominant. The soil, having lost its protective mantle of grass, is damaged by trampling and is increasingly susceptible to erosion.

In some woodlands, particularly Rosewood/ Belah *Alectryon oleifolius*/*Casuarina pauper*, and in mallee, consumption of seedling trees and shrubs by livestock and feral grazers may suppress recruitment and lead to a longer term loss of dominant woody species (Chesterfield and Parsons 1985).

In chenopod shrublands, degradation is equated with shrub loss. Saltbush *Atriplex* spp. and Bluebush *Maireana* spp. are useful livestock drought reserves — they are eaten once the herbage layer has gone — but their prime value in the long term is as protective cover for potentially unstable soils. Loss of these perennial shrubs because of heavy grazing inevitably leads to replacement by shorter-lived species, often other chenopods or grasses which, being less persistent, have less value for soil protection in poor seasons (Graetz and Wilson 1984). Noble (1984) does not identify any grazing-induced degradation in mallee, although he expresses concern for the impact of marginal cropping activities, which are beyond the scope of this paper.

The fauna of the semi-arid rangelands of New South Wales has been profoundly altered in the time since European settlement. At least 38 per cent of species of mammals, and 2 per cent of species of birds that formerly occurred in the arid and semi-arid rangelands of New South Wales are now extinct, either locally or globally. The geographic range of many other species of mammals, birds, and reptiles has also decreased over the same time period. Dingoes have been virtually eliminated and densities of kangaroos have increased.

Sources of change

At least five grazing mammals have a marked impact in the rangelands. Sheep, cattle, kangaroos, feral goats and rabbits all have the potential to bring about within- and between-community change. Landsberg *et al.* (1992) found that less than half the total grazing pressure in two semi-arid woodland paddocks in north-west New South Wales was due to sheep. Even when sheep were stocked conservatively, their low numbers could be counterbalanced by high goat and kangaroo densities.

The spatial distribution of both feral and native grazers does not necessarily match that of the commercial livestock. Goats and kangaroos are not restrained by fences and can migrate from one watering point to another. Rabbits are not dependent on watering points. Instead, their distribution is controlled by forage quality and suitable habitat for warren sites, and their impact is concentrated around their warrens (Wilson and Harrington 1984).

The effect that grazing mammals have is a product of their density and distribution, their particular dietary preferences and behaviours like trampling or digging. Goats eat more browse than sheep, kangaroos eat more grass and rabbits favour green feed of any kind as well as having a penchant for the bark of woody species (Wilson and Harrington 1984). However, these are simply preferences, and other feed types will be eaten as supply dwindles. As a consequence, the differences in impact at the herbaceous species level are small. On the other hand sheep, goats and rabbits in sufficient numbers are all capable of inducing structural change in the long term either by reducing the vigor of the pasture layer, to the benefit of woody species or, alternatively, eliminating the seedlings of trees and shrubs in some rangeland types (Harrington *et al.* 1984b; Chesterfield and Parsons 1985).

The extent of competition among grazing animals for forage depends on the land type and seasonal conditions (Harrington *et al.* 1984b; Noble 1984). When forage is abundant, competition for food is unlikely (Edwards 1990) but, more often, non-domesticated grazers have a significant impact on pastoral production and on the habitat (Wilson 1991). Any attempt to manage for biodiversity in pastoral lands will need to include control of feral and native grazing animals as well as livestock.

Fire was once an integral part of many rangeland landscapes, initiated either by Aborigines or lightning, but it has often been suppressed by European landholders. Very likely, not only the incidence of fires was higher in the past, but the size and intensity of fires were different too. This probably means that prior to the intervention of Europeans there was patchiness at various spatial and temporal scales, and resource quality within some landscapes, which is now gone.

Grazing and fire are interdependent in some landscapes. In the semi-arid woodlands, removal of perennial grasses by heavy grazing virtually eliminates the opportunity for burning. The use of fire as an inexpensive management tool to reduce the density of unpalatable perennial shrubs is therefore eliminated too.

The causes of changes in faunal species are not easily disentangled. The impacts of habitat change wrought by domestic and other grazers and by altered fire regimes are complicated by feral predators such as foxes and cats and, in some landscapes, by radical tree clearing. The dramatic decline of native mammal species in the rangelands has been attributed to all factors but the last (Morton 1990; Dickman *et al.* 1993). It is suggested that the decline was not steady, but underwent a spurt at critical times and places, when drought refuges were occupied by both competitors and predators.

Grazing by domestic stock has been implicated by many authors as a major cause for the extinction of faunal species in New South Wales (Krefft 1866; Dickman *et al.* 1993; Sadlier and Pressey 1994; Smith *et al.* 1994). While this assessment may be correct, there is no direct evidence linking grazing with loss of biodiversity because of confounding factors such as rabbit plagues, habitat clearing and predation by foxes and cats.

Whereas the distribution and abundance of many native mammalian species have declined through the indirect impact of European settlement, other species have increased. The densities of Red, Eastern and Western Grey Kangaroos and of Emus are higher now than 150 years ago (Newsome 1975, 1990; Caughley *et al.* 1980). The presumed mechanism for this increase is the extermination of dingoes "inside" the dingo fence (Caughley *et al.* 1980), and the provision of large numbers of watering points across western New South Wales.

Only a few studies have specifically examined the effect of grazing on native fauna in Australia. Greenslade (1992) reported a significant increase in the number of species and abundance of invertebrates inside a long-term grazing exclosure, compared with outside, in chenopod shrublands in South Australia. In contrast, Abensperg-Traun (1992) found no changes in diversity or abundance of termite faunas in relation to grazing intensity in the wheatbelt in Western Australia. As far as we can determine, the only research aimed at quantifying the impact of grazing on native fauna in the arid and semi-arid rangelands is a project currently being conducted by one of us (C. D. James). Preliminary studies using fence-line contrasts of high versus low grazing intensity have indicated that diurnal reptiles, birds, and some taxa of invertebrates in heavily grazed areas are significantly less abundant and species rich than in lightly grazed areas (C. D. James, unpubl. data). At this stage, there are not enough data for a synthesis of diversity patterns at community or landscape scales.

Are changes reversible?

Grazing affects biota both directly and indirectly. For example, in the mulga woodlands, many perennial grasses will die during drought regardless of grazing, but the death of palatable species is disproportionately large and dependent on the degree of grazing pressure (Hodgkinson 1993). Seed production is also suppressed disproportionately while recruitment is less sensitive. This is a direct grazing effect which presumably is reversible until the seed bank is exhausted or the productive potential of the soil is damaged. In chenopod shrublands, the same is true — the shrub populations will recover if the topsoil and seed supply are in good condition.

Excessive establishment of shrubs is also reversible while the perennial grass layer remains capable of sufficient growth to support fire and while shrubs are small enough to succumb. Once the shrubs are large and capable of capturing resources in competition with the pasture layer, recovery is likely to depend on rare opportunities for burning, possibly decades apart, or on major management inputs. These will be expensive in terms of income foregone or financial outlay.

Where grazing acts indirectly through soil change, recovery through simple grazing control becomes even less likely. Small changes to the soil surface of mulga woodlands may alter hydrological processes markedly (Tongway and Ludwig 1990), and starve the vegetation of water. As well, nutrients are concentrated in the topmost layers of these soils (Tongway and Ludwig 1990) so that the loss of a very few centimetres would seriously deplete the nutrient capital.

Restoration depends on the re-creation of spatial and temporal processes which conserve resources. Tongway (1990) suggests that the re-establishment of mulga log mounds on degraded woodland slopes will begin the processes of resource entrapment needed to halt and reverse degradation there. In chenopod shrublands where the bush has gone and the soil is scalded, recovery depends on mechanical treatment (e.g., Cunningham 1987) to trap scarce soil particles, organic matter, water and seeds.

Regarding the reversal of faunal losses, we can only assume that rehabilitation of vegetation will lead to improved faunal habitat. If faunal species are only locally extinct, they may re-invade unaided or be re-introduced. The greater likelihood is that, without effective management of predators and competitors and the restoration of processes controlling the supply of food, shelter and breeding sites, the recovery of even the most well-studied species will be tenuous.

MODELS RELATING GRAZING AND DIVERSITY

The study of biodiversity currently suffers from a dearth of theoretical paradigms against which to test and refine hypotheses for land management. The need for such models at a general ecological level is evident, given the growing worldwide interest in the impact of human activity on biodiversity. Figure 2 illustrates four models of potential changes in biodiversity in response to a hypothetical disturbance gradient, of which grazing could be an example. These models suggest a variety of responses, including thresholds at which declines in biodiversity could be rapid. The models are discussed in more detail below.

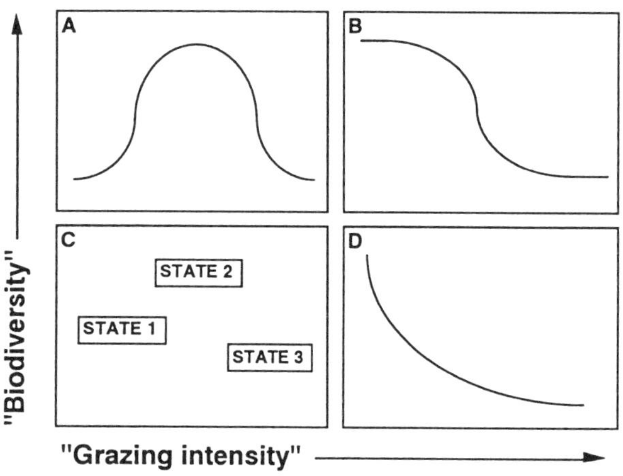

Fig. 2. Models of potential change in biodiversity along a hypothetical gradient of disturbance. A — Intermediate disturbance model; B — Island biogeographic model; C — State-and-transition model; D — Suggested relationship between species richness of vertebrate animals and an index of grazing intensity, for semi-arid woodland sites in northwestern New South Wales.

The intermediate disturbance hypothesis (Connell 1978; Milchunas *et al.* 1988) predicts that at high or frequent levels of disturbance, and low or infrequent levels of disturbance, diversity will be lower than at some intermediate level of disturbance (Fig. 2A). Studies of the impact of grazing on the diversity of plant communities in other countries are consistent with the predictions of this hypothesis, although the trends are often weak (Andresen *et al.* 1990; Archer *et al.* 1987; Wilcox *et al.* 1987). A test of this hypothesis in the temperate rangelands of New South Wales found a pattern consistent with the predictions but the trend was similarly weak (S. McIntyre, pers. comm.).

In mulga woodlands, grazing at high levels increases the homogeneity of the landscape by increasing the scale of the grove-intergrove patchiness described above (D. J. Tongway, pers. comm.). Thus, although moderate grazing may result in higher diversity within a patch, diversity at larger scales (among habitat patches) may decline as a result of a change in the diversity of habitats (Chaneton and Facelli 1991).

It is possible that unreliable climate and fire regimes of the arid and semi-arid rangelands mean that levels of disturbance are never truly low, even when grazing disturbance is minimal or absent. No species may have all the appropriate attributes to dominate, whatever the vagaries of climate and fire. What constitutes a high or a low disturbance regime needs clarifying.

Island biogeographic theory predicts a relationship between species richness and the size of islands (Fig. 2B — MacArthur and Wilson 1967). In recent years this theory has been modified to predict a relationship between species richness and the size of habitat fragments resulting from the clearing of native vegetation (Diamond 1975; Gilpin and Diamond 1980). In this situation, habitat fragments are considered islands because they are surrounded by a landscape that is hostile to the species inhabiting the fragment. This model predicts a dramatic decline in the diversity of species in an area at some critical threshold of fragmentation. Grazing may have the effect of fragmenting habitats because of the tendency for grazing animals to focus on high quality patches where productivity is high. The model may apply to grazing impacts for some taxa although it has not yet been tested.

The state-and-transition model (Fig. 2C — Westoby *et al.* 1989) differs from the previous two models in that it predicts multiple stable states of a response variable, rather than continuous variation. Each state is relatively stable over a range of disturbance (e.g., grazing intensities), and it may require a large disturbance to push a system from one state to the next. This model does not predict the diversity of a given state, and may be compatible with predictions of the previous models. However, it may indicate what states are to be avoided and suggest what processes lead to deleterious states.

Very few studies have examined the impact of different levels of grazing on faunal communities, and no published studies exist for New South Wales. Most studies in the international literature have used a two-treatment experiment — presence and absence of grazing — to examine the effect of grazing on a variety of taxa. Results from these studies have usually reported lower abundance, biomass and species richness of grazed sites compared with ungrazed sites (e.g., Jones 1981; Bowland and Perrin 1989; Jepson-Innes and Bock 1989; Heske and Campbell 1991; but see Clark *et al.* 1989; and Jameson 1989). It is also clear from these studies that changes in abundance and community structure are often complex (Quinn and Walgenbach 1990), and biodiversity will not be altered in any one consistent direction as a consequence of grazing. Testing theoretical models such as the intermediate disturbance hypothesis will require more rigorous hypothesis testing than has occurred in past studies.

It is critical to know what levels of grazing may have severe effects on biodiversity so that appropriate management can be developed. If a model such as the island biogeographic hypothesis (Fig. 2B) is supported, then management could be modified to operate at grazing levels below the threshold which would lead to a severe decline in biodiversity. Preliminary data relating species richness of vertebrates to an index of grazing intensity suggests that a model such as Figure 2D may be appropriate for the vertebrate fauna of the semi-arid woodlands of northwestern New South Wales (C. D. James, unpubl. data). However, these results relate to only one sample of a few habitats within the landscape. We do not yet have enough information about spatial or temporal variation in the distribution and abundance of species to be confident of the shape of the function illustrated in Figure 2D. This result may simply be a subset of the island biogeographic model. Further work covering the full range of grazing intensities across the landscape will be needed to test these models.

MONITORING FOR MANAGEMENT

The ways of managing grazed native pastures for their biodiversity as well as commercial production are limited. Land areas are large and economic returns are low. Potential actions include controlling numbers of livestock as well as non-domestic grazers and predators, judicious placing of fencing and watering points, and rehabilitating degraded land. The difficulty facing the landholder is that these actions are costly, while at the same time there are no guarantees as to the outcomes because of the innate variability of landscape and climate.

In this circumstance, management has to be adaptive. In other words, when a particular action is taken in a relatively unknown situation, the action must be designed to provide the manager with the maximum possible information for future use. Consequently, its outcome has to be monitored and, if it is not as good as it could be, then the action is modified. Monitoring is an essential part of refining management, with the added benefit that, if enough information accumulates, there is an increasing potential for predicting what the outcome will be.

What should be monitored?

Noss (1990) argues that three primary attributes are important at all levels of organization, from genetic to landscape-region: composition, structure and function. We provide a reduced list (Table 1) of his proposed indicator variables, but recommend that readers examine the complete version in Noss (1990). The variables are not exclusive to a particular level and nor are the techniques for assessing them. The techniques include remote sensing, via satellite and aircraft-borne imaging, geographic information system technology, ground-based resource inventories and censuses for data gathering, and data analysis across both time and space.

It is important that "charismatic fauna" like our small marsupials should not be the focus of monitoring, despite their natural appeal. Damage to habitat may be well advanced before deleterious change to native fauna becomes apparent and so monitoring of habitat attributes is likely to provide a better early warning of decline. In addition, common faunal groups such as ants and birds are likely to be better indicators of deteriorating ecological processes than small marsupials, for which trapping success is often very low.

Assessing "biodiversity"

How does information about species composition and abundance in different habitat patches relate to the concept of biodiversity at community or landscape scales?

Linking measurement of species composition and abundance at sample sites with the concept of biodiversity at regional scales is a major methodological stumbling block. There is no single index of diversity that can be used to indicate biodiversity, and even previously accepted measures of diversity such as the Shannon-Weaver information statistic, have now fallen out of favour. However, a potentially fruitful avenue of investigation is to link site-based samples with regional habitat pattern using "mosaic diversity" indices (e.g., Turner 1989; Scheiner 1992). These indices attempt to combine information about species richness, abundance, and dominance among communities along an environmental gradient. Such techniques are potentially useful, particularly in comparative studies, but will require careful calibration and application of the index to ensure that real biological patterns are being reflected.

Recent approaches to practical assessment

Some specific indicators of the status of habitats, appropriate to the scale of native pasture management in Australia, have been explored recently. Most existing methodologies are oriented towards pastoral production and depend on ground-based measures of community-ecosystem level attributes (Friedel 1993). Typically, plant data, often biomass or species abundance, are presented as an average. Recently, Tongway

Table 1. Some indicator variables for monitoring biodiversity in terms of composition, structure and function at three levels of organization, for grazed native pastures; modified from Noss (1990).

	Composition	Structure	Function
Landscape-region	Identity, distribution, richness and proportion of patch (habitat) types; collective patterns of species distributions	Heterogeneity; connectivity; contrast; grain size; patch size frequency distribution; perimeter-area ratio	Disturbance processes (e.g., areal extent), frequency, intensity and seasonality; nutrient cycling rates; energy flow rates; patch turnover rates; rates of erosion, and geomorphic and hydrologic processes
Community-ecosystem	Identity, relative abundance, frequency, richness, evenness and diversity of species and guilds; proportions of, e.g., endemic and exotic species; dominance-diversity curves; life-form proportions	Soil variables; vegetation biomass and physiognomy; horizontal patchiness; abundance, density and distribution of key physical features (e.g., sinks), and structural elements (e.g., log mounds)	Biomass and resouce productivity; herbivory and predation rates; colonization and local extinction rates; patch dynamics (fine scale); nutrient cycling rates
Species-population	Absolute or relative abundance; frequency; importance or cover value; biomass; density	Dispersion; range; population structure; habitat variables (see above)	Demographic processes (e.g., fertility, recruitment rate, survivorship and mortality); life history; phenology

(1990) proposed a system of soil and landscape analysis to determine the nature and scale of patchiness. This approach integrates vegetation and soil survey up to watershed scale and is directed towards restoration of landscape processes. Stafford Smith and Pickup (1990) have outlined a sequence of steps for interpreting vegetation patterns generated by spatial processes, including grazing. They point out that complex ecological patterns involve changes in variances, as well as means, of attributes across a landscape. Landscape analysis, they suggest, might include initial stratification from satellite imagery and subsequent assessment of variance amongst replicates at different distances from watering points.

Temporal variability is a potentially useful monitoring tool, too. Data from grazing trials in Western Australian stony halophyte pastures (Alec Holm, Western Australian Department of Agriculture, pers. comm.) show that variance over time in shrub biomass could indicate grazing effects not apparent from mean values (Friedel 1993). As well, differences in shrub biomass on upslope areas and on drainage lines were detected both in the mean values over time and the variances, reflecting likely differences in hydrological processes.

Whatever is monitored, it is essential that appropriate information is available to land managers so that they can assess the outcomes of their actions.

IS GRAZING CONSISTENT WITH CONSERVING BIOLOGICAL DIVERSITY?

The loss of fauna and the destruction of habitat since Europeans occupied the grazing lands has lead to some grim forecasts. "We cannot guarantee the long-term survival of the semi-arid ecosystems under the present regime of domestic and feral herbivores and feral carnivores" (Pickard 1994). The continued loss of Australia's mammal species and the accelerated loss of bird species will be followed by the decline of reptiles and amphibians (Recher and Lim 1990). As well, the semi-arid woodland habitat may be quite incompatible with grazing. "Continuous grazing by sheep, even at low densities, appears to degrade this wooded grassland" (Hodgkinson 1993).

Is it possible to sustain both pastoral production and biodiversity on native pastures, and should it be attempted? We believe there is no real choice. In essence, existing parks and reserves are unable to encompass all the diversity of the grazing lands. The community cannot afford to purchase and manage all the land required for the purpose, and it may not be able to support the economic and social cost of retiring land from commercial grazing. The only option is to manage for both production and biodiversity, as recent publications attest (Mott and Bridgewater 1992; Pickard 1994).

There are no blueprints for success, but several prospects are worth considering. The first step is to recognize that economic production is the aim of individual enterprises, whereas conservation of biodiversity is a regional goal. It is not realistic to expect every species on every part of a grazing enterprise to be conserved. We accept that grazing will inevitably bring about change. Managing for conservation as well as grazing means that changes should be reversible if grazing is removed. In practical terms, this means maintaining basic ecological processes so that soils remain in good health, and maintaining vigorous seed-bearing plants within dispersal range to restore the complement of species. Presently, so little is known about the requirements of fauna that we must simply assume that high quality habitat is a surrogate for healthy faunal communities.

Some encouraging examples are available from the grazing industry. Low stocking rates coupled with careful management of land and livestock has allowed perennial forage to regenerate and persist, and produced satisfactory profits, for innovative landholders in various regions of arid Australia (Lange *et al.* 1984; Purvis 1986; Morrissey and O'Connor 1988). Research by Hodgkinson (1993) has shown that, while continuous grazing at any stocking rate may degrade the semi-arid woodlands, tactical grazing which protects perennials during drought and at seed-set after rains, will prevent adverse change. We assume that the recovery of the perennial component of native pastures indicates the gradual recovery of ecological processes and, hence, of at least some components of biodiversity.

At a broader scale, a review of the winter rainfall area of Western Australia's arid zone (Curry and Hacker 1990) concluded that various ecological processes and the genetic diversity of flora and fauna as indicated by species richness had, by and large, not been seriously affected by pastoralism. However, the authors were at pains to point out that, while the overall incidence of severe damage was very low, it was disproportionately large in areas of the landscape of especial importance to native biota as well as livestock. Clearly, management solutions which operate at this landscape-region level are necessary, and they need to be carefully planned and implemented (Reid and Puckridge 1992).

At the regional level, as we have just indicated, some landscape elements will be more important than others. They may, for instance, be particularly species rich or they may comprise a network of seasonal breeding grounds. For practical conservation of biodiversity, some means of integrating conservation with grazing management must be devised at the regional scale. As Noss and Harris (1986) explain, the linking together of "high-quality" areas should enhance the survival of wide-ranging species, facilitate gene flow and help perpetuate large-scale ecological processes.

With this in mind, Morton *et al.* (1995) advocate a hierarchical system of land use for arid Australia: (1) full inclusion into National Parks, (2) small excisions of resource-rich habitat from existing land holdings to enhance networks and incorporate variation, (3) restricted use of locally resource-rich habitats within existing land holdings at critical times (breeding seasons, droughts) and (4) general sustainable use for the majority, i.e., those areas not covered by (1), (2) or (3). Such a system requires careful setting of land use priorities, followed by objective, quantitative analysis of environmental patterns, before changes are made to existing land uses. Once revised uses are in place, their impacts will need continuing management and monitoring. Realistic rewards for land managers and appropriate social and political change will also be integral to the success of the proposal.

Australia's native pastures are unique, in that they have not been subjected to thousands of years of livestock grazing. We still have the option of conserving their biodiversity in something like their original state. The task is substantial but worth the challenge.

ACKNOWLEDGEMENTS

We thank Jill Landsberg, Sue McIntyre, John Pickard, Mark Stafford Smith, David Tongway and an anonymous referee who contributed either their unpublished information or their time, and sometimes both, to improve this paper.

REFERENCES

Abensperg-Traun, M., 1992. The effects of sheep grazing on the subterranean termite fauna (Isoptera) of the Western Australian wheatbelt. *Aust. J. Ecol.* **17:** 425–32.

Andresen, H., Bakker, J. P., Brongers, M., Heydemann, B. and Irmler, U., 1990. Long term changes of salt marsh communities by cattle grazing. *Vegetatio* **89:** 137–48.

Archer, S., Garrett, M. G. and Detling, J. K., 1987. Rates of vegetation change associated with prairie dog (*Cynomys ludovicianus*) grazing in North American mixed grass prairie. *Vegetatio* **72:** 159–66.

Atlas of Australian Resources, Third Series., 1980. *Soils and land use, volume 1* ed by T. Plumb. Division of National Mapping: Canberra.

Atlas of Australian Resources, Third Series., 1990. *Vegetation, volume 6.* Department of Administrative Services: Canberra.

Bowland, A. E. and Perrin, M. R., 1989. The effect of overgrazing on the small mammals in Umfolozi Game Reserve (South Africa). *Zeitschrift Fuer Saeugetierkunde* **54:** 251–60.

Caughley, G. C., Grigg, G. C., Caughley, J. and Hill, G. J. E., 1980. Does dingo predation control the densities of kangaroos and emus? *Aust. Wildl. Res.* **7:** 1–12.

Chaneton, E. J. and Facelli, J. M., 1991. Disturbance effects on plant community diversity: spatial scales and dominance hierarchies. *Vegetatio* **93:** 143–56.

Chesterfield, C. J. and Parsons, R. F., 1985. Regeneration of three tree species in arid south-eastern Australia. *Aust. J. Bot.* **33:** 715–32.

Clark, B. K., Kaufman, D. W., Finck, E. J. and Kaufman, G. A., 1989. Small mammals in tall grass prairie: patterns associated with grazing and burning. *Prairie Nat.* **21:** 177–84.

Connell, J. H., 1978. Diversity in tropical rainforests and coral reefs. *Science* **199:** 1302–310.

Cunningham, G. M., 1987. Reclamation of scalded land in western New South Wales — a review. *J. Soil Conserv. Service NSW* **43:** 52–61.

Curry, P. J. and Hacker, R. B., 1990. Can pastoral grazing management satisfy endorsed conservation objectives in arid Western Australia? *J. Environ. Manage.* **30:** 295–320.

Diamond, J. M., 1975. The island dilemma: lessons of modern biogeographic studies for the design of natural reserves. *Biol. Conserv.* **7:** 129–46.

Dickman, C. R., Pressey, R. L., Lim, L. and Parnaby, H. E., 1993. Mammals of particular conservation concern in the Western Division of New South Wales. *Biol. Conserv* **65:** 219–48.

Edwards, G. P., 1990. Competition between red kangaroos and sheep in arid New South Wales. Ph. D. Thesis, University of New South Wales.

Friedel, M. H., 1993. Spatial and temporal heterogeneity — a question of scale. Pp. 43–47 in *Plant dynamics and interpretation in rangeland ecosystems.* Western Australian Department of Agriculture Miscellaneous Publication 27/93.

Gilpin, M. E. and Diamond, J. M., 1980. Subdivision of nature reserves and the maintenance of species diversity. *Nature* **285:** 567–68.

Graetz, R. D. and Wilson, A. D., 1984. Saltbush and bluebush. Pp. 209–22 in *Management of Australia's rangelands* ed by G. N. Harrington, A. D. Wilson and M. D. Young. CSIRO Australia: Melbourne.

Greene, R. S. B., 1992. Soil physical properties of three geomorphic zones in a semi-arid mulga woodland. *Aust. J. Soil. Res.* **30:** 55–69.

Greenslade, P., 1992. Conserving invertebrate diversity in agricultural, forestry and natural ecosystems in Australia. *Agric. Ecosystems Environ.* **40:** 297–312.

Harrington, G. N., Mills, D. M. D., Pressland, A. J. and Hodgkinson, K. C., 1984a. Semi-arid woodlands. Pp. 189–207 in *Management of Australia's rangelands* ed by G. N. Harrington, A. D. Wilson and M. D. Young. CSIRO Australia: Melbourne.

Harrington, G. N., Friedel, M. H., Hodgkinson, K. C. and Noble, J. C., 1984b. Vegetation ecology and management. Pp. 41–61 in *Management of Australia's rangelands* ed by G. N. Harrington, A. D. Wilson and M. D. Young. CSIRO Australia: Melbourne.

Heske, E. J. and Campbell, M., 1991. Effects of an 11-year livestock exclosure on rodent and ant numbers in the Chihuahuan Desert, southeastern Arizona. *Southwest. Nat.* **36:** 89–93.

Hodgkinson, K. C., 1993. Tactical grazing can help maintain stability of semi-arid wooded grasslands. Pp. 75–76 in *Proc. 17th Int. Grassland Congress,* New Zealand and Australia, February 1993.

Jameson, M. L., 1989. Diversity of coprophagous Scarabaeidae (Coleoptera) in grazed versus ungrazed sandhills prairie in western Nebraska (United States of America). *Trans. Nebraska Acad. Sci.* **17:** 29–35.

Jepson-Innes, K. and Bock, C. E., 1989. Response of grasshoppers (Orthoptera: Acrididae) to livestock grazing in southeastern Arizona: differences between seasons and subfamilies. *Oecologia* **78:** 430–31.

Jones, K. B., 1981. Effects of grazing on lizard abundance and diversity in western Arizona. *Southwest. Nat.* **26:** 107–15.

Krefft, G., 1866. On the vertebrated animals of the lower Murray and Darling, their habits, economy, and geographical distribution. *Trans. Phil. Soc. NSW* **1862–1865:** 1–33.

Landsberg, J., Stol, J., Stafford Smith, M. and Hodgkinson, K., 1992. Distribution of sheep, goats and kangaroos in semi-arid woodland paddocks. Pp. 117–21 in *Conference papers, Australian rangeland society 7th biennial Conference, Cobar New South Wales, 5–8 October, 1992.* The Australian Rangeland Society: Cobar.

Lange, R. T., Nicolson, A. D. and Nicolson, D. A., 1984. Vegetation management of chenopod shrublands in South Australia. *Aust. Rangel. J.* **6:** 46–54.

Lynch, J. J., Hinch, G. N. and Adams, D. B., 1992. *The behaviour of sheep.* CSIRO Australia: Melbourne.

MacArthur, R. H. and Wilson, E. O., 1967. *The theory of island biogeography.* Princeton University Press: Princeton, United States of America.

Milchunas, D. G., Sala, O. E. and Lauenroth, W. K., 1988. A generalized model of the effects of grazing by large herbivores on grassland community structure. *Am. Nat.* **132:** 87–106.

Morrissey, J. G. and O'Connor, R. E. Y., 1988. Twenty-eight years of station management: Fair use and a fair go. In *Conference papers, Australian rangeland society 5th biennial conference, Longreach, Queensland, 14–17 June, 1988.* The Australian Rangeland Society: Longreach.

Morton, S. R., 1990. The impact of European settlement on the vertebrate animals of arid Australia: a conceptual model. *Proc. Ecol. Soc. Aust.* **16:** 201–13.

Morton, S. R., Stafford Smith, D. M., Friedel, M. H., Griffin, G. F. and Pickup, G., 1995. The stewardship of arid Australia's ecology and landscape management. *J. Environ. Manage.,* in press.

Mott, J. J. and Bridgewater, P. B., 1992. Biodiversity conservation and ecologically sustainable development. *Search* **23:** 284–87.

Newsome, A. E., 1975. An ecological comparison of the two arid-zone kangaroos of Australia and their anomalous prosperity since the introduction of ruminant stock to their environment. *Q. Rev. Biol.* **50:** 389–424.

Newsome, A. E., 1990. The control of vertebrate pests by vertebrate predators. *Trends Ecol. Evol.* **5:** 187–91.

Noble, J. C., 1984. Mallee. Pp. 223–40 in *Management of Australia's rangelands* ed by G. N. Harrington, A. D. Wilson and M. D. Young. CSIRO Australia: Melbourne.

Noss, R. F., 1990. Indicators for monitoring biodiversity: a hierarchical approach. *Conserv. Biol.* **4:** 355–64.

Noss, R. F. and Harris, L. D., 1986. Nodes, networks and MUMs: preserving diversity at all scales. *Environ. Manage.* **10:** 299–309.

Pickard, J., 1994. Land degradation and land conservation in the arid zone of Australia: grazing is the problem . . . and the cure. Pp. 131–37 in *Conservation biology in Australia and Oceania* ed by C. Moritz and J. Kikkawa. Surrey Beatty & Sons: Chipping Norton.

Purvis, J. R., 1986. Nurture the land: my philosophies of pastoral management in central Australia. *Aust. Rangel. J.* **8:** 110–17.

Quinn, M. A. and Walgenbach, D. D., 1990. Influence of grazing history on the community structure of grasshoppers of a mixed-grass prairie. *Environ. Entomol.* **19:** 1756–766.

Recher, H. F. and Lim, L., 1990. A review of current ideas of the extinction, conservation and management of Australia's terrestrial vertebrate fauna. *Proc. Ecol. Soc. Aust.* **16:** 287–301.

Reid, J. R. W. and Puckridge, J. T., 1992. Innamincka Regional Reserve — a mis-managed multiple landuse reserve. Pp. 370–73 in *Conference papers, Australian rangeland society 7th biennial conference, Cobar, New South Wales, 5–8 October, 1992.* The Australian Rangeland Society: Cobar.

Sadlier, R. and Pressey, R. L., 1994. Reptiles and amphibians of particular conservation concern in the Western Division of New South Wales. *Biol. Conserv.* **69:** 41–54.

Scheiner, S. M., 1992 Measuring pattern diversity. *Ecology* **73:** 1860–867.

Smith, P. J., Pressey, R. L. and Smith, J. E., 1994. Birds of particular conservation concern in the Western Division of New South Wales. *Biol. Conserv.* **69:** 315–38.

Stafford Smith, D. M. and Morton, S. R., 1990. A framework for the ecology of arid Australia. *J. Arid Environ.* **18:** 255–78.

Stafford Smith, D. M. and Pickup, G., 1990. Pattern and production in arid lands. *Proc. Ecol. Soc. Aust.* **16:** 195–200.

Tongway, D. J., 1990. Soil and landscape processes in the restoration of rangelands. *Aust. Rangel. J.* **12:** 54–57.

Tongway, D. J. and Ludwig, J. A., 1990. Vegetation and soil patterning in semi-arid mulga lands of Eastern Australia. *Aust. J. Ecol.* **15:** 23–34.

Turner, M. G., 1989. Landscape ecology: the effect of pattern on process. *Ann. Rev. Ecol. Syst.* **20:** 171–97.

West, N. E., 1993. Biodiversity of rangelands. *J. Range Manage.* **46:** 2–13.

Westoby, M., Walker, B. and Noy-Meir, I., 1989. Opportunistic management for rangelands not at equilibrium. *J. Range Manage.* **42:** 266–74.

Wilcox, B. P., Bryant, F. C. and Belaun Fraga, V., 1987. An evaluation of range condition on one range site in the Andes of central Peru. *J. Range Manage.* **40:** 41–45.

Williams, R. J., 1990. Grazing cattle within subalpine heathland and grassland communities on the Bogong High Plains: disturbance, regeneration and the shrub-grass balance. *Proc. Ecol Soc. Aust.* **16:** 255–65.

Wilson, A. D., 1990. The effect of grazing on Australian ecosystems. *Proc. Ecol. Soc. Aust.* **16:** 235–44.

Wilson, A. D., 1991. The influence of kangaroos and forage supply on sheep productivity in the semi-arid woodlands. *Rangel. J.* **13:** 69–80.

Wilson, A. D. and Harrington, G. N., 1984. Grazing ecology and animal production. Pp. 63–77 in *Management of Australia's rangelands* ed by G. N. Harrington, A. D. Wilson and M. D. Young. CSIRO Australia: Melbourne.

Harvesting wildlife: kangaroos and waterfowl

S. C. CAIRNS[1] and R. T. KINGSFORD[2]

Large numbers of wildlife have been harvested in Australia, sometimes almost to the point of extinction. Today, only two wildlife groups are still harvested in any great numbers on a broad scale: kangaroos and waterfowl. This paper examines the management practices for these two groups and the effects of harvesting on their conservation; concentrating on New South Wales.

Kangaroos and waterfowl are harvested legally in most states and territories of Australia for pest mitigation, products and recreation. Harvest quotas for kangaroos require Commonwealth approval even though harvest offtakes rarely reach the quotas set. No species of kangaroo has ever been over-harvested. No species of waterfowl or non-target waterbird has been endangered by the activities of hunters.

The futures of kangaroo and waterfowl harvesting are canvassed as are their associated conservation problems and some possible approaches that could be taken to deal with them. Kangaroo harvesting could increase if demand expands with human consumption. Waterfowl harvesting could decline as public attitudes change. Either way, management of both these groups may require consideration of theoretical aspects of harvesting and the use of predictive models, and experimentation with different harvesting regimes. These should be used with current aerial surveys of populations.

INTRODUCTION

THE harvesting of wildlife is an activity as old as humankind itself (Leakey and Lewin 1977). In its most basic form, it is the hunting and gathering of food. At more sophisticated levels, it is cropping free-ranging animals for some sustainable yield, culling them to reduce populations that either compete with domestic livestock or threatening crops, or hunting them for recreation.

Initially, wildlife harvesting in Australia involved the hunter-gatherers who arrived on the continent over 40 000 years ago. They were supplanted by agriculturists who arrived about 200 years ago. Aborigines probably harvested all available forms of animal life. The only vertebrates that were not hunted by the Aborigines were the large whales, although stranded carcasses were no doubt utilized (Bryden 1978). It is difficult to assess the impact of Aboriginal hunting on wildlife populations, but it was probably low because the Aboriginal population was not large (Hughes 1987). European settlers harvested the wildlife *ad libitum* for recreation, subsistence, commercial gain and pest mitigation (Frith 1973). Commercial harvesting ranged across all major groups, reptiles, birds and mammals. Culling for pest mitigation involved mainly birds and mammals.

Marine turtles were harvested for meat and shell, crocodiles were harvested for skins, while other species of reptile were traded alive (Table 1). In the 1920s, an industry based on turtle products was established in Queensland. In a decade, the Green Turtle *Chelonia mydas* was severely over-exploited by canneries on North-West and Heron Islands (Worrell 1966). Today, the only form of marine turtle harvesting allowed is Aboriginal hunting and egg collecting (Bustard 1972). The history of crocodile harvesting in the Northern Territory (Webb *et al.* 1987), Queensland (Taplin 1987) and Western Australia (Burbidge 1987) was similarly short and sharp. The industry almost wiped out its resource between the 1940s and 1960s (Letts 1987); hunters switching to Freshwater Crocodiles *Crocodylus johnstoni* as Saltwater Crocodile *C. porosus* numbers dwindled (Webb *et al.* 1987). Today, only crocodile eggs are harvested for stocking crocodile farms. Although all reptiles are now protected, with the collection and trading of live specimens being banned in 1962 (Worrell 1966), there still remains an illegal trade (Callister and Williams, this volume).

Birds have been harvested for food, leather, oil and feathers, hunted for recreation, culled as pests and traded alive (Table 1). The Emu *Dromaius novaehollandiae*, once hunted for its meat and oil as well as being culled as an agricultural pest (Marshall 1966a), is now only culled as a pest, mainly in Western Australia. Waterfowl, Quail *Coturnix* spp. and the Mutton Bird or Short-tailed Shearwater *Puffinus tenuirostris* are today harvested-for food and recreation. Before prohibition in 1959, there was a flourishing export trade in live birds (Marshall 1966a). An illegal trade still persists (Callister and Williams, this volume).

Mammals constitute a major chapter of the history of wildlife harvesting in Australia (Table 1). By the mid 19th Century, every terrestrial mammal with marketable fur was hunted (Marshall 1966b). Marine mammals were harvested indiscriminately (Moorehead 1966; Hughes 1987). By 1820, some seal colonies had been exterminated; the rest were remnants (Busch 1985). Shore-based whaling stations closed with the collapse of the Humpback Whale *Megaptera novaeangliae* "fishery" in the early 1960s (Frost 1978).

[1]Department of Zoology, University of New England, Armidale, New South Wales, Australia 2351.
[2]New South Wales National Parks and Wildlife Service, P.O. Box 1967, Hurstville, New South Wales, Australia 2220.

Pages 260–72 *in* CONSERVING BIODIVERSITY: THREATS AND SOLUTIONS, ed by R. A. Bradstock, T. D. Auld, D. A. Keith, R. T. Kingsford, D. Lunney and D. P. Sivertsen. Surrey Beatty & Sons. 1995.

Table 1. Wildlife harvested in Australia for either the products they offered, recreation or pest mitigation.

Species	Purpose	Period of harvest	Reference
Reptiles			
Marine Turtles *Chelonia* spp.[a]	food and shell	1920–72	Worrell (1966); Bustard (1972)
Saltwater Crocodile *Crocodylus porosus*[a]	skins	1940–72	Letts (1987); Webb *et al.* (1987)
Freshwater Crocodile *C. johnstoni*[a]	skins	1960–72	Webb *et al.* (1987)
Other reptiles (lizard, snakes and turtles)	pet trade and collections	pre-1960s	Worrell (1966)
Birds			
Emu *Dromaius novaehollandiae*	pest mitigation, meat, leather and oil	pre-1960	Marshall (1966a); Frith (1973)
Albatrosses *Diomedea* spp.	feathers	1890–1910	Marshall (1966a)
Short-tailed Shearwater *Puffinus tenuirostris*	food and oil	current	Naarding (1980)
Cormorants *Phalacrocora* spp.	pest mitigation	current	—
Egrets *Egretta* spp.[a]	feathers	1890–1910	Marshall (1966a)
Waterfowl (Anatidae)	pest mitigation, food and recreation	current	—
Cape Barren Geese *Cereopsis novaehollandiae*[a]	pest mitigation, food and recreation	pre-1965	Marshall (1966a)
Wedgetail Eagle *Aquila audax*[a]	pest mitigation	?	Marshall (1966a)
Quail *Coturnix* spp.	food and recreation	current	—
Snipe *Gallinago* spp.	food and recreation	?	—
Pigeons (Columbidae)[a]	food and recreation	pre-1920	Frith (1982b)
Parrots (Psittaciformes)	pest mitigation and pet trade	current	Frith (1973)
Lyrebirds *Menura* spp.	feathers	pre-1920	Marshall (1966a)
Honeyeaters (Meliphagidae)	pest mitigation	current	Frith (1973)
Finches (Estrildidae)	pet trade and collections	?	—
Bowerbirds (Philonorhynchidae)	feathers	pre-1920	Marshall (1966a)
Riflebirds *Ptiloris* spp.	feathers	pre-1920	Marshall (1966a)
Crows and Ravens *Corvus* spp.	pest mitigation	current	—
Mammals			
Thylacine *Thylacinus cyanocephalus*[b]	pest mitigation	pre-1930	Strahan (1987)
Koala *Phascolarctos cinereus*[a]	skins	pre-1927	Marshall (1966b)
Wombats *Lasiorhinus* spp. and *Vombatus* spp.	pest mitigation	?	—
Possums *Trichosurus* spp. and *Pseudocheirus* spp.	skins	current	Marshall (1966b); Frith (1973)
Kangaroos and Wallabies (Macropoidae)	food, skins and pest mitigation	current	Frith (1973)
Flying Foxes *Pteropus* spp.	pest mitigation	current	—
Water Rats *Hydromys* spp.	skins	?	Strahan (1987)
Dugong *Dugong dugon*	food and oil	pre-1977	Bertram & Bertram (1973)
Fur Seals *Arctocephalus pusillus* and *A. fosteri*[a]	skins	1790–1820	Busch (1985)
Australian Sea-lion *Neophoca cinerea*[a]	skins	1790–1820	Busch (1985)
Southern Elephant Seal *Mirounga leonina*[b]	oil	1790–1820	Busch (1985)
Whales *Megaptera novaeangliae*, *Physeter catodon* and *Eubalaena australis*[a]	oil	1790–1962	Frost (1978)

[a]Known to have been harvested to dangerously low population levels.
[b]Extinct, probably almost certainly the result of harvesting.

Only a few groups of wildlife are now harvested legally in Australia, mostly for pest mitigation. There are some species that are only harvested in one locality such as mutton birds in Tasmania (Naarding 1980), or are only harvested in small numbers, such as quail. There are two groups, however, that are harvested in large numbers in many parts of Australia. These are kangaroos and waterfowl. The aim of this paper is to review current management and harvesting practices associated with these groups and to examine the effects of harvesting on the conservation of species within them. Although the emphasis is on New South Wales, the underlying principles of what is discussed are essentially the same throughout Australia. The critical conservation issue is whether current harvesting levels and practices will contribute to the list of harvesting disasters that have occurred in Australia since European settlement.

KANGAROOS

From the beginning of European settlement, kangaroos and wallabies were exploited for meat and skins (Barker and Caughley 1990, 1992,

1994). The expansion of agriculture and pastoralism in the 19th Century saw their status change from that of resource to pest. In the 1860s, huge numbers of kangaroos were destroyed in kangaroo drives (battues) in New South Wales and Victoria (Frith 1973; Barker and Caughley 1992). By the 1880s, legislation in the eastern colonies required that kangaroos be destroyed under a bounty system. Between 1881 and 1889, the number of large kangaroos in New South Wales declined from about six million to one million with payment of about eight million bounties, while the number of wallabies declined from about four million to three million with payment of four million bounties (Robertshaw and Harden 1989). In Queensland, bounty payments were made on both kangaroos and wallabies from the mid 1870s until 1906, and on wallabies alone until 1917. By 1917, about 27 million bounty payments had been made (Robertshaw and Harden 1989).

With the end of the bounty system came the development of a skin-based kangaroo harvesting industry (Poole 1984) which, by the late 1950s, had led to the development of a kangaroo meat industry involving an export trade, principally with Germany. This trade peaked in 1961, but by 1964, following a series of scandals and scares, it had all but collapsed (Marshall 1966b).

The haphazard development of the kangaroo industry was based almost entirely on the Red Kangaroo *Macropus rufus* and the Eastern Grey Kangaroo *M. giganteus*. By the mid 1960s, about half a million kangaroos were being taken each year by the industry in eastern Australia (Marshall 1966b). Initially uncontrolled, harvesting later became regulated through the registration of professional shooters and processors. In the 1970s, scientifically based methods of population monitoring were introduced as part of the regulatory process.

Kangaroo harvesting in practice

Large numbers of kangaroos and some species of wallaby are present in Australia today. Whether there are more now than at the time of European settlement is debated. However, it is generally conceded that in the pastoral regions of Australia (the arid and semi-arid rangelands), the establishment of bores and dams, modification of the landscape through grazing by domestic herbivores, and effective elimination of dingos with the erection of the dog fence have resulted in an increase in kangaroo numbers in modern times (Shepherd and Caughley 1987).

Today, kangaroos are regarded as agricultural and pastoral pests. According to Gibson and Young (1988), a single kangaroo reduces farm net income by about $9 annually; equivalent to a 1.4 per cent net loss from the gross value of agricultural production for Australia ($8,082 million). Further to this, it is contended that, at current population levels, kangaroos are responsible for reducing stocking rates by some 8.65 million dry sheep equivalents (Gibson and Young 1988). About 69 per cent of landholders in the sheep/wheat regions and 80 per cent of landholders in the pastoral regions use professional kangaroo shooters as part of their property management programmes (Gibson and Young 1988).

Kangaroos are harvested both commercially and non-commercially; commercially in all states except Victoria. The main target species for commercial harvesting are the Red Kangaroo *M. rufus*, the Western Grey Kangaroo *M. fuliginosus* and the Eastern Grey Kangaroo *M. giganteus*. Wallaroos or Euros *M. robustus* are also taken in the mainland states and Whiptail or Pretty-faced Wallabies *M. parryi* are taken in Queensland. Bennetts Wallabies *M. rufogriseus rufogriseus* and Red-bellied Pademelons *Thylogale billardierii* are taken in Tasmania on a mainly non-commercial, recreational basis. In Tasmania, the commercial harvesting of wallabies and pademelons is almost non-existent; the great majority of harvesting being conducted on a non-commercial basis by recreational shooters (Driessen 1993).

Kangaroo management is administered by the states and territories. Although the relevant acts and regulations differ in detail, they have essentially the same objectives. Despite the key role played by commercial harvesting, central to the various management programmes is the commitment to "the maintenance of self-sustaining populations of each species throughout its preferred range rather than the maintenance of populations capable of sustained yield for the benefit of commercial operators" (Poole 1984).

For kangaroo management and harvesting, New South Wales is divided into 15 zones, with the eastern one closed to commercial operations. Commercial chillers are located throughout the zones to ensure an even harvest (J. Everleigh, pers. comm.). All harvested kangaroos must carry a licence tag and, if required, meet the size provisions of the commercial trade (Poole 1984). The whole operation is administered by the NSW National Parks and Wildlife Service (NSWNPWS). The administration of kangaroo management in the other mainland states that support commercial harvesting is conducted on a similar basis.

Population monitoring

Those kangaroo populations subjected to harvesting are monitored on a regular basis, mainly by aerial survey (see Caughley *et al.* 1976).

Aerial surveys are relatively cheap, can cover large areas and provide baseline information on trends in kangaroo numbers. They are particularly useful in the rangelands of New South Wales (J. Caughley *et al.* 1984), South Australia (Cairns and Grigg 1993) and Western Australia. Despite this, there has been criticism of the method. Landholders have argued that, even with the use of correction factors, kangaroo numbers estimated from fixed-wing aerial surveys are conservative. Opponents of the commercial harvesting of kangaroos have argued that the estimates are inflated.

In Queensland and Western Australia, kangaroo populations have mainly been monitored through the harvest (Poole 1984; Nance 1985), with more recent aerial surveys conducted by Australian Nature Conservation Agency (ANCA). Fixed-wing aerial surveys have not been favoured in Queensland because of possible sightability problems in a landscape more heterogeneous than the rangelands of western New South Wales and the South Australian pastoral zone (Hill *et al.* 1985). Recent attempts to resolve this problem have involved comparison of different fixed-wing aircraft survey methods with helicopter surveys and ground surveys (Pople *et al.* 1993).

In Tasmania, population trends of Bennetts Wallabies and Red-bellied Pademelons are monitored by spotlight counts, surveys of road kills, hunter surveys and analysis of fur trade statistics (Poole 1984; Driessen 1993).

Some effort has also been made to model changes in kangaroo numbers in relation to resource availability; usually indexed by rainfall (J. Caughley *et al.* 1984; Bayliss 1985, 1987; Cairns and Grigg 1993). Although not altogether representative of the dynamics of kangaroo populations (see Cairns and Grigg 1993), some use has nevertheless been made of these models to make management decisions (P. Alexander, pers. comm.). Also, although not yet validated, a model of sex ratio change in relation to harvest offtake has been developed by Nance (1985).

Table 2. Triennial estimates of numbers, approved harvest quotas and actual harvest offtakes for Red, Eastern Grey and Western Grey Kangaroos in the 14 kangaroo management zones in New South Wales, 1981–90. Proportions of harvest quota to estimated numbers (Q/N), harvest offtake to estimated numbers (H/N) and harvest offtake to harvest quota (H/Q) are also presented. Source: Australian Nature Conservation Agency (formerly Australian National Parks and Wildlife Service) (unpubl. data).

	Numbers (×1 000)	Quota (×1 000)	Harvest (×1 000)	Q/N (%)	H/N (%)	H/Q (%)
Red Kangaroo						
1981	3 837	333	—	8.7	—	—
1984	1 663	399	158	24.0	9.5	39.6
1987	2 770	313	270	11.3	9.7	86.3
1990	4 499	626	377	13.9	8.4	60.2
Grey Kangaroo[a]						
1981	2 886	357	—	12.4	—	—
1984	1 171	100	64	8.5	5.5	64.0
Eastern Grey Kangaroo						
1987	1 824	189	140	10.4	7.7	74.1
1990	2 755	271	171	9.8	6.2	63.1
Western Grey Kangaroo						
1987	806	75	63	9.3	7.8	84.0
1990	1 296	105	84	8.1	6.5	80.0

[a]Eastern and Western Grey Kangaroos are not easily distinguished from survey aircraft. The separation of Eastern and Western Grey Kangaroos for 1987 and 1990 is based upon the relative proportions given by Caughley *et al.* (1984) of each of these species within individual survey blocks.

The commercial kangaroo harvest

Each year, the states set quotas for commercial kangaroo harvests based upon estimates of the numbers of kangaroos in their agricultural and pastoral zones. Under the 1982 *Federal Wildlife Protection (Regulations of Exports and Imports) Act,* these quotas require Commonwealth approval. For mainland states, when setting these quotas the level of pest mitigation in relation to the impact of "total" grazing pressure by domestic livestock, kangaroos and feral pests (e.g., rabbits, goats, etc.), and the maintenance of a viable kangaroo products industry are considered in association with the principal conservation objective of "the maintenance of self-sustaining populations of each species . . ." (see Poole 1984). Although of secondary importance, harvesting for a sustainable yield is also a consideration. The level of harvesting that will result in a sustainable yield from a kangaroo population has yet to be determined. Yield for Red Kangaroos could be maximized over an extended period if 10–15 per cent of the population were harvested each year, reducing the population to 60–70 per cent of its unharvested size (Caughley 1987). However, harvesting Eastern Grey Kangaroos at levels >15 per cent over an extended period could be detrimental to the long-term viability of a population (Nance 1985).

Over the 1981–1990 decade, triennial harvest quotas in New South Wales for kangaroos were 8–14 per cent of the estimated populations of Red or Grey Kangaroos in the commercial management zones (Table 2). The one exception was in 1984, when the quota for Red Kangaroos was set at 24 per cent. This was almost certainly an aberration associated with the 1982–83 drought (J. Caughley *et al.* 1984); the quota having initially been set in relation to the estimated population for 1983. Harvest offtakes were 5–10 per cent of estimated populations, never reaching the level of the quota (Table 2). Quotas are probably never filled because the industry, which is carcass-based, is predominantly demand-driven and relatively stable in terms of shooter and processor numbers.

Relationships between estimated populations, harvest quotas and actual harvest offtakes in South Australia and Western Australia, with carcass-based kangaroo industries, are similar to those in New South Wales. In Queensland, where the industry is predominantly skin-based, there is a greater propensity for the quotas to be filled.

Some Wallaroos are also taken as part of the commercial harvest in the mainland states. Quotas for Wallaroos are around 10 per cent of those for large kangaroos. In New South Wales, the quotas for Wallaroos between 1981 and 1990 have increased from 5 000 to 20 000 (ANCA, unpubl. data), but harvest offtakes have only been 10–20 per cent of the quota. Relative to quotas for large kangaroos, quotas for Wallaroos in South Australia and Western Australia are similar to those in New South Wales, but harvest offtakes are about 50 per cent of the quota (ANCA, unpubl. data). The quota for Wallaroos in Queensland is much higher than in the other states. Harvest offtakes for Wallaroos in Queensland are near to the quota, which is a consequence of the skin-based industry.

Whiptail Wallabies are harvested commercially in Queensland. Relatively small harvest quotas (25 000–75 000) are set for this species, with harvest offtakes of >50 per cent of the quota (ANCA, unpubl. data). In recent years, however, the quota has been low (25 000), with the actual harvest offtake being only 10–20 per cent of the quota.

Large numbers of Bennetts Wallabies and Red-bellied Pademelons (about 650 000) are harvested each year in Tasmania, mainly by recreational shooters (Driessen 1993). There has been no Commonwealth approved macropod management programme for Tasmania since 1986, prior to which only about one-third of relatively small harvest quotas were taken commercially. Quotas set for Bennetts Wallabies and Red-bellied Pademelons before 1987 were about 140–200 000 and 100–125 000, respectively (ANCA, unpubl. data).

Non-commercial harvesting of kangaroos and wallabies also occurs in the mainland states and territories, but not on the same scale as Tasmania. In most states, the non-commercial harvest represents a small proportion of the total harvest, but it can involve culling some species of macropods not harvested commercially (e.g., Red-necked Wallabies *M. rufogriseus banksianus* and Swamp Wallabies *Wallabia bicolor*). In New South Wales, the non-commercial harvest was 5.5 per cent of the total harvest for 1990, comprising mainly Eastern Grey Kangaroos culled in the eastern management zone closed to commercial harvesting (Table 3).

Table 3. The 1990 commercial and non-commercial harvests of kangaroos and wallabies in New South Wales. Source: Everleigh (1991).

Species	Commerical harvest	Non-commercial harvest
Red Kangaroo	377 155	4 155
Eastern Grey Kangaroo	170 766	24 889
Western Grey Kangaroo	83 708	6 165
Wallaroo	1 967	1 527
Red-necked Wallaby	—	135
Swamp Wallaby	—	266
Total	633 596	37 137

Effects of harvesting on macropod conservation

Although macropods in general, and kangaroos and wallabies in particular, have been harvested since people first arrived in Australia, there is little evidence that this activity alone has had a detrimental effect on macropod distribution, abundance and biodiversity. Limitations on distributions, declines in abundance and reductions in biodiversity can usually be attributed to activities related to land use rather than harvesting *per se*.

Red and Grey Kangaroos, and the Wallaroo are integral components of the pastoral landscape. Modern pastoral practices have created an environment to which these species are well adapted (Shepherd and Caughley 1987). Intensive agriculture has often created a sub-optimal environment (Short and Grigg 1982). In the pastoral and less intensively managed agricultural areas where most kangaroo harvesting takes place, climate affects kangaroo populations more than does harvesting (Bayliss 1987; Cairns and Grigg 1993). Red and Grey Kangaroo populations in eastern Australia crashed in the 1982–83 drought (Caughley *et al.* 1985). Despite being harvested at around 10 per cent before, during and after the drought, these populations recovered by the end of the 1980s to the levels of abundance that existed before the drought (NSWNPWS unpubl. data; Cairns and Grigg 1993). The impact of harvesting on population dynamics is such that under good conditions harvest mortality is probably additive in relation to natural mortality, while under conditions of drought it is compensatory. That is, during drought harvest mortality is not additional to any natural, drought-induced mortality, but is thought to be a substitute for part of that mortality. Those animals taken as part of a harvest could well be animals that would later succumb to the effects of the drought (Bayliss 1987).

Today, although harvested in large numbers, the Red, Eastern Grey and Western Grey Kangaroos are all widespread and abundant throughout their ranges (Caughley *et al.* 1984; Cairns and Coombs 1992). The distribution of the Red Kangaroo has not changed markedly for at least several decades. The sharply delineated eastern boundary of its distribution in New South Wales is set primarily by climatic constraints (Caughley *et al.* 1987) and land use such as the growing of food and fibre crops (Cairns and Coombs 1992).

There have been no discernible shifts in the distributional boundaries of the Western Grey Kangaroo over the last several decades (Caughley *et al.* 1984; Cairns and Coombs 1992). These boundaries are essentially determined by climatic constraints (Caughley *et al.* 1987).

Despite being harvested in large numbers in New South Wales and Queensland, the Eastern Grey Kangaroo has increased in numbers and expanded its range westward over the last several decades. The Darling River was thought of as the western limit of its distribution in New South Wales. However, in the early 1970s, Eastern Grey Kangaroos were reported occurring west of the Darling River (Denny 1975) and then later west of the Barrier Ranges and just into South Australia (W. E. Poole, cited in Caughley *et al.* 1984). Caughley *et al.* (1984) confirmed this "westward spread" and suggested that it was, until 1982, occurring at a rate of 3–4 km year^{-1}. This species now extends beyond the western boundary previously defined for it by Caughley *et al.* (1984) (see Cairns and Coombs 1992).

The extensive recreational harvesting of Bennetts Wallabies and Red-bellied Pademelons in Tasmania does not appear to have any detrimental effect on the distributions and abundances of these two species (Driessen 1993).

Apart from the Toolache Wallaby *M. greyi*, which may have been hunted to extinction as an animal of both sporting value and with a fine pelt (Frith 1973), no kangaroo or wallaby has been driven to extinction or become endangered through harvesting. Agricultural and pastoral development have resulted in the extinction or endangered status of a number of the smaller species of macropod, as has the introduction of associated predators such as the fox. Even the Toolache Wallaby, with a high level of harvesting, was also affected by massive alteration of the habitat of its very limited range in southwestern Victoria and southeastern South Australia (Frith 1973). Agricultural and pastoral development have resulted in a decline of the Bridled Nail-tail Wallaby *Onychogalea fraenata* and the Whiptail Wallaby. Next to the Toolache Wallaby, the Bridled Nail-tail Wallaby, once widely distributed throughout eastern Australia, is probably the most spectacular example of the impact of pastoralism and agriculture on a macropod species. It is officially an endangered species that is now found only in remnant populations on the properties "Taunton" and "Red Hill" in central Queensland, near the edge of its range (Lavery and Tierney 1985). Another example of a species whose distribution has also contracted is the Whiptail Wallaby. Although abundant and widespread in southeastern Queensland where it has been harvested, it has declined in New South Wales where it is not harvested. Once common in northeastern New South Wales, it is now found only in low densities in the far northeastern corner of the state (Cairns and Coombs 1992; C. J. Southwell, unpubl. data). The decline of this species in New South Wales is probably due to habitat alteration and disturbance associated with agricultural development.

WATERFOWL

Aboriginal Australians hunted adult waterfowl (ducks, geese and swans) and collected eggs from nesting sites (McGilp 1923). Today, they continue to hunt some species and harvest the eggs of Magpie Geese *Anseranas semipalmata* in the Northern Territory (Dexter and Bayliss 1991). European colonists also hunted waterfowl for food. In the late 19th century, wild ducks were sold through the Sydney and Melbourne markets; up to 159 000 pairs of ducks passing through the Melbourne market each year (Norman 1983).

The regulation of waterfowl harvesting was instigated relatively early in colonial history. Victoria introduced "declared" open seasons for hunting in 1862; progressively reducing the length of the season from 244 days (early 1860s) to 92 days (1946–53) (Norman 1983). With the declaration of these open seasons came regulation of the species to be taken and the bag sizes. In New South Wales, the 1901 *Bird Protection Act,* and all other subsequent acts (including the 1974 *National Parks and Wildlife Act*) protected all waterfowl within the state, outside proclaimed open seasons. Despite its introduction, the effectiveness of legislative protection has probably always been contingent upon prevailing community attitude toward the harvesting of waterfowl (Norman and Young 1980).

Waterfowl harvesting in practice

Today, waterfowl are no longer harvested commercially in Australia. Instead, they are either culled for damage mitigation, usually in relation to irrigated crops such as rice, or they are hunted recreationally during declared open seasons.

Damage mitigation

To prevent damage to rice crops, ducks can be culled under permit outside declared open seasons. The extent of the damage caused by ducks can be considerable; rice growers estimated that ducks caused $1.23–4.07 million of damage to the 1987–88 rice crop in New South Wales alone, with up to a 90 per cent loss of yield from some farms (CSIRO 1988). About 4 300 licences (600 to farmers and 3 700 to hunters) were sold in 1992–93 for the destruction of waterfowl on rice. Licences issued to farmers have a bag limit of 100 ducks, while those issued to hunters have a bag limit of 25 ducks. Most of the ducks shot under these licences (>95%) are the Grey Teal *Anas gracilis*, the Maned Duck *Chenonetta jubata* and the Pacific Black Duck *Anas superciliosa* (S. V. Briggs, unpubl. data; 1983–84).

Open-season recreational hunting

Hunters shoot ducks during declared open seasons. The birds taken are usually eaten. Management of the waterfowl harvest through declaration of open seasons and legal game species is conducted at the state and territory level. There is, however, co-operation between the relevant authorities, particularly among southeastern states.

The timing of open seasons and the composition of game species in each state or territory (Table 4) is based upon abundance, distribution and timing of breeding and moulting (flightless stage). Open seasons generally fall outside breeding and moulting periods. Although different numbers of game species are declared in the states or territories, there is considerable overlap among the species involved. Currently, seven species can be shot legally within New South Wales: Australian Shelduck *Tadorna tadornoides*, Maned Duck *Chenonetta jubata*, Pink-eared Duck *Malcorhynchus membranceus*, Grey Teal *Anas gracilis*, Chestnut Teal *Anas castanea*, Pacific Black Duck *A. superciliosa* and Hardhead *Aythya australis*. The distributions of these seven species are given in Marchant and Higgins (1990).

Table 4. Opening dates, lengths, bag sizes, and numbers of declared game species for the 1992 shooting seasons in those states and territories that permit the recreational hunting of waterfowl.[a]

State or territory	Opening date	Length (weeks)	Bag size[b] (hunter^{-1} day^{-1})	Number of game species[c]
NSW	21 Mar.	8	10	7
NT	1 Sept.	12	10[d]	8
Qld	7 July	11	12	6
SA	15 Feb.	16	12[e]	8
Tas.	7 Mar.	12	10[f]	4
Vic.	21 Mar.	8	10e	8

[a]The ACT does not declare an open season and the Western Australian Government banned duck shooting in that state in 1992.
[b]Number of birds.
[c]Game species in each state or territory as indicated by the superscript following species name (N, New South Wales; O, Northern Territory; Q, Queensland; S, South Australia; T, Tasmania; V, Victoria):
Magpie Goose, *Anseranus semipalmata*[O]; Plumed Whistling-Duck *Dendrocygna eytonii*[O,Q]; Wandering Whistling-Duck *Dendrocygna arcuata*[O,Q]; Australian Shelduck *Tadorna tardornoides*[N,S,T,V,]; Maned Duck *Chenonetta jubata*[N,O,Q,S,V]; Pink-eared Duck *Malcorhynchus membranceus*[N,O,S,V]; Grey Teal *Anas gracilis*[N,O,Q,S,T,V]; Chestnut Teal *Anas castanea*[N,S,T,V]; Pacific Black Duck *Anas superciliosa*[N,O,Q,S,T,V]; Australasian Shoveler *Anas rhynchotis*[V,S]; Hardhead *Aythya australis*[N,O,Q,S,V].
[d]or seven Magpie Geese and seven Ducks.
[e]no more than two Australasian Shovelers.
[f]no more than four Australian Shelducks.

The protocol for the declaration of open seasons on particular game species of waterfowl is similar for all states and territories that permit duck hunting. In New South Wales, in accordance with the *National Parks and Wildlife Act*, the Minister for the Environment declares open seasons on game species of waterfowl if their conservation is not threatened. This decision is based upon the occurrence of rainfall in preceding years, game species population indices, the distribution of water throughout arid Australia and catch per unit effort data collected as indices of local populations.

Rainfall — The bag size of game species (an index of abundance) in any given year is positively correlated with the amount of rainfall recorded in previous years (Briggs and Holmes 1988). Game species are known to breed in response to good rainfall and flooding (Maher and Carpenter 1984; Crome 1986; Kingsford 1989). Hence, after widespread rains, populations of game species are high. Conversely, there is poor breeding and reproductive success during drought years (Kingsford 1989); the abundance of game species being low in dry periods (Norman and Nichols 1991). Following the widespread drought of the early 1980s, the 1983 game season was not declared in southeastern Australia.

Population indices — The distributions of all game species transcend state boundaries, although some species are restricted to either the southern or northern parts of the Australian continent (Marchant and Higgins 1990). Individuals of some species sometimes move extensive distances to and from arid Australia (Blakers *et al.* 1984; Frith 1982a; Lawler *et al.* 1993). Hence, waterfowl management is not simply a state responsibility (Frith 1982a). In 1983, CSIRO and the wildlife authorities of four states (Queensland, New South Wales, Victoria and South Australia) set up a project to investigate effects of duck hunting on game species of waterfowl throughout eastern Australia (Braithwaite *et al.* 1986). The basis of this programme is an extensive annual aerial survey that produces population indices for game species in eastern Australia (Kingsford *et al.* 1992). All wetlands >1 ha in 10 survey bands (30 km wide), extending from the Northern Territory border to the east coast, are surveyed for waterbirds in October of each year. These surveys also provide information on more than 40 other species of wetland birds. Concern over the population index of the Australasian Shoveler *Anas rhynchotis* prompted its removal from the game species list in New South Wales in 1987. Analyses investigating the effect of duck hunting on indices of game species in Australia have been completed with climate identified as being most related to game species abundance (Wong 1993).

Distribution of water — Because popular shooting sites can attract high mortality, sufficient refuges must exist away from these sites for game species of waterfowl to survive (Briggs *et al.* 1983; Halse *et al.* 1993). These refuges for game species are important for waterfowl management, particularly in New South Wales where relatively few significant wetlands are reserved for conservation (Braithwaite *et al.* 1986; Maher and Braithwaite 1992). In New South Wales, hunting is concentrated in the south-west of the state (Briggs *et al.* 1985). This guarantees that some wetlands, not officially gazetted as reserves, inadvertently act as refuges because they are not vulnerable to hunting. Data on wetland availability, assessed during the annual aerial survey of waterbirds in eastern Australia (Kingsford *et al.* 1992), are combined with meteorological and satellite data to determine the distribution and extent of wetlands in eastern Australia at the time of the incoming open season for the hunting of declared game species. Impetus to close the 1983 season came from a survey that showed that few wetlands in arid New South Wales had water in them in the summer of 1982–83 (M. T. Maher, unpubl. data).

Catch per unit effort — Survey teams from NSW National Parks and Wildlife Service collect catch per unit effort data from hunters during the morning of opening weekend each year. These track the populations of game species taken from the two most intensively shot wetlands in New South Wales, Barrenbox Swamp and Lake Cowal. Seasonal offtakes are also measured by mail card surveys sent to a random sample of hunters (Briggs *et al.* 1985).

Effects of harvesting on waterfowl conservation

Duck shooting in New South Wales, as in other states and territories, is concentrated mainly on the first weekend of the declared open season (Norman and Powell 1981; Briggs *et al.* 1985) and usually at a few popular sites where the mortality of game species can be particularly high (Briggs *et al.* 1983; Halse *et al.* 1993). Outside intensively hunted areas, hunting mortality appears to be quite low. Halse *et al.* (1993) estimated that, although hunting mortality in Pacific Black Duck and Grey Teal in south-western Australia could be as high as 60 per cent in intensively hunted areas, overall it was <25 per cent of mortality. During the open season in New South Wales, each hunter is estimated to shoot about 23 ducks over a period of about three and one-half days (Briggs *et al.* 1985). Of the seven species currently declared as game species in New South Wales, the Maned Duck, the Grey Teal and the Pacific Black Duck constitute nearly 90 per cent of all ducks shot during the open season (Briggs *et al.* 1985).

Although local hunting mortality can be high during a declared open season, its overall impact is thought to be moderate to low. At Lake Cowal, between Forbes and West Wyalong, and Barrenbox Swamp, near Griffith, the numbers of game species present have remained relatively stable over the 20 year period, 1972–1992; as have mean bag sizes for the first morning of the open season (Briggs *et al.* 1993). Similarly, hunting had no detectable effect upon either Pacific Black Duck populations or Grey Teal populations in southwestern Australia (Halse *et al.* 1993).

Significantly, there has probably been a steady decline in the numbers of game species shot over the last ten years in New South Wales because the number of game licences sold has declined from about 14 000 in 1975 to about 4 000 in 1992, with a substantial fall between 1987 and 1988 (Briggs *et al.* 1993). Sales of duck licences increased for a period after 1969 (Caughley and Briggs 1983), stabilized between 1972 and 1987, and then declined (Briggs *et al.* 1993). The numbers of duck licences sold in Victoria (65 000 in 1987 and 23 000 in 1992) and South Australia (21 000 in 1975 and 7 000 in 1988–89; R. Storr, pers. comm.) have also declined dramatically.

Each year in New South Wales, a number of non-game species are shot illegally, particularly on the opening day. Those shot during the opening weekend on Barrenbox Swamp and Lake Cowal over the three years, 1990–92, and collected by volunteers from animal welfare groups are given in Table 5. Concern about these birds, in particular, the Freckled Duck *Stictonetta naevosa*, which has been declared vulnerable and rare saw the introduction of a waterfowl identification test in New South Wales in 1990. The Freckled Duck was listed by Frith (1982a) as one of a number of species likely to be vulnerable to hunting, being thought to be mistaken by hunters for Pacific Black Duck.

The introduction of this test has increased the knowledge and involvement of duck shooters in waterfowl conservation in New South Wales. However, its impact on reducing the numbers of non-target species shot is unclear because search effort is not standardized and there is no record of the numbers of hunters shooting on any particular wetland. Nevertheless, there has been a decline in recent years in the range of species illegally shot during opening weekend

Table 5. Non-game species of waterbirds shot during the first weekend of the 1990–93 open seasons at Lake Cowal and Barrenbox Swamp, New South Wales.

Waterbird species[a]	1990	1991	1992	1993
Australasian Grebe	—	2	—	—
Hoary-headed Grebe	1	1	—	1
Great Crested Grebe	2	4	—	—
Australian Pelican	—	—	—	1
Darter	1	—	—	—
Greater Cormorant	5	2	1	1
Pied Cormorant	1	1	—	2
Little Black Cormorant	—	2	1	1
Little Pied Cormorant	—	1	1	1
Pacific Heron	—	—	—	1
Australian White Ibis	—	—	1	—
Straw-necked Ibis	1	1	—	—
Royal Spoonbill	1	—	—	—
Yellow-billed Spoonbill	—	—	—	1
Plumed Whistling-Duck	—	—	—	1
Musk Duck	—	—	1	—
Freckled Duck	4	7	5	9
Black Swan	—	1	—	—
Australasian Shoveler	2	5	3	8
Black-tailed Native-hen	—	—	—	1
Purple Swamphen	1	—	1	—
Eurasian Coot	7	10	5	3
Black-winged Stilt	—	—	3	2
Silver Gull	—	2	—	—
Total	26	39	22	33

[a]Common names follow Marchant and Higgins (1990).

on Lake Cowal and Barrenbox Swamp (R. T. Kingsford and S. V. Briggs, pers. obs.). Waterfowl identification tests, first introduced in Victoria, are now in use in all eastern states.

Even though some non-target species are shot, the numbers do not seriously affect the conservation of any of these species, even the Freckled Duck. Most populations of Freckled Duck probably occur on wetlands far removed from areas of high duck hunting activity (Norman *et al.* 1994; Kingsford and Porter 1994). Also, because of the remoteness and the unpredictable nature of its environment, the species probably is much more abundant than official estimates (19 000; Garnett 1992). NSWNPWS routinely flies aerial surveys before each open season to determine if any large flocks of Freckled Duck are present on wetlands that will be subjected to hunting. During the open seasons of 1991 and 1992, Tullakool evaporation basin (near Deniliquin) was closed to hunting because of the presence of significant numbers of Freckled Ducks.

In contrast to the direct impact of hunting on waterfowl species, both game and non-game, the indirect effect of resultant dumping of lead into wetlands is potentially a serious problem. This mostly comes from lead shot, although some wetlands have increased lead concentrations because of road run-off (Lund *et al.* 1991). As a result of duck shooting, about 350 tonnes of lead shot are deposited into wetlands throughout Australia each year (Sharley *et al.* 1992). This shot can be mistakenly ingested by waterfowl as grit, which they ingest naturally to aid digestion and provide minerals (Trost 1981). Once ingested, lead shot can be ground down by the muscular action of the gizzard and absorbed as lead salts that are detrimental to the brain and the immune system (Dieter and Finley 1979; Trust *et al.* 1990; Rocke and Samuel 1991). The ingestion of lead shot may cause death.

Serious problems with lead shot have been detected on three wetlands in Australia: Howards Springs in the Northern Territory (Whitehead and Tschirner 1991), Bool Lagoon in South Australia (Koh and Harper 1988; Harper and Hindmarsh 1990) and Lake Buloke in Victoria (Wickson *et al.* 1992). At present, waterfowl shot on New South Wales wetlands do not appear to be as affected by the ingestion of lead as they are in other states (Kingsford *et al.* 1989; Kingsford *et al.* 1994). This has been attributed to low hunter density in New South Wales (Kingsford *et al.* 1994). Nevertheless, long-term (>100 years) decisions on use of lead shot will need to be made because of the cumulative nature of its impact and the likely long-term breakdown of lead shot in wetlands into more mobile compounds. Apart from its direct impact on bottom-feeding waterfowl, lead shot may also indirectly affect other species such as raptors that ingest the lead shot found in waterbirds (Pain and Amiardtriquet 1993). There is also increasing evidence that lead inhibits growth in aquatic biota (Greger and Kautsky 1991; Timmermans *et al.* 1992).

Alternatives to lead shot exist in the form of steel shot and bismuth shot. Both are considered to be non-toxic. Steel shot is used in North America and in South Australia. Its use has been in force on Bool Lagoon since 1987, and on all other South Australian game reserves since 1992 (L. Best, pers. comm.). A state-wide ban on the use of lead shot came into force in South Australia in 1994 (L. Best, pers. comm.). A problem does exist with steel shot, however, as it causes damage to some gun barrels (Brister 1992). The other alternative to lead shot, bismuth shot, is not yet readily available in Australia. It appears to be non-toxic (Ringelman *et al.* 1993).

The recreational harvesting of waterfowl has produced what could probably be some benefits from the perspective of maintaining biodiversity in wetland communities. Field and game organizations in different states actively manage wetlands for game species. Nest boxes (>25 000 in Victoria) have often been erected with some success (Norman and Riggert 1977; Ogden 1992). In South Australia, extensive areas of wetlands that were previously drained for agriculture have been reclaimed by hunting groups through the placement of a series of levee banks to hold water. Such works not only benefit the game species, but also many of the other waterbird species that inhabit wetlands. Hunting groups also actively lobby against the destruction of wetlands. Levies on licences (38% of the $40 in New South Wales) have supported research that has helped to understand the ecology of wetlands, to identify those wetlands of critical importance for the conservation of all waterbirds, and to determine the effects of other potentially threatening processes (e.g., habitat loss, river management, cropping) upon both wetlands and waterfowl.

CONCLUSIONS, FUTURE PROBLEMS AND FUTURE SOLUTIONS

Because kangaroos and waterfowl are harvested for quite different reasons, the futures of the harvesting of these two groups could take quite different paths. The level of kangaroo harvesting could increase, while the level of waterfowl harvesting could decline both as a response to changing public attitudes. Whether this happens or not, the harvesting of both these groups still presents problems relevant to the conservation of the constituent species of these groups.

Most commercial kangaroo harvesting is for a carcass industry. In all states except South Australia, where the sale of kangaroo meat for human consumption has been legal since 1980, the meat from trade carcasses has been used mainly for pet food, with a small proportion being exported for human consumption. With New South Wales, Queensland and Western Australia now following South Australia in legalizing the sale of kangaroo meat for domestic human consumption, the demand for trade carcasses could increase with increased prices for kangaroo meat. This would increase the value of kangaroos as rangeland producers. If this happened, it could result in a substantial shift in land use in the arid and semi-arid rangelands, with kangaroos replacing sheep, in part at least, as a source of income for landholders as well as for kangaroo shooters and processors (Grigg 1988, 1989). Within the context of land care, the implications of this are potentially quite significant. A reduction of "total" grazing pressure achieved through a reduction in the number of cloven-hoofed grazers (sheep) and an increase in kangaroo harvest offtakes coupled with effective control of feral grazers could slow or even reverse the rate of rangeland degradation.

Waterfowl populations have been thought to be in decline as a result of wetland loss, destruction of nesting habitat by domestic livestock and the hunter (Frith 1982a). However, despite having been subjected to sometimes high levels of hunting since the early days of colonial Australia, no game species of waterfowl has become endangered or extinct as a result of harvesting alone. The declaration of the Freckled Duck as rare and vulnerable and the removal of the Australasian Shoveler from the list of game species were pre-emptive conservation measures and not measures taken in response to hunting pressure. Alteration and loss of wetland habitat are processes far more threatening to waterfowl conservation (Kingsford, this volume) than harvesting. Thus, waterfowl conservation requires effort to be directed towards preventing or reversing the processes that destroy critical habitat. On a positive note, although the Maned Duck has always been hunted, its numbers have actually increased in recent times (Blakers *et al.* 1984; Briggs *et al.* 1985; Kingsford 1992), having benefitted from the creation of suitable habitat through the digging of farm dams in rural areas (Kingsford 1992) and the establishment of greenbelt areas with water (e.g., golf courses) within the urban landscape.

To ensure conservation of those species subjected to harvesting, management practices such as the long-term monitoring of populations are essential. Although such monitoring (e.g., Cairns and Grigg 1993; Kingsford *et al.* 1992) can reveal the status of a population over time, it is often difficult to establish conclusively the underlying causes of observed changes in numbers unless other information is available.

If harvesting levels were likely to increase, then more attention would need to be paid to the theoretical aspects of harvesting and the construction of models that could be used to predict the responses of populations to different levels of harvesting. Attempts have been made at this for kangaroos by Nance (1985), Caughley (1987) and Cairns (1989). However, a need to move beyond the stages reached with these models could well arise if the demand for meat or other kangaroo products replaces pest mitigation as the main reason for kangaroo harvesting.

Coupled with this and the practice of long-term population monitoring is the need, where possible, for experiments aimed specifically at assessing the response of populations to different levels of harvesting. Nichols (1991) advocates more use of experimentation to determine the causes of changes in waterfowl numbers. For example, different areas could operate under different hunting regimes, with a control of no hunting being enforced in some of them (Nichols and Johnson 1989). Banding studies conducted at the same time could be used to measure the survival of waterfowl and therefore the extent of hunting mortality. Although such an experiment might be possible in North America where hunting impact is reasonably uniform, its application in Australia would be difficult because of the patchy distribution of hunting activity. Perhaps a "natural" experiment could be conducted instead, with local effects being examined by pre-season banding and the monitoring of assumed control and impacted sites. Which ever approach is taken, sites need to be separated sufficiently for independence.

Taking an experimental approach to the management of harvested wildlife, be they kangaroos or waterfowl, offers great opportunities for examining the responses of the populations to harvesting. However, taking an experimental approach can be expensive sometimes prohibitively so. By way of a compromise, properly organized population surveys, conducted on a continuous basis over relatively long periods and in conjunction with the recording of harvest offtakes, could well prove to be an adequate and cost-effective substitute for the more expensive experimental approach to the formulation of proper management procedures for harvesting kangaroos, waterfowl or any other wildlife.

REFERENCES

Barker, R. D. and Caughley, G., 1990. Distribution and abundance of kangaroos (Marsupialia: Macropodidae) at the time of European contact: Tasmania. *Aust. Mammal.* **13:** 157–66.

Barker, R. D. and Caughley, G., 1992. Distribution and abundance of kangaroos (Marsupialia: Macropodidae) at the time of European contact: Victoria. *Aust. Mammal.* **15:** 81–88.

Barker, R. D. and Caughley, G., 1994. Distribution and abundance of kangaroos (Marsupialia: Macropodidae) at the time of European contact: South Australia. *Aust. Mammal.* **17:** 73–83.

Bayliss, P., 1985. The population dynamics of red and western grey kangaroos in arid New South Wales, Australia. II. The numerical response function. *J. Anim. Ecol.* **54:** 127–35.

Bayliss, P., 1987. Kangaroo dynamics. Pp. 119–34 in *Kangaroos: their ecology and management in the sheep rangelands of Australia* ed by G. Caughley, N. Shepherd and J. Short. Cambridge University Press: Cambridge.

Bertram, G. C. L. and Bertram, C. K. R., 1973. The modern Sirenia: their distribution and status. *Biol. J. Linn. Soc.* **5:** 297–338.

Blakers, M., Davis, S. J. J. F. and Reilly, P. N., 1984. *The atlas of Australian birds.* Royal Australasian Ornithologists Union, Melbourne University Press: Melbourne.

Braithwaite, L. W., Maher, M., Briggs, S. V. and Parker, B. S., 1986. An aerial survey of three game species of waterfowl (Family Anatidae) in eastern Australia. *Aust. Wildl. Res.* **13:** 213–23.

Briggs, S. V. and Holmes, J. E., 1988. Bag sizes of waterfowl in New South Wales and their relation to antecedent rainfall. *Aust. Wildl. Res.* **15:** 459–68.

Briggs, S. V., Maher, M. T. and Davey, C. C., 1985. Hunter activity and waterfowl harvests in New South Wales, 1977–82. *Aust. Wildl. Res.* **12:** 515–22.

Briggs, S. V., Brickhill, J. G., Kingsford, R. T. and Hodgson, P. F., 1993. Ducks, hunters and rainfall at two sites in New South Wales. *Wildl. Res.* **20:** 759–69.

Briggs, S. V., Maher, M. T., Brickhill, J. G. and Kingsford, R. T., 1983. Mortality of the maned duck and grey teal during the 1982 open season at Barrenbox Swamp, New South Wales. *Aust. Wildl. Res.* **10:** 537–41.

Brister, B., 1992. Steel shot: ballistics and gunbarrel effects. Pp. 26–38 in *International waterfowl and wetlands research bureau special publication No. 16.* Slimbridge: United Kingdom.

Bryden, M. M., 1978. Whales and whaling in Queensland waters. *Proc. Roy. Soc. Qld* **88:** 5–18.

Burbidge, A. A., 1987. The management of crocodiles in Western Australia. Pp. 125–27 in *Wildlife management: crocociles and alligators* ed by. G. J. W. Webb, S. C. Manolis and P. J. Whitehead. Surrey Beatty & Sons: Chipping Norton.

Busch, B. C., 1985. *The war against the seals: a history of the North American seal fishery.* McGill-Queen's University Press: Kingston and Montreal.

Bustard, H. R., 1972. *Sea turtles: natural history and conservation.* Collins: London.

Cairns, S. C., 1989. Models of macropodid populations. Pp. 695–704 in *Kangaroos, wallabies and rat-kangaroos* ed by G. Grigg, P. Jarman and I. Hume. Surrey Beatty & Sons: Chipping Norton.

Cairns, S. C. and Coombs, M. T., 1992. The monitoring of the distributions of commercially harvested species of macropod in New South Wales. Unpublished report to the Australian National Parks and Wildlife Service. 69 pp.

Cairns, S. C. and Grigg, G. C., 1993. Population dynamics of red kangaroos *(Macropus rufus)* in relation to rainfall in the South Australian pastoral zone. *J. Appl. Ecol.* **30:** 444–58.

Callister, D. J. and Williams, L. M. (this volume). Australia's native wildlife trade: scale, trends and impacts. In *Conserving biodiversity: threats and solutions* ed by R. A. Bradstock, T. D. Auld, D. A. Keith, R. T. Kingsford, D. Lunney and D. P. Sivertsen. Surrey Beatty & Sons: Chipping Norton.

Caughley, G., 1987. Ecological relationships. Pp. 159–87 in *Kangaroos: their ecology and management in the sheep rangelands of Australia* ed by G. Caughley, N. Shepherd and J. Short. Cambridge University Press: Cambridge.

Caughley, G. and Briggs, S. V., 1983. Management of waterfowl. Pp. 68–72 in *Parks and wildlife — wetlands* ed by C. Haigh. National Parks and Wildlife Service: Sydney.

Caughley, J., Bayliss, P. and Giles, J., 1984. Trends in kangaroo numbers in western New South Wales and their relation to rainfall. *Aust. Wildl. Res.* **11:** 415–22.

Caughley, G., Grigg, G. C. and Smith, L., 1985. The effect of drought on kangaroo populations. *J. Wildl. Manage.* **49:** 679–85.

Caughley, G., Sinclair, R. and Scott-Kemmis, D., 1976. Experiments in aerial survey. *J. Wildl. Manage.* **43:** 775–77.

Caughley, G., Brown, B., Dostine, P. and Grice, D., 1984. The grey kangaroo overlap zone. *Aust. Wildl. Res.* **11:** 1–10.

Caughley, G., Short, J., Grigg, G. C. and Nix, H., 1987. Kangaroos and climate: an analysis of distribution. *J. Anim. Ecol.* **56:** 751–61.

Crome, F. H. J., 1986. Australian waterfowl do not necessarily breed on rising water level. *Aust. Wildl. Res.* **13:** 461–80.

CSIRO and Bird Pest Research Group, 1988. The duck problem on rice fields in New South Wales: a feasibility study. Unpublished report to to the Rice Research Committee.

Denny, M. J. S., 1975. The occurrence of the eastern grey kangaroo (*Macropus giganteus*, Shaw) west of the Darling River. *Search* **6:** 89–90.

Dexter, N. and Bayliss, P., 1991. The effect of experimental clutch harvest on numbers of magpie geese nests and juvenile recruitment. *Wildl. Res.* **18:** 533–38.

Dieter, M. P and Finley, M. T., 1979. ALAD enzyme activity in blood, brain and liver of lead dosed ducks. *Environ. Res.* **19:** 127–35.

Driessen, M. M., 1993. Effects of hunting and rainfall on Bennetts wallaby and Tasmanian pademelon populations. Unpublished M.Sc. thesis, University of Tasmania.

Everleigh, J., 1991. New South Wales 1990 Kangaroo Management Programme Annual Report. Unpublished report to Australian National Parks and Wildlife Service. 3 pp.

Frith, H. J., 1973. *Wildlife conservation.* Angus and Robertson: Sydney.

Frith, H. J., 1982a. *Waterfowl in Australia.* 2nd edition. Angus and Robertson: Sydney.

Frith, H. J., 1982b. *Pigeons and doves of Australia.* Rigby: Adelaide.

Frost, S., 1978. Whales and Whaling. Report of the Independent Inquiry conducted by the Hon. Sir Sydney Frost. Australian Government Publishing Service: Canberra.

Garnett, S., 1992. Threatened and extinct birds in Australia. RAOU Report Number 82. York Press: Melbourne.

Gibson, L. M. and Young, M. D., 1988. Kangaroos: Counting the Cost. Report to Australian National Parks and Wildlife Service. Project Report No. 4. CSIRO: Melbourne. 74 pp.

Greger, M. and Kautsky, L., 1991. Effects of Cu, Pb and Zn on two *Potamogeton* species grown under field conditions. *Vegatatio* **97:** 173–84.

Grigg, G. C., 1988. Kangaroo harvesting and the conservation of the sheep rangelands. *Aust. Zool.* **24:** 124–28.

Grigg, G. C., 1989. Kangaroo harvesting and the conservation of arid and semi-arid rangelands. *Biol. Conserv.* **3:** 194–97.

Halse, S. A., James, I. R., Fitzgerald, P. E., Diepeveen, D. A. and Munro, D. R., 1993. Survival and hunting mortality of Pacific black ducks and grey teal. *J. Wildl. Manage.* **57:** 42–48.

Harper, M. J. and Hindmarsh, M., 1990. Lead poisoning in magpie geese, *Anseranas semipalmata,* from ingesting lead pellet at Bool Lagoon Game Reserve (South Australia). *Aust. Wildl. Res.* **17:** 141–45.

Hill, G. J. E., Barnes, A. and Wilson, G. R., 1985. Time of day and aerial counts of grey kangaroos. *J. Wildl. Manage.* **49:** 843–49.

Hughes, R., 1987. *The fatal shore.* Collins Harvill: London.

Kingsford, R. T., 1989. The effect of drought on duckling survival of maned ducks. *Aust. Wildl. Res.* **16:** 405–12.

Kingsford, R. T., 1992. Maned ducks and farm dams: a success story. *Emu* **92:** 163–69.

Kingsford, R. T., (this volume). Ecological effects of river management in New South Wales. In *Conserving biodiversity: threats and solutions* ed by R. A. Bradstock, T. D. Auld, D. A. Keith, R. T. Kingsford, D. Lunney and D. P. Sivertsen. Surrey Beatty & Sons: Chipping Norton.

Kingsford, R. T. and Porter, J. L., 1994. Waterbirds on an adjacent freshwater lake and salt lake in arid Australia. *Biol. Conserv.* **69:** 219–28.

Kingsford, R. T., Flanjak, J. and Black, S., 1989. Lead shot and ducks on Lake Cowal. *Aust. Wildl. Res.* **16:** 167–72.

Kingsford, R. T., Kacprzak, J. L. and Ziaziaris, J., 1994. Lead in livers and gizzards of waterfowl shot in New South Wales. *Environ. Pollut.* **85:** 329–35.

Kingsford, R. T., Porter, J. L. and Ferster Levy, R., 1992. An aerial survey of wetland birds in eastern Australia — October, 1991. New South Wales National Parks and Wildlife Service Occasional Paper 12.

Koh, T. S. and Harper, M. J., 1988. Lead-poisoning in black swans, *Cygnus atratus,* exposed to spent lead shot on Bool Lagoon Game Reserve, South Australia. *Aust. Wildl. Res.* **15:** 395–403.

Lavery, H. J. and Tierney, P. J., 1985. Scarcity and extinction. Pp. 25–54 in *The kangaroo keepers* ed by H. J. Lavery. University of Queensland Press: St Lucia.

Lawler, W., Kingsford, R. T., Briggs, S. V. and Milkovits, G., 1993. Movement of grey teal *Anas gracilis* from drying, arid zone wetland. *Corella* **17:** 58–60.

Leakey, R. E. and Lewin, R., 1977. *Origins.* E. P. Dutton: New York.

Letts, G. A., 1987. The management of Crocodilians in Australia — introductory comments. Pp. 103–05 in *Wildlife management: crocodiles and alligators* ed by G. J. W. Webb, S. C. Manolis and P. J. Whitehead. Surrey Beatty & Sons: Chipping Norton.

Lund, M., Davis, J. and Murray, F., 1991. The fate of lead from duck shooting and road runoff in three Western Australian wetlands. *Aust. J. Mar. Freshw. Res.* **42:** 139–49.

Maher, M. T. and Braithwaite, L. W., 1992. Patterns of waterbird use in wetlands of the Paroo: a river system of inland Australia. *Rangel. J.* **14:** 128–42.

Maher, M. T. and Carpenter, S. M., 1984. Benthic studies of waterfowl breeding habitat in south-western New South Wales. II. Chironomid populations. *Aust. J. Mar. Freshw. Res.* **35:** 97–110.

Marchant, S. and Higgins, P. J., 1990. *Handbook of Australian, New Zealand and Antarctic birds. Vol. I, ratites to ducks.* Oxford University Press: Melbourne.

Marshall, A. J., 1966a. Of the plume trade, and emu oil. Pp. 43–74 in *The great extermination* ed by A. J. Marshall. Heinemann: Melbourne.

Marshall, A. J., 1966b. On the disadvantages of wearing fur. Pp. 9–42 in *The great extermination* ed by A. J. Marshall. Heinemann: Melbourne.

McGilp, J. N., 1923. Birds of the Lake Frome district, South Australia. *Emu* **23:** 237–44.

Moorehead, A., 1966. *The fatal impact: the invasion of the South Pacific, 1767–1840.* Hamish Hamilton: London.

Naarding, J. A., 1980. Study of the short-tailed shearwater, *Puffinus tenuirostris,* in Tasmania. Unpublished report to National Parks and Wildlife Service, Tasmania. 78 pp.

Nance, C. M., 1985. Macropodid management of cereal and pastoral regions of Queensland. Unpublished report prepared for the Reserve Bank of Australia. 45 pp.

Nichols, J. D., 1991. Extensive monitory programmes viewed as long-term population studies — the case of North American waterfowl. *Ibis* **133:** 89–98.

Nichols, J. D. and Johnson, F. A., 1989. Evaluation and experimentation with duck management strategies. Pp. 566–93 in *Transactions of the 54th American wildlife and natural resources conference.*

Norman, F. I., 1983. Protection of native waterfowl. *Roy. Hist. Soc. Vict.* **54:** 33–43.

Norman, F. I. and Nicholls, N., 1991. The southern oscillation and variations in waterfowl abundance in south-eastern Australia. *Aust. J. Ecol.* **16:** 485–90.

Norman F. I. and Powell, D. G. M., 1981. Rates of recovery of bands, harvest patterns, and estimates for black duck, chestnut teal, grey teal, and mountain duck shot during Victorian open seasons, 1953–77. *Aust. Wildl. Res.* **8:** 659–64.

Norman F. I. and Riggert, T. L., 1977. Nest boxes as nest sites for Australian waterfowl. *J. Wildl. Manage.* **41:** 643–49.

Norman, F. I. and Young, A. D., 1980. Short-sighted and doubly short-sighted are they . . . Game laws of Victoria, 1858–1958. *J. Aust. Studies* **7:** 2–24.

Norman, F. I., Kingsford, R. T. and Briggs, S. V., 1994. The freckled duck *Stictonetta naevosa* as a threatened taxon. International Wetlands Research Bureau (IWRB) *Waterfowl Research Group Newsletter* **5:** 11–13.

Ogden, R., 1992. *Modern hunting and conservation.* The Field Game Federation: McMillan Printing Group.

Pain, D. J. and Amiardtriquet, C., 1993. Lead poisoning of raptors in France and elsewhere. *Ecotoxicol. Environ. Safety* **25:** 183–92.

Poole, W., 1984. Management of kangaroo harvesting in Australia. Australian National Parks and Wildlife Service Occasional Paper No. 9. 25 pp.

Pople, T., Cairns, S., Clancy, T., Grigg, G. and Beard, L., 1993. Accuracy of kangaroo surveys in Queensland using fixed-wing aircraft. Unpublished report to the Australian National Parks and Wildlife Service. 75 pp.

Ringelman, J. K., Miller, M. W. and Andelt, W. F., 1993. Effects of ingested tungsten-bismuth-tin shot in captive mallards. *J. Wildl. Manage.* **57:** 725–32.

Robertshaw, J. D. and Harden, R. W., 1989. Predation on Macropodoidea: a review. Pp. 735–53 in *Kangaroos, wallabies and rat-kangaroos* ed by G. Grigg, P. Jarman and I. Hume. Surry Beatty & Sons: Chipping Norton.

Rocke, T. E. and Samuel, M. D., 1991. Effects of lead shot ingestion on selected cells of the mallard immune system. *J. Wildl. Diseases* **27:** 1–9.

Sharley, A. J., Best, L. W., Lane, J. and Whitehead, P., 1992. An overview of lead poisoning in Australian waterfowl and implications for management. *IWRB Special Publ.* **16:** 73–77.

Shepherd, N. and Caughley, G., 1987. Options for the management of kangaroos. Pp. 188–219 in *Kangaroos: their ecology and management in the sheep rangelands of Australia* ed by G. Caughley, N. Shepherd and J. Short. Cambridge University Press: Cambridge.

Short, J. and Grigg, G. C., 1982. The abundance of kangaroos in suboptimal habitats: wheat, intensive pastoral and mallee. *Aust. Wildl. Res.* **9:** 221–27.

Strahan, R. (ed), 1983. *Complete book of Australian mammals.* Angus and Robertson: Sydney.

Taplin, L. E., 1987. The management of crocodiles in Queensland. Pp. 129–40 in *Wildlife management: crocociles and alligators* ed by G. J. W. Webb, S. C. Manolis and P. J. Whitehead. Surrey Beatty & Sons: Chipping Norton.

Timmermans, K. R., Peeters, W. and Tonkes, W., 1992. Cadmium, zinc, lead and copper in *Chiromus riparius* (Meigen) larvae (Diptera, Chironomidae) — uptake and effects. *Hydrobiol.* **241:** 119–34.

Trost, R. E., 1981. Dynamics of grit selection and retention in captive mallards. *J. Wildl. Manage.* **45:** 64–73.

Trust, K. A., Miller, M. W., Ringelman, J. K. and Orme, I. M., 1990. Effects of ingested lead on antibody production in mallards *(Anas platyrhynchos). J. Wildl. Disease* **26:** 316–22.

Webb, G. J. W., Whitehead, P. J. and Manolis, S. C., 1987. Crocodile management in the Northern Territory of Australia. Pp. 107–24 in *Wildlife management: crocodiles and alligators* ed by G. J. W. Webb, S. C. Manolis and P. J. Whitehead. Surrey Beatty & Sons: Chipping Norton.

Whitehead, P. J. and Tschirner, K., 1991. Lead shot ingestion and lead poisoning of magpie geese, *Anseranas semipalmata,* foraging in a northern Australian hunting reserve. *Biol. Conserv.* **58:** 99–118.

Wickson, R. J., Norman, F. I., Bacher, G. J. and Garnham, J. S., 1992. Concentrations of lead in bone and other tissues of Victorian waterfowl. *Wildl. Res.* **19:** 221–32.

Wong, P., 1993. The analysis of ten years of aerial survey data to determine the effects of hunting on game species of waterfowl. Unpublished M.Sc. thesis, University of Sydney. 74 pp.

Worrell, E., 1966. The unpopular ones. Pp. 75–94 in *The great extermination* ed by A. J. Marshall. Heinemann: Melbourne.

Australia's native wildlife trade: scale, trends and impacts

D. J. CALLISTER[1] and L. M. WILLIAMS[2]

Commercial use of Australian wildlife by Europeans began soon after the establishment of the first colony and for many years was largely unregulated and unmonitored. While it is unlikely that any extinctions resulted from this early exploitation, a number of species were reduced to dangerously low levels before harvesting limits or bans were imposed. Wild harvest of native flora and fauna is currently controlled by state and territory conservation agencies; controls over fauna harvest are generally stricter and better managed than those for flora. Federal involvement is limited to controlling export trade and approving management plans for exported species. In recent years, with some notable exceptions (e.g., kangaroos, possums), captive breeding and cultivation has become increasingly important. Currently the legal domestic and export trade in Australia's native wildlife is worth around A$100 million annually, and growing, with more than half of this coming from the kangaroo industry. The extent of illegal trade cannot be quantified. Data on the impact of wildlife trade are incomplete for many taxa, however, they suggest that it is unlikely to be detrimental for most populations, although localized impacts are possible. Current resources for controlling and managing wildlife trade are inadequate and, with any expansion in trade, are likely to become more so. There is continuing debate as to the future character, size and direction of this country's expanding wildlife trade. The time now seems opportune for an open consultative process to address the growing number of issues being generated by this debate.

INTRODUCTION

IT has been estimated that the global trade in wild animals, plants and their products is worth up to US$20 billion per year (Russell and Mills 1993). Up to a quarter of this trade is believed to be illegal. Australia is not a major player in the global wildlife trade. We do, however, consume some of our own and other countries' wildlife and we do export an increasing amount of our own wildlife — both legally and illegally. The annual value of the Australian industry in wild animals and related products has been put at $100 million (Wilson *et al.* 1992) to $132–156 million (Ramsay 1994). These figures include trade in wild populations of both native and introduced species. Australia also has a healthy domestic and export trade in wild and cultivated flora and exports captive bred specimens, or their products, of a number of native fauna species.

Broad trends in Australia's wildlife trade show fairly extensive commercial exploitation of native species following European settlement, with trade levels declining as restrictions were progressively introduced. Trade in some species hunted in these early days of European settlement persisted and new trades developed. Interestingly, quite recently there has been an upsurge in interest in commercial utilization of Australian wildlife, with some increasing trade activity as a result. This has primarily centred around flora, but there is increasing interest in commencing trades in certain vertebrate fauna such as Magpie Geese *Anseranas semipalmata*, Flying Foxes Pteropodidae spp. and Cockatoos Cacatuinae spp. On the other hand, some long-standing trades such as Short-tailed Shearwaters *Puffinus tenuirostris* are in decline.

The issue of sustainable utilization of wild species is reaching new levels of prominence in the international conservation debate. Views on the relative merits of sustainable utilization as a conservation tool range across the full spectrum from the "use it or lose it" advocates through to those who view any use of wildlife by humans as immoral. This issue is increasingly coming to the fore in Australia. In early 1994 a major symposium was held on the theme of conservation through sustainable use of wildlife, and a number of state and territory government departments are attempting to re-define policy towards a more "pro-use" stance.

This paper gives an overview of the history and current status of trade within and from Australia in native Australian flora and fauna. It does not cover the trade in introduced species, nor does it look at forestry or commercial food fisheries. The latter two are extensive trades in their own right and deserve detailed individual attention elsewhere; trade in wild introduced species is covered in Ramsay (1994).

HISTORY

Table 1 lists many of the vertebrate species that have been legally traded in and from Australia. In some cases the numbers traded were quite high. By far the most commonly traded commodity was skins. In a number of cases wildlife trade was linked in with bounty or pest

[1]TRAFFIC Oceania, P.O. Box R594, Royal Exchange, Sydney, New South Wales, Australia 2000.
[2]426 Queens Parade, North Fitzroy, Victoria, Australia 3068.

Pages 273–91 *in* CONSERVING BIODIVERSITY: THREATS AND SOLUTIONS, ed by R. A. Bradstock, T. D. Auld, D. A. Keith, R. T. Kingsford, D. Lunney and D. P. Sivertsen. Surrey Beatty & Sons. 1995.

Table 1. Australian vertebrate taxa which have entered legal wildlife trade since European settlement. Adapted, with additions, from Thomson *et al.* (1987), Table 10.1, p. 230. Sources of status information: threatened marsupials and monotremes — Kennedy (1992); all crocodiles — Thorbjarnarson (1992); other threatened species — Groombridge (1993); all other species — Thomson *et al.* (1987).

Scientific name	Common name	Status
Ornithorhynchus anatinus	Platypus	Potentially vulnerable
Notoryctes typhlops	Marsupial Mole	Potentially vulnerable
Macrotis lagotis	Bilby	Vulnerable
M. leucura	Lesser Bilby	Extinct
Phascolarctos cinereus	Koala	Potentially vulnerable
Psuedocheirus peregrinus	Common Ringtail Possum	Common
Trichosurus caninus	Mountain Brushtail Possum	Potentially vulnerable
T. vulpecula	Common Brushtail Possum	Potentially vulnerable
Potorous tridactylus	Long-nosed Potoroo	Potentially vulnerable
Bettongia penicillata	Brush-tailed Bettong	Endangered
Petrogale xanthopus	Yellow-footed Rock-wallaby	Potentially vulnerable
Thylogale billardierii	Tasmanian Pademelon	Abundant, range reduced
Macropus fuliginosus	Western Grey Kangaroo	Abundant
M. giganteus	Eastern Grey Kangaroo	Abundant
M. greyi	Toolache Wallaby	Extinct
M. parryi	Whiptail Wallaby	Common
M. robustus	Common Wallaroo	Abundant
M. rufus	Red Kangaroo	Abundant
Wallabia bicolor	Swamp Wallaby	Common
Thylacinus cynocephalus	Thylacine	Extinct
Hydromys chrysogaster	Water Rat	Sparse to common
Physeter catodon	Sperm Whale	Insufficiently known
Eubalaena australis	Southern Right Whale	Vulnerable
Megaptera novaeangliae	Humpback Whale	Vulnerable
Neophoca cinerea	Australian Sea Lion	Rare
Arctocephalus pusillus	Australian Fur Seal	Abundant
Mirounga leonina	Southern Elephant Seal	Abundant
Dugong dugon	Dugong	Vulnerable
Dromaius novaehollandiae	Emu	Common
Puffinus tenuirostris	Short-tailed Shearwater	Common
Psittaciformes spp.	Parrots, Cockatoos, etc.	Not applicable
Ploceidae spp.	Finches, Mannikins, etc.	Not applicable
Crocodylus johnstoni	Freshwater Crocodile	Not depleted
C. porosus	Saltwater Crocodile	Depleted/Vulnerable

control schemes and it is generally difficult to ascertain whether it was these schemes, or the economic benefits from trade, which was driving the harvest. While some of the species shown in Table 1 are now extinct or threatened, with the exception of some of the marine mammals it is unlikely that wildlife trade played a major part in reducing their population. Some of the more common species were subject to heavy exploitation and consequently suffered population declines; in general, these species have recovered following the introduction of harvesting restrictions.

The commercial exploitation by Europeans of Australian wildlife probably began with the seal trade. Nine-thousand skins of the Australian Fur Seal *Arctocephalus pusillus* were harvested in Bass Strait during a voyage of the vessel *Nautilis* in the late 1790s (King 1989). An export trade commenced and sealing vessels came to Australia from America, England and India. By 1820, seal herds were very much reduced and the size of the industry subsequently declined. Seals were given protection in Victorian waters in 1891, however, sealing continued in Tasmania until at least 1923. Fur seal populations had recovered to current levels by around the mid-1940s (King 1989; Warneke 1983). Colonies of Elephant Seals *Mirounga leonina* in Tasmania were believed to have been hunted to extinction in the early 19th Century (Bryden 1983; King 1989; Strahan 1983). Reports on whether this was a breeding population are conflicting. Bass Strait populations of the Australian Sea Lion *Neophoca cinerea* are also reported to have been wiped out by sealers (Marlow 1983).

The history of whaling in Australian waters is given in Bannister (1989). Sperm Whales *Physeter catodon* were heavily harvested in the early part of the last century for their oil; this century there were two phases of hunting, both off Albany, Western Australia, during 1912–15 (small catches) and 1955–78 (large catches). The Southern Right Whale *Eubalaena australis* comprised most of the catch of "bay" whaling which commenced in Tasmania in 1806, with harvests of this species reaching a peak in the 1830s and 1840s. The Humpback Whale *Megaptera novaeangliae* was also taken in Tasmanian "bay" whaling operations but only became commercially important in the Southern Ocean at the turn of this century with the advent of modern whaling techniques. Off the west coast four phases of hunting occurred: 1912–16; 1925–28; 1936–38; and 1949–63. Between 1934 and 1939, 7 244 Humpback Whales were taken off the cost of Western Australia, with a further 12 000 harvested between 1949 and 1963. Off eastern Australia, significant numbers were taken from 1949 onwards; between 1950 and 1962 around 7 000 animals were taken from these waters. This hunting had a very negative impact on Humpback Whale stocks; from estimated unfished stock levels of 12–17 000 and 10 000, the Western Australian and Eastern Coast stocks were believed to have declined to around 800 and 500 animals respectively by 1962. The Australian whaling industry ceased entirely in 1979 with the adoption of a non-whaling policy.

Commercial exploitation of macropods in Australia began as a bounty payment on kangaroo and wallaby scalps (Poole 1984; Thornback and Jenkins 1982). For most species this was done mainly as a pest control measure. In Queensland, between 1877 and 1917, bounties and government subsidies were paid on some 26 million macropod scalps (Poole 1984). The Toolache Wallaby *Macropus greyi* and Yellow-footed Rock Wallaby *Petrogale xanthopus*, however, had attractive pelts that were much in demand (Russell *et al.* 1989; Thornback and Jenkins 1982). Large numbers of Toolache Wallaby skins were traded in Melbourne around the turn of the century (Thornback and Jenkins 1982). By the 1920s the species was presumed extinct. While habitat destruction probably played a part in the demise of this species, hunting for trade and sport were undoubtedly contributing factors. In the mid 1800s the Long-nosed Potoroo *Potorous tridactylus* was sold for game (Wheelwright 1861, cited in Seebeck and Rose 1989) and around 1900 the Brush-tailed Bettong *Bettongia penicillata* was sold live for coursing (Thomson *et al.* 1987). In the 1910s the skins of other macropods began to be traded. In Queensland 300 000 to 400 000 macropod skins were processed annually, increasing to 450 000 in the 1950s (Poole 1984). A thriving trade in macropod meat, skins and leather continues to this day (see below).

Taken for its pelt, the Koala *Phascolarctos cinereus* was hunted from the late 19th Century until the 1920s in numbers large enough to bring it close to extinction (in combination with the effects of habitat destruction and epidemic) (Thomson *et al.* 1987). Millions of skins were traded, with many exported; about 2 million in 1924 alone. In the last open season, in Queensland in 1927, 584 738 Koalas were killed.

A market for Brushtail Possums, *Trichosurus vulpecula* and *T. caninus*, was established by the early 1900s. Both species were taken. In 1906 four million skins were marketed in London and New York, and heavy harvests were at least partly responsible for the elimination of the Common Brushtail Possum *T. vulpecula* from much of its range in Queensland (Frith 1973; Thomson *et al.* 1987; Winter 1979). The last open season on the mainland was in 1963 in Victoria, when 90 295 animals were taken (Winter 1979). A commercial trade in Common Brushtail Possum skins persists in Tasmania (see below).

Common Ringtail Possums *Pseudocheirus peregrinus* began to be traded for their fur before the 1920s. Peak numbers were hunted between 1929 and 1938; of the 3 975 700 skins traded in this period, 3.5 million were from Tasmania. The last open season was in 1959, in Victoria, when 2 500 skins were traded (Thomson *et al.* 1987).

Other native mammals traded included: Dugong *Dugong dugon* for its oil (Lack 1968, cited in Marsh 1989); Platypus *Ornithorhynchus anatinus* for its pelt in what was reported to be primarily a domestic trade (Grant 1989); Thylacine *Thylacinus cynocephalus*, where 3 482 skins were exported to the United Kingdom during 1878–93 to be made into waistcoats (Dixon 1989); Marsupial Mole *Notoryctes typhlops* which were sold by Aborigines to Europeans and Afghan camel drivers who then sold them at markets, for what final use is not known (Johnson and Walton 1989); and Bilby *Macrotis* spp., including *M. lagotis*, specimens of which were killed in large numbers for their skins in South Australia early this century (Wood-Jones 1923–25, cited in Thornback and Jenkins 1982).

W at was to become a large commercial enterprise utilizing the meat, oil, eggs, fat and feathers of the Short-tailed Shearwater, was begun on Bass Strait islands in the 1830s by sealers and aboriginals (Backhouse 1843; Begg and Begg 1979). Prior to the prohibition of the taking of eggs in 1902, an estimated 100 000 Short-tailed Shearwater eggs were harvested annually (Cott 1953). Most of the harvest, however, consisted of birds (principally chicks). Early this century approximately one million birds per annum were being harvested (Skira 1987), the catch declining thereafter to around 300 000 birds by the early 1980s (Callister 1991b). A commercial trade persists to this day, but appears to be in decline (see below).

Trapping of birds for the pet trade, both legally and illegally, occurred at various stages in Australia's European history. The principal target species were parrots, cockatoos and the like (Psittaciformes spp.) and finches and allied species (Ploceidae spp.). Comprehensive data on the nature and extent of this historical trade are lacking.

Commercial trade in crocodiles began in the 1920s but these endeavours petered out (Webb *et al.* 1987). In 1945 commercial hunting of the Saltwater Crocodile *Crocodylus porosus* in the Northern Territory commenced. Between 1945 and 1972 some 270 000 to 330 000 animals were believed to have been killed across the country (Anon. 1985). Due to its skin being commercially inferior, the Freshwater Crocodile *C. johnstoni* was largely ignored before 1958, when advances in tanning technology made the skin of this species economically marketable (Webb *et al.* 1987). By 1964, possibly 60 000 *C. johnstoni* skins had been traded. *C. johnstoni* was protected in 1962 in Western Australia, 1964 in the Northern Territory and 1974 in Queensland; protection followed for *C. porosus* in 1971 (Western Australia), 1972 (Northern Territory) and 1974 (Queensland), and export of this species was banned in 1972 (Groombridge 1982).

CURRENT AND RECENT TRADE

A wide and increasing variety of Australian flora and fauna enter trade domestically and internationally, both legally and illegally. Specimens and species traded range from well known wildlife trade items such as birds, orchids and kangaroos, through to the more obscure such as Red-back Spider *Latrodectus hasselti* venom and Southern Bull-kelp *Durvillaea potatorum.*

Management of wildlife trade

In general, responsibility for the management of exploited species rests with states and territories, normally within the department responsible for conservation management, and is governed by the relevant policies and nature conservation legislation of that state or territory. Where the trade includes farming or slaughter for meat, there is usually some involvement of other government departments such as primary industries or health. Federal involvement only becomes relevant for taxa which are subject to export trade. Exported species fall under the purview of the *Wildlife Protection (Regulation of Exports and Imports) Act* 1982 *(Wildlife Protection Act)*, which contains specific management provisions and requirements for native species subject to export. This legislation is administered by the Australian Nature Conservation Agency (ANCA). Wild-harvested native taxa which meet the requirements of the *Wildlife Protection Act* can be approved for export under the "approved management programme" or the less stringent "controlled specimens" provisions of the Act. To a large extent, however, management occurs at the state level, with the federal government's role being to ensure management programmes comply with the requirements of federal legislation and approving those that do.

The norm for management of wild-harvested species in Australia is that greater management effort is expended on animals rather than plants. For some trades such as the kangaroo industry, considerable effort and resources are provided to monitor populations and regulate the harvest and trade. For others, especially some flora harvesting operations, field monitoring and management is virtually non-existent. Partial management costs would be recouped from licence fees, permits and the like. Comprehensive data are unavailable but it seems highly unlikely that these fees would cover existing management costs, let alone the costs that would be involved if management was raised to appropriate levels across all exploited species. This is certainly the case for ANCA, where permit fees amount to only 10 per cent of the total cost of associated administrative activities (Ley 1992). If the trade in some species was made truly "user pays" in terms of covering harvest and population monitoring and management costs, the trade would probably become uneconomic.

One consequence of these apparent management shortcomings is that exploitation levels are based on incomplete data. This could lead to either over- or under-harvesting; the former undesirable from a conservation perspective, and the latter from an economic one. Another problem is that in a number of cases harvest and trade data are generated, but lack of resources preclude any in-depth scrutiny or analysis of these data.

Considering all these factors, it appears that for some traded species, especially plants, the level of knowledge of the population, ecology and so on is insufficient to draw scientifically rigorous conclusions about the likely or actual impacts of harvesting and trade. At the moment, for many species, the best that can be achieved with current resources is to make decisions based on existing knowledge, try to introduce some safeguards, and hope for the best.

Current legal trade

Macropods

In terms of both economic value and number of animals involved, Australia's macropod trade is among the world's biggest based on the harvest of a wild native mammal. The species of macropod subject to commercial harvest and the number of animals killed are shown in Table 2. Since 1986, harvest levels have stood at between 2 and 3 million animals annually. In recent years, *Macropus rufus* has overtaken *M. giganteus* as the most heavily harvested species; these two account for around 90 per cent of the kangaroos killed in the last eight years. Queensland harvests the most animals, followed by New South Wales, with the South Australian kill overtaking that of Western Australia since 1991.

Table 2. Commercial kangaroo kill by state and species, 1979–1993. Blanks indicate that no data are available. Source: Australian National Conservation Agency. *Victorian figures include both commercial and non-commercial authorized kills.

	1979	1980	1981	1982	1983	1984	1985	1986	1987	1988	1989	1990	1991	1992	1993
New South Wales															
Macropus fuliginosis									62 926	72 786	67 252	83 708	106 629	117 994	129 378
M. fuliginosis/giganteus				263 168	134 900	63 754	111 968	181 463							
M. giganteus									140 061	130 335	126 072	170 766	253 791	264 447	284 344
M. robustus				2 066	714	632	763	0	0	0	97	1 967	1 378	1 377	1 678
M. rufus				399 108	264 863	157 629	213 297	263 046	270 467	218 086	297 029	277 155	495 986	412 189	359 820
TOTAL	**520 000**	**619 023**	**488 647**	**664 342**	**400 477**	**222 015**	**326 028**	**444 509**	**473 454**	**421 207**	**490 450**	**533 596**	**857 784**	**796 007**	**775 220**
Queensland															
M. agilis	6 935	1 712	100	649	0	0	0	0	0	0	0	0	0	0	0
M. dorsalis	262	253	0	238	19	0	0	0	0	0	0	0	0	0	0
M. dorsalis/rufogriseus	0	0	0	0	0	0	0	0	0	0	0	0	0	0	0
M. fuliginosis/giganteus	866 395	649 534	398 860	682 583	520 948	543 430	744 966								
M. giganteus								957 621	1 231 889	1 292 196	1 143 314	1 097 890	1 017 086	919 234	989 578
M. parryi	53 908	29 431	31 981	33 823	21 130	12 707	10 727	26 455	9 498	24 296	13 456	14 954	4 226	2 071	2 737
M. robustus	34 054	22 005	27 937	67 191	66 149	77 595	72 927	69 919	65 441	67 048	93 452	76 745	98 052	122 967	143 841
M. rufogriseus	9 714	2 337	241	197	9	0	0	0	0	0	0	0	0	0	0
M. rufus	342 622	237 282	220 216	352 117	268 904	226 212	297 496	289 526	365 138	359 985	473 985	476 636	471 643	570 885	595 488
Wallabia bicolor	370	545	49	0	109	0	0	0	0	0	0	0	0	0	0
Wallaby spp.	0	0	0	0	726	0	0	0	0	0	0	0	0	0	0
TOTAL	**1 314 260**	**943 099**	**679 384**	**1 136 798**	**877 268**	**960 670**	**1 126 116**	**1 343 520**	**1 671 966**	**1 743 525**	**1 724 207**	**1 666 225**	**1 591 007**	**1 615 157**	**1 731 644**
South Australia															
M. fuliginosis						9 898	9 043	9 987	14 849	13 778	11 546	18 593	14 533	18 999	32 798
M. robustus						5 907	7 043	6 147	9 167	8 892	5 756	7 014	8 966	11 252	18 436
M. rufus						83 469	98 557	112 789	100 507	118 232	124 173	172 793	213 628	219 338	227 056
TOTAL	**93 234**	**148 329**	**193 326**	**229 453**	**155 785**	**99 274**	**114 643**	**128 923**	**124 523**	**140 902**	**141 475**	**198 400**	**237 127**	**249 589**	**278 290**
Tasmania															
M. rufogriseus			77 000	49 000	64 000	71 000	55 000	33 000							
Thylogale billardierii			31 000	28 000	46 000	47 000	37 000	27 000							
TOTAL	**244 918**	**220 226**	**108 000**	**77 000**	**110 000**	**118 000**	**92 000**	**60 000**	**0**	**0**	**0**	**0**	**0**	**0**	**0**
Victoria*															
M. fuliginosis/giganteus			23 347	30 885											
M. rufogriseus				473											
M. rufus				25											
W. bicolor				365											
TOTAL			**23 347**	**31 748**	**0**	**0**	**0**	**0**	**0**	**0**	**0**	**0**	**0**	**0**	**0**
Western Australia															
M. fuliginosis	40 000	58 000	34 000	38 793	35 775	37 000	37 024	37 019	40 092	29 019	23 912	36 800	37 847	46 694	47 070
M. robustus	4 500	4 930	5 000	3 916	4 365	5 500	6 244	3 955	0	819	4 164	5 191	5 399	4 434	8 477
M. rufus	151 000	100 000	136 000	134 827	162 262	136 000	161 057	153 655	150 462	215 771	173 663	223 801	185 117	107 605	139 833
TOTAL	**195 500**	**162 930**	**175 000**	**177 536**	**202 402**	**178 500**	**204 325**	**194 629**	**190 554**	**245 609**	**201 739**	**265 792**	**228 363**	**158 733**	**195 380**
TOTALS ALL STATES	**2 367 912**	**2 093 607**	**1 667 704**	**2 316 877**	**1 745 932**	**1 578 459**	**1 863 112**	**2 171 581**	**2 460 497**	**2 551 243**	**2 557 871**	**2 664 013**	**2 914 281**	**2 819 486**	**2 980 534**

The value of the kangaroo industry has been put at some $50–60 million annually and increasing (Ramsay 1994). There is an extensive domestic trade in kangaroo products such as fur and leather souvenirs, meat for the pet food trade and an increasing trade in meat for human consumption. Comprehensive trade data (volume and value) are, however, only available for the export trade (Table 3). Using Ramsay's figures on the annual value of the trade, this suggests that the export trade accounts for just over half of the income derived from the harvest. The products exported are meat (both fresh, and unfit for human consumption), skins, leather and their manufactured products. The value of the export trade of all goods totalled remains high ($32 335 653 in 1992/93), continuing on from a marked jump in 1989–90.

The countries importing kangaroo products fluctuate, especially for goods such as meat. Over recent years the major importers of kangaroo products have included: skins — Germany, Italy, Japan; fresh meat — Germany, Japan; meat unfit for human consumption — Indonesia, Macau; and leather — France, Germany, Japan, United States of America.

Considerable resources and effort are expended at the state (in particular) and federal level towards management of the kangaroo harvest. Detailed management plans are followed in those states which harvest kangaroos commercially and these are scrutinized by the Commonwealth Government as part of the process of export approval. While kangaroo management programmes are probably not perfect (indeed, it would be difficult to find any management programme for the exploitation of a wild species which was), they are generally regarded as the model for the large-scale exploitation of wild species within Australia.

Brushtail Possum

The Common Brushtail Possum has been commercially traded since at least the turn of the century; Tasmania is the only state where the trade persists. Trade in Brushtail Possums has been under government control in Tasmania since 1918 (Callister 1991a; Ramsay 1994), in what is essentially a trade based on culling for pest control purposes. Current management occurs under a management programme administered by the Tasmanian Department of Environment and Land Management, and approved under the *Federal Wildlife Protection Act.* In Tasmania, the Common Brushtail Possum is classified as "partly protected", which allows for regulated harvest and trade.

Most of the trade is in skins, which are used for garment manufacture or to line shuttles used in textile manufacture (Callister 1991a; Ramsay 1994). There has also been a small amount of trade in possum fur (Callister 1991a) and the meat is sold in Tasmania as pet food (Ramsay 1994). The industry is very much subject to the vagaries of prevailing skin prices, which affect both the number of animals shot in a season and the number of hunters that will apply for permits in the following season (Callister 1991a).

Trade and harvest data are shown in Table 4. Comparison of Tasmanian and Commonwealth sources of data on the trade indicate discrepancies which could be caused by either statistical errors, skin stockpiling or illegal trade (see also Callister 1991a; Ramsay 1994). This is a matter of some concern in that it appears that illegal skins could enter the trade either at the point of export from Tasmania, or from mainland Australia, without the trade statistics alerting the authorities to the fact that this is happening. The most regular importers of possum skins from Australia are Japan, the United States of America, United Kingdom and Italy. Italy was formerly the biggest importer of Australia's possum skins but has ceased importing in recent years. Examination of Australian Bureau of Statistics export data indicates that exports have remained at low levels during the 1990s (1990/91 — 14 000 skins, value $39,571; 1991/92 — 12 750 skins, value $66,579; 1992/93 — 35 920 skins, value $209,142).

The primary population monitoring tool used is annual spotlight surveys. These periodically show significant increases and decreases in the population (see e.g., Hocking 1988, 1989), but no trends which would indicate that harvesting is having a detrimental impact on the species. Some suggestions for improvements to the Tasmanian management programme for this species were given in Callister (1991a), however, she concluded that overall the management was adequate.

Emus

The Emu *Dromaius novaehollandiae* is fully protected in all states other than Western Australia and Tasmania (the Emu is not native to Tasmania) (Wilson *et al.* 1992). In Western Australia it is protected under the *Wildlife Conservation Act* 1950, but it is also a declared animal under the *Agriculture and Related Resources Protection Act* 1976, meaning that pest animals can be killed under permit.

Over 50 Emu farms in Australia are currently engaged in commercial captive breeding of Emus, in what is pretty much a fledgling industry (Anon. 1994a). Approval for the commercial slaughter of farmed Emus was first given in 1987, in Western Australia (Wilson *et al.* 1992). This state still accounts for the bulk of production with 38 farms holding the majority of the national stock of 30 000 birds (Ramsay 1994).

Table 3. Annual quantity and value of kangaroo meat and skins exported for the period 1963/64 to 1992/93. (Blank cells indicate that figures were not available for that item. Totals for years where all exports are not available should therefore be regarded as minimum figures.) Qty = quantity; kg = kilograms; A$ = Australian dollars; no. = number; sq.m = square metres. Source: Australian Bureau of Statistics.

	MEAT						SKINS						LEATHER		
	Fresh		Unfit for human use		TOTAL		Raw		Pickled		TOTAL				Minimum total export
YEAR	Qty (kg)	Value (A$)	Qty (kg)	Value (A$)	Qty (kg)	Value (A$)	Qty (no.)	Value (A$)	Qty (no.)	Value (A$)	Qty (no.)	Value (A$)	Qty (sq.m)	Value (A$)	value ($A)
1963/64					2 432 564	683 000	1 117 131	1 511 000							2 194 000
1964/65					4 913 251	1 392 000	1 485 157	1 611 000							3 003 000
1965/66					4 927 399	1 429 000	1 275 590	1 266 000							2 695 000
1966/67					3 870 476	1 331 000	1 078 830	1 426 000							2 757 000
1967/68					1 834 677	590 000	860 933	1 180 000							1 770 000
1968/69					1 009 577	343 000	817 655	1 209 000							1 552 000
1969/70					775 869	236 000	835 279	1 568 000							1 804 000
1970/71					232 901	77 000	847 896	1 847 000							1 924 000
1971/72					424 940	120 000	551 475	1 355 000							1 475 000
1972/73*					206 153	65 000	898 387	2 771 000							2 836 000
1973/74*					0	0	0	0	13 599	10 000	13 599	10 000			10 000
1974/75*					0	0	11 001	24 000	1 040	1 000	12 041	25 000			25 000
1975/76*					16 058	6 000	163 406	515 000	181 241	608 000	344 647	1 123 000			1 129 000
1976/77					41 875	16 000	192 052	649 000	346 594	1 585 000	538 646	2 234 000			2 250 000
1977/78					122 931	48 000	227 079	867 000	232 033	1 000 000	459 112	1 867 000			1 915 000
1978/79					257 138	92 000	248 484	1 312 000	712 931	3 756 000	961 415	5 068 000			5 160 000
1979/80					1 272 638	715 000	708 159	4 868 000	1 212 854	7 905 000	1 921 013	12 773 000			13 488 000
1980/81					1 671 445	3 247 000	282 294	1 404 000	844 896	3 334 000	1 127 190	4 738 000			7 985 000
1981/82					1 520 837	2 081 000	214 682	1 161 000	1 352 925	6 703 000	1 567 607	7 864 000			9 945 000
1982/83	597 044	942 000	678 024	455 000	1 275 068	1 397 000	205 865	1 328 000	1 565 397	8 344 000	1 771 262	9 672 000			11 069 000
1983/84	225 202	347 827	261 658	145 076	486 860	492 903	68 917	475 383	993 598	5 687 447	1 062 515	6 162 830		841 618	7 497 351
1984/85**	183 090	225 848	333 667	210 934	516 757	436 782	129 246	1 003 578	1 199 883	10 911 206	1 329 129	11 914 784	26 271	205 100	12 556 666
1985/86	291 529	393 562	380 706	252 082	672 235	645 644	90 730	817 507	929 540	8 998 083	1 020 270	9 815 590	72 108	1 584 096	12 045 330
1986/87	209 967	309 420	481 251	374 180	691 218	683 600	82 135	674 909	1 056 976	7 926 654	1 139 111	8 601 563	122 566	2 587 873	11 873 036
1987/88	137 131	150 819	466 729	350 162	603 860	500 981	98 454	853 118	1 310 931	9 147 104	1 409 385	10 000 222	69 200	1 497 087	11 998 290
1988/89	150 793	205 581	336 443	231 623	487 236	437 204					736 864	5 709 935	122 514	2 244 268	8 391 407
1989/90	48 076	138 763	349 887	263 739	397 963	402 502					1 592 303	14 092 994	344 535	6 498 154	20 993 650
1990/91	175 405	175 817	192 330	146 006	367 735	321 823					1 774 566	23 919 444	478 827	9 714 475	33 955 742
1991/92	444 820	498 953	276 070	220 072	720 890	719 025					1 612 444	15 459 478	542 998	11 707 612	27 886 115
1992/93	427 098	769 820	319 860	297 164	746 958	1 066 984					1 992 669	18 832 357	625 810	12 436 312	32 335 653

*export ban applied March 1973; lifted New South Wales and South Australia February 1985, Queensland and Western Australia 1975, Tasmania June 1976.
**In addition, there was a re-export of 521 raw skins, value $1 672. The figure for skins included values for November to June only.

Table 4. Number of Brushtail Possums harvested and traded, 1975/76–1989/90. (Blanks indicate no data. ABS = Australian Bureau of Statistics export data; ANCA = Australian Nature Conservation Agency export data). Data on animals shot, royalties paid and skins sold are from Tasmanian State Government agencies. Source: Adapted from Callister 1991, Tables 5 and 6; additional data from ANCA.

Year	Animals shot	Royalties paid	Skins sold	ABS exports	ANCA exports
1975/76		120 072	138 758	110 718	
1976/77		226 571	311 388	242 121	
1977/78		239 643	305 622	201 502	
1978/79		298 019	366 211	294 046	
1979/80	405 578	291 848	302 808	272 853	
1980/81	373 400	270 421	303 039	257 649	
1981/82	277 578	159 395	171 724	196 299	
1982/83	161 275	97 281	116 361	101 347	
1983/84	217 900	136 546		94,005	
1984/85	118 300	57 743	78 802	87 229	82 206
1985/86	115 233	59 922	62 599	25 996	100 562
1986/87	106 734	61 799	62 821	57 677	71 442
1987/88	252 086	181 089	213 724	91 371	204 407
1988/89	177 808	93 255	101 894	76 477	88 259
1989/90	59 731	34 047	43 923	31 024	38 562

A number of other states are in the process of developing Emu industries and enacting legislation to allow Emu farming. It has been projected that by 1995, there will be around 100 000 birds on farms throughout Australia (Anon. 1994a). Farm stock is primarily sourced from captive-bred birds, although some wild harvest has been allowed to establish stocks (Anon. 1994a; Wilson *et al.* 1992).

Domestically, there is not much demand for Emu products, however, small quantities (20–40 kg/year) of meat are sold to restaurants (Anon. 1994a). Export trade in products derived from Emus has been, most significantly, in skin, meat and oil (Table 5), with a lesser trade in a range of other items including eggs, feathers, live animals and manufactured goods such as handbags, keyholders and purses. France and the United States have, to date, been the major markets, between them importing 87 per cent of meat, 74–98 per cent of skins (depending on whether by number or weight) and 95 per cent of oil exported from 1989–1993. A promising recent development for the industry was the securing of a contract worth $3 million for Western Australia to supply Emu products to France over three years (Anon. 1994a). Table 5 clearly shows that the export trade has expanded considerably in recent years.

Short-tailed Shearwaters (Muttonbirds)

The Short-tailed Shearwater is fully protected in all states other than Tasmania where it is classified as "partly protected wildlife" under the *Wildlife Regulations* 1971 of the *National Parks and Wildlife Act* 1970. This classification permits the taking of species under licence in declared open seasons, and the trading of products of such species, by licensed commercial operators.

Commercial harvesting of the Short-tailed Shearwater is subject to a management plan, prepared by the existing state conservation agency (currently the Department of Environment and Land Management). Of 83 reserved colonies, commercial harvesting is permitted in seven (585 ha.), non-commercial harvesting in 24 (206 ha.), and no harvesting is allowed at the remaining 52 colonies (273 ha.). A further 84 colonies (481 ha.) are mostly on privately owned land and are harvested for both the commercial and non-commercial harvest (Anon. 1994b). The total number of Short-tailed Shearwater burrows in Tasmania is estimated to be 11.4 million (Skira *et al.* 1986). Birds can only be taken in an open season which runs for about two (non-commercial) to five (commercial) weeks in March and April.

Harvesting is solely of the chicks (adults were allowed to be taken up to 1976 (Skira *et al.* 1986)). The products are: whole birds — fresh, salted or frozen; feathers; and oil. These are mainly sold in Tasmania or exported to New Zealand (Table 6); less than 10 000 birds are sold interstate (Anon. 1987a). In 1985 the industry was worth $328,000 (Skira 1987). Although there has been a waxing in some years, the commercial harvest figures show an overall decline over the last 13 years, with the

Table 5. Export of Emu products from Australia, 1989–1993. Source: Australian Nature Conservation Agency.

	1989	1990	1991	1992	1993	TOTAL
Skins* (number)	50	384	322	183	7 740	8 679
Meat (kilograms)	0	7	1 657	7 103	11 950	20 717
Oil (litres)	2	152	1 051	175 909	18 339	195 453
Eggs** (number)	43	159	78	798	92	1 170
Feathers (grams)	0	0	20	470	2 050	2 540
Manufactured items†	4	0	0	83	63	150
Products†† — litres	0	0	38	221	3 224	3 483
— kilograms	0	0	940	3 179	170	4 289
— number	0	0	675	1 010	1 925	3 610

*includes skin pieces and leg skins; excludes exports of 9 kg (1992) and 423 kg (1993) of skins.
**includes souvenir eggs and egg shells.
†includes garments, handbags, key holders, purses, belts and desk accessories.
††probably oil (litres), meat (kilograms) and skins (number).

Table 6. Commercial muttonbird statistics, 1981–1993. (Blank indicates no data available; no. = number; kg = kilograms.) Source: Tasmanian Department of Environment and Land Management management programme 1994 (Anon. 1994b).

Year	Number of birds taken	Exports			Number of operators	Number of catchers
		Birds (no.)	Oil (litres)	Feathers (kg)		
1981	369 085	183 025	4 654	4 968	15	69
1982	359 305	186 120	4 842	7 344	15	64
1983	142 645	229 159	4 781	6 102	15	70
1984	367 219	163 590	4 805	5 535	16	69
1985	324 579	116 985	2 945	5 233	16	64
1986	249 014	123 865	3 255	4 212	14	50
1987	235 890	79 500	1 210	3 276	13	41
1988	310 336	120 800	2 351	5 523	14	43
1989	217 638	57 660	4 112	4 104	15	54
1990	202 824	68 284	2 430	3 543	17	42
1991	160 997	20 166	2 857	3 202	11	34
1992	137 059	20 040	1 660		9	32
1993	203 705	51 680	2 418	1 188	9	35

number of catchers and operators active in the industry also dwindling (Table 6). The decline has been attributed to a number of factors: diminished interest in the muttonbird as a foodstuff; decreasing economic viability; and a lessening interest in the tradition of "muttonbirding" (Skira 1987).

The Department of Environment and Land Management manages the harvest using a combination of regulation, field surveys and collection and analysis of harvest statistics. These data enable them to calculate an estimated Maximum Sustained Yield (MSY) and compare it with harvest levels. MSY has been exceeded on rare occasions on some commercial rookeries (Callister 1991b), but harvest is normally well below MSY (Anon. 1994b). The non-commercial harvest has been more problematic. Following indications of ongoing harvests at some non-commercial rookeries of over 90 per cent of chicks (Anon. 1987a; Skira *et al.* 1986), in 1987 harvesting was restricted to Bass Strait islands and seven colonies on the west coast (Callister 1991b). It appears that harvesting at some non-commercial rookeries is still well in excess of MSY (Anon. 1994b; Callister 1991b), although this is occurring at too few rookeries to be likely to be having a major impact on the overall population.

Cage birds

The export of native birds has been banned since 1960 (Antram 1990). Some exceptions are currently made for scientific and zoological purposes and for exports of pets of a few species owned by people permanently departing the country. There is, however, an extensive domestic bird trade, primarily in captive-bred birds, but with some trade in both legally and illegally acquired wild stock. Trade occurs at a major commercial scale and also at a more informal level between the many thousands of aviculturists throughout the country. Estimates of the scale and value of this domestic trade are not available; meaningful data may be impossible to obtain.

All states and territories have legislation to control the keeping and sale of native birds; some common species are exempted from control provisions. Lack of uniformity from state to state hampers the effective implementation of these laws and can create loopholes which facilitate illegal trade (see Halstead (1992) and Robinson and Brouwer (1989) for more discussion of this). New South Wales, in particular, has been singled out in this regard for its so-called "nineteen bird rule". Under this provision a permit is not required to keep less than 19 birds. This facilitates the laundering of illegal stocks of birds by keeping part of the trade outside any form of record keeping system that allows for scrutiny of the numbers and species of birds kept (Halstead 1992).

The most popular native aviary birds are parrots and finches. Western Australia allowed the commercial trapping of finches until the last open season in 1986 (Anon. 1987b; Antram, 1990). Over 260 000 finches were trapped over the 12 years to 1985 (Table 7), including 22 328 Gouldian Finches *Erythrura gouldiae,* a species now considered to be endangered. While trapping has been suggested as a possible cause of the decline of *E. gouldiae,* others feel that this is unlikely (see discussion in Garnett 1992). Nevertheless, trapping continued until 1981, despite declining trapping returns over the previous decade (Blakers *et al.* 1984).

Some states still allow wild-trapped parrots to enter trade. Precise data on numbers are not available for most states; general indications of the scale of this trade in 1990 are summarized in Table 8. Current levels of trapping for trade are not known, although in New South Wales numbers of wild-caught birds legally entering trade have declined from the levels shown in

Table 7. Number and species of finches trapped for aviculture in Western Australia, 1974–1985. Source: Western Australian Department of Conservation and Land Management, in Anon. 1987b.

Scientific name	Common name	Number
Emblema picta	Painted Firetail	78
*Erythrura gouldiae**	Gouldian Finch	22 328
Lonchura castaneothorax	Chestnut-breasted Mannikin	25 164
*Lonchura flaviprymna***	Yellow-rumped Mannikin	129
Lonchura pectoralis	Pictorella Mannikin	23 071
Neochmia phaeton	Crimson Finch	12 310
Neochmia ruficauda	Star Finch	51 015
Poephila acuticauda	Long-tailed Finch	75 135
Poephila bichenovii	Double-barred Finch	24 561
Poephila personata	Masked Finch	30 669
Finch spp.	Unknown	1 500
TOTAL		265 960

*excluded from trapping in 1982.
**excluded from trapping in 1976.

Table 8 in recent years (J. Cook, pers. comm.). This trade seems ostensibly to be a by-product of pest control measures. Without a better indication of the number and location of birds taken, it is difficult to draw any conclusions about the impact this trade has on these "pest" populations. Certainly any moves to significantly expand this trade from wild-harvested birds should be accompanied by the development of more thorough management measures, including population monitoring and record keeping, by the relevant state conservation agencies.

Crocodiles

Crocodylus johnstoni is listed on Appendix II of the Convention on International Trade in Endangered Species of Wild Fauna and Flora (CITES), which allows for commercial international trade subject to the meeting of certain safeguards and the issuing of the appropriate permits. In 1985, the Australian population of *C. porosus* was downlisted from Appendix I of CITES (which prohibits commercial international trade) to Appendix II, pursuant to Resolution Conf. 3.15 on ranching. This means that trade from Australia in this species must come from either captive bred animals, or those which were taken as eggs or hatchlings from the wild and raised in a controlled environment (so-called "ranching"). At the time of writing, ANCA is considering presenting a proposal to the November 1994 meeting of the Conference of the Parties to CITES, which, if approved, would place *C. porosus* on Appendix II without the ranching restriction. This would open the way for the export of skins from adult animals killed in the wild.

Table 8. Estimated number and species of parrots and cockatoos wild-trapped annually for the trade as of 1990. Source: Antram 1990.

	State			
Species	NSW	SA	Vic.	WA*
Barnardius zonarius				2 860
Cacatua galerita	2 000			?
Cacatua roseicapilla	2 000	2 000	?	60
Cacatua sanguinea		2 000		?
Cacatua tenuirostris			??	
Platycercus elegans			?	
Platycercus eximus			?	
Platycercus icterotis				1 999
Purpureicephalus spurius				1 474
Trichoglossus haematodus				?

*data for 1989 and 1990.
?wild-trapped birds believed to enter trade — volume unknown.
??wild-trapped birds possibly entering trade — volume unknown.

The history of commercial crocodile farming and ranching in Australia goes back little more than 10 years. The Northern Territory is presently (by far) the largest supplier of crocodiles for commercial exploitation, with an industry still largely based on ranching. It is expected, however, that the production of captive-bred animals will increase over coming years with improved breeding techniques (Anon. 1992). The first skins from the Northern Territory were sold in 1987 and by 1992, production stood at around 3 000 skins of *C. porosus* (about 15% of total world trade in this species for that year).

The scale of the industry in the other two states involved in crocodile trade — Western Australia and Queensland — is considerably lower. Queensland currently has four crocodile farms, all operating as closed-cycle captive-breeding operations. Some wild-caught animals are transferred to farms as breeding stock, including problem animals, especially in Queensland. Only farm-bred animals are allowed to be slaughtered for trade (Anon. 1989). There are currently three crocodile farms in Western Australia; the first of these was established in 1989. Population levels of *C. porosus* in Western Australia are lower than those of the Northern Territory. Out of an estimated current adult population of a few thousand, between 1983 and 1992, 208 adults and 365 eggs/hatchlings were taken from the wild, mostly from the Cambridge Gulf river systems (Anon., undated). Surveys indicate that populations are coping with these levels of authorized collection. At the end of 1992, stock levels on all Western Australian crocodile farms combined were (hatchlings through to adults): 1 943 *C. porosus;* 2 421 *C. johnstoni.*

The domestic market for crocodile products is small; most items are exported. The major export item is skins (Table 9), with 74 per cent of skin exports between 1984 and 1992 coming from ranched (management plan) animals. Japan has imported 57 per cent of *C. porosus* skins and France 33 per cent over this period;

Table 9. Exports of crocodile products from Australia, 1984–1992 (MP = Management Programme/ranged; CB = captive bred). Source: Australian CITES Annual Reports.

		Crocodylus porosus				*Crocodylus johnstoni*				**Total both species**			
Year	Source	Skins — whole (no.)	Skins— parts*	Meat (kg)	M/factured articles**	Skins — whole (no.)	Skins — parts*	Meat (kg)	M/factured articles**	Skins — whole (no.)	Skins — parts*	Meat (kg)	M/factured articles**
1984	MP	0	0	0	0	0	0	0	0	0	0	0	0
	CB	100	0	0	0	0	0	0	0	100	0	0	0
	TOTAL	**100**	**0**	**0**	**0**	**0**	**0**	**0**	**0**	**100**	**0**	**0**	**0**
1985	MP	0	0	0	0	0	0	0	0	0	0	0	0
	CB	98	0	0	0	0	0	0	0	98	0	0	0
	TOTAL	**98**	**0**	**0**	**0**	**0**	**0**	**0**	**0**	**98**	**0**	**0**	**0**
1986	MP	0	0	0	0	0	0	0	0	0	0	0	0
	CB	400	0	0	0	0	0	0	0	400	0	0	0
	TOTAL	**400**	**0**	**0**	**0**	**0**	**0**	**0**	**0**	**400**	**0**	**0**	**0**
1987	MP	299	0	0	0	219	0	0	0	518	0	0	0
	CB	200	0	150	0	0	0	0	0	200	0	150	0
	TOTAL	**499**	**0**	**150**	**0**	**219**	**0**	**0**	**0**	**718**	**0**	**150**	**0**
1988	MP	1 048	0	0	0	1 240	0	0	0	2 288	0	0	0
	CB	200	0	0	0	0	0	0	0	200	0	0	0
	TOTAL	**1 248**	**0**	**0**	**0**	**1 240**	**0**	**0**	**0**	**2 488**	**0**	**0**	**0**
1989	MP	1 766	0	0	0	262	0	0	0	2 028	0	0	0
	CB	493	0	0	0	557	0	0	0	1 050	0	0	0
	TOTAL	**2 259**	**0**	**0**	**0**	**819**	**0**	**0**	**0**	**3 078**	**0**	**0**	**0**
1990	MP	1 367	0	160†	24	838	200	35	21	2 205	200	195†	45
	CB	1 463	50	20	11	0	0	0	0	1 463	50	20	11
	TOTAL	**2 830**	**50**	**180**	**35**	**838**	**200**	**35**	**21**	**3 668**	**250**	**215**	**56**
1991	MP	2 043	34	90	142	878	4	0	11	2 921	38	90	153
	CB	573	27	0	84	0	0	0	0	573	27	0	84
	TOTAL	**2 616**	**61**	**90**	**226**	**878**	**4**	**0**	**11**	**3 494**	**65**	**90**	**237**
1992	MP	2 260	205	591	173	1 495	16	15	0	3 755	221	606	173
	CB	1 129	0	180	46	159	180	0	0	1 288	180	180	46
	TOTAL	**3 389**	**205**	**771**	**219**	**1 654**	**196**	**15**	**0**	**5 043**	**401**	**786**	**219**
TOTAL	**MP**	**8 783**	**239**	**841**	**339**	**4 932**	**220**	**50**	**32**	**13 715**	**459**	**891**	**371**
	CB	**4 656**	**77**	**350**	**141**	**716**	**180**	**0**	**0**	**5 372**	**257**	**350**	**141**
	TOTAL	**13 439**	**316**	**1 191**	**480**	**5 648**	**400**	**50**	**32**	**19 087**	**716**	**1 241**	**512**

*includes back skins, horny backs, back straps and skin pieces, by number.
**includes handbags, belts, wallets, watch straps, key holders, purses, bags, hat accessories, cases, collars, boxes and jewellery, by number.
†includes 25 kilograms of unidentified origin.

for *C. johnstoni* skins, Singapore takes 63 per cent of exports and Japan 34 per cent. The volume of skin exports is gradually increasing, and in recent years an increasing number of other crocodile products were entering export trade, particularly manufactured goods such as belts and wallets.

Comprehensive data on the value of Australia's crocodile trade are not available. Data from the Australian Bureau of Statistics on the value of exports of reptile leather, hides and skins of Australian origin (in theory this should only include crocodile products) puts the value of the export trade at around $977,000 in 1990/91, increasing to $1,624,000 in 1991/92 and $1,723,000 in 1992/93. Other sources put the value of output from the six farms in the Northern Territory at $1.3 million in 1990/91, increasing to around $1.9 million in 1991/92 (Anon. 1992).

Flora — Western Australia

Australia's native cut flower industry is an expanding multi-million dollar one. Export value data for the last three years of available figures for wild-harvested plants (Table 10), largely illustrate this expansion. The largest importer of Australian cut wildflowers is Japan (Anon. 1993a).

Western Australia is by far the largest player in this industry, constituting 68 per cent of the country's total production (Anon. 1993a). The industry has undergone considerable expansion over the last 15 years. In 1981/82, cut flower exports from Western Australia (almost all of native species) were worth $1.5 million; current exports have grown to around $15 million annually (Anon. 1993a, 1993b).

In recent years there has been a trend in Western Australia — in some species at least — from harvesting of wild stands to cultivated plants. Currently, 200 growers cultivate some 570 ha planted with native species and cultivated stocks now make up 65 per cent of total export of cut wildflowers from the State (Anon. 1993a, 1993b). The Waxflower *Chamaelaucium uncinatum* is perhaps the prime example of the switch to cultivation: 132 000 single stems were wild-harvested in 1981 (Anon. 1993a); in 1992 3 962 single stems and 12 670 bunches of stems were wild-harvested, while 89 872 bunches of stems and 1 010 single stems were cultivated (Table 11).

Some 150 species of Western Australia flora are still wild-harvested (some of these are also cultivated). Table 11 shows the volume of harvest for some of those taxa taken in the greatest amounts. Reasonable data on the extent of the harvest are only available for 1992 and 1993. Pickers are required to provide returns to the Department of Conservation and Land Management but data for previous years has not been compiled. Obviously this is not an ideal situation if the trade is to be properly monitored; the recent trend to compilation of harvest data should be continued. Management plans for the Western Australia flora industry indicate that population assessment and monitoring occurs. Species specific data are not given in the management plan and we cannot therefore comment on the adequacy of these management measures.

Flora — other

Approvals for export of wild-collected Sphagnum Moss *Sphagnum* spp. were granted under the "exceptional circumstances" clause of the *Wildlife Protection Act* from 1986 to 1991 and, more recently, under the "controlled specimens" provisions of the Act. In 1989 exports stood at 10 590 kg, increasing to 14 071 kg in 1990, then dropping to only 7 296 kg in 1991; in 1992 there were no exports and only 12 kg in 1993 (ANCA data). (Note — there were no export approvals operative for all of 1992 and much of 1993). Details of harvest for domestic markets are not known. All export harvesting operations are in Victoria or Tasmania. Alpine bog communities, of which *Sphagnum* spp. is a key component, are listed as a threatened community under the Victorian *Flora and Fauna Guarantee Act* 1988. This notwithstanding, the Victorian Government endorsed the commencement of a commercial harvest and export operation for sphagnum. According to Whinam (1992) there is a growing interest in the commercial exploitation of Sphagnum Moss. But with as few as 300 ha of peatlands in Tasmania and probably less in other states, she raises concerns about the

Table 10. Amounts and values of wild-harvested cut flower exports, 1990/91–1992/93 (wt = weight; kg = kilograms; qty = quantity; no. = number) Source: Australian Bureau of Statistics.

	Wild-picked flowers and buds			Foliage, branches and grasses		Other wild-picked plants		
Year	Wt (kg)	Qty (no.)	Value ($)	Wt (kg)	Value ($)	Wt (kg)	Qty (no.)	Value ($)
1990/91	1 044 000	4 803 716	6 644 000	305 000	2 103 000	1 000	2 005	6 000
1991/92	1 063 000	4 735 649	5 834 000	233 000	1 275 000	51 000	8 580	73 000
1992/93	1 157 000	9 213 839	6 074 000	214 000	987 000	173 000	86 052	157 000
Total	3 264 000	18 753 204	18 225 000	750 000	4 365 000	225 000	96 637	236 000

impact of sphagnum harvesting and advocates the adoption of a more integrated and cautious management approach.

Australia is the world's fourth largest exporter of live tree ferns (Oldfield, undated); by far the bulk of Australian exports are of the Soft Tree Fern *Dicksonia antarctica.* Between 1987 and 1992 7 521 wild-collected live plants and 1 159 wild-collected stems of *D. antarctica* were exported, mainly to the Netherlands (Australian CITES Annual Reports). The export trade, however, pales in comparison to the domestic trade. In 1991 it was estimated that 20 000–40 000 specimens of *D. antarctica* were taken from private land in Victoria, while around a further 200 000 were imported from Tasmania (M. White, pers. comm.). Up to 50 000 propagated specimens (mostly the Rough Tree Fern *Cyathea australis* — this species is easily cultivated from spores and, unlike *Dicksonia,* does not transplant well) were also released to trade. It is estimated that some two million tree ferns a year are taken through salvage operations in logging areas in Tasmania (Braggins 1991; cited in Oldfield, undated). Although much smaller than the trade in live plants, tree fern stems are also utilized as blocks, poles, containers or shredded for growth media. In Victoria a system for tagging specimens taken from private land has been in place since 1977; declines in the number of tags issued has been construed as representing a depletion in tree fern numbers on private land (M. White, pers. comm.). There have been problems with issuing tags without checking on either the number of tree ferns on the property or the number harvested. Unused tags are rarely returned and consequently can be attached to illegally taken specimens.

Australia has a healthy domestic and export trade in native orchids. Most of this is in artificially propagated specimens, however, there is domestic trade in both legally and illegally collected wild specimens; there is currently no legal commercial export of wild-collected native orchids. The extent of the domestic trade in native orchids is not known. However, in the late 1980s there were 170 orchid societies around the country, 14 devoted exclusively to native orchid species (Jones 1988), which indicates the level of interest in keeping and growing native orchids within Australia. From 1990 to 1992 some 13 085 native orchid plants and seedlings were exported from Australia. These were of 163 different species from 31 genera; among those exported in the largest quantities were specimens of *Dendrobium* spp., *Diuris* spp., *Pterostylis* spp., *Sarcochilus* spp. and *Thelymitra* spp.

In addition to orchids, there are devotees to the horticulture of many other Australian plant types including ferns, cycads, palms and carnivorous plants. Details of the volume and value of these trades and the extent and impact of wild-collection are not readily available.

Table 11. Size of harvests of selected plant species in Western Australia in 1992 and 1993 (part), cultivated plants and wild harvested from Crown land and private property. (Quantity in bunches of stems and single stems). Source: Department of Conservation and Land Management, Western Australia.

Year		*Chamelaucium uncinatum*[1]			*Stirlingia latifolia*[2]			*Banksia* spp.[3]			*Agonis* spp.[4]		
		cultivated	wild — Crown land	wild — private land	cultivated	wild — Crown land	wild — private land	cultivated	wild — Crown land	wild — private land	cultivated	wild — Crown land	wild — private land
1992	bunches	89 872	2 917	9 753	1 160	73 542	38 584	65 198	30 255	23 248	1 431	118 373	112 165
	singles	1 010	345	3 617	0	3 400	0	126 216	1 350 024	526 756	0	7 480	1 386
1993*	bunches	49 758	0	11 700	0	88 756	18 158	0	71 170	64 819	16 211	155 043	0
	singles	2 520	0	0	0	10 070	0	171 477	1 523 061	1 300 403	0	1 470	170 777

*first three-quarters of the year only.
[1]plus 10 bags of stem from Crown land in 1993
[2]plus 500 bunches of unknown provenance in 1992
[3]plus 698 single stems of unknown provenance in 1993
[4]plus 590 bags of stems on unknown provenance in 1992; 3 000 kg of stems from private property in 1992; 0.45 kg of stems from Crown land in 1992; 1 500 bags of stems of unknown provenance in 1993

Other flora harvesting operations for overseas markets are expanding around the country, if the number of operations applying for export approval is any indication. Species harvested from the wild include *Adiantum formosum, Caustis blakei, Craspedia globosa, Nothofagus cunninghamii, Pteridium esculentum* and *Xanthorrhoea* spp. Table 12 shows the export volumes for some of these. Management and monitoring effort for flora harvests varies; for some it is virtually non-existent. It seems clear that conservation agencies will allow wild harvests of flora based on less information, monitoring and regulation, than what would be deemed acceptable for fauna. Having said this, however, some of the species being harvested are very widespread and common and are subject to ongoing clearing operations (e.g., *P. esculentum*). For such species it seems highly unlikely that harvest for trade at current levels will be negatively impacting populations.

Other

There is some legal domestic trade in live reptiles and amphibians; commercial export of live reptiles and amphibians is prohibited. Most domestic trade probably occurs between private herpetologists and would involve swapping and barter of specimens in addition to straight cash transactions. To the outside observer, the world of reptiles and amphibian keeping in Australia is something of a mysterious one, with data on the scale and nature of the trade difficult to obtain. As with birds, there is an ongoing illegal demand for live animals both domestically and overseas. Ehmann and Cogger (1985) estimate that 18 000 reptiles and amphibians are taken from the wild by amateur herpetologists each year — 1 700 to keep and 16 300 to feed captive animals. They do not consider this scale of collection to be having any significant ecological impact.

Freshwater fish are the only native vertebrate species for which live wild-collected specimens can be exported commercially. Between 1984 and 1989, approval was given for the export of a number of species of wild-collected freshwater fish from the Northern Territory under the exceptional circumstances provision of the *Wildlife Protection Act.* Over this period, export permits were issued for 161 759 fish of at least 49 different species — both wild-caught and captive bred (ANCA data). Exports dropped sharply between 1988 (30 179 fish) and 1989 (4 260), rising dramatically to 1 950 535 fish in 1991, 2 241 429 in 1992 and back down to 58 025 in 1993. Most of this increase was accounted for by captive-bred Silver Perch *Bidyanus bidyanus* 1991–1993: 4 092 266 fish). Other major export species were Golden Perch *Macquaria ambigua* and Australian Bass *M. novemaculeata* (1991–1993: 150 950) and Saratoga

Table 12. Scale of exports of certain wild-collected plant species, 1989–1993. (Extract in kilograms, all other figures by number). Source: ANCA unpublished statistics.

Year	*Xanthorrhoea* spp.				*Caustis* spp.			*Pteridium esculentum*			*Adiantum formosum*		
	Stems*	Live plants	Flowers	Extract	Stems*	Live plants	Flowers	Stems*	Live plants	Flowers	Stems	Fronds	Leaves
1989	23 071	0	0	0	574 129	0	0	230 364	13	0	30 000	160 000	0
1990	127 316	39 386	0	0	397 028	90 450	0	1 143	0	0	32 280	120 000	0
1991	796 191	0	1 016	0	221 777	0	4 415	7 912	0	390	40 000	63 750	0
1992	3 078 680	33	24	18 600	710 526	0	4 045	106 456	0	95	0	280 000	0
1993	7 433 179	25	300	302 337	1 413 705	0	50	35 040	0	0	30 000	40 000	40 000
TOTAL	11 458 437	39 444	1 340	320 937	3 317 165	90 450	8 510	380 915	13	485	132 280	663 750	40 000

*includes leaves and fronds.

Scleropages leichardti (1989–1993: 7 951). All these fish were captive-bred. Value of the trade stood at $550,000 in 1990–91, rising to $703,000 in 1991–92 and $1,216,000 in 1992–93 (Australian Bureau of Statistics (ABS) data). It should be noted that there are considerable differences between the volume of exports given by ABS and ANCA, therefore, these figures should be viewed as indicative only. Data on the scale of domestic trade are not available. In 1992 the Northern Territory Government applied to have a "controlled specimens" management plan approved for the export of wild freshwater fish; approval is still pending. According to this application, in 1991 there were 30 people licensed to take wild freshwater fish in the Northern Territory but only two of these were harvesting to any great extent (Department of Primary Industry and Fisheries, undated).

Only a few insect taxa are specifically protected by state legislation, therefore collection outside protected areas, and domestic sale, are to a large extent unregulated. Export of insects is covered by the provisions of the *Wildlife Protection Act* and there is a small legal commercial export trade, primarily in captive bred butterflies (Lepidoptera). A study by Hawkeswood *et al.* (1991) identified an escalating illegal export trade in Australian insects. It identified six species of Australian Lepidoptera and 82 species of Beetles (Coleoptera) for sale overseas, most of which appeared to have been illegally exported. These insects were being sold for prices ranging from a few dollars to over $1,000 a pair. It was suggested that while the trade did not appear to be of conservation concern for most of the species involved, it did appear to be increasing and warranted further monitoring.

Trade in Australian shells has been estimated to be worth $2–$2.5 million annually; 80 per cent of this is accounted for by the specimen shell trade and the remainder by the ornamental and other trades (Malacological Society of Australia, undated). Most of this trade is for export (Anon. 1994c). In 1991, permits were issued for the export of 36 799 shell specimens (Ley 1992). There are currently 48 approved shell exporters, mainly based in Western Australia (54%), Queensland (19%) and South Australia (15%). Since the introduction of the *Wildlife Protection Act* in 1984, shell exports have been authorized under the "exceptional circumstances" clause (Section 44) of the Act. Use of this clause to permit an ongoing trade is a clear misuse of this section of the legislation. Difficulty in developing management plans for marine molluscs has been the reason given for not bringing specimen shell exports under the same form of management programme provisions required for other wildlife exports. Moves to rectify this situation are currently under way. A management programme for Western Australia (Barrington and Stewart 1994) was approved by the State Fisheries Minister on 17 May 1994 (F. Antram, in litt.) and this plan and a draft plan for South Australia (Anon. 1994c) are currently being considered for approval by ANCA under the *Wildlife Protection Act.*

Australia also trades in live Giant Clams *Tridacnidae* spp. and Giant Clam products. These are all sourced from aquaculture operations and present no threat to wild populations. Indeed, captive-bred clams have been used to restock areas where wild populations are depleted. Of more conservation concern is an apparent ongoing illegal import trade in wild-taken Giant Clam shells from the Philippines into Australia (see Wells, in prep.).

Harvesting of the seaweeds *Durvillaea potatorum* and *Gelidium asperum* and Seagrass *Posidonia australis* occurs in Tasmania and South Australia respectively. Harvesting is of beach-cast or detached specimens only (Antram, in litt.; Cheshire 1991; Weston, in litt.). This is a multi-million dollar export industry, which is growing in value (1990/91 — $1,679,000; 1991/92 — $1,781,000; 1992/93 — $2,112,000 (ABS data)). The bulk of the export trade is in *D. potatorum* (1989–1993: 89 436 litres and 2 520 kg of extract; and 13 558 470 kg and 215 000 fronds of processed and unprocessed seaweed), followed by *P. australis* (1992–1993: 134 880 kg of leaves and stems; and 100 live plants), and *G. asperum* (1989–1993: 21 094 kg of fronds) (ANCA data). Harvest and management information for *P. australis* indicates that 100 per cent of available material will be harvested (Weston, in litt.). The impact of this level of harvesting on communities reliant on decomposing seagrass is not known.

Some aquatic taxa such as beche-de-mer *Holothuroidea* spp. and jellyfish *Cnidaria* spp. have their exports controlled under the *Wildlife Protection Act.* However, as these fall more logically under the category of fisheries they will not be considered here. Other wildlife items are exported in small quantities including sponges, snake venom, spider venom, locusts and so on.

Illegal trade

Attempts have been made to quantify the value of the illegal trade in Australian wildlife (e.g., $40 million annually — Bottom 1985), however, the basis for deriving such figures has never been clear. Indeed, given the nature of the trade — that is, illegal — it is impossible to put a figure on its scale that is anything more than a guess, or an educated guess at best.

Australia has a healthy, and perhaps even thriving, illicit trade in wildlife, both for domestic and international markets. A combination of factors has led to an ongoing illegal trade in Australian wildlife including: generally strict policies regarding wildlife trade, particularly for export of live wildlife; the high degree of endemicity of Australia's flora and fauna; and the difficulty of breeding certain species in captivity. Scarcity of supply, coupled with continuing demand, has resulted in some Australian species commanding high prices on domestic and international markets. This in turn has continued to make the illegal trade a financially attractive option for those prepared to take the risk.

The highest profile of these illegal trades is in birds and reptiles. Australian bird species which attract the most attention of illegal trappers, nest robbers and smugglers are cockatoos and to a lesser extent some parrot and finch species. Markets are both domestic and overseas. In Western Australia, for example, DNA-testing of blood samples from 30 Baudin's White-tailed Black Cockatoos *Calyptorhynchus baudinii* and six Red-tailed Black Cockatoos *C. magnificus* owned by nine aviculturists, indicated that only one breeder had actually bred the birds in captivity as claimed (Mell and Wetherall 1992). Other black cockatoos were also seized from Western Australian aviculturists after this action when they admitted taking them from the wild. In 1993, 21 Pink Cockatoos *Cacatua leadbeateri* and 10 Gang-Gang Cockatoos *Callocephalon fimbriatum* were seized in New Zealand, smuggled in on a light plane originating from Australia. In 1993, government authorities in the Netherlands seized 80 black cockatoos (van Kreveld, in litt.) (this probably also included gang-gangs); work by TRAFFIC indicated that there were no Gang-gang Cockatoos and only very few black cockatoos held in that country in 1991 (Antram and Salisbury 1991). Very recently, six people were charged in the United States of America over an alleged conspiracy to smuggle wild-collected eggs of Galahs *Cacatua roseicapilla,* Red-tailed Black Cockatoos, Pink Cockatoos and Long-billed Corellas *C. tenuirostris.* The operation was alleged to have involved several hundred eggs worth over US$1 million over a 10 year period (US Department of the Interior Press Release, 29 April 1994).

So what is the impact of this illegal trade? Morris (1994) suggests that illegal trade could be the principal factor in distribution contractions and population declines for the Eclectus Parrot *Eclectus roratus macgillivrayi,* Superb Parrot *Polytelis swainsonii,* Naretha Blue Bonnet *Northiella haematogaster narethae,* Golden-shouldered Parrot *Psephotus chrysopterygius,* Swift Parrot *Lathamus discolor,* Palm Cockatoo *Probosciger aterrimus aterrimus,* eastern subspecies of the Glossy Black Cockatoo *Calyptorhynchus lathami lathami* and Carnaby's Black Cockatoo *C. latirostris.* Other bird species for which illegal trapping has been suggested to pose a threat, at least for isolated populations, include the Black-breasted Button Quail *Turnix melanogaster* (Holmes 1987, cited in Garnett 1992), and subspecies of the Black-throated Finch *Poephila cincta cincta* and Crimson Finch *Neochmia phaeton evengelinae* (Garnett 1992). The difficulty is that baseline data on both population trends and the extent of illegal take, which would conclusively support or refute such claims, are generally lacking.

Australian reptiles are in demand for overseas markets and illegal exports are ongoing. There is also continuing illegal take from the wild for the domestic market, although this may be primarily for personal collections. Animals can survive being sent overseas using the postal system, a *modus operandi* that makes detection of illegal reptile trade difficult. Cases which have been detected indicate that much of the illegal reptile trade is in common species such as Shingleback Lizards *Trachydosaurus rugosus.* The Action Plan for Australian Reptiles (Cogger *et al.* 1993) lists two species for which one of the threatening processes is, or has been, illegal wild collection — Mary River Tortoise *Gen. nov. sp. nov.* (Endangered) and Broad-headed Snake *Hoplocephalus bungaroides* (Vulnerable). Illegal collecting of some snake and amphibian species may be responsible for local population declines in parts of the Queensland Wet Tropics Region (Werren 1993).

Other animals are also subject to unauthorized trade. There are ongoing rumours regarding smuggling of marsupial species such as Koalas — evidence to corroborate such claims is lacking. The thriving illegal export trade in insects has already been discussed. Recently a man was prosecuted for attempting to illegally export 32 specimen shells, including rare species absent from Australian museum collections (Hider 1994).

Plants are also the target of illegal collectors. Unauthorized harvesting of the seeds of Cape Melville Palms *Wodyetia bifurcata,* a species endemic to Cape Melville National Park in Queensland, has recently attracted considerable publicity. Tens of thousands of seeds of the palm have been seized over recent years. Illegal operations are believed to involve organized gangs who sell the seeds both interstate and overseas (Anon. 1993c). Unauthorized collecting of orchids, ferns, palms, cycads and Ant Plants *Myrmecodia* sp. in the Queensland Wet Tropics World Heritage Area is believed to be threatening populations of some species (Nias *et al.* 1993). These include *Lycopodium dalhousianum, Oberonia attenuata, Phalaenopsis amabilis* and *Vrydagzynea paludosa;* other threatened plants from the region likely to be of interest to collectors include *Gardenia actinocarpa, Macadamia* sp. "Woopen Creek", *Toechima pterocarpum* and *Xanthostemon formosus.*

The major conservation impacts of illegal collecting from the wild are likely to be localized. Participants in such activities are known to revisit collection sites year after year. For example, bird smugglers returning to the same set of nesting trees, or seed harvesters to the same mature plants. If such activities continue over a long period of time, there would appear to be potential for localized impacts on population recruitment. However, our knowledge about both populations of targeted species and the scale of illegal harvest are, in most cases, insufficient to make any firm judgements about the conservation impact of illegal trade. Nevertheless, illegal trade has been shown to have a detrimental effect on the populations of many species outside of Australia, therefore the potential for the same in Australia cannot and should not be discounted.

CONCLUSIONS

The trade in wild Australian flora and fauna is a multi-million dollar business. Available (incomplete) data indicate a legal domestic and international trade in wild plants and animals and their products approaching $100 million in value annually. The true value is higher — how much higher is difficult to determine. Information on the size of the domestic trade is especially scarce. Export data are available but are not comprehensive and quite probably also not accurate. On top of this is an ongoing illegal trade — both domestic and international — which cannot be quantified.

The trade is increasing. The number of applications for export received by ANCA grew by 54 per cent over the two years July 1989 to June 1991 (Ley 1992). Interest in commercial exploitation of wild flora and fauna is growing around the country. Certain state and territory governments such as Queensland and the Northern Territory are examining their options for increasing utilization of wild species, including both wild harvests and farming. A recent meeting of the Australian and New Zealand Environment and Conservation Council agreed on a more active role for its Task Force on Sustainable Use of Wildlife.

While on the one hand there is the pro-utilization lobby, on the other there is an increasingly vocal band of (mostly) non-government organizations who take a fairly tough anti-trade line. This lobby has mainly been active overseas, however stirrings of it are starting to become evident in Australia. If Australia continues the push towards greater utilization of wildlife and especially if this involves mammal, bird or reptile species, it is likely that the debate will escalate and probably draw the attention of animal welfare interests from outside the country. The power and influence of these organizations should not be underestimated and they have mounted very successful campaigns — against the wild-caught bird trade for example — in countries such as the United Kingdom and the United States of America.

So what are the implications for Australia if existing trends of increasing interest in utilizing native wildlife for commercial gain, coupled with increasing debate on the merits of such activities, continue? Debate in and of itself is unlikely to have much impact on biological diversity conservation unless it results in policy changes. However, increasing trade will lead to an increased burden of administration, monitoring, research and enforcement if the trade is to be controlled at levels which ensure no negative conservation impact. To date, expansion of the trade does not seem to have been accompanied by any increase in management resources. This trend needs to be watched carefully to ensure that sufficient resources are directed towards managing the trade to ensure that harvests remain within safe limits. Part of this burden should fall on the trade itself. Those benefiting financially from the wildlife trade should be made more accountable in helping to provide the necessary resources to ensure that Australia's wildlife trade is carried out on a legal and sustainable basis.

The current reality is that firm conclusions about the sustainability or otherwise of Australia's legal wildlife trade can only be made for a few of the species exploited. These are the long-standing harvests such as kangaroos, possums and muttonbirds, which have received considerable management investment over the years and also been subject to external scrutiny and examination. With many of the other harvests it appears that sufficient data are not available to make scientifically justifiable determinations about their sustainability. For many, however, the scale of the trade and the nature of the taxa involved suggest that utilization is unlikely to be having a detrimental effect on the overall population, although localized impacts are possible.

Illegal trade is more problematic. The fact that by its very nature it is uncontrolled means that conservation safeguards can never be incorporated into illegal harvests. Another concerning hallmark of illegal trade is that there is always a demand for that which is hard to obtain. With Australia's strict laws this sometimes leads to a demand for common species. However, there is nearly always an underlying market for less common species, a demand which increases once markets for common species are filled. With some species, the rarer it becomes, the greater is the demand, a pattern which can be disastrous, especially when species are also battling other threatening processes.

At the moment, however, there is no information to suggest that any Australian species is threatened with extinction due to illegal trade.

Given the recent interest in expanding wildlife trade within and from Australia, plus a general trend in that direction anyway, it would seem to be an opportune time for Australia to thoroughly overview its policies and practices in this area. Current limited resources mean that important, and increasingly potentially controversial, decisions about the utilization of Australia's wildlife are being made by the small number of government officials charged with monitoring and controlling the trade. A thorough and open review would, hopefully: make the management of wildlife trade more transparent; foster more public debate on the issue; and help answer some of the many unanswered questions about the true nature, extent and future of wildlife trade in and from Australia.

ACKNOWLEDGEMENTS

Completion of this paper would not have been possible without the patience and perseverance of David Keith, NSW National Parks and Wildlife Service. Extensive data on the Western Australian wildflower trade was kindly provided by Sarah McEvoy, Department of Conservation and Land Management. Robert Moore (ANCA) forwarded a considerable volume of ANCA export data in a very timely fashion and Cindy Steensby (ANCA) assisted with missing kangaroo harvest statistics. Peter Dickinson, Murray Donaldson, Mike Sverns, Ron Waters and Matt White (Victorian Department of Conservation and Natural Resources), and John Cook (NSWNPWS) also provided useful advice and information. A financial contribution to support the completion of this paper was provided by the NSW National Parks and Wildlife Service and is gratefully acknowledged.

REFERENCES

Anonymous (undated). Management Program for the Saltwater Crocodile *Crocodylus porosus* and the Freshwater Crocodile *Crododylus johnstoni* in Western Australia. Western Australian Department of Conservation and Land Management, unpublished paper.

Anonymous, 1985. Australian Submission to C.I.T.E.S. on *Crocodylus porosus*. Proposal submitted to Resolution Conference 3.15 on Ranching, fifth meeting of the Conference of the Parties to CITES.

Anonymous, 1987a. Management of the Short-tailed Shearwater *(Puffinus tenuirostris)* in Tasmania. Department of Lands, Parks and Wildlife, unpublished paper.

Anonymous, 1987b. Finch trapping in W. Australia to cease? *Traffic Bulletin* **VIII**(4): 71.

Anonymous, 1989. Crocodile Management Plan 1989. Queensland National Parks and Wildlife Service, unpublished paper.

Anonymous, 1992. Northern Territory Crocodile Industry Strategy. Unpublished paper.

Anonymous, 1993a. Floriculture a blooming business. *WA J. Agri.* **34:** 87–91.

Anonymous, 1993b. Commercial Flora Harvesting. CALM Policy Statement No. 13, Department of Conservation and Land Management: Como.

Anonymous, 1993c. Seizures and prosecutions — State, Queensland. *TRAFFIC Bulletin,* **13**(3): 109.

Anonymous, 1994a. The Farming of Protected Wildlife in Queensland. Unpublished policy options paper. Queensland Department of Primary Industries.

Anonymous, 1994b. The Short-tailed Shearwater Management Program in Tasmania. Parks and Wildlife Service, Department of Environment and Land Management, unpublished paper.

Anonymous, 1994c. Specimen Shell Fishery Management Plan. Department of Primary Industries South Australia — Fisheries, draft unpublished report.

Antram, F., 1990. Bird trade politics in Australia. Paper presented at the ICBP XX World Conference, November 1990, unpublished.

Antram, F. and Salisbury, R. (eds), 1991. International Trade in Australian Parrots. Consultancy report to the Australian National Parks and Wildlife Service, unpublished report.

Backhouse, J., 1843. *A narrative of a visit to the Australian colonies.* Hamilton, Adams and Co.:, London.

Bannister, J. L., 1989. Balaenidae. Pp. 988–92 in *Fauna of Australia. Volume 1B. Mammalia* ed by D. W. Walton and B. J. Richardson. Australian Government Publishing Service: Canberra.

Barrington, J. H. S. and Stewart, G., 1994. Management Arrangements for Specimen Shell Collection in Western Australia. Fisheries management paper No. 62. Fisheries Department of Western Australia: Perth.

Begg, A. C. and Begg, N. C., 1979. *The world of John Boultbee.* Whitcoulls: Christchurch.

Blakers, M., Davies, S. J. J. F. and Reilly, P. N. (eds), 1984. *The atlas of Australian birds.* Melbourne University Press: Carlton.

Bottom, B., 1985. *Connections.* Sun Books: South Melbourne.

Bryden, M. M., 1983. Southern Elephant Seal. Pp. 467–68 in *Complete book of Australian mammals* ed by R. Strahan. Angus and Robertson: Sydney.

Callister, D. J., 1991a. A review of the Tasmanian brushtail possum industry. *TRAFFIC Bulletin* **12**(3): 49–58.

Callister, D. J., 1991b. Exploitation of the Short-tailed Shearwater in Tasmania. *TRAFFIC Bulletin* **12**(1/2): 5–11.

Cheshire, A., 1991. A Management Program under the Wildlife Protection (Regulation of Exports and Imports) Act 1982 for the Harvesting of Macro-algae from Coastal Locations on King Island and North-west Tasmania. Draft unpublished paper.

Cogger, H. G., Cameron, E. E., Sadlier, R. A. and Eggler, P., 1993. The Action Plan for Australian Reptiles. Australian Nature Conservation Agency: Canberra.

Cott, H. B., 1953. The exploitation of wild birds for their eggs. *Ibis* **95:** 409–49.

Department of Primary Industry and Fisheries (undated). Management of Freshwater Fishes in the Northern Territory. Application for the listing of certain species of Northern Territory freshwater fish as controlled specimens under the ANPWS Wildlife Protection (Regulation of Imports and Exports) Act, unpublished.

Dixon, J. M., 1989. Thylacinidae. Pp. 549–59 in *Fauna of Australia. Volume 1B. Mammalia* ed by D. W. Walton and B. J. Richardson. Australian Government Publishing Service: Canberra.

Ehmann, H. and Cogger, H., 1985. Australia's endangered herpetofauna: a review of criteria and policies. Pp. 435–47 in *Biology of Australasian frogs and reptiles* ed by G. Grigg, R. Shine and H. Ehmann. Surrey Beatty & Sons: Chipping Norton.

Frith, H. J., 1973. *Wildlife conservation.* Angus and Robertson: Melbourne.

Garnett, S. (ed), 1992. Threatened and Extinct Birds of Australia. RAOU Report 82, Royal Australasian Ornithologists Union, Moonee Ponds and Australian National Parks and Wildlife Service: Canberra.

Grant, T. R., 1989. Ornithorhynchidae. Pp. 436–50 in *Fauna of Australia. Volume 1B. Mammalia* ed by D. W. Walton and B. J. Richardson. Australian Government Publishing Service: Canberra.

Groombridge, B., 1982. *The IUCN amphibia-reptilia red data book. Part 1: Testudines, Crocodylia, Rhynchocephalia.* IUCN: Gland.

Groombridge, B. (ed), 1993. *1994 IUCN red list of threatened animals.* IUCN: Gland and Cambridge.

Halstead, B., 1992. Traffic in Flora and Fauna. Trends and Issues in Crime and Criminal Justice No. 41, Australian Institute of Criminology: Canberra.

Hawkeswood, T., Callister, D. J. and Antram, F., 1991. Collection and export of Australian insects: an analysis of legislative protection and trade to Europe. *TRAFFIC Bulletin* **12**(3): 41–48.

Hider, J., 1994. Fisherman fined for shell smuggling. *The City Messenger,* p.3.

Hocking, G. J., 1988. Management Programme for Brush Possum *Trichosurus vulpecula* (Kerr) in Tasmania 1989. Department of Lands Parks and Wildlife, Hobart, unpublished.

Hocking, G. J., 1989. Management Programme for Brush Possum *Trichosurus vulpecula* (Kerr) in Tasmania 1990. Department of Parks, Wildlife and Heritage, Hobart, unpublished.

Johnson, K. A. and Walton, D. W., 1989. Notoryctidae. Pp. 591–602 in *Fauna of Australia. Volume 1B. Mammalia* ed by D. W. Walton and B. J. Richardson. Australian Government Publishing Service: Canberra.

Jones, D. L., 1988. *Native orchids of Australia.* Reed Books: Frenchs Forest.

Kennedy, M., 1992. *Australasian marsupials and monotremes: an action plan for their conservation.* IUCN: Gland.

King, J. E., 1989. Otariidae and Phocidae. Pp. 1012–19 in *Fauna of Australia. Volume 1B. Mammalia* ed by D. W. Walton and B. J. Richardson. Australian Government Publishing Service: Canberra.

Ley, J. F., 1992. *Australia's protection and conservation of wildlife.* Australian Government Publishing Service: Canberra.

Malacological Society of Australia. (undated). The Sea Shell Trade in Australia. Unpublished report.

Marlow, B. J., 1983. Australian Sea-lion. Pp. 460 in *Complete book of Australian mammals* ed by R. Strahan. Angus and Robertson: Sydney.

Marsh, H., 1989. Dugongidae. Pp. 1030–38 in *Fauna of Australia. Volume 1B. Mammalia* ed by D. W. Walton and B. J. Richardson. Australian Government Publishing Service: Canberra.

Mell, D. and Wetherall, J., 1992. To catch a thief. *Landscope* **7**(4): 28–32.

Morris, C., 1994. Trafficking in Native Birds. RAOU Conservation Notes 14, Royal Australasian Ornithologists Union: Moonee Ponds.

Nias, R. C., Kennedy, M., Reynolds, C. and Habel, S., 1993. A Regional Action Plan for the Conservation of Threatened Species and Ecological Communities in the Wet Tropics of Queensland World Heritage Area. Unpublished report to the Wet Tropics Management Authority.

Oldfield, S. (undated). Significant Trade in CITES Appendix II Plants: Tree Ferns. Draft unpublished report prepared for the CITES Plants Committee meeting, 26–28 October 1992, Chiang Mai, Thailand.

Poole, W. E., 1984. Management of Kangaroo Harvesting in Australia (1984). Occasional Paper No. 9, Australian National Parks and Wildlife Service: Canberra.

Ramsay, B. J., 1994. Commercial Use of Wild Animals in Australia. Australian Government Publishing Service: Canberra.

Robinson, D. and Brouwer, J., 1989. Legislative Protection to Birds in Australia and its Territories. RAOU Report No. 62, Royal Australasian Ornithologists Union: Moonee Ponds.

Russell, E. M., Lee, A. K. and Wilson, G. R., 1989. Natural history of the Metatheria. Pp. 505–26 in *Fauna of Australia. Volume 1B. Mammalia* ed by D. W. Walton and B. J. Richardson. Australian Government Publishing Service: Canberra.

Russell, S. and Mills, J., 1993. *TRAFFIC project portfolio.* WWF International: Gland.

Seebeck, J. H. and Rose, R. W., 1989. Potoroidae. Pp. 716–39 in *Fauna of Australia. Volume 1B. Mammalia* ed by D. W. Walton and B. J. Richardson. Australian Government Publishing Service: Canberra.

Skira, I. J., 1987. Socio-economic aspects of muttonbirding in Tasmania, Australia. *ICBP Tech. Publ.* **6:** 63–75.

Skira, I. J., Wapstra, J. E., Towney, G. N. and Naarding, J. A., 1986. Conservation of the short-tailed shearwater, *Puffinus tenuirostris,* in Tasmania, Australia. *Biol. Conserv.* **37:** 225–36.

Strahan, R. (ed), 1983. *Complete book of Australian mammals.* Angus and Robertson: Sydney.

Thomson, J. M., Long, J. L. and Horton, D. R., 1987. Human exploitation of and introductions to the Australian fauna. Pp. 227–49 in *Fauna of Australia. Volume 1A. General articles* ed by G. R. Dyne and D. W. Walton. Australian Government Publishing Service: Canberra.

Thorbjarnarson, J., 1992. *Crocodiles: an action plan for their conservation.* IUCN: Gland.

Thornback, J. and Jenkins, M. (eds), 1982. *The IUCN mammal red data book. Part 1: threatened mammalian taxa of the Americas and the Australasian zoogeographic region (excluding Cetacea).* IUCN: Gland.

Warneke, R. M., 1983. Australian Fur-seal. Pp. 462 in *Complete book of Australian mammals* ed by R. Strahan. Angus and Robertson: Sydney.

Webb, G. J. W., Whitehead, P. J. and Manolis, S. C., 1987. Crocodile management in the Northern Territory of Australia. Pp. 107–24 in *Wildlife management: crocodiles and alligators* ed by G. J. W. Webb, S. C. Manolis and P. J. Whitehead. Surrey Beatty & Sons: Chipping Norton.

Wells, S. M. (in prep.). Giant Clams: Status, Trade and Mariculture, with Special Reference to the Role of CITES in Management. IUCN/Species Survival Commission: Gland and Cambridge.

Werren, G. L., 1993. Conservation strategies for rare and threatened vertebrates of Australia's wet tropics region. *Mem. Qld Mus.* **34**(1): 229–41.

Whinam, J., 1992. Tasmanian sphagnum peatlands. *Aust. Ranger* **25:** 17–18.

Wilson, G., McNee, A. and Platts, P., 1992. *Wild animal resources: their Use by Aboriginal communities.* Australian Government Publishing Service: Canberra.

Winter, J. W., 1979. The status of endangered Australian Phalangeridae, Petauridae, Burramyidae, Tarsipedidae, and the koala. Pp. 45–59 in *The status of endangered Australasian wildlife* ed by M. J. Tyler. Royal Zoological Society of South Australia: Adelaide.

The practical application of population viability analysis for conservation planning

H. P. POSSINGHAM[1]

Population Viability Analysis (PVA) is the only tool for quantitatively assessing the effect of land management and planning on the persistence of populations. There are many misconceptions about the utility and application of PVA. This confusion leads some wildlife managers to avoid using this as a tool for conservation and land management.

In this paper I do not review the field of PVA, instead I illustrate its application with three examples and discuss its utility and some of the ideas concerned with the viability of populations. I emphasize the use of PVA for species that are not in imminent danger of extinction, the kinds of data required to carry out PVA, and the educational value of PVA.

INTRODUCTION

MANY individuals, groups and government inquiries have called for a national commitment to the preservation of biodiversity (Biodiversity Advisory Committee 1992). Population Viability Analysis (PVA) is a tool that is playing an increasing role in moves to preserve biodiversity (Resource Assessment Commission 1992). Despite an increasing global awareness of PVA there are many misconceptions about what it can and cannot contribute to the sustainable management of natural resources. In this paper I discuss some of the crucial concepts, like population viability and minimum viable population, and illustrate the application of PVA using brief studies of Leadbeater's Possum *Gymnobelidius leadbeateri*, Southern Brown Bandicoot *Isoodon obesulus* and Greater Glider *Petaurus volans*. The examples are intended to demystify the PVA approach. In particular I argue that:

i) PVA does not necessarily require complex computer simulation models — various methods can be used, including experiments, observation and by analogy with similar species.

ii) PVA can be applied to any population in danger of **local** extinction; it is not only for rare and threatened species.

iii) PVA can be used to quantify the impact of resource use and habitat modification on the viability of populations. Environmental Impact Statements that are intended to estimate the impact of a development on a vulnerable population should use PVA; there is no other way for assessing such an impact.

iv) It is not necessary to have an enormous amount of data to assess the viability of a population. Although population viability analysis can always be improved with more data, analyses with even minimal information (including a sensitivity analysis) are educational and useful for guiding management and future research.

v) Some mathematical models can be solved analytically to deduce the viability of populations, and in particular minimum viable populations (e.g., Goodman 1987). These have been educational and are of great theoretical interest but they are not useful for specific application. Viability analyses will be different for every species and every situation. Only after many studies have been done may it be possible to make generalizations about the viability of populations. Simple rules like "500 is a minimum viable population size" have served to show that we need at least hundreds of individuals of every species, but they are of limited applied value (Soulé 1987).

Numerous authors have recently reviewed PVA (Burgman *et al.* 1988; Shaffer 1990; Possingham 1991; Akçakaya 1992; Boyce 1992; Burgman *et al.* 1993; Lindenmayer *et al.* 1993b). Therefore, the purpose of this paper is not to review PVA but to discuss some of the misconceptions about the process and broaden our perspectives on where this tool can be useful.

WHAT IS A VIABLE POPULATION?

The idea of a viable population (or minimum viable population) can be dangerous because it implies that such populations will not become extinct. A minimum viable population is merely the smallest population that is viable, although it is worth noting that the viability of a population will depend on more than its size (Nunney and Campbell 1992). Numerous properties of a population, like age structure, spatial distribution and genetic make-up, affect population viability (Burgman *et al.* 1993).

[1]Department of Applied Mathematics, The University of Adelaide, Adelaide, South Australia, Australia 5005.

Pages 292–99 *in* CONSERVING BIODIVERSITY: THREATS AND SOLUTIONS, ed by R. A. Bradstock, T. D. Auld, D. A. Keith, R. T. Kingsford, D. Lunney and D. P. Sivertsen. Surrey Beatty & Sons. 1995.

If a wildlife manager claims that a certain park contains a viable population of a particular species then we may be lulled into believing that we do not have to worry about conserving other populations of that species. This sense of security is false because there is no such thing as a population that will never become extinct, and the value of each population needs to be seen in the context of not just population viability, but also the "long-term evolutionary potential of a species". (The long-term viability of a population reflects its ability to evolve in response to environmental changes like, new diseases and climate change, and also its ability to evolve into new species.) Indeed most local populations of animals are vulnerable to catastrophic and rare events, which by their very nature are difficult to predict. Multiple viable populations are usually essential for long-term persistence.

It is worth remembering that even in the very short term every population has a small but finite probability of becoming extinct. Consider for example the human population. Suppose we each have a one in a million chance (1 in 10^6) of dying an independent death in the next 24 hours, and suppose there are about 5.5 billion people in the world, then there is a one in $10^{33\,000\,000\,000}$ chance there will be no people alive tomorrow. Although this probability is small, it is finite.

The term viable population is used too freely. Wildlife managers and conservation biologists frequently refer to whether or not populations are viable. For example conservation agencies and groups often state that their objective is to maintain viable populations of all species, while groups concerned with using resources (forestry, mining, agriculture) may argue that their operations have no impact on the viability of a species. But the term, viable population, has no meaning outside the context of a time frame and a probability. Shaffer's (1981) original definition of a viable population is a population that has a 99 per cent chance of persisting for over 1 000 years. Other authors use different definitions. Regardless of the definition most statements about whether a population is viable or not are not substantiated but are merely subjective guesses made without the framework of an explicit definition.

Misconceptions concerning the term "viable population" could be eliminated if we made use of the term in a relative sense. For example, it is always true that killing some animals in a population or removing suitable habitat decreases the viability of that population. These statements can be made without knowing the absolute probability of population extinction. Consequently the objective of a group responsible for a threatened species should be to "increase or maintain the viability" of a population, not "create a viable population". The impact of resource use or habitat modification on a population should be measured in terms of its relative effect on a population's viability, i.e., does it result in a significant increase in the probability of extinction and is that increase acceptable?

WHAT IS POPULATION VIABILITY ANALYSIS AND WHY SHOULD WE DO IT?

PVA is a process in which the likelihood that a population will become extinct is assessed, within a specified time and under particular circumstances (Shaffer 1981). This process usually involves the use of mathematical models that are explored using a computer simulation. However, there are other methods for assessing population viability.

Many authors have used data on habitat patch or island occupancy, invariably using bird species, to make inferences about the viability of populations (for example Pimm *et al.* 1988; Gibbs and Faaborg 1990; Cutler 1991). For species with short generation times we can use an experimental approach to assessing population viability by constructing isolated populations and monitoring their properties (see for example Borlase *et al.* 1993). Although these observational and experimental studies usually refer to common species, inferences can be made about the viability of populations of more threatened species that have similar life-history traits. For example Walter (1990) found that a small (15–20 pairs) stable population of the Red-tailed Hawk *Buteo jamaicensis socorrensis* has survived on Socorro Island for at least 50 years, probably considerably longer. By analogy it is tempting to infer that other raptor species with a similar life-history need only this number of pairs to form a population that has a reasonably low probability of extinction.

PVA by analogy with empirical observations of another species has advantages and disadvantages. Empirical data on population viability are more compelling in some sense, as they do not include all the assumptions of a simulation model, however differences between species and locations may make the analogy tenuous, while none of these studies enable us to make inferences about the viability of entire metapopulations.

Regardless of how we do a PVA, the assumptions behind an assessment of population viability must be explicit, and it is only responsible to highlight circumstances or conditions under which the analysis is poor.

For detailed management or assessment of impacts rough analyses are inadequate. With a relatively small input of time and money, one can use an existing mathematical simulation model (e.g., VORTEX, Lacy 1993; RAMAS-Space,

Akçakaya and Ferson 1992; ALEX, Possingham *et al.* 1992) or in special circumstances, construct a model specific to a population and location. The last option is preferable but can take several months of full time work. In the examples below I have used an existing model (ALEX, Possingham *et al.* 1992) that is suited to modelling vertebrates in spatially complex environments where catastrophes, like fire, are important. The parameter estimation and simulation process takes about a week in each case. Collaboration between the mathematical modeller and wildlife experts is essential in all cases.

Although PVA has probably been carried out on over 100 populations (Boyce 1992), the details of many studies are published in literature that is generally not accessible, while many other studies remain unpublished. Some overseas examples of the use of PVA are: work by Shaffer (1983) on the Grizzly Bear *Urus arctos horribilis*, several papers and reports on the Northern Spotted Owl *Strix occidentalis caurina* (e.g., Marcot and Holthausen 1987; Lande 1988; Doak 1989; Thomas *et al.* 1990) and many reports and papers that use VORTEX to assess the viability of small populations (Lacy 1993). In Australia PVA is being increasingly used (Lacy and Clark 1990; Burgman and Lamont 1992; Possingham and Noble 1991; Lindenmayer and Lacy 1993; Lindenmayer *et al.* 1993a).

Despite all these studies of population viability, only a few explicitly use PVA to choose between different management options (two recent examples are Armbruster and Lande 1993; Haig *et al.* 1993). Determining the probability that a species will become extinct is an interesting academic exercise that is useful for ranking threatened species lists and thereby directing conservation effort, but has little other value for resource management and applied conservation biology. The most celebrated example of using PVA for guiding resource use decisions is that of the Northern Spotted Owl. Thomas *et al.* (1990) use PVA, to choose a number of habitat conservation areas that support up to 20 pairs each, and ensure landscape management practices that reduce dispersal mortality (a significant threatening process for this population).

It is a common misconception that PVA is only intended for threatened species. Although the most widely used package, VORTEX (Lacy 1993), focuses on those species that are on the brink of extinction, PVA can be applied to any population, whether that population represents all the individuals of a species, or whether the population is simply one of many populations for that species. In two of the examples below we apply PVA to species that are locally threatened, but are not in immediate danger of global extinction. Used in this way PVA can help us to maintain species throughout their normal range, not just the last one or two populations. This is critical to the long-term objectives of preserving both species diversity and within species diversity.

Finally we need to remind ourselves that PVA gives estimates of the *probability* of extinction, it does not tell us when the species will become extinct. For example if a PVA suggests that a population has a 90 per cent chance of becoming extinct in the next 50 years, then there is still a 10 per cent chance that it will be around in 50 years time and its persistence over 50 years is not evidence that the model has failed. No model is exactly correct, it is only a tool that educates us, adds rigour to our thinking, helps us to choose between management options, and allows us to extrapolate our knowledge over large spatial and temporal scales. During those 50 intervening years we should be continually upgrading the model as further research identifies new processes and refines estimates of key parameters.

EXAMPLES OF THE APPLICATION OF PVA

The examples presented here are intended to highlight some of the questions that PVA can be used to answer.

Assessing the value of parcels of habitat

A very useful exercise for exploring management options for a species is to determine the relationship between size of a patch of habitat and the probability of extinction in that single patch. Although most species of interest occur in many patches of habitat, (they exist as a metapopulation) this relationship provides a rough answer to the question — What is the size of an area of protected habitat that makes a useful contribution to the viability of the species?

In Figure 1 the relationship between patch size and probability of extinction over 200 years for Leadbeater's Possum *Gymnobelideus leadbeateri* in old growth forest in the absence of fire is shown. A detailed description and justification for the parameters and model used to generate this graph is in Lindenmayer and Possingham (1995) and will not concern us further. The crucial issue is that the sigmoidal shape of the curve is typical. This means that there tends to be a relatively narrow range of patch sizes over which the probability of extinction changes from being almost certain, to being relatively small.

Interpreting Figure 1, first note that the minimum size of habitat necessary for a viable population (using Shaffer's definition of 99% chance of persisting 1 000 years) is about 250 ha. Second, there is a rapid decline in viability around 50 ha. Habitat patches bigger than 100 ha

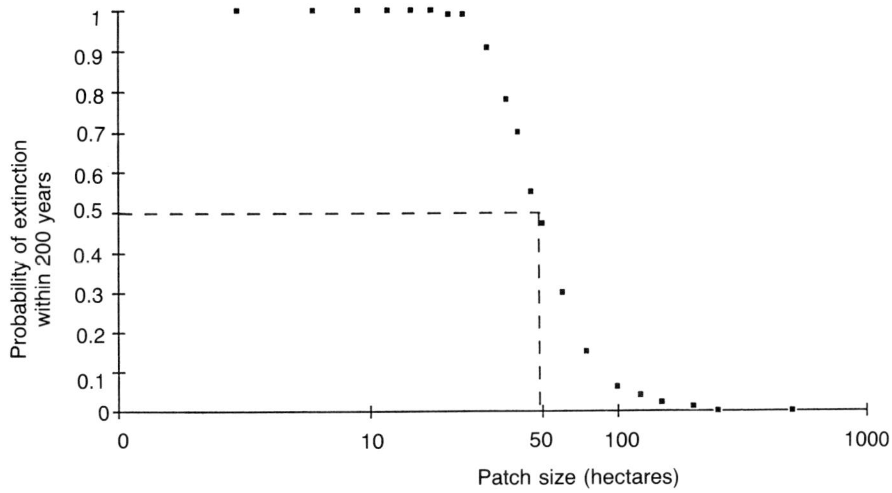

Fig. 1. The relationship between the area of a single patch of old-growth forest and the probability of extinction over 200 years of a population of Leadbeater's Possum *Gymnobelideus leadbeateri* in the absence of fire.

are likely to sustain a population for a significant period of time, and in association with other patches, would make an important contribution to population viability. Patches less than 20 ha make little contribution and might only be useful if associated with numerous other nearby patches. Because wildfire makes habitat unsuitable and kills possums, for long-term population viability we need a number of patches of habitat, each with low probabilities of extinction in the absence of fire. Such patches must be distributed so that the probability all patches are burnt simultaneously is extremely small. Once patches are burnt there needs to be the possibility of natural recolonization.

This kind of analysis might lead us to recommend that a conservation strategy for Leadbeater's Possum should focus on preserving and/or restoring a large number of old-growth forest patches that are bigger than 100 ha. Although this analysis is preliminary, it already provides insights that are new to the conservation of this threatened species.

Fire management for a mid-successional species

The Southern Brown Bandicoot *Isoodon obesulus* appears to favour mid-successional habitat in southeastern South Australia. Although fires kill some individuals, and habitat immediately after fire is not suitable, habitat that is not burnt for a long time has low quality. Only at the mid-successional stage, when the heath understorey is dense, does habitat quality reach its peak (Heinsohn 1966, Paull 1992). Paull's (1992) work suggests that patch quality after a fire is low, from where it increases to a peak between five and seven years, after which it slowly declines to a low level (Fig. 2).

Many of the remaining patches of land in which the species persists in south-east South Australia are managed by the Woods and Forests Department. They control burn some of the habitat and are interested to find the fire interval that minimizes the likelihood of local bandicoot extinctions.

Although the Southern Brown Bandicoot is not globally endangered, the species has disappeared or declined over much of its range. In the south-east of South Australia it occurs in about 30 patches of remnant vegetation ranging from 5 ha to 1 500 ha (Paull 1992).

One of the most critical patches of habitat for the species is Honans forest block. The total area of 1 000 hectares is divided into about 25 blocks each of about 40 ha. Three of these are reserved from control burns while the remainder are burnt on a staggered ten year rotation, i.e., each block is burnt every ten years but adjacent blocks are burnt in different years. This information was entered into the PVA model, ALEX, and I explored the relationship between control burn interval and the probability of extinction (see Possingham and Gepp 1993, for details). The result in Figure 3 shows that the probability of

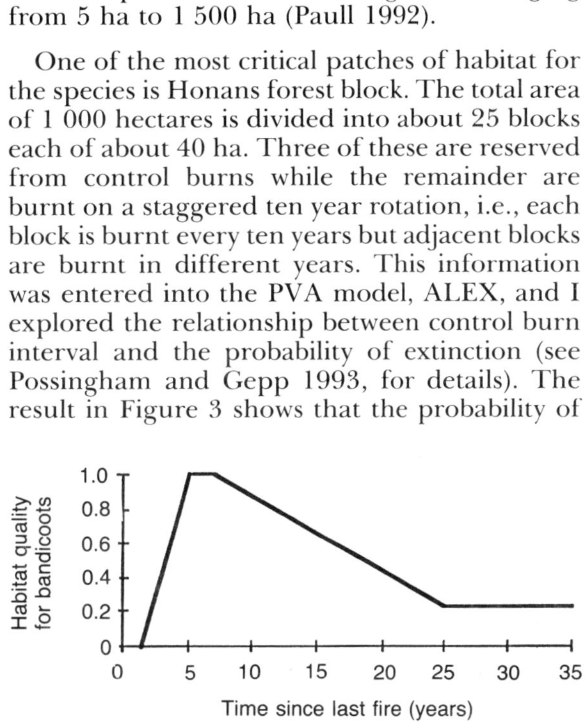

Fig. 2. The relationship between habitat quality and the time since fire for the Southern Brown Bandicoot in south-east South Australia used in the PVA model ALEX for simulations on this species.

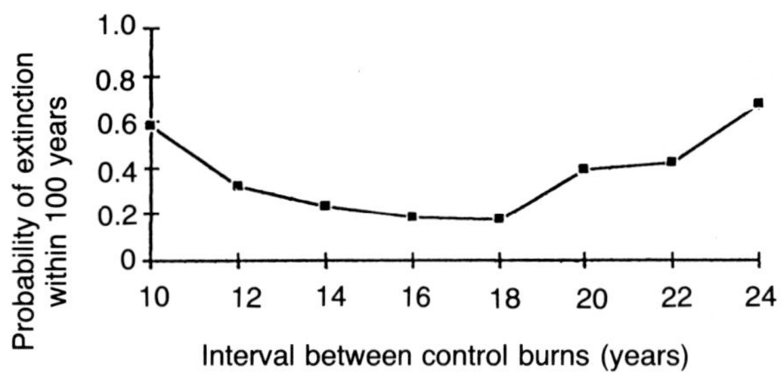

Fig. 3. Extinction probability over 100 years of the Southern Brown Bandicoot *Isoodon obesulus* in a 1 000 ha patch of scrub (Honans) in the south-east of South Australia as a function of the probability that adult females move between adjacent forest compartments in any year.

extinction over 100 years is minimized by choosing a control burn interval of 16–18 years which is significantly longer than current practice and the point at which habitat quality is at its highest.

This counter-intuitive result needs some explanation. First note that, ignoring the impact of the burn, we want to have a high average habitat quality, not maximize the amount of time at the highest habitat quality, which means the burn needs to be after the peak in habitat quality. Second, a control burn causes a dramatic local decline in bandicoot density because we assume that most of the bandicoots are killed during the control burn and habitat is very poor for one or two years after a fire. By burning well after the peak in habitat quality fluctuations in population size are reduced while the average habitat quality is still reasonable.

In this example I use PVA to explore a specific management question and arrive at a strategy that maximizes bandicoot population viability. The question remains, how sensitive are these results to changes in the parameters? By changing different parameters, like dispersal rates and birth and death rates, the absolute probability of extinction varies significantly but the minimum extinction probability always occurs with a control burn interval around 15–20 years. Conversely, varying the effect of fire on habitat quality and bandicoot survival alters the optimal strategy significantly. For example, if the habitat quality of unburnt habitat is a lot higher, then the optimum strategy may be never to burn. This sensitivity analysis shows that there needs to be more research into the response of various life-history parameters to the successional stage of the vegetation and the direct impact of fire. One general rule does emerge, control burns should always be well after the peak in habitat quality.

This example illustrates two points.

i) The modelling process has generated the rule that is of wider educational value: burn well after a peak in habitat quality.

ii) Sensitivity analysis of a PVA result can help to isolate those parameters and processes that we need to target for future research, and give us an idea of the robustness of a conclusion.

Assessing timber harvesting strategies

Timber harvesting in Australia has a significant local impact on many species of arboreal marsupial. The Greater Glider *Petauroides volans* is the most abundant arboreal marsupial (by number and/or biomass) in many of the wetter forests of Victoria, New South Wales and Queensland. There is no reason to expect it to become extinct in the near future, however, because it can be thought of as an indicator species for hollow-dependent fauna, forest managers are interested in maintaining local populations in areas used for timber harvesting.

Possingham and Noble (1991) briefly explored the impact of timber harvesting on the greater glider; this is an extension of that previous work. Here I use ALEX with the parameters introduced by Norton and Possingham (1991).

Consider eight 80 ha coupes of high quality Greater Glider habitat. Assume that 72 ha of each coupe is normally available for harvesting, leaving 8 ha in permanent reserves (buffer zones, steep areas, stream-side reserves etc.). Also assume that after harvesting the quality of habitat in a block is negligible, but recovers slowly to the maximum potential after 100 years. In this example I explore the impact of different harvesting strategies on the probability the Greater Glider persists in this forest block (assuming a closed population). The results are only indicative of the kinds of resource-use management questions that PVA can help us to answer.

In Table 1 the probability that the glider population persists in the 640 ha block is shown for seven different scenarios.

Table 1. Forest management scenarios and their effect of the viability of local Greater Glider population.

Scenario	Chance of persisting 200 years	Description of scenario
1	100%	No harvesting.
2	0%	Forest harvested on a 100-year cycle. All coupes cut simultaneously. Each coupe has 8 ha uncut.
3	3%	As for 2 with 16 ha reserves per coupe.
4	55%	As for 2 but with one coupe unharvested.
5	3%	As for 2 with a 200-year cycle.
6	0%	As for 2 but harvesting is staggered, four coupes every 50 years.
7	63%	Forest harvested on a 200-year cycle, four coupes every 100 years.

In the absence of harvesting the population is very likely to persist and is viable by most definitions. When 72 ha of each coupe is simultaneously harvested (scenario 2) every 100 years the probability of the population persisting 100 years falls to zero! What strategy can be used to raise this probability while still harvesting the forest?

Comparing scenarios 2, 3 and 4 suggests that reserving slightly larger areas of forest scattered through the block is of little conservation, value but reserving one large block of forest, an entire coupe, significantly improves population viability.

Lengthening the cutting cycle from 100 to 200 years does not increase viability much (compare scenarios 2 and 5) while harvesting the coupes asynchronously also has little effect when the rotation length is 100 years (compare scenarios 2 and 6). However an increase in the rotation time to 200 years coupled with an asynchronous harvesting strategy results in a population with a 63 per cent chance of persisting.

This example shows how PVA can be used to explore the consequences of, and possibly modify, resource use strategies. This sort of analysis needs detailed information on the response of a species to habitat modification, in particular changes in fecundity, mortality and home range size. The example here is only intended to show the capabilities of the process, not to provide general prescriptions.

DISCUSSION

Sensitivity analysis

A sensitivity analysis of the results is essential. This needs to be carried out by repeating the simulations with parameters varied around their expected value. For some parameters the estimate of the probability of extinction may vary markedly. This often disturbs people, especially given uncertainty about many parameters and processes in the model. However, the primary purpose of PVA is not to quantify extinction risk, but to choose between management options and explore impacts. In this context it is the *ranking* of management scenarios that is important, in particular, which management option is associated with lowest probability of extinction. The ranking of options is usually much less sensitive to changes in parameter values.

Sensitivity analysis has one further advantage. Where two or three of the parameters are found to significantly impact upon extinction probability and/or the ranking of management options, we can direct our research effort into refining estimates of those parameters. We often find that the value of many parameters has little impact on our conclusion so that further information about these parameters is not essential to the PVA process. This is illustrated by the bandicoot example above, where the optimal control burn interval is relatively insensitive to many parameters. A framework for the application of PVA for choosing management options for threatened populations, including the incorporation of sensitivity analysis, is discussed by Possingham *et al.* (1993).

Data requirements — how much data is enough

More data is always good, however, as indicated above, good estimates of many parameters are not essential. There are a variety of tactics by which we can use PVA where there is crude or even missing information about a species.

Where a species is very rare or cryptic we invariably have poor data on fecundity and mortality. In these circumstances it is possible to estimate parameters from a species with a similar life-history strategy and size. For example, it is reasonable to assume that all large owls have low fecundity and long life spans. In the future we may be able to find relationships between habitat area and extinction probability (Fig. 1) for critical groups like the arboreal marsupials and parrots without assessing every species in detail.

The frequency and effect of catastrophes are usually hard to estimate because of their rarity. It is possible to use historical records to estimate the frequency of catastrophes like fire, hurricane and flood. However, estimating the impact of these events on population and habitat usually requires intelligent estimation. It is important to factor in these rare events, as they may be the only processes that threaten the persistence of a species. The likelihood that a species recovers following a catastrophe, and the frequency of that catastrophe, is critical to estimating population viability. Because of these uncertainties sensitivity analyses of parameters associated with catastrophes are particularly important.

The key issue about estimating parameters is that we take care to explicitly describe where our estimates come from. Clear statements of assumptions in deducing parameters are often more important than the actual parameter estimate because they allow future scientists to extend what has already been done. By taking such care, work can be continually upgraded as new information comes to light.

Further pros and cons of PVA

The PVA process has a number of benefits aside from helping us to choose between management strategies. It helps to bring experts together and focuses them on the parameters and processes that influence population viability. The results from the model, and the dynamics of the population can be viewed and checked. By simulating the dynamics of a population over long times, wildlife biologists gain some appreciation of the fluctuating nature of the dynamics of populations. The educational spin-offs of building a model of population dynamics, and having to state explicitly the assumptions that underlie that model cannot be over-estimated. Placing research in wildlife biology in the context of understanding population dynamics often helps managers and biologists to focus on key issues, like estimating basic demographic rates.

Despite these numerous advantages, we need to think carefully before investing finite resources in PVA for particular species.

i) What is the value of PVA when there is no possibility for improving the viability of the species? In some cases PVA may provide no more than an estimate of the likelihood of species extinction, which in the absence of practical management options, may be of only academic interest.

ii) Should we do a PVA for every species? Clearly this is impossible. The species that need to be targeted for PVA need to be carefully chosen (Possingham 1991). Carrying out a PVA on every species for which we have a little data will be a waste of time. Candidates for PVA should have at least one of the following attributes: species of high cultural or economic value, keystone species, indicator species, species for which we have a lot of information, and threatened species.

iii) How much time will a PVA take and who should be involved? PVA should not be thought of as an "extra" task in the threatened species management or resource use processes. Once target species are chosen, and it is decided PVA is a useful tool, the PVA process should proceed in harmony with research, management and data collection. Ideally the PVA process should involve all concerned individuals, including the wildlife expert and modeller. Currently it is not possible to use one of the generic PVA packages and produce an adequate PVA without some knowledge of the mathematics behind the model. As more case studies are published it may eventually be possible to do reasonable PVAs of different populations without involving experts in a particular model. Just as research into wildlife management is an ongoing process, the PVA task has no finite lifetime.

In the future PVA will be used increasingly for Environmental Impact Assessment and threatened species management. The greatest danger is that we blindly follow simple rules (for example, that 500 is a minimum viable population for all species). A useful PVA requires some effort and cannot be carried out in a couple of days. Ideally PVA requires the collaboration of wildlife biologists and modellers. Given this effort and care, PVA will make a significant contribution to preserving biodiversity.

ACKNOWLEDGEMENTS

The ideas and results presented in this paper would not be possible without the continuing support of, and collaboration with, Tony Norton, David Lindenmayer, Ian Davies, Ian Noble and Brian Gepp. I am particularly grateful to David Lindenmayer for commenting on the manuscript. Mark Burgman and an anonymous referee provided valuable improvements. Work on the Southern Brown Bandicoot was partially supported by a National Estates grant to Brian Gepp, the preliminary results for Leadbeater's Possum was supported by a grant to David Lindenmayer from the Endangered Species Unit, Australian Nature Conservancy Agency.

REFERENCES

Akçakaya, H. R., 1992. Population viability analysis and risk assessment. In *Proceedings of wildlife 2001: populations* ed by D. R. McCullough. Elsevier: Amsterdam.

Akçakaya, H. R. and Ferson, S., 1992. RAMAS/Space. Applied Biomathematics. Setauket: New York.

Armbruster, P. and Lande, R., 1993. A population viability analysis for African elephant *Loxodonta africana:* how big should reserves be? *Conserv. Biol.* **7:** 602–10.

Biodiversity Advisory Committee, 1992. A national strategy for the conservation of Australia's biodiversity. Australian Government Publishing Service: Canberra.

Borlase, S. C., Loebel, D. A., Frankham, R., Nurthen, R. K., Briscoe, D. A. and Daggard, G. E., 1993. Modeling problems in conservation genetics using captive *Drosophila populations:* consequences of equalization of family sizes. *Conserv. Biol.* **7:** 122–31.

Boyce, M. S., 1992. Population viability analysis. *Ann. Rev. Ecol. System.* **23:** 481–506.

Burgman, M. L. and Lamont, B., 1992. A stochastic model for the viability of *Banksia cuneata* populations: environmental, demographic and genetic effects. *J. Appl. Ecol.* **29:** 719–27.

Burgman, M. A., Akçakaya, H. R. and Loew, S. S., 1988. The use of extinction models for species conservation. *Biol. Conserv.* **43:** 9–25.

Burgman, M. A., Ferson, S. and Akçakaya, H. R., 1993. Risk assessment in conservation biology. Chapman and Hall: London.

Cutler, A., 1991. Nested faunas and extinction in fragmented habitats. *Conserv. Biol.* **5:** 496–505.

Doak, D., 1989. Spotted owls and old growth logging in the Pacific Northwest. *Conserv. Biol.* **3:** 389–96.

Gibbs, J. P. and Faaborg, J., 1990. Estimating the viability of ovenbird and kentucky warbler populations in forest fragments. *Conserv. Biol.* **4:** 193–96.

Goodman, D., 1987. Considerations of stochastic demography in the design and management of nature reserves. *Nat. Res. Model.* **1:** 205–34.

Haig, S. M., Belthoff, J. R. and Allen, D. H., 1993. Population viability analysis for a small population of red-cockaded woodpeckers and an evaluation of enhancement strategies. *Conserv. Biol.* **7:** 289–301.

Heinhson, G. E., 1966. Ecology and reproduction of the Tasmanian bandicoots (*Perameles gunnii* and *Isoodon obesulus*). *Univ. Calif. Pub. Zool.* **80:** 1–107.

Lacy, R. C., 1993. VORTEX: a computer simulation model for population viability analysis. *Wildl. Res.* **20:** 45–66.

Lacy, R. C. and Clark, T. W., 1990. Population viability assessment of the Eastern Barred Bandicoot in Victoria. Pp. 131–35 in *Management and conservation of small populations* ed by T. W. Clark and J. H. Seebeck. Chicago Zoological Society: Chicago.

Lande, R., 1988. Demographic models of the northern spotted owl *Strix occidentalis caurina*. *Oecologia* **75:** 601–07.

Lindenmayer, D. B. and Lacy, R. C., 1993. Predictions of the impacts of changes in population size and environmental variability on Leadbeater's Possum *Gymnobelideus leadbeateri* McCoy (marsupalia: Petauroidae) using population viability analysis: an application of the computer model VORTEX. *Wildl. Res.* **20:** 67–86.

Lindenmayer, D. B. and Possingham, H. P., 1995. The risk of extinction: Ranking management options for leadbeater's possum using population viability analysis. Centre for Resource and Env. Studies, ANU, Canberra.

Lindenmayer, D. B., Norton T. W. and Possingham, H. P., 1993a. An approach for determining wildlife metapopulation viability using GIS to couple habitat models and forest resource data. Pp. 429–39 in *Conference proceedings GIS'93*. Vancouver, British Columbia, Canada.

Lindenmayer, D. B., Clark, T. W., Lacy, R. C. and Thomas, V. C., 1993b. Population viability analysis as a tool in wildlife management: a review with reference to Australia. *Environ. Manage.* **17:** 745–58.

Marcot, B. G. and Holthausen, R., 1987. Analyzing population viability of the spotted owl in the Pacific Northwest. *Trans. Nth Amer. Wildl. Nat. Res. Conf.* **52:** 333–47.

Norton, T. W. and Possingham, H. P., 1991. Modelling the population viability of forest biota: a case study using the Australian arboreal marsupial folivore, *Petauroides volans* (Kerr). *Proceedings of the ninth biennial conference on modelling and simulation.* Pp. 442–47.

Nunney, L. and Campbell, K., 1992. Assessing minimum viable population size: demography meets population genetics. *Trends Ecol. Evol.* **8:** 234–39.

Paull, D. J., 1992. The distribution, ecology and conservation of the southern brown bandicoot *Isoodon obesulus obesulus* in South Australia. MA Thesis, Geography Department, The University of Adelaide.

Pimm, S. I., Jones, H. L. and Diamond, J., 1988. On the risk of extinction. *Amer. Natur.* **132:** 757–85.

Possingham, H. P., 1991. The role of population viability analysis in forest management. Pp. 35–40 in *Conservation of Australia's forest fauna* ed by D. Lunney. The Royal Zoological Society of New South Wales: Sydney.

Possingham, H. P. and Gepp, B., 1993. The application of population viability analysis to assess management options for the southern brown bandicoot *Isoodon obesulus* in the SE of South Australia. *Proceedings of the International Simulation Society Conference.* Pp. 633–38: Perth.

Possingham, H. P. and Noble, I. R., 1991. An evaluation of population viability analysis for assessing the risk of extinction. Research consultancy for the Resource Assessment Commission, Forest and Timber Inquiry: Canberra, Australia.

Possingham, H. P., Davies, I., Noble, I. R. and Norton, T. W., 1992. A metapopulation simulation model for assessing the likelihood of plant and animal extinctions. *Math. Comput. Simul.* **33:** 367–72.

Possingham, H. P., Lindenmayer, D. B. and Norton, T. W., 1993. A framework for improved threatened species management. *Pac. Conserv. Biol.* **1:** 39–45.

Resource Assessment Commission 1992. Forest and Timber Inquiry. Volume 1. Australian Government Printing Office: Canberra.

Shaffer, M. L., 1981. Minimum population size for species conservation. *BioScience* **31:** 131–34.

Shaffer, M. L., 1983. Determining minimum viable population sizes for the grizzly bear. *Internat. Confer. Bear Manag.* **5:** 133–39.

Shaffer, M. L., 1990. Population viability analysis. *Conserv. Biol.* **4:** 39–40.

Soulé, M. E. (ed), 1987. *Viable populations for conservation.* Cambridge University Press: Cambridge.

Thomas, J. W., Forsman, E. D., Lint, J. B., Meslow, E. C., Noon, B. R. and Verner, J., 1990. A conservation strategy for the Northern Spotted Owl. United States Government Printing Office: Portland.

Walter, H. S., 1990. Small viable population: the red-tailed hawk of Socorro Island. *Conserv. Biol.* **4:** 441–43.

Biodiversity and fisheries management: the implications of extracting maximum yields from interactive ecosystems

R. E. KEARNEY[1]

Harvesting fish changes ecosystems. The degree of impact is usually related to the level of exploitation and the nature of the method used to capture fish. Very few, if any, fish species have been driven to extinction by fishing but the balance of many ecosystems have been changed by direct and indirect effects of harvesting. Species richness has been minimally impacted by fishing but community structure has suffered.

It is argued that human activities, other than fishing, pose a greater threat to the existence of species, particularly in freshwater systems, and have changed ecosystems more dramatically.

There are several million fishers, commercial and recreational, in Australia and all have at least some interest in the long-term availability of the resource they exploit. Properly harnessed they represent a potent force for the conservation of aquatic biodiversity.

INTRODUCTION

Is fishing a serious threat to the conservation of biodiversity? Does fisheries management, including aquaculture, represent a solution to potential loss of diversity?

Harvesting fish by commercial or recreational fishing does change ecosystems. The degree of change is largely determined by the proportion of the total population of the species that is exploited, by the role the target species play in the particular ecosystem and by the impact of the gear on the ecosystem.

There is no doubt that fishing has caused dramatic declines in some fish populations; the Peruvian Anchovetta, North Sea Herring, Atlantic Cod and Southern Bluefin Tuna are world-renowned examples. Local coastal examples include Abalone, Eastern Rock Lobsters and Snapper.

Even though all of the above species are unquestionably overfished and no longer produce maximum sustainable yields (with due respect to fisheries purists who wish to argue terminology, such as Larkin 1977) none of them is, as yet, classified as endangered, and there is optimism that with careful management, all can be returned to optimum productivity.

It is argued by the world's commercial fishing interests that commercial fishing has not led to the extinction of a single fish species. The disappearance of several species of clams from a number of South Pacific islands suggests that invertebrates may not have been so fortunate. If fishing has obliterated any fish species then the number of such species would be small. The threat of fishing is more real to several marine mammals, such as Blue Whales, whose future still hangs in the balance. But is whaling really fishing? Or should the exploitation of marine mammals be considered more as hunting? Regardless of the classification given to their exploitation most countries have special arrangements for the management of marine mammals and my comments which follow are restricted to more classical fish.

Even if one or two species of fish have been fished to extinction this would represent a small fraction of the thousands of species that are fished commercially and recreationally throughout the world. It would certainly be much less than the number of species threatened, or even driven to extinction, by human activities, other than fishing. If the primary criterion for determining biodiversity is the number of species in existence then one must conclude that fishing, at least by conventional methods, is not a major threat.

If, on the other hand, broader, and more commonly used interpretations of biodiversity are accepted, especially those that relate to the abundance of species in ecosystems, such as community composition, then fishing obviously is a significant factor. Any definition of biodiversity that gave weight to the total abundance of species, or even presence or absence in localized areas, would identify fishing as a problem. The species richness versus community composition debate surfaces again. The question of whether changes to ecosystems are always bad for biodiversity also cannot be ignored.

---

[1]New South Wales Fisheries, Fisheries Research Institute, 202 Nicholson Parade, Cronulla, New South Wales, Australia 2230.

Pages 300–05 *in* CONSERVING BIODIVERSITY: THREATS AND SOLUTIONS, ed by R. A. Bradstock, T. D. Auld, D. A. Keith, R. T. Kingsford, D. Lunney and D. P. Sivertsen. Surrey Beatty & Sons. 1995.

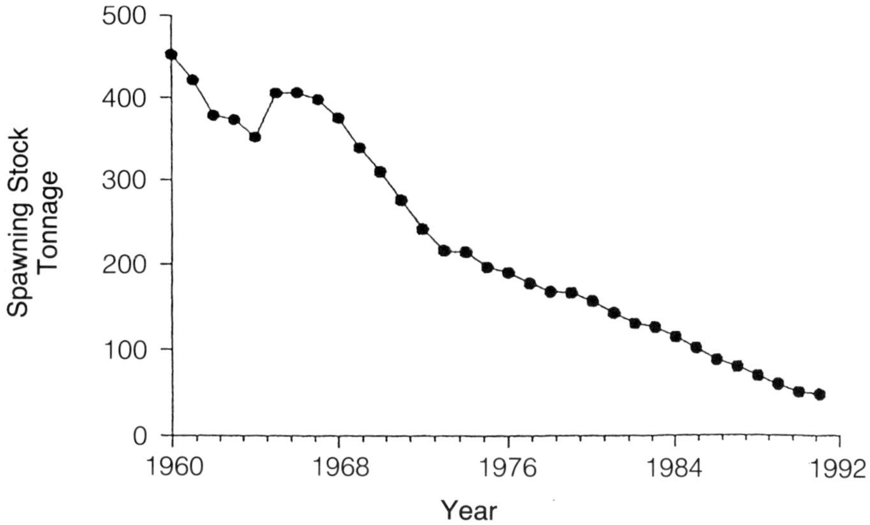

Fig. 1. Decline in size of the spawning stock of Southern Bluefin Tuna (from Polacheck, Sainsbury and Klaer 1992).

FISHING AND ECOLOGICALLY SUSTAINABLE DEVELOPMENT

Fishing, as an exploitive activity, alters the balance of ecosystems. It does so in more ways than just by changing the abundance of target species. Large scale commercial fishing almost always causes changes in the size composition of fish populations and in their distribution, and has secondary impacts on other species. The existence of these changes confirms that fishing changes ecology and is therefore by the strictest definitions not ecologically sustainable. On the other hand the impact of fishing on ecology is much less than that of many other human activities, such as agriculture which, by definition, cultures organisms and manipulates ecosystems, often to the detriment of biodiversity regardless of how it is defined. There are few more obvious examples of deliberately changing ecology than the farming of introduced species, both animal and plant, in Australia.

MARINE FISHERIES

Abundance

Many Australian marine fish species have been fished to the extent that population levels are now below those which would support maximum sustainable yields. Southern Bluefin Tuna is a good example of a species whose abundance is well down on virgin levels (Fig. 1). While the remaining population of Southern Bluefin is of the order of tens of thousands of tonnes (Polachek and Sainsbury, in press) it could be argued there is not currently any real threat of extinction. On the other hand, the species is still being fished, and the impact of such fishing remains unclear.

Stock and recruitment relationships

Fish, unlike most terrestrial megafauna, show extremely unpredictable relationships between stock size and subsequent recruitment to the population. Immense fecundity, often exceeding one million eggs per female per season, and for some species as high as 200 million eggs per female per year, and the wide and variable distribution of eggs and juvenile life stages provide for immense variability in survival, even from relatively stable parental populations.

Environmental variability and density dependent compensation can often result in even depleted parental populations producing abnormally high numbers of progeny in some years. Alternatively, juvenile survival can be low in years when the parental population is great.

Species compensation

In a complex ecosystem with an area of tens of millions of square kilometres, no definite lateral boundaries and a truly three dimensional nature, such as inhabited by Southern Bluefin Tuna and Gemfish, it is extremely difficult to determine the impact of the diminution, or even total removal, of one apex predator. It is, however, probable that the declines in Southern Bluefin Tuna and Gemfish populations have led to increases in the availability of food for other large pelagic species. Increases in abundance of competitors with Southern Bluefin and Gemfish could be anticipated. It is possible that the apparent increased catches of Albacore off eastern Australia in recent years (Kailola *et. al.* 1993) may be a reflection of the decline in Southern Bluefin Tuna.

Compensatory changes in populations of marine pelagic species have been documented in a number of fisheries, for example the

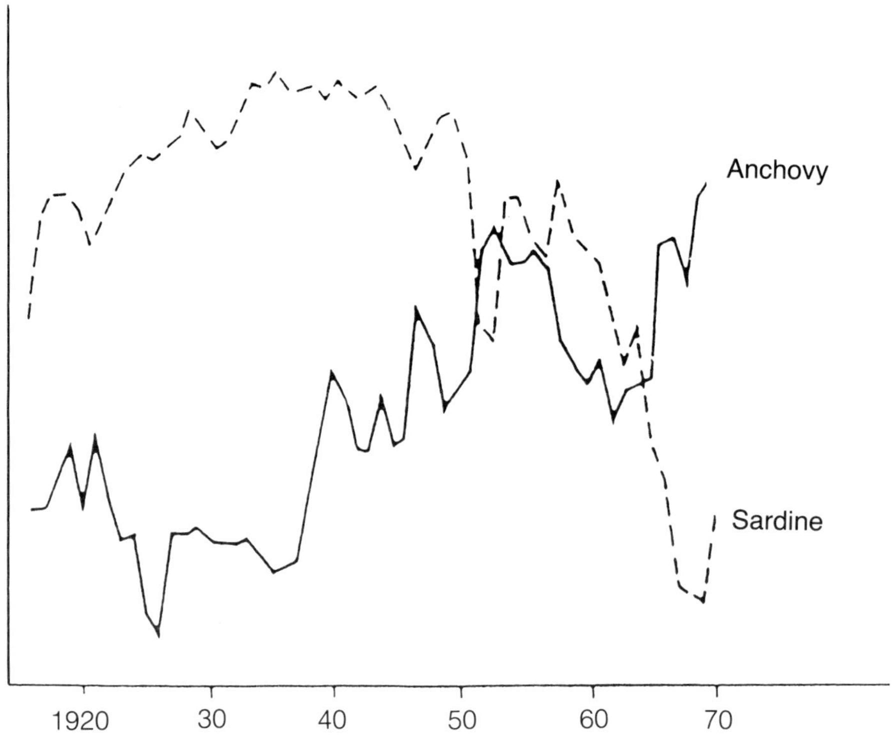

Fig. 2. Fluctuation in production from the California Sardine and Anchovy fishery (from Caddy and Gulland 1983).

Californian Sardine and Anchovy (Fig. 2). When the abundance of one of these species is up, the other is down. Other components of related ecosystems obviously change as a result. While fishing is often blamed for recent changes in the balance of these particular species, scale deposits in sediments confirm a long history of fluctuating dominance, well before human intervention (Soutar and Isaacs 1969). Human intervention has, however, probably increased the frequency of fluctuations.

The impact of fishing on fish habitats has also led to changes in the composition of fish communities in many areas. While it is extremely unlikely that the offshore, oceanic environment of species such as Southern Bluefin Tuna have been demonstrably affected by fishing, bottom trawling is not so benign. Damage by trawlers to epibenthic communities can be substantial and this in turn can contribute significantly to major changes in relative abundance and species composition, including the replacement of commercially important species by those of relatively little value. A recent review of this subject concludes that decades may be required for recovery (Jones 1992).

Predator-prey relationships

With the declining abundance of an apex predator, such as Southern Bluefin Tuna, it is probable that there would be at least a short-term increase in the abundance of the species on which the apex predator preys. In the longer term, increased abundance of another apex predator, such as Albacore, could be anticipated to take advantage of the increased food at this trophic level, as already discussed.

If the removal of a predator leads to the increase of a prey species, then pressure on life forms lower in the food chain would be expected to increase. Species such as Southern Bluefin Tuna have shown at least one order of magnitude change in abundance and changes of at least the same magnitude in the abundances of species lower in the food chain could well have occurred.

Even though gross changes in abundance of prey species unquestionably do occur, the total oceanic ecosystem appears to be remarkably resilient. Many depleted species appear to be able to re-establish some form of equilibrium, even though this may be at population levels considerably lower than those which result in maximum productivity of the species. Again, assessment of the impact on biodiversity is primarily influenced by the definition of diversity.

Changing fish behaviour

There is no doubt that changes in abundance brought about by fishing affects the overall behaviour of fish populations. The average size of fish in a population usually changes with exploitation, for example Gemfish off eastern Australia (Fig. 3).

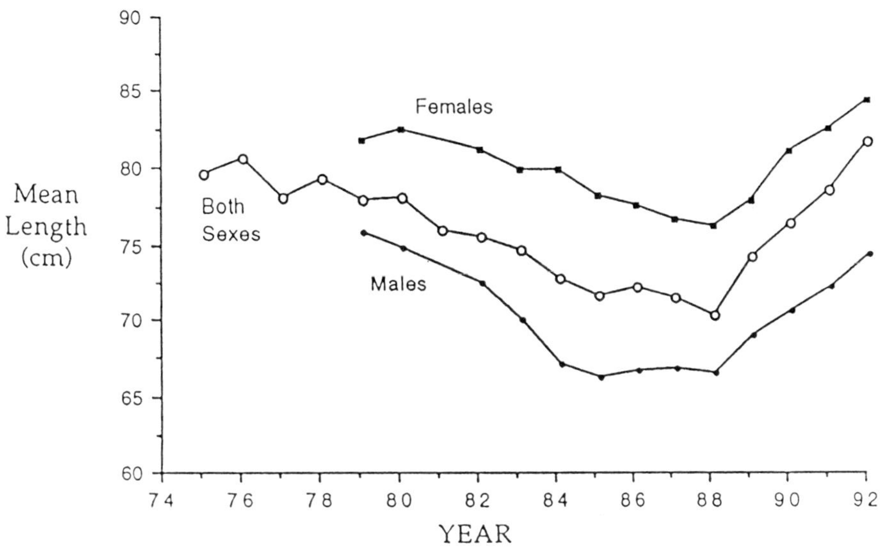

Fig. 3. Mean lengths of spawning-run Gemfish, 1975 to 1992 (K. Rowling, unpublished, Fisheries Research Institute, Cronulla).

In general, the distribution of a species contracts as its population declines. For example, Southern Bluefin tuna disappeared off the coast of New South Wales in the mid 1980s as a direct result of fishing, and Australian Salmon are no longer abundant in the estuaries of the southern parts of this state, where prior to the 1940s they supported a major canning industry.

Declines in the geographic distribution of a species unquestionably change the ecosystem, and decrease the diversity, at least in the areas of the original distribution that the species no longer inhabits. But because the species still exists in other areas, total global biodiversity, as determined by the number of species, has not declined. On the other hand when distribution is diminished it is difficult, or impossible, to determine if a homogenous population has merely reduced its area in order to maintain some density dependent relationship, or if some genetically unique populations have been lost. It could certainly be argued that the genetic diversity of the species must in the long term, i.e., geological time scale, be diminished if the geographic range of the species, and hence its exposure to variations in climate and environment, is decreased. Over time the likelihood of evolutionary change within a species would be expected to be reduced at decreased population levels. But what about compensatory changes in abundance or distribution of other species to their advantage? Fishing probably reduces diversity within species in the short term but its long-term impact on total evolutionary change is debatable.

FRESHWATER FISHERIES

Many species of Australia's fish have been classified as threatened. Almost all of these are freshwater species (Table 1). In all cases the primary cause of the threat has come from activities other than the exploitation of the population for maximum yields, i.e., managed fishing. The many dams and weirs that humans have built on our inland river systems, other forms of habitat destruction and water degradation brought about by activities such as farming and sewage disposal, have led to predictable declines in native fish populations. However biodiversity is measured there are several river reaches where declines have been precipitous even though the species still exist elsewhere.

Table 1. Conservation status of Australian freshwater fish 1985–1992. (From Jackson 1992.)

Classification	1985	1988	1989	1990	1991	1992
Extinct	0	0	0	0	0	0
Endangered	4	5	6	8	9	11
Vulnerable	4	5	5	6	6	5
Potentially threatened	5	3	4	5	7	10
Indeterminate	2	2	2	2	2	2
Restricted	29	33	31	30	32	30
Uncertain status	15	16	16	13	13	14
Total	59	64	65	64	69	72

Total population levels of probably all of our native inland fish species are below those which would produce maximum sustainable yields. Total ecosystems have been dramatically altered by tremendous changes to the physical and chemical environment and by introduced species, such as Carp, often better suited to the environments humans have designed, or inadvertently created.

Certainly the ecosystems of most of our inland freshwater rivers and lakes have been changed. The larger river systems, such as the Murray-Darling, no longer represent a continuum through which a number of species migrate freely. The many sections of this mighty system created by dams and weirs represent great change to the virgin environment. Peter Cullen (Cullen and Lake, this volume) described the current system as "a series of little eutrophic lakes". On the other hand we have created a number of semi-discrete ecosystems which may have, or may develop, their own individuality. But who would like to argue that this increase in the number of ecosystems represent a gain for biodiversity? Certainly we have not conserved the original ecosystem and the subdivisions have decreased the total abundance and increased the vulnerability of many species. But this conference is, after all, about conserving biodiversity, which is slightly different to, but not necessarily more worthy than, conserving ecosystems and is considerably different to preserving virgin habitats.

AQUACULTURE

I have already conceded that agriculture, by definition, changes ecosystems and therefore impacts biodiversity. Aquaculture is in many ways more similar to agriculture than it is to capture fisheries.

In many parts of South East Asia degradation of coastal habitats in the interests of aquaculture (Fast 1992) have paralleled the massive destruction of terrestrial habitats by other human activities. Even in Australia productive habitats such as mangrove swamps were converted to aquaculture ponds or, more often, destroyed in the interests of alternative development. Such gross habitat destruction under the guise of aquaculture development is no longer a reality in this country (Maguire and Allan 1992). Threats to biodiversity from aquaculture now come more from the misuse of chemicals, increased incidence and new types of disease, disposal of waste, and the translocation and escape of cultured species. While all of these problems pose threats in their own right, their current impacts are, on average, less than from agriculture, with the possible exception of escaped species. However, large-scale increases in aquaculture are anticipated in the next decade. New South Wales currently imports more than 60 per cent of the fish consumed here and our wild fisheries resources are already declining.

Aquaculture obviously has the potential to alter biodiversity, particularly in localized environments. For example, it has already prevented the extinction of the Lake Eacham Rainbow Fish, which no longer occurs in the wild, and has enabled the re-introduction of the Eastern Freshwater Cod to the Richmond River, from where it disappeared approximately 40 years ago. The species richness scorecard would have fisheries leading at least 1: nil.

Genetic "improvement", through selective breeding programmes and, in the extreme, by directly altering or replacing genes, are also factors. Superficially these would appear to offer considerable potential for increasing biodiversity, but they highlight the need to decide whether increased diversity for its own sake is desirable.

Selective breeding is already carried out in almost all intensive aquaculture systems. Even for extensive aquaculture, such as for edible oysters, human intervention in natural selection has played its part. It can be argued that 100 years of selection for slower growing Sydney rock oysters, through harvesting the larger oysters and allowing slow growers to breed at least one more season, has already had a negative impact on the growth rate of the average oyster. Properly designed and executed breeding programmes could reverse this trend and eventually lead to improved strains (stocks or sub-populations) better suited to our current, or even past, conditions. The degree to which such strains would change, or even restore ecosystems, and possibly even increase biodiversity, is debatable. But again this conference is about conserving biodiversity, not increasing it.

CONCLUSIONS

It appears that fishing on its own has caused the extinction of few, if any, fish species. A few species of sedentary invertebrates, such as clams, may not have been so fortunate. Localized depletions of fish stocks have certainly resulted from fishing, causing unknown, or at best poorly described, secondary effects. Whether or not these are fully reversible remains largely unknown. However, the threat to species, and the relative intensity of ecosystem change resulting from fishing is, in most cases, much less than that resulting from human activities other than fishing.

The interactive nature of marine ecosystems, where recruitment potential is immense, species eat species, adults eat juveniles including their own, juveniles eat larvae and eggs, and predator and prey density at all life stages influences survival by orders of magnitude, appears the reason for our inability to eliminate species by fishing. Millions of people have tried.

Fisheries management is improving globally and there is reason to be confident that many currently overexploited species will be assisted to recover in the years ahead, even if the motive is to return to maximum yields rather than species, ecosystem or biodiversity conservation *per se*.

Our freshwater ecosystems have, on average, suffered more acute damage than their marine counterparts. But, while fishing has played its part in reducing several freshwater species to dangerously low population levels, major responsibility for damage can be attributed to activities other than fishing.

While it is acknowledged that fishing exploits certain species it must also be noted that the goal of fisheries managers is to have maximum, or optimum, long-term yields; in fisheries jargon, maximum sustainable yields. This goal is completely complementary to sustainable development and much closer to ecologically sustainable development than is true for many other human activities. When it is also noted that one of the primary tools for achieving maximum sustainable yields is the protection, and often total preservation, of key, or representative fish breeding, nursery or production areas, fisheries management is even closer to the true intent of ESD. The conservation of fish species and their habitats (biodiversity) is a core function of most fisheries management agencies.

When, in his opening address to this Conference, Professor Beattie argued that profit was not only a legitimate excuse for conserving biodiversity, but one of three primary reasons for doing so, he was criticized, albeit politely, for being too biased towards economic considerations. Why, he was asked, did he not give more support for the right of a species to exist? In supporting the right of each and every species I would also stress that we do have to face the political and social realities of our time. In so doing we must acknowledge that if there is profit in the process then support for furthering the process, or at least conserving it, is more likely. The more people, or the more influential the sum of the people, who share the profit, the greater the potential support, and hence the greater the power of the lobby for the process (or ecosystem) to continue to exist. Fishing, both commercial and recreational, should be regarded as part of the process of taking profit from our ecosystems. The real challenge is to manage fisheries for long-term benefit and prevent over-exploitation by short-term profit takers.

To most Australians, at least all those who eat or catch fish, one of the great values of the total ecosystem is its yield of fish. It is also true that consumers, both diners and anglers, benefit from the diversity of the fruits of the sea. There is unquestionably a broad spectrum of preferences for fish as food, or to hunt.

This immense interest in harvesting a variety of aquatic life could, on the one hand, be regarded as a threat to some forms of biodiversity. But I would hope I have convinced at least some of you that it is a relatively minor threat.

On the other hand the strong support for taking maximum, but sustainable yields from our oceans, estuaries, rivers and lakes, expressed by the many commercial and recreational fishers, has already led to the conservation of several of our most fragile fish nursery areas, such as mangrove swamps and sea-grass beds. This same group of increasingly conservation-conscious profit-takers, i.e., the fishing community, who number approximately two million in New South Wales alone, represent one of the greatest potential forces for the conservation of aquatic biodiversity.

REFERENCES

Caddy, J. F. and Gulland, J. A., 1983. Historical patterns of fish stocks. *Marine Policy* **7:** 269.

Cullen, P. and Lake, P. S., (this volume). Water resources and biodiversity: past, present and future problems and solutions. In *Conserving biodiversity: threats and solutions* ed by R. A. Bradstock, T. D. Auld, D. A. Keith, R. T. Kingsford, D. Lunney and D. P. Sivertsen. Surrey Beatty & Sons: Chipping Norton.

Fast, A. W., 1992. An overview of Asian marine shrimp culture. Pp. 623–40 in *Marine shrimp culture: principles and practices* ed by A. W. Fast and L. J. Lester. Elsevier: New York.

Jackson, P., 1992. Australian threatened fishes: 1992 supplement. *Aust. Soc. Fish Biol. Newslet.* **22**(2)**:** 18–21.

Jones, J. B., 1992. Environmental impact of trawling on the seabed: a review. *NZ J. Mar. Freshw. Res.* **26:** 59–68.

Kailola, P. J., Williams, M. J., Stewart, P. C., Reichelt, R. E., McNee, A. and Grieve, C., 1993. Australian fisheries resources. Bureau of Resource Sciences: Canberra.

Larkin, P. A., 1977. An epitaph for the concept of maximum sustained yield. *Trans. Amer. Fish. Soc.* **106:** 1–11.

Maguire, G. B. and Allan, G. L., 1992. Shrimp culture industry in Australia. Pp. 757–69 in *Marine shrimp culture: principles and practices* ed by A. W. Fast and L. J. Lester. Elsevier: New York.

Polacheck, T. and Sainsbury, K. J., (in press). Age structure assessment with multiple and conflict use information: a case study based on southern bluefin tuna. Bulletin. Bureau of Resource Sciences: Canberra.

Polacheck, T., Sainsbury, K. J. and Klaer, N., 1992. Assessment of the status of the Southern Bluefin Tuna stock using virtual population analysis — 1992. 11th Trilateral Meeting of Australian, Japanese and New Zealand Tuna Scientists, Shimizu, Japan, 5–10 October 1992. Background document SBFWC/92/WP7.

Soutar, A. and Isaacs, J. D., 1969. History of fish populations inferred from fish scales in anaerobic sediments off California. California Cooperative Oceanic Fisheries Investigations. Reports. CalCOFI **13:** 63–70.

PLATE 18 *(above)*. There are more than 30 000 duck hunters in southeastern Australia who may potentially shoot up to 10 birds per day. Such a harvest may have an impact on the survival of game species of waterfowl. *(Photograph by Wayne Lawler.)*

PLATE 19 *(below)*. The Pacific Black *Anas superciliosa* is one of the three more common species shot by hunters in southeastern Australia. Populations have declined but this has been attributed mainly to habitat loss. *(Photograph by Wayne Lawler.)*

Changes to Fire Regimes

PLATES 20 *(above)* and 21 *(below)*. Contrasting intensities of fire in a semi-arid mallee shrubland. High intensity fire *(above)* results in the combustion of crowns of the eucalypt overstorey. Low intensity fire *(below)* in contrast consumes less plant biomass with resultant differences in post-fire conditions and regeneration. Fire intensity is an integral component of any fire regime: the consequences of intensity must be considered when framing fire regime objectives for conservation. *(Photographs by Tony Auld and Ross Bradstock.)*

Extinction of biota by fires

A. M. GILL[1] and R. BRADSTOCK[2]

Legislative imperatives to conserve biota aim to prevent extinctions. Extinctions can be local or nationwide, temporary or permanent. Here, examples of fire-caused extinctions of plants and animals in Australia are reviewed using simple models.

Historically, many plant species have become locally extinct due to too-frequent fires. Typically, these species have fire-sensitive adults and rely on seed for their re-establishment after fire ("obligate seeder species"). Fire-sensitive species may become rare and become confined to "fire shadows" in the landscape.

Breeding populations of small mammal species become locally extinct (temporarily) after waves of population growth at various times after fires in eastern Australian forests and heaths. Recolonization may occur in these apparently facultative habitats from core areas where the species is found always, perhaps independent of fire occurrence.

A suite of small mammalian herbivores and omnivores have become extinct in arid Australia but the role of fires in these extinctions needs clarification. Changes in the patchinesses of fires have been implicated. The sizes, shapes and distributions of burned areas for the fire management of reserves need to be determined.

Derivation of simple rules to guide management is possible based on existing knowledge. Fire managers faced with the possibility of extinctions from reserves where conservation is important can use monitoring systems to assist in implementing such rules. Indeed, monitoring can be based on the sensitivities of plant species to fires in order to guide decision making relevant to the frequency of planned fires and the suppression of unplanned ("wild") fires.

INTRODUCTION

FROM Kosciusko to Kimberley, from Cairns to Karridale, from Arnhemland to the Nullarbor Plain, fires are a major factor in influencing terrestrial plant communities across Australia. Fires are easy to ignite and can spread widely. They can be a cheap management tool and a costly reality. Fires can be considered both "natural" and "artificial". They impact on human values and ecosystem characteristics while their influences can be dramatic or enigmatic. It is no wonder that fires and their effects have been, and continue to be, controversial topics. With feelings often running high on conservation issues involving perceived or possible extinctions of plant and animal species, and the interest in species rarity, biodiversity and ecologically sustainable development, the controversy is exacerbated. In the account which follows, we attempt to present factual data and informed opinion but, often, we have had to rely on circumstantial evidence, although published. That this is the case should not be surprising as there is no telling where extinction may occur; appropriate documentation is therefore very difficult to obtain.

WHAT IS EXTINCTION?

Examination of the meaning of "extinction" has more than academic interest. With increasing demand for species conservation in areas managed for multiple uses, definition becomes imperative because the time and area bases for the definition have operational significance.

"Extinction" comes from "extinguish" which means "to put out" (a fire, for example) or "to destroy" or "to wipe out". "Extinction" can refer to the destruction of plants and animals as either individuals, populations or species but, here, we are referring to the species level of extinction which also encompasses losses of individuals and populations.

Extinction is time based. For Australian plants, Leigh and Briggs (1992) define species as "presumed extinct" if:

(i) they have not been collected or otherwise verified over the past 50 years despite thorough searching; or,

(ii) "all known wild populations have been destroyed more recently."

Burbidge and McKenzie (1989) provided a similar definition but for Australian animals: a species is considered extinct if there have been "reasonable searches but no substantiated records since 1955".

Plant species may appear to be extinct because no material is obvious, yet be present as dormant propagules in the soil: "*Leptorhynchos gatesii* was rediscovered by Mary White in 1984 near Lorne, Victoria following its resurrection after a severe wildfire in 1983. The species had not been seen since 1922" (Briggs and Leigh 1988). A similar intriguing case is that of a number of species on Middle Island of the Recherche Archipelago. There, botanical explorations have been few and brief but Weston's (1985) observations suggest that a number of plants on the island,

[1]CSIRO Division of Plant Industry, G.P.O. Box 1600, Canberra, Australian Capital Territory, Australia 2601.
[2]NSW National Parks and Wildlife Service, P.O. Box 1967, Hurstville, New South Wales, Australia 2220.

Pages 309–22 *in* CONSERVING BIODIVERSITY: THREATS AND SOLUTIONS, ed by R. A. Bradstock, T. D. Auld, D. A. Keith, R. T. Kingsford, D. Lunney and D. P. Sivertsen. Surrey Beatty & Sons. 1995.

apparently unseen for 150 years, reappeared in very conspicuous populations after fires in 1972–73. Furthermore, Weston (1985) suggested that a fire on the island just prior to the visit by Robert Brown in 1802 was the cause of these species — apparently short-lived perennials (species of *Alyogyne, Scaevola, Solanum* and *Villarsia*) — being visible and collectable.

The time base chosen for studies of extinction is one of convenience and will vary with purpose. Here, of **no** concern are the megafaunal extinctions of former geologic epochs. Extinctions in the period of European history are of most concern here. Also of interest are landscape patterns which imply that extinctions have occurred in certain areas; whether the patterns observed first arose in European, Aboriginal or earlier periods is often an unresolved matter.

Extinction is not only a time-based phenomenon, it is also a spatially-based phenomenon. Over what area should extinction be considered? Briggs and Leigh (1988) were concerned with the whole continent of Australia; State Departments are concerned with extinctions and threats within their political boundaries; Park managers are concerned with losses and threats within park limits. Thus, the area of concern can be an outcome of the area of jurisdiction. On the other hand, the area of concern can be decided on the basis of a process that, if continued even for a presently common species, would lead to the elimination of the species.

There have been numerous extinctions in areas now urban or agricultural; the species concerned may persist elsewhere in the landscape but they are now extinct from the cities and farms. In an area set aside for conservation, there may be a legislative imperative to maintain all species there; such an imperative may mean management will have to be sensitive to species which occupy sites for only particular times and make sure that these particular habitats are ever present.

Space and time may both require consideration from a management point of view. Consider the case of a mistletoe species, and its associated Mistletoe Bird, within a fire-prone conservation reserve. If the Mistletoe is killed by fire and there are no propagules for restoration of the population from within the burned area (Gill 1981), propagules need to come from areas where mature plants exist. Without areas of mature plants, there will be no restoration; without a suitable period of regrowth, the burned area will not be suited to re-establishment of Mistletoe. Thus, areas of potential habitat (where Mistletoe has been eliminated by fire) need to be considered along with areas of current habitat (where mature plants are present), the risk of losing mature plants to fire in areas of current habitat, and the time necessary for potential habitat to become suitable habitat. A species absence may be temporary. Such an argument applies to few plants but to many animals (see discussion on mosaic models, below).

LOCAL EXTINCTION; A NATIONAL SPREAD OF PLANT EXAMPLES

Examples of fire-induced local extinctions of native plants in Australia span the continent. The list which follows should not be considered exhaustive although it does reflect a wide cross section of literature on the subject. Generally, authors do not state over what area these local "extinctions" have taken place. Nomenclature is that of the quoted authors. Examples include:

Gymnospermae

- *Callitris* spp., common small trees in many parts of Australia, including the Pilbara in Western Australia (Burbidge 1985) and parts of Central Australia (Bowman and Latz 1993).
- *Diselma archeri,* a heathland shrub of Tasmania's mountains (Kirkpatrick 1984).
- *Athrotaxis* spp., woodland conifers of Tasmania's mountains (Kirkpatrick 1984).
- *Lagarostrobus franklinii,* a woody Tasmanian endemic found mostly in riparian sites (Gibson *et al.* 1991).

Angiospermae

Casuarinaceae

- *Casuarina distyla,* a heathland shrub associated with *Banksia ericifolia* (Siddiqi *et al.* 1976).
- *Casuarina pusilla,* a heathland shrub (Specht *et al.* 1958).

Chenopodiaceae

- *Atriplex vesicaria* (salt bush), a shrub of the Western Australian goldfields (Mitchell 1986).

Euphorbiaceae

- *Phyllanthus* sp., a small shrub from the Perth area (Baird 1977).

Mimosaceae

- *Acacia aneura* (mulga), a widespread arid zone dominant (Start 1986).
- *Acacia pulchella,* a woodland shrub from southwestern Western Australia (Baird 1977).

Myrtaceae

- *Eucalyptus regnans* (mountain ash), a dominant of tall eucalypt forests in Victoria and Tasmania (Ashton 1981).

Fabaceae

- *Gompholobium tomentosum,* a woodland shrub from Perth, Western Australia (Baird 1977).
- *Hovea* sp., a woodland shrub from the Perth area (Baird 1977).

Proteaceae

- *Banksia ericifolia,* a dominant species of heathlands or woodland understoreys, especially on the central coast of New South Wales (Siddiqi *et al.* 1976).
- *Banksia ornata,* a heathland dominant in South Australia and Victoria (Specht *et al.* 1958; McMahon 1984a).
- *Grevillea nematophylla,* an arid-zone shrub of South Australia (Lay 1986).
- *Hakea teretifolia,* a shrub associated with *Banksia ericifolia* (Siddiqi *et al.* 1976).

Rhamnaceae

- *Pomaderris aspera,* a major understorey shrub or small tree of tall eucalypt forests in Victoria and Tasmania (Ashton 1981).

Thymelaeaceae

- *Pimelea* sp., a small shrub from Perth (Baird 1977).

The causes of these local extinctions is examined below in the section on models. However, it can be noted here that these species come from a diverse array of families and genera, and from a wide range of habitats. Extinction is not the exclusive province of particular plant families, genera or habitats. All the listed species are woody perennials; the absence of species from open-forests of eucalypts in the list may be significant.

While one fire is responsible for the local extinction in the final analysis, the timing of the previous fire is usually critical in the extinction process; fire intensity may also be critical, especially for trees. To generalize, fire effects, including extinctions, depend on the fire regime — a specific set of fire frequencies, fire intensities, seasons of burning and types of fires (Gill 1975, 1981). Leigh and Briggs (1992) list 19 species as being threatened with extinction at state or federal level due to the inappropriateness of current fire regimes.

UNDERSTANDING EXTINCTIONS THROUGH MODELLING

Models are a valuable guide to prediction. However, models are abstractions of reality, caricatures, and do not emulate the real world. Models focus on attributes and events that explain or predict as much variation as possible while keeping their structure as simple as possible. Models use a minimum number of variables for the maximum result. Models may be designed for particular purposes and have various degrees of sophistication. They provide us with a framework around which complications can be considered. There is an art to the development of models exhibited by the method of model construction and the choice of variables; and, there is an art to the use of models because the right model must be chosen for the problem at hand. Below, we consider a number of simple models, broadly defined, which have been used to explain the behaviour of species in relation to fire. We use the word "model", somewhat liberally, to include concepts. Models may be formulations of hypotheses or be explanatory in nature.

Plant models

1. Life-history-and-attribute models

Life history models use only the characteristics of species to predict the outcome of particular fire regimes. These models usually take no account of any effects of weather, fire intensities, or herbivory; they do not make explicit recognition of seed production, probabilistic mortality nor variable rates of establishment. They do consider general sensitivity of plants to fire, location of propagules for replenishment after disturbance, times to set seed, seed longevities and plant longevities. The most comprehensive and normalized model relevant to fire is that of Noble and Slatyer (1981).

Let us consider two simple examples:

(1) If a species vegetatively regenerates after fire, it may be considered to be immune to fires, a circumstance frustrating to managers trying to eliminate some species of woody "weeds" from pasture lands (Wilson and Mulham 1979). However, while immunity may indeed occur in many cases, deliberate setting of fires may cause population depletion over time (e.g., Noble 1989), and the possibility of local extinction. Species in this group are often called "sprouters". Sprouters can be regarded as less vulnerable to fires than the class depicted in the next example, the "seeders".

(2) Many plants are fire sensitive in that any fire that kills the crown of the plant, kills the whole plant. An indicator of crown death is complete leaf death, or "scorch", of the crown, and this has been used as a biological indicator of fire intensity for classifying species' responses (Gill 1981; Gill and Bradstock 1992). While variations in fire intensities in the field may affect responses for woody plants, the assumption of the model would normally be that the plant was killed by any fire. The model is conservative in this respect. Because "seeders" are "killed" by fire, they rely on seed for

regeneration whether the seed is dormant in the soil, held in woody fruits on the plant, or must immigrate from unburnt areas if re-establishment is to occur. The Achilles heel of these species is that if they do not have a suitable seed store when they are burnt a second time, they are eliminated from the site; seed stores in the soil are generally more persistent through fire events than seed stores in plant canopies although considerable variation occurs.

Some plants may exceed their longevity before there is an opportunity for regeneration so the species is potentially lost from the area by too few fires. Sometimes, a community of plants, like a *Banksia ornata* heath, may change character completely upon death of the dominants from over-age, yet the dominant species may survive as seedlings at much reduced densities (McMahon 1984b). Another circumstance is that plants, even vegetatively regenerating ones, may be out-competed by other plants if the interval between fires is long, such that the species loses density or, at the extreme, is eliminated.

Nearly all the plant species mentioned as examples of extinction in the section of examples above are woody, fire-sensitive perennials. There are two exceptions to the fire sensitivity of the examples:

(i) *Leptospermum myrsinoides,* noted by Specht *et al.* (1958) as having "seed and regrowth regeneration . . . equally important"; and,

(ii) the "less vigorously regenerating shrubs" *Hovea, Pimelea* and *Phyllanthus* of a Western Australian woodland (Baird 1977).

As a general rule, the species more vulnerable to extinction through fire may be considered to be those perennials which have plants normally killed by fire, whether the source of propagules for replenishment of their populations after the fire are seeds in the soil (e.g., *Pomaderris*) or seeds in woody fruits on the plant canopy (e.g., *Banksia, Actinostrobus, Leptospermum*) (i.e., category 8 of Gill and Bradstock 1992). Annuals may be considered less vulnerable than perennials to extinction by frequent fires because they produce and shed seed (usually into a soil-seed store) in less than a year while fire intervals are normally a year or longer. For most Australian species, fire responses have not yet been recorded; an attempt is being made to rectify this situation (Gill and Bradstock 1992).

Some rare herbaceous species seem to be endangered due to the inappropriateness of current fire regimes for their persistence. We use two examples from the Canberra area as described by Leigh and Briggs (1992) — *Swainsona recta* and *Rutidosis leptorrhynchoides*. *S. recta* is a perennial herb with colourful pea flowers; it grows among grasses of woodland or open forest. Only 1 200 plants are known. Unlike most of the woody examples, the problem here seems to be the rarity of fire. Burning reduces competition from the grasses, and, under suitable conditions, stimulates seed to germinate. The plants already established are fire resistant. Similarly, *R. leptorrhynchoides* is a perennial herb found among grasses; while the total known population is an order of magnitude greater than that of the *Swainsona,* the absence of burning seems to allow other plants to overwhelm it. These examples could be predicted from life-history-attribute models if the plant longevities and seed longevities were known with and without grass competition.

2. Fire-shadows and "refuges"

Consider a landscape, vegetated with a variety of seeders and sprouters and uniform except for the presence of streams and occasional boulder outcrops or "breakaways". Now impose fires across this landscape. As fires jump the streams in the form of burning brands, there are portions of the downwind stream bank missed by the fire; the same may occur near boulder outcrops. The areas that miss being burned may be called "fire-shadow areas" (by analogy with "rain shadows"). If the intervals between fires on the broad sweep of the landscape decline, we might expect that some seeders with relatively long juvenile periods will become locally extinct but survive as remnants in the fire shadow areas. If this sort of model has any credibility, we would expect to find that fire-shadow areas would have more seeder species than the surrounds and would include species with relatively long juvenile periods, possibly trees. Below, we consider a number of examples of plant distributions in relation to this model. Note that the same model may be invoked if the sites described were simply less flammable remnants rather than areas missed by fire because of "shadowing".

On the southern Tablelands of New South Wales, *Callitris* is often found on the less fire-prone sites in the landscape, such as rocky river valleys. However, there are isolated individuals and populations of the genus on the intervening *Eucalyptus* woodland areas. In these areas native grasses and eucalypt litter may have carried fires much more frequently than the sparser, more disrupted fuels in the valleys. It seems likely that *Callitris* has been eliminated from the interfluves by burning. In similar vein, Lacey (1973) noted that *Callitris* may be excluded from mixtures with hardwoods "under conditions of regular burning". However, grazing by domestic stock and rabbits could also be involved, a situation observed in Victoria (Cheal *et al.* 1979) and southwestern Queensland (Johnston 1969). It is highly likely that numbers of species of the genus *Callitris,* a genus found in all states of the nation, have become extinct in many local areas because of burning practices (Bowman and Latz 1993), perhaps combined with grazing by rabbits or domestic stock.

Grevillea rivularis is a rare shrub that occurs along a restricted section of Kangaroo River and Douglas Creek near Carrington Falls, New South Wales. Briggs and Leigh (1990) summarizing available information on this rare species found it to be absent from areas of similar habitat elsewhere. Plants are fire sensitive and take more than five years to reach reproductive maturity, let alone produce a substantial seed bank. On the basis of species characteristics, the species was considered by Briggs and Leigh (1990) to be threatened by fire (see discussion of this under life histories, below). Whether or not the location along the riverbank was a fire shadow enabling the species to avoid fires on occasion was not discussed.

Muir (1985) argued that rare species may be found in areas where fires are rare. Of the 100 species gazetted as rare in Western Australia, Muir found that 34 were "found on monadnocks, rocky outcrops and breakaways, yet these habitat types [occupied] only a very small proportion of the total land area of the south-west." Pate and Hopper (1993) observed that a large number of gazetted woody species in the south-west were seeders. Furthermore, all endangered species in the family Restionaceae in southwestern Australia were seeders while species in other categories of rarity were mixtures of seeders and sprouters. These observations fit the model suggested at the outset of this section but there has been no test of this model-hypothesis. Rarity of habitat could explain the occurrence of rare species, and susceptibility to drought by seedlings of seeders could be involved (Burgman and Lamont 1992).

3. Edges

Edges are often forgotten because most attention is focused on sweeps of landscapes. However, edges can be the critical dimension of fire-mediated vegetation change in some cases. If a rainforest surrounded by eucalypt forest, for example, is eliminated from an area by the attrition of its edges through repeated fire occurrences, all the rainforest species become locally extinct. Similarly, edge phenomena may be important to the dynamics of *Callitris* communities in some cases: "Crown flaring of dense regeneration occurs on the edges of the stands in contact with the fire . . . and the stems within the stand are unaffected" (Lacey 1973). Our simplest model here would be that the difference in fire regime to which the vegetations, or species, were exposed is not due to "fire shadows" but to differences in intrinsic flammabilities of the vegetations.

In the Northern Territory patches of rainforest known as monsoon forests (Russell-Smith and Dunlop 1987) occur. Monsoon forests are the home of the mound building jungle fowl. Finding mounds outside the forests in eucalypt communities suggests that the monsoon forest has retreated and been replaced (Stocker 1971; Russell-Smith 1985). The same phenomenon occurs in Queensland (Lavarack and Godwin 1987) and Western Australia (Johnstone and Burbidge 1991). The monsoon communities appear often to have been subject to attrition at the margins from fires in the adjacent savannah woodlands "as evidenced by the burnt out shells and standing remnants of monsoon vine forest species" (Russell-Smith and Dunlop 1987). Recent observations in Queensland suggest that rare severe frosts may dispose the edge of the rainforest to greater flammability (Duff and Stocker 1989). While fire may pass right through monsoon forests on occasion, fires in the surrounding vegetation, with flames able to scorch and kill trees on the edges, appear to be more common and able, eventually, to cause extinction at the margins even though a number of species resprout after the first damaging fire (Russell-Smith and Dunlop 1987; Bowman 1991). On the other hand, if fires become rarer and of reduced intensities, rainforest species may encroach into the eucalypt community.

Rainforests of many types occur from the Kimberley in the north-west, across the north of Australia to Cape York Peninsula, and south along the east coast to Victoria and Tasmania (Winter *et al.* 1987). Many of these rainforests occur in what appear to be fire-free havens — such as wet valleys and broken rocky areas — or occur on better soil types than surrounding eucalypt communities. A great deal of comment has been advanced, and controversy generated, regarding the place of fire in controlling rainforest boundaries throughout their distribution (Kahn and Lawrie 1987; Lavarack and Godwin 1987; Russell-Smith and Dunlop 1987; Young and McDonald 1987; Ash 1988; Melick 1990). It seems likely that there are core habitats in which the rainforest species are more-or-less immune to the occurrence of fire because of the moist microclimate, low fuel loadings (sometimes discontinuous as well) and lack of grass, and there are adjacent areas which are facultatively occupied by rainforest. The same idea of core habitat and facultative habitat could apply to other apparently remnant communities.

As an addendum to this discussion it may be pertinent to mention that, just as extinctions are occurring in some places, distributions of communities like rainforests are expanding locally (e.g., Kahn and Lawrie 1987), and densities of species like *Callitris* spp. have increased locally in some areas (Wilson and Mulham 1979). The causes of both the extinctions and expansions may be the same, viz. changed fire regimes.

Animal models

1. *Population models*

Fire has been implicated in the local extinctions of a number of bird species, both rare and common, in Australia. Among the rare bird species, Bristlebirds *Dasyornis broadbentii* and Noisy Scrub Birds *Atrichornis clamosus* in south-western Western Australia have been influenced by changed fire regimes (Smith 1977) while in eastern Australia the Ground Parrot *Pezoporus wallicus* may provide a similar example (Meredith *et al.* 1984). The Mallee Fowl *Leiopoa ocellata*, however, is a relatively common bird which has suffered substantial range and population reductions in Australia (e.g., Burbidge and McKenzie 1989, for Western Australia). Its population dynamics in north-western Victoria have been modelled in relation to fire by Benshemesh (1990).

Mallee Fowl are relatively large, ground dwelling birds remarkable for their use of decomposing litter and soil scratched up into large mounds to control the incubation temperatures of their eggs. Extinction of this species can occur if the long-unburned habitat, in which they may flourish in the absence of feral predators, is burned relatively frequently. Reproduction of mallee fowl may not begin until the vegetation has been unburned for 10 years. Fires at 10 year intervals or less will thus cause local extinction. After the first decade, an increasing proportion of the Mallee Fowl population may begin breeding but it may be 40 years before maximum breeding density occurs. Because of the long period before breeding begins and the graded nature of the return to full breeding potential, fire frequencies of one in 60 years still only lead to a 54 per cent potential of maximum production of offspring according to the model. With fire exclusion, which may be deemed the optimum regime for Mallee Fowl from the models, vegetation changes with unknown implications for Mallee Fowl may take place. Successful fire exclusion, however, is an unlikely possibility in practice. This example raises the question: "can the optimization of performance of one species lead to the local extinction of another?"

2. *Mosaic models*

Mosaic models may be regarded as spatially explicit models. For the "mosaic" to be significant, we may expect that the edges of the patches, or the context of the patches, is important to the result. However, these factors are not always explicit. While "mosaics" and "patches" consisting of elements burned at different times are sometimes advocated as desirable aims of management (see Gill 1986), greater definition is needed in terms of patch sizes, patch shapes, patch densities and patch histories if sensible dialogue is to be conducted.

We can distinguish three types of effects due to patch burning:

(i) The effect of the patch is through the creation of edges. At the edge, shelter may be provided by the unburned vegetation, while the burned area may be more productive of food. This effect would depend on the overlap of home ranges and the fire-caused patch. A good example is that of the rare Rufous Hare Wallaby *Lagorchestes hirsutus* in Central Australia (Bolton and Latz 1978). As home ranges vary in size according to the size of the animal species, and as requirements for shelter and food vary widely, the results of such patchiness in burning may be diverse.

(ii) It is possible that the changed habitat represented by a burned patch larger than the home range of the animal allows the multiplication of the population.

(iii) A third possibility concerns the burning of large areas in which the significance lies in the patches, or "remnants", that did not burn. "Remnants" may be defined as areas where life continues as it would independent of burning in the surrounding area while "refuges" may be defined as areas where animals have temporarily migrated because of the burning of the surrounding area.

All three of these possibilities may occur in association with a large fire.

Small burned patches within a home range become the focus of herbivore activity because of the increased palatability of regrowth. This increases the effective stocking rate of the area and the chances of local extinction of the plants. The multiplier for the maximum possible stocking rate on the burned patch is equal to the inverse of the proportion of home range area burned; if one hundreth of the area burns, then the maximum possible stocking rate on the burned area is 100 times that before any burning. The scale of the phenomenon depends on the size of the animal and the location of the patch in relation to the locations of animal home ranges. In Tasmania, Mount (1976) found that increasing logged-burned areas from 10–20 to 300–400 ha reduced the browsing of tree seedlings to below critical values for tree crop re-establishment. The model can be summarized as follows: the chances of local extinction of plants increases with stocking rate; the stocking rate can be increased locally by increasing palatability and productivity of the vegetation; palatability and productivity often increase due to regrowth after burning; and, if burned patches are small relative to the home range, stocking rates are maximized. Experiments in subalpine woodland have demonstrated that grazing by rabbits after fire can cause the local loss of palatable species (Leigh *et al.* 1987).

An interesting example of the possible effects of fires of various sizes and intensities causing extinctions is that of koalas in eastern Australia (T. W. Norton and A. M. Gill, pers. comm.). As experiments were socially undesirable, a modelling approach using all available information was appropriate. Life history data were essential to the modelling process described here as well as spatial information. The assumptions adopted by the researchers were that: koalas died when the trees in which they lived were completely crown scorched (i.e., all leaves were killed by the fire); koalas reached reproductive age at year two; koalas raised one young per year; koalas moved into burned areas from unburned edges after one year's tree recovery and when one year old; they moved at a rate of 100 m per year; there was no shortage of young of the right sexes at the recolonizing edge due to habitat factors; and, koalas depended solely on edge populations for restocking of burned areas. These considerations, which require amplification and verification, and only advanced here by way of example, suggest a rate of recolonization of burned forest of 50 m per year. If the tree-scorched area was circular and 1 km in diameter, the time for full recolonization of the burned site would be 10 years. Scorching fires at the centre of the burned area every nine years or less would mean that koalas were made locally extinct there. If the burned area was the same but the shape was more elliptical, the time for recolonization would be less. If fire intensities were insufficient to scorch the trees then no losses of animals would occur according to the assumptions made. This type of modelling is not confined in its use to koalas. Mistletoes provide a plant example.

Patch dynamics caused by fires were hypothesized by Burbidge (1985) to have had an important role in the persistence or extinction of small to medium sized marsupials in hummock grasslands of the Australian arid zone. The theory was that with the changed patterns of burning associated with migrations and changing sociologies of Aboriginal communities, various animal species in the areas they occupied have become extinct. According to this theory, as Aboriginal burning decreased, patch size of fires increased (Burrows and Christensen 1990) and the animals became locally extinct. Detailed mechanisms of extinction have not been advanced although the timing of extinction events appears to correspond with the timing of changed burning practices.

Burbidge and McKenzie (1989) reviewed the data for the causes of extinction of native animals in Australia and highlighted the associations of extinction with climate (arid areas more than mesic), animal size (35 to 5 500 g weight, reflecting relatively high energy needs), diet (herbivores and omnivores compared with carnivores) and locomotion (ground dwelling rather than flying animals). These authors saw changed fire regimes in the arid zone reducing vegetative cover "leading to increased evaporation, loss of litter fauna and disrupted nutrient cycling" (p.183), and increased "environmental homogeneity" (p.185).

Bolton and Latz (1978) have provided some detail as to how changed fire regimes may have affected the rare Western Hare-wallaby *Lagorchestes hirsutus* in Central Australia. They emphasized the importance of the juxtaposition of shelter (in spinifex at least four years since fire) and the presence of nearby food (stimulated by recent fire). The ideal "tight mosaic of vegetation in various stages of fire succession" generated by "consistent winter burning" by Aborigines was replaced in most areas by "uncontrolled hot summer fires" started by lightning. This model suggests that the animals owe their existence to the presence of Aborigines and that different sized animals, with different home-range areas, may respond differently to the "mosaic".

Morton (1990), in his studies of extinctions of small mammals in Central Australia, reached similar conclusions to Burbidge and McKenzie (1989). He hypothesized that "refuges" (core areas), which formerly provided sources of animals to surrounding country for the replenishment of populations after severe drought, were degraded to such an extent by overstocking with introduced herbivores that extinctions in times of severe drought resulted. Documenting similar extinctions in South Australia, Tunbridge (1991), in a study independent of Morton's, observed coincidences between extinctions, droughts and occupation of the landscapes by domestic grazing stock. There seems to be no single factor explanation; fires, feral animals, domestic stock, droughts and "refuges" may all be involved.

Burning may alter the quality of habitat dramatically for certain animals such that, with time since fires, remarkable changes in abundance may occur. One well-studied example is that of animal communities in the heaths and forests at Nadgee Nature Reserve in southeastern

New South Wales (Newsome and Catling 1983). For the first few years after fire, the introduced House Mouse *Mus domesticus* flourished. (*Pseudomys* provides a native analogue but was not present in the area of study and may have become locally extinct — Catling 1986.) *Rattus* then became prominent, then *Antechinus* spp. After each population wave, a decline followed by very low populations occurred; the low populations that were present were probably transients rather than breeding animals (Catling, pers. comm.).

An important question for understanding the way in which fire affects small mammal communities and how it could cause more permanent local extinctions is: what is the source of small mammals for each wave of population increase? Catling (pers. comm.) believes that certain sites are core areas for species and these sites are unaffected or little affected by fire; house mice were found on seaward dunes and in boulder fields at the foot of coastal cliffs, while native small mammals are believed to persist in core areas such as *Gahnia* creek lines — apparently unaffected by fire for long periods in the Nadgee area. A critical factor in management then is the identification and care of core areas and the prevention of the complete burning of isolated reserves — whether by high or low intensity fire. (Note that the animals may persist whether or not the non-core areas are patch burned.) If the core area is destroyed (by roading, erosion, sedimentation, etc.) and the whole reserve burns, then there is no source of animals to replenish populations made locally extinct by the change in habitat. A corollary is that if a reserve is declared which has no core area, and is completely burned, then the reserve may be depleted of its extant small mammal representation. This example is reminiscent of postulates for the extinctions of small to medium sized mammals of the arid zone although the circumstances are somewhat different. A suggestion that combines elements of various theories proposed for extinctions in the arid zone, and mimics the suggestions for the heaths and forests, would be that extinctions were caused by the destruction of the core areas by drought and domestic grazing animals (Morton 1990), along with large-area fires in the surrounding country (Burbidge 1985; Burrows and Christensen 1990).

Fox (1982) studied small mammals in the Myall Lakes area and compared his data with similar data from Victoria. From a correlation between diversity of small mammal species and habitat diversity (variation in the vegetation), Fox (1982) reasoned that habitat diversity created by different times since fire would show the same correlation. Thus, to maintain maximum diversity of small mammals — and presumably minimize the chances of their extinctions — he advocated maintaining a patchiness in vegetation caused by a diversity of times since fires.

The consideration of edges, remnants, fire-intensity distributions within fire events, and arrangements, sizes, ages and densities of burned and remnant patches, is most important to management. All these factors, and any interactions between them, could influence the population densities and species richnesses of animals in reserves. We need to know more of the locations and relative flammabilities of core areas and how fire regimes affect their biota.

3. Habitat models

Habitat models are particularly useful for the design of management systems. The importance of hollows for many species of Australian birds and arboreal mammals is now well known, and this aspect of habitat can be limiting for population size and the possibility of local extinction (e.g., Lindenmeyer 1992). In forestry, the dynamics of hollows becomes particularly important as logging and burning can have immediate and dramatic impact. We do not further explore the subject of hollows here but rather concentrate on broader issues of habitat. In our example we use the habitat score of Newsome and Catling (1979) because of its sensitivity to fires.

Newsome and Catling (1979) related the number of terrestrial vertebrate species in a habitat to the characteristics of that habitat. A score was calculated on the basis of tree cover, shrub cover, herbaceous cover, presence of hollow logs, water supply and site rockiness. Because the organic components of this index are related to the occurrences of fires, fire regimes (Gill 1975) change the score and, consequently, the numbers of species — either up or down. If the numbers go down, then some species have become locally extinct. (Note that we are not referring here to changes taking place due to plant development after fire, although the index encompasses such changes.) Catling (1991) used the score, and the types of changes in habitat likely to have taken place as the result of particular fire regimes, to suggest possible management strategies for the maintenance of species diversity (and the avoidance of extinction). Some might postulate that the scenario most likely to lead to extinction of native animals is in a reserve, completely isolated from all other suitable habitat (thereby restricting immigration of native animals but maximizing opportunity for feral predators) and long unburned (when populations of many species would be at a very low ebb), which is then completely burnt by crown consuming fires (making maximum changes to habitat and creating maximum exposure to predators). However, in model comparisons between rare high intensity spring fires and frequent autumn low intensity fires (the preferred forest management scenario for litter reduction in southeastern Australian forests), extinctions were postulated as being more likely under the latter scenario. Under the regime of frequent low intensity fires many introduced species were postulated to be advantaged together with an equal number of native species; twice as many native species, and only one introduced species, would be disadvantaged by this regime.

Catling suggested that dense shrubby understories to forests result from severe spring fires at low frequencies. However, most fire managers would baulk at the idea of deliberately lighting intense fires for management purposes because of the chances of damage to timber values or the chances of fire escape. Possibly, low intensity fires on dry soils could achieve a similar result for shrubby legumes although the chances of fire escape may be considered by managers to be too great under these circumstances as well. There is a need to examine the circumstances under which dense shrubby understories can be formed as the result of particular fire regimes and species attributes.

CHANGED FIRE REGIMES

Changes to fire regimes, or the effects of different fire regimes, have been a theme in the discussion of the models introduced above. Aborigines are likely to have changed fire regimes when they entered the country thousands of years ago but the way they changed fire regimes is often obscure. When Europeans arrived 200 years ago, further changes were wrought which varied from burning as often as possible with maximum intensity possible (Wakefield 1970) to attempted elimination of fire. In the arid zone, average fire size increased by an order of magnitude on the cessation of traditional aboriginal burning (Burrows and Christensen 1990). Changes are still occurring as new species become naturalized, as fire management policies change, and as land uses change.

In many places there have been invasions of exotic plants which may have had an impact on survival of species through the changes they allow in fire regimes and/or the effects they may have on seedling competition after fire. The number of examples of these effects is increasing. Perhaps the first to receive exposure in the Australian literature was the example of exotic grass invasion into King's Park, Western Australia through "fires and firebreaks" (Baird 1977). Though not specific about any changes the grass may have on fire regimes (Baird 1977) noted that the effect of the introduced grass "has been to change the whole cycle of regeneration after fires". Examples like this are multiplying in southwestern Australia where disturbances and fertilizers are allowing invasions of exotic plants into native communities (Hobbs and Atkins 1988). Many of these communities contain fire sensitive shrubs with canopy stored seed and may have become vulnerable to potential increases in fire frequency. Invasions of exotics into mallee from road verges are occurring in New South Wales and Victoria so fire frequencies there may have increased also.

Exotic plants often affect fire regimes in various ways. Stocker and Mott (1981) suggest that *Lantana* can block grass invasion in disturbed rainforests near Cairns and provide a buffer to low intensity fires in the early dry season; on the other hand, *Melinus*, a grass which is a rapid colonizer with vigorous growth, can support high intensity fires. Also in the tropics, *Pennisetum* (in the Darwin area), may alter the intensities and times of year at which fires occur.

At rainforest edges in the Kimberley, fire intensities may have been exacerbated — and botanical erosion of patches enhanced — by the establishment of grasses. Grass establishment, assisted by seed dispersal through domestic cattle, allows fires to penetrate further into rainforest than they would have in the absence of cattle (Kendrick and Rolfe 1991; McKenzie and Belbin 1991).

Logging practices affect fire regimes in many ways, sometimes creating significant artificial-fuel arrangements (Williams *et al.* 1994). In wetter eucalypt forests, logging practices (probably including the effects of fires), may cause sharply reduced numbers of mature plants of sensitive species like *Dicksonia antarctica* (and their dependent epiphytes), possibly to the extent of causing local extinction (Ough and Ross 1992; Mueck and Peacock 1992).

In Tasmanian alpine areas, Kirkpatrick and Dickinson (1984) suggest that repeated fires may have their main impact on species composition through soil erosion.

MANAGEMENT SCENARIOS

In this section, we list some possible aims of management for the conservation of species and assess the consequences of those aims for the avoidance of extinctions. For brevity and convenience, we do not consider any interaction of these aims with other aims such as protection of human life and buildings. We are concerned with the intervals between fires and the area over which burning might take place. Other aspects of fire regimes are important but are neglected here. Concerning the area burned, we neglect important questions relating to optimal shapes and densities of areas to be burned. We address three major questions, albeit briefly.

- What are the consequences to species survival if we prescribe fire in order to: maximize performance of a single species; maximize plant-species richness; maximize variance in fire intervals; or "keep the bush healthy"?
- What are the consequences to species survival if we prescribe fire to encompass the whole area of the reserve or, burn only a fraction of any species' distribution, community, ecosystem or reserve?
- Can we prescribe fire to avoid extinctions?

Consequences to species survival

1. Maximize performance of a single species

It may be tempting to centre one's management around the performance of a single species, perhaps a dominant plant species, an endangered species or a "cuddly and furry" animal. While our knowledge of any one community is insufficient to explore the consequences of this, we touch on some recent ideas.

In Western Australian heathlands, Burgman and Lamont (1992) suggested that a fire regime to maximize performance of the obligate seeder *Banksia cuneata,* a rare species, does **not** minimize the probability of its own extinction. They assumed that the probability of extinction of the species was high in the young seedling stage because the chances of drought death in the first year after fire were high. Thus, every fire put the species into a potentially extinction-prone position. Even so, if fire was completely excluded (a theoretical possibility, if not actual), the longevity of the species would be exceeded and extinction would appear inevitable. The species has existed presumably for hundreds of generations so, to postulate that drought death of all seedlings is possible or likely now, some climatic or other change must be invoked (as the authors do). Many assumptions were made in modelling this situation and the authors stress the need for further data.

Maximizing the cover of dominant plant species may minimize the apparent species richness of the community (Specht and Specht 1989) but not necessarily affect extinction probability of non-dominant species. Indeed, in heathlands, the probability of the extinction of species by fires appears greatest for the dominant — often an obligate seeder such as *B. ornata* or *B. ericifolia.*

The models described for Mallee Fowl suggested that the longer the fire-free interval, the greater the success of the bird (Benshemesh 1990). Taken to extremes, this suggests that fire exclusion would be the best policy. However, fire exclusion is likely to lead to the loss of some plant species, even the dominant mallee eucalypts. Extrapolation of results of the models is unwise.

2. Maximize plant species richness

The plotting of the number of plant species with time since fire has become a standard way of illustrating change. In many plant communities, species richnesses may decrease with time after fire: e.g., in hummock grasslands of arid areas (Allan and Baker 1990; Burrows and Christensen 1990); or in heathlands of south-eastern Australia (Specht *et al.* 1958; McMahon 1984b; Cheal *et al.* 1979). Frequent burning may seem desirable to maintain the floral diversity in such cases but, although the *apparent* species richness may be greater on average over time, *total* diversity (including species present as soil seed) may have been reduced through elimination of species, even the dominants in some cases.

3. Maximize variance in fire intervals

Concern has been expressed at the monotonic intervals considered appropriate to prescribed burning in some Australian plant communities. Varying the fire interval seems more "natural" than having a uniform fire interval. Maximizing variance is the opposite extreme; it may represent what has happened historically in some plant communities where, in one place, maximum frequencies may have been considered desirable, whereas in others, absence of fire has been the aim. In this discussion, maximizing variance would be the aim through time, rather than in space. Maximizing variance would undoubtedly cause the extinction of some plant species because maximization of variance would include some short interval fires — which appear to be the prime circumstances for local extinctions of woody plants in Australia — and would include some very long intervals — which appear to pose a problem for the survival of some herbaceous species also.

4. Keep the bush healthy

The remarkable recuperative power of the bush in Australia is often noted by the public with the assumption that it has returned just the way it was. Vigour has been equated with composition. With species in some *forests,* this could be a reasonable assumption because species richness is constant with time after fire within certain limits (Purdie and Slatyer 1976; Venning 1978; Gill *et al.* 1991). An abundance of foliage should not be taken as an indicator of species richness, however; comparisons before and after fire are necessary for any change, including local extinction, to be detected.

5. Burn the whole reserve?

It could be argued that there is an optimum solution to the problem of the avoidance of extinction such that a whole reserve, community or ecosystem be treated uniformly. However, such arguments do not consider the likely occurrence of an unexpected fire which may, if it burns the whole area, eliminate many species like some obligate seeders. A consequence of this could be the elimination of animals dependent on "seeders", e.g., the Glossy Black Cockatoo *Calyptorhynchus lathami* which appears to depend on the fruit of an obligate reseeding *Casuarina* in the Pilliga area of New South Wales (observations of R.A.B.). If only a portion of the reserve is burnt by prescription, then the impact of the unexpected premature fire that burns the whole reserve is reduced. If portion of the reserve only is to be burned, which portion of which species distribution, community, ecosystem or reserve area should be chosen? This question is unanswerable at the moment but, as a general principle for plants, we suggest that the reserve be managed for the most sensitive species and that not all the reserve be burned at any one time. For small mammals on the north coast of New South Wales, Fox (1982) advocated patch burning after observing a correlation between habitat diversity and small mammal species diversity; a direct test of this idea is needed. There may be circumstances where management of reserves for mobile animals, like nectar-feeding birds, involves the consideration of fire management across regional sets of reserves, rather than for individual reserves independently.

Can we prescribe fire to avoid extinctions?

In experiments (involving scientific monitoring), the imposition of a new, or changing, fire regime may take one cycle of the new regime to indicate change or may take many cycles before a new "equilibrium" is established, the latter case being particularly pertinent with vegetatively regenerating plants such as those Western Australian shrubs noted by Baird (1977) and the New South Wales shrubs modelled by Bradstock and Myerscough (1988); several cycles may be needed for habitat change for fauna (Catling 1991). Therefore, anticipation of the impacts of changes in fire regimes with respect to extinctions may take long periods of observation over a number of "cycles" if trends are to be ascertained. Concluding that no species will go extinct after observing the restoration of species numbers after one fire may be unrealistic in the long term. The same point is pertinent when fire regimes are changing historically. In nature, it is possible that there is no equilibrium at all and that the vegetation is always in a state of flux, sometimes too subtle to measure in the short term. Monitoring of some sort is necessary if we are to know what is actually happening to species.

A practical system of species monitoring as a guide to prescribing fire provides the most effective way of avoiding plant extinctions at present. The system devised by Gill and Nicholls (1989) uses relatively fire-sensitive species as site indicators of plant condition. After all species have flowered following a fire, a period of time is left for seed build up to occur, then the site is considered to be "safe" from extinctions should a fire be prescribed or occur unexpectedly. Currently, the system is only operative for "higher" plant species, not animal species (including invertebrates) or non-vascular plants.

CONCLUSIONS

"Extinction is forever" only when the last plant and seed of a species, or the last living animal, has died. However, extinction can also be a local phenomenon so that a reversal is possible. Fires have caused the local extinctions of many species of fire sensitive plants, even dominants, across a wide variety of Australian habitats. Some native animals have become extinct in modern times in the arid zone but the place of fires in these extinctions is uncertain at present. Models based on the life history characteristics, spatially significant processes and habitat modifications help explain why species have, and may yet, become locally extinct. Certain management aims may flow from modelled situations but care and insight are necessary if inferences are to be accurate and appropriate; particular caution is needed if the use of a model developed in one region is considered for application in another.

ACKNOWLEDGEMENTS

This paper has benefited from the comments and suggestions made by Messrs John Briggs and Peter Catling, and Drs Gordon Friend and David Keith.

REFERENCES

Allan, G. and Baker, L., 1990. Uluru (Ayers Rock-Mt Olga) National Park: an assessment of a fire management programme. *Proc. Ecol. Soc. Aust.* **16:** 215–20.

Ash, J., 1988. The location and stability of rainforest boundaries in north-eastern Queensland, Australia. *J. Biogeog.* **15:** 619–30.

Ashton, D. H., 1981. Fire in tall open-forests (wet sclerophyll forests). Pp. 339–66 in *Fire and the Australian biota* ed by A. M. Gill, R. H. Groves and I. R. Noble. Australian Academy of Science: Canberra.

Baird, A. M., 1977. Regeneration after fire in King's Park, Western Australia. *J. Roy. Soc. West. Aust.* **60:** 1–22.

Benshemesh, J., 1990. Management of mallee fowl — with regard to fire. Pp. 206–11 in *The mallee lands. A conservation perspective* ed by J. C. Noble, P. J. Joss and G. K. Jones. CSIRO: Melbourne.

Bolton, B. L. and Latz, P. K., 1978. The Western Hare-wallaby *Lagorchestes hirsutus* (Gould) (Macropidae), in the Tanami Desert. *Aust. Wildl. Res.* **5:** 285–93.

Bowman, D. M. J., 1991. Recovery of some northern Australian monsoon forest tree species following fire. *Proc. Roy. Soc. Qld* **101:** 21–5.

Bowman, D. M. J. and Latz, P. K., 1993. Ecology of *Callitris glaucophylla* (Cupressaceae) on the MacDonnell Ranges, Central Australia. *Aust. J. Bot.* **41:** 217–25.

Bradstock, R. A. and Myerscough, P. J., 1988. The survival and population response to frequent fires of two woody resprouters *Banksia serrata* and *Isopogon anemonifolius*. *Aust. J. Bot.* **36:** 415–31.

Briggs, J. D. and Leigh, J. H., 1988. Rare or threatened Australian plants. 1988 revised list. Australian National Parks and Wildlife Service Special Publication 14.

Briggs, J. D. and Leigh, J. H., 1990. Delineation of important habitats of threatened plant species in south eastern New South Wales. Research Report to the Australian Heritage Commission.

Burbidge, A. A., 1985. Fire and mammals in the hummock grasses of the arid zone. Pp. 91–94 in *Fire ecology and fire management in Western Australian ecosystems* ed by J. R. Ford. Western Australian Institute of Technology Environmental Studies Group Report 14.

Burbidge, A. A. and McKenzie, N. L., 1989. Patterns in the modern decline of Western Australia's vertebrate fauna: causes and conservation implications. *Biol. Cons.* **50:** 143–98.

Burgman, M. A. and Lamont, B. B., 1992. A stochastic model for the viability of *Banksia cuneata* populations: environmental, demographic and genetic effects. *J. Appl. Ecol.* **29:** 719–27.

Burrows, N. D. and Christensen, P. E. S., 1990. A survey of aboriginal fire patterns in the Western Desert of Australia. USDA Forest Service General Technical Report SE-69, pp. 297–305.

Catling, P. C., 1986. *Rattus lutreolus,* colonizer of heathland after fire in the absence of *Pseudomys* species? *Aust. Wildl. Res.* **13:** 127–39.

Catling, P. C., 1991. Ecological effects of prescribed burning practices on the mammals of southeastern Australia. Pp. 354–63 in *Conservation of Australia's Forest Fauna* ed by D. Lunney. Royal Zoological Society of New South Wales: Mosman.

Cheal, P. D., Day, J. D. and Meredith, C. W., 1979. Fire in the national parks of north-west Victoria. Victorian National Parks Service.

Duff, G. A. and Stocker, G. C., 1989. The effects of frosts on rainforest/open forest ecotones in the highlands of north Queensland. *Proc. Roy. Soc. Qld* **100:** 49–54.

Fox, B. J., 1982. The influence of disturbance (fire, mining) on ant and small mammal species diversity in Australian heathland. USDA Forest Service General Technical Report PSW-58, pp. 213–19.

Gibson, N., Davies, J. and Brown, M. J., 1991. The ecology of *Lagarostrobos franklinii* (Hook.f) Quinn (Podocarpaceae) in Tasmania. 1. Distribution, floristics and environmental correlates. *Aust. J. Ecol.* **16:** 215–22.

Gill, A. M., 1975. Fire and the Australian flora: a review. *Aust. For.* **38:** 4–25.

Gill, A. M., 1981. Adaptive responses of Australian vascular plant species to fires. Pp. 243–71 in *Fire and the Australian biota* ed by A. M. Gill, R. H. Groves and I. R. Noble. Australian Academy of Science: Canberra.

Gill, A. M., 1986. Research for the fire management of Western Australian state forests and conservation reserves. Western Australian Department of Conservation and Land Management Technical Report 12. 82 pp.

Gill, A. M. and Bradstock, R. A., 1992. A national register for the fire responses of plant species. *Cunninghamia* **2:** 653–60.

Gill, A. M. and Nicholls, A. O., 1989. Monitoring fire-prone flora in reserves for nature conservation. In *Fire management on nature conservation lands* ed by N. Burrows, L. McCaw, and G. Friend. Western Australian Department of Conservation and Land Management Occasional Paper 1/89.

Gill, A. M., Hoare, J. R. L. and Cheney, N. P., 1991. Fires and their effects in the wet-dry tropics of Australia. Pp. 159–78 in *Fire in the tropical biota* ed by J. Goldammer. Springer: Berlin.

Hobbs, R. J. and Atkins, L., 1988. Effect of disturbance and nutrient addition on native and introduced annuals in plant communities in the Western Australian wheat belt. *Aust. J. Ecol.* **13:** 171–79.

Johnston, T. N., 1969. The effect of sheep and rabbit grazing on regeneration of white cypress pine. *Aust. For. Res.* **4**(2): 3–12.

Johnstone, R. E. and Burbidge, A. H., 1991. The avifauna of Kimberley rainforests. Pp. 361–91 in *Kimberley rainforests of Australia* ed by N. L. McKenzie, R. B. Johnson and P. G. Kendrick. Surrey Beatty & Sons: Chipping Norton.

Kahn, T. P. and Lawrie, B. C., 1987. Vine thickets of the inland Townsville region. Pp. 159–99 in *The rainforest legacy, Australian national rainforests study I.* Australian Heritage Commission Special Australian Heritage Publication Series 1.

Kendrick, P. G. and Rolfe, J. K., 1991. The reptiles and amphibians of Kimberley rainforests. In *Kimberley rainforests of Australia* ed by N. L. McKenzie, R. B. Johnson and P. G. Kendrick. Surrey Beatty & Sons: Chipping Norton.

Kirkpatrick, J. B., 1984. Tasmanian high mountain vegetation II — Rocky Hill and Pyramid Mountain. *Pap. Proc. Roy. Soc. Tas.* **118:** 5–20.

Kirkpatrick, J. B. and Dickinson, K. J. M., 1984. The impact of fire on Tasmanian alpine vegetation and soils. *Aust. J. Bot.* **32:** 613–29.

Lacey, C. J., 1973. Silvicultural characteristics of white cypress pine. Forestry Commission of New South Wales Research Note 26.

Lavarack, P. S. and Godwin, M., 1987. Rainforests of northern Cape York Peninsula. Pp. 201–22 in *The rainforest legacy, Australian national rainforests study I.* Australian Heritage Commission Special Australian Heritage Publication Series 1.

Lay, B. G., 1986. The significance of fire in arid rangelands of South Australia. P. 604 in *Rangelands: a resource under siege* ed by P. J. Joss, P. W. Lynch and O. B. Williams. Australian Academy of Science: Canberra.

Leigh, J. H. and Briggs, J. D. (eds), 1992. *Threatened Australian plants: overview and case studies.* Australian National Parks and Wildlife Service: Canberra.

Leigh, J. H., Wimbush, D. J., Wood, D. H., Holgate, M. D., Slee, A. V., Stanger, M. G. and Forrester, R. I., 1987. Effects of rabbit grazing and fire on a subalpine environment. 1. Herbaceous and shrubby vegetation. *Aust. J. Bot.* **35:** 433–64.

Lindenmeyer, D., 1992. The ecology and habitat requirements of arboreal marsupials in the montane ash forests of the central highlands of Victoria: a summary of studies. Victorian Department of Conservation and Environment. VSP Internal Report 6. 15 pp.

McKenzie, N. L. and Belbin, L., 1991. Kimberley rainforest communities: reserve recommendations and management considerations. Pp. 453–80 in *Kimberley rainforests of Australia* ed by N. L. McKenzie, R. B. Johnson and P. G. Kendrick. Surrey Beatty & Sons: Chipping Norton.

McMahon, A., 1984a. The effects of fire regime components on heathlands in the Little Desert, N.W. Victoria, Australia. Pp.101–02 in *MEDECOS IV. Proceedings of the 4th international conference on mediterranean ecosystems* ed by B. Dell. Botany Department, University of Western Australia: Nedlands.

McMahon, A., 1984b. The effects of time since fire on heathlands in the Little Desert, N.W. Victoria, Australia. Pp. 99–100 in *MEDECOS IV. Proceedings of the 4th international conference on mediterranean ecosystems* ed by B. Dell. Botany Department, University of Western Australia: Nedlands.

Melick, D. R., 1990. Ecology of rainforest and sclerophyllous communities in the Mitchell River National Park, Gippsland, Victoria. *Proc. Roy. Soc. Vic.* **102:** 71–87.

Meredith, C. W., Gilmore, A. M. and Isles, A. C., 1984. The ground parrot (*Pezoporus wallicus Kerr*) in south-eastern Australia: a fire adapted species? *Aust. J. Ecol.* **9:** 367–80.

Mitchell, A. A., 1986. The effects of fire on a bluebush *(Maireana sedifolia)* shrubland in the Western Australian goldfields. Pp. 606–07 in *Rangelands: a resource under siege* ed by P. J. Joss, P. W. Lynch and O. B. Williams. Australian Academy of Science: Canberra.

Morton, S. R., 1990. The impact of European settlement on the vertebrate animals of arid Australia: a conceptual model. *Proc. Ecol. Soc. Aust.* **16:** 201–13.

Mount, A. B., 1976. Some problems associated with small coupes. *Search* **7:** 462.

Mueck, S. G. and Peacock, R. J., 1992. Impacts of intensive timber harvesting on the forests of East Gippsland, Victoria. Victorian Department of Conservation and Environment VSP Technical Report 15. 107 pp.

Muir, B. G., 1985. Fire exclusion: a baseline for change? Pp.119–28 in *Fire ecology and fire management in Western Australian ecosystems* ed by J. R. Ford. Western Australian Institute of Technology Environmental Studies Group Report 14.

Newsome, A. E. and Catling, P. C., 1979. Habitat preferences of mammals inhabiting heathlands of warm temperate coastal, montane and alpine regions of south-eastern Australia. Pp. 301–16 in *Ecosystems of the World. 9A. Heathlands and related shrublands* ed by R. L. Specht. Elsevier: Amsterdam.

Newsome, A. E. and Catling, P. C., 1983. Animal demography in relation to fire and shortage of food: some indicative models. *Ecol. Stud.* **43:** 490–505.

Noble, I. R. and Slatyer, R. O., 1981. Concepts and models of succession in vascular plant communities subject to recurrent fire. Pp. 311–35 in *Fire and the Australian biota* ed by A. M. Gill, R. H. Groves and I. R. Noble. Australian Academy of Science: Canberra.

Noble, J. C., 1989. Fire regimes and their influence on herbage and mallee coppice dynamics. Pp. 168–80 in *Mediterranean landscapes in Australia. Mallee ecosystems and their management* ed by J. C. Noble and R. A. Bradstock. CSIRO: Melbourne.

Ough, K. and Ross, J., 1992. Floristics, fire and clearfelling in wet forests of the Central Highlands, Victoria. Victorian Department of Conservation and Environment VSP Technical Report 11. 49 pp.

Pate, J. S. and Hopper, S. D., 1993. Rare and common plants in ecosystems, with special reference to the south-west Australian flora. *Ecol. Stud.: Anal. Synth.* **99:** 293–325.

Purdie, R. W. and Slatyer, R. O., 1976. Vegetation succession after fire in sclerophyll woodland communities in south-eastern Australia. *Aust. J. Ecol.* **1:** 223–36.

Russell-Smith, J. J., 1985. Records of change: studies of vegetation history on the South Alligator River, Northern Territory. *Proc. Ecol. Soc. Aust.* **13:** 191–202.

Russell-Smith, J. and Dunlop, C., 1987. The status of monsoon vine forests in the Northern Territory: a perspective. Pp. 227–88 in *The rainforest legacy, Australian national rainforests study I.* Australian Heritage Commission Special Australian Heritage Publication Series 1.

Siddiqi, M. Y., Carolin, R. C. and Myerscough, P. J., 1976. Studies in the ecology of coastal heath in New South Wales. III. Regrowth of vegetation after fire. *Proc. Linn. Soc. NSW* **101:** 53–63.

Smith, G. T., 1977. The effect of environmental change on six rare birds. *The Emu* **77:** 173–79.

Specht, R. L., Rayson, P. and Jackman, M. E., 1958. Dark Island heath (Ninety-Mile Plain, South Australia) VI. Pyric succession: changes in composition, coverage, dry weight and mineral nutrient status. *Aust. J. Bot.* **6:** 59–88.

Specht, R. L. and Specht, A., 1989. Species richness of sclerophyll (heathy) plant communities in Australia – the influence of overstorey cover. *Aust. J. Bot.* **37:** 337–50.

Start, A. N., 1986. Status and management of mulga in the Pilbara region of Western Australia. Pp. 136–38 in *The Mulga lands* ed by P. S. Sattler. Royal Society of Queensland: Brisbane.

Stocker, G. C., 1971. The age of charcoal from old jungle fowl nests and vegetation change on Melville Island. *Search* **2:** 28–30.

Stocker, G. C. and Mott, J. J., 1981. Fire in the tropical forests and woodlands of northern Australia. Pp. 425–39 in *Fire and the Australian biota* ed by A. M. Gill, R. H. Groves and I. R. Noble. Australian Academy of Science: Canberra.

Tunbridge, D., 1991. *The story of the Flinders Ranges.* Kangaroo Press: Kenthurst, New South Wales.

Venning, J., 1978. Post-fire responses of a *Eucalyptus baxteri* woodland near Penola in South Australia. Aust. For. **41:** 192–206.

Wakefield, N. A., 1970. Bushfire frequency and vegetation change in south-eastern Australian forests. *Vic. Nat.* **87:** 152–58.

Weston, A. S., 1985. Fire — and persistence of the flora of Middle Island, a southwestern Australian offshore island. Pp. 111–18 in *Fire ecology and management in Western Australian ecosystems* ed by J. R. Ford. Western Australian Institute of Technology Environmental Studies Group Report 14.

Williams, J. E., Whelan, R. and Gill, A. M., 1994. Fire and environmental heterogeneity in southern temperate forest ecosystems: implications for management. *Aust. J. Bot.* **42:** 125–37.

Wilson, A. D. and Mulham, W. E., 1979. A survey of the regeneration of some problem shrubs and trees after wildfire in western New south Wales. *Aust. Range. J.* **1:** 363–68.

Winter, J. W., Atherton, R. G., Bell, F. C. and Pahl, L. I., 1987. An introduction to Australian rainforests. Pp. 1–7 in *The rainforest legacy, Australian national rainforests study I.* Australian Heritage Commission Special Australian Heritage Publication Series 1.

Young, P. A. R. and McDonald, W. J. F., 1987. The distribution, composition and status of the rainforests of southern Queensland. Pp. 119–41 in *The rainforest legacy, Australian national rainforests study I.* Australian Heritage Commission Special Australian Heritage Publication Series 1.

Fire and conservation: imperatives and constraints on managing for diversity

R. A. BRADSTOCK,[1] D. A. KEITH[1] and T. D. AULD[1]

Management of fire to achieve biodiversity conservation is constrained by a number of problems including inadequate knowledge, fragmentation of natural systems and the concurrent need to achieve other objectives such as protection of human life. These problems have contributed to a lack of flexibility of fire management in many instances. We discuss examples which deal with the derivation of target fire regimes to achieve conservation goals, emphasizing the inflexible and restrictive nature of these target regimes. Examples are drawn from fire-prone communities in southeastern Australia. A more flexible approach when setting fire management objectives may be achieved by using principles of conservation biology, such as population extinction risk as an alternative to density maximization. We illustrate how a flexible approach may operate in practice using the example of shrubland communities. In these communities, recent research indicates that the variance of fire frequency is positively related to plant species diversity. Thus it is impossible to set a fixed fire regime target for conservation management. Rather, management there must be flexible and continually take into account the history of fire. Short-term actions can be tailored to meet the objective of providing fire regime variability using the concept that there is a threshold in fire regime variability that demarcates a change from high to low species diversity. An extension of this concept, to deal with the problem of managing fire concurrently for conservation and protection, is outlined using some conceptual models. These models indicate the way in which the use of prescribed fire can be quantitatively related to protection and conservation on an explicit landscape scale.

INTRODUCTION

FIRE is a conservation issue of national significance (ANZECC 1993). Fire is important because of its influence on species and the ecological processes that mediate the persistence of species in communities and ecosystems. Fire is also a landscape process that people manipulate.

Conservation must deal with the problem of maintaining assemblages of species and the ecological processes associated with those species (Western 1992). Within that context, fire management depends on the formulation of clear objectives that can be addressed through the manipulation of fire regimes. Aside from managing fire to achieve conservation, other objectives such as human protection and protection or enhancement of natural resources harvested or exploited by humans (e.g., timber, water, pastures, etc.) are important. Particular fire regimes may be needed to meet each of these objectives and we contend that the appropriate regimes for each objective may overlap to some degree (Fig. 1). Fire management solutions for a particular combination of objectives (field of overlap) may be different. The way in which conservation objectives for fire management are framed and achieved may vary according to the particular mix of objectives that must be met in a landscape or area. The overlap between the fire regimes needed to achieve protection from wildfires and the regimes necessary for conservation is of great importance in many remaining areas of natural and semi-natural bushland. A major imperative is to define this overlap and ensure that management targets the fire regimes within it.

Despite much research, the ability to define the fire regimes suitable for conservation in any particular place or system is limited. A number of major constraints to knowledge and the way in which knowledge is applied to the management of fire stand in the way. Current approaches to fire management often seek to impose inflexible fire regimes on natural systems. Inflexibility can result from a combination of limited knowledge and bureaucratic and political pressures. Inflexibility may unnnecessarily restrict the fire regime options needed to conserve biota and thus restrict opportunites to meet other concurrent management objectives. Other problems contribute to limit the effectiveness of management. These include:

(1) Native species and their habitats now exist in a fragmented state (Saunders *et al.* 1987). Maintaining populations of species and their habitats is difficult because of the reduced contiguity of populations (Soulé and Gilpin 1991; Burgman *et al.* 1993). Turning back the clock to manage for putative fire regimes prior to fragmentation is therefore unlikely to succeed, even where the nature of such regimes can be properly defined (Gill 1977).

(2) Knowledge of the ecological effects of fire is limited in its coverage of species and interaction with life history processes. Often, what we know amounts to a loose set of "cautionary tales" (Whelan and Muston 1991) that describe changes to species at points on the ground but which are difficult to extrapolate to populations spread across landscapes. Given the nature of fire as a landscape process and the problem of habitat fragmentation, this is a significant limitation.

---

[1]NSW National Parks and Wildlife Service, Box 1967, Hurstville, New South Wales, Australia 2220.

Pages 323–33 *in* CONSERVING BIODIVERSITY: THREATS AND SOLUTIONS, ed by R. A. Bradstock, T. D. Auld, D. A. Keith, R. T. Kingsford, D. Lunney and D. P. Sivertsen. Surrey Beatty & Sons. 1995.

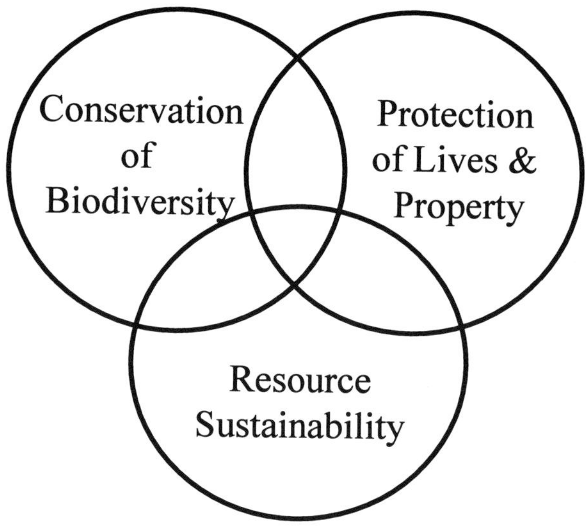

Fig. 1. A conceptual representation of the overlapping sets of fire regimes needed to achieve varying land management objectives.

(3) Our knowledge of fire as a landscape process is limited. Notwithstanding the ability of people to use fire in a controlled manner, we know little about how patterns of fire regimes are generated within landscapes and about how people may alter these patterns.

Here we examine several syndromes of fire management for conservation that illustrate these problems and show how these problems have arisen because of limitations in knowledge and its application. We point the way to a flexible approach to fire management that may more fully encompass the fire regimes necessary for conservation. In doing so we suggest future directions for research and the framework of management necessary to apply knowledge on the ground. We argue that an approach that puts emphasis on defining the risk of extinction (Burgman *et al.* 1993) attached to particular fire regimes may be necessary. Such an approach, when generalized to groups of species, may define a threshold that demarcates desirable from undesirable fire regimes in a conservation sense.

Our bias is toward studies drawn from some relatively well studied, southeastern Australian communities. The south-east represents a significant challenge because it is fire-prone, densely populated and contains significant areas of natural bushland which are managed for conservation. Human protection from fire poses a major problem for managers of these areas. Fire management for protection, through suppression and various forms of hazard reduction, influences fire regimes. The influence of protection measures on conservation has yet to be formally evaluated at the scale of landscapes.

MANAGING FIRE: THE INFLUENCE OF DIFFERENT ECOLOGICAL PERSPECTIVES

Ecological goals, the choice of favoured fire regimes and fire management actions have been influenced by the type of knowledge available concerning the effects of fire on species, communities and habitats. Ecological knowledge about the effects of fire varies depending on whether fire is seen as either a single event followed by a process of regeneration (succession); or as a regime of recurring events in which frequency, season, intensity and spatial extent determine ecological response (Gill 1981).

Plant succession and demography

Research in fire-prone shrublands (e.g., Gill and Groves 1981; Kruger 1983; Zedler and Zammit 1989) has produced a detailed understanding of plant population and vegetation dynamics, particularly for species of woody plants that give these communities their characteristic structure. When fire is viewed as a single disturbance the initial composition model of succession (i.e., post-fire changes reflect differing rates of growth and senescence in the species initially present) applies in many plant communities (Noble and Slatyer 1981), including southeastern Australian heaths and shrublands (e.g., Posamentier *et al.* 1981). When fire is viewed as a recurrent disturbance, demographic knowledge of woody species (Gill 1977, 1981) indicates that high frequency fire may eliminate species because such fires interrupt key life cycle processes such as maturation and development of fire resistant organs. Much of the plant ecological perspective on fire is therefore concerned with fire frequency, although other aspects of the fire regime may also influence ecological change.

Animal succession, refugia and mosaics

Studies of animal succession after a single fire (e.g., Fox and McKay 1981; Newsome and Catling 1983) have shown that there is a replacement series of species, with some species peaking in presence and abundance soon after fire and others later (i.e., a relay successional model, Noble and Slatyer 1981). Friend (1993) noted that post-fire changes in species abundance were related to simple habitat models, i.e., changes in the availability of shelter and food correspond to changes in vegetation structure and composition. An important distinction is between those animal species that persist through fire and those which disappear at some stage following fire and must recolonize a site or area through dispersal.The successional view of fire and animal dynamics has resulted in an emphasis on the importance spatial aspects of fire for the conservation of animal populations. The management imperative has been the "mosaic" paradigm which identifies patchiness of fires within landscapes (provision of "unburnt" refugia or adjacent areas of contrasting fire history) as desirable for animal conservation (Fox 1983; Fox and Fox 1987).

There is a paucity of research on the effects of recurring fires on animals. The few relevant studies (e.g., Rowley and Brooker 1987; Catling 1991; York 1994) draw a parrallel conclusion to plant studies: that high frequency fires cause elimination of species. The mechanism of loss is through change in habitat (decline in vegetation composition and structure).

DETERMINISTIC FIRE MANAGEMENT AND ITS ADEQUACY FOR CONSERVATION

The idea of a target fire regime, whether in time or space, carries with it the notion of an ideal state or at least an ideal compromise that suits the requirements of particular species within the area of concern. The fields of fire ecology briefly described above have influenced fire-management of heath/shrublands in the south-east in a number of ways, involving the definition of target regimes.

Single species management

Target fire regimes are sometimes prescribed for the conservation of single species or small groups of species, including those that are rare, dominant or high in the public profile. A simple example stems from the plant demographic approach which has resulted in fire strategies that define and deal with the juvenile periods of fire sensitive woody plants as determinants of minimum fire-free intervals or maximum fire frequency in conservation reserves. This approach has been extended so that the recommended fire-free interval makes allowance for the build-up of seedbank to a level sufficient for populations to remain stable or increase in density (Gill and Nicholls 1989). Typically on the east coast, heath/shrubland communities are dominated by tall shrubs which are fire sensitive and store their seeds in woody fruits, e.g., *Banksia ericifolia, Allocasuarina distyla, Hakea teretifolia*. Population models developed for *B. ericifolia* (Bradstock and O'Connell 1988) predict that the density of these shrubs will be maximum when fire frequency is moderate (13–25 years). Shorter intervals between fire (6–12 years) may lead to the decline of populations and very short intervals <6 years may lead to local extinction because of the time required for maturation. Management for fire free periods in excess of 10 years has been specified in plans for conservation reserves based on this knowledge (e.g., fire management plans for Ku-ring-gai Chase and Royal National Parks near Sydney; Conroy 1988; Harrigan and Conroy 1990).

Recent research indicates that rigid adherence to a narrow range of fire-free intervals based on the requirements of overstorey shrubs only, may lead to a lowering of plant species diversity in these heathlands. The work of Cary (1992) and Morrison *et al.* (1995) shows a positive relationship between variance in fire frequency and species diversity. Other co-habiting plants in these communities vary in life-history but are smaller in stature than the dominant shrubs. Maintenance of a dense overstorey thicket may lead to the exclusion of these understorey species, especially slower growing woody species (Keith and Bradstock 1994). Historical evidence derived from aerial photography suggests that the overstorey thicket can be labile in time and space as a function of fire patterns (Keith 1995). Thus a fire regime directed solely at maintaining dominant shrub species at high densities could be detrimental to conservation by leading to the exclusion of co-habiting plants which make up the bulk of floristic diversity at a site. In this case, fire frequency needs to vary for species diversity to be maintained. There is no single target value of fire frequency or inter-fire interval that is likely to favour all species but rather a range within which conservation of full diversity is possible.

Mosaic fire management

The mosaic paradigm has had an important influence on fire management for conservation. It has led to detailed and highly interventionist management strategies which define a target regime of fire aimed at maximizing the density of a single species, e.g., a rare and/or prominent animal species. Often this involves dividing landscapes into blocks and subjecting them to a burning regime, usually of a fixed fire frequency. Ignition of different blocks is staggered through time to provide the "mosaic" of different post-fire age classes. Post-fire trends in abundance of the target species are used to select a fire frequency: the target frequency is equated to the time since fire at which maximium abundance occurs. The size, arrangement and burning order of blocks are often determined by logistic factors rather than ecological information.

Deterministic burning strategies aimed at maximizing density of a target species are attractive to managers faced with the problem of grappling with complex and diverse systems, because they provide a simple, regular formula which is compatible with bureaucratic requirements for accountability. These strategies invoke enthusiasm in people who want to "get in and manage the place". Such enthusiasm is often heightened by external political pressures. This style of management is often regarded as good for protection because it breaks up the pattern of fuel (as a function of time since fire) in a landscape. Thus rotational block burning is viewed as a good compromise between conservation and protection objectives.

Problems with this style of management include: (i) lack of attention to the requirements of non-target species; (ii) unexpected outcomes for the target species due to interactions with non-target species (e.g., elimination or decline of food sources) and; (iii) the difficulty of imposing target regimes of fire in landscapes prone to stochastic events of unplanned fire. The latter is a significant but often underestimated problem (Whelan and Muston 1991). Use of planned fire could result in an overall increase in fire frequency with the decline or elimination of some species as a possible outcome, particularly in heath/shrubland systems where fuel accumulation is rapid (Conroy 1993). When ignitions are chronic and positively related to fire danger, such as on the fringe of urban areas (NSW National Parks and Wildlife Service, unpubl. records), the problem is exacerbated. It becomes difficult to predict the fire regimes that will be generated in each block over time.

These problems are well illustrated by a case study of Barren Grounds Nature Reserve south of Wollongong, where a rotational block-burning strategy was devised in the 1970s explicitly for conservation of the ground parrot *(Pezoporus wallicus)*, a rare bird species. Heathland was divided into blocks (<50 ha) to be burnt every 5–7 years on a staggered basis. However, prescriptions were curtailed because wildfires burned through treated blocks in the early 1980s and disrupted schedules (Baker and Clarke 1991). Uncertainty over the consequences of imposing further burning treatments has prevailed. Objections to the planned fire frequency were raised on the basis that a simple time since fire correlation with target animal density may be misleading and took no account of habitat (vegetation) dynamics. Burning at the time of peak density may alter the habitat in a way that is actually detrimental to the species by lowering the density of plants (particularly obligate seeder shrub species) which it uses for food and shelter (Meredith *et al.* 1984). There may be demographic or other reasons why burning target populations at the apparent time of peak density is not an optimal strategy for minimizing the probability of extinction (Burgman and Lamont 1992; Possingham, this volume). The prescribed burning frequency may also be detrimental to mammals which require older heath as habitat (e.g., Newsome and Catling 1983).

Another manifestation of the mosaic paradigm is the "balanced age-class" approach to fire management. This approach seeks to accomodate the needs of animal species in general. It is based on the notion that refugia and different successional stages are necessary to conserve a range of animal species — a translocation of the temporal patterns of relay-type succession of the type documented by Fox and McKay (1981) into space. Fire management is targeted to produce a range of age or time since fire classes, up to a limit, within a particular vegetation type and/or in an area as a whole (e.g., specified in fire management plans for Ku-ring-gai Chase National Park; Conroy 1988 and Grampians National Park, Wouters 1993). It is important to note that this approach is often implemented without specific data. It is a bet hedging strategy: the better the balance the more chances are provided to species with unknown requirements.

One of the chief problems with the balanced age-class approach is that achieving it in an area or landscape is not easy. Unplanned fires can quickly negate the effects of careful planning. Because it is focused on time since fire, there is no way of reconciling effects of habitat (vegetation) dynamics that may be generated by sequences of fires. There is little cognisance of the way in which size, shape and juxtaposition of patches of different ages influence the abundance of species. Commonly, certain patches are held in the same post-fire age class over time, because they are subject to arson or are continually burnt by prescribed fire (close to human assets) or else excluded from fire. The consequences for vegetation and habitat dynamics, which affect animal populations, are obvious. This re-emphasizes the general problem with the mosaic paradigm: animal habitat is considered to be static and fire is seen as a single event rather than a periodic regime of disturbance with consequences for habitat quantity and quality. Reconciliation of temporal and spatial components of fire regimes is lacking (Gill and Bradstock, this volume).

TOWARD FLEXIBLE MANAGEMENT

In the ground parrot example, the goal of density maximization and the consequent fixed burning strategy are central to the problems raised. A solution is to jettison the goal of density maximization and, instead, aim to avoid the extinction of populations of species. Density may then fluctuate within limits, the limits being shaped by the level of risk of extinction and the extent (e.g., area or range) that we may choose to tolerate any extinctions. Fire regimes can then vary accordingly: management is more flexible and intervention occurs when appropriate, not on a deterministic timetable. Opportunities then arise to use contemporary principles of conservation biology in fire management, such as assessment of risk of population extinction (e.g., Burgman *et al.* 1993).

Under such a scenario how would we know when to intervene? For example, is there a safe lower limit of ground parrot density that should not be crossed? Does it matter if fire regimes temporarily do not suit the species and it disappears from the reserve? Will dispersal allow recolonization? These issues all affect the limits to flexibility and the level of risk we may wish to carry in setting those limits. For example, in the absence of knowledge about long-range dispersal we may adopt a conservative approach that avoids extinction of a species such as the ground parrot in any single reserve. This would result in a different limit to fire regimes rather than acceptance of possible short-term reserve level extinction followed by recolonization through dispersal. It is important to emphasize the fundamental nature of this distinction, particularly when dealing with species in fragmented habitats and the ensuing problematic nature of dispersal (Hobbs 1992). A conservative approach to fire management that places emphasis on avoidance of local extinction within remaining tracts of bushland (irrespective of their size) may be well warranted.

If we return to the deterministic scenario of managing for a balanced range of fire history or successional classes in a landscape, we can ask how this is affected by a more flexible approach to management. If such a balanced mosaic of fire-classes cannot be achieved what is the margin of error before species are lost? If the goal is to avoid major extinctions of plants or animals within that landscape, we suggest that it becomes a problem of understanding the limits to spatial heterogeneity of fire in terms of species diversity. In other words what is the range of fire history combinations that are acceptable within a landscape? Providing an answer to this question is a challenging problem because of the differing scales at which individual species may be viewed.

If an animal species is unable to survive the combustion and post-fire shock phases (Friend 1993), it is axiomatic that its persistence will depend on: i) unburnt habitat with individuals present, or ii) alternative burnt habitats where it does survive, and iii) its dispersal capability. If dispersal is short relative to the spatial extent of fires and proximity of any alternative habitat that may offer refuge, then intervention may be necessary to provide a mosaic of unburnt refuges. A systematic approach to define the limits to patchiness, in terms of the size, shape and configuration needed to maintain animal diversity will require an improved understanding of habitat availability and dispersal (Gill and Bradstock, this volume). For animal species it requires viewing and understanding fire as a recurrent disturbance, particularly in relation to habitat change, as attempted by Catling (1991). Meta-population viability models offer one possible means of solving problems about the variability of fire in space, by integrating information about survival, as a function of habitat type, with data on dispersal (Possingham *et al.* 1992; Possingham, this volume). Such an approach could show whether existing or putative reserves are adequate for conservation of particular species or groups of species, in terms of habitat and dispersal, given existing or altered patterns of fire regimes.

RELATING SPECIES DIVERSITY TO FIRE REGIME VARIABILITY: A CASE STUDY

Imposition of a fixed spatial and temporal regime of fire may be questioned at a more fundamental level. Does a fixed disturbance regime favour lower species diversity than a more variable regime? It is recognized that the co-existence of species is maintained by environmental variability, imposed in part by periodic disturbance (Connell 1978; Petraitis *et al.* 1989). Landscape fires cause repeated disturbances in many ecosystems and communities. Variability both in species responses and in fire regimes has been recognized as a factor that allows persistence and resilience of plant and animal communities (Fox 1983; Fox and Fox 1987; Keeley *et al.* 1989; Lamont *et al.* 1993). Particular emphasis has been placed on spatial variability of fires, but less emphasis has been given to recognizing the role that temporal variability may play in maintaining species diversity. Hobbs and Hueneke (1992) have suggested that floristic diversity may be maximized by promoting intervals between fire that are intermediate between the time taken for plants to mature and the time of senescence, based on the intermediate disturbance regime hypothesis (Connell 1978).

The co-existence of species with different life history strategies provides the key to understanding how varied disturbances will affect species diversity (Noble and Slatyer 1980; Pickett and White 1985). Returning to the example of coastal heath/shrublands, it can be shown how full diversity of plant species may only be maintained by variable fire regimes (Keith and Bradstock 1994). Patches exposed to a short interval fire will lose overstorey species temporarily until seed dispersal after subsequent fires enables recolonization from adjacent areas. Rates of recolonization by dispersal are poorly known, but will be related to the area thus affected (Keith 1994): large areas exposed to uniform fire regimes would recover more slowly than smaller areas. Flux of the overstorey in this manner is necessary to maintain the full complement of species over time at any particular point on the ground, because of the competitive effects on understorey plants described above. Fire intervals of varying length (high variance in fire frequency) are needed to achieve this. It follows that repeated sequences of the same fire interval length, whether short or long (i.e., low variance in frequency variance), may lead to species declines and losses.

Conceptually, this example can be used to outline a relationship between species diversity and variability of fire regimes (Fig. 2). This model proposes that there is a threshold of temporal (fire frequency) variability that demarcates a reduction in species diversity due to fire regime uniformity (low variability). To conserve full diversity, management needs to promote fire frequencies on the "variable" side of the threshold. Note that a wide range of variable fire regimes is postulated to be compatible with conservation. In practice, some concrete definition of the threshold is needed. For the heathland example such a definition is attempted in Table 1. The definition summarizes critical aspects of fire regimes that demarcate uniformity from variability. It deals mainly with fire in a temporal sense but also with fuel consumption, a component of fire intensity which affects seedling recruitment through germination (Auld and O'Connell 1991; Bradstock and Auld 1995).

The threshold concept relates fire regimes to changes in species abundance within functional groups. It is important to emphasize that it does not deal with how, or to what extent fire regime

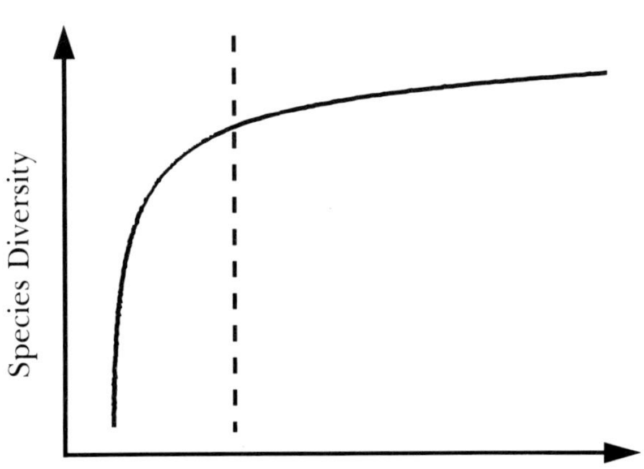

Fig. 2. A model proposing that diversity of plant species at a site is a function of fire regime variability (in this case frequency and intensity — see Table 1) in fire-prone shrub/woodlands of Sydney. Fire regime variability exceeding a threshold (dashed line), will maintain species numbers at or near maximum. Note that high levels of variability indicated here could lead to a decline in diversity.

variability is generated *in situ*, but merely indicates the responses of flora if exposed to particular regimes. We suggest that there is sufficient ecological knowledge available to derive thresholds similar to those defined in Table 1 for a broad range of plant communities, particularly if functional classifications are used to lump the responses of species into a small number of categories whose competitive relationships are known (Moore and Noble 1990; Keith and Bradstock 1994).

Table 1. A definition of the fire regimes that are likely to lead to a decline in plant species density and diversity in coastal heaths, and associated shrublands and woodlands, in a patch of about 1 ha or less (see text). Such fire regimes may be considered undesirable for conservation.

A decline in populations of plant species can be expected when:

- there are more than two consecutive fires less than 6–8 years apart (fire-sensitive shrubs decline);
- intervals between fires exceed 30 years (herbs and shrubs with short-lived individuals and seedbanks decline);
- three or more consecutive fires occur at intervals of 15–30 years (sub-dominant herbs and shrubs decline);
- more than two consecutive fires occur which consume less than 8–10 tonnes ha^{-1} of surface fuel (species with heat-stimulated seedbanks in the soil decline).

FLEXIBLE FIRE MANAGEMENT IN PRACTICE

The threshold defined in Table 1 can be thought of as a set of guidelines that could direct fire management in heath/shrublands to achieve conservation of floristic diversity. Acceptance of this approach means that fire management cannot be carried out by rote. Flexible management requires the state of fire regimes at any point in time to be evaluated in relation to some set of ecological guidelines or criteria such as those shown in Table 1. This is a necessary pre-requisite to the formulation of strategies and actions. Strategies and actions will change over time in line with fire regimes as a consequence of ongoing re-evaluation. Using our example, under a flexible management system, active use of fire is discretionary, being required when variability tends to uniformity (Fig. 2; Table 1).

The guidelines in Table 1, can be applied in practice to points on the ground. Application is more difficult on a broader scale. How do we evaluate the significance of fire regimes, in relation to species diversity over a landscape? Following our example of a heathland, what is the maximum area over which extinctions (reduction in species diversity) should be tolerated? Does species diversity have to be maximized over the entire landscape or do we draw the line at some lesser proportion? Can we allow local extinctions if recolonization is probable in the future? Conceivably the scale of

Noble, I. R. and Slatyer, R. O., 1981. Concepts and models of succession in vascular plant communities subject to recurrent fire. Pp. 311–35 in *Fire and the Australian biota* ed by A. M. Gill, R. H. Groves and I. R. Noble. Aust. Acad. Sci.: Canberra.

Petraitis, P. S., Latham, R. E. and Niesenbaum, R. A., 1989. The maintenance of species diversity by disturbance. *Quart. Rev. Biol.* **64:** 393–418.

Pickett, S. T. A. and White, P. S., 1985. *The ecology of natural disturbance and patch dynamics.* Academic Press: Orlando.

Posamentier, H. G., Clark, S. S., Hain, D. L. and Recher, H. F., 1981. Succession following wildfire in coastal heathland (Nadgee Nature Reserve New South Wales). *Aust. J. Ecol.* **6:** 165–75.

Possingham, H. P., this volume. The practical application of population viability analysis for conservation planning. In *Conserving biodiversity: threats and solutions* ed by R. A. Bradstock, T. D. Auld, D. A. Keith, R. T. Kingsford, D. Lunney and D. P. Sivertsen. Surrey Beatty & Sons: Chipping Norton.

Possingham, H. P., Davies, I., Noble, I. R. and Norton, T. W., 1992. A metapopulation simulation model for assessing the likelihood of plant and animal extinctions. *Math. Comp. Simul.* **33:** 367–72.

Rowley, I. and Brooker, M., 1987. The response of a small insectivorous bird to fire in heathlands. Pp. 211–18 in *Nature conservation: the role of remnants of native vegetation* ed by D. A. Saunders, G. W. Arnold, A. A. Burbidge and A. J. M. Hopkins. Surrey Beaty & Sons: Chipping Norton.

Saunders, D. A., Arnold, G. W., Burbidge, A. A. and Hopkins, A. J. M., 1987. *Nature conservation: the role of remnants of native vegetation.* Surrey Beatty & Sons: Chipping Norton.

Soulé, M. E. and Gilpin, M. E., 1991. The theory of wildlife corridor capability. Pp. 3–8 in *Nature conservation 2: the role of corridors* ed by D. A. Saunders and R. J. Hobbs. Surrey Beatty & Sons: Chipping Norton.

Walker, B., 1989. Diversity and stability in ecosystem conservation. Pp. 121–30 in *Conservation for the twenty-first century* ed by D. Western and M. Pearl. Oxford University Press: New York.

Western, D., 1992. The biodiversity crisis — a challenge for biology. *Oikos* **63**(1)**:** 29–38.

Whelan, R. J. and Muston, R. M., 1991. Fire regimes and management in southeastern Australia. Pp. 235–58 in *Proceedings of the Tall Timbers Fire Ecology Conference.* **15**.

Wouters, M. A., 1993. Developing fire management planning in Victoria: a case study from the Grampians. In *Victoria Department of Conservation and Environment, Fire Management Branch* Research Report No. 39. 54 pp.

York, A., 1994. The long-term effects of fire on forest ant communities: management implications for conservation of biodiversity. *Mem. Qld Mus.* (in press).

Zedler, P. H. and Zammit, C. A., 1989. A population-based critique of concepts of change in the Chaparral. Pp. 73–83 in *The California chaparral: paradigms re-examined* ed by S. C. Keeley. No.34 Science series, Natural History Museum of Los Angeles County.

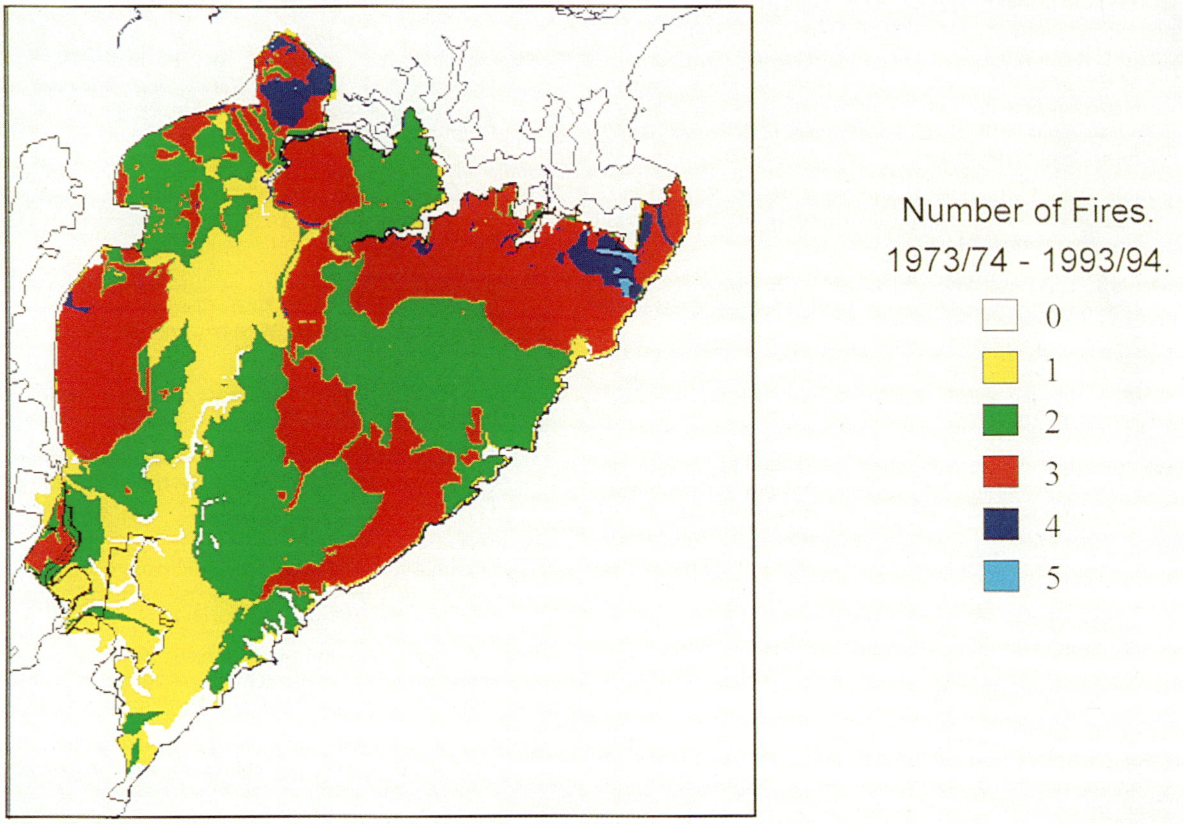

PLATE 22 *(above).* A map of the number of fires (fire frequency) that have occurred within the period 1975–95 in the Royal National Park near Sydney. Records of fire form an essential resource in targeting the management of fire regimes to achieve conservation objectives in fire-prone landscapes. *(Photograph by Ross Bradstock.)*

PLATE 23 *(below).* Houses set among bush in the northern suburbs of Sydney. Manipulation of fire regimes in adjacent bushland, to provide protection to residents is the paramount objective. How compatible are these regimes with those needed to conserve biodiversity? A scientific approach that enables quantification and comparison of different options is needed to provide the best solution to this problem. *(Photograph by Ross Bradstock.)*

Can Governments Solve the Problems?

PLATE 24 *(above)*. Remnants of natural vegetation among farmland in the Bega valley of southeastern New South Wales. These remnants serve as a reminder of the responsibility for conservation of biodiversity which private individuals and landholders bear. *(Photograph by Dan Lunney.)*

PLATE 25 *(below)*. Lobby groups and activists have pressured governments to take action in many controversies concerning development and conservation. Government intervention is, however, only one of the many options that society may use to bring about effective conservation of biodiversity.

Policy instruments for conserving biodiversity on private land[1]

D. FARRIER*

There are increasing international pressures on the Commonwealth Government to address the issue of biodiversity conservation on privately owned land. Traditionally, the states have exercised land management responsibilities, and the Intergovernmental Agreement on the Environment assumes that the Commonwealth will be able to satisfy its international obligations by substantially delegating implementation to the states. Currently the states employ a variety of policy instruments having a bearing on biodiversity conservation on private land. These range from entirely voluntary arrangements between government agencies and landholders at one extreme to command and control regulation, with no provision for any financial recompense, at the other. The focus is usually exclusively negative, neglecting the need for ongoing management of preserved vegetation and habitat.

Unlike the position in the United States of America, there are no constitutional limitations on government regulation of land use by private landholders in Australia. In practice, the ideology of private property within rural communities is sufficiently strong to undermine command and control initiatives. So, for example, vegetation clearance permit systems will usually operate so that the focus is on what conditions should be attached to permits rather than whether or not a permit should be granted in the first place. While this may cater for land degradation concerns, it will frequently be an inadequate response where the conservation of biodiversity is at issue and a precautionary approach is appropriate. Where, on the other hand, regulation rests initially on inducing landholders to enter into voluntary agreements, experience suggests that few agreements will be reached because government will rarely be prepared to offer adequate inducements.

Command and control regulation cannot by itself address the issue of ongoing management of native vegetation and wildlife habitat on private land. Even where regulatory agencies are prepared to go further than symbolic command, and instigate effective controls, the reality is that alienated landholders will make poor land managers. Regulation must be combined with adequate financial incentives. These should take the form of forward-looking payments for management rather than backward-looking compensation. Possible sources of funding for such a strategy include the National Landcare Program and the Rural Adjustment Scheme. The priorities of these initiatives need to be adjusted to ensure that greater emphasis is placed on the retention of existing native vegetation rather than replanting, and that restructuring in rural communities is not based exclusively on short-term productivity concerns, but also takes account of the much longer term economic interest that the community has in conserving biodiversity.

INTERNATIONAL OBLIGATIONS

THE United Nations Convention on Biological Diversity, signed by Australia at the United Nations Conference on Environment and Development in Rio de Janeiro on June 5, 1992,[2] imposes upon the Commonwealth Government a number of international obligations in relation to the conservation of Australia's biological diversity, defined to include not only diversity between species, but also genetic and ecosystem diversity.[3] It will be committed "as far as possible and appropriate" to:

- "regulate or manage biological resources (including genetic resources and populations) important for the conservation of biological diversity whether within or outside protected areas, with a view to ensuring their conservation and sustainable use";
- "control or eradicate those alien species which threaten ecosystems, habitats or species";
- develop necessary legislation and/or other regulatory provisions for the protection of not only threatened species, but also threatened populations;
- identify types of activities likely to have significant adverse impacts on the conservation of biodiversity, monitor the effects of these activities and "regulate or manage" them;
- adopt "economically and socially sound measures that act as incentives" for the conservation of biodiversity.[4]

There are a number of important points to make about these commitments. In the first place, they relate not only to endangered species, but also to genetic and ecosystem diversity. Legislative initiatives which focus on endangered species will not suffice.[5] Secondly, they require action to be taken not only through the creation and management of protected areas, such as national parks and nature reserves, but also in relation to privately owned land. The conservation of genetic and ecosystem diversity on privately owned land is perhaps the most intractable environmental problem currently faced by government. The two major threats to the conservation of biodiversity on privately owned land in Australia today are those posed by habitat destruction from

See Notes at end of paper.

*Faculty of Law, University of Wollongong, Northfields Avenue, Wollongong, New South Wales, Australia 2522.

Pages 337–59 *in* CONSERVING BIODIVERSITY: THREATS AND SOLUTIONS, ed by R. A. Bradstock, T. D. Auld, D. A. Keith, R. T. Kingsford, D. Lunney and D. P. Sivertsen. Surrey Beatty & Sons. 1995.

agricultural land clearing, private forestry, residential development and the invasion of exotic species, on the one hand, and direct destruction of native species by feral animals on the other. Thirdly, the Commonwealth Government simply cannot avoid identifying agricultural land clearing and private forestry as activities having significant adverse impacts on the conservation of biodiversity. The obligation is then to "regulate or manage" both the activities and the biological resources which they destroy. The argument advanced in this paper is that the twin threats of habitat destruction and feral animals can only be adequately confronted by a regulatory strategy which ensures that existing native vegetation/wildlife habitat is retained and at the same time adequately managed on a continuing basis through the provision of incentives to private landholders. The disjunction which the Convention draws between regulation and management cannot be sustained at a practical level.

THE CONSTITUTION AND THE INTERGOVERNMENTAL AGREEMENT

The formal legal position is that the Commonwealth Parliament has power under section 51(xxix) of the Constitution (the external affairs power) at least to enact legislation giving effect to Australia's obligations under an international convention.[6] The Convention on Biological Diversity imposes binding obligations flowing from the use of the mandatory word "shall", in spite of the substantial qualification stemming from the addition of "as far as possible and as appropriate". However, while the Commonwealth will be under a qualified *duty* under international law to take steps to protect Australia's biodiversity, it simply has the *power* under the Constitution. There is no constitutional duty to take action. The approach that the Commonwealth Government will take in practice is spelt out in the Intergovernmental Agreement on the Environment which it concluded with the states, territories and the Australian Local Government Association in May 1992. The Agreement attempts to demarcate the environmental policy and management responsibilities of each level of government, although the treatment of local government responsibilities is cursory.[7] In general terms, it represents a substantial retreat by the current Keating Government from the assertive approach to the use of Commonwealth constitutional powers taken by its predecessor. It begins by recognizing that "environmental concerns and impacts respect neither physical nor political boundaries and are increasingly taking on interjurisdictional, international and global significance in a way that was not contemplated by those who framed the Australian Constitution".[8] This does not, however, lead to the conclusion that the Commonwealth should use its broad constitutional powers to enact legislation which addresses environmental issues on a national basis, and ensures a consistent and uniform approach throughout the states and territories. The Agreement clearly acknowledges only two areas of environmental policy development for which the Commonwealth is primarily responsible, one of them being foreign policy, particularly entering into international conventions and ensuring that international obligations are met.[9] Even here, the interests of the states and territories are recognized.[10] There are clear indications that the Commonwealth's international responsibilities are to be pursued through forums in which the states and territories are heavily represented, i.e., the Biological Diversity Advisory Committee and the Australian and New Zealand Environment and Conservation Council (ANZECC). So, for example, in Schedule 6, covering Biological Diversity, while the parties note the Commonwealth's responsibility for ensuring implementation of the Convention, it is ANZECC, in consultation with other Ministerial Councils, which is to provide advice on the "manner in which implementation of the proposed Convention may be undertaken".[11]

When it comes to national environmental issues, the emphasis is placed specifically on the development of policy through co-operation between the Commonwealth, states and territories. The Commonwealth's role is simply to facilitate the co-operative development of national standards and guidelines in the limited circumstances where the need for these are recognized elsewhere in the Agreement.[12] This is to be formalized through the creation of a National Environment Protection Authority (NEPA), consisting of Ministerial representatives from each of the states and territories, as well as the Commonwealth. NEPA will be able to make binding decisions on the basis of a two-thirds majority, but only on a limited range of issues. These are confined to aspects of pollution control and waste management, excluding, for example, endangered species.[13] Nevertheless, there is still agreement that:

> a national approach should be taken to rare, vulnerable and endangered species given that the distribution of these species and their habitats is not confined or determined by state or Commonwealth borders and that a national approach is desirable to avoid duplication of effort, to ensure appropriate outcomes and to maximize the effectiveness of available resources.[14]

This is to be achieved, however, purely through consensus, without even the limited benefit available to NEPA of decision-making by a two-thirds majority. By May 1992, ANZECC, which is given the overall co-ordinating role, was supposed to have reported back on a national strategy, including a discussion of options for the protection of species and habitats on land in private ownership, taking into account the work of the Endangered Species Advisory Committee.[15] The most worrying scenario is that agreement will ultimately be reached on a national strategy, but that it will take the form of the lowest common denominator, with the terms essentially dictated by those states least inclined to face the difficult decisions needed if adequate steps are to be taken in relation to privately owned land to protect species at risk.

This attempt to develop a national approach is restricted to rare, vulnerable and endangered species. In line with Australia's obligations under the Convention on Biological Diversity, the Agreement also recognizes the broader dimensions of biodiversity conservation, although it deals with these under the heading of nature conservation in Schedule 9. The Agreement reflects the provisions of the Convention by acknowledging that the demand for nature conservation will not be adequately met through the creation of a representative reserve system, and that this must be supplemented by off-reserve protection and management.[16] There are, however, strong indications that in addressing these issues, the states are to be left substantially to their own devices.

Some confusion stems from the fact that "biological diversity" and "nature conservation" are dealt with quite separately in different schedules to the Agreement. As we have seen, Schedule 6, dealing with biological diversity, recognizes the Commonwealth's responsibility for implementing the Convention, and this is reinforced in Schedule 9, dealing with nature conservation, although the reference here is to commitments under "international nature conservation treaties", rather than to the Convention on Biological Diversity specifically. The Commonwealth's "particular interest in facilitating the effective and efficient co-ordination of nature conservation across all jurisdictions" is also acknowledged.[17] On the other hand, it is a fundamental principle of the whole Agreement that the development of regulatory strategies is to be left to individual jurisdictions:

> Each state has responsibility for the policy, legislative and administrative framework within which living and non living resources are managed within the state.[18]

This is reinforced in Schedule 9, which recognizes the primary responsibility of the states and territories in relation to nature conservation,[19] leaving them "to take whatever action they deem appropriate within their respective jurisdictions to protect any native species and habitats which they consider requires [*sic*] specific action".[20] This is a *de facto* legitimization of the present situation, with the states and territories employing a diversity of policy instruments to address the core issue for biodiversity conservation of retention of native vegetation on privately owned land.[21]

The Convention on Biological Diversity, with its unclear and ambiguous requirement to "regulate or manage" biodiversity, is flexible enough to allow Australia to meet its obligations while still contemplating a diversity of regulatory responses by the states and territories. At the same time these obligations will not be met by regulatory regimes which are purely symbolic or which are historically rooted in addressing concerns, such as the prevention of soil erosion, which bear little relationship to biodiversity conservation objectives. One of the objectives of the present paper is to survey the various initiatives which have so far been taken in the states and territories with a view to inducing landholders to conserve native vegetation on privately owned land, and to assess their effectiveness, both in light of the little empirical material which is available, and concerns derived from broad theoretical considerations.

CONSERVING NATIVE VEGETATION ON PRIVATE LAND

The issues of how much native vegetation needs to be conserved on private land, what kind and where, are better left to be dealt with by others, although answers to these questions necessarily have significant implications for the design of an appropriate regulatory strategy. If, for example, we are only talking about a limited area in the hands of a small group of landholders, then it may be less socially and economically disruptive simply for the state to purchase the land or to compensate for the restrictions imposed. The remainder of this paper rests on the assumption that the demand for conservation of biodiversity cannot be so readily met and dealt with so painlessly. It assumes that the reserve system will fall considerably short of meeting legitimate demands for the conservation of biodiversity.[22]

Where native vegetation is located on land in public ownership, such as state forest or national park, the only issue to be resolved is the purpose for which the land is to be managed. The issues are much more complex where private land is involved. The land use and management options cover a wider range. In the case of publicly owned land, the contest is essentially between what has traditionally been called nature conservation and forestry. It has never been suggested that the state should consider agriculture, or the sowing of improved pastures,[23] preceded by clearance of native vegetation, as a primary use for land which is to remain within public ownership, although on occasions rough grazing may be tolerated. Yet, agriculture is at present viewed by most private rural landholders as the most attractive development option. Up until recently the management of privately owned native forests for *sustainable* timber-getting has rarely been seen by landholders as a viable economic option, although many have financed agricultural land clearing operations by a once-off timber-getting exercise. The recent work of the Resource Assessment Commission has emphasized that there are considerable quantities of merchantable timber on privately owned land.[24] An obvious but important point to make here is that sustainable timber-getting is a more sympathetic land use than agriculture in terms of biodiversity conservation and may prove to be an acceptable compromise in some circumstances. Agricultural land clearing is totally incompatible with both nature conservation and sustainable timber-getting and, to this extent, those advocating these land uses have a common interest in inducing or requiring landholders to abort this activity.

The other reason why the issues are more complex where privately owned land is involved is that the state must also determine a regulatory strategy for achieving its policy objectives on appropriate land use, in a context where few of the options are going to be totally palatable to the landholder. The forms of regulation available range from financial inducements to persuade landholders to enter into voluntary agreements, at one extreme, to control through enforcement of criminal law, at the other. Careful attention needs to be given at the stage of regulatory design to how the regulatory form will be implemented in practice. Ostensibly identical forms of regulation might be implemented in different ways, depending on the culture and resources of enforcement agencies and subtle differences in regulatory form. Economists categorizing traditional legal regulatory strategies into the single pigeon-hole of "command and control" are at considerable risk of oversimplifying what is in practice a highly complex phenomenon.

Simple retention of native vegetation is by no means the only issue which must be addressed by policy instruments. The land must be managed on an ongoing basis. Many of the government initiatives discussed below have treated the retention of native vegetation not simply as a starting point for in-situ conservation, but as the ultimate goal, quite neglecting the need for continuing management. Management for the purposes of biodiversity conservation includes not only management of processes for which current activities of landholders are directly responsible, such as stock invasion and run-off, but also feral animal and plant control, which represent a broader community problem. The issue for policy-makers is what is the most appropriate means for ensuring both retention of native vegetation and continuing management.

The debate at a national level about appropriate regulatory strategies is characterized by confusion, disagreement and caution, reflecting the divergent approaches currently taken by the states and territories, and even within jurisdictions. Some have been unhelpfully circumspect. The ESD Forest Use Working Group recommended that forest practice codes and environmental prescriptions should be applied to private forests. When it comes to the issue of how we go about ensuring that private forests are not converted to agricultural or pastoral land use, the guarded response is that *appropriate* legislative controls over the permanent clearing of native forests should be adopted by state, territory and local governments.[25]

Others have opted for a kitchen-sink approach which employs every possible strategy which has ever been suggested. The draft National Strategy for the Conservation of Australia's Biological Diversity recommends "introduction of, or continuing support for, controls on further clearance of native vegetation". It then goes on to suggest the provision of financial assistance and other incentives for conservation, and the use of heritage agreements, including agreed property and management plans.[26] More recently, the National Forest Policy Statement has indicated that:

> Sustainable management of private forests will be encouraged through a combination of measures that may include dissemination of information about and technical support for forest management, education programmes, conservation incentives, land-clearing controls, harvesting controls, and codes of forest practice.[27]

There is no sensitivity here to the conflicts which might arise if some landholders are to be gently persuaded by financial inducements to enter into voluntary arrangements, while others are to be coerced.

The ESD Working Group on Agriculture, on the other hand, appears at one stage to opt firmly for a policy of vigorous enforcement. It recommends that state and territory governments review regulatory procedures in relation to the protection of remnant vegetation to ensure strict application of the criteria by requiring authorizations for clearing; that these measures be underpinned by appropriate sanctions, and that these sanctions be applied rigorously. It believes that this is an area where the use of penalties can be effective.[28] This stance is the more remarkable for the fact that it was supported by all members of the Working Group, including representatives of the National Farmers Federation and state government Departments of Agriculture, and that it was prefaced by a discussion which revealed an underlying scepticism about the efficacy of direct regulation.[29] The only area of disagreement was over whether an immediate moratorium on clearing should be declared in all states and territories with the exception of South Australia.[30] On the other hand, there is evidence of vacillation when it comes to the Working Group's response to the issue of enforcement. The suggestion is that landholders are increasingly prepared to dob in their neighbours and that therefore the role of controls is "in underpinning the operations of community-based groups". Clearing restrictions do "not necessarily need to be policed and enforced by government officers".[31] The problem with this is that controls which rely upon command and control have traditionally been implemented so as only to have any significant impact upon those landholders who are particularly insensitive to what are usually quite limited land conservation requirements. Moreover, these can be justified as being in their direct economic interests, or those of the broader community. We can expect much less co-operation if licensing systems are to be tightened so as to impact upon a much greater proportion of landholders and become more explicitly concerned with the conservation of biodiversity rather than the prevention of land degradation, in which landholders have a direct economic interest.

CURRENT INITIATIVES IN THE STATES

Species protection legislation

The Convention on Biological Diversity has redefined the problem to be addressed to include not simply endangered species, but also genetic and ecosystem diversity. From one perspective, this broadened concern could simply be seen as stemming from a precautionary approach, with the ultimate objective still being species conservation. This would be quite wrong. The Convention quite clearly regards genetic and ecosystem diversity as having value to human beings in their own right, quite apart from the role which they play in conserving species. To this extent, legislation which requires us to spend considerable time and resources *identifying* endangered or vulnerable species, without, at the same time, creating ground rules to protect endangered genetic stock and ecosystems is nonsense. In fact it is positively mischievous insofar as it allows governments to present themselves as addressing the problem, when they are simply wrongly defining it.

Modern endangered species legislation is usually not quite so crude.[32] The Victorian *Flora and Fauna Guarantee Act* 1988, for example, enables the listing of communities of flora and fauna, as well as species.[33] However, in practice, communities are likely to be the poor relations: as at May 1993, only 14 had been listed under the Act.[34] Interim conservation orders under the *Flora and Fauna Guarantee Act* cannot be made to protect communities, as distinct from taxa, on private land.[35] The recent Commonwealth *Endangered Species Protection Act* 1992[36] is even more dismissive: although "ecological communities" can be listed, the definition has not even been finalized at this stage.

Even if the *Flora and Fauna Guarantee Act* does attempt to a limited extent to grapple with the broader issue of biodiversity, it contravenes what has become another fundamental tenet of sustainable development, the precautionary principle. A version of this appears in the Preamble[37] to the Convention on Biological Diversity, where the parties "note" that:

> where there is a threat of significant reduction or loss of biological diversity, lack of full scientific certainty should not be used as a reason for postponing measures to avoid or minimize such a threat.

What this means is that once the threshold test[38] has been satisfied by showing the existence of a *threat* of significant reduction or loss of biological diversity, the argument that action needs to be taken cannot be resisted by claims that there is no scientific proof that there will in fact be a significant reduction or loss.[39] In essence, the burden of proof is reversed, so that those arguing that there will *not* be any significant reduction or loss of biological diversity must prove it. Yet even in relation to the preliminary issue of defining the circumstances when a species is to be *listed*, the *Flora and Fauna Guarantee Act* places a far heavier burden of proof on those advocating listing than is envisaged by the threshold test set out in the precautionary principle:

> A taxon or community of flora or fauna is eligible to be listed if it is in a *demonstrable* state of decline which is *likely* to result in extinction or if it is *significantly* prone to future threats which are *likely* to result in extinction.[40]

The legislation contemplates that a recommendation from the Scientific Advisory Committee may take up to three years from the date a nomination is made.[41] Only after listing is a management plan (an action statement) mandatory ("as soon as possible").[42] Otherwise such plans are optional.[43] Regulatory provisions requiring a licence to be obtained before protected flora is disturbed only constrain owners of private land after the "critical habitat" of a species has been identified.[44] Although interim conservation orders can be used as soon as a nomination for listing has been made,[45] none has ever been made in practice, even in relation to listed species. They are regarded as a last resort.[46] This looks much more like a risk-taking than a precautionary approach.

At least at a symbolic level, the current position in New South Wales is quite different. Under the *National Parks and Wildlife Act* 1974, it is an offence to "take" not only *endangered*,[47] but *protected*[48] fauna, anywhere in New South Wales, without first obtaining a licence from the National Parks and Wildlife Service. The Court of Appeal, in *Forestry Commission of New South Wales v Corkill*,[49] breathed fresh life into this legislation when it held that the destruction of habitat by a logging programme authorized by the Forestry Commission in a state forest amounted to a "taking" even though this was not the objective of the Commission. Previously, it had been assumed that "taking" only covered direct human interference. The decision in *Corkill* has been confirmed in amendments to the legislation by the *Endangered Fauna (Interim Protection) Act* 1991, as a result of which "take" is specifically defined to include "significant modification of the habitat of the fauna which is likely to adversely affect its essential behavioural patterns".[50]

The result is an impressive command and control regulatory façade, because the definition of protected fauna is such that it currently covers all native fauna apart from a limited number of native birds.[51] The *Corkill* decision and the confirming amendments have effectively imposed broadacre clearing controls over all private land which is the habitat of protected species.[52]

This more clearly reflects a precautionary approach than the Victorian provisions in that, at least in theory, it forces those wishing to interfere with fauna habitat to make out a case. Landholders must first decide whether their land does provide such habitat in order to work out whether they have to submit a licence application. Otherwise they risk prosecution. In practice, however, the National Parks and Wildlife Service does not have the resources to process applications for licences relating to *protected* species which do not fall into the endangered category, and as a result this limb of the regulatory system does not operate. Even where *endangered* species are concerned, early figures indicate that the focus of the National Parks and Wildlife Service will be on granting conditional licences rather than outlawing activities altogether.[53] In this context, its bargaining position may look strong in theory but is likely to be weak in practice.

Apart from this, the overall approach under the New South Wales legislation is reactive and *ad hoc*. There is no provision for forward planning or ongoing management. The burden of gathering basic data on endangered species is cast on to the developer, who is ordinarily required to submit a fauna impact statement with applications to take endangered, as distinct from protected species. There is a strong argument that basic resource inventory work, as distinct from assessing the impact of a particular development, should be carried out by the state, and not individual developers. Finally, fauna only covers mammals, birds, reptiles and amphibians, not arthropoda, including insects and arachnids. Endangered flora is not covered at all by these provisions.

Consensus strategies

Until recently the regulatory tradition has been to seek nature conservation on private land through voluntary agreement with landholders rather than resorting to command and control strategies. The earliest attempts simply involved contractual arrangements between landholder and government agency, which could be revoked at will and could not bind later purchasers of the land even if this was the wish of the original parties.[54] Recent initiatives provide for agreements to "run with the land", so that later purchasers are bound.[55] While attractive to government agencies and to those landholders who possess a conservation ethic, this may serve only to discourage those concerned about the potential impact on land value.

Such strategies assume that landholders can do as they please with the land unless they are persuaded not to do so in a voluntary transaction. Rather than having to go to the conservation agency cap in hand to obtain a licence, as they would if a regulatory system was in operation, the agency must come to them. This amounts to a fundamental shift in bargaining power. It is arguable, however, that such a strategy, provided that it is effectively put into practice, can still satisfy the precautionary principle, which does not contemplate the implementation of any particular types of measure, provided that measures taken are designed to avoid or minimize any threat to biodiversity. Unlike the polluter-pays principle, there is nothing in the precautionary principle which requires those interfering with the environment to pay for any damage which they cause.

The evidence suggests that consensus strategies will never have more than a marginal impact on vegetation conservation. Prior to the enactment of the *Native Vegetation Management Act* 1985 (South Australia), heritage agreements were the main tool employed in South Australia in an attempt to persuade private landholders to conserve native vegetation. Fowler has commented:

> . . . it had become clear by 1983 that the concept was attractive only to those landholders with an existing, strong commitment to vegetation retention. Landholders who proposed to clear land . . . proved unwilling in the vast majority of cases to enter into a heritage agreement or to modify or abandon their proposals in any way.[56]

The crucial issue here was the inadequacy of the inducements offered to landholders to enter into what are entirely voluntary agreements. These can oblige the landholder to carry out positive management activities as well as imposing land use restrictions.[57] There is no provision in the legislation for payment for management activities or compensation for lost productivity or decline in land value, although the Minister can offer relief from rates and taxes[58] and assistance with stock-proof fencing.[59]

In other situations, where inducements are attractive to landholders, constraints may be imposed by governments. The Remnant Vegetation Protection Scheme in Western Australia provides a subsidy of 50 per cent of the total cost of materials and labour for the construction of a stockproof fence, to assist private landholders to protect and manage areas of native vegetation. The Scheme protected a total of 7 000 ha in 1988/89 and was anticipated to cover a further 10 000 ha in 1989/90.[60] But demand exceeds supply: only 150 out of 350 applicants were successful in 1988/89.

Command and control through the planning system

Recognizing the limitations of a strategy based on heritage agreements, South Australia turned initially to the planning system, and to a regulatory strategy resting on very different assumptions. Instead of landholders having a right to clear native vegetation which they could bargain away if they wished, the state government attempted to deny them the right to clear under threat of criminal sanction, but to allow it as a privilege in particular cases through a system of individual approvals. In South Australia, this approach foundered on the shoals of existing use protection,[61] but the lesson was well learnt in New South Wales and Victoria, where regulation through the environmental planning system continues to be the primary response to destruction of vegetation on private land.

The position in Victoria, under the State Section of all planning schemes, is that a permit must be obtained to remove, destroy or lop native vegetation on areas of at least 0.4 hectares, unless the proposal falls within one of a number of exemptions. All applications to clear areas of 10 hectares and above must be referred for comment to the Director-General of Conservation and Environment.[62] If the Director-General objects to the grant of a permit then the legislation requires the municipal council to reject the application.[63] The operation of these provisions can be excluded by provisions in the regional and local section of the planning scheme. This may, for example, incorporate more restrictive provisions. Councils are in fact to be encouraged to develop their own controls to replace the state-wide provisions.[64] A number of councils have incorporated more stringent provisions into their schemes.

In New South Wales, the approach has generally been *ad hoc*, with an increasing number of local environmental plans including a requirement to obtain development consent for clearing in particular zones. One of the more radical examples is found in the Wollongong LEP 1990, where there is a general provision requiring development consent for "clearing or partial clearing of vegetation", although none of these terms is defined. Under the Queanbeyan LEP 1991, clearing in all rural and environmental protection zones requires development consent except where it is authorized under other legislation. A number of recent LEPs take a more cautious approach, providing for the identification of areas of environmentally sensitive land and imposing a development consent requirement where trees, or certain species of trees, or more than a certain area of trees, are to be destroyed.

There have been two significant initiatives in the context of biodiversity conservation at the state level in New South Wales.[65] Under State Environmental Planning Policy 14, coastal wetlands have been mapped in a number of local government areas, and development consent is required for land clearing, land draining, land filling and the construction of levees. These are classed as designated development, and consequently an environmental impact statement must be prepared and third-party appeals against decisions to grant consent are available. The concurrence of the Director of Planning is required before consent can be given. State Environmental Planning Policy 26 applies to certain mapped areas of littoral rainforest as well as a buffer zone of 100 metres surrounding them. Residential areas are excluded from the buffer zone. A range of activities in these areas requires the consent of the council and the concurrence of the Director of Planning (or the Minister where development by public authorities is involved). This includes disturbance of native vegetation and, in the areas of rainforest themselves, use of land for any purpose, although this will be subject to the provisions protecting the continuation of existing lawful uses, such as grazing. Within the littoral rainforest, these activities are designated development.

Both Victoria and New South Wales have taken advantage of the fact that planning instruments can be made under delegated powers so as to come into force immediately, without public consultation, let alone debate by the legislature. This is an important consideration in the present context where there is a danger of panic clearing if advance warning is given. In an ideal world, environmental planning instruments can also be targeted at specific areas for specific reasons after careful assessment of the needs of particular environments. This places an onus on planners to assess the resource at a broad level, rather than leaving it up to those seeking approvals to produce information about specific parts of it on an *ad hoc* basis. Ideally such assessments should be carried out before controls are introduced, but application of the precautionary principle, in the face of the massive research operation frequently required, may dictate intervention on the basis of preliminary evidence. SEPP 14 wetlands, for example, were initially identified through aerial photographs.[66]

Interferences with so-called "private property rights" imposed through planning systems can lay claim to a degree of legitimacy stemming from a long-established tradition of intervention in urban areas. The movement from town to environmental planning can be presented as a change of emphasis rather than a radical new initiative in which the rural landholder has been singled out for separate treatment. As a result, opponents are placed in the position of having to defend continued non-intervention in rural areas now that previously hidden land use problems have now come to light. At the same time, planning systems are not completely careless of the claims of private property holders. Existing uses, such as current grazing practices, are protected from new regulatory initiatives, whether in the form of prohibitions or requirements to obtain development approval, although in some states the precise ambit of these rights are currently being renegotiated. What this effectively means is that a government which wishes to terminate a non-conforming use must purchase it in the open market. From one perspective, it is no more than reasonable that existing operations, as distinct from new initiatives, should be allowed to continue, but what this means in the present context is that "passive clearing" caused by existing grazing regimes will ordinarily be exempt from clearing controls imposed through planning systems.[67]

Ultimately, however, the success of any system which, though symbolically prohibitive, in practice simply delegates to agencies the power to give approvals on a case-by-case basis, lies in its implementation. There is some evidence that at this level, the interests of biodiversity are not being met. One of the fundamental problems is that planning systems are not focused but attempt to cater for a wide range of competing considerations. This includes not only the protection of fauna and flora habitat and genetic diversity, but also landscape conservation and the prevention of land degradation.[68] Some of these objectives can be met more easily than can the exacting demands of biodiversity conservation, clearing the way for a compromise decision in which some degree of interference is allowed and the symbolic prohibition relaxed. More fundamentally, the New South Wales courts have held that the interests of property owners must be taken into account in deciding whether or not to grant development consent, including the fact that the current use may not be economically viable,[69] with the inference that landholders should be given the opportunity of changing it to one which is. This has significant implications for proposals to clear land currently used for rough grazing to allow intensive agriculture.

The way in which discretionary powers to grant approvals are usually formulated in legislation or plans means that the weight to be given to individual factors lies with the particular decision-maker, which can quite legitimately allow its own values to prevail, without giving the courts grounds for interference. The legal requirement is that factors set out in legislation must simply be considered, not be determinative. A council can quite legitimately make "development-at-all-costs" decisions, provided that the environmental impact is at least considered, before, perhaps, being dismissed. As Mr Justice Moffitt has put it, before an inference can be drawn that a council failed to take relevant factors into account:

> [T]here must be displaced for example a view that the decision, even an unwise one, was come to on the basis that the merits of the development and the burden which would

be placed on the developer if required to mitigate the harm to the environment were considered to outweight [*sic*] protecting the environment, after the relevant s. 90(1) matters had been taken into consideration.[70]

The South Australian experience of using the planning system to regulate the removal of native vegetation in the period prior to the enactment of the *Native Vegetation Management Act* 1985, suggests that in these circumstances decision-makers are reluctant to refuse applications outright, as distinct from granting approval subject to conditions. In the two years following the introduction of controls under the *Planning Act* 1982, over 80 per cent of the applications dealt with were approved, although usually under an arrangement whereby some vegetation was to be retained.[71] Early evidence from Victoria showed that of the 129 applications referred to the Department of Conservation and Environment in the first six months of operation of the interim controls, nearly 70 per cent were approved. Of 10 444 ha for which applications were received, only 2 577 ha were protected by outright refusal.[72] This appears to be reinforced by more recent data from eight shires, covering the period November 1989–June 1993: 86 per cent of applications were granted conditionally, with only 11 per cent refused altogether and a further 3 per cent withdrawn. There are, however, large differences between shires, with clearance permitted of only 23 per cent of the total area applied for in Mildura Shire, for example, compared with 81 per cent in Orbost Shire.[73] More research needs to be carried out to identify the factors leading to large areas being conserved, and the extent to which this is being done by outright refusal. There is enough evidence at this stage, however, to support the hypothesis that there is indeed a culture of consent among decision-makers.[74]

One argument is that a precautionary approach should be built into statutory formulations of decision-making powers, curtailing the free play of the decision-maker's values.[75] One approach is to require the decision-maker *to be satisfied* of certain things before it can make a particular decision. The Tumbarumba Local Environmental Plan 1988, for example, provides that in environmental protection areas, consent should not be given to land clearing unless council is satisfied either that:

- clearing is necessary for the reasonable economic use of the land or the provision of utility services; or
- clearing is necessary to reduce the risk of bushfires.

Here the onus is on the applicant to show, for example, that the current land use is not economically viable before consent can be given. A similar provision in a draft Tumut Local Environmental Plan would have applied across a number of rural zones. It took an even more restrictive approach than the Tumbarumba LEP by providing that consent should not be given unless council was satisfied that clearing would not adversely affect wildlife. This is a clear application of the precautionary principle: in light of the scientific uncertainty associated with land clearing, the burden of disproving the ostensibly obvious threat to biodiversity is placed on the landholder. Clauses such as these represent an attempt to ensure that regulatory provisions which are symbolically prohibitive, but allow for individualized exceptions, do not in practice result in a situation where the exception becomes the rule.[76]

Both the initiatives in New South Wales and Victoria at the state level retain some degree of state government supervision over council decisions.[77] In the past, planning systems have focused their energies on urban as distinct from rural land use and consequently local councils have no established expertise in the field of nature conservation.[78] Apart from this, they are inevitably closer to their constituencies than state government agencies, making hard decisions even more difficult. Durkin's survey of the implementation of clearing provisions by certain Victorian local councils does not paint an optimistic picture of their ability to hold the line when it comes to the issue of biodiversity conservation. Phrases such as "ecological balance" and "biological integrity", which appeared in the list of factors to be considered when making decisions on permit applications, were interpreted in different ways by planners and councillors[79]. The emphasis was on negotiation and compromise over permit conditions on a case by case basis.[80] There was evidence of a community view that land clearing was an acceptable part of farming. When it came to questions of enforcement, Durkin found that although a dob-in system worked well where trees were concerned, there were major enforcement problems in relation to shrubby understorey, the conservation value of which was not appreciated. Breaches which were detected were generally dealt with by negotiation. Illegal clearing was often legitimated by the retrospective granting of a permit.[81] Where permits were granted, it frequently proved difficult to monitor and enforce conditions.[82]

Fook's work[83] on the implementation of SEPP 14 in New South Wales indicates that regulatory failure may occur even where there is supervision by the state government. She refers to a number of occasions when *retrospective* consent from the council and concurrence from the

Director of Planning were given to illegal activities not subject to prosecution. She also raises the possibility that it is elected councillors, rather than council officers, who are responsible for this approach, indicating that the level of political interference is a crucial variable in determining policies pursued by enforcement agencies.

The final point to make about using the planning system as a vehicle for clearing controls is that it has traditionally been negative in orientation. Its primary focus has been on restricting inappropriate rather than promoting appropriate development.[84] To this extent, it is ill-adapted to providing the two-pronged strategy of retention and management needed to address the issue of biodiversity conservation. Any suggestion that refusals of clearing approval should be associated with some form of payment for management would be vigorously resisted by policy-makers because of the fear that payments would be perceived as compensation, and the potential flow-on effects that this might have in relation to urban development.

An alternative way of securing some commitment to ongoing management is to require this as part of a bargain in which the grant of approval is used as a bargaining chip.[85] At one stage, for example, the Victorian provisions provided that those who were initially given permission to clear pursuant to a management plan or a works programme would require no further permits in the future to clear land in accordance with the plan or programme. There was a built-in incentive to take advantage of these provisions, stemming from the fact that many of the specific exemptions (including grazing by domestic stock and clearing of regrowth less than ten years old to re-establish cultivation or pasture) ceased to operate after five years.[86] The difficulty with this bargaining approach is that it assumes a level of permitted land clearing which, while justifiable where the objective is land conservation and the prevention of land degradation, may not be appropriate where the concern is the conservation of biodiversity.

Command and control through land conservation regimes

Another response to pressures to adopt a more aggressive response to matters of nature conservation, generated by concerns over biodiversity and ecological integrity, has been to graft nature conservation objectives on to existing regulatory regimes which started out life with an exclusive concern for land conservation.

Under section 21C of the *Soil Conservation Act* 1938 (New South Wales), trees on "protected land" must not be destroyed, removed or injured without an authority. "Tree" is said to include both "shrub" and "scrub"[87], so that other native flora is probably not protected. "Protected land" includes:[88]

- the area within 20 metres of the bed or bank of specified rivers and lakes: in practice, all major watercourses and a large number of other streams and lakes are covered (Sixth Schedule);
- steeply sloping land in notified catchment areas and land within 20 metres of watercourses which is identified on maps prepared by the Soil Conservation Service: most of the Eastern sea board has been mapped under this provision, including a great deal of timbered country;
- land mapped by the Soil Conservation Commissioner as being "environmentally sensitive or affected or liable to be affected by soil erosion, siltation or land degradation". Environmentally sensitive land can include areas such as arid or semi-arid lands, saline areas, land containing rare or endangered fauna or flora, land containing areas of archaeological or historical interest, land containing bird breeding grounds, wetlands and areas of scenic beauty.

In their original form, the protected land provisions, administered by a body known as the Catchment Areas Protection Board, were primarily concerned with the prevention of certain forms of land degradation resulting from vegetation destruction on steeply sloping land and along watercourses. This focus was reflected in the Board's policy with regard to the issue of authorities. Initial discussions with soil conservation field officers led to some proposals being dropped. Of the applications made between 1 January 1980 and 30 June 1984, the last date for which figures are available, the vast majority were granted: 730 permits out of 747 applications involving land along prescribed watercourses, and 1 176 out of 1 210 applications for steeply sloping land in catchment areas.[89] During the year ended 30 June 1984, only one permit was refused on the ground of environmental factors. The emphasis was placed rather on the attachment of conditions as the means of providing environmental protection. The Soil Conservation Service's submission to the RAC *Forest and Timber Inquiry* indicated that "in most cases, the Service has little or no ecological information at its disposal on a site" and that therefore each proposal "can only be considered on its merits or on an *ad hoc* basis".[90]

More recently, however, a number of legal and administrative changes are resulting in a broadening of the objectives pursued under the protected land provisions, beyond the previous focus on land degradation.[91] The introduction of Part V of the *Environmental Planning and Assessment Act* 1979 means that the Soil Conservation Commissioner (who has taken over decision-making responsibility from the Catchment Areas Protection Board) must comply with the general duty under section 111 to take into account to the fullest extent reasonably practicable[92] all matters affecting or likely to affect the environment before granting an authority. If the proposal is likely to significantly affect the environment, an environmental impact statement must be prepared.[93] As a result of amendments to Part V made in 1991, the Soil Conservation Commissioner, as a determining authority, must also consider whether there is likely to be a significant effect on the environment of endangered fauna, and if there is, the applicant must prepare a fauna impact statement. Finally, the provisions allowing the mapping of environmentally sensitive land, which were introduced in 1986, gave clear indication that the role of protected land regulation was no longer to be confined to narrow land degradation concerns.

The potential impact of the environmentally sensitive land provisions could be dramatic. However, everything depends on how successful the Commissioner is in mapping land as environmentally sensitive and whether, in relation to these areas, the Commissioner is prepared to reject applications for authorities outright, on nature conservation grounds. The limited amount of land mapped so far as environmentally sensitive includes endangered bird habitat, remnant rainforest communities outside dedicated conservation reserves, significant Aboriginal sites and areas of visual amenity. Apart from this, section 111 only requires a broader range of environmental factors to be *taken into account.* While these include any endangering of any species of fauna and flora and any impact on the habitat of protected and endangered fauna,[94] they do not have to be given particular weight, or indeed any weight at all. Legally, it is quite acceptable for the Soil Conservation Commissioner to give more weight to the prevention of land degradation, rather than the preservation of wildlife habitat and vegetation communities. The legislation gives no guidance on the question of priorities. Nevertheless, in practice, there has recently been a very real attempt to set up procedures, to develop expertise and to gather information which will allow decisions sensitive to these considerations to be made.

Ultimately, however, it will be the values of the Soil Conservation Commissioner and the Service which will, quite legitimately within the framework set by the law, determine the sensitivity of the protected land regime to matters of biodiversity conservation. Unlike local councils, they are not made directly accountable through the electoral process, although the Commissioner is subject to the directions of an elected Minister.[95] In these circumstances, the expectations of their landholder clientele will have a vital influence when it comes to determining priorities. Soil conservation agencies have traditionally perceived themselves, and been perceived, as educators rather than regulators, as being primarily concerned with assisting landholders to conserve the land resource in their own best interests, as well as those of the wider community. Retention of native vegetation is only one approach to addressing the problem of land degradation. Alternatives may simply require the retention of vegetation in strategic places, such as on slopes and along rivers, or sowing pasture to counter the threat of soil erosion. Consequently, the needs of land conservation can frequently be met by attaching appropriate conditions to licences rather than refusing applications outright. This allows for bargains to be made, behind the ostensibly coercive facade, whereby the landholder gives and receives concessions. Viewed in this light, land conservation is not an end in itself but a restraint on the pursuit of development in the form of agriculture or forestry. Put another way, land conservation is not, like agriculture and forestry, itself a land use, but a restriction on other land uses.

When it comes to the conservation of biodiversity, there is no obvious short-term economic self-interest on the part of the landholder to which appeals can be made. On top of this, in many situations, demands for biodiversity conservation cannot be met by arrangements such as those which currently exist in the Western Division of New South Wales under the maximum allowable area policy, whereby each lessee is, in effect, allowed to clear up to a maximum area, depending on the zone in which the land is situated, the purpose of the lease, the purpose of the clearing and the size of the property.[96] Biodiversity does not distribute itself so conveniently. A regulatory system orientated towards conditional approvals rather than outright prohibitions cannot ordinarily satisfy the demands of nature conservation.[97] Moreover, if landholders were to be completely denied permission to clear, the state then has nothing to offer to them, in terms of limited clearing, in return for the positive management of the land which biodiversity conservation requires.

Land conservation provisions in Western Australia have also come under some pressure to incorporate nature conservation values. Under the Soil and Land Conservation Regulations 1985, landholders must give notice of intent to clear land in excess of one hectare where the clearing would result in a change in the use of the land and the Commissioner then has the opportunity of intervening. The notice must be given at least 90 days before the clearing begins and must include a plan showing the position of the land to be cleared. If the Commissioner takes no action within this period, the landholder has satisfied his or her legal obligations and can go ahead and clear. "Land degradation" is defined broadly in the enabling legislation to include "the removal or deterioration of natural or introduced vegetation, that may be detrimental to the present or future use of the land".[98] So far, however, the Commissioner's apparent interest in broadening the scope of his role to incorporate nature conservation concerns has been baulked by legal advice to the effect that the legislation does not allow for the retention of native vegetation solely on nature conservation grounds.[99]

One option open to the Commissioner in response to a notice of intention to clear is to issue a soil conservation notice prohibiting clearing. Yet in spite of this coercive façade, this system has traditionally operated through a process of bargaining and compromise which allows some level of clearing to go ahead. In nearly all cases it is possible to reach agreement on the clearing plan, providing for areas to be retained.[100] Amendments to the legislation in 1990 now formalize this bargaining process by providing for agreements to preserve native vegetation to be registered on the land title, either as "conservation covenants" (where expressed to be irrevocable) or "agreements to reserve".

Another feature of the Western Australian approach sets it apart from the New South Wales protected land provisions, even though both are ultimately based on command and control regulation. Agency directions forbidding activities, such as soil conservation notices, present as exceptional prohibitions in the context of what is otherwise a general permission allowing clearing subject to notification. They require the regulatory agency to take the initiative by introducing what may appear to be an exceptional prohibition. To enforcement agencies, notices look like weapons of last resort rather than a typical response. In essence, there is a presumption against coercive regulation. The New South Wales protected land provisions, on the other hand, place the onus upon those who are subject to a general prohibition against land clearing to seek exemption. This general prohibition can claim greater legitimacy from its status as legislation than an individualized direction. Permits can be depicted as departures from a statement formally condemning the general activity in question. They can be presented as the exception rather than the rule.

These differences in emphasis should not distract us from the real function of coercive instruments in the hands of enforcement agencies in this context, either in the form of a general criminal prohibition or an individualized direction. These instruments are not intended to be used as a typical response. They are not an end in themselves, but a means to the end of inducing an *agreed* solution. Their real function is to reweight the balance of bargaining power in favour of the agency and to ensure some degree of co-operation from landholders in the negotiating process. However, regulatory systems focusing on land conservation ultimately depend on agencies being able to offer something to landholders, by way of restricted land clearing. If a focus on the conservation of biodiversity prevents agencies from offering very much, and they have to rely on their coercive powers as a matter of routine, while the regulatory system may retain the same form, it begins to operate under a completely different set of assumptions.

NATIVE VEGETATION PROTECTION IN SOUTH AUSTRALIA[101]

The distinctive feature of the *Native Vegetation Management Act* 1985 (South Australia), repealed in 1991, was its attempt to combine a symbolically coercive system of regulation, similar to those discussed in the previous section, with the offer of a financial package to those refused permission to clear. The result was a significant tightening up in terms of the granting of approvals in comparison with the prior position under which land clearing was a regulated development under the planning system.[102] Of the total area for which applications were made between 1986 and 1989 involving broadacre clearing, about 94 per cent was protected by an outright refusal.[103]

The legislation prohibited land clearing and woodcutting without consent from the Native Vegetation Authority, on both private and Crown leasehold land, subject to a number of exemptions spelt out in Regulations. These included grazing by domestic stock and clearance of regrowth and shrub invasion in certain carefully defined circumstances. Where an activity was not exempt, owners of land who were given a conditional approval or were refused consent, could generally insist on the Minister entering into a heritage agreement under the *South Australian Heritage Act*.[104] Once such an agreement had been concluded, the landholder was entitled to the payment of "a sum of money"[105] based on "the diminution (if any) in the market value of the land resulting from the Authority's decision".[106]

One of the basic problems which arose was how this financial payment was to be characterized: was it to be seen as backward-looking compensation or a fee for land management services to be rendered in the future? The government representatives on the Review Committee which considered the early stages of the operation of the legislation were anxious to emphasize that the payment of the money was related not to the mere refusal of consent but to entry into a heritage agreement. It was financial assistance, not compensation. They also emphasized the need for long-term management: refusal of consent did not guarantee this.[107] The fact remains, however, that the amount paid was related to loss in market value and therefore looked much more like compensation than a management fee.[108] The landholder could require the Minister to enter into an agreement on refusal of consent. *Once the agreement was concluded,* the landholder was *entitled* to the payment of a sum of money based on a set formula. This confronted the government with massive obstacles when it came to attempts to link the financial payment to management obligations on the part of the landholder. There was, for example, no provision for varying the amount paid according to the degree of management needed. Under the terms of the heritage agreement itself, all that the landholder was provided with was relief from payment of rates and taxes and the construction, but not the maintenance of fences[109] in return for simply *dedicating* the land to the conservation of native vegetation and native fauna. The way in which landholders viewed the arrangement was no more clearly evident than in the argument put by United Farmers and Stockowners' representatives that farmers were being expected to manage areas at no cost to the government.[110] At this stage, it was agreed by the Committee that a separate management fee should be considered by the Authority on the merits of each case.[111]

Even with this combination of coercive restrictions on land use and an apparently generous financial package, many landholders were reluctant to enter into heritage agreements.[112] In its 1987–88 Annual Report, the Native Vegetation Authority reported that only just over 44 000 hectares were protected by heritage agreements stemming from a refusal to grant approval to clear, but in that financial year alone refusals covered over 68 000 hectares.[113] In response to this, administrative arrangements were made to make the financial payment more generous. The new arrangements appeared to have some effect because in 1988–89 a further 60 agreements were concluded with landholders, covering an area of 52 487 hectares.[114]

These new arrangements discriminated between the conservation of biodiversity, on the one hand, and land conservation on the other.[115] The discount factor, which in essence meant that the landholder was not compensated for vegetation preserved on a proportion of the holding, was now abandoned where the objective was biodiversity conservation, as distinct from land conservation. The community from now on was expected to pay for all of its biodiversity conservation. Restrictions on clearing for these purposes were defined solely in terms of the provision of a public benefit rather than the prevention of a public harm. More surprisingly, this scheme assumed that it was quite appropriate to compensate landholders to a substantial degree where the objective was to prevent them from degrading the land in their own best interests or to protect the broader community from polluting spill-overs caused by land degradation associated with vegetation clearance. This runs totally against the polluter-pays principle.

The *Native Vegetation Management Act* 1985 has been repealed by the *Native Vegetation Act* 1991. Applications for consent to clear must now be accompanied by a native vegetation management plan.[116] In addition, an attempt has been made to restrict the discretion of what is now the Native Vegetation Council. It must not make a decision which is "seriously at variance with" the principles of clearance of native vegetation. These are set out in Schedule 1 of the Act, delineating a number of situations in which native vegetation *should not be cleared.* They include both biodiversity conservation (e.g., the vegetation comprises a high level of diversity of plant species) and land conservation (e.g., clearance is likely to contribute to soil erosion or salinity in an area in which appreciable soil erosion or salinity has already occurred). This strategy is quite different from simply requiring a decision-maker to take certain factors into account before reaching a decision. It embodies a precautionary approach by not only making it clear that the norm is *not* to develop, thereby creating a presumption against development, but also taking the further step of forbidding a decision which deviates significantly from this presumption in favour of the *status quo.*[117] Even though the Council is required to have regard to the applicant's desire to operate the business as efficiently as possible, this is only to operate as a factor within these overall limits,[118] representing a significant downgrading of traditional economic considerations in the decision-making process.

Under the new decision-making procedure outlined above, very few consents should be granted. The new legislation attenuates the connection between the refusal of permission to clear and the payment of compensation. Those refused can no longer demand that the Minister enter into a heritage agreement in order to claim a financial payment based on the decline in the value of the land arising from the refusal of permission to clear. But the spectre of compensation is not completely expunged. As an inducement to enter into a heritage agreement, an amount can be offered representing the loss in land value *resulting from the heritage agreement,* as distinct from the land use restriction imposed. In recognition of the fact that this amount may not be significant, the Minister is also empowered to offer an amount "as an incentive to enter into a heritage agreement".[119] What the landholder offers in return is an extension of land use restrictions to include activities exempted from the clearing controls, notably grazing,[120] rather than any management commitments.[121]

At the same time, the 1991 Act does show a much greater commitment on its face than its predecessor to management of vegetation following the imposition of restrictions.[122] Listed as the first of its objects is "the provision of incentives and assistance to landowners in relation to the preservation, enhancement and management of native vegetation".[123] The Council has broad powers to grant financial or other assistance in managing the land and preserving, managing, and enhancing native vegetation, in accordance with management guidelines[124] adopted after public comment has been considered.[125] However, these management arrangements are not part of the initial heritage agreement: applications can only be made *after* the landholder has entered into such an agreement. The heritage agreement can then be varied.[126] Management obligations are not to be part of the original bargain in the form of the heritage agreement, which appears to be primarily orientated towards inducing the abandonment of existing uses, such as grazing. They must be purchased separately.

AN ECONOMIC PERSPECTIVE

In the field of industrial pollution control, we are witnessing an increasing movement away from command and control regulation towards the use of fiscal instruments, in particular, charges on emissions and tradeable pollution permits.[127]

In the industrial pollution context, the argument is that fiscal instruments are more flexible than uniform emission standards which industrial units are required to comply with by law. Fiscal instruments allow each individual unit to make decisions which are cost-effective from its point of view. The theory is that if it is cheaper for a particular industrial unit to use different inputs, for example an alternative fuel, or to install pollution control equipment than to pay a tax or purchase additional tradeable permits, then it will be driven by market forces to make these changes. Those who do not find it cheaper will make a commercial decision to pay the charge/purchase the tradeable permits and to continue to pollute. In comparison with individualized emission standards fixed by conditions attached to licences, fiscal instruments have the advantage of leaving decisions to be made by the industrial units themselves. These units are in a stronger position, in terms of information and resources, than regulatory agencies, to make decisions about how far each of them should go in terms of pollution abatement. Apart from this, regulatory agencies issuing licences are under considerable pressure to treat everybody as equally as possible.

Fiscal instruments in the form of charges and tradeable permits, as distinct from subsidies, are also attractive because in addition to this efficiency dimension, they also have the strong moral element of making the polluter pay. This can be seen as to some extent making up for abandoning condemnation through the criminal sanction, which is an intrinsic part of command and control regulatory strategies.

The first obstacle we meet when we try to translate this approach to biodiversity conservation is the ideology of private property, with its claim that landholders can do what they like with their land. We are reluctant to treat landholders as being akin to polluters where there are no physical spillovers *from* their land,[128] even though the loss of biodiversity does involve public costs which ought to be taken into account in the economic calculus. Indeed, historically, even where there have been diffuse physical spillovers stemming from land degradation, in the form of salinization or siltation, for example, we have been reluctant to treat agricultural landholders in the same way as point-source industrial polluters, frequently preferring to subsidize remedial work rather than apply the polluter-pays principle.

Another problem is that, when it comes to the destruction of biodiversity, there are no measurable outputs to be taxed. The spillover for which the "polluter" should be required to pay is loss of biodiversity. This is impossible to measure in our current state of knowledge. In response to equivalent difficulties experienced in measuring diffuse agricultural run-off, those concerned with preventing land degradation have suggested either clearing taxes/"conversion charges",[129] perhaps based on the proportion of vegetation cleared, the area of land cleared and its susceptibility to erosion, etc., or alternatively, the use of tradeable clearing permits.[130] This last would involve governments determining an acceptable level of land clearing in a particular area and then issuing permits which, taken together, would cover this total amount. These could be freely traded in the market-place. The problem here is that there is not necessarily any equivalence between the damage caused by clearing in one area and clearing in another. This becomes even more obvious when our concern is with loss of biodiversity, as distinct from diffuse agricultural pollution. As its name implies, biodiversity does not have the common currency contemplated by a uniform tax or a trade in equivalents.

It is also important to remember that fiscal instruments are not designed simply to make the polluter pay, but to regulate the behaviour in question within acceptable limits through market mechanisms. Because of the difficulties of determining what is an *optimum* level of biodiversity loss, it becomes essential from an economic perspective to determine what is an *acceptable* level of loss of biodiversity and to fix a charge which will ensure that this level is not exceeded. Given our totally inadequate information about what is out there, our even greater ignorance of its potential utility to humankind, and the irreversibility of its destruction, the notion of an acceptable level of loss is nonsense.

Finally, while some have attempted to extrapolate to the land degradation context[131] the argument that industrial polluters are in a better position than regulatory agencies to work out least-cost approaches to pollution abatement, this cannot be sustained in the context of the conservation of biodiversity. Economists acknowledge that one of the situations when society is likely to place constraints on individual sovereignty is "where individuals have incomplete or erroneous information or for various reasons are unable to process the available information properly".[132]

THE REGULATORY ALTERNATIVE

Writing in the context of the prevention of land degradation, Chisholm argues that "outright prohibition" of clearance is unlikely to prove an optimal policy in the long term, but he concedes that it can be "justified as a short-term emergency measure to provide time for more information to be gathered and analysis undertaken".[133] The argument which I wish to advance here is that there is basically no alternative to some form of regulation once our objective is to prevent irreversible loss of biodiversity. The primary issue becomes whether land use restrictions should be accompanied by financial inducements, not whether they should be imposed at all. We simply cannot afford to allow the question of whether or not a particular species or ecosystem should be irreversibly destroyed or damaged to be resolved by negotiations between landholder and government agency, where the former is in a position to name his or her price. Command and control regulation of land clearing does not necessarily mean vigorous prosecution. Its crucial role is to change the ground rules so that bargaining power is reweighted in favour of the state rather than leaving it in the hands of individual landholders.[134]

We need, however, to go further than simply setting up symbolic constraints, behind the facade of which the issue becomes what conditions should be attached to approvals. If we are to continue to allow exceptions to be made through a licensing/approval system, the precautionary approach must be carried through into that system to ensure that the individually approved exception does not become the rule. The licensing system must be insulated against the danger that the land degradation horse, with its strong historical roots and the attraction of compromise which it holds out, will pull the biodiversity cart. The conservation of biodiversity must be identified as the pre-eminent objective. In most situations, by satisfying the demands of biodiversity conservation, we will automatically meet those of land conservation and landscape protection.

The values of decision-makers must not be given the free play contemplated by the legislative formula which simply requires them to "take into account" a list of considerations, with biodiversity one of many. I have already provided examples of how the precautionary principle can be built into decision-making processes through formulae requiring that decisions not be "seriously at variance with" enunciated principles or that decisions to permit clearing will not "adversely affect" the conservation of biodiversity.[135]

A precautionary approach must also escape the tyranny of *ad hoc* decision-making on a case-by-case basis. If we are to convince landholders that their private benefits in clearing land have to give way to the public interest in conserving biodiversity, governments must at least be prepared to make a substantial commitment to biodiversity assessment and regional planning.[136] It is quite inappropriate to cast the burden of proof on landholders who wish to clear land, where this necessarily includes the task of generating most of the data required to satisfy that burden. In the medium term effective regional planning, which identifies significant areas and justifies their significance, will serve to insulate decision-makers from the pressures arising from having to deal with individual applicants, in a context where they cannot offer convincing justifications for frustrating landholder plans. Landholders are more likely to accept decisions made well in advance of particular proposals than those which take the shape of last minute reactions.

COMPENSATION?

Use of regulatory instruments without some form of financial recompense will alienate the very landholders on whom we must continue to rely to manage land not taken into public ownership. Moreover, the evidence strongly suggests that where there is some kind of trade-off, regulatory agencies are happier about making the hard decisions on individual permit applications. However, the precise form that this payment takes is crucial. Framed in terms of compensation, it looks backwards and offers a palliative to those whose plans have been frustrated. Compensation for losses caused by land use restrictions "does not really create an incentive to change production in the desired direction".[137] Demands for compensation rest on the assumption that landholders have some form of private property right to destroy the very material on which the survival of life on this planet depends. Such a statement only needs to be made to be dismissed. It may have been legitimate so long as vegetation conservation was defined in terms of "nature conservation". For the broader community, nature conservation has generally been seen as being about matters of aesthetics and recreation — as a public benefit for which it was legitimate to expect society to pay where it was provided at the expense of individual landholders. The nature of the debate has changed with its redefinition in terms of the conservation of biodiversity and the growing realization that the issue is no longer about *life-style* but *life-support.* Attention is increasingly drawn to the dramatic spillover costs resulting from the destruction or degradation of nature rather than the public benefits resulting from the decision to retain it. Viewed in this light, the destruction of vegetation becomes a polluting externality and compensation for preventing harm is tantamount to paying the aspiring polluter.

There is nothing in any of the State Constitutions which guarantees compensation for landholders, even in situations where they are totally excluded from their land by state action. The payment of compensation where land is resumed for public purposes is purely a matter of convention. Section 51(xxxi) of the Commonwealth Constitution does provide that any "acquisition" of property by instrumentalities of the Commonwealth Government must be made on just terms. However, in the *Tasmanian Dam* case, three of the four members of the High Court who dealt with the issue made it clear that even the severe restrictions on land use imposed under the *World Heritage Properties Conservation Act* 1983 (Cth) did not constitute an "acquisition" requiring the payment of compensation under section 51(xxxi). According to Mason J, the Commonwealth had acquired no proprietary interest in the land and therefore did not have to pay compensation, even though in terms of its potential use the property was sterilized in the same way as a dedicated park, subject only to the power of the Minister to consent to development on a case by case basis.[138]

There is, in fact, a long tradition of land use restrictions being imposed on private landholders without compensation being paid.[139] The common law, developed by the courts, including in particular the law of nuisance, recognized that some restrictions were inevitable if private property holders were to be protected even from each other. If each was allowed to develop land according to their own wishes, there would inevitably be conflicts arising from inconsistent uses which had spill-over effects on neighbouring land, such as factories situated next to residential areas. Town planning laws took this process a stage further, introducing a preventive dimension. Environmental laws are simply part of this heritage, stemming from the growing realization that human beings are not only detrimentally affected by those who conduct activities in their immediate neighbourhood. We are now coming to recognize that spill-over effects encompass not only pollution but also the destruction and degradation of biodiversity, and that they have an impact not only on close neighbours but on the broader community, as well as on future generations.

This is simply another way of saying that the concept of property is continually evolving with our increasing knowledge of the way in which our surroundings impact upon us. Macpherson has argued:

> The meaning of property is not constant. The actual institution, and the way people see it, and hence the meaning they give to the word, all change over time. We shall see that they are changing now. The changes are related to changes in the purposes which society or the dominant class in society expect the institution of property to serve.[140]

Any institution of property needs to be justified by "some more basic human or social purpose" and property in land stands in more need of justification than property in consumables.[141] In face of our growing awareness of the dramatic repercussions which destruction of vegetation can have in terms of loss of biological diversity and ecological integrity, quite apart from obvious and immediate polluting spillovers into common property resources, any concept of private property which gives landholders an unfettered right to use their land as they see fit can no longer be sustained by any of the justifications which have traditionally underpinned it. Sax has pointed to the incompatibility of assumptions underlying traditional conceptions of private land ownership with the interconnected web presented to us by the science of ecology:

> Land is not seen as part of a common trust, like a spaceship in which we are all travelling, or even as an ecosystem, but as a collection of essentially separate and separable entities driven by notions of decentralization and autonomy, individual dominion and private aggrandizement. The evocative language of this conception is brought to mind by such familiar phrases as "good fences make good neighbours" or "keep out, private property".[142]

To argue that there is no right to compensation stemming from some historical notion of private property rights is not to deny that compensation can be justified in particular circumstances. Chisholm has argued that secure expectations are vital to the efficient functioning of the property market and that changes in environmental policy which produce unexpected losses or gains are disruptive. At the same time, he suggests that from an equity perspective, compensation should only be payable in special circumstances, given that government policy produces such gains and losses in all spheres of economic activity.[143] If we were to start closing down *existing* operations without compensation, such as the existing use of land for grazing, this would be both socially disruptive and inequitable. To require compensation in such circumstances is a different matter from requiring it where what has been lost is some nebulous expectation of being able to expand an existing operation to increase profitability or to provide for family members. The requirement to pay compensation may need to be taken a step further. The distinction to be drawn must be between legitimate expectation and speculation. Much will depend on the viability of the existing use. In some situations, genuine hardship will result if land use restrictions prevent a planned expansion of the area of land under productive use to make it into an economically viable holding. Hardship payments may have to be made in these circumstances, although this should in part depend on the extent to which payments for managing the land for the purposes of biodiversity conservation provide an alternative pathway to economic viability.

PAYMENTS FOR MANAGEMENT

Paying private landholders to manage land in the interests of conserving biological diversity and protecting ecological integrity is a different mechanism to paying compensation to those who are required to forego an activity which will degrade or destroy the basic resource itself.[144] While compensation is backward-looking, management payments look to the future and provide a legitimate subsidy for the production of a commodity[145] — biological diversity — for which there is currently only a very restricted market. Sax has in fact argued that our legal institution of private property needs to be taken beyond its traditional concern with negative obligations to embrace affirmative duties.[146] In other words, there should be a duty on landholders to manage land in an appropriate manner, without having to be paid to do so. Yet, while it may be quite appropriate to argue that the community should not have to pay for the management of land to avert land degradation, it is a much greater leap to extend such an obligation to carry out positive management of land which the landholder has been denied permission to develop, in the long-term community interest of conserving biodiversity, where the potential private rewards are both distant and nebulous.

The level of management that I have in mind goes far beyond fencing off areas of land to keep off domestic stock. Australia's biodiversity and ecosystems are not only threatened by the clearance of native vegetation but by feral animals and invasive exotic plants.[147] Fire management is also crucial in some contexts. Beyond this, it may well be possible to train some landholders, at least, to carry out the work involved in making a preliminary rapid assessment and identification of the organisms which are present in and on their land.[148]

There are two potential funding sources for such a proposal already in existence — the National Landcare Program and the Rural Adjustment Scheme — but it would necessitate a substantial reordering of priorities. In 1992, the House of Representatives Standing Committee on Environment, Recreation and the Arts reported on its inquiry into the contribution made to biodiversity conservation by the four different limbs of the National Landcare Program (Save the Bush Program, One Billion Trees Program, the National Soil Conservation Program and the Murray-Darling Basin Natural Resources Management Strategy Program). All of the evidence on the Save the Bush Program viewed it as the only one of the four to focus specifically on biodiversity.[149] The Committee concluded that the current level of funding was "desultory".[150] The National Soil Conservation Program received almost twenty times the funding, but its focus was on land degradation and the projects it supported were mostly reactive.[151] Many submissions pointed to the contradictions inherent in allowing clearing to continue while providing funding to protect and manage remnant native vegetation or to support revegetation projects.[152] The Committee was particularly sceptical about the spin-off effects of on-farm demonstration projects and suggested that in some cases it might be more appropriate to put a large pool of resources into direct action to ameliorate a major regional environmental problem.[153] Since the Committee's report, the National Landcare Program has undergone substantial changes.[154] The present proposal, however, would demand a radical reorientation of the basis on which funds are allocated.

This is also the position in relation to the Rural Adjustment Scheme. In 1992–93, a total of $166.46 million was allocated to the Scheme.[155] The Scheme was reviewed and reshaped in 1992, but it still falls far short of representing a real attempt to integrate rural adjustment and environmental protection. The Review rejected the use of the Scheme as an instrument of environmental policy, and in particular, the idea that financial support should be conditional upon the adoption of specified farming practices.[156] In other words, while mouthing the platitude that "sustainable agricultural systems require economically strong farms",[157] it rejected the idea that the Scheme should be used as a policy instrument to counter land degradation, let alone loss of biodiversity.

There has been some attempt, in the significant changes made to the Scheme following the Report, to make it at least appear more sensitive to land degradation considerations which impact directly on the profitability of the individual farm enterprise. However, there is no indication that it is to be concerned with preventing degradation associated with particular farm management practices from spilling over on to other land (e.g., dryland salinity resulting from inappropriate land clearing), let alone with advancing the broad social interest in the conservation of biodiversity. In his second reading speech on the Rural Adjustment Bill 1992, the Minister emphasized the movement away from the focus on *assistance* in the former Scheme to a new emphasis on structural adjustment, which was necessary to address "the formidable economic and environmental problems facing the sector". This involved enhancing the productivity, *sustainability* and profitability of farm businesses.[158] In a press release, the Minister took this a step further by emphasizing the need for agricultural production systems to "take account of Australia's eco-systems, to ensure sustainable development and long-term profitability".[159] One of the objectives of the *Rural Adjustment Act* 1992 is "to improve the competitiveness of the farm sector in a sustainable manner".[160] However, when it comes to spelling out the assessment criteria to be applied by state bodies making decisions on whether to grant interest subsidies and loans, the emphasis is not on *ecological sustainability* but on improving "the productivity of the farming enterprise, leading to the *sustainable long-term profitability* of the enterprise". The ability to meet financial commitments relating to investment in sustainable farming systems becomes only one of a number of indicators of future profitability.[161] Another limb of the Scheme allows grants and loans to be given for training and professional advice, but the focus here is on the improvement of managerial, technical, financial and business knowledge and skills.[162]

CONCLUSION

While conservation reserves may satisfy community needs traditionally associated with nature conservation values, they fall far short of meeting the demand from the international community for the conservation of biodiversity as a resource, with the strong economic overtones associated with this. This squarely raises the question of biodiversity conservation on privately owned land, and the policy instruments available for inducing landholders to act against what they see as being in their immediate short-term economic interests. The strategy adopted must address not only the question of retention, but also that of ongoing management. I suggest that this is best achieved through a combination of command and control regulation and ongoing payments to landholders for managing the land in the interests of biodiversity conservation. This can be linked to the movement towards rural adjustment: economic restructuring in rural communities can no longer be based exclusively on short-term productivity concerns but must also take account of the much longer term economic interest that the community has in conserving biodiversity. Such an approach serves the broader educational function of sensitizing rural communities to the significant differences between land and biodiversity conservation, while offering the reassurance that governments are prepared to make financial investments in something which will often appear to be remote and nebulous, with no market pay-offs in the short term. It will prove far more attractive in political terms, especially during rural recession, than will providing funding to enhance the management capabilities of conservation agencies. Besides, the reality is that these agencies are unlikely to purchase privately owned land against the wishes of landholders, so that there really is no alternative to supporting its management in private hands where it has significance for the conservation of biodiversity.

NOTES

[1]It is the convention among legal writers to guarantee that their discussion of the law is up to date as of a particular time. Insofar as this is relevant, the law discussed is as it was known to me in May 1993. The aim of this paper is not, however, to provide a comprehensive survey of Australian legislation relevant to the conservation of biodiversity on privately owned land, but to discuss examples of some of the different legal strategies used in this context, with a view to developing improved policy responses.

[2]The Convention entered into force on 29 December 1993 on ratification by thirty countries, including Australia, in accordance with Article 36(1).

[3]United Nations Convention on Biological Diversity Article 2.

[4]*Ibid.* Articles 8(C), 8(H), 8(K), 7(C), 8(L), 11 and 20(1). Parties to the Convention will be required to submit annual reports on the measures for implementation taken and their effectiveness in meeting the Convention's objectives: Article 26.

[5]See, for example, the narrow definitions of endangered and vulnerable species proposed in the unsuccessful Endangered and Other Threatened Species Conservation Bill 1993 (New South Wales). The general position proposed was that a species should only be classified as "endangered" if it was likely to become extinct in New South Wales in 20 years, and as "vulnerable" if it was likely to become extinct in 100 years.

[6]*Commonwealth* v *Tasmania* (1983) 46 ALR 625; *Richardson* v *Forestry Commission* (1988) 77 ALR 237.

[7]Intergovernmental Agreement on the Environment cl. 2.4.

[8]*Ibid.* p. 2.

[9]*Ibid.* cl. 2.2.1. The other involves preventing situations where the policies and practices of one state have significant adverse effects in another jurisdiction.

[10]*Ibid.* cl. 2.3.3.

[11]*Ibid.* Schedule 6 cll. 3, 7.

[12]*Ibid.* cl. 2.2.1(iii).

[13]*Ibid.* Schedule 4.

[14]*Ibid.* Schedule 9 cl. 4.

[15]*Ibid.* Schedule 9 cll. 7–8.

[16]*Ibid.* Schedule 9 cl. 15.

[17]*Ibid.* Schedule 9 cll. 3, 10.

[18]*Ibid.* cl. 2.3.2.

[19]*Ibid.* Schedule 9 cl. 2.

[20]*Ibid.* Schedule 9 cl. 6.

[21]For present purposes, privately owned land is taken to include Crown leasehold, because in spite of the massive divide at a theoretical level between Crown land held under lease and freehold land, the practicalities are very different. This is especially true in a situation such as that currently faced in New South Wales, where the State Government is set to alienate to private ownership all Crown leasehold land in the Eastern and Central Divisions currently reserved from sale on nature conservation grounds. J. Benson and M. Doherty, "The Implications for Nature Conservation of Selling off New South Wales Leasehold Crown Lands," (1993) 37(4) *National Parks Journal* 15.

[22]Pressey has argued that conservation reserves in New South Wales have been set aside on an opportunistic basis, and as a result are not representative of biodiversity: R. L. Pressey, "Opportunism in Acquiring Land for Reserves," (1992) 36(4) *National Parks Journal* 19.

[23]Throughout the remainder of this paper, I take agriculture to cover pasture improvement.

[24]Out of 43 million hectares of forested land, 11.3 million are located on private land. Of this, 7.6 million are technically available for logging: RAC Forest and Timber Inquiry, *Final Report,* Vol 1, paras 1.21–1.23.

[25]Ecologically Sustainable Development Working Groups, *Final Report — Forest Use,* Recommendation 4.

[26]DASETT, 1992, paras 4.1.1, 4.1.4, 5.3.4.

[27]Commonwealth of Australia, *National Forest Policy Statement: A New Focus for Australia's Forests* (1992), p. 13. See also p. 27. This differs from an earlier draft, which committed state and territory governments to establishing a system of land clearing controls to maintain native forests on private land where removal would compromise regional conservation objectives: The Future of Australia's Forests, Draft Policy Statement by the Commonwealth, State and Territory Governments, 4 October 1992, p. 14.

[28]Ecologically Sustainable Development Working Groups, *Final Report — Agriculture,* pp. 159–62.

[29]*Ibid.* pp. 128–29.

[30]The Ecologically Sustainable Development Working Group Chairs, *Intersectoral Issues Report* (1992) is more circumspect, recommending that "[o]ff-reserve management arrangements, which include involvement by resource users, should be strengthened" (p. 34). ANZEC, as it then was, was committed to supporting "existing state, territory and Commonwealth measures to protect remnant native vegetation and investigate additional financial incentives for its protection": *Towards a National Greenhouse Strategy* (1990), p. 27; cf pp. 41, 61. The RAC Forest and Timber Inquiry evades the issue: *op. cit.* n. 24, paras 8.77–8.86; 8.110–8.115. See also *Ecologically Sustainable Development,* A Commonwealth Discussion Paper, June 1990, where it is accepted that there is a need for a response which goes beyond governments buying up or setting aside land, "because of the costs involved and because the environment needs to be treated with care whether or not it is of sufficient quality to justify a special classification" (para 127). This is then followed by a reference to the need to find "other ways" of protecting and enhancing our biological resources outside special areas.

[31]*Op. cit.* n. 28, p. 161.

[32]But see the proposals in the unsuccessful Endangered and Other Threatened Species Conservation Bill 1993 (New South Wales), with its narrow definitions of "endangered" and "vulnerable" species, and its failure to deal with threatened communities.

[33]Section 11.

[34]*Flora Protected under the Flora and Fauna Guarantee Act* 1988 (6 May 1993).

[35]Section 26(1)(c),(d). They can be made in relation to communities on Crown land: s. 26(1)(a),(b).

[36]Sections 4(1), 159(1)(c). This legislation is not discussed further because, apart from co-operative arrangements and the possible provision of financial assistance, it is largely irrelevant to issues of species conservation on private land because its application is restricted to Commonwealth land.

[37]Consequently it does not give rise to any international obligations. Article 14(1)(b) of the Convention only requires that "the environmental consequences of its programmes and policies that are likely to have significant adverse impacts on biological diversity" are *taken into account,* which is hardly precautionary. However, the Intergovernmental Agreement on the Environment specifically espouses the precautionary principle, although the version which appears in that document is predicated more narrowly on there being proof of "threats of serious or irreversible damage".

[38]Lawyers would refer to this as the evidentiary burden of proof. Once it has been met, the ultimate probative burden passes to those arguing that there will be no significant reduction or loss.

[39]This does not appear, however, to outlaw arguments based on economics rather than science. For example, it could be argued that action should not be taken to avoid the threat to biodiversity because the economic costs outweigh the benefits. The problem with this attempt to bypass the precautionary principle is that it is usually impossible to assess the real costs of destruction of biodiversity because of scientific uncertainty about the ways in which it might in the future benefit human beings. See J. V. Krutilla, "Conservation Reconsidered," [1967] *American Economic Review* 777.

[40]Section 11(1), emphasis supplied. See also the Flora and Fauna Guarantee Regulations 1990 Schedule 1, which spells out more detailed sub-criteria: Flora and Fauna Guarantee Information Paper No 2, *Guide to the Flora and Fauna Guarantee Regulations* 1990 (February 1991).

[41]Section 15(2).

[42]Section 19.

[43]Section 21.

[44]Section 47(2)(b). "Take" is specifically defined to include "disturb" (section 3(1)) and therefore prohibits interference with habitat. Forty-five taxa (approximately 680 species), formerly protected under the *Wildflowers and Native Plants Protection Act* 1958 have been declared to be protected plants by order of the Governor in Council dated 25 September 1988: *op. cit.* n. 34. Identification of critical habitat, however, only requires a determination of the Director-General (s. 20). Note also that these provisions only cover protected *plants.* There are provisions in the *Wildlife Act* 1975 prohibiting the unlicensed taking of protected, notable and endangered wildlife (ss. 41(1), 42(1), 43(1), 44(2)). However, here the concept of "take" may not cover disturbance of habitat, as distinct from direct human interference, which the legislation specifically includes (s. 68(4)).

[45]Section 26(1)(d).

[46]Department of Conservation, Forests and Lands (Victoria), *Questions and Answers on the Flora and Fauna Guarantee Act* 1988, p. 12; N. Oates, "Flora and Fauna Guarantee Act" (1988) 30(2) *Trees and Natural Resources* 11, at p. 13; G. Dent and T. Beadle, *Flora and Fauna Guarantee Act* 1988 (Victoria), paper presented to the 11th National Environmental Law Conference, Perth, 20–22 September 1992, p. 11.

[47]Section 99.

[48]Section 98.

[49](1991) 73 LGRA 247.

[50]Section 5(1).

[51]Endangered fauna are listed in Schedule 12 of the Act, the list being broken down into "threatened" and "vulnerable and rare" species. Schedule 12 was revised at the beginning of 1992. Protected fauna are not specifically listed. There is a list of unprotected fauna in Schedule 11, and protected fauna are simply defined as any fauna not included in this list (ss. 5(1), 93). Endangered fauna are classified as a special type of protected fauna.

[52]Except where the taking of protected, but not endangered, fauna is "essential for the carrying out of" a project which requires some other kind of approval from a government body: s. 98(5).

[53]All 62 licence applications to take or kill endangered fauna, received after the enactment of the *Endangered Fauna (Interim Protection) Act* 1991, have been granted: *Report to Parliament Pursuant to the Timber Industry (Interim Protection) Act,* 1992, July 1993, PP 119 of 1993.

[54]For example, wildlife refuges under section 68 of the *National Parks and Wildlife Act* 1974 (New South Wales); agreements under section 74 of the *Territory Parks and Wildlife Conservation Act* (Northern Territory), prior to amendments made by the *Territory Parks and Wildlife Conservation Amendment Act* 1992; sanctuaries under section 44 of the *National Parks and Wildlife Act* 1972 (South Australia).

[55]See, for example, conservation agreements entered into under section 69C of the *National Parks and Wildlife Act* 1974, discussed by D. Farrier, *Environmental Law Handbook: Planning and Land Use in New South Wales* (2nd ed., 1993), pp. 271–72. In practice, very few agreements have been reached under this provision.

[56]R. J. Fowler, "Vegetation Clearance Controls in South Australia — A Change of Course," (1986) 3 *Environmental and Planning Law Journal* 48 at p. 49.

[57]*South Australian Heritage Act* 1978, s. 16b(1)(a).

[58]*Ibid.* s. 16b(c).

[59]I. R. McPhail and T. Dendy, "Biological Conservation in South Australia," (1989) 2(2) *Australian Biologist* 18.

[60]Government of Western Australia submission to Resource Assessment Commission, *Forest and Timber Inquiry,* May 1990, p. 23.

[61]See D. Farrier, "Vegetation Conservation: The Planning System as a Vehicle for the Regulation of Broadacre Agricultural Land Clearing," (1991) 18 *Melbourne University Law Review* 26.

[62]*Planning and Environment Act* 1987, s. 55.

[63]*Ibid.* s. 61(2).

[64]Ministry for Planning and Environment (Victoria), *Planning Guide for Native Vegetation Retention* (1989), p. 3.

[65]See also State Environmental Planning Policy 19, which gives some protection to natural bushland, but only applies to local government areas listed in the Schedule. Currently this includes those within the Sydney Region, along with Gosford. It covers areas zoned or reserved for public open space purposes, and these will usually be under the control and management of local councils, leaving bushland on a great deal of private land unprotected. Development consent must ordinarily be obtained before bushland is "disturbed" in these areas. Disturbance is defined broadly to include changes in the natural ecology resulting in degradation, for example changes in drainage patterns. Development proposals must be advertized, but they do not fall within the definition of designated development and therefore an environmental impact statement is not required by law.

[66]The disadvantage of targeting particular areas is that this may result in sterile disputes about whether the area in question falls within the protected status, for example, a "wetland" protected by SEPP 14. See P. Adam, "Wetlands and Wetland Boundaries; Problems, Expectations, Perceptions and Reality," (1992) 11 *Wetlands (Australia)* 60.

[67]For a general discussion of the use of the planning system in this area, with special reference to problems of existing use and compensation, see D. Farrier, *op. cit.* n. 61. The assumption in Victoria is that grazing is not protected by existing use protection: there is a special exemption for grazing by domestic stock from the requirement to obtain a permit.

[68]See, for example, cl. 7–4.2 of Amendment S16 to all planning schemes in Victoria, 5 June 1991.

[69]*Bauer Holdings Pty Ltd* v *Sydney C C* (1981) 48 LGRA 356.

[70]*Parramatta C C* v *Hale* (1982) 47 LGRA 319 at pp. 344–45.

[71]*Op. cit.* n. 59, p. 23; *op. cit.* n. 56, p. 50. Fowler argues that, at least in the short term, the controls actually accelerated land clearance.

[72]*The Submission by the Department of Planning and Urban Growth to the Panel on Amendment S5 To All Planning Schemes,* Victoria (August 1990), p. 9.

[73]Personal communication from Fiona Fraser, Office of the Environment, Victoria.

[74]There were 19 applications under SEPP 14 in 1991: 14 were approved, two were rejected, one was withdrawn and two were still being considered. In 1992, there were 13 applications: five were approved, three were rejected and five were still being considered. See K. Fook, "Protect Wetlands! The Disastrous Tale of SEPP 14," (1993) 37(4) *National Parks Journal* 11.

[75]See generally on the relevance of the precautionary principle to the way in which administrative discretions are structured, D. Farrier and L. Fisher, "Reconstituting Decision Making Processes and Structures in Light of the Precautionary Principle," *Precautionary Principle Conference,* University of New South Wales, 20–21 September 1993; J. Whitehouse, "Will the Precautionary Principle Affect Environmental Decision Making and Impact Assessment," *Precautionary Principle Conference,* University of New South Wales, 20–21 September 1993.

[76]See also s. 30(1) of the *Australian Heritage Commission Act* 1975, as interpreted by Wilcox J in *Yates Security Services Pty Ltd* v *Keating* (1990) 77 LGRA 165, discussed by D. Farrier and L. Fisher, *ibid.* See also State Environmental Planning Policy No 19, which provides that before granting consent to the disturbance of native bushland in the areas covered, councils must:

- make an assessment of the need to protect and preserve the bushland, taking into account certain objectives spelt out in the policy;
- be satisfied that the disturbance of the bushland is essential for a purpose in the public interest and no reasonable alternative is available;
- be satisfied that the amount of bushland to be disturbed is as little as possible and, where the disturbance is to allow construction work to be carried out, the bushland will be reinstated as far as possible.

[77]*Planning and Environment Act* 1987 (Victoria) ss. 55, 61(2). The New South Wales SEPPs require the concurrence of the Director of Planning.

[78]G. Mather, *Rural Tree Preservation* (TEC, 1991), p. 25.

[79]P. Durkin, *Good intentions and lost opportunities: Review of local planning schemes and procedures to achieve vegetation retention at the local government level,* Conservation Council of Victoria (1990), p. 32. Durkin's findings discussed here are based on open-ended interviews with planning staff at seven shires: Werribee, Healesville, Sherbrooke, Cranbourne, Flinders, Hastings and Lilydale. Note that these observations are not confined to controls directed specifically to broadacre land clearing. He examined all provisions which had a potential bearing on vegetation conservation, including the consideration of vegetation retention as an aspect of the consideration of development applications in general, including those involving urban development. See also the discussion of the difficulties faced by local councils in *The Submission by the Department of Planning and Urban Growth to the Panel on Amendment S5 To All Planning Schemes,* Victoria (August 1990), pp. 41–42.

[80]*Ibid.* pp. 37–38, 42, 57.

[81]*Ibid.* pp. 42, 56.

[82]*Ibid.* p. 41.

[83]See K. Fook. *op. cit.* n. 74.

[84]Neutze has characterized it as "mostly responsive rather than positive and directive": M. Neutze, *Urban Development in Australia* (1981), pp. 237–39.

[85]For example, allowing the landholder to clear part of the area in return for incurring fencing/management obligations in relation to the remainder: Durkin, *op. cit.* n. 79, p. 32.

[86]This followed a recommendation in the Report of a Panel pursuant to Section 153 of the Planning and Environment Act, *Native Vegetation Clearance Controls* (1991). The five year time limit on exemptions was later removed, thus taking away the immediate incentive to carry out whole farm planning exercises: see Amendments S15 and S16.

[87]Section 3.

[88]Sections 21AB, 21B.

[89]*Report of the Catchment Areas Protection Board* for the period 1 January 1980 to 30 June 1984.

[90]Submission 90/075.

[91]N. Abraham, I. Hannam and P. Spiers, *Protected Land and its role in land management,* 7th International Soil Conservation Organization Conference, 1992.

[92]*Guthega Development Pty Ltd* v *Minister Administering the National Parks and Wildlife Act* [1986] 7 NSWLR 353 at p. 366.

[93]Section 112.

[94]Environmental Planning and Assessment Regulation 1980, cl. 56(e1),(f).

[95]*Soil Conservation Act* 1938 s. 4(1).

[96]D. Farrier, *op. cit.* n. 55, p. 281.

[97]Although conditions can be of some help. So, for example, under the New South Wales protected land provisions, 50 per cent of larger trees in river red gum stands have to be retained and a minimum of one such tree every 15 metres along each side of the watercourse. Trees used for nesting by rare or endangered fauna cannot be destroyed: D. Farrier, "Regulation of Rural Land: Coercion or Consensus," (1990) 2(1) *Current Issues in Criminal Justice* 95.

[98]*Soil and Land Conservation Act* 1945–90, s. 4. See also the *Soil Conservation and Land Care Act* 1989 (South Australia) in which degradation is defined to include a decline in the quality of vegetation resulting from over-grazing, over-clearing or any other human activity (s. 3) and soil conservation orders can be made requiring the landholder to refrain from destroying vegetation and to submit a property plan detailing the proposed management of the land (s. 38). Contrast the *Soil Conservation and Land Utilization Act* 1958 (Victoria), under which the Director-General can only give directions relating to the use of land for the purpose of preventing soil erosion (s. 17).

[99]In South Australia, even though the original controls in the *Soil Conservation Act* 1939 required written notice of *an intention to clear land where substantially the whole of the natural vegetation remained,* and compliance with ministerial directions relating to the protection of trees, in practice, the concern was solely with the prevention of soil erosion: R. J. Fowler, *op. cit.* n. 56, p. 49.

[100]Commissioner of Soil Conservation and the Division of Resource Management of the Department of Agriculture (WA), *Annual Report 1989–90,* p. 39.

[101]For a discussion of the role that the South Australian approach to conservation of native vegetation plays in the conservation of biodiversity, see J. Bradsen, "Biodiversity Legislation: Species, Vegetation and Habitat," (1992) 9 *Environmental and Planning Law Journal* 175.

[102]D. Farrier, *op. cit.* n. 61. This regulatory regime eventually foundered on the shoals of existing use: *Dorrestijn* v *South Australia Planning Commission* (1984) 49 ALJR 104.

[103]Native Vegetation Authority, *Annual Report* 1987–88, Appendix 2, and 1988–89, Appendix 2.

[104]Section 16a(3).

[105]Section 27.

[106]Section 28(1).

[107]South Australia, *Report of the Review into the Native Vegetation Management Act* 1985 (1986), pp. 4, 11–12, 20.

[108]This was acknowledged by the Minister for Environment and Planning in his second reading speech on the Native Vegetation Bill: "[*The Native Vegetation Management Act*] recognized the need for payment of a level of financial assistance to landholders for what in effect was a partial loss of property rights and access to land which otherwise may have been available for development purposes": Legislative Council Debates, SA, 13 February 1991, p. 2890.

[109]*Op. cit.* n.107, p. 29; Native Vegetation Authority, *Annual Report* 1988–89, p. 6.

[110]*Ibid.* p. 20.

[111]*Ibid.* p. 32.

[112]*Op. cit.* n. 59, p. 24.

[113]Native Vegetation Authority, *Annual Report* 1987–88, p. 11 and Table 3.

[114]Native Vegetation Authority, *Annual Report* 1988–89, p. 11.

[115]Previously, the formal position under the legislation was that if the area of land affected by restrictions was 12.5 per cent or less of the total area of the holding, the Minister was not bound to make any payment (s. 26(6)). If the area exceeded 12.5 per cent, the payment due was to be calculated in such a way as to take into account only the area in excess of the 12.5 per cent (s. 28(2)). Under the new arrangements, this discount was abandoned in certain circumstances. Where the area of vegetation to be protected was greater than 12.5 per cent of the holding, compensation was to be given for the *whole area* if clearance was refused solely on biological grounds or a landholder suffered extreme economic hardship because of the clearance controls. The Minister would purchase properties which were made non-viable (Native Vegetation Authority, *Annual Report* 1988–89, p. 6). Where the native vegetation was less than 12.5 per cent of the holding, compensation was now to be paid if the vegetation was of outstanding conservation significance (Native Vegetation Authority Fact Sheets 1 and 2/1987; Native Vegetation Authority, *Annual Report* 1988–89, p. 6).

[116]Section 28(3).

[117]The only qualification to this is that the Council does not have to follow this very restrictive approach where "isolated plants" have an economically detrimental impact upon primary production. Even here, a condition must be attached requiring the establishment of native vegetation on other land specified by the Council and the Council must be satisfied that the environmental benefits provided by this will outweigh those now provided by the vegetation to be cleared.

[118]Section 29(3).

[119]Section 23(3).

[120]There is a specific exemption from the clearing controls for "grazing of domestic stock in a manner and at a rate that is consistent with the manner in which, and the rate at which, the land has been grazed by domestic stock of the same species during the previous 10 years": Regulations under the *Native Vegetation Act* 1991, cl. 3(1)(r).

[121]The standard heritage agreement clauses make it clear that the landholder's negative obligations extend to any damaging activities, including grazing and "any other activity that in the opinion of the Minister, is likely to damage, injure or endanger the native vegetation or native fauna on the subject land". They have little to say about positive management, other than requiring the landholder to maintain any fences supplied by the government and reminding landholders of their obligations under existing pest eradication legislation: Department of Environment and Planning, South Australia, *Guide to the Native Vegetation Act* (1991), Appendix III.

[122]"This Bill . . . is about land management as distinct from land development": Minister for Environment and Planning, second reading speech, Legislative Council Debates, South Australia, 13 February 1991, p. 2891.

[123]Section 6(a).

[124]Section 24.

[125]Section 25.

[126]Section 24(4).

[127]Tradeable permits are already being used in relation to fisheries regulation and extraction of water from inland rivers for irrigation purposes. On the latter, see D. Farrier, *op. cit.* n. 55, pp. 334–36.

[128]Cf. *Palos Verdes Estates Pty Ltd* v *Carbon* (1991) 72 LGRA 414, where it was held that the clearing of vegetation to make a track did not constitute "pollution" within section 3(1) of the *Environmental Protection Act* 1986 (Western Australia), even though pollution was specifically defined to include "direct or indirect alteration of the environment . . . to its detriment or degradation".

[129]OECD, *Guidelines for the Application of Economic Instruments in Environmental Policy,* Background Paper No 1, Environment Committee Meeting at Ministerial Level, Paris, January 1991, p. 54.

[130]A. Chisholm, "Abatement of land degradation: regulations versus economic incentives," in A. Chisholm and R. Dumsday, *Land Degradation: Problems and Policies* (1987), pp. 235–36.

[131]*Ibid.* p. 232.

[132]*Ibid.* p. 343.

[133]*Ibid.* p. 235.

[134]Where it is not regulated, the landholder is in essence given a "right to pollute", which must be purchased by anybody who wishes to avoid the effects of pollution. Given the significant transaction costs and freeloader problems associated with any private initiatives to conserve biodiversity, it falls to the state to carry out the negotiations on behalf of the community. This is the rationale underlying legislative provision for heritage or conservation agreements. One question is whether this allocation of rights in favour of landholders is economically efficient. Although the Coase theorem tells us that in the absence of transaction costs, the initial allocation of property rights will not affect the efficiency of outcomes, this is otherwise where these costs are present. Their existence may result in desirable transactions not being negotiated. It then becomes important that rights are allocated to the party to whom they are most valuable: G. Edwards, "Commentary" in A. Chisholm and R. Dumsday, *Land Degradation: Problems and Policies* (1987), p. 253. Apart from this, the initial allocation of rights has very significant implications in terms of the distribution of resources in society.

[135]In decisions by the New Zealand courts, two pieces of legislation have been interpreted so as to require decision-makers to give greater weight to factors concerned with conservation of the natural environment: *Ashburton Acclimatisation Society* v *Federated Farmers of New Zealand Inc* [1988] 1 NZLR 78 and *Environmental Defence Society Inc* v *Mangonui County Council* [1989] 3 NZLR 257. In relation to the water conservation legislation under consideration in the Ashburton Acclimatisation Society case, Cooke P concluded: ". . . as a general working rule or guideline preservation of the natural state, either as fully as possible or to the extent of protection of outstanding characteristics or features, is to be aimed at unless clear and clearly sufficient reason is shown to the contrary. The ultimate criterion must be the public interest. The presumption is in favour of conservation. A strong, really compelling case is needed to displace it" (p. 88).

[136]See House of Representatives Standing Committee on Environment, Recreation and the Arts, *Biodiversity: The Contribution of Community-Based Programs* (1992), Chapters 3 and 4; *Draft National Strategy for the Conservation of Australia's Biological Diversity,* DASETT, 1992, para 4.1.2.

[137]*Op. cit.* n. 129, p. 59. The Resource Assessment Commission has categorized compensation payments as economic instruments indirectly affecting resource use: Forest and Timber Inquiry, *Final Report* Volume 1, March 1992, para 16.57.

[138]*Commonwealth* v *Tasmania* (1983) 158 CLR 1 at pp. 145–46.

[139]J. Bradsen and R. Fowler, "Land Degradation: Legal Issues and Institutional Constraints," in A. Chisholm and R. Dumsday, *Land Degradation: Problems and Policies* (1987), p. 130.

[140]C. B. Macpherson, *Property: Mainstream and Critical Positions* (1978), p. 1.

[141]*Ibid.* pp. 11–12; J. S. Mill, *Principles of Political Economy,* Longmans Green (1904), pp. 140–44.

[142]J. L. Sax, "The Law of a Liveable Planet," Paper presented at the International Conference on Environmental Law, Sydney, 15 June 1989, pp. 3–4.

[143]*Op. cit.* n. 130, pp. 242–43. See also OECD, *Op. cit.* n. 129, p. 55, where it is argued that compensation type subsidies should only be allowed for a limited period in cases where severe difficulties occur.

[144]See the discussion by the RAC Forest and Timber Inquiry in its *Final Report,* Vol 1, paras 8.77–8.105.

[145]OECD, *op. cit.* n. 129, p. 53. See also p. 56, where it is argued that output-related subsidies such as those related to the number of endangered species found on farmers' fields, will provide "a much greater incentive to study the ecological interrelationships and to further improve their management than if they only get compensation payments for the maintenance of certain traditional practices".

[146]*Op. cit.* n. 142, pp. 4–7: "Wc have given parts of the machine into the hands of these owners and yet we have not imbued them with the responsibility for assuring the continued good functioning of those parts. This is an oversight of primordial importance" (p. 5).

[147]House of Representatives Standing Committee on Environment, Recreation and the Arts, *Biodiversity: The Contribution of Community-Based Programs* (1992), para 5.20.

[148]See the discussion of Rapid Biodiversity Assessment, *ibid.* para 4.7.

[149]*Ibid.* para 2.8.

[150]*Ibid.* para 2.16.

[151]*Ibid.* para 2.77.

[152]*Ibid.* paras 5.7–5.19.

[153]*Ibid.* para 2.75.

[154]It is now supported by legislation — the *Natural Resources Management (Financial Assistance) Act* 1992 (Cth), which allows the Commonwealth to enter into partnership agreements with the states and territories, covering such things as the funding of community Landcare projects (s. 5) and direct funding agreements with research and community organizations for projects of national significance (s. 6). The legislation establishes a Natural Resources Management Fund, from which payments are to be made under agreements with research and community organizations (s. 11). It also sets up the National Landcare Advisory Committee to make recommendations to the Minister for the Arts, Sport, the Environment and Territories on natural resource management, including priorities and strategies (s. 13). This covers use, development and conservation of soil, water and vegetation (s. 4(1)). The Committee is made up of government and community representatives, with the latter forming a majority.

[155]Media Release, Minister for Primary Industries and Energy, 21 September 1992.

[156]Synapse Consulting (Aust) Pty Ltd, *Report of the Review of the Rural Adjustment Scheme,* DPIE (1992), paras 2.3.4; 4.3.2; 4.3.4. At the same time, it recommended that applicants for support should have to produce a comprehensive farm plan covering financial, physical and social factors, and that efforts should be made to ensure that these should be based on a "best practice" approach to farming.

[157]*Ibid.* para 2.3.4.

[158]Hansard, House of Representatives, 3 November 1992, pp. 2412–413.

[159]*Op. cit.* n. 155.

[160]Section 3(1)(b).

[161]Minister for Primary Industries and Energy, *Rural Adjustment Act* 1992: *Policy Guidelines Applying to Interest Subsidies and Loans for Enhancing Farm Productivity,* 25 April 1993 (emphasis supplied). The New South Wales guidelines specifically indicate that support can assist with "increasing resource intensity (including capital, labour and/or *land*)": New South Wales Rural Assistance Authority, *Information Bulletin No: 2/93* (emphasis supplied). The ACF's conclusion that the new Scheme "adds ecological criteria" is not supported by the analysis here: see J. Alexandra, *Conservation Movement Input into Federal Government Soil Conservation Initiatives,* ACF, March 1993, p. 14.

[162]*Ibid.*

Some socio-economic and ecological implications of alternative biological diversity conservation strategy options

M. D. YOUNG[1]

At the state level it is impossible to value all the elements of a biological-diversity conservation strategy. It is possible, however, to quantify the trade-offs between conservation and production in both economic and ecological terms. Moreover, for any stated ecological objective or policy target it is possible to find the most cost-effective course of action across space and through time. It is also possible to use a variety of policy instruments to realize biological-diversity conservation objectives. If all ways lead to similar recommendations for reserve acquisition then the problem is academic. However, if alternative specifications lead to different recommendations then careful analysis is necessary. Consideration is given to the trade-offs between off-reserve conservation and reserve-based conservation. Off-reserve conservation can be implemented via the use of conservation easements and may be more effective when budgets are limited. In some cases conservation easements offer an effective transitional arrangement. There is also opportunity also to combine expenditure on biodiversity conservation with that on the prevention of land degradation so that both objectives can be achieved at less cost.

ECONOMICS AND ECOLOGY

MOST economic activity depends upon the maintenance of the ecological processes that underpin them and the integrity of ecological systems economics and ecology are inextricably linked. In short, there is an ecological shadow to most economic activity (Fig. 1). Most markets, however, do not recognize this. Acting in their own self interest and surrounded by market pressures, consumers and producers tend to ignore the consequences of failing to conserve biological diversity. Often the impact of any one transaction on future options is trivial. However, the cumulative impact of many small transactions and decisions can be catastrophic.

Recognition of the failure of markets recognize many biodiversity values has led to calls for resource accounting systems that quantify the value of biological diversity and for governments to change market signals so that resource management policy and land use practice ensures maintenance of biological diversity.

VALUATION

Responding to the challenge, economists have developed methods and to new conceptual frameworks for valuing non-market considerations. Recognizing use values and non-use values, economists are trying to estimate total economic value of conditionally-renewable natural resources (Fig. 2).

The valuation of non-market values, like the protection of a forest that contains an endangered species has become a major growth industry. Around the world, many economists are using contingent valuation techniques to estimate how much people are willing to pay for policy decisions like the prevention of logging in a native forest. Demand for such studies is high as they provide decision-makers with a platform from which they can argue that the value of an undeveloped resource is greater than its value in alternative uses like timber harvesting. Contingent valuation can provide a useful political and analytical tool. However, that the value has meaning only within the context of a decision. Moreover, for issues like biological diversity the contingent valuation methodologies being used are controversial.

On theoretical grounds some economists argue that the contingent-valuation methodologies are flawed. Sagoff (1988) reasons that value is a consumer concept and those who apply it to issues like biological diversity make a "category mistake". Sagoff claims that it is wrong to equate consumer preferences, which are driven by ability to pay, with citizen choices about moral and ethical issues. Consumer choices cannot be used to reveal community preferences. Others argue that contingent-valuation techniques are useful for the analysis of near-market decisions, like those associated with recreation and tourism, but that as one moves further from the market place, prices lose their relativity and the techniques become less and less reliable. The best Australian critique of contingent value is by Blamey and Common (1992). One of the best lay explanations of non-market valuation techniques can be found in Pearce *et al's*. (1989) book "Blueprint for a green economy".

There is also a problem of whether to use a willingness-to-pay or a willingness-to-accept

[1]CSIRO Division of Wildlife and Ecology, P.O. Box 84, Lyneham, Canberra, Australian Capital Territory, Australia 2602.

Pages 360–64 *in* CONSERVING BIODIVERSITY: THREATS AND SOLUTIONS, ed by R. A. Bradstock, T. D. Auld, D. A. Keith, R. T. Kingsford, D. Lunney and D. P. Sivertsen. Surrey Beatty & Sons. 1995.

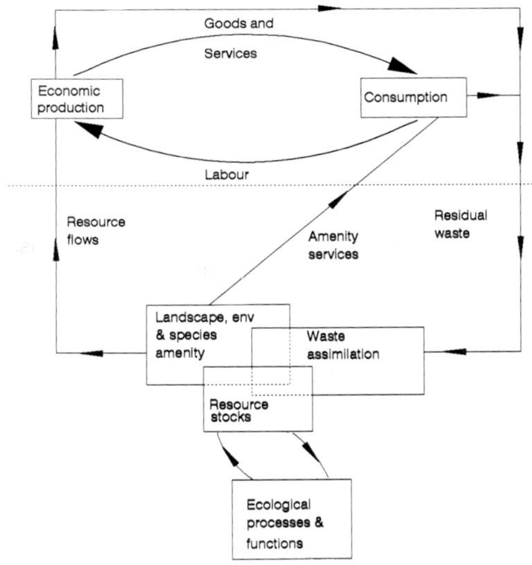

Fig. 1. The ecological shadow of economic activity.

method. Most analysts prefer a willingness-to-pay method because they give "more reliable" estimates. Implicitly, the willingness-to-pay methodology assumes that those interested in biodiversity must be willing to buy the right to protect it. That is, it is assumed that a prior decision has already been made to reduce biological diversity and that people must pay to reverse that decision. Conversely, the willingness-to-accept methodology asks how much each person would accept as compensation for a decision to reduce biodiversity. The assumption is that this money can then be used to buy substitutes of value equivalent to the value of the satisfaction and services that the biodiversity gave them. Note that the people who express these preferences rarely receive a compensation payment. In most cases, it is assumed that the benefits will trickle down. Note also that the contingent valuation methodology also depends upon an assumption that it is reasonable to analyse biodiversity decisions in a partial framework. That is, that any change in biodiversity will have no noticeable impact on other ecosystems or other parts of the economy. Each decision is regarded as a small one of trivial consequence for other ecosystems and for people not directly affected by it. No ecological surprises are expected. Another problem is that the techniques do not involve actual transactions within an available budget. As respondents do not have to choose among options like protecting Australia's southeastern forests, Coronation Hill, the receipt of unemployment benefits, improvements to Medicare and an overseas trip they may be inclined to indicate that they are willing to pay more than they actually would pay if they had to make such a decision.

There are also a considerable number of empirical problems with attempts to value biological diversity. While it is possible to add up all the answers to questions about willingness-to-pay to protect an entire ecosystem, no one has yet worked out how to reliably partition these estimates amongst all the vertebrate, invertebrate, plant and soils that form that ecosystem and then distribute these estimates across space. That is, while it may be relatively easy to value a forest or a nature reserve it is still impossible to use these estimates for the design of protected-area networks which include National Parks, Nature Reserves and a range of "off-reserve" measures like the declaration of wildlife sanctuaries. Thus, while politically useful, contingent valuation studies are limited in their capacity to assist those people responsible for the development and maintenance of Australia's national parks and reserves. They do not allow one to ask questions about whether or not it is better to:

- set targets and conserve all areas equally; or
- conserve only those areas threatened by current land-use practices; or
- focus on those threatened elements that can be protected at little cost.

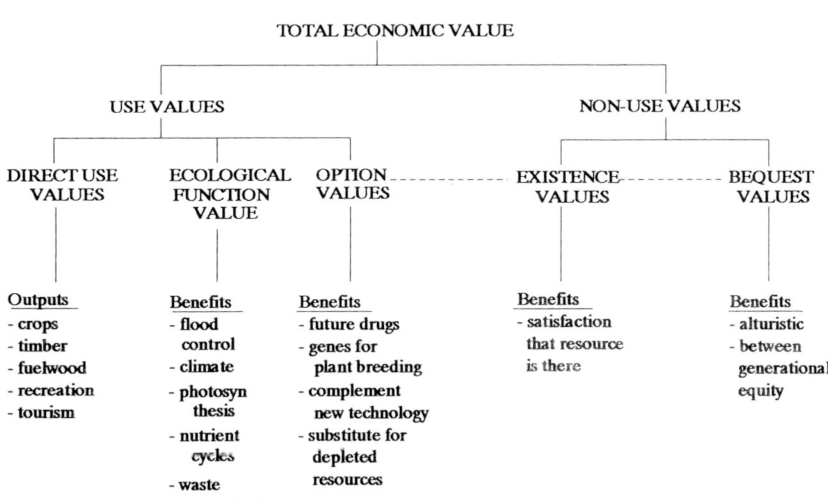

Fig. 2. The nature of total economic value (Young 1992, after Barbier 1991).

DESIGNING PROTECTED-AREA NETWORKS

There are two broad approaches to the design of protected-area networks. The first approach focuses upon policy targets and the second on the recognition of trade-offs between specified objectives within pragmatic constraints. Of these two approaches, most conservation agencies only recognize the first — the *"target approach"*. Targets usually take the form of a recommendation that a fixed proportion of each ecosystem should be placed in a reserve. In most cases, however, the target selected is an arbitrary one. The methodologies used to choose from the available areas assumes that the target will be achieved. Usually rare, endangered and threatened species are dealt with separately and areas sufficient to protect these species are included in the final solution. Important concepts like complementarity, rarity, representativeness, flexibility and efficiency are built into the algorithms (Margules *et al.* 1988; Pressey 1990; Kirkpatrick 1983) but it is always assumed that the target provides the optimum outcome.

The alternative *"trade-off approach"* draws upon standard land-use planning techniques. Linear and goal programming techniques, which are available in packages like LUPIS[1], are used to choose between competing objectives. This latter approach makes trade-offs transparent. The technique, for example, forces one to consider which species are of greater anthropocentric value, which species have the greatest probability of survival and what is the value of land for other purposes. Examples of operational objectives include:

- maximize the number of environmental domains and functional groups represented in reserves at a level of at least 10 per cent of their total area;
- maximize the number of threatened species that have a viable population within a reserve;
- maximize the biodiversity represented within a protected area network;
- place within a reserve system at least three viable populations of all those species likely to decline under existing land-use practices;
- as far as possible, locate the protected area network so that it complements other social and economic objectives; and finally, but by no means least
- maximize the value of economic activity within each region.

Note that within this framework, reserves are a subset of a protected area network which could also include off-reserve conservation. Note that the "trade-off methodology" also builds in a number of safe minimum standards. Examples of safe minimum standards include a commitment to

- place at least 2.5 per cent of every environmental domain and every functional type within a reserve;
- place at least 10 per cent of each environmental domain in the protected area network; and
- ensure that the reserve system is capable of sustaining a viable population of each endangered species.

Under the trade-off approach, preference rules can also be introduced. For example greater preference could be given to species at risk or those with significant economic potential. Similar preference rules can be set for wildlife corridors, buffer areas and so forth. Once this has been done, with the assistance of a geographic information system and decision-support systems like LUPIS and CODA[2], alternative strategies can be costed out, trade-offs assessed and the most cost-effective options identified. Using a grant from the Department of Environment, Sports and Territories, we in the CSIRO Division of Wildlife and Ecology have just begun such a study for the state of New South Wales (CSIRO 1993).

POLICY CONSIDERATIONS

Biological diversity can be conserved both on-reserve and off-reserve. Moreover, in some cases, equivalent amounts of conservation (perhaps measured in terms of reduction in risk to the element of interest) may be achieved at much less cost off-reserve than it can be achieved on reserves. Off-reserve conservation can be achieved by restricting land-use practices and changing the economic incentives presented to land and resource users. Opportunities to use economic incentives to improve biodiversity are both direct and indirect.

Indirect measures

In a perverse way, government policies targeted at other objectives can undermine biodiversity objectives. Failure to recover irrigation infrastructure and supply costs, for example,

[1]LUPIS is a Land Use Planning and Information System available from the CSIRO Division of Wildlife and Ecology.

[2]CODA is a "Conservation Options and Decisions Analysis package" developed by Michael Bedward, Bob Pressey and David Kerr within the NSW National Parks and Wildlife Service.

can increase salinity problems and result in the destruction of unique habitats. Failure to charge for the full cost of waste disposal has similar consequences — especially in marine ecosystems. Thus economists recommend that resource users be made to:

- pay for the full cost of making government services and infrastructure available to then *(the User-Pays Principle);* and
- pay for the costs of pollution control and prevention plus the cost of any damage they impose on others *(the Polluter-Pays Principle).*

Policy makers recognize these policy prescriptions but, because of institutional and government failure, it is rare to find regions where they are fully implemented. Some government policy however, is moving in this direction. Drought subsidies are being removed, water charges are being increased and income taxation provisions are being made more commensurate with biodiversity objectives. As policy moves in this direction, the threat that many forms of land use pose for the conservation of biological diversity decreases. Note also that the likely effectiveness of off-reserve conservation strategies increases as the Polluter-Pays and User-Pays Principles are implemented.

Direct measures

The converse of the User-Pays Principle and the Polluter-Pays Principle is a *Beneficiary-Compensates Principle* (Young 1992). If it is ethically right to make users and polluters pay, it is also right to reward those who provide non-market services of value to others. The Beneficiary-Compensates Principle requires that any costs associated with the provision of positive non-market benefits, and not expected of all resource users, be compensated. Note, however, that such compensation should be limited to the reimbursement of costs incurred in enhancing and maintaining non-market benefits. Where the issue is one of preventing land clearance property rights should be changed in a way that does not enable resource users to hold the environment to ransom (Young 1992). During the transitional period when property rights are changed, however, an equity based case can be made for the payment of compensation. One example of such an approach is the acquisition of conservation easements which, for example, prevent the drainage of a significant wetland or clearance of an important vegetation remnant. These easements attach to a land title and, whenever that title is sold the obligation. A one-off compensation payment is made when the easement is first attached to the title but thereafter no further payments are made.

Conservation easements

When money has been made available for the acquisition of conservation easements, participation rates have been high. These, and other economic instruments, recognize that while markets may be poor masters, they can be excellent servants. Learning from the conservation reserve scheme that operates within the United States, but not making the policy mistakes associated with it, it is not difficult to imagine a scheme whereby farmers could be offered the opportunity to place their land under a permanent conservation easement. Land within these easements could not be cleared, fertilized, cropped nor grazed intensively. Weed and vermin control would still be required. Participation would be voluntary and begin with a call for tenders. Using a geographical information system, land would be selected according to its perceived biological value, the price offered and the nature of land already in the protected area network. Prior information about the type of land preferred for inclusion in the scheme would be advertised widely. Note that careful attention must be paid to the acceptance price. Experience in the United States suggests that, in the initial stage of such a programme, the acceptance price should be quite low. Moreover, it is critical that the conservation easements attach to a land title in perpetuity so that liability for default transfers from owner to owner (Young 1992).

Note also that such schemes may enable a much more precautionary approach to be adopted. Consistent with the Precautionary Principle (Young 1993), until much more is known about some ecosystems, it may be wiser to place 5 per cent of each ecosystem in a reserve and use conservation easements to protect a further 20 per cent until more information becomes available. Policy studies that examine such questions may find that this approach is effective. Conservation easements may be a useful way to construct fer zones around key areas and build habitat corridors between areas of biological significance.

A warning about conservation easements is necessary. It must be realized that the introduction of such systems is not easy. Issues such as whether or not mortgagees have to consent to their creation, what penalties apply for default and whether or not one easement can be swapped for another requires careful consideration.

Management agreements

Consistent with the Beneficiary-Compensates Principle, another option to consider is the use of management agreements. Taking the form of a contract between a resource user and the government, or a non-government organization, users are compensated for the cost to them of providing community benefits that they can not sell. The best known example is that between Ducks Unlimited in the United States of America and the owners of wetlands. Under these agreements Ducks Unlimited pays farmers for the cost of maintaining wetlands in prime duck breeding condition so that their members have more ducks to shoot. Other examples include compensation to farmers for the cost of maintaining traditional landscapes in Austria and environmentally sensitive areas in the United Kingdom (Young 1991). In Australia, such arrangements may be particularly well suited for the maintenance of unique habitats that are too small to place in a national park. Note that with few exceptions, management agreements, which are negotiated with resource users, are most effectively superimposed upon a conservation easement that attaches to a land title or other form of resource right.

Stretching the available budget

Finally, there are opportunities to stretch biological diversity conservation budgets through a search for complementary land-use activities. Most dryland salinity strategies, for example, recommend the retention of native vegetation and extensive tree-planting schemes. With astute co-ordination, biodiversity objectives might be achievable without the use of money allocated for reserve acquisition or the development of a protected area network. If opportunities to do this are significant, and casual observation would suggest they are, it may be possible to conserve much more biological diversity within the available budget.

CONCLUDING COMMENTS

Examining the interface between economics and ecology within the context of conserving biodiversity suggests that effort should shift from valuation of ecosystems to the evaluation of trade-offs. Effort should also involve the development of frameworks that allow the pursuit of "optimum" conservation strategies in a way that recognizes the full spectrum of social and economic demands placed on Australia's resources. In many cases, biological diversity may be conserved more cost-effectively via the simultaneous development of on-reserve and off-reserve strategies. It may also be appropriate to step beyond the safe minimum standard or target approach and begin to specify policy objectives so account is formally given to trade-offs and budget constraints.

REFERENCES

Blamey, R. and Common, M., 1992. Sustainability and the limits to pseudo market valuation. Unpublished paper presented to the second meeting of the International Society for Ecological Economics, Stockholm, August, 1992. (Copies can be obtained from Blamey or Common at CRES, ANU.)

CSIRO, 1993 Definition of methodology for a study on the socio-economic implications of alternative biological diversity conservation options. First report to Department of Environment, Sports and Territories, CSIRO Division of Wildlife and Ecology, Canberra.

Kirkpatrick, J. B., 1983. An iterative method for establishing priorities for the selection of nature reserves: an example from Tasmania. *Biol. Cons.* **25:** 127–34.

Margules, C. R., Nicholls, A. O. and Pressey, R. L., 1988. Selecting networks of reserves to maximize biological diversity. *Biol. Cons.* **43:** 63–76.

Pearce, D., Markandya, A. and Barbier, E. B., 1989. *Blueprint for a green economy.* Earthscan: London.

Pressey, R. L., 1990 Reserve selection in New South Wales: where to from here? *Aust. Zool.* **26:** 70–75.

Sagoff, M., 1988. *The economy of the earth.* Cambridge University Press: Cambridge.

Young, M. D., 1993. For our children's children: Some practical implications of intergenerational equity, the precautionary principle, maintenance of natural capital and the discount rate. CSIRO Division of Wildlife and Ecology AMNRS Program Working Document 93/5.

Young, M. D. (ed), 1991. *Towards sustainable agricultural development.* Belhaven Press: London.

Young, M. D., 1992. *Sustainable investment and resource use: Equity, environmental integrity and economic efficiency.* Parthenon Press: Carnforth and Unesco, Paris.

How well does the EIA process protect biodiversity?

R. BUCKLEY[1]

The EIA process in Australia provides very little protection for biodiversity either in law or in practice. Only in New South Wales, and to a lesser extent Queensland, is there an enforceable legal requirement to prepare EIS's for developments which would have a significant effect on the environment, including plant and animal species. Even in New South Wales, EIA laws do not contain any requirement to protect biodiversity, but merely to consider it. The EIA process does, however, encourage site-specific flora and fauna surveys, and provide a formal public avenue for government agencies, research organizations and community groups to raise concerns over impacts on biodiversity.

SCOPE AND FORMAT

THIS contribution, as requested by the Convenors of this Conference, is rather specific and limited in its focus. It examines only:

- protection of biological diversity, not the environment in general;
- by the EIA process, not the overall legislation and administrative framework;
- in Australia, and particularly New South Wales, not globally.

In particular, it does not consider the protection of biodiversity through endangered species, nature conservation or national parks legislation.

Definitions and salient features of biodiversity are canvassed by others at the Conference. EIA processes, however, are not. This contribution therefore starts with a brief review of environmental planning legislation in Australia, and of relevant litigation in New South Wales. This is drawn largely from Buckley (1991a).

The limitations of EIA processes as regards conservation of biodiversity are then summarized in four main categories: institutional frameworks, information, decision criteria, and onus of proof. The issue of risk and uncertainty is particularly significant and is considered in somewhat more detail, including a case study of the Mt Todd gold mine in the Northern Territory and its potential impacts on the Gouldian Finch.

EIA IN AUSTRALIA

EIA legislation

Each state and territory, and the Commonwealth, has different legislation for planning and EIA (Fowler 1985; Bates 1987; Buckley 1991a). The intentions are similar, but the mechanisms are different. General patterns have been summarized by Martyn *et al.* (1990), Preston (1990), Buckley (1991a), Fitzgerald (1991), Fowler (1991), and Hibbert (1991). Particularly significant differences include the following.

(1) Whether the Minister or agency responsible for EIA has the power to refuse planning or development approval or to impose conditions on consent, or whether another Minister or agency makes the final decision.

(2) Whether planning and regulatory authorities are required only to "have regard to" or "take account of" environmental factors along with other competing factors, with an implicit trade-off between factors, or whether there is a legal requirement for environmental protection.

(3) Whether there is a strict requirement for particular activities or developments to be subject to EIA, or whether the requirement for EIA is ultimately at the discretion of the responsible Minister or agency; and if the former, whether the requirement applies to particular designated types of development, or development in particular designated areas, or activities which satisfy a defined legal test such as being "likely to significantly affect the environment".

(4) Whether or not there is a requirement for EIA to address a predefined list of issues or aspects.

(5) Whether or not there are defined criteria to determine whether defined issues have been addressed adequately, or whether the EIS as a whole is adequate.

(6) Whether third parties have the right or standing to challenge decisions of a planning authority or Minister through the courts if any of these requirements, especially those under (3) to (5) above, are not met.

---

[1]Griffith University, P.M.B. 50, Gold Coast, Queensland, Australia 4217.

Pages 365–74 *in* CONSERVING BIODIVERSITY: THREATS AND SOLUTIONS, ed by R. A. Bradstock, T. D. Auld, D. A. Keith, R. T. Kingsford, D. Lunney and D. P. Sivertsen. Surrey Beatty & Sons. 1995.

When is EIA required?

Under the *Environment Protection Act (Impact of Proposals) Act* 1974 (Cth), the Commonwealth Minister for the Environment can decide whether an EIS is required and what it should contain, but the Commonwealth Act applies only to activities which, for Constitutional reasons, require Commonwealth consent.

In Western Australia, the *Environment Protection Act* 1986 provides for three levels of environmental impact assessment, at the discretion of the Environment Protection Authority or responsible Minister. In the Northern Territory, the requirement for an EIS under the *Environmental Assessment Act* 1982 is at Ministerial discretion. In South Australia, the *Planning Act* 1982 provides for EISs for projects of "major social, economic or environmental importance" (s. 49(1)), but leaves the application of these criteria to Ministerial discretion. Tasmania has no formal EIA legislation, though EIA is carried out under administrative guidelines pursuant to the *Environmental Protection Act* 1973.

In Victoria, the *Environmental Effects Act* 1978 applies to public works, Ministerial decisions and works specified by the Minister responsible for planning, if they "could reasonably be considered to have or be capable of having a significant effect upon the environment". Administrative guidelines issued by the responsible agency specify subsidiary tests, including (a) nature, scale, intensity or complexity; (b) location near sensitive, valued or significant environments; and (c) social impacts or controversy. The guidelines also list examples of types of development commonly requiring an Environmental Effects Statement (EES), but this does not have the force of a list of designated developments, and the requirement for an EES is ultimately left to Ministerial discretion.

In Queensland, EIA may be required at the discretion of the responsible Minister or statutory authority under the *State Development and Public Works Organization Act* 1971 (a range of Acts regulating mining, tourist resorts, canals, electricity generation, and water quality), or the *Local Government (Planning and Environment) Act* 1990. This last Act has a formal requirement that an EIS be prepared for designated developments and for developments in designated areas, defined by regulation. It has significant deficiencies, however, as regards public participation and appeal rights. In particular, there are avenues of appeal to the Minister for Housing, Local Government and Planning; but not the Minister for Environment and Heritage.

In New South Wales, the *Environmental Planning and Assessment Act* 1979 requires that an EIS is required for any development of a type designated in Pt IV of the Act, or for any "activity likely to significantly affect the environment", under Pt V. If an EIS is not prepared for such a development proposal, or is inadequate, the planning decision or development consent is open to challenge in the Land and Environment Court. The implications and effects of these provisions have been reviewed by Preston (1986a,b), Gibbs (1991) and Hibbert (1991). Contrary to the initial fears of some, there has not been a flood of vexatious or frivolous litigation (Pain 1989). Indeed, the provisions seem to have contributed quite effectively to improving the quality of environmental planning and impact assessment in New South Wales, and could provide a model which other states might do well to emulate. Note, however, that Gibbs (1991) concluded that the Act "does not work effectively" in relation to at least one class of development, namely electricity transmission lines. This may well apply equally to development by other statutory authorities and government agencies, where the proponent and the consent authority, under Part V of the Act, are different arms of the same government.

What does EIA include?

Different governments and legislation emphasize different functions for their EIA documents. Some specify the contents of EIA documents only in terms of the outcome they are intended to achieve, others in terms of the issues they must address or the specific questions they must answer. In Victoria, the contents of an EES are suggested in administrative guidelines rather than required by regulation. In Queensland and South Australia, planning agencies provide case-specific guidelines for the contents of EIA documents, in response to a proponent's initial notification or request. The *Local Government (Planning and Environment) Act* 1990 (Queensland) specifies formal requirements for the contents of an EIS, but these are very general: a description of the proposal, a statement of potential impacts, and information to substantiate the latter. In New South Wales, the required contents of an EIS are defined quite precisely by Regs. 34, 56 and 57 under the *Environmental Planning and Assessment Act* 1979.

Litigation in New South Wales

Test cases in the Land and Environment Court have established a set of principles for the application of EIA legislation in New South Wales (see Buckley 1991a for case citations). In regard to biodiversity conservation, the following are the most significant.

(1) In the test phrase, "activity likely to significantly affect the environment", "likely" means "possible", rather than "more probable than not".

(2) In the same test, a "significant" effect means one which is "important", "notable", "weighty", or "more than ordinary". This is "a matter of fact and degree"; and it must be judged in the light of prior impacts on the environment concerned.

(3) "Effects" include impacts on flora and fauna, communities, ecosystems and scientific values, and include risks, long-term and cumulative effects, and increased demands on scarce resources. Community concern alone, however, does not satisfy the test.

(4) For the purposes of this test, the environment is defined "to include all aspects of the surroundings of man". Its spatial extent for any proposed activity is a question of fact, but includes "the geographic location in which it is to be carried out and the area of which it is physically a part".

(5) An EIS which does not satisfy all the requirements of Regs. 34 or 56 and 57 of the Act is inadequate, and an inadequate EIS fails the requirements of the Act as if no EIS had been prepared.

The New South Wales forestry cases

Many of the test cases referred to in the preceding section were actions brought against the New South Wales Forestry Commission, notably the *Prineas, Jarasius* and *Bailey* cases (for citations see Buckley 1991a). Essentially, these cases forced the Commission to prepare EIS's for logging commencing or extending into new areas. One of the ways in which logging could significantly affect the forest environment is through its impacts on biodiversity. Because of the format of EIA legislation, however, these cases did not force the Commission to protect biodiversity, but merely to expose the impacts of logging to public review and comment. The subsequent *Chaelundi* action,[1] which did make a very significant contribution to protecting biodiversity in the forests of northern New South Wales, was brought under the *National Parks and Wildlife Act* 1974 (NSW) (McDonald and Buckley 1993; McDonald 1993; Stewart 1993; McDonald and Munchenberg 1995).

As a result of the *Chaelundi* cases, the New South Wales Government first tried to exempt the Forestry Commission from the *National Parks and Wildlife Act* 1974 (NSW) regulations. When this failed, the New South Wales Parliament[2] passed the *Endangered Fauna (Interim Protection) Act* 1991 (NSW) and the *Endangered Fauna (Interim Protection) Amendment Act* 1992 (NSW), which effectively specified that any activities likely to have a significant impact on the environment of a protected species would automatically trigger the EIA requirements under Part IV or Part V of the *Environmental Planning and Assessment Act* 1979 (NSW). At the same time, however, the New South Wales Government promulgated the *Timber Industry (Interim Protection) Act* (NSW), which effectively legalized the Forestry Commission's breaches of the *Environmental Planning and Assessment Act*, but required it to prepare EIS's or *Fauna Impact Statements* in future (McDonald 1993).

Under s112(1B) of the *Environmental Planning and Assessment Act* (NSW) as amended, a *Fauna Impact Statement* must be prepared in accordance with s92D of the *National Parks and Wildlife Act* before carrying out any activity "likely to have a significant effect on the environment" of endangered fauna, as specified in Schedule 12 of the Act. Guidelines under s4A of the amended *EPA Act* specify that the test must consider: habitat removal and modification; species sensitivity to habitat disturbance; regeneration time for habitat critical to species survival; species recovery capabilities; any mitigation plans; any wilderness assessment under the *NPW Act*; and "any adverse effect on the survival of . . . species . . . or . . . populations." This phrase does not specify whether a change in the probability of species extinction, local extinction or population reduction will be considered an adverse effect. This will therefore be left to the courts to determine. There are precedents from other countries and other areas of the law to suggest that it might, but the outcome is by no means certain. Even if the test is satisfied, however, it only requires preparation of an FIS, and a FIS simply describes the species concerned, their habitat, their regional distribution and conservation status, and the likely impacts of the proposed activity. It does not in itself provide any protection for endangered species or other components of biodiversity. Note also that plant species are not considered, and that FIS's are only required if animal species listed in Schedule 12 are likely to be affected.

As an example, the EIS for the Wingham Forest Management Area (Forestry Commission of New South Wales 1992a) incorporates an FIS (Denny 1992) which identified 25 endangered

[1]Corkill *v.* Forestry Commission of New South Wales (1991) 73 L.G.R.A. 126; Forestry Commission of New South Wales *v.* Corkill (1991) 73 L.G.R.A. 247.

[2]The *Endangered Fauna (Interim Protection) Act* 1991 went through the New South Wales Lower House promoted by three independent MPs and supported by the Labour Party opposition, who together outnumbered the Coalition minority government. It passed the Upper House with the support of an independent MLA. The Act was thus forced on the government by Parliament. The Act amends the *National Parks and Wildlife Act* 1974 (New South Wales). The *Timber Industry (Interim Protection) Act* 1992 was the Coalition Government's counter to the *Endangered Fauna (Interim Protection) Act*, (D. Lunney, pers. comm.).

animal species in the Area; but the only practical measure proposed to protect them is "ensuring that a minimum of 5 per cent of the area covered by each of the 22 forest types in the WMA are preserved in an unlogged condition . . ." (Forestry Commission of New South Wales 1992a, p.11). Similar comments apply for the Dorrigo Management Area (Forestry Commission of New South Wales 1992b). Put bluntly, therefore, the changes to legislation may assist in identifying adverse impacts, but do little to prevent them. Their main effect is to draw public attention to the issues, and to provide an opportunity for public comments, which must be considered in the assessment of the EIS.

LIMITATIONS OF EIA FRAMEWORKS

No substantive provisions

Endangered-species legislation (ESL), such as that in the United States of America, generally contains substantive provisions forbidding actions which would further endanger listed species or critical habitat. EIA law, however, generally contains only procedural provisions for the preparation, exhibition and assessment of EISs, and in some cases substantive provisions for the application, content and adequacy of EISs. EIA law generally does not set substantive criteria for granting or refusing development consent on the basis of impacts predicted, including impacts on biodiversity.

Multiple aims

The conservation of biodiversity is only one of many aims for the EIA process. Other aims (Buckley 1990a-c, 1991b,c, 1992a,b) are to: identify environmental issues; predict environmental impacts; inform the public; enable government assessment; demonstrate proponent competence; make environmental protection commitments; assist in plant and process design; assist in setting lease conditions and discharge standards; and provide a basis for continuing environmental management. In practice, impacts on biodiversity may receive less attention than other potential impacts such as those on air and water quality, the social environment, and so on.

Advisory role

An EIS is only one input into planning processes, and essentially serves only an advisory function to the decision-making authority concerned. In particular, environmental costs outlined in an EIS are commonly weighed against potential economic benefits if development approval were to be granted (Buckley 1990a,c, 1992a).

Time constraints

Most EISs are prepared, exhibited and assessed under tight time constraints, which limit the accuracy and completeness of their information content, and limit the opportunity for comment, debate and further research if required (see Buckley 1991b,c and case study below). This is of particular concern in seasonal or variable environments, and in remote or other relatively unstudied areas.

Limited public participation

Planning law commonly limits the formal involvement of the public to a single relatively brief period for comments, i.e., consultation rather than participation. In some States, public exhibition is at the discretion of the relevant authority (Buckley 1990a, 1991a, 1992b).

TECHNICAL INFORMATION AND VALUES

Technical information

Baseline surveys are part of almost any EIA process, and most draft EISs or equivalent documents contain lists of plant and animal species recorded or expected at the proposed project site and its immediate vicinity. For major projects, especially in greenfields or other remote or rural areas, these lists are commonly prepared from field surveys carried out by specialist consultants, and to that extent may represent new information on biodiversity generated by the EIA process.

Even though the individual consultants may be highly qualified, however, and employ state-of-the-art survey techniques, they are rarely granted the time or funding for an exhaustive survey. All too often, such surveys involve a few days field work in a single season of a single year, and cover only the more conspicuous, easily collected and easily identified plant and animal taxa. Typically they survey only vertebrates, or only some vertebrate families.[3] They have only a slim chance of encountering ephemeral or uncommon plants or hibernating, cryptic, small or rare animals. In addition, such information is often very site-specific and not always set in context, often because the taxonomy, biogeography, ecology and population genetics of the species concerned are simply not known (see Buckley 1991d and case study below for examples).

---

[3]One reason for this, in New South Wales at least, is that whilst EIA legislation does not in itself define fauna, other legislation such as the *National Parks and Wildlife Act* fails to acknowledge invertebrates as fauna.

Because of this, and because of limitations in process descriptions and prediction techniques, there is often substantial uncertainty as to the types, magnitudes, probabilities, and accuracies of impacts on biodiversity as on other components of the environment (Buckley 1991c, 1992b; Buckley and McDonald 1991). The ability of project proponents or government agencies to mitigate environmental impacts, whether predicted or not, and to encourage the recovery of species populations affected either directly or indirectly, is often also uncertain (Buckley 1991d).

Values

Values attached to the conservation of biodiversity (Buckley 1992c), relative to other environmental and economic goals, are rarely specified explicitly in EIA documents. In addition, the relative significance or value of populations, subspecies, assemblages and different taxa is rarely defined.

UNCERTAINTY AND RISK

General issues

There are no predefined criteria for an acceptable level of uncertainty over baseline data, impact predictions, or mitigation and recovery plans, and in practice these are rarely defined, and differ widely between projects and impacts.

Current planning procedures assume that we can predict environmental impacts with sufficient confidence to make an informed judgement as to whether the economic benefits of a proposed activity would outweigh its environmental costs (Buckley 1992a; Brown 1992; Stewart 1993). They assume that even if we cannot predict definite impacts, we can identify potential impacts and calculate their probabilities and consequences. But this is not always possible in practice. Predicting probabilities and consequences is difficult enough even for engineering risks. For ecological risks, predictions are even more imprecise.

For example, how can we calculate the degree to which a particular activity would change the probability that a particular species, known to be rare or vulnerable in some way, might become extinct? We might be able to predict the gross physical impacts of the project proposed, and estimate overall effects on vegetation and animal communities, but we are unlikely to know what effect those impacts would have on individual plants or animals. The chances are that many other factors would contribute to the population dynamics of the species concerned, but that we would not know how, or even what those factors were (Shaffer 1987). Even if it was a conspicuous and well-known species, the chances are that we wouldn't even know its current population size or its habitat requirements. We might not know if it was a distinct species, or a subspecies or form.

How should planning procedures treat such issues, where the probability of a particular environmental impact is unknown, but its ecological consequences are potentially severe? The so-called precautionary principle, namely that if we can not predict the consequences of an activity we should not undertake it, is generally accepted as a principle, but there is considerable argument in practice as to (a) how reliable our predictions are, (b) how reliable they need to be, and (c) acceptable risk levels. As a general issue, this may be stated as follows: "how should planning procedures handle potential impacts where the best estimate is a negligible impact, but the worst scenario is a very serious and irremediable one?". The same question applies to industrial and transport risks, but the level of uncertainty is generally greater for risks to species' survival.

Gouldian Finches at Mt Todd

Issues of risk and uncertainty are well illustrated by the potential impacts of a proposed gold mine at Mt Todd, Northern Territory, on Gouldian Finches, *Erythrura gouldiae*. Details are set out in the case study report below, which was written during public review of the DEIS. I make no comment on whether the Mt Todd mine should or should not have received development approval. The critical issue here is that even for a single conspicuous and publicly well-known species, and a relatively localized and well-defined development, there was inadequate baseline information on the distribution, habitat requirements and population ecology of the species, and on the design and operation of the project, to predict the impacts of the project on the species' population size and survival probability, but the project was approved nevertheless.

ONUS OF PROOF

The overall effect of the above is that under the EIA process, if someone is concerned that a particular development poses a threat to biodiversity, it is up to that person to prove the existence and significance of the threat and publicize it sufficiently for politicians to take into account. Under ESL systems, in contrast, the onus is on development proponents to prove that their proposal will not endanger any listed species or critical habitats.

CONCLUSIONS

Requirements for EIA differ between states, territories and the Commonwealth. Most rely ultimately on the discretion of the responsible Minister or planning agency to determine whether EIA is required for a particular development or activity, what form it should take, what issues should be covered and in what detail, the degree of public information and participation, and the weight accorded to EIA in the final approval or rejection of a planning or development application. Only New South Wales and Queensland have legal requirements for EIA for a defined class of activities. The New South Wales legislation relies on a list of designated types of development or those required under an environmental instrument, such as SEPP14 on Coastal Wetlands (Lunney, pers. comm.) and a threshold test, namely "activities likely to significantly affect the environment". The Queensland legislation lists designated types of development and designated geographical areas. It is only in New South Wales that there is a significant body of past litigation challenging the absence or adequacy of EIS's and defining the threshold test.

The EIS process does, however, encourage site-specific flora and fauna surveys, and provides a formal public opportunity for community groups, research organizations and government agencies to signal concerns and provide information on biodiversity to development proponents and planning authorities.

Overall, therefore, EIA processes are valuable and perhaps necessary in conserving biodiversity, but far from sufficient.

CASE STUDY: GOULDIAN FINCHES AND THE MOUNT TODD GOLD MINE

Written after the Draft Environmental Impact Statement (DEIS) was released for public comment and reflecting publicly available information on the project and the species at that time. After that date an agreement was reached between the proponent (Zapopan NL), the Northern Territory Government, and the Jawoyn Association Aboriginal Corporation (Anon 1993). The agreement refers principally to native title in the light of the Mabo decision[4] *, but also covers environmental issues (Anon 1993). It is not clear, however, whether that agreement addresses the issues raised below, and in any event, the agreement was reached outside the EIA process. For an industry view and discussion see Strapp (1993) and Buckley (1993).*

1. This proposal and its assessment will test whether current planning and impact assessment law provides adequate protection for biodiversity in Australia.
2. If the Gouldian Finch were a listed species under the *US Endangered Species Act*, there is no doubt that the proposed mine would be prohibited, and that damage already caused to Gouldian Finch nest trees during exploration would be actionable.
3. Although industry criticism of the *US ESA* has centred on economic issues, the Mt. Todd EIS makes no attempt to assign economic values to possible changes in Gouldian Finch populations, either at the proposed minesite at Yinberrie Hills, other areas of the Northern Territory, Western Australia or Queensland, or in captivity. The DEIS simply argues that the proposed Mt Todd project would pose no significant threat to the Yinberrie Hills population. Implicitly, it appears to accept that if there were such a threat, the project should not proceed.
4. The proponent's argument that the project would have no significant long-term effect on the Yinberrie population of Gouldian Finches, relies on the following claims:
    - that clearing vegetation for the mine pit would destroy only a small proportion of Gouldian Finch nest hollows, namely 0.2 per cent, and that individual birds affected will be able to nest successfully in other parts of the Yinberrie Hills without reducing the breeding success of the remainder of the Gouldian Finch population;
    - that clearance of vegetation for other components of the mining operation would have no effect on the Gouldian Finch population
    - that noise, blast, shock, lights, etc. would have no effect on the Gouldian Finch population;
    - that cyanide in impounded water will not affect Gouldian Finches;
    - that draw-down of groundwater, consumption of surface water, and modifications to water quality would have no detrimental impacts on vegetation in Finch habitat, e.g., death of nest trees;
    - that any changes in fire regime associated with the project would have no detrimental impact on the finches' food supply, notably *Sorghum* species;
    - that there would be no other adverse effects of the project on Gouldian Finches, e.g., an increase in population of feral cats scavenging in the project area;

[4]*Mabo* v. *State of Queensland* (No. 2) 1992 176 CLR 1.

- that the Yinberrie Hills population of Gouldian Finches is sufficiently large, genetically diverse and healthy, to recover from any reduction caused during project construction or operation;
- that any reduction in the Yinberrie Hills population would have no significant effect on the population of wild Gouldian Finches as a whole, or on its genetic diversity or long-term viability;
- that if the Yinberrie Hills population should start to decrease, effective recovery measures will be possible and will in fact be taken in sufficient time to prevent any further reduction;
- that the population of Gouldian Finches will be monitored in sufficient detail to detect any reduction in time to put such a recovery plan in action.

5. The Guidelines for the DEIS specifically required the proponents to substantiate some of the above claims. In particular, the Guidelines required the proponents to demonstrate that noise, blast and dust would not reduce the Gouldian Finch population at Yinberrie Hills. This has not been done.
6. Currently available evidence for each of the claims or assumptions listed above, as presented in the DEIS and its Appendices, is summarized below. Most of them may or may not be correct; few if any are substantiated.
7. The area shown in the DEIS is different to that shown in previous documentation. The changes increase the total area and decrease the proportion occupied by the proposed mine. The habitat area as shown in the DEIS appears to have been drawn on the basis of vegetation type rather than Gouldian Finch records. For the area shown as Gouldian Finch habitat, the favoured nest tree, *Eucalyptus tintinnans*, is dominant, whereas in the adjacent vegetation type, it is subdominant. The distinction between these two vegetation types is probably somewhat arbitrary. In addition, whilst Gouldian Finches may be more abundant where *Eucalyptus tintinnans* is dominant, they are presumably still also present at lower frequency in adjacent areas. This suggests that they are probably present in other parts of the proposed mining operation, as well as the main pit. Records of Gouldian Finch nests mapped in the DEIS show no reduction in density towards the southeast boundary of the vegetation type with *E. tintinnans* dominant. This suggests strongly that Gouldian Finches also use, and probably nest in, the area proposed for other components of the mining operation. The Guidelines for the DEIS specifically require surveys of the entire mine area, but these have not been provided.
8. The DEIS estimates that there were 3 668 Gouldian Finch nests in 5 350 ha of prime Gouldian Finch habitat at Yinberrie Hills in 1990, with 95 per cent confidence limits of 2 312–5 024 nests. It estimates that there were 1 026 Gouldian Finches breeding in 1990, and that the average number of Gouldian Finches present at Yinberrie Hills in the middle of the dry season from 1986–88, including both adults and juveniles, was 1 368 individuals. It appears from Appendix 11 of the DEIS (Collins and McNee 1992) that a total of 90 person-weeks' searching actually found only three active Gouldian Finch nests at Yinberrie Hills (Collins and McNee 1992, p. 21), and 163 Gouldian Finch nests overall (Collins and McNee 1992, Table 8, p. 15). The estimates quoted above were extrapolated from these figures according to a method described in an unpublished consultant's report to the proponent (Lane 1991). Given the reliance on that report, it should have been included as an Appendix to the DEIS, but it was not.
9. According to the DEIS (Collins and McNee 1992, Table 8, p. 15), Gouldian Finches have been found at 14 other sites in the Northern Territory, with nests at six of these: 140 at Newry, 15 at Eva Valley, 40 at Timber Creek, six at Manbulloo, eight at Kidman, and five at Thymanan. The DEIS does not say how many of these nests were active. Excerpts from Lane's report, however, published in a consultant report to the Ecologically Sustainable Development Working Group on Mining (Buckley 1991d) quote 25 active nests at Yinberrie Hills, 20 at Newry, three at Eva Valley, and two at Timber Creek. According to the DEIS (Collins and McNee 1992, Table 7, p. 15, quoted from Lane 1991) the total numbers of Gouldian Finches recorded from these six sites in 1991 were respectively 2, 1, 12, 1, 5 and 2.
10. Gouldian Finches have also been recorded from the Kimberley in Western Australia. In 1990 up to 37 Finches were recorded at Pumpkin Springs, and a single count of 57 birds at Stewart River Spring 3 (Collins and McNee 1992, pp. 4–5). In 1991 a maximum of 11 Gouldian Finches were recorded on any one day on King River Road, 3 on Valentine Creek, 14 at Pumpkin Spring, and 34 at Granite Creek. Counts at Valentine and Granite Creeks included immatures. In 1992, a flock of 223 Gouldian

Finches, with 81 adults and 84 immatures, was recorded at Granite Creek on one particular day, but on most other days only 10 to 20 birds were recorded. A maximum of 89 Gouldian Finches (29 adults, 60 immatures) were recorded at Doon Doon, 98 (43 + 55) at Wilson River, and 12 (8 + 4) at Ellenbrae. Intensive searches for nests (Collins and McNee 1992, p. 12) found 2 active nests with chicks, and seven used Gouldian Finch nests, in the entire East Kimberley. Of these six were at Doon Doon, one at Granite Creek and 1 at Wilson River. The age of the used nests was not estimated. A total of 759 trees were searched to find these nine nests. As noted by Collins and McNee (1992), "... these data provide the only direct proof of breeding by Gouldian Finches in the Kimberley".

11. A range of different methods, techniques and people were used in searching for Gouldian Finches at different sites and in different years. Some, but not all, of the searches were carried out by trained ornithologists who had previous experience with Gouldian Finches. Interestingly, no attempt seems to have been made to enlist the expertise of commercial Gouldian Finch trappers, though one might imagine that such trappers would already have known the locations of many nests.

12. Though searches for Gouldian Finches in the Kimberley continued into 1992, it is not clear what has been done to improve knowledge of the Yinberrie Hills population since the unpublished work of Lane in 1991. The DEIS quotes 163 Gouldian Finch nests at Yinberrie Hills, whilst in October 1991 only 44 had been identified positively, according to statements by the proponent published in Buckley (1991d, p. 15). Excerpts from Lane's report quoted in Buckley (1991d) indicate that as of June 1991 the number of breeding Gouldian Finches at Yinberrie Hills had been estimated as between 16 and 930 (95% confidence interval), with a best estimate of 167; and the total number of Gouldian Finch nests had been estimated at 991–2 176. It is not clear, however, whether the differences between the figures quoted to the ESDMWG and those in the DEIS are due to further surveys at Yinberrie Hills, or to reinterpretation of Lane's results. Other than the main pit area, areas proposed for mining operations do not seem to have been surveyed for Gouldian Finches. In addition, the uncertainty of population estimates is so high that monitoring during operations would be unable to detect population change due to mining.

13. The DEIS states at several points that if Gouldian Finch populations at Yinberrie Hills were reduced by mining, the proponent would implement a recovery plan in conjunction with the Conservation Commission of the Northern Territory (CCNT). It gives no indication, however, of what this plan might be. Indeed, given that Gouldian Finches are already endangered, then if such a recovery plan is so straightforward, why has it not already been implemented?

14. The DEIS claims that the probability of extinction of the Gouldian Finch as a result of mining at Mt. Todd would be nil. Presumably, this claim relies on the substantial number of birds in captivity, estimated in the DEIS at 50 000. No information is given, however, on the genetic diversity of the captive population relative to that in the wild. It is probable that even though the numbers are larger, genetic diversity is much lower since the entire captive population has probably been bred from a relatively small number of captured birds. Besides making the captive population potentially more susceptible to disease, lack of genetic diversity may reduce the chance that captive birds reintroduced to the wild would survive and breed successfully and Yinberrie Hills.

15. Appendix 16 of the DEIS (Faith and Wood 1992) points out statistical difficulties associated with the proposed monitoring programme. The authors suggest that the monitoring programme should be designed so as to be able to detect a 10 per cent reduction in the Yinberrie Hills population of Gouldian Finches. Given the enormous uncertainty surrounding current population estimates, this would in practice be quite impossible under the construction schedule outlined in the DEIS, no matter how great the sampling effort during operations. In the BACI (Before-After/Control-Impact) terminology of Faith and Wood (1992), if parameters BC (Before-Control) and BI (Before-Impact) are highly uncertain, then comparisons BA (Before-After) will lack statistical power. According to the Northern Territory Department of Mines and Energy (NTDME) and the Conservation Commission of the Northern Territory (CCNT), quoted in Buckley (1991d, p. 19), it may be impossible to test whether EIS predictions are correct or not, because: the frequencies of actual nesting, per nest hole or per tree, are very low; current estimates of Gouldian Finch numbers have very wide confidence limits; and hence only very large changes in

Gouldian Finch populations or nesting would be statistically demonstrable.

16. As stated by Faith and Wood (1992), therefore, it is essential that at least one and preferably two years' detailed monitoring of Gouldian Finch populations both at Yinberrie Hills and at other sites be carried out before any construction activity commences at Mt Todd, if the impacts of mining on Gouldian Finch populations are in fact to be detectable.

17. In addition, development approval if granted should define precisely what action is to be taken if the Gouldian Finch population does decline. It should specify the time scale for such action, financial responsibilities, and penalties for noncompliance. The proportional reduction in Gouldian Finch numbers, or the absolute population size, at which such actions would be triggered must also be specified, together with a measure of probability.

18. The Australian mining industry has repeatedly rejected the so-called precautionary principle espoused by the international Business Council for Sustainable Development, albeit without specifying its reasons (e.g., Farnell 1992). Instead, the industry favours so-called adaptive management, where environmental impacts are addressed as they arise. Two of the essential features of adaptive management are firstly, the ability to detect impacts and secondly, a response strategy. For Gouldian Finches at Yinberrie Hills, the DEIS contains neither. The DEIS asserts that any impacts on Gouldian Finches from the proposed mining operation would be minimal. This may or may not be correct, but the DEIS contains neither the evidence to substantiate the assertion, nor any concrete or realistic proposals for action should it prove incorrect.

19. From the evidence available in the DEIS, it appears that:

    - the area cleared for the mine-pit will indeed be only about 1 per cent of the total area of the vegetation type preferred by Gouldian Finches at Yinberrie Hills;
    - it may, however, represent a significantly larger proportion of the area actually used for nesting by Gouldian Finches at Yinberrie Hills; this is unknown;
    - in addition, it is highly likely that Gouldian Finches also occur on areas proposed for other components of the mining operation, albeit probably at lower density; again, this is unknown;
    - the effects of noise, blasting and dust will extend across at least half the total area of prime Gouldian Finch habitat at Yinberrie Hills. As noted by Collins and McNee (1992, Summary, Item 8) "little is known regarding the influence that noise, dust and blasting associated with mining activity are likely to have on Gouldian Finches in the Yinberrie Hills . . ."; it seems most improbable, however, that these effects will be insignificant.

20. Without substantially more accurate information on baseline populations of Gouldian Finches at the proposed mine site and elsewhere, and substantially better information on the impacts of noise, blast and dust on Gouldian Finches, the impacts of the proposed mine on Gouldian Finches will be neither predictable nor detectable, and there is no guarantee that they would be reversible. In that case, it is clear that existing planning and EIA legislation would fail to provide adequate protection for this endangered species.

REFERENCES

Anon, 1993. Mt Todd project agreement. *Aust. Environ. Rev.* **8**(4): 9.

Bates, G. M., 1987. *Environmental law in Australia.* (2nd ed.) Butterworths: Sydney.

Brown, A. J., 1992. Prayers of sense and reason: mining, environmental risk assessment and the politics of objectivity. *Environ. Plan. Law J.* **9:** 387–410.

Buckley, R. C., 1990a. What's wrong with EIA? *Search* **20:** 146–47.

Buckley, R. C., 1990b. Environmental science and environmental management. *Search* **21:** 14–16.

Buckley, R. C., 1990c. Shortcomings in current institutional frameworks for environmental planning and management. *Canb. Bull. Publ. Admin* **62:** 50–56.

Buckley, R. C., 1991a. Environmental planning legislation: court backup better than ministerial discretion. *Environ. Plan. Law J.* **8:** 250–56.

Buckley, R. C., 1991b. How accurate are environmental impact predictions? *Ambio* **20:** 161–62.

Buckley, R. C., 1991c. Auditing the precision and accuracy of environmental impact predictions in Australia. *Environ. Monitor. Assess.* **18:** 1–23.

Buckley, R. C., 1991d. Sustainable development in the Australian mining and petroleum industries: case studies and general issues. *Griffith Environment Monographs* **1:** 1–81. Report to Ecologically Sustainable Development Working Group on Mining. Griffith University, Gold Coast. ISBN 0 86857 442 2.

Buckley, R. C., 1992a. Triggers, thresholds and tests for EIA: policy issues. *Environ. Pol. Law* **22:** 146–49.

Buckley, R. C., 1992b. Environmental impact predictions in science and planning. *Ambio* **21:** 323–24.

Buckley, R. C., 1992c. Economic issues in conserving biodiversity. In *Biological diversity* AIC/Euromoney: Sydney.

Buckley, R. C., 1993. Biodiversity and EIA: the Mt Todd case study. *Aust. Environ. Law News* **3/93:** 46–47.

Buckley, R. C. and McDonald, J., 1991. Science and law: the nature of evidence. *Search* **22:** 94–95.

Collins, B. G. and McNee, S. A., 1992. Impact of the proposed Mt. Todd gold mine and other environmental factors on the Gouldian Finch in Northern Australia. *Mt. Todd Gold Project Draft Environmental Impact Statement, Appendix 11.* 28 pp. NSR Environmental Consultants P/L, Hawthorn, Victoria, for Zapopan NL.

Denny, M., 1992. Wingham Management Area fauna impact statement for proposed forestry operations. In Forestry Commission of New South Wales 1992a, Appendix A. 85 pp.

Faith, D. P. and Wood, J., 1992. Monitoring of the indirect effects of the Mt Todd gold project on the Gouldian Finch *(Erythrura gouldiae). Mt. Todd Gold Project Draft Environmental Impact Statement, Appendix 16.* 35 Pages. CSIRO, Canberra, for NSR Environmental Consultants P/L and Zapopan NL.

Farnell, G., 1992. Mining industry perspective. In *Biological diversity.* AIC/Euromoney: Sydney.

Fitzgerald, G. E., 1991. *Commission of Inquiry into the conservation, management and use of Fraser Island and the Great Sandy Region: Report, Appendix 18.* Government Printer: Brisbane.

Forestry Commission of New South Wales, 1992a. *Environmental Impact Statement, Wingham Management Area Proposed Forest Management.* Truyard P/L: Crows Nest.

Forestry Commission of New South Wales, 1992b. *Dorrigo Management Area proposed forestry operations environmental impact statement.* Sinclair Knight: Sydney.

Fowler, R., 1985. Legislative bases for environmental impact assessment. *Environ. Plan. Law J.* **2:** 200.

Fowler, R. J., 1991. *Proposal for a Federal Environment Protection Agency.* Report to Australian Conservation Foundation and Greenpeace Australia Ltd. University of Adelaide.

Gibbs, Harry (Sir), 1991. *Inquiry into community needs and high voltage transmission lines development.* Government Printer: Sydney.

Hibbert, S., 1991. What does an EIS require? — a changing area. In *Duties, liabilities and opportunities for directors under environmental law.* IIR Conference: Sydney.

Lane, B. A., 1991. *Gouldian Finch investigations, Mt. Todd Gold Project, Northern Territory.* Consultant Report to Billiton Gold Pty Ltd.

Martyn, A., Morris, M. L. and Downing, F., 1990. *Environmental impact assessment process in Australia: discussion paper.* Environment Institute of Australia: Canberra.

McDonald, J., 1993. Environmental law. Pp. 275–314 in *An annual survey of Australian law 1992* ed by R. Baxt and A. Moore. Adelaide Law Review: Adelaide.

McDonald, J. and Buckley, R. C., 1993. The taking offence and lawful activity defence under the *Nature Conservation Act* 1992 (Queensland): when is habitat disturbance a taking? *Environ. Plan. Law J.* **10:** 198–204.

McDonald, J. and Munchenberg, S., 1995. Public interest environmental litigation: chipping away at procedural obstacles. *Environ. Plan. Law J.* (in press).

Pain, N., 1989. Third party rights: public participation under the *Environmental Planning and Assessment Act* 1979 (New South Wales.): do the floodgates need opening or closing? *Environ. Plan. Law J.* **6:** 26.

Preston, B. J., 1986a. Adequacy of environmental impact statements in New South Wales. Environ. *Plan. Law J.* **3:** 194.

Preston, B. J., 1986b. Third party appeals in environmental matters in New South Wales. *Aust. Law J.* **60:** 215.

Preston, B., 1990. The EIS threshold test: when is an activity likely to affect the environment? *Environ. Plan. Law J.* **7:** 147–62.

Shaffer, M., 1987. Minimum viable populations: coping with uncertainty. In *Viable populations for conservation* ed by M. Soulé. Cambridge University Press.

Stewart, A., 1993. Environmental risk assessment: the divergent methodologies of economists, lawyers and scientists. *Environ. Plan. Law J.* **10:** 10–17.

Strapp, T., 1993. Letter to the editor. *Aust. Environ. Law News.* **3/93:** 45–46.

ADDED IN PROOF

NOTE: Material on legislation and litigation current to conference date, not publication date.

The role of non-government organizations in the development of international and national policies for the conservation of biological diversity

R. C. NIAS[1]

Non-government conservation organizations (NGOs) have developed in parallel to government and arise from a public demand that cannot be adequately met from other institutions. NGOs are critical in all stages of biodiversity conservation — they have instigated most of the important biodiversity policies and are active in most of the implementation and practice.

NGOs work effectively through national and international networks, and the best are responsive to community concerns, practical in their outlook, and are politically sophisticated.

NGOs provide a variety of services to the community — at extremely cost-effective rates. The NGO movement is not static and the survival of any group depends on developing practical solutions and marketing these effectively to the consumer.

Conservation biologists, practitioners, and policy makers are an essential part of the NGO process.

INTRODUCTION

NON-GOVERNMENT conservation organizations have existed since the early part of this century. The Royal Australasian Ornithologists Union, which claims to be the oldest conservation group in Australia, was founded in 1901. In other words, NGOs have been developing for as long as Australia's system of Federal Government. It was primarily in the 1970s, however, in line with increasing concern for the environment, that the NGO movement expanded to the size and importance that it now occupies.

There are now literally hundreds if not thousands of individual NGOs active in Australia. From very specific and localized groups of residents to the larger global organizations of WWF and Greenpeace.

NGOs AND BIODIVERSITY

NGOs have always played a major role in the conservation of biological diversity. As community-based organizations they have been able to reflect and affect community attitudes on biodiversity and to translate these into policies and programmes. During the 1980s in particular, conservation groups promoted a range of biodiversity concerns that have since entered into the mainstream of national and international politics. In order to illustrate the effectiveness of the NGO movement, two related examples will be considered of how biodiversity concerns have grown from rather obscure beginnings to the forefront of the environmental agenda. The first is the development of an international convention on biological diversity.

THE INTERNATIONAL BIODIVERSITY CONVENTION

At the 1984 IUCN General Assembly in Madrid, a number of Australian environmentalists got together with scientists concerned about the loss of agriculturally important species to propose the need for an international convention to conserve biological diversity. The Convention was eventually signed in 1992 at the Rio Earth Summit by a record number of countries. The convention has major implications for the conservation of biological resources and draws attention to many areas not traditionally considered central to the environment debate, such as international trade relations and property rights. Through the ability to form networks nationally and internationally, considerable public interest, concern, and technical input was bought to bear on the development of this convention.

In Australia, an International Biodiversity Coalition was formed in 1992 with the aim of promoting the final development of the convention and its ratification. Based on the successful Antarctic and Southern Oceans Coalition model, a network of interested parties was kept informed of key events and opinions relating to the convention. These international networks continue to operate effectively in promoting global issues, and are very much part of international NGO operations. Similar networks on issues from biotechnology, tropical rainforests, and even opposition to golf-courses are flourishing, thanks in particular to the widespread use of electronic mail services.

[1]World Wide Fund for Nature (WWF) Australia, G.P.O. Box 528, Sydney, New South Wales, 2001.

Pages 375–79 *in* CONSERVING BIODIVERSITY: THREATS AND SOLUTIONS, ed by R. A. Bradstock, T. D. Auld, D. A. Keith, R. T. Kingsford, D. Lunney and D. P. Sivertsen. Surrey Beatty & Sons. 1995.

By 1993, the Biodiversity Coalition network encompassed some 50 active core members, including a number of government agencies, and a further 120 interested groups in 32 countries on all continents. The coalition played a critically important role in promoting an informed debate.

THE NATIONAL STRATEGY FOR THE CONSERVATION OF BIOLOGICAL DIVERSITY

The second example is the development in Australia of a National Strategy for the Conservation of Biological Diversity. The development of this yet to be finalized strategy, was initiated by NGOs in 1988 and became a major feature of the Prime Minister's 1989 environment statement "Our Country, Our Future." Since then the strategy has been the subject of intense development and debate.

One of the eight guiding principles for the strategy is that "co-operative involvement of conservation groups, resource users, indigenous peoples and the community in general is critical to the conservation of biological diversity."

Major public input to the development of the strategy came during and after the 1992 Fenner Conference on the Environment. NGOs played a major part in shaping the agenda for this document and have been influential in many of its key recommendations (Department of the Environment, Sport and Territories 1993a,b). The strategy identifies many specific actions that could be carried out by NGOs.

Obviously the first step for NGOs will be to ensure the development of a final strategy document and to obtain the necessary political and public will to have its recommendations implemented. The strategy identifies a number of important areas of opportunity for participation by NGOs.

1. Identification of biodiversity components

There are several NGOs with the ability and scientific credibility to carry out the identification of particular elements of biodiversity requiring conservation. For example, two major taxonomic overviews of threatened species, the Marsupial and Monotreme, and the Bird Action Plans, were both compiled and edited primarily by NGOs (Garnett 1992; IUCN 1992). In many cases NGOs can bridge the gap between different agencies in such projects, even between government departments that do not communicate with each other.

Other regional or taxa specific studies may also be carried out by NGOs. In September 1992 WWF brought together over 60 experts on tropical biota for a workshop that led to a comprehensive overview of biodiversity conservation, together with an action plan, for the Queensland Wet Tropics (Werren 1993). Similarly there are many studies carried out by national or state NGOs on matters involving the National Estate.

2. Developing strategies for the conservation of biological diversity

Many NGOs have been involved in the development of specific strategies for the conservation of terrestrial and marine ecosystems. NGOs would expect to expand their input in the development of regional plans for biodiversity. Furthermore the implementation of such plans will clearly require a high level of community input.

The promotion and implementation of conservation strategies at a local community level is one of the most important areas of NGO activity. For example, the Australian Conservation Foundation and National Farmers' Federation initiated the development of the Landcare approach to the problems of land degradation. Community groups such as Greening Australia are playing a major role in implementing this approach. The Australian Littoral Society is currently developing a national community-based network to promote marine and coastal conservation. The National Threatened Species Network (a co-operative venture between the Federal Government's Endangered Species Program, WWF, and state and territory conservation groups) is actively involved in promoting the conservation of threatened species across Australia.

These are successful models and ones which should grow in the future. The combination of a central co-operating body working through state-based community networks is well suited to our federal system of government.

The House of Representatives Standing Committee Report on Biodiversity — the contribution of community-based programmes, provides many examples of how government funded programmes can supplement and give valuable direction and support to community-based efforts (HORSCERA 1992). Similarly, the involvement of NGO groups in government programmes such as Save the Bush provide good evidence of how the effectiveness of these programmes can be magnified.

3. Wildlife conservation

NGOs have been instrumental in the development of threatened species programmes at both Commonwealth and state levels. In the courts, particularly in New South Wales, NGOs have been successful in forcing government departments to carry out basic environmental studies and impact assessments in relation to threatened species. NGOs are also involved in planning species conservation work and the Endangered Species Advisory Committee established under the recently proclaimed Federal *Endangered Species Protection Act* 1992 includes three conservation NGOs and three industry representatives.

The legal and illegal trade in wildlife products imported and exported from Australia is also monitored and analysed by an NGO — TRAFFIC (Trade Records Analysis of Flora and Fauna in Commerce). This organization provides important perspective on the implementation of international treaties such as CITES, as well as acting as an independent source of information on the extent and nature of the legal and illegal trade in wildlife.

4. Protected areas

NGOs have been major driving forces behind the establishment of Protected Areas ever since the development of the concept. Some of the most important protected areas in Australia such as the Great Barrier Reef Marine Park, Kakadu National Park, east coast temperate rainforests, Queensland Wet Tropics, and many others have been the subject of long-term NGO campaigns.

NGOs continue to pursue the development of a biologically representative, well managed system of protected areas as one of their major concerns. The development of appropriate land management practices, rehabilitation of degraded ecosystems, and a whole landscape oriented approach toward the conservation of biological diversity are new areas of concern suited to greater NGO participation.

There are few successful models of how these concerns can be co-ordinated at a community level. A key to the success of these more holistic approaches to the environment will be the development of appropriate regional community structures. Along lines such as Landcare Groups and Total Catchment Management Committees, these new groups will need to consider the whole range of environmental issues within their region in an integrated fashion. Since the majority of community groups at present tend to be either single issue or campaign based, or active within a single town or suburb, such change will not happen overnight.

5. Major biodiversity users

A substantial proportion of Australia's national income is derived from the exploitation of biological diversity. This is particularly true for industries such as pastoralism, forestry, fisheries, and tourism. The existence of many others such as mining and manufacturing is fundamentally dependent on the capacity of the environment to assimilate waste and pollution.

NGOs are clearly concerned with how such economic pursuits can be accomplished without degrading the biological resources upon which they depend. This is an extremely difficult task, and one which governments frequently appear to be reluctant to provide clear direction or solutions.

This is an area that NGOs will need to continue to develop expertise and influence. NGOs can provide the pressure to conserve resources that is often lacking in the market place. By presenting challenges to inefficient and wasteful industries, NGOs assist the development of demand, and new markets for, more sustainably-produced commodities.

It is also worth responding to the frequent suggestions about the need for local action and community education, particularly of farmers, in relation to the use of biological diversity. Although action by local groups in the rural community is vital to the success of conservation efforts, the enormous influence wielded by industry associations, with backing from aggressive and powerful pro-development government departments should not be underestimated.

In our dealings we have found that many individual farmers, corporations and other major users of biodiversity are already convinced of the need to conserve this resource. Indeed, we cannot keep up with the demand for practical suggestions for ways they can improve their operations.

However, it is often the city-based industry associations and their lobbyists that set policy, and in turn, pursue this vigorously with government ministers and officials. A great number of professional lobbyists are former public servants or ministerial advisors, giving them excellent understanding of government process and access to decision makers.

In order to address the enormous impact that such access brings, NGOs need to be similarly equipped with political nous and contacts at the highest levels, and even on the odd occasion, with a demonstration of old-fashioned civil disobedience. This often leads to criticism, in the media and elsewhere, of NGOs being too political. This view neglects the fact that virtually all major interests in the community, from blue-collar workers to the doctors to the major mining companies spend millions of dollars each year on professional lobbyists and advertising campaigns, working the corridors of power, and by a variety of other means exercising the economic and political muscle required to obtain the output they want. Although this is all an accepted part of industrial relations and corporate behaviour, it is frequently seen as unseemly or undesirable if the cause being promoted is ecological rather than economic.

There are too many examples of how, even with all the best research, education, community support, and goodwill from environment and conservation agencies, the environment can be threatened by decisions taken by the major political players in isolation from community concerns. Without a politically sophisticated and effective NGO movement, local concerns and the public interest will inevitably suffer.

NGOs PROVIDE A SERVICE TO THE COMMUNITY

NGOs provide a large part of the total input that goes into policy development, education, research and practical implementation of biodiversity conservation. In a report to the Commonwealth Government in 1992, for example, it was estimated that the major national and State NGOs contributed nearly 3 million hours of community conservation work at an estimated value of about $55 million (Henry and Olson 1992). This figure is about 40 per cent of the total Commonwealth Environment Program expenditure in 1991–92. All this would be expected to cost a considerable amount, and much has been made, for example, about the size of grants to conservation organizations by the Federal Government. However, this review also pointed out that of the estimated $2 billion of Federal funds directed to community programmes across all areas of government activity, such as the arts, sport, education, primary industry, and so on, less than one-tenth of 1 per cent went to environment groups.

Governments have come to rely heavily on input from NGOs to deal with the increasing expectation by the Australian community to be informed, and have a proper influence, in the management of their environment. NGOs are required by their memberships to be involved in practically every environmental decision that is made. Most NGOs have representatives on a variety of committees and contribute to numerous inquiries, environmental impact assessments, and other documents requiring public input. In many cases NGOs send representatives, at their own expense, on official delegations to international meetings and treaty negotiations.

THE FUTURE FOR NGOs

The future for the environmental NGO sector is bright, so long as the community is concerned to have an independent voice acting in their interest. However, this does not mean that the situation is static. As issues and economies wax and wane, various groups will come and go, fulfilling Andy Warhol's prediction that every one will be famous for at least 15 minutes.

In order to survive these changes, NGOs will need to match the increasing complexity of the environment debate. New issues such as biodiversity, climate change, ecologically sustainable development, the integration of economics and the environment, are providing major challenges to NGOs to become increasingly sophisticated in their approach and to make greater use of knowledge that resides within government, academia and the private sector. Furthermore, the messages used by NGOs will have to compete with the information and sensory overload that already overwhelms us. The difference between attention grabbing hysteria and carefully articulated opinion is one that is too subtle for most public relations agencies.

Most of all, however, NGOs will increasingly need to provide practical and economically valid solutions to global problems. As we near the end of the century, there will almost certainly be a trend to reflecting on the damage to the environment from 2000 years of unbridled human population growth and resource development. Similarly there will be a great desire for new visions and solutions for the coming century.

NGOs operate in the market place. The community, government and industry is demanding solutions to pressing environmental problems. To be successful in this market place, NGOs will need to develop practical, achievable solutions and market these effectively. In this context NGOs will rely increasingly on the work of people such as yourself. In large part it will be people like you that must develop these solutions.

REFERENCES

Department of the Environment, Sport and Territories, 1993a. Biological diversity. Its future conservation in Australia. Proceedings of the Fenner Conference 1992. Commonwealth of Australia 1993.

Department of the Environment, Sport and Territories, 1993b. A national strategy for the conservation of Australia's biological diversity. Synopsis of Public Submissions. Commonwealth of Australia 1993.

Garnett S., 1992. *The action plan for Australian birds.* Australian National Parks and Wildlife Service: Canberra.

Henry, D. and Olson, M., 1992. Supporting community conservation action in Australia. Report on the Ministerial Review of Commonwealth Grants to Voluntary Conservation Organizations.

House of Representatives Standing Committee on Environment, Recreation and the Arts, 1992. *Biodiversity. The contribution of community-based programmes.* Australian Government Publishing Service: Canberra.

IUCN, 1992. *Australasian marsupials and monotremes. An action plan for their conservation.* IUCN: Gland, Switzerland.

Werren, G. L., 1993. Conservation strategies for rare and threatened vertebrates of Australia's Wet Tropics region. *Mem. Queensland. Museum* **34**(1): 229–41.

The role of activists

R. BRECKWOLDT[1]

The title of "activist" is often applied in a perjorative sense and used as a basis for stereotyping. However, people involved in the voluntary conservation movement are crucial advocates for environmental conservation because they are unencumbered by some of the constraints placed upon those employed by natural resource management agencies. It is this freedom to speak out, question and criticize that delineates activists from other groups. An informed and jealous public is essential in protecting biodiversity.

While such advocates may feel free to speak out on issues they are, of course, limited in numerous other ways. The most common limitation is that of financial support. Conservation campaigns are now complicated, tightly contested and highly politicized. Experience in the south-east forest campaign shows that there is far from a level playing field in environmental debates and some suggestions about how to develop a more equitable framework for environmental debates will be made.

The media has brought environmental conflicts in distant regions into the loungerooms of urban Australia. There are obvious gains and some losses. One of the losses is that the confrontation of protest or blockade has become a benchmark by which to measure whether there is an environmental campaign for a particular area or resource. Protests occur when all other avenues have failed and the urban television viewer or newspaper reader is unaware of the countless lobbying trips to Canberra or the relevant state parliaments, the endless escorting of politicians and bureaucrats on inspections, the review of myriads of EISs and government reports and the constant preparation of media releases and interviews that are the main aspect of the contemporary conservation campaign.

Those who become a spokesperson for a particular conservation campaign will undoubtedly become widely known in their local area. This will result in some vivid messages on their phone answering machine. It is also possible to use the publicity chosen or thrust upon them, by standing for election to local government. Success at election will result in having a legitimate forum to debate important environmental planning and resources management issues.

INTRODUCTION

"GOVERNMENTS are like turtles: they operate best when their neck is out". Anon.

People who are active in the conservation movement come from all walks of life and represent a great diversity of age groups, occupations, political leanings or affiliations. Working together with such a great variety of people on a focussed campaign such as the south-east forests is one of the great rewards of being involved in the process of social and environmental change.

Added to the diversity of people involved in environmental activism is the fact that every conservation campaign has different physical characteristics and political nuances. For example, the south-east forest campaign has been terribly handicapped by the vast travelling distances involved in a woodchip concession area that covers 279 000 hectares. On the political front it has been equally burdened by falling within the marginal Federal Seat of Eden Monaro which carries the adage that if this seat changes so does the government. Consequently, Federal Government paralysis is endemic in the south-east forests.

Whilst the details of every campaign will vary, the general role of activists remains much the same. That role is another necessary task in the full spectrum of conserving biological diversity. It is also quite possible, and hopefully it will become more widely accepted, to be an activist in a particular campaign while fully involved in one's occupation, social commitments and community.

Why can't one be scientist, educator, resource manager and activist? Why should a scientist not speak out and protest when the values which he or she knows and understands so well are being destroyed? The desire to conserve biological diversity surely stems from knowledge and experience of what it is that ought to be protected. Those of us who so often use the need to consider natural systems as interlinked and interdependent can themselves demonstrate a blind spot and be equally afflicted with the tendency to stereotype people and strive, inadvertently or deliberately, to keep them confined in boxes.

Thus people involved in the demand for social change provide a target for perjorative stereotypes and dismissive comments such as only "representing a noisy minority". Stereotyping is a time-honoured method of avoiding the issues involved. As far as the minority group tag goes — it is simply axiomatic. There is no need to vigorously pursue change if you are in the majority.

---

[1]"Hilltop", Candelo, New South Wales, Australia, 2550.

Pages 380–86 *in* CONSERVING BIODIVERSITY: THREATS AND SOLUTIONS, ed by R. A. Bradstock, T. D. Auld, D. A. Keith, R. T. Kingsford, D. Lunney and D. P. Sivertsen. Surrey Beatty & Sons. 1995.

However, the gatekeepers of the *status quo* are also very active. But one does not hear spokespersons for the timber industry, for example, being referred to as radical activists. We frequently overlook the fact that our subtle but powerful selection of a particular form of language, used to describe our perception of what is right and wrong, tends to support the already dominant groups in society.

Whilst it is often observed that today's radical cause becomes tomorrows conservative dogma it is of no comfort to those who see threats to biodiversity accelerating. So, ordinary people in the course of their lives ask themselves what they can contribute and start to look at the full range of options that lay before them. What follows is an account of what many people involved in the voluntary conservation movement do to protect and conserve biological diversity.

JEALOUS GUARDIANS OF THEIR LOCAL AREA

Many people who become actively involved in conservation campaigns do so from a close association with an area under threat. It is easy to dismiss such people by saying that altruism is overridden by a parochial or even selfish concern for their immediate environs. I reject such a view on two grounds: firstly, if local people do not care about their area then who will?, and why should others do their bidding?; secondly, the importance of the issue will emerge over time and what may at first appear to be of local significance can turn out to have regional, state, national and international ramifications.

Living close to an area under threat means being in a position to provide surveillance and work in concert with the large urban-based voluntary conservation organizations. The greatest progress towards conservation of biological diversity in the south-east forests of New South Wales has occurred since the formation of a number of active groups whose membership is drawn from around the areas under threat.

SYSTEM MAINTENANCE

The brief confrontation of protest seen on television, or the quick news grab on the midday news, can promote the impression that that is all that activists do with their time. The reality for a major campaign is very different. For every second on the media there are countless hours, days, weeks and even years of background work. Most of the work of activists consists of building and maintaining an effective organization. In other words, meetings, meetings and more meetings. Making meetings interesting, educational and involving for a broad range of members is one of the great challenges of a long-running campaign.

An enormous amount of time and energy goes into maintaining an effective local campaign group. This ground work is a continuing process and must take place before and during any actual campaigning. Lack of funds is a perennial problem and fund raising is a time consuming task that always needs addressing. Inevitably, voluntary activists find themselves in opposition to large business corporations, well funded industry lobby groups and the government bureaucracies directed at encouraging the exploitation of the resource at issue.

PROPAGANDA UNDER THE MICROSCOPE

Because of their proximity and commitment, activists are in a position to expose the myths, rhetoric and propaganda spun by resource management agencies and/or industries that want access to the area being contested. There are many examples of this that I could relate just from the south-east forest campaign. One of the most graphic is the case of the koala in Tantawangalo State Forest. In summary, after 20 years of woodchipping in the south-east forests the Forestry Commission, now State Forests of New South Wales, kept on denying persistent and reliable reports of an important koala population.

Finally, it was the members of the Tantawangalo Catchment Protection Association who went out on weekends, in their own time and at their own expense, who put the issue of the existence of koalas at rest once and for all. In November 1990, during an organized search on a grid pattern, in an area where Anne Breckwoldt had recently found koala scats while assisting on a botanical survey, Kim Taysom of the Association found a female with a pouch young and accompanied by another juvenile.

That is now history which is undeniable but has yet to be properly and decently acknowledged by State Forests, if indeed it ever will be. Nor does the NSW National Parks and Wildlife Service have a great deal to be proud of in this episode where a species of major importance was ignored by the organization charged with the responsibility of protecting native fauna. Moreover, the fact that a species as large as the koala was ignored for so long merely points out the problems faced by the invertebrates and all the other small life forms that make up the forest ecosystem.

A postscript to this is that both the National Parks and Wildlife Service and State Forests have begun to address the koala issue in the south-east forests with a regional survey and some follow up studies. Additionally, the *South East Timber Protection Act* 1994, passed to let State Forests avoid preparing an environmental impact statement on further logging in areas of high conservation value, places all known koala areas in a moratorium. However, these small gains were a direct result of the work of the voluntary conservation movement.

EDUCATION AND PUBLIC AWARENESS

A major consequence of questioning the way a natural resource such as a forest is managed is the discovery of new information and acquisition of knowledge. It is disturbing that the dissemination of such information invariably falls upon voluntary conservation groups.

In the south-east forests there prevails a situation where State Forests staff speak publicly without any apparent restraint on any and every topic of even the faintest relevance to their perspective on forest management. However, other resource agencies who are central to the debate are silenced. Indeed, an officer of the National Parks and Wildlife Service speaking out candidly without fear or favour as an advocate for the conservation of biological diversity is unheard of.

Even the most timid and circumspect points of view expressed by the Service in the south-east forest debate are not without the approval of the Director and/or Minister. To do otherwise is to risk sanctions that jeopardize career options. Other agencies such as the Soil Conservation Service and the Department of Water Resources continue to make abstract statements on the value of Total Catchment Management while the very catchments of greatest significance are ripped apart from limb to limb.

This creates a steeply sloping playing field on which the voluntary conservation movement has to compete. First, we have to continually wear the criticism, unfounded as it is, that our spokespersons could not possibly know as much about forest management as a "scientifically trained forester"? Secondly, there is the cost factor. The local campaigners, strapped for time and cash, are left to counter the combined propaganda campaigns of the State Forests and the timber industry while the other resource agencies are sitting on their cheque books and turning the other cheek.

LOBBYING, CONDUCTING INSPECTION TOURS, REPORT WRITING

An incredible amount of time is spent taking politicians, academics and members of government committees on inspections and conducted tours. In the south-east forest dispute there has been an ongoing demand for such inspections since 1987. This puts a high burden on the financial and personal resources of activists.

The phone rings and it is yet another VIP who wants to hear and or see the conservation point of view. Naturally, such opportunities are welcomed and we are the first to cry foul if some such identity comes to the area and only listens to the timber industry. However, when providing one of these guided tours it frequently dawns upon you that out of all the people present that day, those in the voluntary conservation movement are the only ones wearing out their own cars and not being paid a wage.

A similar degree of effort is required to write written responses to environmental impact statements and government inquiries that are held during major environmental disputes. An additional demand is to go on lobbying visits to the Federal and State Parliament. Many days, indeed weeks, are spent tramping the corridors of government seeking out and using every opportunity to persuade and convince. It is through these diverse, time consuming and expensive activities that activists put forward a view that can be very different from government departments and the industry concerned. This role is crucial in the entire overall debate in conserving biodiversity.

LEGAL ACTION

With the increase in the amount and complexity of legislation put in place to assist in the conservation of biodiversity comes the extra burden of policing it and taking appropriate action in the Land and Environment Court when there are breaches. Again the preparation of a solid legal case is costly, time consuming and can be emotionally draining on those concerned.

The Judgement handed down in Jarasius vs Forestry Commission of New South Wales and others in 1988 was a landmark decision that forced the State Forests to prepare environmental impact statements for the Coolangubra State Forest and other areas in the south-east forests. The sustained effort put in by the members of the Towamba Valley Catchment Protection Association over a two year period exacted a high price from the key people involved.

In 1993, the South-East Forest Conservation Council comprising local forest campaigners took action in the Land and Environment Court to appeal against the decision of the Director of the National Parks and Wildlife Service to issue a licence to State Forests to log a particular compartment in the Cathcart Special Prescription Area of Tantawangalo State Forest. The result of that action was that State Forests withdrew from the area under challenge before the matter was heard. However, in awarding costs against both the Service and State forests, Justice Stein stated that the Service had attempted to prevent the licence being issued but that "Ministerial will" had "prevailed" over that of the Minister for Environment, the Hon Chris Hartcher.

Here is a case where a particular compartment, known by both the National Parks and Wildlife Service and State Forests to contain endangered species, would have been logged but for the local voluntary conservation movement. The costs awarded are invariably, and at best, direct party to party costs and never reflect the time and effort of unpaid volunteers in mounting a successful action.

It is important that legislators of environmental law remember that it is unlikely to be a wealthy multinational corporation that will take legal action against a breach of an Act of Parliament designed to conserve biodiversity. Those who are most likely to apply the legislation will be ordinary members of the public with limited financial capacity. And when these individuals risk so much to take the matter to court they face resource agencies or industry bodies and lobby groups that have such a line up of barristers and Queens Counsel on the payroll that you would swear they grew on trees.

PROTEST

People who feel strongly about an issue may make a conscious decision to make a public statement in the form of protest. It is invariably a last resort. It is the sort of thing you might do if all your life you have been taught to "Protect Your Forests" but find out that when you want a small say in how "your" forests are managed then every door is slammed tightly shut. The campaign for the Tantawangalo and Coolangubra began in earnest in 1981. The first protest action did not take place until 1989. No protest took place until every single opportunity for negotiation, deferral or compromise had been exhausted.

People who judge protest action from the safety of dinner party conservation and get no closer to it than the television set may not appreciate the hardship involved. The notion that people do it for fun or excitement is sheer nonsense. Protests can be intimidating and frightening and involve extreme physical hardship. It certainly involves a high risk of being arrested, sometimes wrongfully. It can also involve being charged with a grossly exaggerated offence that results in a heavy fine, community service or a gaol sentence. These are not issues to be taken lightly. People who have never had so much as a traffic fine can find themselves in a difficult position when gaining a security clearance for a senior government position or a visa to travel to certain overseas destinations.

Who should stand before the court facing a charge of malicious damage? — the protester chained to a peg in a small hole dug in a half finished road that is being bulldozed through an old growth forest, or State Forests for using its vast physical, financial and political resources to push the Wog Way through the heart of Coolangubra one of the great and beautiful wilderness areas of the south-east!

Some protests are designed to gain publicity either by creative attention-grabbing actions or the display of sheer numbers at rallies. Other protests and blockades are designed to actually stop work on the resource under threat. Many of the blockades seen on television are carried out by people who are deeply committed in their belief that what they are attempting to stop is wrong. That they fail to stop the activity in question by blockade or protest alone is understandable once you have been involved.

The full weight of the law applied to protect the timber industry in the south-east is great in its scope and intensity and its capacity to endure is surprising. There was an overwhelming police presence sent by the New South Wales Government in "Operation Redgum" to protect the interests of Harris-Daishowa and Boral in the south-east. The justification for that course of action by the government of the day and the involvement of timber workers themselves in the "Greens costs jobs" campaign is part of the social fabric that makes the conservation of biological diversity as political as it is scientific if not more so.

DEVELOPING ALTERNATIVE INDUSTRY STRATEGIES

The voluntary conservation movement is subjected to considerable moral coercion to provide alternative management and restructuring packages. Activists are portrayed by their opponents as a negative element of society. It is hard to ignore this pressure and the South East Forest Alliance adapted to it by developing alternative industry strategies in the form of a plantation programme and a full restructuring and employment plan for the south-east timber industry.

Although I became part of this process I also retain some scepticism about it. This is because if the government of the day really wanted a timber industry restructuring package, then the Premier and the Cabinet would simply demand one from the Minister responsible for forestry. However, strategies for conserving biodiversity do not tend to drift from the Cabinet room to the populace at large. One of the lessons learnt from lobbying many politicians is the high number of Ministers who vigorously defend an industry constituency and end up articulating that narrow perspective in their policies and, presumably, their debate in Cabinet. As a consequence of this narrow focus, complex strategies for change are often developed by the voluntary conservation movement.

SELF PRESERVATION

Working towards the conservation of biodiversity also entails some self preservation. Only a small minority of people involved in the voluntary conservation movement are paid a wage. These are the staff of the major conservation organizations and their wages and conditions are poor enough. These people are essential for their organizational and administrative skills and it is sad that there are not more of them. There is a phenomenal imbalance in the amount of money devoted to ensuring the exploitation of a resource compared to that available to the voluntary conservation movement.

The majority of people who contribute to conservation do it in their own time and at their own expense. It angers me greatly to hear people say that because a protest is held during a working day it therefore follows that all the protesters must have nothing better to do with their time. A protest aimed at stopping the destruction of a forest has to be held during working days because that is when it is happening. It is not a matter of choice.

Because of my involvement it is highly likely that I will know most of the people at a protest in the south-east forests. They are people who have sacrificed their wages and or recreation leave or left their farm or business to be there. Many will be targeted from that day on and their future employment or business opportunities will be affected. If a proportion are unfortunate enough to be unemployed then it probably reflects the national average brought about by government policies over which they have no control. If they are out of work, well good on them for putting their time and effort into something as socially constructive as conserving biodiversity.

Additional to the financial sacrifice, people who have lived relatively peaceful lives can suddenly find themselves unpopular among a certain section of the community. People worry about whether their children will be victimized at school. In various ways, a long-running campaign can be hard on the people involved and their families are dragged in whether they like it or not. People in business may be very supportive but feel unable to speak out or join a campaign because people with an opposing view will take their custom elsewhere. For a variety of such reasons, there is no denying that to be a long-time campaigner on a conflict over resource management that is forever boiling over on your doorstep can be hard going.

I am aware of choice and free will and that people volunteer their help for conservation campaigns. However, no discussion on the role of activists is complete without a rounded picture of what it is actually like in the bush, on the ground. It is also just a simple way of saying that even activists have to do all the things that others do to maintain their personal lives and income. There always remains this nagging conflict between the aims and objectives for the campaign and keeping the other important aspects of life in sufficient balance and harmony. There are so many talented and dedicated campaigners and there is never any shortage of ideas or even knowing exactly what needs to be done to achieve a certain goal. It is just finding the time and money to do it that is the problem.

GENERAL REMARKS WITH A WISH LIST

Society is characterized by a great inertia when it come to social change. In many ways this is an essential human survival mechanism. To be exposed to rapid change would put society at the mercy of the moral entrepreneurs or charismatic leaders who prey on collective fear and vulnerability. Notwithstanding this need or desire for continuity and stability, there are some impediments to social and environmental change that require attention if we are serious about conserving biological diversity and perhaps even the life support systems of the planet itself.

(a) *More constructive ways of resolving conflict.* While conflict can be useful and lead to social improvement there does need to be a system of arbitrating environmental conflicts. The resource agency concerned with exploiting the resource under threat should not be the only advocate in deciding which areas are conserved.

Independent land use or environmental tribunals need to be established to resolve major disputes. Agencies such as the State Forests should be allowed to manage public assets only after they are allocated by such tribunals. Periodical review is also essential and this means that access to resources such as native forests for resource exploitation is never guaranteed in perpetuity.

(b) *Separation of scientific research from the agency that has been established to exploit the resource.* It is inappropriate that a resource agency whose charter and industry allegiances are directed at exploiting resources of land or water is also given responsibility for determining the level or consequences of its activities on conserving biodiversity. For example, public confidence in the future of biodiversity in the south-east forests will not be high as long as State Forests is in charge of biological research and consequent conservation strategies. Similarly, it is wrong for State Forests to be conducting hydrological research into whether the woodchip operation has any impact on water quality and quantity in the Tantawangalo catchment. Caesar judging

Caesar and putting the fox in charge of the chook house are not flippant comments on this serious issue. Clearly, efforts must be made to reduce the values that impinge on scientific research in the conservation of biodiversity. One way might be through an independent organization that has absolutely no vested or pecuniary interest in the outcome.

(c) *Freedom of speech for knowledge.* It is acknowledged that someone who is employed as a bureaucrat or political adviser by a government led by a political party of a particular persuasion might be duty bound to maintain a public air of neutrality or be an advocate for that government. However, it is far more difficult to defend a scientist employed by a natural resource management agency or supposedly independent research institution who refrains from speaking freely or disseminating scientific knowledge. Security of employment for scientists or research grants must not be dependant on silence or circumscribed and qualified opinion.

Science as it relates to the environment, and scientists as they relate to society, can continually examine the ethical basis that is imposed on them by the practice of science. There is a certain inconsistency in having the comfort and security of being employed in a government department and claiming the prestige associated with being a scientist but refusing to speak freely on the substance of that science.

The fine line is, of course, when is it knowledge that should be in the public domain and when is it value judgement or policy and how do the two interact? These are the issues that have to be negotiated. With the trend towards contract employment and enterprise bargaining the time is ripe for scientists to demand of their employers the fundamental right to speak and publish freely and openly on the matters of knowledge that they address in their work. Where are the activists for the freedom of science?

(d) *Freedom of information.* The *Freedom of Information Act* as it stands in New South Wales has not been much help in the south-east forest debate. Invariably, the information required was arbitrarily classified as commercially or politically sensitive. Either that or the cost of obtaining copies of the documentation are prohibitive. An example of the former is the application under the Act that was made in mid 1993 to the Director of the National Parks and Wildlife Service requesting a copy of Saxon M. J., Shepherd T. G., 1993 *An assessment of the significance of current and proposed integrated harvesting in Tantawangalo State Forest.* The Director refused the application on the grounds that it was an internal document and its release was not in the public interest.

(e) *Public participation.* The opportunities for genuine and constructive public participation in conflicts over natural resource management issues are limited. Public participation in forest management in New South Wales too often consists of making submissions on environmental impact statements that State Forests or the Department of Planning can choose to ignore if they wish. Public participation at this level is tokenism and only constitutes the very first rung in the ladder towards genuine involvement and power sharing.

Given the high level of government control over resource management in Australia it is unlikely that power sharing with community groups will ever catch on in a big way. There are, however, various mechanisms short of this that could be used to head off major confrontations before they inevitably result in protest action. This is not the place to elaborate on the various constructive alternatives for public participation in resource management for that is a paper in itself. It is, however, appropriate to emphasize that there is an urgent need for these options to be developed and instituted in a way that blends resource management with the demands of an increasingly well educated and informed public.

(f) *Social justice and equity.* Being involved in a very active campaign to bring about social and environmental change exposed a great number of examples where social justice and equity were lacking. Sometimes these incidents caused people great distress that should have been taken up and followed through such avenues as the Ombudsman or the Council for Civil Liberties. Inevitably, it didn't happen because of limited resources and the need to focus on the task at hand which was the forest.

The New South Wales Police and the Australian Capital Territory Police did not apply the law to the drivers of timber trucks that blockaded roads or Parliament House the same way as they did to protesters as they travelled to and from protests. Then there were frequent arrangements between police and State Forests regarding mass arrests. On one occasion, for example, it was possible to make a very strong statement because many hundreds of people willing to be arrested all arrived on the same day. To overcome any such impact or perhaps the court system not being able to cope with such pressure, State Forests simply declared the previously prohibited area open for the day. The next day, after all the protesters had returned to their jobs, the forest was closed again.

If a farmer carted cattle to the local saleyard under the same conditions as people were transported in prison vans to Eden Police Station, sometimes over 70 kilometres away and much of it on rough unsealed roads, then he or she would be fined by the RSPCA many thousands of dollars. These are just two issues out of many I can remember where the law and its application are left wanting. The wish — that the law is applied equally and not manipulated to favour the government of the day or the resource agency being criticized, and that all prison vans such as those used in "Operation Redgum" are either scrapped or converted into vehicles appropriate to the carriage of living things.

CONCLUSION

These are some of the impressions gained during the south-east forest campaign when I was directly involved as an elected office bearer in both the Tantawangalo Catchment Protection Association and the South East Forest Conservation Council as well as a frequent delegate to the South East Forest Alliance, the umbrella group based in Sydney. This period also coincided with a term as an elected Councillor on the Australian Conservation Foundation.

Consequently, I like to think that the above discussion is fairly representative of the role of activists in various conservation campaigns and I truly hope that I have done them full justice. Any misconceptions and oversights are no ones fault but my own and I hope that others have the opportunity to put them right. What I really wanted to convey in all the above is the importance of the voluntary conservation movement in bringing about the social changes on which conserving biological diversity must inevitably be based.

People who are spokespersons for a particular campaign will become widely known in their local area. One way to consolidate the gains made in the campaign is to stand for election to local government. Success at election will provide a recognized and legitimate forum to debate the ideas about development and conservation that propelled you into public life in the first instance. This was the path I chose in the September 1991 New South Wales Local Government elections and subsequently took my place as one of the 12 Councillors on Bega Valley Shire Council. Again, it is really interesting and you can achieve some really worthwhile things, but that is another story.

Conserving biodiversity — a viewpoint on the role of the landholder

M. TOOTH[1]

For most landowners (farmers) who are serious about the sustainability of their asset, conservation in all its connotations is an ever present reality. Security of title, past, present and future government policies, markets, weather, the condition of soil and water, costs of inputs all have an impact on the sustainability of agriculture and have intended or unintended consequences on conserving biodiversity.

THE ROLE OF LANDOWNERS

A LANDOWNER has many responsibilities. They include the need to make a profit, to increase the intrinsic value of the asset, to nurture the soil, vegetation and water, to produce a product that "the market" wants, the need to embrace new technology and research, to pay for and provide clean, pure, nutritious and healthy food for "the market", to control noxious weeds and feral animals, and how to conserve biodiversity. These needs have become increasingly difficult to fulfil as costs go up and prices for commodities go down. The value of net farm production (Australian Bureau of Agricultural Resource Economics 1993), which is an indirect measure of farm income, is now at extremely low levels (Fig. 1).

A further dilemma we face when making policies for the future is how we will cope with the pressure of an increasing world's population. In 1990, the world's population was 5.2 billion. By 2050 it is estimated the world's population will be about 10 billion (Sadik 1990). While the consequences of population pressure poses problems for all sections of society, consider the extent of the challenge faced by agriculture. Worldwide we are now using 90 per cent of all the potentially arable land and the United Nations Food and Agriculture Organization has forecast that all the world's stock of potentially arable land will be fully utilized by the end of the decade. Furthermore, the stock of potentially and presently available arable land is declining as a result of population growth, urbanization, rising incomes leading to higher food consumption, land degradation, and increasing concern to protect parts of the natural environment in its original state.

Partly because of population changes, but mainly because of income growth, the demand for agricultural products is expected to grow strongly. Farmers worldwide will be under pressure and Australia will not be insulated or isolated from this pressure.

CONSERVING BIODIVERSITY

If we are serious about the sustainability of land use and further embracing the conservation ethic, then the most efficient and long-lasting method of achieving this (a desirable outcome for further Australians) is through security of land titles and adding to these titles the encouragement to the ownership and capacity to market both flora and fauna, domestically and internationally.

Just as unfettered land tenure will be important to meet the consumption needs of future generations, there are appearing some new thoughts about how to combat the diminishing species of flora and fauna. Sixty-one per cent of Australia's total land area of 768 million hectares is grazed by sheep and cattle and 3 per cent is used for broad area cropping. Only 6 per cent of the grazing lands have been improved by the introduction of exotic plant species and the use of fertilizers (Hutchinson 1992). This suggests that Australia did not lose its wildlife through farming, grazing, mining or land clearance. Australia lost its wildlife through the introduction of foxes, cats and other feral animals.

In support of this argument are the experiences of John Wamsley, a noted conservationist from South Australia. He has initiated an alternative experiment for protecting our biodiversity. He started the Warrawong Sanctuary in 1969 with a cat and fox proof fence and has had phenomenal success. All introduced rare species have thrived and multiplied.

Other sanctuaries like Warrawong have been established and many are planned that have extensive multiple land uses. Some include farms, some mines and some towns. The important factor that emerges from this success is that there are no land uses which contradicts the concept of a sanctuary provided you remove the known threat to fauna — the cats, foxes and, I would add, wild dogs.

---

[1]"Sevenoaks", P.O. Box 449, Armidale, New South Wales, Australia 2350.

Pages 387–90 *in* CONSERVING BIODIVERSITY: THREATS AND SOLUTIONS, ed by R. A. Bradstock, T. D. Auld, D. A. Keith, R. T. Kingsford, D. Lunney and D. P. Sivertsen. Surrey Beatty & Sons. 1995.

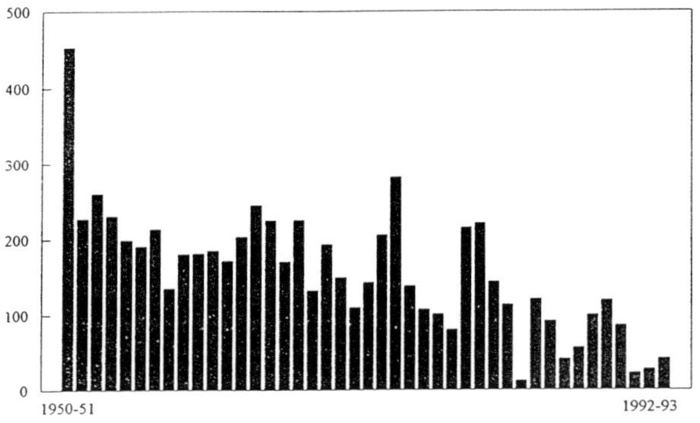

Fig. 1. Real net value farm production. Source: ABARE 1993 (Index 1980 = 100).

Politicians are now paying heed to the problem. Bob Carr, the current Leader of the Opposition in New South Wales and former Minister for the Environment in past State Governments, focused recent media attention on the problem when he pinpointed the fact that feral animals, including rabbits, foxes, cats, dogs, goats and pigs were responsible for the destruction of our native flora and fauna. Among the measures Mr Carr highlighted as necessary to solve the problem were bounties for destruction of feral animals, better pest management in protected areas and investigation of the role in which markets and property rights (e.g., farming of native species for overseas sale) may have in protecting rare and endangered species.

We have an environment corrupted by the introduction of exotic species with no natural predators. It would be nice to be able to turn the clock back on their introduction, but unfortunately we cannot. Our strategies must confront the reality of the situation. I would argue that no successful capacity to conserve biodiversity has been demonstrated within the portion of Australia's land mass that is managed by governments (about one-third).

The approaches we have adopted for conserving our diminishing stock of endangered species have been predicated upon a regulated, imposed view of the environmental world from governments. The amount of legislation in this area is skyrocketing. Extra controls and edicts are forever being promulgated. New layers of bureaucrats are being added and further restrictions on the rights of landholders are forever being enacted. The creation of new reserves and National Parks will not solve the problem, given that current resources are inadequate to eradicate pests within existing Parks, let alone new additions. To achieve a genuine conservation success we must target the most effective source of maintaining biodiversity. Most land in Australia is in non-government hands. Therefore biodiversity conservation will only be possible when an effective system of property rights exists which will allow the landholders of Australia to profit from biodiversity, rather than being penalized for encouraging it.

Barns *et al.* (1993), have stated: "To achieve more satisfactory outcomes, governments need to change conservation policy from a reliance on public management, towards an approach that employs market mechanisms based on secure property-rights". Privatizing control over an asset, whether it be land, wildlife or housing generates considerable incentives for its preservation.

If we wish to reduce the impact of our own activities on some natural rate of extinction then property rights, markets and voluntary action offer useful approaches. By substituting compulsion for willingness, government intervention will often serve to frustrate the effectiveness of markets.

Allowing landholders the right to export native fauna will give them a value and an incentive to maintain the existence of endangered species. They would be protected from the criminal smugglers who are not interested in sustainable yields. The earnings from these exports would help to cover the costs of ensuring their survival.

The answer to conserving biodiversity is not more restrictions on the titles of those whom own the land, it is not more legislation. The answer is the security of title of the land already owned and privatizing or making flora and fauna a marketable commodity.

EXPENDITURE ON CONSERVATION

The majority of farmers who by financial necessity rely on leaving their land in better condition than when they arrived will embrace new technology and profitable options, as evidenced in the rapid uptake of Landcare groups in the quest of economic sustainable production.

The rapid expansion of Landcare groups and the embracing of the concept by farmers is one of the most significant recent changes in rural Australia. It shows that despite the hardships of falling incomes owners are prepared to work together to overcome the most pressing of the perceived problems of their particular areas.

There are now 478 Landcare groups in New South Wales and nearly all of New South Wales is covered by Total Catchment Management committees. A recent survey by the Australian Bureau of Agricultural Resource Economics (ABARE) (Ockerby and Roper 1994) has examined farm expenditure on land conservation (Table 1). The same survey showed the importance of farm profitability to such expenditure (Table 2).

Table 1. Results of an ABARE survey of expenditure by farmers and conservation (Ockerby and Roper 1994).

(a) Conservation-land care expenditure

	All farmers	Wheat	Wool	Beef	Cereal grains	Sheep meat
Farmers making significant expenditure in recent years on conservation-land care	52%	51%	55%	52%	51%	52%
Average expenditure among those farmers	$19 034	$18 613	$17 250	$22 455	$14 740	$18 532

(b) Details of the expenditure by farmers on conservation-land care

	To the 52% of farmers with significant expenditure %
Tree planting. Revegetation	42
Fencing. Fencing off salt, unstable areas	28
Contour banks. Building, maintaining	21
Water conservation. Dams, catchments, recycling	15
Soil conservation. Erosion, water runoff	15
Weed, regrowth control. Eradication	12
Water control. Stabilization. Flood mitigation	11
Pasture improvement. Reseeding, planting	8
Fertilizer treatment. Super	7
Land/laser levelling. Grading	6
Salinity control. Saline drainage	5

Other expenditures include: various specialist machinery, bores and water recirculation, vermin control, earthworks, land clearing and reclamation, soil improvement, access roads.

Table 2. Results of ABARE survey showing impact of decline in real net farm income on expenditure by farmers on conservation (Ockerby and Roper 1994).

(a) 1991 plans for expenditure on conservation-land care

Effects on income prospects on plans for conservation expenditure	All farmers %	Wheat %	Wool %	Beef %	Cereal grains %	Sheep meat %
No effect	38	37	35	42	41	34
Uncertain	9	8	8	10	7	11
Expenditure plans affected	53	55	57	48	52	55

(b) Details of the effect of changes to income prospects on 1991 conservation expenditure plans

What will not be done that was planned?	To the 53% of farmers who have amended plans %
Limit tree planting. None, less, reduction	35
No expenditure at all. All areas	32
Fencing curtailed. Reduced, stopped	24
Contour banks not completed	15
Fertilizer treatments reduced. Super, lime	9
Soil conservation cut. Reduced	9
Water conservation postponed. Dams deferred	8
No pasture improvement. Reduced seeding	6
Weed control reduced. Limited	6
Water control discontinued. Drainage, levies	5

Other areas where planned expenditure will be affected are: land levelling, salinity control and wisalt banks, vermin control, irrigation, land and earthworks, essential maintenance only.

It is also apparent from the ABARE survey that members of Landcare groups were highly informed and well understood the nature of landcare tax concessions. Existing tax concessions include 100 per cent deductions for work done to control land movement, weed control and planting trees, and expenditure for water storage.

With farm incomes likely to be low or fluctuating in the foreseeable future, future tax incentives could include rebates, if greater than tax payable in the year concerned as a lump sum similar to a tax refund. An alternative approach would be to create a tax credit in a taxless period drawable in later years when the Landcare rebate is smaller than the amount payable as tax.

The National Farmers' Federation recommends the eligibility for tax deductions or optimal rebate/credits to apply to:

- expenditure on new equipment with a use primarily and principally for the purpose of combating land degradation,
- expenditure to alter equipment to satisfy conservation purposes,
- expenditure incurred in applying minerals to soil to assist development of soil structure,
- expenditure relating to in-kind or cash support of landcare activities when the group is duly formed under National Soil Conservation Program guidelines,
- expenditure on use of water storage and delivery systems for land conservation or water management, and
- fencing relating to water distribution.

However, greatest in incentive for increased expenditure on Landcare projects or programmes would be an increase in gross farm incomes.

In conclusion farmers will take up the conservation of biodiversity, both flora and fauna, if given the right encouragement.

REFERENCES

Australian Bureau of Agricultural and Resource Economics (ABARE), 1993. Commodities Statistical Bulletin. Canberra.

Barns, G., Brown, K., Chisolm, A. G., Moran, A. J. and Richardson, C., 1993. Concluding comments. Pp. 233–31 in *The price of preservation* ed by A. G. Chisolm and A. J. Moran. Tasman Institute: Melbourne.

Hutchinson, K. J., 1992. The grazing resource. Pp. 54–60 in *Looking back — planning ahead*. Proceedings of the 6th Australian Society of Agronomy Conference, Armidale.

Ockerby, J. and Roper, H., 1994. Survey of landcare and land management practices 1992–93. Australian Bureau of Agricultural and Resource Economics Research Report No. 94.6.

Sadik, N., 1990. *The state of world population 1990*. United Nations Population Fund: New York.

Biodiversity conservation in New South Wales: the role of the National Parks and Wildlife Service

D. J. PAPPS[1] and P. D. WILSON[1]

New South Wales has undergone dramatic modification of its natural environments, particularly since European colonization more than 200 years ago. Biodiversity loss has been considerable and is continuing. The conservation of biodiversity is now a central policy objective of Australian governments following their endorsement of a National Strategy for Ecologically Sustainable Development in 1992. Protection of biodiversity and maintenance of "essential ecological processes and life-support systems" is one of the Strategy's core objectives. The National Parks and Wildlife Service is the state's principal biodiversity conservation agency, a responsibility which has been compromised over the past two decades by changes in its operating environment. As part of a broader Natural Heritage Conservation Strategy, the Service has drafted its first comprehensive biodiversity conservation strategy, which has six objectives:

1. to manage the threats to the biodiversity of New South Wales;
2. to manage the protected area system to conserve biodiversity at the ecosystem, species and genetic levels;
3. to increase knowledge and understanding of biodiversity through the acquisition and dissemination of scientific information;
4. to limit any further decline in genetic and species diversity by increasing understanding of their requirements and applying this in management programmes;
5. to increase community understanding of and support for conserving biodiversity; and,
6. to ensure that there is a suitable legislative basis for conserving biodiversity.

Establishment of a representative protected areas system and promotion and implementation of community-based biodiversity conservation across all land tenures and land uses are critical to achieving these objectives.

INTRODUCTION

BIOLOGICAL diversity (biodiversity) can be defined as:

> *the variety of all life forms — the different plants, animals and micro-organisms, the genes they contain and the ecosystems of which they form a part. It is not a fixed entity, but constantly changing; it is increased by genetic change and by evolutionary processes and reduced by extinction and habitat degradation. The concept emphasizes the interrelatedness of the biological world. It covers the terrestrial, aquatic and marine environments* (BDAC 1992).

The draft "A National Strategy for the Conservation of Australia's Biodiversity" and the "Global Biodiversity Strategy" recognize three levels of biodiversity: genetic diversity, species diversity and ecosystem diversity (BDAC 1992; Courrier 1992). Genetic diversity is the total genetic information contained in the genes of biota. Genetic diversity occurs within and between populations of species. Species diversity is the variety of living organisms on the earth. Ecosystem diversity relates to the variety of habitats, biotic communities and ecological processes in the biosphere. It is the diversity of the different types of communities formed by living things, the relationships between them and ecological processes (Courrier 1992).

Australia is one of 12 megadiverse (i.e., extensive total biodiversity) countries that together account for 75 per cent of total global biodiversity (Mott and Bridgewater 1992). Australia is the only complete continent represented and it and Antarctica are the only continents that lie completely within the southern hemisphere.

Human-induced loss of biodiversity (including extinction of species, the loss of local populations and natural system dysfunction) is global, considerable and continuing (BDAC 1992). Australia has suffered extinctions and severe declines in total biodiversity during its period of human occupation (at least 60 000 years) but particularly since colonization by a technologically advanced society more than 200 years ago (Mott and Bridgewater 1992). New South Wales is a microcosm of Australia and its natural environments have been modified dramatically (Papps 1992).

The New South Wales National Parks and Wildlife Service exists to conserve, protect and manage the state's natural heritage, key elements of the state's cultural heritage, and to promote its appreciation and appropriate use. The Service's natural heritage programme has three components: biodiversity, wilderness and natural phenomena (e.g., geological landscapes). The Service is New South Wales' principal biodiversity conservation agency.

---

[1]NSW National Parks and Wildlife Service, P.O. Box 1967, Hurstville, New South Wales, Australia 2220.

Pages 391–409 *in* CONSERVING BIODIVERSITY: THREATS AND SOLUTIONS, ed by R. A. Bradstock, T. D. Auld, D. A. Keith, R. T. Kingsford, D. Lunney and D. P. Sivertsen. Surrey Beatty & Sons. 1995.

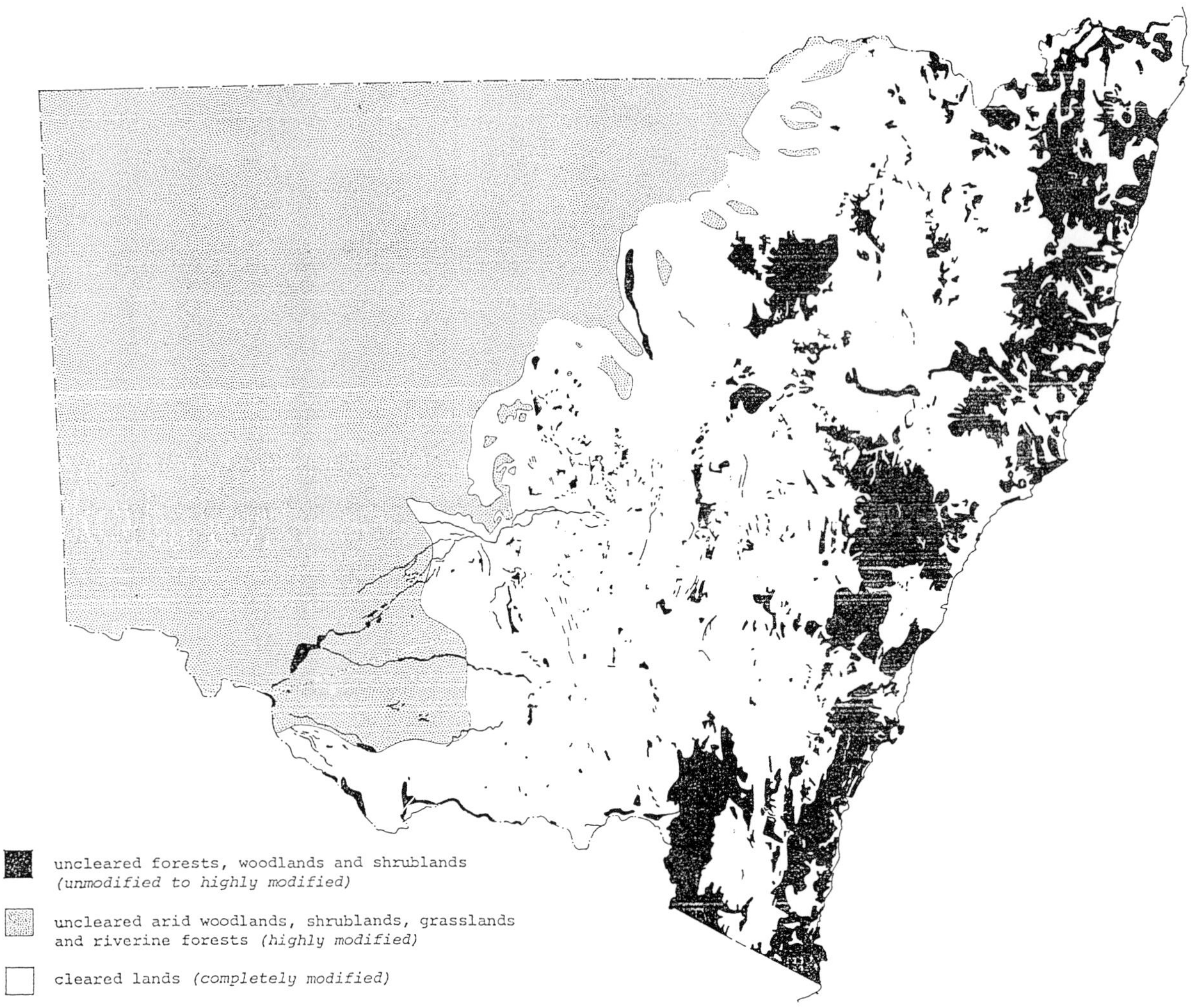

Fig. 1. Modification of the native vegetation of New South Wales (Interpreted from 1: 1 000 000 1982–86 LANDSTAT imagery).

This paper identifies the Service's role in biodiversity conservation in New South Wales, puts that role into the context of State Government policies and objectives, and provides an outline of the strategies to be utilized by the Service in meeting its responsibilities.

HUMAN IMPACT ON THE BIODIVERSITY OF NEW SOUTH WALES

Information on the current conservation status of the state's biodiversity is not complete and what is known can be drawn from the literature (NPWS 1992). However, it is relevant to this discussion to note the extent, severity and causes of the decline in biodiversity in New South Wales as this has a direct bearing on the nature of the response and the urgency with which remedial activities are undertaken.

Over the past 206 years of European settlement the natural environment of New South Wales has been modified dramatically. The degree of modification has not been uniform across the State (see Fig. 1) (NPWS 1992).

These impacts have been greater on certain groups of biota. An often-used indicator of biodiversity decline has been the number of extinct and endangered species of native fauna and flora. There is some variation in these lists according to the criteria used. The most recent New South Wales study was undertaken in 1992 by an independent, statutory Scientific Committee using criteria defined in the *Endangered Fauna (Interim Protection) Act* 1991. Schedule 12 of the *National Parks and Wildlife Act* 1974 now consists of three parts with Part 1 (Threatened) and Part 2 (Vulnerable and Rare) of relevance here. In deciding whether to place a species of fauna in Part 1, the Committee has to have regard to:

(a) whether the population of a species has been reduced to a critical level;

(b) whether habitat of a species has been drastically reduced or modified;

(c) whether a species may be in danger of extinction; and,

(d) whether a species may now be considered extinct but has been seen in the wild in the last 50 years.

For Part 2, the criteria are:

(a) whether the population of a species is decreasing because of over-exploitation, extensive destruction of habitat or other environmental disturbance;

(b) whether the population of a species has been seriously depleted and its ultimate security has not yet been assured;

(c) whether the population of a species is still abundant but is under threat from severe adverse factors throughout its range; and

(d) whether a species has a small population contained in restricted areas or habitats or thinly scattered over a more extensive area.

The approach taken by the Committee was modelled on Millsap *et al.* (1990) and drew extensively on the expertise of scientists (Lunney *et al.*, in press).

As a group, mammals, and the small ground-dwelling marsupials in particular, have suffered the greatest loss. Of the 130 mammal species that occurred in New South Wales at the time of European settlement, 27 species (or 21%) are now presumed extinct and a further 50 species (or 38.5%) are threatened (i.e., included on Part 1 or 2 of Schedule 12). Only 40.8 per cent of the original complement of mammals are not considered either threatened or vulnerable and rare.

After terrestrial mammals, birds are the next most affected group with 12 species (or 2.2%) presumed extinct and 98 species (21%) considered threatened. However, Recher and Lim (1990) argue that the avifauna has yet to reach an equilibrium with the altered landscape and predict further extinctions. There is also some evidence to suggest that birds are in decline worldwide (Noon and Young 1991).

No species of New South Wales amphibian is considered to be extinct but nine species (27%) are considered threatened. One species of reptile is currently known to be extinct. Twenty-six species (12%) are threatened. However, these figures are likely to be misleading as there is far less information on these two groups in comparison to mammals and birds. In recent years scientists have documented a global decline in amphibians (Rabb 1990; Wyman 1990) and there is some evidence to suggest that this is occurring in Australia (M. Mahoney, pers. comm.; Ferraro and Burgin 1993a,b; Richards *et al.* 1993).

Of the 883 vertebrate species known to occur in New South Wales, 233 (26%) have been identified as endangered and are listed on Schedule 12 of the *National Parks and Wildlife Act* 1974, hereafter referred to as the *NP&W Act.* Of these 40 are extinct (Lunney *et al.*, in press).

Compared to terrestrial fauna, the rate of extinction among vascular plant species has not been as great. Leigh and Briggs (1992) note only 15 extinct species from New South Wales representing 0.76 per cent of the original number. However, if grazing pressure in some parts of the State (particularly the Western Division) continues at current levels then there is likely to be a significant increase in extinction of higher plants due to a lack of regeneration.

Due to the paucity of information on "lower" plants and invertebrates, accurate estimates of extinction or decline are not possible. One species of insect has probably become extinct in New South Wales in historical times, and that is the Lord Howe Island Phasmid or stick insect, *Dryococelus australis*. Several other New South Wales invertebrate species are considered, on the basis of available knowledge, to be under some threat (Hill and Michaelis 1988; Willan 1977; Williams 1990; Yen and Butcher 1994). In view of the important role that invertebrates play in the functioning of natural systems (New 1984; Williams 1990), the Service regards improving the information base and therefore the conservation management of this group as a critical issue to be addressed in any strategy to conserve biodiversity even though legislative protection of invertebrates in New South Wales is inadequate (NPWS 1992). Both New (1984, 1991) and Greenslade and New (1991) provide a very sound basis for devising conservation strategies for Australian invertebrates, and Ponder (1992) has issued a spirited challenge to agencies such as the Service to devise such strategies thereby correcting the current imbalance in favour of vertebrates.

There are several main factors (or "threatening processes") involved in the loss of biodiversity and extinction of species and communities. Threatening processes are "processes that have or can threaten the survival, abundance or evolutionary development of any species or ecological community" (ESAC 1992).

The following summary of the major threatening processes operating in New South Wales is based on the Service's Draft *Natural Heritage Conservation Strategy — Biodiversity Component* (NPWS 1992).

Habitat loss and fragmentation

Habitat loss is considered the most important influence on species extinction rates for all biota. Most species cannot exist outside a natural or semi-natural area. Habitat loss also represents a loss of biodiversity at the ecosystem level. Two hundred years of European occupation in New South Wales has resulted in enormous losses of habitat. For example, between 1893 and 1921, 25.7 million hectares or about 32 per cent of

New South Wales was ringbarked and partially cleared (Reed 1991). A very significant proportion of this habitat destruction occurred before the establishment of a protected area system and a truly representative system is now impossible (Reed 1991).

Clearing for agriculture continues to pose a threat to many species in New South Wales. The rare plant Ooline *Cadellia pentastylis* is restricted to a few localities on the north-west slopes of New South Wales and Central Queensland and is threatened primarily by land clearance for agriculture (Benson 1993a). The Regent Parrot *Polytelis anthopeplus anthopeplus* and the Superb Parrot *Polytelis swainsonii* are threatened by continuing habitat loss and the effects of habitat fragmentation (Webster and Ahern 1992; Webster 1991).

While less devastating than complete habitat loss, fragmentation of habitat can lead to extinction of a species when numbers in isolated patches fall below a minimum viable population (Soulé 1987). When populations are small they face an increased risk of extinction due to a combination of the following factors:

(a) Environmental variation (including randomness and stochasticity) — temporal variation in the net rate of population growth due to variation in external factors such as weather or food supply;

(b) Genetic processes e.g., loss of fitness by genetic drift or inbreeding;

(c) Catastrophe — infrequent severe environmental events which cause populations to fluctuate (e.g., droughts, fires); and,

(d) Demographic factors — changes in population size and structure due to chance variation (see, for example, Soulé 1987; Shaffer 1987; Gilpin 1987; Harrison 1991; Possingham 1991).

Habitat degradation

In many cases habitat is not totally destroyed nor significantly fragmented but is changed in some critical way so as to make it unsuitable for endemic fauna or flora. Processes leading to these types of changes include modified fire regimes, erosion, wetland drainage, damming of rivers and salinization. The New South Wales Soil Conservation Service found that 72 per cent of the State was affected by some form of land degradation as a direct result of European land-use (SCS 1989). One of the most significant forms of habitat degradation occurs when human-altered fire regimes change the composition of the plant species of an area (Bradstock 1993). This results in changes to the habitat of a vast assemblage of vertebrate and invertebrate animals and plants.

Species introduction

Exotic species can threaten native flora and fauna directly by predation or competition or indirectly by altering natural habitat. Between 1840 and 1880, more than 60 species of exotic vertebrates were released in Australia. Many occur in New South Wales and have caused, or have the potential to cause extinction of species. These include rabbits, foxes, goats, cattle, horses, pigs and rats. The Bilby *Macrotis lagotis*, a species of bandicoot, is now extinct in New South Wales, partly as a result of competition with rabbits. Two exotic species, cats and foxes, have substantially contributed to the endangerment of several mammals and ground-dwelling birds (Dickman and Read 1992; Dickman 1993; Dickman *et al.* 1993; Smith *et al.* 1994).

Exotic plants can pose a serious threat to native species and communities (Humphries *et al.* 1991) and are invading, or have invaded, large areas of the state. Such invasive weeds are often referred to as "environmental weeds". Boneseed *Chrysanthemoides monilifera monilifera*, Bitou Bush *C. m. rotunda* and Scotch Broom *Cytisus scoparius*, for example, have displaced native plants and may also contribute to declines in native animals not adapted to the modified habitats created by these foreign species (Humphries *et al.* 1991; ANPWS 1991; Benson 1992; Love and Dyason 1985). Ironically, native animals often aid the spread of invasive weeds by spreading seeds (Bass 1990a,b; Loyn and French 1991).

The Lord Howe Island Phasmid or stick insect *Dryococelus australis* is thought to have been eliminated by introduced rats (Williams 1990).

Direct exploitation

Hunting and poaching are now not as significant in New South Wales as other factors but have contributed to the loss of some species. For example, Southern Right Whales *Eubalaena australis* were exploited to the edge of extinction (Ling 1988). The Black-Throated Finch *Poephila cincta* is now locally extinct throughout much of its former range in central Queensland and New South Wales partly, it is thought, as a result of excessive collection (Garnett 1992; Parker and Reid 1978). This has resulted in its listing on Part 1 (Threatened) of Schedule 12 of the *NP&W Act*. The impact of illegal collection for the bird trade has been implicated in decline of several other bird species in New South Wales, but definitive evidence on the importance of that threat is lacking (Garnett 1992). The Greenhood Orchid *Pterostylis gibbosa* has disappeared from many areas of its former range in the state because of illegal collecting coupled with habitat destruction (Leigh and Briggs 1992). Concern has been expressed over the continued exploitation of a number of Australian freshwater fish occurring in New South Wales (Wager and Jackson 1993). It seems likely that some species have suffered through exploitation and have thus been made more vulnerable to habitat destruction or other threatening processes. Furthermore, there continues to be an illegal trade in native wildlife which may threaten a number of species, particularly amongst birds.

Pollution

Both air and water pollution may adversely affect ecosystem function and reduce populations of sensitive species. Several alpine aquatic insects whose streams are subjected to pollution from ski resort development are considered by Kennedy (1990: Appendix II) to be threatened. The Dwarf Mountain Pine *Microstrobus fitzgeraldii*, endemic to a few locations in the Blue Mountains west of Sydney, is threatened principally by pollution of mountain streams (Jones 1994).

Cumulative effects

Extinction is generally due to multiple causes. The cumulative effect of various disturbances can seriously threaten a region's biodiversity. Some disturbances may be synergistic rather than merely additive — for example, habitat fragmentation plus climate change. Land degradation, in all its forms, is often the result of cumulative impacts. For example, unsustainable agricultural practices where no investment has been made in soil conservation can severely disrupt biological productivity and lead to a self-reinforcing cycle of further degradation (ESAC 1992). The Mallee Fowl *Leipoa ocellata*, once widely distributed across southern and central Australia, is now endangered in New South Wales mainly due to the combined effects of the destruction of its mallee habitat and predation of chicks by foxes (Priddel 1990). The Pigmy Cypress Pine *Callitris oblonga* is threatened by a combination of changed fire regimes, clearing for agriculture and competition from weeds (Nadolny and Benson 1993). Habitat disturbance from ski resort development coupled with the likely adverse impacts of climate change could combine to threaten the Mountain Pygmy-possum *Burramys parvus* (Mansergh and Broome 1994).

WHY CONSERVE BIODIVERSITY?

The reasons for acting to conserve biodiversity have been well documented (see, for example, McNeely *et al.* 1992; Courrier 1992; BDAC 1992; DEST 1993). It is important to identify where biodiversity conservation falls within the range of government objectives; without government commitment it will remain a peripheral, and therefore largely fruitless, exercise.

The emergence of biodiversity conservation as an issue in New South Wales reflects international and national trends. In 1992, the Global Biodiversity Strategy was published as a complementary initiative to the development of an international Convention on Biological Diversity in order to act "as a basis for the practical action that should be taken while the Convention is being ratified and entering into force" (Courrier 1992).

In July 1989, the then Prime Minister of Australia, Bob Hawke, announced that the Australian Government would prepare a National Strategy for the Conservation of Biological Diversity. The first stage was the appointment of a Biological Diversity Advisory Committee (BDAC) to advise the Commonwealth Minister for the Environment, Sport and Territories on the status of, and threats to, Australia's biological diversity and the adequacy of existing conservation and legislative mechanisms.

The Committee was also charged with developing the national strategy which, when the first draft was released in 1992, generated a great deal of controversy and, in particular, opposition from industry groups. Consequently, the BDAC was expanded to include representatives from the mining, fishing, tourism, forestry and agriculture industries.

A revised version of the strategy, representing the majority view of BDAC, was prepared and is currently under consideration by the Commonwealth Minister. Following adoption of an "Intergovernmental Agreement on the Environment" (IGAE) by the Federal and State Governments in May 1993, a Working Group of agencies from the Australian and New Zealand Environment and Conservation Council's Standing Committee on Conservation is currently preparing a draft National Strategy for the Conservation of Biodiversity. This will eventually be considered by the Council of Australian Governments. The preparation and adoption of such a strategy is one of Australia's obligations under the International Convention on Biological Diversity.

An initiative potentially more capable of driving biodiversity conservation programmes and policies into mainstream economic and planning decision-making is the adoption of the principles of ecologically sustainable development (ESD).

In December 1992, the Council of Australian Governments endorsed the National Strategy for Ecologically Sustainable Development (Commonwealth of Australia 1992a). The strategy will play a critical role in setting the scene for the broad changes in direction and approach that governments will take to try to ensure that Australia's development is ecologically sustainable. In particular, the Council agreed that the future development of all relevant policies and programmes, particularly those which are national in character, should take place within the framework of the ESD strategy and the IGAE.

ESD is defined as "using, conserving and enhancing the community's resources so that ecological processes, on which life depends, are maintained, and the total quality of life, now and in the future, can be increased" (Commonwealth of Australia 1992a). Simply, ESD is development which aims to meet the needs of today's Australians while conserving ecosystems for the benefit of future generations. There are two main features which distinguish an ecologically sustainable approach to development (Commonwealth of Australia 1992a):

- the need to consider, in an integrated way, the wider economic, social and environmental implications of our decisions and actions for Australia, the international community and the biosphere; and,
- the need to take a long-term rather than a short-term view when taking those decisions and actions (i.e., operating in biological rather than political time-frames).

The ESD strategy has three core objectives:

- to enhance individual and community well-being and welfare by following a path of economic development that safeguards the welfare of future generations;
- to provide for equity within and between generations; and,
- to protect biological diversity and maintain essential ecological processes and life-support systems.

Biodiversity conservation, therefore, now formally occupies a position as a central plank of Commonwealth Government policy. It sits comfortably with the State Government's general policy objective of fostering economic development while maintaining rigorous environmental standards.

THE NEW SOUTH WALES NATIONAL PARKS AND WILDLIFE SERVICE

The National Parks and Wildlife Service was established by the *National Parks and Wildlife Act* 1967 which was replaced by the *National Parks and Wildlife Act* 1974 (that is, the *NP&W Act*). The Service does not have corporate status; rather it is a government department under the control of the Director-General of National Parks and Wildlife who is subject to the control and direction of the Minister for the Environment.

The functions of the Director-General of National Parks and Wildlife are specified in the Act and the *Wilderness Act* 1987. In summary, these functions are:

- to investigate and acquire land as national parks and nature reserves so as to conserve a complete range of the State's natural environments;
- to acquire and preserve as historic sites, buildings, objects, monuments and other sites of special cultural significance throughout the state;
- to preserve and protect Aboriginal sites, buildings, objects, monuments and other sites of historical significance;
- to investigate and acquire land as state recreation areas;
- to manage these resources so that they can be enjoyed by present generations and conserved for future generations;
- to develop facilities for visitors without impairing the values of areas managed;
- to protect, manage and research native flora and fauna throughout the state;
- to identify, protect and preserve wilderness; and,
- to promote awareness, understanding and appreciation of wildlife, national parks, Aboriginal and historic sites, wilderness and conservation.

Under the *NP&W Act*, five categories of land — national parks, nature reserves, historic sites, Aboriginal areas and state game reserves — are vested in the Director-General and managed by the Service. The *NP&W Act* also provides for the vesting of state recreation areas in trusts, and where there are no trusts, the care, control and management of these areas devolve to the Minister. The Minister has delegated this responsibility to the Director-General.

The *NP&W Act* provides very limited protection of plant species and, outside land vested in the Director-General, no protection for native fish or invertebrates.

As at 30 June 1994, 3 955 318 hectares were reserved under the *NP&W Act*. This represents approximately 4.94 per cent of the land area of New South Wales.

National parks and historic sites are managed in a way that will preserve their natural and cultural features while still allowing for their use and enjoyment by visitors. State recreation areas are managed to maximize their recreation potential while preserving and protecting their natural and cultural features. Because nature reserves and Aboriginal sites are preserved for their scientific and cultural value and are usually small in area, public access to them is generally limited.

The Service is responsible for six other types of areas which are not vested in the Director-General: Aboriginal places, protected archaeological areas, wildlife refuges, wildlife management areas, conservation agreement areas and wilderness protection agreement areas.

The Service makes submissions under the *Environmental Planning and Assessment Act* 1979 for activities that:

- will have a detrimental effect on Service estate;
- will have a detrimental effect on areas being examined for addition to Service estate;
- will reduce the conservation value of areas in which the Service has an interest (because of their heritage values); and,
- will have a detrimental effect on native flora and fauna.

The Director-General is the authority for the protection and care of native fauna in New South Wales. As the Director-General's responsibility for fauna covers all land, any development which will affect fauna or its habitat is a matter of concern for the Service. Under amendments to the *NP&W Act* made by the *Endangered Fauna (Interim Protection) Act* 1991 the Director-General is responsible for the control of activities likely to "take or kill" endangered fauna where take or kill is defined to include significant habitat disturbance. Within the meaning of the amended *NP&W Act,* however, native fish, invertebrates, micro-organisms and plants are not given the same degree of protection.

The Service controls development and building on its own estate. For places such as Thredbo this means control over entire towns, including services such as sewerage and garbage removal.

The Service is the principal management agency for two of the three areas in New South Wales listed under the World Heritage Convention (i.e., the Lord Howe Island Group and the Australian East Coast Temperate and Sub-Tropical Rainforests). It has a major role in the management of the third area, the Willandra Lakes Region, as it includes Mungo National Park.

TRENDS IN THE OPERATING ENVIRONMENT

In the 27 years since the establishment of the Service there has emerged a number of significant changes in the environment in which it operates. These changes, or more accurately in some cases, the Service's failure or inability to respond to them, have compromised its performance in meeting its responsibilities in biodiversity conservation. In this the Service is not unique as similar problems have beset other agencies in Australia and internationally, for example, in the United States and New Zealand.

Rapid increase in land management responsibilities

At the time of the passage of the *NP&W Act*, New South Wales' 20 national and state parks, 47 nature reserves and six historic sites between them covered just over 860 000 hectares — less than 1 per cent of the state was securely protected. Land acquisition was the highest conservation priority of successive governments over the next two decades.

The accepted wisdom of the time was that a minimum of five per cent of the state's land needed to be reserved. This belief, and the rapid economic growth of the time with its concomitant changes to the natural and cultural landscapes of New South Wales, drove the process of acquisition. Between 1967 and 1982 the area of lands under the Service's control underwent a fourfold increase from some 0.8 million hectares to 3.2 million hectares (and currently is 3.9 million hectares).

By any measure, the resourcing of park management has never kept pace with land acquisition. As in most states, the majority of the Service's funding for biodiversity conservation, including management of its protected area system, is provided from consolidated revenue. The Service generates nearly 15 per cent of its income, principally from park user fees and concessions. Without the revenue from winter recreational use of Kosciusko National Park, this figure would be considerably less.

The Resource Assessment Commission summarized relevant expenditure and revenue data for 1990–91 as published in the annual reports of state and territory conservation agencies (RAC 1992a:220). On these figures, the Service expends about $22.90 per hectare of reserved land. This compares favourably with a national average of $12.70 per hectare and the lowest figure of $1.74 in South Australia. However, the Service's expenditure includes administration and corporate services, statewide flora and fauna conservation programmes and some statewide cultural heritage programmes. The New South Wales figure is therefore a gross overestimate and the comparisons between states are relatively meaningless as agency arrangements for financial reporting vary widely (RAC 1992a). A more realistic figure for New South Wales park management is $13.90 per hectare in that year. (This is calculated from figures published in the 1990–91 Annual Report, namely, that approximately $40 million was expended on essential operations and works and the area managed was 3.8 million hectares as at 30 June 1991.)

There is no national or international benchmark against which to measure this performance. However, there can be no doubt, considering surveys within the Service and by State Government management specialists, that expenditure on park management, let alone broader biodiversity conservation, is inadequate. The Service therefore has never had the people or funds necessary to maintain its estate. In July 1994 the New South Wales Government announced a major enhancement to the Service's budget. This will contribute to closing the gap between the resources required to adequately discharge the Service's statutory responsibilities and those provided.

Major shifts in scientific knowledge and theory

As in all other fields of scientific endeavour, there have been dramatic changes in the scientific knowledge and theory which governs the best practice of heritage conservation. This is most evident in the global shift to an emphasis on biological diversity (or biodiversity) conservation, replacing the old wildlife management approach of nature conservation agencies. This new synthesis has been termed "conservation biology" (Soulé 1986; Soulé and Wilcox 1980; Western and Pearl 1989). However, examination of the scientific principles underpinning conservation biology is continuing (see, for example, Caughley 1994), as are the implications of applying the concepts underlying conservation biology in the Australian context (Walker 1992; Walker and Nix 1993).

One of the most significant insights arising from conservation biology has been the realization that achievement of biodiversity conservation through reservation of land in a protected area system is, and always has been, a limited option (Recher 1990; Pressey 1990; Mott and Bridgewater 1992). The Service has accepted this maxim for some time but has failed to develop an adequate suite of "off-park" or community-based conservation programmes. Practice has not reflected best available knowledge.

Emphasis on recreation

While difficult to demonstrate, it is our contention that for much of its history the Service has focused on its role as a provider of outdoor recreational opportunities. This has been to the detriment of its biodiversity conservation work, although there have been some notable successes (e.g., the recovery of the Lord Howe Island Woodhen) in this field. The Service has a legitimate responsibility to provide for appropriate (i.e., consistent with protection of heritage values) public use of its protected area system and this will become an increasingly important means of providing for public awareness and education on biodiversity conservation. However, this is a secondary objective for the Service and resource allocation must reflect this.

Poor or nonexistent information and limited analyses

It is widely acknowledged that the information base on which the Service operates is either poor or nonexistent for many species, ecosystems or geographical areas. This situation is not unique to the Service. The RAC Forest and Timber Inquiry noted, for example, that:

> *. . . a major impediment to assessing the representativeness of the existing reserve system is poor knowledge of the distribution of forest types and species in Australia . . . In many cases, very little is known about the distribution of species in conservation reserves or production forests* (RAC 1992a).

In recent years the Service has expended many millions of dollars in rectifying this deficiency but the need is far greater than the Service's resources can satisfy. The Service's *Draft Natural Heritage Conservation Strategy — Biodiversity Component* does address this problem and assigns priorities for survey work. This area requires a considerable investment of resources.

An additional problem faced by the Service is the effort required to collate and analyse existing data or information on the biological diversity of the state. Vast quantities of data remain isolated in innumerable reports, studies, papers or notebooks. Until that data is collated into a single accessible form, it cannot be used to guide conservation assessment and planning. The Service recognizes the importance of drawing on this body of knowledge. For example, the Service is continuing the development of the *Atlas of New South Wales Wildlife* (Ellis 1992) and, together with financial support from the Natural Resources Audit Council, is accelerating the collation of locality records for fauna and flora from a wide range of sources.

Changes in public administration and agency focus

A number of government agencies in New South Wales now have a major role in the field of environmental protection. These include the Department of Planning, the Environment Protection Authority and the Department of Conservation and Land Management. In other departments there has been a shift in focus to increase emphasis on conservation even though this remains a secondary function, most notably in State Forests (formerly the Forestry Commission) of New South Wales and New South Wales Fisheries.

Since 1988 there have also been massive policy, operational and cultural changes in public administration in New South Wales with, among other things, greater emphasis on organizational outcomes, commercial objectives and the contributions of agency services to core business.

Government initiatives, such as Total Catchment Management (TCM), which are designed to address land degradation and are consistent with ESD principles, have also been set in legislation. TCM incorporates mechanisms for extensive community involvement. Government policy also requires land management agencies such as the Service to participate in community-based activities such as Landcare groups.

Nonetheless, there is a need to better define the boundaries between government agency responsibilities to maximize efficiency (i.e., best use of public resources) and minimize conflict. Interagency conflict is a major impediment to biodiversity conservation in New South Wales, as it appears to be elsewhere (Grumbine 1991, 1992; Connor *et al.* 1993; Landres 1992).

Increased national and international responsibilities in nature conservation

Nature conservation is now a major national and international issue. For example, Australia is party to nine international conventions relevant to biodiversity conservation. These include:

- The recently promulgated Convention on the Conservation of Biodiversity (the Rio Convention) which came into force on the 29th December 1993;
- The Convention Concerning the Protection of the World Cultural and Natural Heritage (World Heritage Convention);
- The Convention on Wetlands of International Importance especially as Waterfowl Habitat (Ramsar Convention);
- The Agreement between the Peoples Republic of China and the Government of Australia for the Protection of Migratory Birds and the Environment (CAMBA);
- The Agreement between the Government of Australia and the Government of Japan for the Protection of Migratory Birds and Birds in Danger of Extinction and the Environment (JAMBA);
- The Convention on International Trade in Endangered Species of Wild Fauna and Flora (CITES);
- The International Convention for the Regulation of Whaling (ICRW);
- The Convention on Conservation of Nature in the South Pacific (Apia Convention); and,
- Convention on the Conservation of Migratory Species of Wild Animals (Bonn Convention).

Increase in legislative responsibilities

The Service has experienced substantial growth in its legislative responsibilities since the passage of its original Act in 1967. The *NP&W Act* was expanded considerably in 1974. Other legislative responsibilities have also been assigned to the Service, for example, the *Wilderness Act* 1987 and the *Endangered Fauna (Interim Protection) Act* 1991. Significant amendments to the *NP&W Act* since 1974 have also augmented the Service's responsibilities, for example, the establishment of the conservation agreement scheme and a capacity to declare wild and scenic rivers in national parks. In most cases, no additional or only very limited resourcing has been provided.

Increased community awareness and expectations

In New South Wales, like the rest of the world, the 1970s saw a "green revolution" and the Service was rapidly embroiled in community debate over conservation versus development. A second wave of environmental concern emerged in the late 1980s following disturbing global environmental impacts: acid rain; tropical deforestation; global warming and a string of environmental disasters — Chernobyl, Bhopal and Exxon Valdez, to name a few.

Australians have become aware that these impacts and issues are not confined to the northern hemisphere. A local example is the recurrence of toxic blue-green algal blooms in the inland rivers of New South Wales and recently in Jervis Bay on the south coast of New South Wales.

Community awareness of environmental issues has continued to increase and with it community expectations of the Service. Recent surveys of public opinion by the Commonwealth Government's Department of Environment, Sport and Territories, and by the Service (Frank Small and Associates 1993) and the New South Wales Environment Protection Authority (New South Wales EPA 1994) have confirmed, even in times of economic difficulty, the importance of environmental protection to the general community. What is also apparent from these surveys is that people and organizations dealing with the Service believe it is unable to meet its responsibilities at current resourcing levels.

A NEW PARADIGM

Conservation biology research has led to a revision of the approach taken by many conservation and management agencies (Grumbine 1991, 1992). This work has emphasized the importance of large, viable and representative reserves and also highlighted the critical need for conservation endeavours outside of such a reserve system. Protection of biodiversity will require:

(a) sufficient habitat for viable populations for all native species;

(b) maintenance of areas large enough to accommodate natural disturbance regimes;

(c) planning over a time line of centuries so that species and ecosystem patterns and processes may continue to evolve; and,

(d) provision for human occupancy and use that does not result in ecological degradation (Grumbine 1991).

There is now consensus that conservation of biodiversity at all levels (genetic, species and ecosystem) is dependent on an integrated system of large, representative protected areas (Grumbine 1991). However, even a large, representative reserve system cannot protect the biodiversity of New South Wales, and as Recher (1990) noted:

> *Wildlife management and conservation, if it is to succeed, must be extended to all land, regardless of tenure, and despite arbitrary political boundaries or bureaucratic divisions.*

The Service recognizes the significance of this axiom and has been undertaking research into new analytical methods for assessing conservation options in the broadest sense (Pressey 1990; Pressey *et al.* 1993). This new paradigm for biodiversity conservation can be more expansively defined as follows:

> *National parks and other protected areas, must indeed play a major role in the conservation of biodiversity, but that role will be compromised if they are inappropriately buffered. Today a mixture of protected areas, highly altered agriculture and pastoral lands, urban and suburban areas and industrial sites, together with relatively natural wildlife linkages and corridors, form the mosaic of the landscape.*
>
> *This landscape oriented ecological framework provides the basic unit for management. More urgent than a precise knowledge of exactly how many species there are is a need for understanding how the different elements of the landscape mosaic interact and how management can target biodiversity conservation for sustainable landscapes/seascapes* (Mott and Bridgewater 1992).

Therefore, while acknowledging the critical role of the protected area system (particularly when it becomes representative and viable), it is incumbent on agencies such as the Service to frame biodiversity conservation strategies within this paradigm. "Off-park" or community-based conservation programmes must go beyond the current standards of wildlife refuges and voluntary conservation agreements (see Papps 1990) to genuine partnerships in conservation endeavours with local communities, landholders and other agencies as described in Campbell (1994) and Wilson (1995).

While awareness of and support for biodiversity conservation within the community is essential, it must be a precursor to participation. Achievement of biodiversity conservation in this way cannot occur in a policy or legislative vacuum, any more than it can occur in the absence of recognition of the critical importance of co-operative endeavour between governments, agencies, communities and so on (Saunders 1990; Connor *et al.* 1993; Craig and Stewart 1994).

Biodiversity conservation agencies will have to increasingly involve themselves in the maelstrom of economics, sociology and politics to ensure appropriate policies (Ehrlich and Wilson 1991; Machlis 1992; Mott and Edwards 1992). Failure to address the economic and social causes of biodiversity decline is an abrogation of responsibility and conservation strategies that do not consider these factors are, as Pickard (1991) suggested, doomed to failure. Similarly, the role of organizational structures and individual behaviour in the success of key conservation programmes (e.g., endangered species conservation) cannot be overlooked and the approach of Clark and others (e.g., Clark 1992; Miller *et al.* 1994) merits further consideration (Smith 1991).

THE SERVICE'S ROLE IN BIODIVERSITY CONSERVATION

The parameters which set the limits within which the Service's role can be defined are:

- the government's objectives (economic, social and environmental);
- the best available theories, paradigms and practices for heritage conservation; and,
- the roles played by other agencies and the general community.

In summary:

- the State Government is committed to the development of a diverse, strong and internationally competitive economy that provides for proper environmental protection and, in particular, the conservation of biodiversity;
- heritage conservation, and especially biodiversity conservation, will only be achieved through community-based, co-operative

endeavours across the full range of landscapes and tenures, built around (but not limited to) a representative protected areas system; and,

- many other government agencies have an important role to play in environmental protection and heritage conservation but all have a different focus of operation with the Service being pre-eminent in heritage conservation.

At the state level, the *NP&W Act* is the principal legislation for achieving biodiversity conservation.

The Service, as the state's leading biodiversity conservation agency, is currently developing a strategy to guide its programmes and projects in this area. The objective or goal of that (draft) strategy is:

> *To conserve in perpetuity the biodiversity of New South Wales at the ecosystem, species and genetic levels.*

Achievement of this goal is predicated on two fundamental actions:

(i) maintaining viable and representative examples (to the extent possible) of the range of natural biodiversity (in protected areas in one form or another); and,

(ii) elsewhere by ecologically sustainable management of the landscape and biota, minimizing the impact of human-induced changes.

This approach reflects consensus among conservation biologists and is a significant change from that traditionally adopted by biodiversity conservation agencies (including the Service) of concentrating, almost exclusively, on "sampling" some components of the biota (usually vertebrates and vascular plants) in national parks and nature reserves.

A BIODIVERSITY CONSERVATION STRATEGY

The Service's *Draft Natural Heritage Conservation Strategy — Biodiversity Component* will aim to:

(i) establish a goal for biodiversity conservation in New South Wales, in which ecologically sustainable management across all lands can ensure that the state's biodiversity is maintained;

(ii) identify the priorities for biodiversity conservation in New South Wales;

(iii) provide the Service with a framework for developing biodiversity conservation programmes; and,

(iv) inform the community and other government agencies of the major threats to the state's biodiversity and the Service's response.

The Service is first and foremost an operational agency. The strategy therefore must be action-oriented or functional. To meet this demand the Strategy is being developed to:

- address the three levels of biodiversity (genetic, species and ecosystem);
- be based on biological principles and processes;
- include a clear procedure for determining conservation issues and priorities (from general at the state level through to specific at the district and regional levels), and a clear statement of what the strategy has identified those issues and priorities to be;
- identify relevant mechanisms (statutory and funding opportunities) to assist in developing conservation programmes;
- establish a clear course of action, based upon the determined priorities;
- be flexible in space, e.g., be able to identify threatening processes operating at different spatial scales;
- be dynamic and responsive to changing conditions, particularly changes to threatening processes or improvements in the information base; and,
- be consistent with and able to be related to other strategies at other levels, or in other areas in the Service (NPWS 1992).

The major principle in constructing the strategy was a focus on threatening processes. While acknowledging the costliness of endangered species recovery programmes and the need for a landscape-oriented approach, the Service will not abandon its commitment to maintaining species diversity.

An evaluation of threatening processes in relation to the distribution patterns of biota is therefore the key step that must precede any attempt to derive actions for conservation (NPWS 1992). From a Service perspective, threatening processes can operate on or off the protected area estate. There are basically three categories of action for dealing with threatening processes that will be applicable to most conservation problems: (a) Threats that exists within its estate and the solution can usually be provided by direct, appropriate Service management; (b) Threats operating beyond the Service estate must be addressed by appropriate management by other land owners or land managers (possibly facilitated by Service expertize or direct assistance); or (c) By Service acquisition of the land followed by appropriate management. (This is illustrated in Fig. 2.)

Within the Service's *Draft Natural Heritage Conservation Strategy — Biodiversity Component*, six fundamental objectives have been formulated

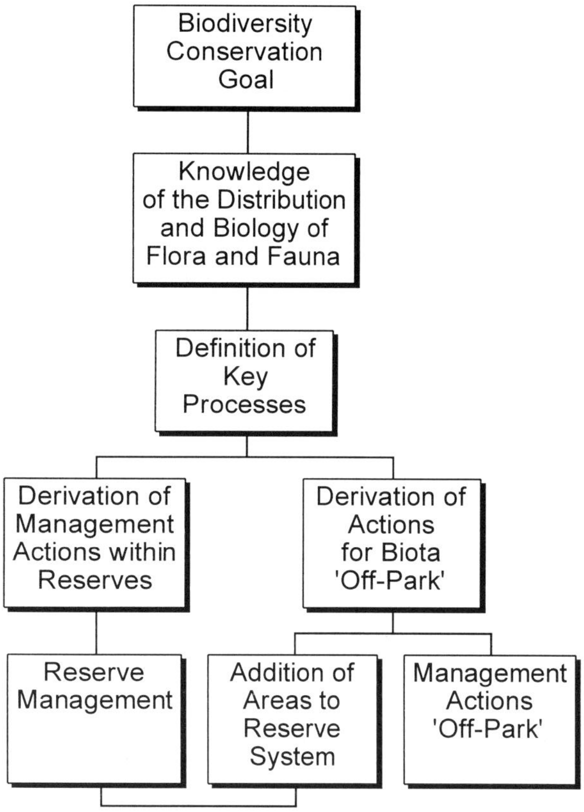

Fig. 2. Conceptual approach to conserving the state's biodiversity.

for conserving biodiversity in New South Wales. These are:

1. to manage the threats to the biodiversity of New South Wales;
2. to manage the NPWS reserve system in accordance with the NPWS's biodiversity conservation goal;
3. to increase knowledge and understanding of biodiversity through the acquisition and dissemination of scientific information;
4. to limit any further decline in genetic and species diversity by increasing understanding of their requirements and applying this in management programmes;
5. to increase community understanding of and support for conserving biodiversity; and,
6. to ensure that there is a suitable legislative basis for conserving biodiversity (NPWS 1992).

A set of programmes has been proposed to give effect to each of these objectives. To achieve the long-term maintenance of biodiversity at all three levels, the Service must focus its limited resources on managing the threats to biodiversity. This involves an evaluation of the effects of current management practices on the natural environment and an assessment of how those practices can be improved to maximize biodiversity conservation (Mott and Bridgewater 1992). It also involves making judgments as to which threatening processes the Service can potentially influence, alter or control.

The two threats to New South Wales' biodiversity that require greatest effort are clearing and grazing. Clearing of native vegetation is still occurring even in areas where the loss has already been substantial. For example, Wells *et al.* (1984) revealed that Central Division of New South Wales, that is in the wheatbelt, between 80 per cent and 95 per cent of the original natural vegetation had been cleared by 1984. In the northern wheatbelt of New South Wales, a study conducted by the Service revealed that between 1973 and 1984, 67 per cent of the remnant native vegetation was removed from the Goondiwindi district, 50.5 per cent was removed in the St George district and 46 per cent from the Moree district (Sivertsen 1994).

The management of total grazing pressure in the Western Division of New South Wales is particularly important (NPWS 1992). (Total grazing pressure refers to the impact of all grazing or browsing species and thus includes feral species (e.g., goats), pest species (e.g., rabbits), stock (e.g., sheep and cattle) and native species such as kangaroos.) Widespread loss of ground cover and impacts on numerous species in this part of the state is well-documented (Benson 1992; Grice and Barchia 1992; Mitchell 1991; Pickard 1991). Programmes addressing other identified dominant threats have also been developed.

Notwithstanding the acknowledged gaps in the current New South Wales reserve system, reserves have a central role to play in conserving the biodiversity of New South Wales. In many cases reserves contain species and communities that are now significantly reduced in range or abundance elsewhere in the state. Reserves are, however, not immune from the impacts of many threatening processes which operate across the entire landscape. The impacts of public visitation can also present problems for the management of reserves. For these reasons, the Service is developing and implementing management practices within reserves to conserve their natural values. Control programmes for introduced weeds and animal pests and ecologically appropriate fire management regimes are also important components of reserve management.

Successful biodiversity conservation, and in particular the establishment and proper management of a representative protected area system, is dependent on high quality, comprehensive information and a good understanding of the science of conservation biology (Benson 1993b). Research and biodiversity survey are fundamental to achieving the Service's conservation objectives. This dependence on research and biodiversity survey poses dilemmas: in many cases conservation managers have neither the time nor the resources to accommodate essential research. Moreover, experience suggests that bureaucracies such as the Service often fail to properly utilize research results. In this regard, Mott and Bridgewater (1992) are probably correct in their assertion that "we are overselling and underutilizing research". And while it is easy to concur that the "main thrust [of biodiversity conservation] will have to be management" rather than an "endless series of independent skirmishes over research priorities" (Mott and Bridgewater 1992), biodiversity conservation agencies will have to be wary of decision-making in the absence of an adequate information base or some reasonable approximation of such data. Confusing activity with progress where it forecloses conservation options is unacceptable.

Western (1991, 1992) noted that the solution to the problem of halting the decline in natural biodiversity lies in the interface between science and management, and recommended two steps to address this problem. First, conservation depends primarily on managers, not biologists. Therefore, scientific ideas or findings must be presented in a way comprehensible to those responsible for initiating action, that is, the managers. Second, the results of research must be developed into useable or workable solutions that the manager can apply directly.

One hope for managers on the horns of this dilemma is technological advance, including computer-assisted conservation planning. Rapid biodiversity assessment or RBA (Oliver and Beattie 1993; Beattie 1993; Cranston and Hillman 1992; Janzen 1991), population viability analysis (Lindenmayer *et al.* 1993a; Possingham 1991), conservation options decision analysis or CODA (Bedward *et al.* 1992), irreplaceability analysis (Pressey *et al.* 1993; Pressey *et al.* 1994), computer-based biodiversity survey design and analysis (Ferrier 1991; Ferrier and Smith 1990), BIOCLIM (Busby 1991) and DOMAIN (Carpenter *et al.* 1993) are examples of tools which can assist decision-making and conservation management. An overview of the Service's response to each of these innovations follows:

Rapid biodiversity assessment (RBA)

Rapid biodiversity assessment appears to offer significant advantages for diverse and taxonomically poorly understood taxa such as insects and micro-organisms (Beattie 1993; Oliver and Beattie 1993; Beattie and Oliver 1994). However, there may be limitations to the use of rapid biodiversity assessment techniques in conservation assessment and planning, particularly when agencies such as the Service must relate the results of rapid assessment surveys to statutory planning or regulatory processes (Wilson 1993).

Population viability analysis

Population viability analysis (PVA) is a generic term for population models that can provide an estimate of the probability of species extinction under a range of management options (Possingham 1991). While PVA has been used extensively in the United States, its capacity to assist in wildlife conservation in Australia has been underutilized. Examples where it has been used in the Australian context are noted in Clark and Seebeck (1990) and Lindenmayer *et al.* (1993b).

However, PVA is time and data intensive and therefore can only be used for a limited number of species. Possingham (1991) suggests:

- species that have been studied intensively for a long time;
- keystone species (i.e., species that are important to a number of other species);
- indicator species (i.e., easily censused species which are sensitive to environmental perturbations);
- species that occur at low population densities;
- species with poor recolonization abilities;
- species of cultural importance.

Conservation options analysis

CODA (Bedward *et al.* 1992) and irreplaceability analysis (Pressey *et al.* 1993; Pressey *et al.* 1994) are significant developments in the field of conservation assessment and planning. They allow interactive manipulation of options and the objective measurement of proximity to reserve system goals to be undertaken by conservation planners and managers. They are flexible (i.e., they can accept a wide range of data types including socio-economic information) and are relatively easy to use. Adaptations of these techniques have now been incorporated into the Service's in-house geographical information system software, E-RMS (Pressey and Ferrier 1994).

Survey design and data analysis

Sound management requires, amongst other things, a sound information base. In pursuit of such a sound information base, the Service has become skilled at the application of geographical information systems to the cost-effective design of biological surveys (Ferrier and Smith 1990) and the application of innovative methods to the storage and analysis of the survey results. The most significant example to date has been the

North-east Forests Biodiversity Survey, a comprehensive survey of major elements of the biodiversity of the area of New South Wales from the Hunter River north to the Queensland border and east of the New England Highway. The results obtained in this study are already being used to develop a comprehensive assessment of conservation needs in the north-east of New South Wales using the techniques for conservation options analysis described above and to respond to environmental impact assessments for forestry operations. A similar broad-based biodiversity survey in the Service's Western Region commenced fieldwork in November 1994. This project is being undertaken in partnership with the Western Region of the New South Wales Department of Conservation and Land Management. Studies are also in the advanced stages of planning for the south-east forest (co-operatively with State Forests of New South Wales), the south-west slopes (co-operatively with many stakeholders), and the central west of the state (again in co-operation with a wide range of stakeholders).

Priority for these surveys has been based first on the level of threat, and second on the adequacy of existing information. Survey data is being used to identify areas of high conservation value which support high densities of critical species or habitats. This is in accordance with the principles set out in the Service's *Draft Natural Heritage Conservation Strategy* (NPWS 1992).

BIOCLIM and DOMAIN

BIOCLIM is a bioclimate analysis and prediction system which can be used to stratify an area on a climatic basis prior to survey and also to predict distributions of species, vegetation types or other biological entities (Busby 1991). The Service has used BIOCLIM in a number of places to guide biodiversity survey and species conservation efforts, most notably in the south-east forests of New South Wales for the Long-footed Potoroo *Potorous longipes* and the Koala *Phascolarctos cinereus*. It has also become an integral part of data validation in the *Atlas of New South Wales Wildlife,* the Service's principal repository of sightings of fauna and flora.

DOMAIN is a more advanced form of analysis that may prove to more effective than BIOCLIM (Carpenter *et al.* 1993).

Much of the potential of these methods is yet to be realized. For conservation managers (and for the governments making land-use decisions on their recommendations) it is essential that reliable and cost-effective methods of survey and analysis are further developed and utilized.

COMMUNITY EDUCATION AND INVOLVEMENT

It is also extremely important for the Service to pursue a programme of public education and broad community involvement in the conservation of biodiversity. Although there is broad support for the work of the NPWS, there is still a considerable gap between acknowledging the need for biodiversity conservation and changing behaviour to reduce impacts on biodiversity. Narrowing that gap is absolutely essential to the success of efforts to conserve biodiversity.

Community education is only useful if it leads to participation in the conservation of biodiversity, either directly or indirectly (Saunders 1990). The chain of events is likely to be awareness leading to understanding; to support; to participation. Community education programmes whose only outcome is agency self-congratulation are a waste of resources.

The Service's Corporate Plan places an increased emphasis on informing and involving the community in all aspects of biodiversity conservation (NPWS 1994).

IMPROVED LEGISLATION

We have noted earlier that a sound legislative base is needed upon which to build programmes for the conservation of the biodiversity of New South Wales. As the lead agency for biodiversity conservation in New South Wales, the Service has a responsibility to advise the government on legislative needs. There is considerable scope to rationalize the provisions of the *NP&W Act* and strengthen them in line with the broad-scale requirements of conserving biodiversity. The same assessment can be made of other New South Wales environmental laws and administrative arrangements.

We consider that the following issues need to be addressed in any review of the *NP&W Act:*

Aboriginal ownership

This important issue remains unresolved in New South Wales at the present time, although the government is currently working to develop a solution in consultation with the Aboriginal community.

The Service has reaffirmed its commitment to playing a leading role in implementing whatever solution the government and the Aboriginal people of New South Wales develop (NPWS 1994). This is particularly important since the Service estate protects many areas of both archaeological and cultural significance.

Whilst acknowledging that progress has been slow in New South Wales, it should also be recognized that the Service has adopted operational policies that seek implement aspects of Aboriginal involvement in reserve management to the fullest extent possible under the current *NP&W Act* (Geering and Roberts 1993; NPWS 1989).

Fortunately, there exist a number of well-studied models in Australia (DeLacy 1994; Chapters in Birckhead *et al.* 1993; Hill and McLeod 1993; Tjamiwa 1991; Toyne and Johnston 1991) and many of the issues that must be dealt with have been the subject of serious consideration throughout Australia (Birckhead *et al.* 1993; Nutting 1994) and recently in New South Wales (Douglas 1994; Johnson 1994; Weatherly and Somerville 1994).

We see a need to balance the legitimate interests and aspirations of Aboriginal people with the requirements of conserving biological diversity in an environment greatly altered in the past 206 years, and thus a need to adjust the *NP&W Act* to provide a legislative basis for realizing that balance.

Invertebrates and plants

The *NP&W Act* gives limited protection to invertebrates and plants which are fully protected only within the Service estate. The *NP&W Act* makes the Director-General of Parks and Wildlife the authority for the protection of native plants in New South Wales. Schedule 13 of the *NP&W Act* lists plant species that are considered in need of protection, but they may be picked or removed from private land with the landowners permission. Trade in protected plants can only occur under licence from the Director-General.

The Service is not empowered by the *NP&W Act* to protect invertebrates outside the Service estate. Modification of the *NP&W Act* to provide adequate protection for invertebrates was recognized by the Service as an important issue in 1980 when appropriate amendments to the *NP&W Act* were proposed. There is thus a long-standing acknowledgement of the problem by the Service with no concrete outcomes.

Effective biodiversity conservation will require a legislative base that deals adequately with all elements of the biota. The biological bias in the *NP&W Act* appears to us to be a major impediment.

Marine areas

The conservation of marine environments is difficult within the current legislative framework in New South Wales. Effective conservation of marine areas requires joint efforts by the Service and New South Wales Fisheries who are responsible for managing species, other than marine mammals and reptiles, within the water column under the provisions of the *Fisheries and Oyster Farms Act* 1948. Only the *NP&W Act* can provide protection of the sea floor.

These problems are not unique to New South Wales (RAC 1992b; Commonwealth of Australia 1992b). Many of the issues relating to the management of the marine environment are the same as those facing agencies working in terrestrial environments (e.g., comprehensiveness, adequacy and representativeness of reserves, ecologically sustainable use). Some, however, are not (e.g., motility of many species, tidal movements) and require responses unique to the marine environment.

The Service is uniquely positioned to bridge the management of protected areas in both terrestrial and marine environments within both the requirements of the *NP&W Act* and the Government's Coastal Policy. The Coastal Policy is currently undergoing a major review (Coastal Committee of New South Wales 1994). A rationalization of the various Acts relating to environmental protection could be beneficial to the conservation of marine biodiversity.

Conservation biology principles

There have been many changes in the principles underpinning conservation action since the *NP&W Act* was given assent in 1974. We have discussed at some length the implications of these principles which have been brought together under the banner of conservation biology.

Many of the principles of conservation biology form the basis for the National Strategy for Ecologically Sustainable Development and the draft National Strategy for the Conservation of Australia's Biological Diversity. These include: the precautionary principle, inter-generational equity, and ecosystem management (i.e., addressing the maintenance of ecological processes and maintenance of ecological integrity).

More recent New South Wales environmental legislation than the *NP&W Act* incorporates some of these principles. For example, the *Protection of the Environment Administration Act* 1991, which sets out the objectives of the New South Wales Environment Protection Authority in section 6 to ensure that ecologically sustainable development is achieved through the implementation of the following key principles and programmes:

(a) the precautionary principle;

(b) intergenerational equity;

(c) conservation of biological diversity and ecological integrity; and,

(d) improved valuation and pricing of environmental resources.

We consider it vital that the *NP&W Act* be revised to bring it into line with current concepts and best practice. Outdated legislation can become an impediment to effective action by agencies such as the Service.

CONCLUSION

In its 27 year history, the Service has both seen and endured many changes in its operating environment. Of profound importance has been the change in the guiding principles for conservation planning and management that has taken place over the last decade. The pace of change appears to be quickening as this new paradigm leads to an ever-increasing body of knowledge.

We believe biodiversity conservation is too important and too difficult to be left to just governments or bureaucracies. It is critical that both are involved. However, it is the support and participation of the community, working to an agreed strategy, that will eventually lead to the next major steps forward in the conservation of biodiversity.

Thus, as it seeks to meet the challenges associated with conserving the remaining biodiversity of New South Wales, the Service will need to quickly adopt new approaches to planning and management. This demands an unprecedented degree of responsiveness for a government bureaucracy. Survival as an agency cannot be regarded as an indicator of success.

ACKNOWLEDGMENTS

The views expressed in this paper do not necessarily represent those of the State Government of New South Wales.

The principal author of the Service's *Draft Natural Heritage Conservation Strategy — Biodiversity Component* was Grant Bywater and major contributions were made by Tom Fox, Dr Ross Bradstock and Jim Starling. A significant part of the strategy was based on regional reports written by John Brickhill, Sue Walker, Jonathan Sanders and Roger Good. The draft strategy formed the basis of this paper and we wish to acknowledge its contribution while accepting responsibility for its interpretation.

Finally, we thank our referees. The paper is much improved for their comments.

REFERENCES

Australian National Parks and Wildlife Service (ANPWS), 1991. Plant invasions: the incidence of environmental weeds. Part One. Plant invasions of Australian ecosystems. Papers selected from the Ninth Australian Weed Conference. Australian National Parks and Wildlife Service: Canberra.

Bass, D. A., 1990a. Dispersal of an introduced shrub *(Crataegus monoguna)* by the brush-tailed possum. *Aust. J. Ecol.* **15:** 227–29.

Bass, D. A., 1990b. Pied currawongs and seed dispersal. *Corella* **14:** 24–27.

Beattie, A. J. (ed.), 1993. *Rapid biodiversity assessment. Proceedings of the biodiversity assessment workshop Macquarie University — 1993.* Research Unit for Biodiversity and Bioresources, Macquarie University: North Ryde.

Beattie, A. J. and Oliver, I., 1994. Taxonomic minimalism. *Trends Ecol. Evol.* **9:** 488–90.

Bedward, M. B., Pressey, R. L. and Keith, D. A., 1992. A new approach for selecting fully representative reserve networks: addressing efficiency, reserve design and land suitability with an iterative analysis. *Biol. Conserv.* **62:** 115–25.

Benson, J., 1992. The effect of 200 years of European settlement on the vegetation and flora of New South Wales. *Cunninghamia* **2:** 343–70.

Benson, J., 1993a. The biology and management of Ooline *(Cadellia pentastylis)* in New South Wales. Species Management Report No. 2. New South Wales National Parks and Wildlife Service: Hurstville.

Benson, J., 1993b. Past, present and future: the role of scientific information in nature conservation. *Nat. Parks J.* **37:** 16–23.

Biological Diversity Advisory Committee (BDAC), 1992. *A national strategy for the conservation of Australia's biological diversity.* Department of the Arts, Sport, the Environment and Territories: Canberra.

Birckhead, J., DeLacy, T. and Smith, L. (eds), 1993. *Aboriginal involvement in parks and protected areas.* Aboriginal Studies Press: Canberra.

Bradstock, R. A., 1993. Management of fire for native plant diversity. In *The burning question: fire management in New South Wales* ed by J. Ross. University of New England: Armidale.

Busby, J. R., 1991. BIOCLIM — A Bioclimate Analysis and Prediction System. In *Nature conservation: cost effective biological surveys and data analysis* ed by C. R. Margules and M. P. Austin. CSIRO: Australia.

Campbell, A., 1994. *Landcare: communities shaping the land and the future.* Allen and Unwin: St Leonards.

Carpenter, G., Gillison, A. N. and Winter, J., 1993. DOMAIN: a flexible modelling procedure for mapping potential distributions of plants and animals. *Biodiv. Conserv.* **2:** 667–80.

Caughley, G., 1994. Directions in conservation biology. *J. Anim. Ecol.* **63:** 215–44.

Clark, T. W., 1992. Practising natural resource management with a policy orientation. *Environ. Manage.* **16:** 423–33.

Clark, T. W. and Seebeck, J. H. (eds), 1990. *Management and conservation of small populations.* Chicago Zoological Society: Brookfield, Illinois.

Coastal Committee of New South Wales, 1994. *Draft revised coastal policy for New South Wales.* Coastal Committee of New South Wales: Sydney.

Commonwealth of Australia, 1992a. *National strategy for ecologically sustainable development.* Australia Government Publishing Service: Canberra.

Commonwealth of Australia, 1992b. *A draft policy for commonwealth responsibility in the coastal zone.* Australia Government Publishing Service: Canberra.

Connor, R., Houlbrooke, R. and Tarihao, F., 1993. Local conservation area ownership and traditional management. In *Fifth South Pacific conference on nature conservation and protected areas, Nuku'alofa, Tonga, 4–8 October 1993, Volume 2: Conference Papers.* SPREP: Apia, Western Samoa.

Courrier, K. (ed.), 1992. *Global biodiversity strategy.* World Resources Institute, The World Conservation Union and the United Nations Environment Program: Washington, D.C.

Craig, J. L. and Stewart, A. M., 1994. Conservation: a starfish without a central disk? *Pac. Conserv. Biol.* **1:** 163–68.

Cranston, P. and Hillman, T., 1992. Rapid assessment of biodiversity using "Biological Diversity Technicians". *Aust. Biol.* **5:** 144–54.

De Lacy, T., 1994. The Uluru/Kakadu model — Anangu Tjukurrpa. 50 000 years of Aboriginal law and land management changing the concept of national parks in Australia. *Soc. Nat. Res.* **7:** 479–98.

Department of Environment, Sport and Territories (DEST), 1993. *Biodiversity and its value.* Biodiversity Unit, Department of Environment, Sport and Territories: Canberra.

Dickman, C. R., 1993. The biology and management of native rodents of the arid zone in New South Wales. Species Management Report No. 12. New South Wales National Parks and Wildlife Service, Hurstville.

Dickman, C. R., Pressey, R. L., Lim, L. and Parnaby, H. E., 1993. Mammals of particular conservation concern in the Western Division of New South Wales. *Biol. Conserv.* **65:** 219–48.

Dickman, C. R. and Read, D. G., 1992. The biology and management of dasyurids of the arid zone in New South Wales. Species Management Report No. 11. New South Wales National Parks and Wildlife Service, Hurstville.

Douglas, G., 1994. Towards a policy Aboriginal ownership of natural areas. *Nat. Parks J.* **38:** 12–13.

Ehrlich, P. R. and Wilson, E. O., 1991. Biodiversity studies: science and policy. *Science* **253:** 758–62.

Ellis, M., 1992. Revision of the Atlas of New South Wildlife. *Aust. Zool.* **28:** 59–64.

Endangered Species Advisory Committee (ESAC), 1992. *An Australian national strategy for the conservation of Australian species threatened with extinction.* Australian National Parks and Wildlife Service: Canberra.

Ferraro, T. J. and Burgin, S., 1993a. Amphibian decline: a case study in western Sydney. In *Herpetology in Australia: a diverse discipline* ed by D. Lunney and D. Ayers. Royal Zoological Society of New South Wales: Mosman.

Ferraro, T. J. and Burgin, S., 1993b. Review of environmental factors influencing the decline of Australian frogs. In *Herpetology in Australia: a diverse discipline* ed by D. Lunney and D. Ayers. Royal Zoological Society of New South Wales: Mosman.

Ferrier, S., 1991. Computer-based spatial extension of forest fauna survey data: current issues, problems and directions. In *Conservation of Australia's forest fauna* ed by D. Lunney. Royal Zoological Society of New South Wales: Mosman.

Ferrier, S. and Smith, A. P., 1990. Using geographical information systems for biological survey design, analysis and extrapolation. *Aust. Biol.* **3:** 105–16.

Frank Small and Associates, 1993. Stakeholder attitudes towards New South Wales National Parks and Wildlife Service. A market research report. Unpublished report to New South Wales National Parks and Wildlife Service.

Garnett, S., 1992. *Action plan for Australian birds.* Australian National Parks and Wildlife Service: Canberra.

Geering, K. and Roberts, C., 1993. Current limitations on Aboriginal involvement in Aboriginal site management in centralwest and northwest New South Wales. In *Aboriginal involvement in parks and protected areas* ed by J. Birckhead, T. De Lacy and L. Smith. Aboriginal Studies Press: Canberra.

Gilpin, M. E., 1987. Spatial structure and population vulnerability. In *Viable populations for conservation* ed by M. E. Soulé. Cambridge University Press: Cambridge.

Greenslade, P. and New, T. R., 1991. Australia: Conservation of a continental insect fauna. In *The conservation of insects and their habitats* ed by N. M. Collins and J. A. Thomas. Academic Press: London.

Grice, A. C. and Barchia, I., 1992. Does grazing reduce survival of indigenous perennial grasses of the semi-arid woodlands of western New South Wales? *Aust. J. Ecol.* **17:** 195–205.

Grumbine, R. E., 1991. Co-operation or conflict? Interagency relationships and the future of biodiversity for US parks and forests. *Environ. Manage.* **15:** 27–37.

Grumbine, R. E., 1992. *Ghost bears: exploring the biodiversity crisis.* Island Press: Washington, D.C.

Harrison, S., 1991. Local extinction in a metapopulation context: an empirical evaluation. *Biol. J. Linn. Soc.* **42:** 73–88.

Hill, L. and Michaelis, F. B., 1988. *Conservation of insects and related fauna.* Occasional Paper No. 13. Australian National Parks and Wildlife Service: Canberra.

Hill, M. and McLeod, K., 1993. Aboriginal and Torres Strait Islander involvement in natural resource management in Australia. In *Fifth South Pacific conference on nature conservation and protected areas, Nuku'alofa, Tonga, 4–8 October 1993, Volume 2: conference papers.* SPREP: Apia, Western Samoa.

Humphries, S. E., Groves, R. H. and Mitchell, D. S., 1991. Plant invasions of Australian ecosystems — a status review and management actions. Report on Endangered Species Program Project No. 58, Australian National Parks and Wildlife Service, Canberra.

Janzen, D. H., 1991. How to save tropical biodiversity. *Amer. Entomol.* **37:** 159–71.

Johnson, C., 1994. Aboriginal management of national parks. *Nat. Parks J.* **38:** 14.

Jones, W., 1994. The biology and management of the Dwarf Mountain Pine *(Microstrobus fitzgeraldii)* in New South Wales. Species Management Report No. 13. New South Wales National Parks and Wildlife Service: Hurstville.

Kennedy, M. (ed.), 1990. *A complete reference to Australia's endangered species.* Simon and Shuster Australia: Brookvale.

Landres, P. B., 1992. Temporal scale perspectives in managing biodiversity. In *Crossroads of conservation: 500 years after Columbus* ed by R. E. McCabe. Transactions of the Fifty-seventh North American Wildlife and Natural Resources Conference, Wildlife Management Institute: Washington, D.C.

Leigh, J. H. and Briggs, J. D. (eds), 1992. *Threatened Australia plants: overview and case studies.* Australian National Parks and Wildlife Service: Canberra.

Lindenmayer, D. B., Clark, T. W., Lacy, R. C. and Thomas, V. C., 1993a. Population viability analysis as a tool in wildlife conservation policy with reference to Australia. *Environ. Manage.* **17:** 745–58.

Lindenmayer, D. B., Lacey, R. C. Thomas, V. C. and Clark, T. W., 1993b. Predictions of the impacts of changes in population size and environmental variability on Leadbeater's Possum, *Gymnobelideus leadbeateri* McCoy (Marsupialia: Petauridae) using Population Viability Analysis: an application of the computer programme VORTEX. *Wildl. Res.* **20:** 67–86.

Ling, J. K., 1988. Overview of recent cetacean research in Australia. In *Marine mammals of Australasia: field biology and captive management* ed by M. L. Augee. Royal Zoological Society of New South Wales: Mosman.

Love, A. and Dyason, R. (eds), 1985. *Bitou bush and boneseed: a national conference on* Crysanthemoides monilifera. Port Macquarie, New South Wales, 8–9th August, 1984. Work proceedings. New South Wales National Parks and Wildlife Service and New South Wales Department of Agriculture: Sydney.

Loyn, R. H. and French, K., 1991. Birds and environmental weeds in south-eastern Australia. *Plant Protect. Quart.* **6:** 137–41.

Lunney, D., Curtin, A., Ayers, D., Cogger, H. G. and Dickman, C. R. (in press). Identifying the endangered fauna of New South Wales: an ecological approach to a systematic evaluation of the status of all species. Submitted to *Pac. Conserv. Biol.*

Machlis, G. E., 1992. The contribution of sociology to biodiversity research and management. *Biol. Conserv.* **62:** 161–70.

Mansergh, I. and Broome, L., 1994. *The mountain pygmy-possum of the Australian alps.* University of New South Wales Press: Kensington.

McNeely, J. A., Miller, K. R., Reid, W. V., Mittermeir, R. A. and Werner, T. B., 1992. *Conserving the world's biological diversity.* National Academy Press: Washington, D.C.

Miller, B., Reading, R., Conway, C., Jackson, J. A., Hutchins, M., Snyder, N., Forrest, S. and Frazier, J., 1994. A model for improving endangered species recovery programmes. *Environ. Manage.* **18:** 637–45.

Millsap, B. A., Gore, J. A., Runde, D. E. and Cerulean, S. I., 1990. Setting priorities for the conservation of fish and wildlife species in Florida. *Wildl. Monogr.* **111:** 1–57.

Mitchell, P. B., 1991. Historical perspectives on some vegetation and soil changes in semi-arid New South Wales. *Vegetatio* **91:** 169–82.

Mott, J. J. and Bridgewater, P. B., 1992. Biodiversity conservation and ecologically sustainable development. *Search* **23:** 284–87.

Mott, J. J. and Edwards, G. P., 1992. Beyond Beef — The Search for a Level Grazing Paddock. *Search* **23:** 223–25.

Nadolny, C. and Benson, J., 1993. The biology and management of the Pigmy Cypress Pine *(Callitris oblonga)* in New South Wales. Species Management Report No. 7. New South Wales National Parks and Wildlife Service: Hurstville.

National Parks and Wildlife Service (NPWS), 1989. *Field management manual.* New South Wales National Parks and Wildlife Service: Hurstville.

National Parks and Wildlife Service (NPWS), 1992. *Draft natural heritage conservation strategy — Biodiversity component.* New South Wales National Parks and Wildlife Service: Hurstville.

National Parks and Wildlife Service (NPWS), 1994. *Corporate plan 1994/95 — 1995/96.* New South Wales National Parks and Wildlife Service: Hurstville.

New, T. R., 1984. *Insect conservation — an Australian view.* Dr W. Junk Publishers: Dordrecht, Netherlands.

New, T. R., 1991. *Butterfly conservation.* Oxford University Press: Melbourne.

Noon, B. R. and Young, K., 1991. Evidence of continuing worldwide declines in bird populations: Insights from an international conference in New Zealand. *Conserv. Biol.* **5:** 141–43.

New South Wales Environment Protection Authority (NSWEPA), 1994. *Who cares about the environment? A benchmark survey of the environmental knowledge, skills, attitudes and behaviour of the people of New South Wales.* New South Wales Environment Protection Authority: Chatswood.

Nutting, M., 1994. Competing interests or common ground? *Habitat* **22:** 28–37.

Oliver, I. and Beattie, A. J., 1993. A possible method for the rapid assessment of biodiversity. *Conserv. Biol.* **7:** 562–68.

Papps, D. J., 1990. The role of voluntary conservation agreements in habitat protection. In Koala summit. Managing koalas in New South Wales ed by D. Lunney, C. A. Urquhart and P. Reed. New South Wales National Parks and Wildlife Service: Hurstville.

Papps, D. J., 1992. Biodiversity conservation and production forestry — the next decade. In *Sustainable forest management* ed by M. Rowland. Board of Environmental studies. Occasional Paper No 18. University of Newcastle: Newcastle.

Parker, S. A. and Reid, N., 1978. Remarks on the status of some Australian passerines. In *The status of endangered Australian wildlife* ed by M. J. Tyler. Proceedings of the Centenary Symposium of the Royal Zoological Society of South Australia, Royal Zoological Society of South Australia: Adelaide.

Pickard, J., 1991. Land management in semi-arid environments of New South Wales. *Vegetatio* **91:** 191–208.

Ponder, W. F., 1992. Bias and biodiversity. *Aust. Zool.* **28:** 47–51.

Possingham, H., 1991. The role of population viability analysis in forest management. In *Conservation of Australia's forest fauna* ed by D. Lunney. Royal Zoological Society of New South Wales: Mosman.

Pressey, R. L., 1990. Reserve selection in New South Wales: where to from here? *Aust. Zool.* **26:** 70–75.

Pressey, R. L. and Ferrier, S., 1994. Adapting a GIS for interactive conservation planning. *Tropical Forest Update* **4:** 3–5.

Pressey, R. L., Humphries, C. J., Margules, C. R., Vane-Wright, R. I. and Williams, P. H., 1993. Beyond opportunism: key principles for systematic reserve selection. *Trends Ecol. Evol.* **8:** 124–28.

Pressey, R. L., Johnson, I. R. and Wilson, P. D., 1994. Shades of irreplaceability: towards a measure of the contribution of sites to a reservation goal. *Biodiver. Conserv.* **3:** 242–62.

Priddel, D., 1990. Conservation of the Malleefowl in New South Wales: an experimental management strategy. In *The mallee lands: a conservation perspective* ed by J. C. Noble, P. J. Joss and G. K. Jones. CSIRO: Melbourne.

Rabb, G. B., 1990. Declining amphibian populations. *Species* **13–14:** 33–34.

Recher, H. F., 1990 Wildlife conservation in Australia: state of the nation. *Aust. Zool.* **26:** 5–11.

Recher, H. F. and Lim, L., 1990. A review of current ideas of the extinction, conservation and management of Australia's terrestrial vertebrate fauna. *Proc. Ecol. Soc. Aust.* **16:** 287–301.

Reed, P., 1991. An historical analysis of the changes to the forests and woodlands of New South Wales. In *Conservation of Australia's forest fauna* ed by D. Lunney. Royal Zoological Society of New South Wales: Mosman.

Resource Assessment Commission (RAC), 1992a. *Forest and timber inquiry. Final report. Volume 1.* Australian Government Publishing Service: Canberra.

Resource Assessment Commission (RAC), 1992b. *Coastal zone enquiry. Government approaches to coastal zone resource management. Information paper No. 1.* Australian Government Publishing Service: Canberra.

Richards, S. R., McDonald, K. R. and Alford, R. A., 1993. Declines in populations of Australia's tropical rainforest frogs. *Pac. Conserv. Biol.* **1:** 66–77.

Saunders, D. A., 1990. The landscape approach to conservation: community involvement, the only practical solution. *Aust. Zool.* **26:** 49–53.

Shaffer, M., 1987. Minimum viable populations: coping with uncertainty. In *Viable populations for conservation* ed by M. E. Soulé. Cambridge University Press: Cambridge, United Kingdom.

Sivertsen, D., 1994. The native vegetation crisis in the wheat belt of New South Wales. *Search* **25:** 5–8.

Smith, A. P., 1991. Forest policy: fostering environmental conflict in the Australian timber industry. In *Conservation of Australia's forest fauna* ed by D. Lunney. Royal Zoological Society of New South Wales: Mosman.

Smith, P. J., Pressey, R. L. and Smith, J. E., 1994. Birds of particular conservation concern in the Western Division of New South Wales. *Biol. Conserv.* **69:** 315–38.

Soil Conservation Service (SCS), 1989. *Land degradation survey: New South Wales 1987–1988.* New South Wales Soil Conservation Service: Chatswood.

Soulé, M. E. (ed), 1986. *Conservation biology: the science of scarcity and diversity.* Sinauer: Sunderland, Massachussets.

Soulé, M. E. (ed), 1987. Introduction. In *Viable populations for conservation.* Cambridge University Press: Cambridge, United Kingdom.

Soulé, M. E. and Wilcox, B. A. (eds), 1980. *Conservation biology: an evolutionary-ecological perspective.* Sinauer: Sunderland, Massachussets.

Tjmamiwa, T., 1991. Nganana wirunya tjunguringkula waakarinyi: We're working well together. *Habitat* **19:** 5–7.

Toyne, P. and Johnston, R., 1991. Reconciliation, or the new dispossession? Aboriginal land rights and nature conservation. *Habitat* **19:** 8–10.

Wager, R. and Jackson, P., 1993. *The action plan for Australian freshwater fish.* Australian Nature Conservation Agency: Canberra.

Walker, B. H., 1992. Biodiversity and ecological redundancy. *Conserv. Biol.* **6:** 18–23.

Walker, B. and Nix, H., 1993. Managing Australia's biological diversity. *Search* **24:** 173–78.

Weatherley, M. and Somerville, J., 1994. Land rights and park proposals — conservation or conflict? *Nat. Parks J.* **38:** 10–12.

Webster, R., 1991. The biology and management of the Regent Parrot *(Polytelis anthopeplus anthopeplus)* in New South Wales. Species Management Report No. 4. New South Wales National Parks and Wildlife Service: Hurstville.

Webster, R. and Ahern, L., 1992. Management for conservation of the Superb Parrot *Polytelis swainsonii* in New South Wales and Victoria. New South Wales National Parks and Wildlife Service and Victorian Department of Conservation and Natural Resources: Melbourne.

Wells, K. F., Wood, N. H. and Laut, P., 1984. Loss of forests and woodlands in Australia: a summary by State, based on rural local government areas. Technical Memorandum 84/4, CSIRO Division of Water and Land Resources, CSIRO: Canberra.

Western, D., 1991. Biology and conservation: making the relevant connection. *Conserv. Biol.* **5:** 431–33.

Western, D., 1992 The biodiversity crisis: a challenge for biology. *Oikos* **63:** 29–38.

Western, D. and Pearl, M. (eds), 1989. *Conservation for the twenty-first century.* Oxford University Press: New York.

Willan, L., 1977. Arthropods. *Parks and Wildlife* **2:** 59–61.

Williams, G., 1990. Invertebrate conservation. In *A complete reference to Australia's endangered species* ed by M. Kennedy. Simon and Shuster Australia: Brookvale.

Wilson, P., 1993. Biodiversity conservation and assessment: A government agency view. In *Rapid biodiversity assessment* ed by A. J. Beattie. Proceedings of the Biodiversity Assessment Workshop Macquarie University — 1993. Research Unit for Biodiversity and Bioresources, Macquarie University: North Ryde.

Wilson, P., 1995. Sustainable development and biodiversity conservation in the South Pacific region: lessons for New South Wales NPWS? Unpublished report, New South Wales National Parks and Wildlife Service: Hurstville.

Wyman, R. L., 1990. What's happening to the amphibians? *Conserv. Biol.* **4:** 350–52.

Yen, A. L. and Butcher, R. J., 1994. An overview of the conservation status of non-marine invertebrates in Australia. Report to the Endangered Species Unit, Australian Nature Conservation Agency: Canberra.

Conserving biodiversity — threats and solutions: overview

R. PURDIE[1]

This paper provides an overview of the conference based predominantly on the spoken presentations, and was presented to conclude the programme. It provides a broad synthesis, for those landscape types discussed during the conference, of the value and location of Australia's biodiversity and the major threats and impediments to its conservation. Major themes are summarized, including some suggested conservation priorities and strategies. A consistent view emerged from the conference that future conservation of Australia's biodiversity will entail multiple use and more active management of protected areas, and developing and implementing specific conservation objectives and management practices for off-reserve land. Individuals will also need to take greater responsibility for promoting knowledge about biodiversity and influencing relevant policies and programmes.

INTRODUCTION

OVER the course of this conference speakers detailed a wide range of threats to and impacts on Australia's biodiversity and possible solutions to its conservation. In a half hour overview it is neither possible nor desirable to repeat this detail. I have seen my task as trying to draw together major themes that emerged over the four days and identify common directions in which we are or should be heading. Of necessity, I will need to generalize. In doing this, it will not be possible to mention every speaker. However, referring only to some is in no way meant to belittle the contribution of others.

BIODIVERSITY — THE TIME CONTEXT

While our attention has been focused on Australia's existing biodiversity, this needs to be put into the wider context (Fig. 1). Many scientists today tend to use as their biodiversity benchmark the suite of species (N) and ecosystems present in Australia c. 200 years ago, as this coincides with the arrival and permanent settlement of non-indigenous Australians, and marks the start of a disturbance regime the impact of which was both high in intensity and widespread in extent, and which presented a major threat to biodiversity.

Based on our knowledge that known introductions have exceeded known extinctions, the recorded biodiversity of today is generally assumed to be slightly higher (N+) than that present 200 years ago, although this assumption may merely reflect our ignorance of total biodiversity then and now. Some 1 000 species of higher plants and 22 species of vertebrates are known to have been lost since 1788 (e.g., Fox). Based on the re-appearance of the Leadbeater's Possum (Gill), there is a possibility that some species currently presumed extinct may still be surviving. There may also be other species that have become extinct but which we are not aware of. Overall however, Australia's extinction rate is a poor record by international standards. Known losses of species have been outweighed by introductions, both deliberate and unintentional (Fox, Panetta). Over 130 introduced vertebrate species and at least 1 400 introduced species of higher plants (Fox) have become naturalized in Australia over the last 200 years.

Perhaps only the foolish would predict the number of species which may be present in 200 years time — certainly no speaker attempted this — but it is likely to be different from today. During the conference, speakers have detailed innumerable threats which currently exist to our biodiversity and may result in at least widespread local extinctions in the future. It is probable that many new species will be introduced and become naturalized, although this may be slowed through tighter regulations and better agency co-ordination (Panetta). There is also a pool of currently "benign" introduced species, such as garden plants, which could become a threat to the native biota should changed environmental conditions release the constraints which today prevent their spread and naturalization (Fox).

Biodiversity as a popular concept is only c. 10 years old, but biodiversity itself obviously is not new. The number of species present in Australia some 200 years ago was the end product of a long period of evolution. As De Decker reminded us, over the last 10 000–18 000 years, species have been subject to a period of considerable environmental variability — substantial fluctuations in sea level and total area of available habitat, and shifting habitats. By these standards, the environmental conditions of today are abnormal, having existed for only c. 15 per cent of this previous period.

[1]Australian Heritage Commission, GPO Box 1567, Canberra, Australian Capital Territory, Australia 2601.

Pages 410–16 *in* CONSERVING BIODIVERSITY: THREATS AND SOLUTIONS, ed by R. A. Bradstock, T. D. Auld, D. A. Keith, R. T. Kingsford, D. Lunney and D. P. Sivertsen. Surrey Beatty & Sons. 1995.

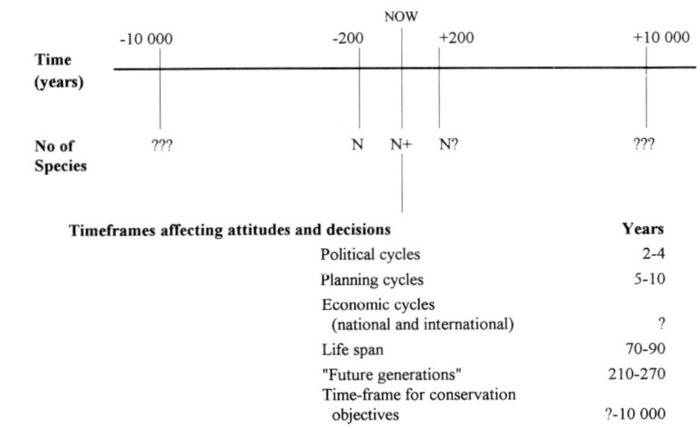

Fig. 1. Long-term perspective of biodiversity, and time-frames relevant to its conservation.

There is no reason to believe that climatic fluctuations over the next 10 000 years will not be comparable to those over the previous 10 000. Hence species diversity will inevitably change in the future, *irrespective* of any changes in diversity resulting from human activities. De Decker was confident that provided their genetic diversity is protected, species will cope with natural climatic changes. However, if "habitat movement" and total available habitat have also been important mechanisms for species' survival, the effectiveness of these mechanisms in the future may be severely reduced by human impacts on the landscape, unless compensatory mechanisms such as the use of habitat corridors are developed (Recher).

The effect of time scales on scientific estimates of biodiversity and the development of conservation objectives was recognized by many speakers. However, the relevance of time scales to the choice of conservation solutions by policy-makers and managers was less well addressed. The time frames which affect human attitudes and decision-making processes (Fig. 1) need to be considered, however, as they may constrain long-term solutions to protecting biodiversity. The short time frames of political and planning cycles for example often make it difficult to adopt and implement policies with conservation objectives spanning decades, let alone "future generations".

Despite the long-term uncertainties in the level of biodiversity described above, many speakers during the conference considered that we should try to conserve as many as possible of today's species knowing, however, that all species are not equally important and undoubtedly cannot all be saved (e.g., Beattie; Kitching).

AN OVERVIEW OF ISSUES AT THE LANDSCAPE LEVEL

If we want to save as much biodiversity as possible, we assume a positive answer to the question: *why save it?* but we need to ask: *where is it?*, *what are the biggest threats?*, and *what are the impediments to saving it?* All except the second of these questions were addressed during the conference.

Consideration of where biodiversity is located, together with an assessment of threats and impediments to its conservation, may assist in determining conservation priorities and strategies. In Table 1, I have attempted to summarize, at a broad scale, some major points relevant to all the above questions for those landscape types which have been the focus of the conference, viz. arid and semi-arid/pastoral, agricultural/wheat and sheep, forest, urban and aquatic. These points are discussed in the following sections.

Where is it? — Relative biodiversity of landscapes

The total area of the above landscape types, or major ecosystems, range from <2 per cent of Australia for the urban environment and major infrastructure such as the connecting road systems (Adam, Woodside), to c. 75 per cent for the arid and semi-arid pastoral landscapes (Friedel). Estimates of the extent of aquatic systems would vary with how they are defined. Although the total area of permanent, natural aquatic systems is small, the total area of natural wetlands and aquatic systems is much higher (although variable due to the ephemeral nature of many), and the total catchment area is immense.

Determining whether there are significant differences in the total biodiversity associated with these landscape types is extremely difficult, since for many groups of taxa we have very little or insufficient information. For species whose taxonomy and broad distribution is reasonably well known, such as vascular plants and birds, some estimate is possible (see Table 1). Although there would be species common to adjacent landscape types these figures suggest that, based on the relative proportion of species present, conservation effort focused in the forest and agricultural landscape types will have a greater overall contribution to conservation of vascular plants and land birds than effort in the urban and arid areas. However, other criteria are also highly relevant to setting conservation priorities. These include the number of restricted endemic species or rare/threatened species or communities per landscape type, the different habitat requirements of species which occur over a range of landscape types (e.g., to protect genetic variation), and the condition of ecosystems and hence conservation potential in the various landscapes types.

Why save it? — Value of biodiversity

Beattie described three main values of biodiversity, viz. beauty in its broadest sense (including the appreciation that comes with

Table 1. Summary of major landscape types and attributes relevant to their biodiversity.

	Landscape type				
	Arid and semi-arid (pastoral)	Agricultural (wheat/sheep)	Forest	Urban (including infrastructure)	Aquatic
Relative area of the continent	75% (Friedel*)	15% (Saunders*)	4–5% (AUSLIG[1])	<2% (Adam*, Woodside*)	?
Relative biodiversity No. of species (% of total in Australia)					
— vascular plants[2]	c. 4 500 (c. 22%) (arid only)	c. 11 200 (c. 55%)	c. 13 500 (c. 66%)	c. 8 100 (c. 40%) (major urban only)	?
— land birds[3]	c. 330 (c. 67%) (arid only)	c. 370 (c. 75%)	c. 460 (c. 93%)	?	?
Value of biodiversity					
— Beauty	low–high	low–high	very high	low–high	low–high
— Functional (Carbon fixation)	?medium	?high	?high	?low	?
— Profit (conventional bioresources)	— food (sheep, cattle) — fibre (sheep) — infrastructure (native plants)	— food (crops, sheep, cattle) — fibre (sheep) — medicine — other	— housing — construction — infrastructure — other	— sense of place — food	— food — (water)
— Spiritual enrichment	+	+	+	+	+
— Intrinsic right	+	+	+	+	+
Main causes of biodiversity loss	1. Invasion 2. Conversion	1. Conversion 2. Fragmentation 3. Invasion	1. Conversion 2. Fragmentation 3. Invasion	1. Conversion 2. Fragmentation 3. Invasion	1. Conversion 2. Fragmentation 3. Invasion
Impediments to biodiversity conservation	— invaders are here to stay — competing uses of land	— invaders are here to stay — > conversion — island effects of fragmentation — political clout — land ethic — competing uses of land	— invaders are here to stay — > conversion of private land — island effects of fragmentation — competing uses of land	— invaders are here to stay — > conversion — island effects of fragmentation — attitudes — NIMBY — competing uses of land	— invaders are here to stay — > conversion — island effects of fragmentation — jurisdictional complexity — whole catchment management — competing uses of land

*Authors in this volume

[1]AUSLIG (1990). *Atlas of Australian Resources. Third series. Volume 6. Vegetation.* Australian Surveying and Land Information Group, Department Administrative Services, Canberra.

[2]Estimates for taxa (including species, subspecies, varieties) from *Census of Australian Vascular Plants* (CAVP), IBIS Database, 23 July, 1993, Australian National Botanic Gardens, Canberra, based on CAVP regions approximating to landscape types; excludes introduced taxa; totals likely to be over-estimates. Total taxa in Australia = 20 245.

[3]Estimates from *RAOU Bird Atlas* Database, Royal Australasian Ornithologists Union, Melbourne, analysed 23 July, 1993 by Wildlife Management Unit, Australian Nature Conservation Agency, based on 1° grid square approximations of landscape types; seabirds, waders and waterbirds excluded from estimates. Total species in Australia = 495.

knowledge), functional value, and profit. During the conference, two other values were mentioned: spiritual enrichment (possibly a sub-set of beauty) and the intrinsic right of species to exist. In Table 1 I have considered the first three values in relation to each landscape type, as an understanding of broad differences in biodiversity value between landscape types may be important in public perceptions of biodiversity and in promoting appreciation and acceptance of its values.

Taking *beauty* in a narrow sense (e.g., the pleasure derived from the appearance of a species or landscape), forests are universally described as being beautiful. No other of the landscape types are probably held in such uniformly high regard in Australia, although aspects of each could variously be ranked as having from low to high beauty.

Using carbon-fixation through photosynthesis as an indicator of one type of *functional value* (Beattie), it could be expected that different landscape types would vary in their intrinsic value, with forests and agricultural areas probably having the highest levels of total C-fixation.

Conventional bioresources may be used as a measure of the *profit* value of biodiversity (Beattie). Using this measure, it is evident that high profit value is associated with all landscape types, with arid and semi-arid/pastoral, agriculture/wheat and sheep, and forests providing the majority of food, fibre, housing, construction and other needs. In the case of food and fibre, it is notable that this profit is largely derived from a small number of introduced species. Although 85 per cent of Australia's population is associated with urban areas (Adam), the conventional bioresources on which they rely to sustain their standard of living are located predominantly in other landscape types. In urban areas, the main profit value of biodiversity is probably associated with a "sense of place": consider how often natural vegetation is featured as a reason to purchase land!

Throughout the conference there was a strong perception that the majority of the Australian community has either little appreciation or understanding of the value of biodiversity, and/or no inclination to protect it (e.g., Burbidge, Noble). Changing these perceptions was seen as a critical element in any strategy to protect biodiversity.

The broad analysis described above supports two strategies suggested during the conference as means of effecting change in community attitudes. These were to focus on those types of biodiversity familiar to the majority of the population (e.g., urban bushland; forests) to promote a broader understanding of biodiversity concepts (Adam), and to link the satisfaction of the community's material needs to the wider landscape types to promote a sense of shared need and responsibility for protecting the biodiversity in these areas (Saunders).

What are the threats? — major causes of biodiversity loss

The notation of Graetz — invasion, fragmentation and conversion — is useful for describing the major causes of biodiversity loss at a landscape (ecosystem/community) level. Although applied by Graetz to terrestrial systems, these terms are equally applicable to aquatic systems. For example, those aquatic systems affected by pollution or salinity, or whose water flow characteristics have been permanently changed by the erection of barriers such as dams or weirs (Cullen, Kingsford), have been "converted" in the sense of Graetz. Speakers have also indicated that aquatic systems have been subject to invasion by introduced species and to fragmentation through loss of riparian vegetation or straightening of natural channels (e.g., Cullen).

All three factors are significant causes of biodiversity loss in agricultural, forest, urban and aquatic systems, with conversion probably being the major cause. Fragmentation does not appear to have been as great a problem in the arid zone, although it is probably an important factor in some semi-arid pastoral areas.

A wide range of threats specific to each landscape type were discussed by speakers. One emerging issue was the commercial use of native species, which is likely to continue and may increase (Callister). Based on our knowledge to date, loss of habitat poses a greater threat to faunal species than commercial use, although the latter may cause significant reductions in abundance and changed distribution (Kearney, Cairns). Although commercialization is seen by some as a means of protecting species, this issue is likely to remain controversial (Callister).

What are the impediments to biodiversity conservation?

A wide range of factors which may impede biodiversity conservation were outlined during the conference. Major factors operating at a landscape level are summarized in Table 1. Recognition of broad-scale impediments is important, as they are realities that must be taken into account when planning for biodiversity conservation at finer scales (regional or local) to avoid attempting the impossible, and to help determine the probability of conservation success.

In arid and semi-arid areas, the continued presence of invading species will remain a major impediment to biodiversity conservation. The invaders are here to stay, either because they are a conventional bioresource (sheep, cattle), or they are naturalized pest species (e.g., rabbits, goats) that will be impossible to eradicate (Bomford). Pest species will similarly be a major impediment in aquatic systems. In agricultural, forest and urban areas, invading species will always be present as either conventional bioresources or as pest species. However, in these landscape types, it may be possible to selectively eradicate pest species or reduce them to a low density in targeted areas of high biodiversity value, using intensive management/control techniques (e.g., baiting foxes — Burbidge; bush regeneration — Buchanan, Fox).

In the agricultural, forest, urban and aquatic landscape types, conversion of ecosystems is a major impediment to biodiversity conservation, through increased clearing for agricultural crops (especially cotton in some areas), clearing of forests on private land, clearing for urban expansion and the draining, infilling or pollution of aquatic habitats. Mining as a competing land use which may result in conversion was not considered to be a major problem for biodiversity conservation (e.g., Noble, Graetz), largely because it affects a small land area (<1% of Australia; Woodside), although its impacts may be long lasting, especially in aquatic systems (Cullen).

In all landscapes affected by fragmentation, the isolation of the native biota in "natural biological islands" was also considered to be an impediment to biological conservation, at least in the long-term (e.g., Bomford, Fox, Possingham). The management of these islands to maintain the ecosystem functions and processes underpinning biodiversity is a major challenge.

Other factors which speakers felt mitigated significantly against biological conservation were the traditional land ethic of the farming community (e.g., Davidson, Graetz) which is paralleled by the general attitude of much of the urban population (e.g., Cullen, Burbidge), and in urban areas, the NIMBY (Not In My Back Yard) syndrome (e.g., Burbidge, Noble). Two additional major impediments were highlighted for aquatic systems (e.g., Crabb, Kingsford, Shepherd): the large number of agencies involved in management of water resources (at least in New South Wales) and the number of jurisdictions involved, and the need for whole catchment management. The latter magnifies the complexity of management and jurisdictional issues and greatly increases the number of stakeholder groups involved.

Underlying all these impediments was people. While the population continues to increase nationally and internationally (Tooth) and continues to require/demand goods and services associated with bioresources, it was noted that land uses associated with their provision will continue to compete directly with land use for biodiversity conservation. The competition will be greatest in landscapes already highly converted and fragmented. This reinforced the need, noted earlier, to educate the wider public about the necessity of conserving Australia's biodiversity. Key aspects for achieving this were considered to be appealing to self interest, i.e., linking the loss of biodiversity to a direct reduction in the quality of people's lives (e.g., Burbidge, Saunders, Kitching), and being able to cite short-term benefits of retaining biodiversity, e.g., the economic gain in using the natural biota of retained vegetation for pest control in agricultural crops rather than chemicals (e.g., Beattie; Davidson).

CONSERVATION OF BIODIVERSITY — HOW CAN WE DO IT BETTER?

A common theme raised by many speakers was the need for new paradigms for the protection of biodiversity.

Protected areas

Speakers noted that there is widespread commitment at a state/territory, national and international level for the development of appropriate systems of protected areas. At a national level, these are embodied in various ESD reports, the Inter-Governmental Agreement on the Environment, the National Forest Policy Statement, the Prime Minister's Environment Statement, Endangered Species Legislation, Draft Biodiversity Strategy, etc.

The development of a comprehensive, adequate and representative network of protected areas will need to be built on the existing reserve system. The recent report of the House of Representatives Standing Committee on Environment, Recreation and Arts, *Biodiversity — the role of protected areas* (January 1993), recommended the expenditure of $150 million over six years for "the acquisition of identified areas to consolidate the establishment of a nationwide system of ecologically representative core protected areas". Given the magnitude of funding required, the time frame which may be required for such resources to be forthcoming, and limits to the amount of land which can be set aside as protected areas, systematic methods of reserve selection are most likely to ensure that networks satisfy the criteria for them to be comprehensive and representative (Pressey), and ensure that the network components show maximum complementarity. Political will is also essential for effective implementation of reserve network recommendations.

It was noted that although values other than biodiversity also need to be considered when selecting reserve areas (e.g., wilderness, heritage, cultural values), objectives such as the protection of all wilderness areas (Jurd) may not be compatible with the objective of a representative network for biodiversity, because of limitations on the amount of land which can be set aside as protected areas. It was also noted that protection of overall biodiversity in reserve networks selected on the knowledge of higher plants and animals may be quite poor. Recher, for example, found that areas appropriate for the protection of invertebrates may not fit conventional reserve models.

While protected areas were identified as being a major element in biodiversity conservation, it was widely acknowledged at the conference that they will not cater for all species, communities or genetic variation within species, and hence other conservation measures will be essential. Because of the likely constraints in time, resources and land in the development of reserve networks, new approaches to protected areas will also be necessary.

New paradigms for protected areas — multiple use; active management

A consistent theme emerging from the conference was the emphasis on protected areas managed for multiple use, rather than managed solely for conservation. While the suggestion that no protected areas should be dedicated to conservation (Tooth) would be unacceptable to many sections of the community, including among conservation biologists, the increasing emphasis on multiple use will require a change in community attitudes. Implementation of this paradigm will require a sophisticated understanding of all values (such as species, communities, wilderness) present in protected areas and their compatibility with various types of land use proposed (Burbidge).

The need for more interventionist management based on models of the biota/ecosystems also emerged during the conference, e.g., the use of fire to maintain species/communities (Bradstock, Gill); specific management to maintain old growth forest (Noble). Speakers noted that conservation objectives for protected areas are often very general (e.g., Bradstock), and stressed the need for more specific objectives related particularly to communities or ecosystem functions (e.g., Burbidge), and also to selected species (Bradstock, Gill). The simple models described by Possingham, Noble, Gill and others will assist in developing options for such management and focusing on conservation objectives.

The long-term view of biodiversity (De Decker, Recher) raised questions about the current focus on protected areas and on threatened/endangered species often located in them. Despite the best methodologies of selection, reserves chosen today may prove to be in the wrong locations to satisfy future conservation needs. Based on this longer-term perspective of natural changes in biodiversity, it could be argued that funds spent today on *in situ* conservation of some species may be better spent by focusing on *ex situ* methods, such as zoos, botanic gardens, aquaculture.

Off-reserve conservation

Throughout the conference, the need for conservation of biodiversity in areas lying outside reserves, and the integration of conservation management with the other land uses, was strongly emphasized. Remnants of natural vegetation and roadside corridors may represent the last opportunity to protect natural biodiversity in essentially converted landscapes such as agricultural areas (Adam). A wide range of measures were discussed which would assist the conservation of biodiversity; some of the more widely applicable ones are summarized in Table 2.

Table 2. Off-reserve management — actions for biodiversity conservation.

- Introduce specific conservation objectives for land management.
- Integrate conservation with other forms of land use.
- Manage on whole catchment/regional basis (for terrestrial and aquatic systems).
- Focus on ecosystems/communities.
- Develop management agreements with land owners.
- Use simple models to provide management options and focus.
- Develop ownership of biodiversity conservation by the community.
- Get community support for and involvement in management for biodiversity conservation.
- Integrate protection measures into normal daily activities.
- Ensure scientists and the community work together.
- Involve local government.
- Involve Aboriginal and Torres Strait Islander communities.
- Improve the liaison and co-ordination between agencies within and between jurisdictions.
- Where necessary, implement changes in administrative structures in government agencies.
- Use legislative incentives, not prohibitions.
- Rely on community education, not legislation.

It was stressed that multiple land use will require the development of specific management objectives for conservation, even in landscape types which are not fragmented (Friedel), and for commercial and non-commercial uses of the land. Models such as those described by Possingham, Noble and others could be used in the development of management options to help resolve the competing and often conflicting demands of biodiversity conservation and other forms of land use.

Monitoring the impacts of land management on biodiversity was seen as essential (Saunders, Bridgewater, Noble, Friedel), using appropriate objective criteria applicable to terrestrial and aquatic systems (Woodside, Graetz, Buckney), with management being modified as necessary, based on the outcomes of monitoring, to meet conservation objectives.

Numerous speakers noted that for effective off-reserve conservation it will be essential to obtain widespread community support for biodiversity conservation and community involvement in developing and implementing appropriate forms of land use. This will require the ownership of, and acceptance of, the need and responsibility for biodiversity conservation by owners and managers of land in pastoral, agricultural and forest regions, and by the urban population whose quality of life is largely dependent on the use of conventional biodiversity resources located in these regions.

In rural areas, Landcare groups were seen to be an important network into which a conservation element could be introduced for biodiversity conservation. In urban areas, bush regeneration groups may play a similar role to help change attitudes and assist in conservation management. Community education and voluntary use of measures to protect biodiversity (Davidson), legislation based on the precautionary principle and legislative incentives (Farrier, Young) were promoted as more effective mechanisms for protecting biodiversity outside reserves than the use of legislated prohibition and regulations. The important role of local government in biodiversity protection was also stressed (Saunders, Bridgewater, Adam), as was the need for more effective environmental impact assessment (Kingsford, Norton, Buckley).

Individual action

Throughout the conference, in both presented papers and during informal discussions out of session, the need for actions, not just more words, was constantly articulated. Despite the large number of solutions suggested during the conference, actions of the type outlined in Table 2 are not those upon which many of us have a direct say or control. Rather than allowing this to become an excuse for lapsing into the LITSE syndrome (Leave It To Someone Else), we need to focus on measures that individually, whether as scientists, land managers, activists or members of non-government organizations, we can take to promote biodiversity conservation.

A number of specific actions were discussed by speakers during the conference (Table 3). These include challenging statements that may gain wide currency in the media which have no scientific foundation and obstruct community awareness of the need for biodiversity conservation or hinder conservation measures (Noble, Bomford).

Table 3. Individual actions to promote biodiversity conservation.

- Challenge tenets and orthodoxies that work against biodiversity conservation.
- Translate scientific knowledge into accessible, practical tools.
- Improve research priorities.
- Help change policies and funding priorities.
- Develop new liaisons for research.
- Become good communicators.
- Help educate the community about the value of biodiversity.
- Communicate the results of the conference, simply.

Speakers highlighted the need for scientists to translate their scientific knowledge into practical tools, e.g., for strategic planning in land management, or land management guidelines for farmers, pastoralists, and miners. The need for scientists to change the focus of their research priorities was also raised, placing greater emphasis on studying the impact of various forms of land use on biodiversity, on research which will provide answers for management, and on research which will meet the needs of land users. The need for taxonomic research, which is fundamental to documenting and understanding Australia's biodiversity, was also emphasized.

Scientists also need to feed their knowledge into policy development and funding priorities. The current focus and level of funding on initiatives related to endangered species and protected areas, for example, has been questioned during the conference in terms of their contribution to the long-term conservation of Australia's biodiversity. Increased support for off-reserve conservation will require communicating this change in emphasis and the reasons why to policy- and decision-makers, and lobbying for a re-order of priorities.

We need to consider new liaisons in our work: conservation biologists working with economists, or with lawyers, or with ethicists; limnologists working with agricultural scientists; land repair technologists from the mining industry sharing ideas with reconstruction technologists in agricultural areas. The cross fertilization of ideas is likely to enhance the development of options and solutions for biodiversity conservation.

As individuals, we each have a role to help educate the wider community — to change community attitudes towards, and understanding of, biodiversity conservation. Burbidge suggested that all scientists should allocate one-tenth of their work time to communication. Communicating the results of this conference, simply, can be a starting priority. We need to become good communicators, to learn to use the media effectively. Communication training should probably be a standard subject in all tertiary science courses.

CONCLUSION

A number of speakers commented that action towards biodiversity conservation commenced only when an ecological crisis point, e.g., algal blooms, salinity, was reached. Speaker after speaker indicated that while there is much that we do not know about Australia's biodiversity, in many areas there is sufficient knowledge to act, and that acting now will be less costly — both biologically and economically — than awaiting a future crisis. The onus is on us all: let us get on with it.

Index

Notes: Scientific names of species and genera are given with a *see* reference to the common name, where there is one. To find an organism, start with the most specific name: e.g., *Eastern Grey Kangaroo* is more specific than *kangaroos,* which is more specific than *macropods,* which is more specific than *marsupials.* Legislation not attributed to specific State Governments is Federal legislation.

This index was prepared by Jonathan Jermey, a registered member of the Australian Society of Indexers.

A

Abalone harvesting 300
Aborigines
  fire regimes of 178, 315, 317
  harvesting of wildlife 260, 265, 275
  role in conservation 19
  site ownership 396–97, 404–05
  weed control by 195–96
Acacia aneura see Mulga
Acacia baileyana see Cootamundra Wattle
Acacia karroo eradication 188
Acacia longifolia 225
Acacia nilotica see Prickly Acacia
Acacia pulchella 310
Acacia sophorae see Coastal Sallow Wattle
Acacia spp. 202–03, 217
acclimatization societies 178
acid rain in Australia 58, 84
acidification of soil 107 *see also* salinization
activism 380–86, 416 *see also* non-government organizations
  against coastal development 62
  against mining 87
  against wildlife harvesting and trade 289
Adiantum formosum 286
aerial surveys *see* surveys
aesthetic impacts 65, 84
African Daisy 194
African Olive 212
Agapetus spp. 133
agriculture *see also* clearing; grazing; pastoral industry
  conflict with conservation 340–41
  damage caused by ducks 265–66
  effect of grazing 249–59
  effect on habitat 39–40
  effect on rivers 119–20
  existing uses protected 344
  extinction due to 310
  introduction of plant species 222–23
  pressures on 387
  water pollution from 133–34, 138
Agriculture and Related Resources Protection Act 1976 (WA) 278
Agrostis avenacea see Blown Grass
air pollution 58, 66, 84
air travel 64–65
Alathyria jacksoni see River Mussel
Albacore harvesting 301
Alectryon oleifolius see Rosewood
ALEX decision support system 293–96
algae
  affected by sedimentation 132
  blooms 116, 118–19, 121, 138–39
  salinization tolerance 127–28
Algal Management Strategy, Murray-Darling Basin 170
alien species *see* introduced species
Allocasuarina littoralis 213, 217
alpine regions 249
Amiscridea edwardsii 128
Ampera xiphoclada 68
amphibians
  decline in numbers 105, 393
  effects of salinization 127
  introduced species 178
  in old growth forests 244
  trade in 286
amphipods 96, 101
Anas castanea see Chestnut Teal
Anas gracilis see Grey Teal
Anas platyrhynchos see Mallard
Anas superciliosa see Pacific Black Duck
Anchovy 302
Andropogon virginicus see Whisky Grass
angiosperm extinction by fire 310–11
Angled Onion 192
angling 120–21 *see also* fishing
animal welfare concerns for feral animals 203
animals *see* fauna
Anredera cordifolia see Madeira Vine
Anseranas semipalmata see Magpie Goose
Ant Plants 288
Antarctic 29, 375
Antechinus 315–16
antibiotic resistance 7
antifertility drugs 181, 203–05
antifouling paint 63
ants as biological resources 6–7
Apia Convention 399
Apis mellifera see Honeybee
aquaculture 304
aquatic environments 115–36, 139
  see also rivers
  effect of urbanization 58–59
  protection for 162, 405
  weed control 224
Arctocephalus pusillus see Australian Fur Seal
arthropods 95–101
Artiodactyla 180
Astacopsis gouldi see Giant Freshwater Crayfish
Asterias amurensis see Pacific Starfish
Astrebla spp. 180
Athel Pine 180, 184
Athrotaxis spp. 310
Atlas of New South Wales Wildlife 398, 404
atmospheric emissions 58, 66, 84
Atrichornis clamosus see Noisy Scrub-Bird
Atriplex vesicaria 310
Attenborough, David 3, 29
Australia
  biodiversity 391, 410–11
  commitment to conservation 337–39, 399
  development of 106
  extinction rates 39
  hydrological variability 139
  introduced species 177
  mining 78–81
  rivers 115–17
  salinity periods 126
  urbanization 55
  wheat-sheep zone 105–06
  wildlife harvesting and trade 273–91
Australian Association of Bush Regenerators 210
Australian Bass 149, 286
Australian Conservation Foundation 376
Australian Fur Seal 274
Australian Grayling 127, 149–50
Australian Littoral Society 376
Australian Mineral Foundation 85
Australian Mining Industry Council workshops 85
Australian National Conservation Association 289
Australian Nature Conservation Agency 276
Australian and New Zealand Environment and Conservation Council 395
  funding weed control 192
  State involvement 338–39
  Task Force on Sustainable Use of Wildlife 289
Australian Pelican 149
Australian Quarantine and Inspection Service 186, 225
Australian Salmon 303
Australian Science and Technology Council 122
Australian Sea Lion 274
Australian Shelduck 266–67
Australian Smelt 132
Australian Water Quality Guidelines 118
Austrolimnius spp. 133
Austrophleboides pusillus 133
Austrotachardia spp. 193
autecology 22, 40
Avon River (Vic) 130
Aythya australis see Hardhead
Azolla spp. 147

B

bacterial oxidation of mine waste 138
balanced age-class approach to fire management 326
ballast water 63, 178, 222
Bandicoot 20, 179, 207, 295–96
Banksia cuneata 318
Banksia ericifolia 311, 325
Banksia ornata 311–12
Banksia spp. 181
Barmah-Millewa forest (NSW) 146, 154, 170
Barren Grounds Nature Reserve (NSW) 326
Barrenbox Swamp (NSW) 267
Bassia quinquecuspis see Black Roly-poly
Batemans Bay (NSW) 242
bats 69, 147–48, 273
beauty of diverse species 3–4
Beche-de-mer harvesting 287
beetles 96–101, 128, 287
  scarab 21, 47, 49
  as bioindicators 139

Belah 252
beneficiary-compensates principle in land management 363
Bennetts Wallaby 262, 264–65
Bettong 21, 275
Bettongia penicillata see Brush-tailed Bettong
Bidyanus bidyanus see Silver Perch
Bilby 179, 275, 394
BIOCLIM decision support system 403–04
biocontrol
  by introduced species 181
  economics of 5–6
  of feral animals 202, 220
  identifying agents 8–9
  of introduced species 181
  of weeds 189–93, 195, 224–25
biodiversity 3–26, 67–68, 177, 255–56, 391
  effect of fish harvesting 300–05
  effect of forestry 245–46
  effect of gardens 56–58
  effect of grazing 249–59
  effect of introduced species 177–83
  effect of urbanization 55–62
  effect of water pollution 139
  effect of weeds 184–201
  on farms 51–52
  and fire regimes 325
  jurisdictional issues 162–65, 170–72
  political aspects 231–39
  population viability analysis 292–99
  on private land 337–59
  quantifying value of 350–51
  reconstruction aimed at 110
  resource harvesting 235–36
  sustainable forestry 240
  water resources 115–25
Biodiversity Convention 76
Biodiversity — the Role of Protected Areas 414
bioindicators 139
  for fire regimes 318–19
  of grazing damage 255
  of river health 118
  of water pollution 84, 141
biological control *see* biocontrol
Biological Diversity Advisory Committee 395
bioresources 7–9
Bird Action Plan 376
Bird Protection Act 1901 (NSW) 265
birds *see also* waterbirds
  aircraft damage by 64
  in cleared and uncleared areas 52
  decline in numbers 105, 393
  domestic trade in 281–82
  effect of sedimentation 131
  Emu harvesting 260, 278–79
  fire extinction of 314
  fish-eating 148–49
  illegal harvesting 288, 395
  international protection 164, 399
  introduced species 178–79
  in old growth forests 244
  population variance 69
Bitou Bush
  biocontrol 184, 195, 224–25
  displaces native plants 394
  herbicide treatment 193
  introduced in ballast 222
  recommended by NSW Soil Conservation Service 87
  replacement with native species 181
  threat to biodiversity 70
Black Box 147–48
Black Rat 69
Black Roly-poly 147
Black-breasted Button Quail 288
Black-Necked Stork 147
Black-Throated Finch 288, 394
Blackberry 120, 193, 213, 216
Blackberry Nightshade 216
Blackwattle 216
Bladey Grass 216
Blown Grass 147
Blue Bell Creeper 193
Blue Couch 213
Blue Gum High Forests (NSW) 30
Blue Rock Dam (Vic.) 134
Boneseed 184, 193–94, 394
Bonn Convention 399
Bool Lagoon (SA) 268
Botany Bay (NSW) 64
bounty system 262, 388
Brachiaria mutica 223
Bridal Creeper 185
Bridled Nail-tail Wallaby 265
Brigalow Belt (NSW) 33
Bristlebird 314
Briza maxima see Quaking Grass
Broad-finned Galaxias 132
Broad-headed Snake 288
Broken Hill (NSW) 89
Brown Rat 69
Brown Trout 127
Brundtland Report 165
Brush-tailed Bettong 21, 275
Brush-tailed Possum 69, 275
bryophytes 58
Buffalo Grass 212
Buffel Grass 185, 194
Bureau of Resource Sciences 203
Burramys parvus see Mountain Pygmy-possum
bush recreation *see* recreation
bushfires *see* fire; fire regimes
bushland 17 *see also* land management
  on farms 45–47, 49–51
  in urban areas 72, 210–19
Buteo jamaicensis socorrensis see Red-tailed Hawk
butterflies 287

C

Cacatua galerita see Sulphur Crested Cockatoo
Cacatua roseicapilla see Galah
Cacatua tenuirostris see Long-billed Corella
Cadellia pentastylis see Ooline
cadmium pollution 141
Californian Sardine 302
Callicoma serratifolia see Blackwattle
Callitris oblonga see Pigmy Cypress Pine
Callitris spp.
  in fire shadows 312–13
  local extinction by fire 310
  local population growth 314
Calluna vulgaris 188
CALM 21, 86, 284
Calyptorhynchus funereus see Carnaby's Black Cockatoo
Calyptorhynchus lathami lathami see Glossy Black Cockatoo
Calyptorhynchus magnificus see Red-Tailed Black Cockatoo
CAMBA 64, 399
Camphor Laurel 212–13
canal estates 62
Canegrass 148
Canis familiaris dingo see Dingo
canopy vines 189–90
Cape Melville Palms 288
Cape Wattle 193, 225
Captains Flat (NSW) 77
Carabidae 96–101
Carduus nutans see Nodding Thistle
Carnaby's Black Cockatoo 39, 288
Carnivora, introduced species 180
Carrington Falls (NSW) 313
Casuarina distyla 310
Casuarina pauper see Belah
Casuarina pusilla 310
Catchment Areas Protection Board (NSW) 346–47
catchment management 61, 124, 165–70 *see also* land management; Murray-Darling Basin; rivers; runoff
  afforestation 116
  aid to farmers 51
  committees for 19, 388–90
  as model for land management 377
  in New South Wales 399
  reducing sedimentation 133–34
  river health 116, 153
Catfish 149
cats, domestic 58 *see also* feral cats
Catsear 213
cattle 249, 251–52 *see also* agriculture; grazing
Caustis blakei 286
Cenchrus ciliaris see Buffel Grass
Centella asiatica 216
Central Division (NSW) 241
Central West Region (NSW) 33
Cercophonius squama 96–101
Chaelundi case 367
Chamaelaucium uncinatum see Waxflower
Chelonia mydas see Green Turtle
chemical control of pests 47, 138, 181, 203–05
Chenonetta jubata see Maned Duck
chenopod shrublands 252–53
Chestnut Teal 266
Cheumatopsyche spp. 128
China Australia Migratory Birds Agreement 64, 399
Chironomidae 139
Chlorophytum cosmosum see Spider Plant
Chondrilla juncea see Skeleton Weed
Christian beliefs conflict with conservation 12
Chromolaena odorata see Siam Weed
Chrysanthemoides monlifera see Bitou Bush
Chuditch 202
Cinnamomum camphora see Camphor Laurel
citadel reserves *see* reserves
CITES 64, 282, 377, 399
citizen choices used to evaluate biodiversity 360
clams 287, 300
clearing 30, 33, 241, 402
  for agriculture 394
  effect on birds 52
  prevention of 341–51
clearing *see also* habitat damage; land management
Clematis vitalba see Old Mans Beard
climatic changes affect river systems 121
Cluster Pine 194
Cnidaria spp. *see* jellyfish
Co-operative Research Centre on the Control of Vertebrate Pest Populations 21
coal mining in New South Wales 79–81
Coast Tea Tree 193–94, 225
coastal development 62

coastal dune revegetation 195, 395
Coastal Sallow Wattle 195
Cobblers Peg 214
Cocatuinae spp. *see* Cockatoo
Cochineal Insect 190
Cockatoo 39, 69, 185, 273, 275, 288, 319
CODA decision support system 362
Codes of Practice for forestry 234
Coleoptera as bioindicators 139
Colletrotichum gloeosporoides 190
Columba livia see Domestic Pigeon
Comb-crested Jacana 147
commercial use of wildlife *see* wildlife harvesting and trade
Common Brushtail Possum 21, 275, 278
Common Galaxias 127, 132
Common Ringtail Possum 275
Commonwealth Constitution on regulation of private land 352
Commonwealth Environment Program 378
communities (ecological) *see* ecosystems
communities (human) 19, 40, 399, 404, 413 *see also* activism
  empowerment 15
  funding of conservation 23
  involvement in mine planning 87–89, 91
  involvement in revegetation 108
  involvement in weed control 193, 195–96
  in Murray-Darling Basin 168–69
  participation in decision-making 385
  participation in environmental impact assessments 368
  support of non-government organizations 375–79
Comostolopsis germana 195
compensation for land management 352–53, 364 *see also* economics
computers 17–18
  aid non-government organizations 375
  ecological modelling on 293–96, 362, 403–04
conservation 11–26, 232–33, 395, 411–13
  activism 380–86
  bargaining approach 346
  best practice 398
  by farmers 43–54, 387–90
  decision support systems 46–47, 293–96, 362, 403–04
  economics of 360–64
  fire management 323–33
  forest reserves 240–41
  grazing lands 256–57
  illegal collecting 289
  land management for 363, 387–88
  National Parks and Wildlife Service 391–409
  non-government organizations 375–79
  off-park strategies 19–20, 400
  personal commitment to 12–15
  population viability analysis 292–99
  professionals in 16
  quantitative values for 369
  river management 151–52
  taxonomy in 68
conservation biology 18–23, 398, 400, 405–06
consumer preferences used to evaluate biodiversity 360
Contingent Valuation Analysis 232–33
contraception *see* reproductive control
Convention Concerning the Protection of the World Cultural and Natural Heritage 399
Convention on the Conservation of Biodiversity 337, 339, 375–76, 395, 399
Convention on the Conservation of Migratory Species of Wild Animals 399
Convention on Conservation of Nature in the South Pacific 399
Convention on International Trade in Endangered Species of Wild Fauna and Flora 64, 282, 377, 399
Convention for the Regulation of Whaling 399
Convention on Wetlands of International Importance 64, 164, 399
Coolabah 147
Cootamundra Wattle 193
copper pollution 141
Coprosma repens see Cobblers Peg
Cord Grass 185
cormorant 148-9
Coronation Hill Joint Venture 86
corridors of bushland 50 *see also* reserves
Cortaderia spp. *see* Pampas Grass
cosmological imperative 12–13
cotton industry 35, 151, 170–71
Coturnix spp. *see* Quail
Couch 212
Craspedia globosa 286
crayfish 121
Crimson Finch 288
Crinia signifera 96, 101
crocodile 260, 276, 282–84
Crocodylus johnstoni see Freshwater Crocodile
Crocodylus porosus see Saltwater Crocodile
Crofton Weed 216
Cryptostegia grandiflora see Rubber Vine
Cubbie Station (Qld) 170–71
Culgoa River (Qld) 146
cultural issues in conservation 12–18
cultural obligations of National Parks 60
Cumberland Plain (NSW) 30
Cumbungi 148–49
cyanobacterial blooms 116
Cyathea australis see Rough Tree Fern
Cynodon dactylon see Couch
Cyrtobagous singularis 224
Cytisus scoparius see English Broom

D

Dactylopius ceylonicus see Cochineal Insect
dams 115, 134, 144, 146, 352
  cause sedimentation 130
  habitat damage due to 118–19
  permanent flooding by 148
  restrict fish migration 149
Daphnia carinata 128
Darling River 170–71 *see also* Murray-Darling Basin
Dartmouth Dam (Vic.) 134
Dasyornis broadbentii see Bristlebird
Dasyurus geoffroii see Chuditch
Dasyurus maculatus see Quoll
data collection *see* surveys
decision support systems 46–47, 293–96, 362, 403–04
  reserve selection 234
  role in conservation 23–24
Department of Conservation and Land Management (WA) 21, 86, 284
Department of Environment and Land Management (Tas.) 281
Department of Primary Industries import regulations 186
Department of Water Resources (NSW) 146, 152–53, 165, 382
deserts, rate of expansion 29
development
  applications for 343–36
  ecologically sustainable 396
  environmental impact assessments for 365–66
  historical attitudes to 115–16
Dicksonia antarctica see Soft Tree Fern
dicotyledons 212–13
dieback 43, 45
Dingo 177, 204, 252–53
dingo-proof fence prevents immigration 205
Dizelma archeri 310
dogs 58, 221
DOMAIN decision support system 403–04
domestic pets as pests 58, 221
Domestic Pigeon 69
Dorrigo Forest Management Area (NSW) 243, 368
Draft Natural Heritage Conservation Strategy — Biodiversity Component (NSW) 401–04
drainage *see* groundwater
dredging causes habitat damage 63–64
drift rate affected by water pollution 139
Dromaius novaehollandiae see Emu
drought effects on macropod population 264
dryland vegetation preservation 171
Dryococelus australis see Lord Howe Island Phasmid
ducks 57, 149, 265–68, 364
Dugong 275
Dugong dugon see Dugong
dune revegetation 195, 395
Durvillaea potatorum 287
Dwarf Mountain Pine 395

E

Earthworm 69
East Finnis River (NT) 84
Eastern Barred Bandicoot 20, 207
Eastern Division (NSW) 241
Eastern Freshwater Cod 304
Eastern Grey Kangaroo 262–65, 276
Eastern Rock Lobster 300
Eclectus Parrot 288
Eclectus roratus macgillivrayi see Eclectus Parrot
ecological modelling 232–39, 403–04 *see also* decision support systems
  effect of fire 311–17, 323–25, 329–31
  effect of grazing on diversity 254–55
  for forestry 316–17
  land-use tradeoffs 362
  population viability analysis 292–96
  uncertainty and risk 369
ecologists 90, 231 *see also* professionals
ecology 233–39, 250
economics 23–24, 337, 360–64, 412–13
  analysis of forestry 232–33
  aquatic flora harvesting 287
  of biodiversity 7–9
  feral animal control 202, 205–07
  fish management 305
  kangaroo management 262, 269, 276–78
  of land management 350–51, 390
  live fish exports 286–87

economics — *continued*
mining industry 78
National Parks 397–98
non-government organizations 378
of pollution 5–6, 342, 349, 363
water use 172
weed control 192, 226
wildlife harvesting and trade 273, 289
ecosystems 4–7, 70–71, 250 *see also* habitat damage; habitat protection and repair
analysis 367
aquatic 115
damage to 38, 70–71, 395
dynamic nature 82
economic valuation 360–64
effect of fire 394
effect of fishing 301
effect of forestry 241
effect of illegal collecting 289
effect of salinization 128–29
estimating biodiversity 255–56
extreme 68
fluctuation in 410–11
introduced species 180
management 300–05, 398
monitoring change 22
recovery of 20–21, 104, 210
required for biocontrol 46
taxonomy required 68
threats to 411
viable population 292–94
Eden Management Area (NSW) 241–42
edge effects 234–35, 313–14
education 16–18, 381–82, 404
about introduced species 182
for revegetation 108
in social and cultural issues 13
for weed control 225–26
Ehrharta calycina see Tussock Grass
Ehrharta erecta 213
EIA *see* environmental impact assessments
Eichhornia crassipes 149, 224
Eichhornia spp. 120
Eleocharis acuta 128
Elephant Seal 274
Elmid beetles 128
Elodia spp. 120
Emex australis see Spiny Emex
empowerment of local communities 15
Emu harvesting 260, 278–79
Endangered Fauna (Interim Protection) Act 1991 (NSW) 367, 392
National Parks and Wildlife Service 397
species protection 243, 342
Endangered Fauna (Interim Protection) Amendment Act 1992 (NSW) 367
Endangered Species Advisory Committee 339, 377
Endangered Species Program 376
Endangered Species Protection Act 1992 341, 377
endangered and threatened species 6, 20–21
classification of risk to 392–93
protection of 339, 367
vulnerable population sizes 292–99, 326–27, 403
endemism in Australian rivers 117
endosulfan 138
English Broom 184
English Ivy 185, 189
EntoloLOGIC decision support system 47
Environment Protection Act 1986 (WA) 366
Environmental Assessment Act 1982 (NT) 366
Environmental Effects Act 1978 (Vic.) 366
environmental impact assessments 347, 365–74
of mining projects 85
not required for introduced plants 223
population viability 298
prompted by activism 383
in river management 152
Environmental Management Overview Strategy (Qld) 87
Environmental Planning and Assessment Act 1979 (NSW) 147, 347, 366, 397
Environmental Protection Act 1973 (Tas.) 366
Environmental Protection Act (Impact of Proposals) 1974 366
Environmental Resource Information Network 86
environmental weeds *see* weeds
Ephemeroptera 139
equilibrium theory of island biogeography 94–95
Equus caballus see horse
eradication *see* feral animals; pest control; weeds
Eragrostis eripodia 251
Erigeron karvinskianus 214
Erigeron spp. *see* Fleabane
erosion
causes sedimentation 130
control on farms 50
due to road construction 66
effect of river management 150
Erythrura gouldiae see Gouldian Finch
ESD Forest Use Working Group 340
ESD Working Group on Agriculture 340–41
ethic of tidiness in farming 43
ethical imperative 13–14
Eubalena australis see Southern Right Whale
eucalypt forests 95, 101, 241, 246 *see also* forestry; forests
Eucalyptus camaldulensis see Red Gum
Eucalyptus largiflorens see Black Box
Eucalyptus microtheca see Coolabah
Eucalyptus regnans 310
Eucalyptus spp. 242–43
Eucalyptus tintinnans 371
European Carp 120
eutrophication 138–39 *see also* algae
evolution 6–8, 293
evolutionary biology 8
exotic species *see* introduced species
expert systems *see* decision support systems
exploration by mining industry 78, 84, 88
exports *see* wildlife harvesting and trade
extension activities 16
extinction 29–33, 38–39, 105–06, 309–10
among marsupials 179
of Bilby 394
by road kills 65–66
due to urbanization 139, 310
due to agriculture 52, 310
due to fire 309–22
fire management to avoid 319, 324–25, 328–29
of freshwater organisms 139
of island species 207
of Koala 315
of Lord Howe Island Phasmid 393
extinction — *continued*
of mammals 315
mining not implicated 83
of plants 69, 185, 310–11
possible survivals from 410
of Potoroo 179
prevention 22
probability of 294–97, 394
in rivers 117–21
secondary 121
of Tasmanian Tiger 179, 275
of Toolache Wallaby 265
viable population size 292–93
weeds involved in 185
within reserves 316–17

F

farmers 43–54, 387–90 *see also* agriculture; pastoral industry
attitudes to conservation 116, 414
incentive to regenerate 53
information needed by 43
need to change practices 107
pest control by 46–47
farming profitability and regeneration 52
farms, wildlife populations on 52
fauna 251–52, 392
as bioindicators 255
decline in numbers 393
effect of development on 105
effect of habitat fragmentation 94–103, 105–06
effects of fire on 314, 316–17, 324–25, 327
harvesting 260–72
introduced species 221, 394
legislative protection 367–68, 392
harvesting and trade 274
Fauna Impact Statements (NSW) 367–68 *see also* environmental impact assessments
Felis catus see cats; feral cats
Fenner Conference on the Environment 376
feral animals 202–09, 220–28
dogs as 58
exclusion of 387
goats as 205, 249, 252
increase with urbanization 58
feral cats
control of 21
effect on isolated population 207
increase with urbanization 58
in rangelands 253
research required 203
Fern Asparagus 212
Ferret 206–07
fertility control *see* reproductive control
finches 275, 281
fire 323–33 *see also* fire regimes
for domestic heating 61
effect on viable population 295
effect on weeds 192–94, 214, 218–19
in monsoon forests 313
regermination of *Acacia* spp. 217
shadow areas 312–13
in urban bushland 61, 211
variation in intensities 311–12
fire regimes 22, 309–33, 370
Aboriginal practices 178, 315, 317
ecological modelling 254, 329–31
ecosystem damage by 394
effect on Southern Brown Bandicoot 295

fire regimes — *continued*
mosaic burning 325–26
prescribed 329–30
suppressed on rangelands 252–53
fish
aquaculture 304
breeding behaviour 147, 150
effect of salinization 127
effect of sedimentation 131–32
effect of water quality 140–41
effect of water temperature 150
harvesting of *see* fishing
introduced species 120, 148–49, 179
ladders for 149, 153
management plan for Murray-Darling 169–70
marine populations 302–03
river populations 117, 119
significant species 139
threatened species 303–04
trade in natives 286
Fisheries and Oyster Farms Act 1948 (NSW) 405
fishing 300–05, 395 *see also* wildlife harvesting and trade
impact on biodiversity 60
overexploitation of 120–21
Fleabane 214
flooding 61, 146–49, 170
Floodplain Mussel 148
floodplains 146–48, 152–53
flora *see* plants
Flora and Fauna Guarantee Act 1988 (Vic.) 284, 341–42
flow regimes 119, 138, 152
Flying Fox 69, 273
foraging resources 245
forestry 231–48, 295–97
activism against 383–84
fire regimes 317
and habitat damage 29
habitat modelling 316
legal requirements 342, 367–68, 383
political issues 380–82
on private land 340
rate of replanting 107
sedimentation due to 130, 133–34
sedimentation from 130
surveys for 384–85, 398
tree decline 43, 45
Forestry Commission of NSW 342, 381–82
Forestry Commission of NSW v Corkill 342
forests, global extent of 240
Fox 21, 58, 204, 253
fragmentation *see* habitat fragmentation
Freckled Duck 267–68
freedom of information 385
Freesia 212
Freshwater Blackfish 132
Freshwater Crocodile 260, 276, 282–84
frogs 96, 101, 131
Fusilade (herbicide) 212

G

Gadopsis marmoratus see River Blackfish
Gahnia 216
Gahnia sieberiana see Gahnia
Galah 21, 39, 69, 288
Galaxias 127, 132
game reserves 396 *see also* reserves
game species *see* wildlife harvesting and trade
Garden Snail 69
Gardenia actinocarpa 288
gardens in urban areas 56–58
Gelidum asperum 287
Gemfish 301–03
genetic diversity 38, 68
genetic manipulation for aquaculture 304
Giant Clam 287
Giant Freshwater Crayfish 121
Giant Sensitive Plant
biocontrol of 195, 225
herbicide treatment 193
impact on wetlands 184
means of introduction 222
supports Water Buffalo 220
Global Network for Anti Golfcourse Action 59
global warming 58, 224
Glossy Black Cockatoo 288, 319
Glyphosphate (herbicide) 212
Goat, as feral animal 205, 249, 252
gold mining 80
Golden Perch 147, 286
Golden-Shouldered Parrot 288
golf courses 59, 62
Gompholobium tomentosum 310
Gouldian Finch 281, 369–73
government 11, 337–59, 365–79, 388
see also legislation; regulation
adoption of industry perspective 383
agencies involved in conservation 398–99
forestry policy 246–47
jurisdictional issues 162–65, 170–72
land management by 24
lobbying for conservation 377–78, 380–86
local 345–46, 386
regulation of wildlife harvesting and trade 276, 278, 280–81, 289
State level 162, 164–65, 345
weed control co-ordination 188
grasses
introduced species 57, 223, 317
weed treatment 213
in woodlands 250–51
grazing 249–59
after fire 315
by kangaroos 262–63, 269
causes sedimentation 130
dieback due to 45
effect on *Callitris* spp. 312–13
extinction due to 52
habitat damage due to 36, 38
land management to control 344, 350, 402
weed invasion due to 189
Greater Glider 243, 296–97
Greater Long-eared Bat 148
Greater Stick-nest Rat 185
Green Turtle 260
Greenhood Orchid 394
greenhouse gases 58, 224
Greening Australia 50–51, 376
Grevillea nematophylla 311
Grevillea rivularis 313
Grey Teal 266–67
Grey-Headed Fruit Bat 69
Grizzly Bear 294
Ground Parrot 314, 326–27
groundwater 126 *see also* runoff; salinization
drainage 119, 129–30
hydrology 66, 253
Murray-Darling Basin 170–71
salinization on farms 50
WA wheat belt 40
Gwydir River 147
Gymnobelideus leadbeateri see Leadbeater's Possum
gymnosperms 310

H

habitat damage 29–42, 55–93, 292, 370–73
by pest infestations 48–49
cumulative impacts 395
due to agriculture 43, 251
due to aquaculture 304
due to fire 310–14, 325
in New South Wales 393–94
penalties for 340–41
on private land 337–38
in rivers 118–20
and species vulnerability 207–08
habitat fragmentation 39, 94–103, 105–06, 413–14
allows introduced species entry 181–82
due to fire 323
due to grazing 254–55
in New South Wales 35
weed invasion 184
habitat protection and repair 77–78, 88, 104–12, 210–19
after fire 318–19, 327–28
after grazing damage 253–54
for arboreal mammals 296–97
of bushland on farms 49–51
by local residents 381
rehabilitation defined 77
river areas 122–23, 152–53
species reintroduction 181
weed control 195
Hakea sericea 190
Hakea teretifolia 311
Haloragodendron lucasii 69
Hardhead 147, 266
Hare Wallaby 314–15
harvest intensity model 235–36
harvesting of wildlife *see* wildlife harvesting and trade
heathlands 190, 325–28
Hedera helix see English Ivy
Helicopsyche spp. 133
Helix aspera see Garden Snail
herbicides 193
environmental effects 121
water pollution by 138
for weed control 212
heritage agreements (SA) 343, 348–49
Himalayan Honeysuckle 189
historical events
attitudes to development 115–16
forestry 241
introduced species brought by First Fleet 178
kangaroo control 262
mining 77–78
river management 144, 163
wildlife harvesting and trade 260, 273–76
hobby farms 60–61
Holothuroidea spp. *see* Beche-de-mer
Honeybee 220
Hoplocephalus bungaroides see Broad-headed Snake
Horse 59–60, 204
horticulture 56–58, 60–61, 226–27
House Mouse 315–16
House of Representatives Standing Committee Report on Biodiversity 376
House Sparrow 69, 178
Hovea spp. 311–12

Hubbard Brook Experimental Forest (USA) 123
humans
  arrival in Australia 177–78
  extinctions caused by 38–39
  population growth 72, 414
Humpback Whale 274–75
hunting *see* wildlife harvesting and trade
hybridization of garden plants 57
hydrology 66, 253 *see also* dams; runoff
  flow regimes 119, 138, 152
Hypolepsis muelleri 216
Hypricum perforatum see St Johns Wort
Hypsela sessiflora 69

I

Ibis 59, 148, 150
Ice Plant 185
Iediparra gallinacea see Comb-crested Jacana
illegal activities 57, 287–90, 394–95
Imperata cylindrica see Bladey Grass
industrial activities 130, 138
infestations *see under* pests
information programmes *see* education
information technology *see* computers
Ink Weed 216
intergenerational equity legislation 405
Intergovernmental Agreement on the Environment 11, 338, 395
intermediate disturbance hypothesis 254–55
International Biodiversity Coalition 375–76
international issues 163–64, 375–76, 378
International Joint Commission (Canada-USA) 164
International Society for Restoration Ecology 85
intraspecies variation 68
introduced species 63–65, 177–83, 185–88, 220–28
  commitment to control 337
  concentrated in urban areas 57
  damage by 21
  due to aquaculture 304
  effect on fire regimes 317
  effects in New South Wales 394
  from elsewhere in Australia 21–22
  habitat repair by 82
  plants in Sydney 70
  river population 120
  threat from 70, 185, 410, 413
  "transformer" functions 180
  of weeds 185
  in wetlands 149
invertebrates
  effect of dams 118
  effect of development 69, 105
  effect of flooding 148
  effect of grazing 253
  effect of habitat fragmentation 96, 101
  effect of sedimentation 132–33
  effect of water pollution 138, 141
  endangered in New South Wales 393
  environmental impact assessments 368
  exploit flooded areas 147
  extinctions of 393–94
  harvesting and trade 287, 300–01, 304
  high river population 117
  introduced species 179
  legislative protection for 121, 405
  salinization tolerance 128–29
Ipomoea indica see Morning Glory
irreplacebility analysis 403
irrigation 149–52 *see also* dams; groundwater; runoff
  contribution to agriculture 144
  drainage water from 119, 129–30
  effect on salinization 126
  effect on wetlands 121
  and flow regimes 119, 138, 152
  pesticides in run-off 138
  for salinization control 129
  water use for 123–24
islands
  biogeography of 94–95, 254–55
  introduced species on 180–81, 207
  role in conservation 22
Isoodon obesulus see Southern Brown Bandicoot
Ivy Gourd 188

J

JAMBA 64, 399
Japan Australia Migratory Birds Agreement 64, 399
Japanese Honeysuckle 212, 217
Jarasius v Forestry Commission of NSW 383
jarrah reforestation after mining 83
jellyfish harvesting 287
Juncus ingens 148

K

Kangaroo Grass 194–95
kangaroos 260–65, 269, 275–78
  advantaged by settlement 39
  damage done by 21, 262
  distribution in woodlands 251
  grazing in alpine regions 249
  increase in rangelands 252–53
  population dynamics 263
Kikuyu 212
Kimberly (WA), invasion of grasses 317
Kinchega National Park 148
Koala
  ecological modelling 404
  fire effects on 315
  protection in forests 242
  trade in skins 275
Kochia scoparia 188
Kooragang Island 71

L

Lagarostrobus franklinii 310
Lagomorpha 180
Lagorchestes hirsutus see Hare Wallaby
Lake Buloke (Vic.) 268
Lake Cowal (NSW) 267
Lake Eacham Rainbow Fish 304
Lake Toolibin Recovery Plan 20
Land Conservation District Committees 19
land damage
  by grazing 255–57
  in New South Wales 394
  prompts weed invasion 189–90
Land and Environment Court (NSW) 366
Land and Environment Plans (NSW) 343–44
land management 40, 337–59, 387–90, 400
  by mining industry 78–79
  compensation for 87, 352–44, 363–64, 388
land management — *continued*
  conflict resolution 384–85
  decision support systems 23–24, 234, 294
  education for 17
  fire regimes 316–19, 328–29
  history of 43
  integrating knowledge 81
  legislation for 15, 348–49
  manuals for 17
  Murray-Darling Basin 169
  National Parks 396–98
  off-reserve 415–16
  organizations for 24
  regulation of 342–43, 388
  reserves 18–19
  uncertainty and risk 369
  urban bush 72
land repair 52, 77–78
  lack of standards for 87–88
  mine sites 77, 83
  monitoring 91
  with native species 181
land repair *see also* habitat protection and repair
land topography 83 *see also* landscapes
Landcare 15, 20, 415–16
  expansion of 388–90
  for farmers 50
  funding for land management 353–54
  implementation in New South Wales 399
  involvement in weed control 195
  vehicle for ecological reconstruction 108–09
landscapes
  habitat protection and repair in 108–10
  synthetic 107
  topographical changes due to mining 83
landscapes *see also* ecosystems
Lantana 213, 216, 317
Lantana camara see Lantana
Large-footed Mouse-eared Bat 147–48
Large-leaf Privet 212
Larus novaehollandiae see Silver Gull
Lathamus discolor see Swift Parrot
Latrobe Valley coal mines 79
law *see* legislation
lead pollution 66, 141, 268
Leadbeater's Possum 294–95
legislation 352–53, 365–74, 404–06, 415–16 *see also* government; illegal activities; regulation
  cage bird trading 281–82
  covering mining industry 86–88
  covering private land 338–39
  due to work by activists 382–83
  Freedom of Information Act inadequate 385
  international law 164–65
  invertebrates not covered by 393
  for land management 15, 348–49
  legal action on environmental issues 376–77, 383–84
  National Parks and Wildlife Service 399
  river management 152
  for social change 15
  to protect species 284, 341–42
  waterbird harvesting 265–66
  weed control 185–88, 222, 225
  wildlife harvesting and trade 276, 278, 280–81, 289

Leipoa ocellata see Mallee Fowl
leisure activities *see* recreation
Leptorhynchos gatesii 309
Leptospermum laevigatum see Coast Tea Tree
Leptospermum myrsinoides 312
Leptospermum spp. 181, 193–94, 225, 312
Lesser Bilby 179
Leucaena leucocephala 223
Leycestaria formosa see Himalayan Honeysuckle
lichens 58
life history models 311–12
lignum killed by flooding 148
Ligustrum lucidum see Large-leaf Privet
Ligustrum sinese see Small-leaf Privet
Ligustrum spp. *see* Privet
Lilium 212
limnology *see* aquatic environments; rivers
Lippia, takeover of wetlands 147
Little Terns 59
livestock improvement with shelter 50
lizards recolonizing minesites 85
local government *see* government
Local Government (Planning and Environment) Act 1990 (Qld) 366
logging *see* forestry
Long-billed Corella 288
Long-footed Potoroo 404
Long-nosed Potoroo 275
Lonicera japonica see Japanese Honeysuckle
Lord Howe Island Phasmid 393–94
Lumbricus spp. *see* Earthworm
LUPIS decision support system 362
Lycopodium dalhousianum 288

M

Macadamia spp. 288
Maccullochella peeli see Murray Cod
Macintyre River 146
Macleay floodplain 147
Macquaria ambigua see Golden Perch
Macquaria australasica see Macquarie Perch
Macquaria novaemaculeata see Australian Bass
Macquarie Harbour 77
Macquarie Marshes (NSW) 146, 152–53, 165
Macquarie Perch 132
macrophytes 127–28, 132
macropods 179, 261–65, 275–78
  see also kangaroos; wallabies; names of individual species
Macropus eugenii see Tammar Wallaby
Macropus fuliginosus see Western Grey Kangaroo
Macropus giganteus see Eastern Grey Kangaroo
Macropus greyi see Toolache Wallaby
Macropus irma see Western Brush Wallaby
Macropus parryi see Whiptail Wallaby
Macropus robustus see Wallaroo
Macropus rufogriseus banksianus see Red-necked Wallaby
Macropus rufogriseus rufogriseus see Bennetts Wallaby
Macropus rufus see Red Kangaroo
Macrotis lagotis see Bilby
Madeira Vine 212
Magpie Goose 265, 273
Malcorhynchus membranaceus see Pink-eared Duck
Mallard 57
Mallee Fowl 208, 314, 395
mallee not damaged by grazing 252
Malurus cyaneus see Superb Blue Wren
mammals 178–80
  core population 316
  decline in numbers 105, 393
  effects of salinization on 127
  extinctions 315
  fire regime for 319
  harvesting of 260
  in old growth forests 244
  recolonizing minesites 85
Managing to Conserve Biodiversity 90
Maned Duck 149, 266–67
mangroves 150
marine environments 405 *see also* aquatic environments
marine turtle harvesting 260
market system fails to recognize ecology 360
Marsupial Mole 275
Marsupial and Monotreme Action Plan 376
marsupials 179–80 *see also* mammals; macropods
Mary River Tortoise 288
median lethal concentrations 140
Mediterranean species adapted to Australia 178
Megaptera novaeangliae see Humpback Whale
Melinus 317
Menindee Lakes 146–48
Mesembryanthum crystallinum see Ice Plant
Mesquite 180
meta-population models 237, 239
metals, water pollution by 66, 121, 138, 140–41, 268
microclimate effect of habitat fragmentation 101–02
Microlaena stipoides 211
Microstrobus fitzgeraldii see Dwarf Mountain Pine
migratory birds *see* birds
Mimosa pigra see Giant Sensitive Plant
Minesite Rehabilitation Research Programme (CSIRO) 85
mining 76–93
  effect on biodiversity 413
  environmental assessment inadequate 370–73
  old sites as tourist attractions 77
  site reconstruction 107
  water pollution from 138
Mirounga leonina see Elephant Seal
Mirror Bush 214
Misgurnus anguillicaudatus see Weather Loach
Mission Grass 194
mistletoe 310
Mistletoe Bird 310
Mitta Mitta River (Vic.) 130
modelling *see* ecological modelling
Moira Grass 147
molluscs 8, 69, 139, 148, 287, 304
Molonglo River (ACT) 121, 138
Monochather paradoxa 251
monocotyledons 212
monsoon forests 313
Mordacia mordax see Short-headed Lamprey
Morning Glory 212
mortality rates of pest populations 48
mosaic fire management 314–16, 325–26
Mountain Pygmy-possum 395
Mount Lofty Range (SA) 194
Mount Lyell 77
Mount Todd (NT), Gouldian Finches at 369–73
Mulga
  local extinction by fire 310
  in woodlands 249, 251, 253
multiple-use areas 18–19, 414–15
Murray Cod 147, 149
Murray River *see* Murray-Darling Basin
Murray-Darling Basin 152–53, 165–70, 172
  damming of 144, 148
  decline of fish species 303–04
  "improvement" of 149
  salinization 126
  water storage areas 145
Murrumbidgee River (NSW) 144
Mus domesticus see House Mouse
Mustela furo see Ferret
Mutton Bird harvesting 260, 273, 275, 280–81
Myall Lakes 316
mycoherbicide biocontrol 190–92
Myotus adversus see Large-footed Mouse-eared Bat
Myriophyllum crispatum 128
Myrmecobius fasciatus see Numbat
Myrmecoida spp. *see* Ant Plant
Myrsiphyllum asparagoides see Bridal Creeper

N

Nadgee Nature Reserve (NSW) 315–16
Namoi River 151
Nannoperca australis see Southern Pygmy Perch
Nardoo destroyed by flooding 148
Naretha Blue Bonnet 288
Narran Lake 170
National Aquatic Reserves 123
National Environment Protection Authority 338–39
National Estate, economic value 232
National Farmers' Federation 376, 390
National Forest Policy 246–47, 340
National Parks 60
  funding for 397–98
  mining in 88
  protection of rivers 153
National Parks and Wildlife Act 1967 (NSW) 396
National Parks and Wildlife Act 1974 (NSW) 367, 396
  endangered and threatened species 393
  protection of fauna 342
  waterbird harvesting 265–66
National Parks and Wildlife Service 391–409
  aid to farmers 51
  denied reports of koala population 381
  kangaroo harvesting 262
  Macquarie Marshes 165
  species protection 342
National Resource Information Centre 86
National Rivers programme 122
National Soil Conservation Program 353
National Strategy for the Conservation of Biological Diversity 340, 375, 395

National Strategy for Ecologically Sustainable Development 11, 395–96
National Threatened Species Network 376
National Weeds Strategy 188
Native Black Duck *see* Pacific Black Duck
native plants *see also* plants
  effect of grazing 249–59
  encouraging germination 211
  illegal collection 57, 288–89
  use in gardens 57
  as weeds 192–93
Native Snow Grass 195
Native Vegetation Act 1991 (SA) 349–50
Native Vegetation Authority (SA) 348–49
Native Vegetation Council (SA) 349–50
Native Vegetation Management Act 1985 (SA) 344–45, 348–49
Natural Resources Management Strategy 169
Nature Conservation Strategy 170
Neochmia phaeton evengelinae see Crimson Finch
Nephoca cinerea see Australian Sea Lion
nesting resources 245
New South Wales
  commitment to biodiversity 400–01
  effect of mining 77
  effect of settlement 391–95
  extinction rates 39
  forestry 240–48, 367
  habitat damage in 30–38
  interdepartmental jurisdictions 165
  kangaroo harvesting 262–64
  land management 249–50, 343–44, 346–48
  legislation for conservation 366–67
  mining industry 79–80, 85
  river management 144–61
  waterbird harvesting 265
New South Wales Forestry Commission 342, 367, 383
Nile Perch 120
nimbyism 14
Nodding Thistle 188
noise pollution 66, 84, 370, 373
Noisy Scrub-bird 20, 314
non-government organizations 375–79 *see also* activism
North American Waterfowl Management Plan 164
North-east Forests Biodiversity Survey 404
Northern Spotted Owl 294
Northern Territory
  crocodile trade 282
  export of live fish 286–87
  Gouldian Finches 369–73
  legislation for conservation 365
  monsoon forests 313
Northiella haemogaster narethae see Naretha Blue Bonnet
Notalina spp. 133
Nothofagus cunninghamii 286
Notonomus spp. 96–101
Notoryctes typhlops see Marsupial Mole
Numbat 21
nutrient pollution 138–39, 190
Nyctophilus timoriensis see Greater Long-eared Bat

O

Oberonia attenuata 288
Ochna 212–13
Ochna serrulata see Ochna
off-park conservation 19–20, 400, 415–16
oil pollution 63–64, 138
old growth forest 240–43, 294–95 *see also* forestry
Old Mans Beard 188
Olea europa var. *africanus see* African Olive
Onchygalea fraeterna see Bridled Nail-tail Wallaby
Onopordum tauricum see Taurean Thistle
Ooline 51–52, 394
Oplismenus spp., vigorous colonization by 211
Opuntia vulgaris see Prickly Pear
oral history in conservation 17
orchid trade 285
Ornithorhynchus anatinus see Platypus
Oryctolagus cuniculus see Rabbit
Our Country, Our Future 376
Oxalis 212
oyster harvesting 304
ozone reducing gases 58

P

Pacific Black Duck 57, 266–67
Pacific Starfish 63, 181
Palm Cockatoo 288
Pampas Grass 188, 212
Paraserianthes lophantha see Cape Wattle
Paratachardina spp. 193
Paratya australiensis 128
Parkcare involvement in weed control 195
Parkinsonia aculeata 225
parrots 281–82, 288, 314, 326–27
Parthenium hysterophorus see Parthenium Weed
Parthenium Weed 188
Paspalum 212
Paspalum dilatatum see Paspalum
Passer domesticus see House Sparrow
pastoral areas *see* grazing
pastoral industry *see also* agriculture; grazing
  destruction of kangaroos 262
  effect on habitat 39–40
  sustainability 256–57
pathogens carried by road transport 65
Paulownia kawakami see Powton Tree
Pelican 149
Pennezitum clandestinum see Kikuyu
Pennizetum polystachion see Mission Grass
Pennizetum spp. 317
Perameles gunii see Eastern Barred Bandicoot
Peramelidae see Bandicoot
Perca fluviatilis see Redfin
Perissodactyla 180
pest control 46–49, 202–09 *see also* biocontrol
  by National Parks and Wildlife Service 402
  chemical 47, 138, 181, 203–05
  costs of 5–6
pests
  domestic pets as 58, 221
  ducks as 265–66
  grasses as 57, 185, 223, 317
  kangaroos as 262
  mortality rates 48
  population density 202–03
  Vertebrate Pests Committee 220–21
pests *see also* weeds
Petauroides volans see Greater Glider
petrochemicals 63–64, 138
Petrogale lateralis see Rock Wallaby
Petrogale rothschildi see Rothschild's Rock-Wallaby
Petrogale xanthopus see Yellow-footed Rock Wallaby
Pezoporus wallicus see Ground Parrot
Phalaenopsis amabilis 288
pharmaceuticals from bioproducts 7
Phascolarctos cinereus see Koala
phosphorus *see* runoff
Phragmites australis 119, 148
Phyla nodiflora see Lippia
Phyllanthus spp. 310, 312
Physeter catodon see Sperm Whale
Phytollacca octandra see Ink Weed
phytophagous fauna associated with weeds 192–93
Phytophthora spp. 21, 65, 67
Pied Currawong 69
Pigmy Cypress Pine 395
Pimelea 311–12
Pindari Dam 146
Pink-eared Duck 266
Pinus pinaster see Cluster Pine
Pittosporum undulatum see Sweet Pittosporum
Planning Act 1982 (SA) 366
plants *see also* native plants; weeds
  biocontrol by 181
  biodiversity in Sydney 70–71
  biomass data as indicator 255
  changes over time 105
  effect of air pollution 58
  effect of disturbance 189
  effect of fire 311–12, 314–15, 318, 324
  effect of flooding 148
  effect of water pollution 141
  endangered in New South Wales 393
  extinction 69, 185, 310–11
  introduced species 57, 149, 220–27, 394
  legislative protection for 396, 405
  on private land 339–41
  seeding of 195, 311–13
  trade in 276, 284–88
Platypus 127, 131, 150, 275
Ploceidae spp. *see* finches
Poa spp. *see* Native Snow Grass
Poaceae 190, 193
Poephila cincta cincta see Black-throated Finch
police attitudes 383, 385–86
politics 23, 231–39, 375, 377–78
polluter-pays principle 342, 349, 363
pollution *see also* water pollution
  of air 58, 66, 84
  damage to ecosystems 395
  economics of 5–6, 342, 349, 363
  from urban areas 58–59
  licences for 350
Polytelis anthopeplus anthopeplus see Regent Parrot
Polytelis swainsonii see Superb Parrot
Pomaderris aspera 311
population viability assessment 292–99, 326–27, 403
populations *see* surveys
port facilities cause habitat damage 63
Posidonia australis see Seagrass
possums 21, 69, 275, 278, 294–95, 395
Potamogeton tricarinatus 128
potoroos 179, 275, 404
Potorous longipes see Long-footed Potoroo
Potorous tridactylus see Long-nosed Potoroo

Powton Tree 182
Prameles gunii see Eastern Barred Bandicoot
precautionary principle 225
in forestry 247
in land management 342–43, 351, 363
legislation for 341, 405
rejected by mining industry 373
reliability of 369
predation 46, 302
prescribed fire 329–30 *see also* fire regimes
Pretty-faced Wallaby *see* Whiptail Wallaby
Prickly Acacia 180, 185, 222
Prickly Pear 190
private land *see* land management
Privet 211–12, 214
Probosciger aterrimus aterrimus see Palm Cockatoo
professionals *see also* ecologists
as activists 385
collaboration by 86, 90, 108, 123, 192
obligations 76, 416
pressure to make decisions 231
tithing of 13, 15
property rights *see* land management
Prosopis spp. *see* Mesquite
Prospect Dam 144
Protasparagus aethiopicus see Fern Asparagus
protected areas *see* reserves
protected species *see* endangered and threatened species
protesting, risks of 383
Prototroctes maraena see Australian Grayling
Pseudemydura umbrina see Western Swamp Tortoise
Pseudocheirus peregrinus see Common Ringtail Possum
Pseudophritis urvilli see Tupong
Pseudoraphis spinescens see Moira Grass
Psittaciformes spp. *see* Cockatoo
Pspehotus chrysopterygius see Golden-Shouldered Parrot
Pteridium esculentum 286
Pteropodidae spp. *see* Flying Fox
Pteropus poliocephalus see Grey-Headed Fruit Bat
Pterostylis gibbosa see Greenhood Orchid
Puffinus tenuirostris see Mutton Bird

Q

quail 260, 288
Quaking Grass 184
Quarantine Act 1908 185–86, 225
quarantine for weed seeds 222
Queensland
crocodile trade 282
kangaroo harvesting 263–64
legislation for conservation 366
Murray-Darling Basin Agreement 167
Queensland Blue Couch 212
Queensland Wet Tropics Action Plan 376
Queenstown (Tas.) 58, 77
Quoll 69
quotas in kangaroo harvesting 263–64

R

Rabbit
damage caused by grazing 249, 252–53
pest population 202–03
success in Australia 180
radioactivity released by mining activity 84
rail corridors 65
rainfall effects on waterbird population 266
rainforests
attrition by eucalypts 313
environmental impact assessments for 343–44
fire regimes in 313–14
rate of destruction 29
weed reduction in 189–90, 211
RAMAS-Space decision support system 293–94
rangelands 249, 251–52, 254
rapid biodiversity assessment 403
rats 69, 127, 131, 185, 315–16
Rattus norvegicus see Brown Rat
Rattus rattus see Black Rat
reafforestation for salinization control 129
Recherche Archipelago 309–10
Recovery Plans 20–21
recreation 59–62
habitat damage by vehicles 66
use of bush 60, 65
use of National Parks 398
use of regenerated mine sites 86
recreational fishing *see* fishing
Red Gum 147–48, 180
Red Kangaroo 262–65, 276
Red-back Spider 276
Red-bellied Pademelon 262
Red-necked Wallaby 264
Red-tailed Black Cockatoo 185
Red-tailed Hawk 293
Redfin 127
reedbeds, flourish in flooded land 148–49
refuges *see* reserves
refuse *see* waste disposal
regeneration *see* habitat protection and repair
Regent Honeyeater 39
Regent Parrot 394
regevetation *see* habitat protection and repair
regional plans for farming 50–51
regulation *see also* government; land management; legislation
of agriculture 340
by consensus 342–43
of environmental impact 365–74
of forestry 246–47
jurisdictional issues 162–65, 170–72
of mining industry 86–88
of pollution 139
of private land 338
of river management 145, 151–52
of wildlife harvesting and trade 276, 278, 280–81, 289
rehabilitation *see* habitat protection and repair
remnant landscapes *see* habitat fragmentation
remnant vegetation 19, 108–09, 170 *see also* reserves
Remnant Vegetation Protection Scheme (WA) 343
reproductive control 181, 203–05
reptiles
decline in numbers 105, 393
effect of salinization 127
effect of sedimentation 131
introduced species 178
in old growth forests 244
recolonizing minesites 85
harvesting and trade 260, 286, 288
research 17–19, 22–23, 122–23, 231, 384–85
directed by sensitivity analysis 297
economic analysis of conservation 232–34
funding for 403
into fire regimes 329
into mining impacts 91
into private land management 351–52
into river management 122, 152
into salinization 127–30
into sedimentation 132
into social attitudes 16
into viable populations 292–94
into water pollution 137, 141
into weed impact 184
into wildlife harvesting and trade 269–70
translating into action 416
reserves 18–19, 361–63, 414
after burning 314–15
exclusion of feral animals 387–88
extinction within 316–17
fires in 319, 326
habitat preservation in 94
inadequate for conservation 339
minimum sizes 294–95
National Parks and Wildlife Service 396–97
non-government organizations 377
selection methods 234
in temperate forests 240–42
for vulnerable population 207–08
waterbird conservation 267
Resource Assessment Commission 231–33
Forest and Timber Inquiry 398
funding for National Parks and Wildlife Service 397–98
population viability analysis 292
on private forestry 340
resource harvesting *see* wildlife harvesting and trade
resource use assessment 292, 360–61, 405
Retropinna semoni see Australian Smelt
revegetation *see* habitat protection and repair
Revegetation Guide to the Central Wheatbelt 108
rice crops, damage caused by ducks 265–66
rights of nature 13–14 *see also* ethical imperative
Rio Convention *see* Convention on the Conservation of Biodiversity
River Blackfish 127
River Murray Commission 166
River Murray Waters Agreement 166
River Mussel 148
Riverina, irrigation of 151
rivers 115–18, 144–73
development of 303
effects of transportation 62
flooding 61, 146–49, 170
habitat damage 413
habitat protection and repair 122
health of 118, 122
"improvement" of 119, 149
riparian zone 119–20, 133–34, 138, 189
roads 65–67, 130, 133–34, 317
Rock Wallaby 21, 202
rodents 177, 180
roosting resources 245
Rosewood 252
Rothschild's Rock-Wallaby 21

Rough Tree Fern 285
Royal Australian Ornithologists' Union 375
Rubber Vine 185, 193, 220
Rubus fruticosus see Blackberry
runoff 66 *see also* hydrology; irrigation
  difficulty in regulating 142
  effect on woodland grasses 251
  from urbanized areas 56–58, 210
  pesticides in 138
  and river health 116
Rural Adjustment Scheme 353–54
Rural Lands Protection Act (Qld) 226
rural subdivisions 60–61
Rutidosis leptorrhynchoides 312

S

St Johns Wort 195
Salinity and Drainage Strategy, Murray-Darling Basin 168–69
salinization 126–30
  effect of river management 150
  groundwater on farms 50
  reversal 168–69
  in rivers 119
  WA wheat belt 40
  in wheat-sheep area 106–07
Salix babylonica see Willows
Salmo trutta see Brown Trout
salmonids 120, 303
salt *see* salinization
Salt-Cedar *see* Athel Pine
saltmarsh 71 *see also* wetlands
Saltmarsh Grass 68
Saltwater Crocodile 260, 276, 282
Salvinia spp. 120, 149, 224
sand mining *see* mining
Saratoga 286–87
Savannah in woodiands 251
Save the Bush Program 353, 376
scarab beetles 21, 47, 49
Schefflera actinophylla see Umbrella Tree
schools *see* education
Scleropages leichardti see Saratoga
sclerophyll forest 211
Sclerotium cepivorum see Angled Onion
scorpions 96, 101
Scotch Broom 394
Seagrass harvesting 287
seal and sealion harvesting 260, 274
seasonality of climate 116, 150
seaweed harvesting 287
sedimentation 130–34
  effect of river management 150
  pesticides in runoff 138
  water pollution by 139
Seed Act 1982 185
seeds, effect of fire on 311–13
selfish imperative 14–15
Senecio phosphorus see African Daisy
Senna 212
Senna pendula see Senna
sensitivity analysis 295–97
sewage, as water pollution 138–39
Sheep, damage caused by grazing 249, 251–52
shells, trade in 287
Shingleback Lizard 288
Shoalhaven River (NSW) 146
Short-headed Lamprey 132
Short-tailed Shearwater *see* Mutton Bird
shrubland fire management 325–26
Siam Weed 224
Silver Gulls 59, 64
Silver Perch 147, 286
silviculture *see* forestry
SIRATAC decision support programme 46–47
Skeleton Weed 195
skin trade *see* wildlife harvesting and trade
Small-leaf Privet 211, 213
smuggling *see* wildlife harvesting and trade
Snapper 300
Snowy Mountain Scheme 145
social attitudes 12–18
  to activism 384–86
  to development 116
  to land reconstruction 107
  to pest eradication 206–07
Society for Growing Australian Plants 51
Soft Tree Fern 285, 317
soil *see also* land management; salinization
  assessment 85, 256
  changes due to grazing 253
  legislation for conservation 346–48, 382
Solanum mauritianum see Wild Tobacco
Solanum nigrum see Blackberry Nightshade
Sollya heterophylla see Blue Bell Creeper
Sonchus oleraceus see Sow Thistle
South Australia 343, 345, 348–50
South Coast Region (NSW) 33
South East Timber Protection Act 1994 (NSW) 381
South Esk River (Tas) 138
Southern Bluefin Tuna 301–03
Southern Brown Bandicoot 295–96
Southern Bull-kelp 276
Southern Pygmy Perch 127
Southern Right Whale 274, 394
Southwest Botanical Province 106
Sow Thistle 214
Sparrow 69, 178
Spartina anglica 71
Spartina spp. *see* Cord Grass
species diversity 9, 68–70
  areas of high richness 117
  effects of habitat damage 38–39
  on farms 52
species-discounting estimates 140
Sperm Whale 274
Sphagnum Moss 284–85
Sphagnum spp. *see* Sphagnum Moss
Spider Plant 212
spiders, source of fibre technology 8
Spinifex sericeus 181
Spiny Emex 185
Sporobolus virginicus see Saltmarsh Grass
Spotted Galaxias 132
Starling 69
State Development and Public Works Organization Act 1971 (Qld) 366
State government *see* government
state-and-transition model 254–55
Statement on the Environment 171–72
Stenotaphrum secondatum see Buffalo Grass
Stictonetta naevosa see Freckled Duck
stock routes 65
stormwater *see* runoff
streams *see* rivers
Strepera graculina see Pied Currawong
Strix occidentalis caurina see Northern Spotted Owl
Sturnus vulgaris see Starling
Sulphur Crested Cockatoo 69
sulphur emissions due to mining 84
Superb Blue Wren 69
Superb Parrot 39, 288, 394
surveys 9, 19–20, 22, 40, 68
  of biodiversity 351–52, 376, 404
  by National Parks and Wildlife Service 402–04
  for ecological studies 122–23
  environmental impact assessments 368
  fauna in New South Wales 392
  for forestry 384–85, 398
  inadequacy of 369–73, 398
  of kangaroo population 262–63
  of waterbird population 266–67
sustainable development 24, 396
  grazing 256–57
  wildlife harvesting and trade 273
Swainsona recta 312
Swamp Wallaby 264
swampland destroyed by flooding 148
Sweet Pittosporum 193–94, 217–18
  invasion by 184
  weed potential 225
Swift Parrot 288
Sydney
  bushland in 214–19
  habitat damage in 30
  plants in 70–71
  water needs 144
  wetlands in 71
Syncarpia glomulifera see Turpentine trees

T

Tallitridae 96, 101
Tallowa Dam 149
Tamarix aphylla see Athel Pine
Tammar Wallaby 21
Tandanus tandanus see Catfish
Tantawangalo State Forest 381
Tasmania 263
Tasmanian Dam case 352
Tasmanian Tiger 179, 275
Taurean Thistle 188
tax concessions for land management 390
taxonomy 22, 68
  aquatic environments 116–17, 122, 139
  of threatened species 376
  of weeds 186–88
technology 8–9, 17, 24, 403–04
  see also computers
  minimising environmental impact 88
  for monitoring grazing damage 255–56
temperate forests *see* forestry
Themeda triandra see Kangaroo Grass
Thomson River (Vic.) 132, 134
threatened species *see* endangered and threatened species
Threskiornis aethiopica see Ibis
Thylacinus cynocephalus see Tasmanian Tiger
Thylogale billardierii see Red-bellied Pademelon
timber harvesting *see* forestry
Timber Industry (Interim Protection) Act 1991 (NSW) 367
tithing of conservation professionals 13, 15
Toechima pterocarpum 288
Toolache Wallaby 265, 275
Total Catchment Management *see* catchment management
tourism *see* recreation
Towards Healthier Rivers 162, 171

Towra Point (NSW) 63–64, 71
toxicity tests 127, 140
Trachydosaurus rugosus see Shingleback Lizard
Trade Records Analysis of Flora and Fauna in Commerce 377
Tradescantia albiflora see Wandering Jew
transportation, effects on biodiversity 62–72
Travel Cost method of valuing reserves 32–33
trawling, damage to epibenthic ecology 302
Tree, Blackberry and Woody Weed Killer 212
trees *see* forestry; plants
tributyl tin 63
Trichglochin procera 128
Trichoptera 128
Trichosurus caninus see Brush-tailed Possum
Trichosurus vulpecula see Common Brushtail Possum
Triclopyr (herbicide) 212
Tridacnidae spp. *see* Giant Clam
tubificid worms 139
Tullakool evaporation basin (NSW) 268
Tumbarumba Local Environment Plan 345
Tumut Local Environment Plan 345
Tupong 127
turbidity *see* sedimentation
Turnix melanogaster see Black-breasted Button Quail
Turpentine trees 63
turtles, effects of salinization 127
Tussock Grass 193–94
Typha spp. *see* Cumbungi

U

Umbrella Tree 225
ungulates, as introduced species 178
United Nations 163, 387
urbanization 55–75
  and attitudes to conservation 414
  extinctions due to 139, 310
  need for water 144
  regeneration of bush 214–19
  water pollution from 137–38, 210
Ursus arctos horribilis see Grizzly Bear
user-pays principle in land management 363

V

vegetation *see* clearing; forestry; plants; remnant vegetation
Velesunio ambiguus see Floodplain Mussel
Vertebrate Pests Committee 220–21
vertebrates *see* fauna
viable population *see* population
Victoria
  environmental impact assessment legislation 366
  private land management 343–45
  river health 118
  salinization in 126
  waterbird harvesting 265
Viney Creek 83
viral immunocontraception of feral animals 204
visual pollution 65, 84
Volterra effect 5
volunteers 23, 380–86
VORTEX decision support system 293–94
Vrydagzynea paludosa 288
Vulpes vulpes see Fox

W

Wallabia bicolor see Swamp Wallaby
wallabies 21, 202, 314–15
  harvesting 260–65, 275
Wallaroo 262, 264
Wandering Jew 212–14
Warrawong Sanctuary (SA) 387
waste disposal 5, 59, 214
water
  availability of 121
  harvesting of 151–52
  managing resources 115–25, 170–72
  storage of 61, 144–46, 148–49
  use in urbanized areas 57
water *see also* groundwater; hydrology; irrigation; rivers; salinization; water pollution
Water Buffalo 220
Water Hyacinth 149
water pollution 137–43, 162
  due to mining 84, 88
  effects 121, 141–42
  from shipwrecks 63–64
  industrial discharges 59
Water Rat 127, 131
water table *see* groundwater
waterbirds 148–50, 260, 265–69
  loss of habitat 64, 147
  management plan for Murray-Darling 170
  significant species 139
  threatened by Cord Grass 185
Watsonia 194, 212
Watsonia pyramidata see Watsonia
Waxflower 284
Weather Loach 120
weeds 21, 184–201, 210–28
  introduced to gardens 57
  spread along roadsides 67
weirs *see* dams
Welcome Reef Dam 146
Western Australia
  crocodile trade 282
  invasion by eastern plants 192
  land management regulations 347–48
  legislation for conservation 365
  wheatbelt revegetation 108
  wildflower trade 284
Western Brush Wallaby 21
Western Division (NSW) 35, 241, 249
Western Grey Kangaroo 262, 264–65
Western Swamp Tortoise 20
wetlands 146–49, 151–53, 267–69
  constructed after mining 83
  effect of urbanization 56, 62–63
wetlands — *continued*
  environmental impact assessments for 343–44
  international importance of 164
  management plan for Murray-Darling 170
  salinization damage to 119
  significant species 139
  in Sydney 71
  threatened by Cord Grass 185
  water availability 121
  weed damage to 184
whale harvesting 260, 274–75, 394
wheat-sheep zone 105–06
wheatbelt regions 40
  habitat damage 33–35
  revegetation 108
Whiptail Wallaby 262, 264–65
Whisky Grass 212
White Rot disease of onion 190–92
Wild Tobacco 190
wilderness, limitations on 414
Wilderness Act 1987 (NSW) 396
wildfires *see* fire; fire regimes
wildflower export 284 *see also* plants; wildlife harvesting and trade
wildlife harvesting and trade 21–22, 231–39, 260–91, 300–03 *see also* fishing
  birds 288, 395
  illegal 221, 287–90, 394–95
  impact on biodiversity 60
  native plants 57, 284
  in New South Wales 394–95
Wildlife Protection (Regulation of Exports and Imports) Act 1982 263, 276, 278, 280
willingness-to-pay 361
Willow 120, 149
Wimmera River (Vic) 126, 129
Wingham Forest Management Area 367–68
wingless grasshoppers 47–49
Wodyetia bifurcata see Cape Melville Palms
Wog Wog (NSW) 95
wood collection for domestic heating 61
woodlands
  ecological modelling 254–55
  effects of fire on 310–11, 324
  native grasses in 250–51
  sustainable grazing in 257
World Heritage Areas 352, 397, 399
Woylies Bettong *see* Brush-tailed Bettong

X

Xanthomyza phyrigia see Regent Honeyeater
Xanthorreah spp. 286
Xanthostemon formosus 288
Xenorhyncus asiaticus see Black-Necked Stork

Y

Yabby, removal through fishing 121
Yellow-footed Rock Wallaby 275
Yinberrie Hills (NT) 372

Z

zinc pollution 138, 141

Common Wombat

Vombatus ursinus

The common wombat is a large burrowing marsupial, which eats roots and leaves. It can be diurnal, but not often found out during the day. Young can be born at any time of the year.

Terrey Hills Grevi

Grevillea Cal

An endangered shru
is restricted to a 6
of the north of Sy
a large spreadi
wide, deeply l
and deep red
flowe

Nephruras vertebralis

Little is known of this small gecko. It lives in the desert, either excavating burrows, or using those made by other animals. It forages for insects at night in open areas.

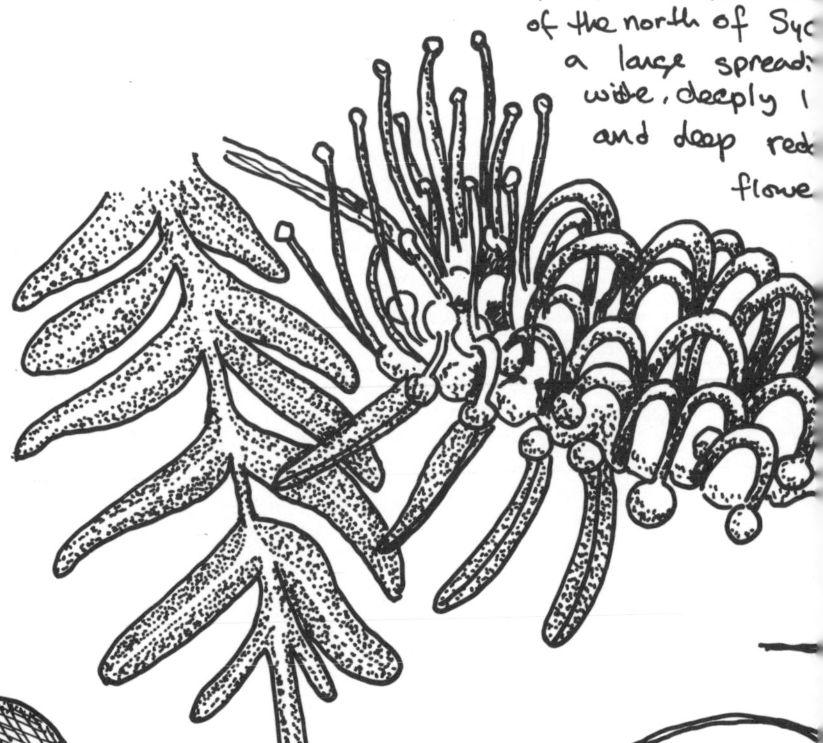